Elementary & Intermediate
Algebra

a combined course

Charles P. McKeague

Cuesta College

BROOKS/COLE

THOMSON LEARNING

Australia • Canada • Mexico • Singapore • Spain
United Kingdom • United States

BROOKS/COLE

★

™

THOMSON LEARNING

Publisher: Emily Barrosse
Executive Editor: Angus McDonald
Developmental Editor: Jay Campbell
Production Manager: Alicia Jackson
Marketing Strategist: Julia Downs-Conover
Production Service: Progressive Publishing Alternatives

Text Designer: Lisa Adamitis
Art Director: Lisa Adamitis
Compositor: Progressive Information Technologies
Printer: Von Hoffman Press—Jefferson City, MO
Cover Image: Vienna, Austria (©Murat Ayranci/SuperStock)

COPYRIGHT © 2000 Thomson Learning, Inc. Thomson Learning™ is a trademark used herein under license.

ALL RIGHTS RESERVED. No part of this work covered by the copyright hereon may be reproduced or used in any form or by any means—graphic, electronic, or mechanical, including but not limited to photocopying, recording, taping, Web distribution, information networks, or information storage and retrieval systems—without the written permission of the publisher.

Printed in the United States of America
4 5 6 7 05 04 03 02 01

For more information about our products, contact us at:
Thomson Learning Academic Resource Center
1-800-423-0563

For permission to use material from this text, contact us by:
Phone: 1-800-730-2214
Fax: 1-800-730-2215
Web: http://www.thomsonrights.com

Asia
Thomson Learning
60 Albert Street, #15-01
Albert Complex
Singapore 189969

Australia
Nelson Thomson Learning
102 Dodds Street
South Melbourne, Victoria 3205
Australia

Canada
Nelson Thomson Learning
1120 Birchmount Road
Toronto, Ontario M1K 5G4
Canada

Europe/Middle East/Africa
Thomson Learning
Berkshire House
168-173 High Holborn
London WC1 V7AA
United Kingdom

Latin America
Thomson Learning
Seneca, 53
Colonia Polanco
11560 Mexico D.F.
Mexico

Spain
Paraninfo Thomson Learning
Calle/Magallanes, 25
28015 Madrid, Spain

Library of Congress Catalog Card Number:
00-100080
ELEMENTARY AND INTERMEDIATE ALGEBRA:
A COMBINED COURSE
ISBN: 0-03-029107-0

CONTENTS

Exponents and Polynomials   *231*

Factoring   *294*

Rational Expressions   *344*

Transitions   *401*

Equations and Inequalities in Two Variables   *468*

Rational Exponents and Roots   *548*

Quadratic Functions   *616*

PREFACE TO THE INSTRUCTOR

Including the appendices, this book contains 92 sections. Half of the sections are in the first six chapters. The material in these chapters is covered at beginning algebra level. The transitions chapter begins the second half of the course. It contains a mixture of levels; some topics are covered at the beginning algebra level, while others are covered at the intermediate algebra level. The topics covered after Chapter 7 are at the intermediate algebra level.

The key to using this book for a two-term course covering both beginning and intermediate algebra is Chapter 7, titled Transitions. It contains a combination of review material and new material. The material in this chapter is organized so that someone entering the course at this point would review all the important topics from the first six chapters. If they work hard on this chapter they will be able to handle the rest of the course.

If you find that you cannot cover all of the first six chapters in your first course, you can interchange Chapters 6 and 7, so that you cover the transitions chapter before the chapter on rational expressions. This way, the transitions chapter still begins your second course. If you cover Chapter 7 before Chapter 6, simply delete the review problems in Sections 7.1 and 7.7. When you are finished with Chapter 7, you can cover Chapter 6 with no loss of continuity.

FEATURES OF THE BOOK

Visualization of Topics This text contains many diagrams, charts, and graphs. The purpose of this is to give students additional information, in visual form, to help them understand the topics we cover. As an example, look in Chapter 1.2 under the "Applying the Concepts" problems.

Tables, Bar Charts, Scatter Diagrams, and Line Graphs Beginning in Chapter 1 and then continued in Chapters 3 and beyond, students are required to analyze information from tables. In addition to simply reading tables, they practice converting data in tabular form to data in graphical form that includes bar charts and scatter diagrams. From there they move on to graphing ordered pairs and linear equations on a rectangular coordinate system.

Early Coverage of Graphing The material on graphing equations in two variables begins in Chapter 3.

Flexible Table of Contents As we mentioned earlier, Chapters 6 and 7 are interchangeable. You can cover Chapter 7 before Chapter 6 simply by deleting the review problems from Sections 7.1 and 7.7. Then cover Chapter 6, and as much of the second half of the table of contents as you like.

Complete Coverage of Functions Functions are introduced early in the second half of the course in Section 8.4. Function notation is covered Section 8.5. Composition of functions and algebra with functions make up Section 8.6. The concept of a function is then carried through to the remaining chapters of the book.

Using Technology   Scattered throughout the second half of the book is material that shows how graphing calculators, spreadsheet programs, and computer graphing programs can be used to enhance the topics being covered. This material is easy to find because it appears under the heading "Using Technology."

Blueprint for Problem Solving   The Blueprint for Problem Solving is a detailed outline of the steps needed to successfully attempt application problems. Intended as a guide to problem solving in general, the Blueprint overlays the solution process to all the application problems in the first few chapters of the book. As students become more familiar with problem solving, the steps in the Blueprint are streamlined.

Facts from Geometry   Many of the important facts from geometry are listed under this heading. In most cases, an example or two accompanies each of the facts to give students a chance to see how topics from geometry are related to the algebra they are learning.

Number Sequences   An introductory coverage of number sequences is integrated throughout Chapter 1. I find that there are many interesting topics I can cover if students have some experience with number sequences.

Unit Analysis   Chapter 6 contains problems requiring students to convert from one unit of measure to another. The method used to accomplish the conversions is the method they will use if they take a chemistry class. Since this is similar to the method used to multiply rational expressions, unit analysis is covered in Section 6.2 with multiplication and division of rational expressions.

Getting Ready for Class   Just before each problem set is a list of four problems under the heading *Getting Ready for Class*. These problems require written responses from students, and can be answered by reading the preceding section. They are to be done before students go to class.

Chapter Openings   Each chapter begins with an introduction in which a real-world application is used to stimulate interest in the chapter. These opening applications are expanded on later in the chapter. Most of them are explained using the rule of four: verbally, numerically, graphically, and algebraically.

Study Skills   Found in the first six chapter openings after the opening application are a list of study skills intended to help students become organized and efficient with their time. The study skills point students in the direction of success.

Organization of the Problem Sets   Six main ideas are incorporated into the problem sets.

1. **Drill**   There are enough problems in each set to ensure student proficiency in the material.

2. **Progressive Difficulty**   The problems increase in difficulty as the problem set progresses.

3. **Odd-Even Similarities**   Each pair of consecutive problems is similar. Since the answers to the odd-numbered problems are listed in the back of the book, the similarity of the odd-even pairs of problems allows your students to check their work on an odd-numbered problem and then to try the similar even-numbered problem.

4. **Applying-the-Concepts Problems**   Students are always curious about how the algebra they are learning can be applied, but at the same time many of them are apprehensive about attempting application problems. I have found that they are more likely to put some time and effort into trying application problems if they do not have to work an overwhelming number of them at one time and if they do not have to work on them every day. For these reasons, I have placed a few application problems toward the end of almost every problem set in the book.

5. **Review Problems**   Each problem set, beginning with Chapter 2, contains a few review problems. Where appropriate, the review problems cover material that will be needed in the next section. Otherwise, they cover material from the previous chapter. That is, the review problems in Chapter 5 cover the important points in Chapter 4. Likewise, the review problems in Chapter 6 review the important material from Chapter 5. If you give tests on two chapters at a time, you will find this to be a timesaving feature. Your students will review one chapter as they study the next chapter.

6. **Extending the Concepts Problems**   In the second half of the book, when students are at the intermediate algebra level, many of the problem sets end with a few problems under this heading. These problems are more challenging than those in the problem sets, or they are problems that extend some of the topics covered in the section.

End of Chapter Retrospective   Each chapter ends with the following items that together give a comprehensive reexamination of the chapter and some of the important problems from previous chapters.

Chapter Summary   These list all main points from the chapter. In the margin, next to each topic being reviewed, is an example that illustrates the type of problem associated with the topic being reviewed.

Chapter Review   An extensive set of problems that review all the main topics in the chapter. The chapter reviews contain more problems than do the chapter tests. Numbers in brackets refer to the section(s) in the text where similar problems can be found.

Chapter Projects   One group project, for students to work on in class, and one research project for students to do outside of class.

Chapter Test   A set of problems representative of all the main points of the chapter.

Cumulative Review   Starting in Chapter 2, each chapter ends with a set of problems that review material from all preceding chapters. The cumulative reviews keep students current with past topics and help them retain the information they study.

SUPPLEMENTS TO THE TEXTBOOK

The first edition of *Elementary and Intermediate Algebra: A Combined Course* is accompanied by a number of useful supplements.

For the Instructor

- ***Instructor's Edition***   An instructor version of the text is available. This version contains everything from the student edition as well as an additional appendix containing the answers to the even-numbered problems in the problem sets.

- ***Printed Test Bank and Prepared Tests***   The test bank consists of multiple-choice and short-answer test items organized by chapter and section. It also includes 15 sample problems per section of the text, labeled by chapter, section, and level of difficulty. The prepared tests comprise fourteen sets of ready-to-copy tests: one set for each chapter and one set for the entire book. Each set comprises two multiple-choice and four show-your-work tests. Answers for every test item are provided.

- ***ESATEST 2000^{TM} Computerized Test Bank***   A flexible, powerful computerized testing system, the ESATEST 2000^{TM} Computerized Test Bank contains all the test bank questions and allows instructors to prepare quizzes and examinations quickly and easily. It offers teachers the ability to select, edit, and create not only test items but algorithms for test items as well. Teachers can tailor tests according to a variety of criteria, scramble the order of test items, and administer tests online, either over a network or via the Web. ESATEST 2000^{TM} also includes full-function gradebook and graphing features. Available in Windows format only.

- ***Instructor's Manual***   The Instructor's Manual contains complete annotated solutions to all even-numbered problems and to every other odd-numbered problem, beginning with problem number 3, in the problem sets.

For the Student

- ***Student Solutions Manual***   The Student Solutions Manual contains complete annotated solutions to every other odd-numbered problem in the problem sets and to all chapter review and chapter test exercises.

- ***MathCue Interactive Tutorial Software for Windows***   This unique computer software is an all-in-one learning tool. It contains Tutorial and Practice modes, enabling students to work any number of problems keyed to chapters of the McKeague text. The Solution Finder tool, an algorithm-based problem generator, provides step-by-step feedback.

- ***Excursions in Real-World Mathematics, by Gina Shankland and Sara Williams of Mt. Hood Community College***   This exciting Web-based activity program will activate the minds of students as they solve problems that they may encounter in their everyday lives. Real data obtained from actual business, industry, and other academic disciplines serve as the raw material used in solving problems. There are more than 80 application situations to increase students' awareness of the power of mathematics; these correspond to topics in McKeague's *Elementary and Intermediate Algebra: A Combined Course.* The applications are organized by level, length, and concept. Professors access an easy-to-use, menu-based system to order exactly the problems they wish to teach. Professors receive an Instructor's Manual with tips and solutions, and students receive the problems in the form of a workbook, which includes a guide to solving these applications and is available in the bookstore. Check out the activities at

http://www.brookscole.com

and contact your sales representative for details and passwords.

Brooks/Cole may provide complimentary instructional aids and supplements or supplement packages to those adopters qualified under our adoption policy. Please contact your sales representative for more information. If as an adopter or potential user you receive supplements you do not need, please return them to your sales representative or send them to:

Attn: Returns Department
Troy Warehouse
465 South Lincoln Drive
Troy, MO 63379

ACKNOWLEDGEMENTS

A project of this size cannot be completed without help from many people. Angus McDonald, my editor at Saunders College Publishing, did an outstanding job of getting all the contract issues settled so that we could proceed with the project. Julia Downs-Conover, Senior Math Marketing Strategist, has provided important feedback from various math departments across the country. Jay Campbell, my developmental editor, worked especially hard to get the book ready for production and to see that I had all the materials I needed to create the manuscript. Lisa Adamitis was responsible for the design of the book. I think you will agree that she did an excellent job. Anne Scanlan-Rohrer, of Two Ravens Editorial Services, Kate Pawlik, and Laura Cornell did an outstanding job of proofreading the book. My son, Patrick, assisted me with this text, and his influence has made this a better book. My thanks to all these people; this book would not have been possible without them. Donna King, of Progressive Publishing Alternatives, implemented all the changes that we needed in the production stage of the project. Working with her has been a very pleasant experience.

Thanks also to Diane McKeague, for her encouragement with all my writing endeavors, and to Amy Jacobs for all her advice and encouragement.

Finally, I am grateful to the following instructors for their suggestions and comments on this text. Some reviewed the entire manuscript, while others were asked to evaluate the development of specific topics, or the overall sequence of topics. My thanks go to the people listed below:

Susan F. Akers, Northeast State Technical Community College
Gerald Allen, Angelo State University
Carol Atnip, University of Louisville
Jon Becker, Indiana University Northwest
Sandra Beken, Horry-Georgetown Technical College
Sandra Belcher, Midwestern State University
Jacqueline "Monty" Briley, Guilford Technical Community College
Beverly Broomell, Suffolk Community College
Karen Sue Cain, Eastern Kentucky State University
Sally Copeland, Johnson County Community College
Jim Corbett, Alvin Community College
Jorge Cossio, Miami-Dade Community College
Barry Cuell, Utah Valley Community College
Margaret F. Donaldson, East Tennessee State University
Elizabeth Farber, Bucks County Community College
Dorothy Gotway, University of Missouri, St. Louis
Carol Hay, Northern Essex Community College
Cathy Hayes, University of Mobile
Betsy Huttenlock, Penn State University — Abington
Mike Iannone, College of New Jersey
Josephine Johansen, Rutgers University
Jamie King-Blair, Orange Coast College
Harriet Kiser, Floyd College
Jillian M. Knowles, Northern Essex Community College
Molly Krajewski, Daytona Beach Community College
William Livingston, Missouri Southern State College
Gene W. Majors, Fullerton College
Aimee Martin, Amarillo College
Mary Lou McDowell, Widener University
Judy Meckley, Joliet Junior College
Beverly Michael, University of Pittsburgh
Ellen Milosheff, Triton College
Peter Moore, Northern Kentucky State University
Linda J. Murphy, Northern Essex Community College
Nancy Nickerson, Northern Essex Community College
Annette Noble, University of Maryland — Eastern Shore
Gene Reid, Central Piedmont Community College
Dennis Riessig, Suffolk Community College
Richard Riggs, New Jersey City University
Gladys Rockind, Oakland Community College
Pat Roux, Delgado Community College
Geoffrey Schulz, Community College of Philadelphia
Margaret Schmid, Black Hawk College
Mark Serebransky, Camden County College

Mark Sigfrids, Kalamazoo Valley Community College
Barbara Jane Sparks, Camden County College
Heidi Staebler, Texas A&M University—Commerce
Katalin Szucs, East Carolina University
Radu Teodorescu, Western Michigan State University
Shirley M. Thompson, North Lake College
Lucy Thrower, Francis Marion University
Kathryn C. Wetzel, Amarillo College
Claude Williams, Central Piedmont Community College
Pete Witt, Glendale Community College
Alice Wong, Miami-Dade Community College
Robert Wynegar, University of Tennessee at Chattanooga

Charles P. McKeague
February 2000

PREFACE TO THE STUDENT

I often find my students asking themselves the question "Why can't I understand this stuff the first time?" The answer is "You're not expected to." Learning a topic in algebra isn't always accomplished the first time around. There are many instances when you will find yourself reading over new material a number of times before you can begin to work problems. That's just the way things are in algebra. If you don't understand a topic the first time you see it, that doesn't mean there is something wrong with you. Understanding algebra takes time. The process of understanding requires reading the book, studying the examples, working problems, and getting your questions answered.

Here are some questions that are often asked by students starting an algebra class.

How much math do I need to know before taking algebra? You should be able to do the four basic operations (addition, subtraction, multiplication, and division) with whole numbers, fractions, and decimals. Most important is your ability to work with whole numbers. If you are a bit weak at working with fractions because you haven't worked with them in a while, don't be too concerned; we will review fractions as we progress through the book. I have had students who eventually did very well in algebra, even though they were initially unsure of themselves when working with fractions.

What is the best way to study? The best way to study is to study consistently. You must work problems every day. A number of my students spend an hour or so in the morning working problems and reading over new material and then spend another hour in the evening working problems. The students of mine who are most successful in algebra are the ones who find a schedule that works for them and then stick to it. They work problems every day.

If I understand everything that goes on in class, can I take it easy on my homework? Not necessarily. There is a big difference between understanding a problem someone else is working and working the same problem yourself. There is no substitute for working problems yourself. The concepts and properties are understandable to you only if you yourself work problems involving them.

HOW TO BE SUCCESSFUL IN ALGEBRA

If you have decided to be successful in algebra, then the following list will be important to you:

1. **If you are in a lecture class, be sure to attend all class sessions on time.** You cannot know exactly what goes on in class unless you are there. Missing class and then expecting to find out what went on from someone else is not the same as being there yourself.

2. **Read the book.**   It is best to read the section that will be covered in class be-forehand. Reading in advance, even if you do not understand everything you read, is still better than going to class with no idea of what will be discussed.

3. **Work problems every day, and check your answers.**   The key to success in mathematics is working problems. The more problems you work, the better you will become at working them. The answers to the odd-numbered problems are given in the back of the book. When you have finished an assignment, be sure to compare your answers with those in the book. If you have made a mistake, find out what it is, and correct it.

4. **Do it on your own.**   Don't be misled into thinking someone else's work is your own. Having someone else show you how to work a problem is not the same as working the same problem yourself. It is okay to get help when you are stuck. As a matter of fact, it is a good idea. Just be sure you do the work yourself.

5. **Review every day.**   After you have finished the problems your instructor has as-signed, take another fifteen minutes and review a section you have already com-pleted. The more you review, the longer you will retain the material you have learned.

6. **Don't expect to understand every new topic the first time you see it.**   Some-times you will understand everything you are doing, and sometimes you won't. That's just the way things are in mathematics. Expecting to understand each new topic the first time you see it can lead to disappointment and frustration. The process of understanding algebra takes time. It requires that you read the book, work problems, and get your questions answered.

7. **Spend as much time as it takes for you to master the material.**   No set for-mula exists for the exact amount of time you need to spend on algebra to master it. You will find out as you go along what is or isn't enough time for you. If you end up spending two or more hours on each section in order to master the mater-ial there, then that's how much time it takes; trying to get by with less will not work.

8. **Relax.**   It's probably not as difficult as you think.

The Basics

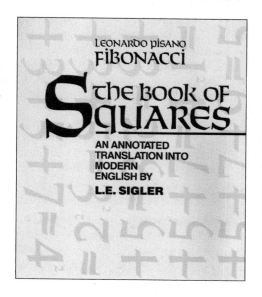

LEONARDO PISANO
FIBONACCI

The Book of
Squares

AN ANNOTATED
TRANSLATION INTO
MODERN
ENGLISH BY
L.E. SIGLER

INTRODUCTION

Much of what we do in mathematics is concerned with recognizing patterns. If you recognize the patterns in the following two sequences, then you can easily extend each sequence.

Sequence of odd numbers = 1, 3, 5, 7, 9, . . .
Sequence of squares = 1, 4, 9, 16, 25, . . .

Once we have classified groups of numbers as to the characteristics they share, we sometimes discover that a relationship exists between the groups. Although it may not be obvious at first, there is a relationship that exists *between* the two sequences shown. The introduction to *The Book of Squares,* written in 1225 by the mathematician known as Fibonacci, begins this way:

I thought about the origin of all square numbers and discovered that they arise out of the increasing sequence of odd numbers.

The relationship that Fibonacci refers to is shown visually in Figure 1.

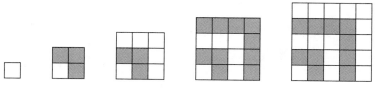

FIGURE 1

Many times we describe a relationship or pattern in a number of different ways. Figure 1 is a visual description of a relationship. In this chapter we will work on describing relationships numerically and verbally, in writing.

STUDY SKILLS

Some of the students enrolled in my algebra classes develop difficulties early in the course. Their difficulties are not associated with their ability to learn mathematics; they all have the potential to pass the course. Students who get off to a poor start do so because they have not developed the study skills necessary to be successful in algebra; they do not put themselves on an effective homework schedule, and when they work problems, they do it their way, not my way. Here is a list of things you can do to begin to develop effective study skills.

1. **Put Yourself on a Schedule** The general rule is that you spend 2 hours on homework for every hour you are in class. Make a schedule for yourself in which you set aside 2 hours each day to work on algebra. Once you make the schedule, stick to it. Don't just complete your assignments and stop. Use all the time you have set aside. If you complete an assignment and have time left over, read the next section in the book, and then work more problems.

2. **Find Your Mistakes and Correct Them** There is more to studying algebra than just working problems. You must always check your answers with the answers in the back of the book. When you have made a mistake, find out what it is and correct it. Making mistakes is part of the process of learning mathematics. In the prologue to *The Book of Squares,* Leonardo Fibonacci (ca. 1170–ca. 1250) had this to say about the content of his book:

   > I have come to request indulgence if in any place it contains something more or less than right or necessary; for to remember everything and be mistaken in nothing is divine rather than human . . .

   Fibonacci knew, as you know, that human beings make mistakes. You cannot learn algebra without making mistakes.

3. **Imitate Success** Your work should look like the work you see in this book and the work your instructor shows. The steps shown in solving problems in this book were written by someone who has been successful in mathematics. The same is true of your instructor. Your work should imitate the work of people who have been successful in mathematics.

Suppose you have a checking account that costs you $15 a month, plus $0.05 for each check you write. If you write 10 checks in a month, then the monthly charge for your checking account will be

$$15 + 10(0.05)$$

Do you add 15 and 10 first and then multiply by 0.05? Or do you multiply 10 and 0.05 first and then add 15? If you don't know the answer to this question, you will after you have read through this section.

Since much of what we do in algebra involves comparison of quantities, we will begin by listing some symbols used to compare mathematical quantities. The comparison symbols fall into two major groups: equality symbols and inequality symbols.

We will let the letters a and b stand for (represent) any two mathematical quantities. When we use letters to represent numbers, as we are doing here, we call the letters *variables*.

VARIABLES: AN INTUITIVE LOOK

When you filled out the application for the school you are attending, there was a space to fill in your first name. "First Name" is a variable quantity, because the value it takes depends on who is filling out the application. For example, if your first name is Manuel, then the value of "First Name" is Manuel. On the other hand, if your first name is Christa, then the value of "First Name" is Christa.

If we denote "First Name" as FN, "Last Name" as LN, and "Whole Name" as WN, then we take the concept of a variable further and write the relationship between the names this way:

$$FN + LN = WN$$

(We use the $+$ symbol loosely here to represent writing the names together with a space between them.) This relationship we have written holds for all people who have only a first name and a last name. For those people who have a middle name, the relationship between the names is

$$FN + MN + LN = WN$$

A similar situation exists in algebra when we let a letter stand for a number or a group of numbers. For instance, if we say "let a and b represent numbers," then a and b are called **variables** because the values they take on vary. We use the variables a and b in the following lists so that the relationships shown there are true for all numbers that we will encounter in this book. By using variables, the following statements are general statements about all numbers, rather than specific statements about only a few numbers.

COMPARISON SYMBOLS

Equality:	$a = b$	a is equal to b (a and b represent the same number)
	$a \neq b$	a is not equal to b
Inequality:	$a < b$	a is less than b
	$a \not< b$	a is not less than b
	$a > b$	a is greater than b
	$a \not> b$	a is not greater than b
	$a \geq b$	a is greater than or equal to b
	$a \leq b$	a is less than or equal to b

The symbols for inequality, $<$ and $>$, always point to the smaller of the two quantities being compared. For example, $3 < x$ means 3 is smaller than x. In this case we can say "3 is less than x" or "x is greater than 3"; both statements are correct. Similarly, the expression $5 > y$ can be read as "5 is greater than y" or as "y is less than 5" because the inequality symbol is pointing to y, meaning y is the smaller of the two quantities.

Next, we consider the symbols used to represent the four basic operations: addition, subtraction, multiplication, and division.

OPERATION SYMBOLS

Addition:	$a + b$	The *sum* of a and b
Subtraction:	$a - b$	The *difference* of a and b
Multiplication:	$a \cdot b, (a)(b), (a)b, (a)b, ab$	The *product* of a and b
Division:	$a \div b, a/b, \dfrac{a}{b}, b\overline{)a}$	The *quotient* of a and b

When we encounter the word *sum,* the implied operation is addition. To find the sum of two numbers, we simply add them. *Difference* implies subtraction, *product* implies multiplication, and *quotient* implies division. Notice also that there is more than one way to write the product or quotient of two numbers.

Note: In the past you may have used the notation 3×5 to denote multiplication. In algebra it is best to avoid this notation if possible, since the multiplication symbol $\times$ can be confused with the variable x when written by hand.

GROUPING SYMBOLS

Parentheses ( ) and brackets [ ] are the symbols used for grouping numbers together. (Occasionally, braces { } are also used for grouping, although they are usually reserved for set notation, as we shall see.)

The following examples show the written language equivalents of the symbols for comparing, operating, and grouping.

EXAMPLES

Mathematical Expression	*Written Equivalent*
1. $4 + 1 = 5$	The sum of 4 and 1 is 5.
2. $8 - 1 < 10$	The difference of 8 and 1 is less than 10.
3. $2(3 + 4) = 14$	Twice the sum of 3 and 4 is 14.
4. $3x \geq 15$	The product of 3 and x is greater than or equal to 15.
5. $\dfrac{y}{2} = y - 2$	The quotient of y and 2 is equal to the difference of y and 2.

EXPONENTS

The last type of notation we need to discuss is the notation that allows us to write repeated multiplications in a more compact form—*exponents*. In the expression 2^3, the 2 is called the *base,* and the 3 is called the *exponent*. The exponent 3 tells us the number of times the base appears in the product. That is,

$$2^3 = 2 \cdot 2 \cdot 2 = 8$$

The expression 2^3 is said to be in exponential form, while $2 \cdot 2 \cdot 2$ is said to be in expanded form. Here are some additional examples of expressions involving exponents.

EXAMPLES   Expand and multiply.

6. $5^2 = 5 \cdot 5 = 25$	Base 5, exponent 2
7. $2^5 = 2 \cdot 2 \cdot 2 \cdot 2 \cdot 2 = 32$	Base 2, exponent 5
8. $10^3 = 10 \cdot 10 \cdot 10 = 1,000$	Base 10, exponent 3

Notation and Vocabulary   Here is how we read expressions containing exponents.

Mathematical Expression	Written Equivalent
5^2	five to the second power
5^3	five to the third power
5^4	five to the fourth power
5^5	five to the fifth power
5^6	five to the sixth power

We have a shorthand vocabulary for second and third powers, because the area of a square with a side of 5 is 5^2, and the volume of a cube with a side of 5 is 5^3.

5^2 can be read "five squared." 5^3 can be read "five cubed."

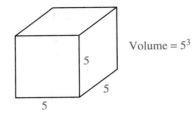

The symbols for comparing, operating, and grouping are to mathematics what punctuation symbols are to English. These symbols are the punctuation symbols for mathematics.

Consider the following sentence:

Paul said John is tall.

It can have two different meanings, depending on how it is punctuated.

1. "Paul," said John, "is tall."
2. Paul said, "John is tall."

Let's take a look at a similar situation in mathematics. Consider the following mathematical statement:

$$5 + 2 \cdot 7$$

If we add the 5 and 2 first and then multiply by 7, we get an answer of 49. On the other hand, if we multiply the 2 and the 7 first and then add 5, we are left with 19. We have a problem that seems to have two different answers, depending on whether we add first or multiply first. We would like to avoid this type of situation. That is, every problem like $5 + 2 \cdot 7$ should have only one answer. Therefore, we have developed the following rule for the order of operations.

RULE (ORDER OF OPERATIONS)

When evaluating a mathematical expression, we will perform the operations in the following order, beginning with the expression in the innermost parentheses or brackets first and working our way out.

1. Simplify all numbers with exponents, working from left to right if more than one of these expressions is present.
2. Then do all multiplications and divisions left to right.
3. Perform all additions and subtractions left to right.

EXAMPLES   Simplify each expression using the rule for order of operations.

9. $5 + 8 \cdot 2 = 5 + 16$ Multiply $8 \cdot 2$ first.

$\quad = 21$

10. $12 \div 4 \cdot 2 = 3 \cdot 2$ Work left to right.

$\quad = 6$

11. $2[5 + 2(6 + 3 \cdot 4)] = 2[5 + 2(6 + 12)] \Big\}$ Simplify within the innermost grouping symbols first.

$\quad = 2[5 + 2(18)]$

$\quad = 2[5 + 36]$ Next, simplify inside the brackets.

$\quad = 2[41]$

$\quad = 82$ Multiply.

12. $10 + 12 \div 4 + 2 \cdot 3 = 10 + 3 + 6$ Multiply and divide left to right.

$\quad = 19$ Add left to right.

13. $2^4 + 3^3 \div 9 - 4^2 = 16 + 27 \div 9 - 16$ Simplify numbers with exponents.

$\quad = 16 + 3 - 16$ Then, divide.

$\quad = 19 - 16 \Big\}$ Finally, add and subtract left to right.

$\quad = 3$

READING TABLES AND BAR CHARTS

The table below shows the average amount of caffeine in a number of beverages. The diagram in Figure 1 is a bar chart. It is a visual presentation of the information in Table 1. The table gives information in numerical form, while the chart gives the same information in a geometric way. In mathematics, it is important to be able to move back and forth between the two forms.

TABLE 1   Caffeine Content of Hot Drinks

Drink (6-ounce cup)	Caffeine (milligrams)
Brewed coffee	100
Instant coffee	70
Tea	50
Cocoa	5
Decaffeinated coffee	4

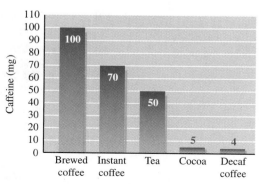

FIGURE 1

EXAMPLE 14   Referring to Table 1 and Figure 1, suppose you have 3 cups of coffee, 1 cup of tea, and 2 cups of decaf in one day. Write an expression that will give the total amount of caffeine in these six drinks, then simplify the expression.

Solution   From the table or the bar chart, we find the number of milligrams of caffeine in each drink; then we write an expression for the total amount of caffeine:

$$3(100) + 50 + 2(4)$$

Using the rule for order of operations, we get 358 total milligrams of caffeine.

NUMBER SEQUENCES AND INDUCTIVE REASONING

Suppose someone asks you to give the next number in the sequence of numbers below. (The dots mean that the sequence continues in the same pattern forever.)

$$2, 5, 8, 11, \ldots$$

If you notice that each number is 3 more than the number before it, you would say the next number in the sequence is 14 because $11 + 3 = 14$. When we reason in this way, we are using what is called *inductive reasoning*. In mathematics we use inductive reasoning when we notice a pattern to a sequence of numbers and then use the pattern to extend the sequence.

EXAMPLE 15   Find the next number in each sequence.

(a) 3, 8, 13, 18, . . .
(b) 2, 10, 50, 250, . . .
(c) 2, 4, 7, 11, . . .

Solution   In order to find the next number in each sequence, we need to look for a pattern or relationship.

(a) For the first sequence, each number is 5 more than the number before it; therefore, the next number will be $18 + 5 = 23$.
(b) For the sequence in part (b), each number is 5 times the number before it; therefore, the next number in the sequence will be $5 \cdot 250 = 1{,}250$.
(c) For the sequence in part (c), there is no number to add each time or multiply by each time. However, the pattern becomes apparent when we look at the differences between consecutive numbers:

Proceeding in the same manner, we would add 5 to the next term, giving us $11 + 5 = 16$.

In the introduction to this chapter, we mentioned the mathematician known as Fibonacci. There is a special sequence in mathematics named for Fibonacci. Here it is:

Fibonacci sequence = 1, 1, 2, 3, 5, 8, . . .

Can you see the relationship among the numbers in this sequence? Start with two 1's, then add two consecutive members of the sequence to get the next number. Here is a diagram.

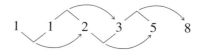

Sometimes we refer to the numbers in a sequence as *terms* of the sequence.

EXAMPLE   16   Write the first ten terms of the Fibonacci sequence.

Solution   The first six terms are given above. We extend the sequence by adding 5 and 8 to obtain the seventh term, 13. Then we add 8 and 13 to obtain 21. Continuing in this manner, the first ten terms in the Fibonacci sequence are

1, 1, 2, 3, 5, 8, 13, 21, 34, 55

Getting Ready for Class

Each section of the book will end with some problems and questions like the ones below. They are for you to answer after you have read through the section, but before you go to class. All of them require that you give written responses, in complete sentences. Writing about mathematics is a valuable exercise. If you write with the intention of explaining and communicating what you know to someone else, you will find that you understand the topic you are writing about even better than you did before you started writing. As with all problems in this course, you want to approach these writing exercises with a positive point of view. You will get better at giving written responses to questions as you progress through the course. Even if you never feel comfortable writing about mathematics, just the process of attempting to do so will increase your understanding and ability in mathematics.

After reading through the preceding section, respond in your own words and in complete sentences.

A. What is a variable?
B. Write the first step in the rule for order of operations.
C. What is inductive reasoning?
D. Explain the relationship between an exponent and its base.

PROBLEM SET 1.1

For each sentence below, write an equivalent expression in symbols.

1. The sum of x and 5 is 14.

2. The difference of x and 4 is 8.

3. The product of 5 and y is less than 30.

4. The product of 8 and y is greater than 16.

5. The product of 3 and y is less than or equal to the sum of y and 6.

6. The product of 5 and y is greater than or equal to the difference of y and 16.

7. The quotient of x and 3 is equal to the sum of x and 2.

8. The quotient of x and 2 is equal to the difference of x and 4.

Expand and multiply.

9. 3^2
10. 4^2
11. 7^2
12. 9^2
13. 2^3
14. 3^3
15. 4^3
16. 5^3
17. 2^4
18. 3^4
19. 10^2
20. 10^4
21. 11^2
22. 111^2

Use the rule for order of operations to simplify each expression as much as possible.

23. $2 \cdot 3 + 5$
24. $8 \cdot 7 + 1$
25. $2(3 + 5)$
26. $8(7 + 1)$
27. $5 + 2 \cdot 6$
28. $8 + 9 \cdot 4$
29. $(5 + 2) \cdot 6$
30. $(8 + 9) \cdot 4$
31. $5 \cdot 4 + 5 \cdot 2$
32. $6 \cdot 8 + 6 \cdot 3$
33. $5(4 + 2)$
34. $6(8 + 3)$
35. $8 + 2(5 + 3)$
36. $7 + 3(8 - 2)$
37. $(8 + 2)(5 + 3)$
38. $(7 + 3)(8 - 2)$
39. $20 + 2(8 - 5) + 1$
40. $10 + 3(7 + 1) + 2$
41. $5 + 2(3 \cdot 4 - 1) + 8$
42. $11 - 2(5 \cdot 3 - 10) + 2$
43. $8 + 10 \div 2$
44. $16 - 8 \div 4$
45. $4 + 8 \div 4 - 2$
46. $6 + 9 \div 3 + 2$
47. $3 + 12 \div 3 + 6 \cdot 5$
48. $18 + 6 \div 2 + 3 \cdot 4$
49. $3 \cdot 8 + 10 \div 2 + 4 \cdot 2$
50. $5 \cdot 9 + 10 \div 2 + 3 \cdot 3$

51. $(5 + 3)(5 - 3)$
52. $(7 + 2)(7 - 2)$
53. $5^2 - 3^2$
54. $7^2 - 2^2$
55. $(4 + 5)^2$
56. $(6 + 3)^2$
57. $4^2 + 5^2$
58. $6^2 + 3^2$
59. $3 \cdot 10^2 + 4 \cdot 10 + 5$
60. $6 \cdot 10^2 + 5 \cdot 10 + 4$
61. $2 \cdot 10^3 + 3 \cdot 10^2 + 4 \cdot 10 + 5$
62. $5 \cdot 10^3 + 6 \cdot 10^2 + 7 \cdot 10 + 8$
63. $10 - 2(4 \cdot 5 - 16)$
64. $15 - 5(3 \cdot 2 - 4)$
65. $4[7 + 3(2 \cdot 9 - 8)]$
66. $5[10 + 2(3 \cdot 6 - 10)]$
67. $5(7 - 3) + 8(6 - 4)$
68. $3(10 - 4) + 6(12 - 10)$
69. $3(4 \cdot 5 - 12) + 6(7 \cdot 6 - 40)$
70. $6(8 \cdot 3 - 4) + 5(7 \cdot 3 - 1)$
71. $3^4 + 4^2 \div 2^3 - 5^2$
72. $2^5 + 6^2 \div 2^2 - 3^2$
73. $5^2 + 3^4 \div 9^2 + 6^2$
74. $6^2 + 2^5 \div 4^2 + 7^2$

Find the next number in each sequence.

75. 1, 2, 3, 4, . . . (The sequence of counting numbers)
76. 0, 1, 2, 3, . . . (The sequence of whole numbers)
77. 2, 4, 6, 8, . . . (The sequence of even numbers)
78. 1, 3, 5, 7, . . . (The sequence of odd numbers)
79. 1, 4, 9, 16, . . . (The sequence of squares)
80. 1, 8, 27, 64, . . . (The sequence of cubes)
81. 2, 2, 4, 6, . . . (A Fibonacci-like sequence)
82. 5, 5, 10, 15, . . . (A Fibonacci-like sequence)

Applying the Concepts

83. **Reading Tables and Charts** The following table and bar chart give the amount of caffeine in five

Caffeine Content in Soft Drinks	
Drink	**Caffeine (milligrams)**
Jolt	100
Tab	47
Coca-Cola	45
Diet Pepsi	36
7UP	0

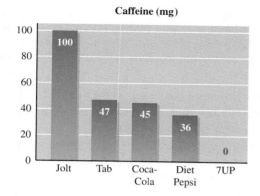

Caffeine (mg)

different soft drinks. How much caffeine is in each of the following?

(a) A 6-pack of Jolt

(b) 2 Coca-Colas plus 3 Tabs

84. **Reading Tables and Charts** The following table and bar chart give the amount of caffeine in five different nonprescription drugs. How much caffeine is in each of the following?

(a) A box of 12 Excedrin

(b) 1 Dexatrim plus 4 Excedrin

Caffeine Content in Nonprescription Drugs	
Nonprescription Drug	**Caffeine (milligrams)**
Dexatrim	200
NoDoz	100
Excedrin	65
Triaminicin tablets	30
Dristan tablets	16

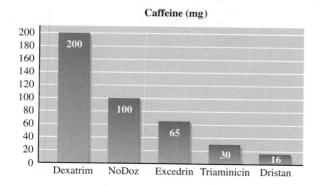

Caffeine (mg)

85. **Reading Tables and Charts** The following bar chart gives the number of calories burned by a 150-pound person during 1 hour of various exercises. The accompanying table should display the same information. Use the bar chart to complete the table.

Calories Burned by 150-Pound Person	
Activity	**Calories Burned in 1 Hour**
Bicycling	374
Bowling	
Handball	
Jogging	
Skiing	

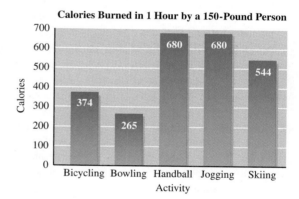

Calories Burned in 1 Hour by a 150-Pound Person

86. **Reading Tables and Charts** The following bar chart gives the number of calories consumed by eating some popular fast foods. The accompanying table should display the same information. Use the bar chart to complete the table.

Calories in Fast Food	
Food	**Calories**
McDonald's Hamburger	270
Burger King Hamburger	
Jack in the Box Hamburger	
McDonald's Big Mac	
Burger King Whopper	

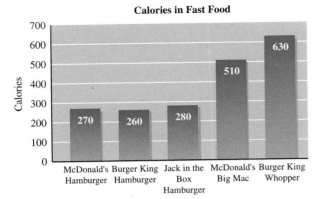

Calories in Fast Food

87. **Stock Market** On Monday Maria buys 10 shares of a certain stock. On Tuesday she buys 4 more shares of the same stock. If the stock splits Wednesday, then she has twice the number of shares she had on Tuesday. Write an expression using parentheses and the numbers 2, 4, and 10 to describe this situation.

88. **Baseball Cards** Patrick has a collection of 25 baseball cards. He then buys 3 packs of gum, and each pack contains 5 baseball cards. Write an expression using the numbers 25, 3, and 5 to describe this situation.

89. **Gambling** A gambler begins an evening in Las Vegas with $50. After an hour she has tripled her money. The next hour she loses $14. Write an expression using the numbers 3, 50, and 14 to describe this situation.

90. **Number of Passengers** A flight from Los Angeles to New York has 128 passengers. The plane stops in Denver, where 50 of the passengers get off and 21 new passengers get on. Write an expression containing the numbers 128, 50, and 21 to describe this situation.

Food Labels In 1993 the government standardized the way in which nutrition information was presented on

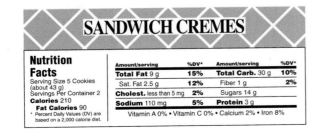

the labels of most packaged food products. The food label shown is from a package of cookies that I ate at lunch the day I was writing this problem set. Use the information on the food label to answer Problems 91–94.

91. How many cookies are in the package?

92. If I paid $0.50 for the package of cookies, how much did each cookie cost?

93. If the "calories" category stands for calories per serving, how many calories did I consume by eating the whole package of cookies?

94. Suppose that, while swimming, I burn 11 calories each minute. If I swim for 20 minutes, will I burn enough calories to cancel out the calories I added by eating 5 cookies?

Food Labels The food label shown below was taken from a bag of corn chips. Use the information on the food label to answer Problems 95–98.

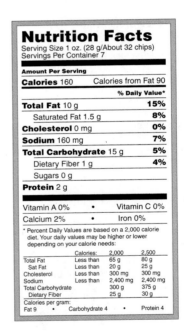

95. Approximately how many chips are in the bag?

96. If the bag of chips costs $0.99, approximately how much does 1 serving of chips cost?

97. The table toward the bottom of the label gives the recommended amount of total fat in grams (g) that should be consumed by a person eating 2,000 calo-

ries per day and by a person eating 2,500 calories per day. Use the numbers in the table to estimate the recommended fat intake for a person eating 3,000 calories per day.

98. Deirdre burns 256 calories per hour by trotting her horse at a constant rate. How long must she ride in order to burn the calories consumed by eating 4 servings of these chips?

1.2   Real Numbers

Table 1 and Figure 1 give the record low temperature, in degrees Fahrenheit, for each month of the year in the city of Jackson, Wyoming. Notice that some of these temperatures are represented by negative numbers.

TABLE 1 Record Low Temperatures for Jackson, Wyoming

Month	Temperature (degrees Fahrenheit)
January	−50
February	−44
March	−32
April	−5
May	12
June	19
July	24
August	18
September	14
October	2
November	−27
December	−49

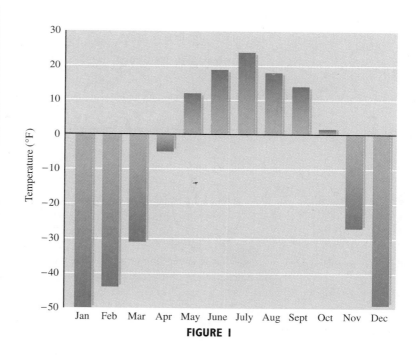

FIGURE 1

In this section we start our work with negative numbers. To represent negative numbers in algebra, we use what is called the **real number line.** Here is how we construct a real number line. We first draw a straight line and label a convenient point on the line with 0. Then we mark off equally spaced distances in both directions from 0. Label the points to the right of 0 with the numbers 1, 2, 3, . . . (the dots mean "and so on"). The points to the left of 0 we label in order, −1, −2, −3, ... Here is what it looks like.

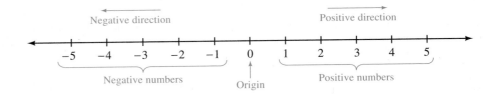

The numbers increase in value going from left to right. If we move to the right, we are moving in the positive direction. If we move to the left, we are moving in the negative direction. When we compare two numbers on the number line, the number on the left is always smaller than the number on the right. For instance, -3 is smaller than -1 since it is to the left of -1 on the number line.

Note: If there is no sign ($+$ or $-$) in front of a number, the number is assumed to be positive ($+$).

E X A M P L E   1   Locate and label the points on the real number line associated with the numbers $-3.5, -1\frac{1}{4}, \frac{1}{2}, \frac{3}{4}, 2.5$.

Solution   We draw a real number line from -4 to 4 and label the points in question.

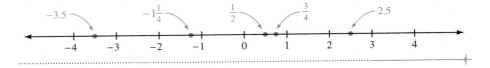

> **DEFINITION**   The number associated with a point on the real number line is called the **coordinate** of that point.

In the preceding example, the numbers $\frac{1}{2}, \frac{3}{4}, 2.5, -3.5$, and $-1\frac{1}{4}$ are the coordinates of the points they represent.

> **DEFINITION**   The numbers that can be represented with points on the real number line are called **real numbers**.

Real numbers include whole numbers, fractions, decimals, and other numbers that are not as familiar to us as these.

FRACTIONS ON THE NUMBER LINE

As we proceed through Chapter 1, from time to time we will review some of the major concepts associated with fractions. To begin, here is the formal definition of a fraction.

DEFINITION   If a and b are real numbers, then the expression

$$\frac{a}{b} \quad (b \neq 0)$$

is called a **fraction.** The top number a is called the **numerator,** and the bottom number b is called the **denominator.** The restriction $b \neq 0$ keeps us from writing an expression that is undefined. (As you will see, division by zero is not allowed.)

The number line can be used to visualize fractions. Recall that for the fraction $\frac{a}{b}$, a is called the numerator and b is called the denominator. The denominator indicates the number of equal parts in the interval from 0 to 1 on the number line. The numerator indicates how many of those parts we have. If we take that part of the number line from 0 to 1 and divide it into *three equal parts,* we say that we have divided it into *thirds* (see Figure 2). Each of the three segments is $\frac{1}{3}$ (one third) of the whole segment from 0 to 1.

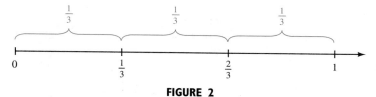

FIGURE 2

Two of these smaller segments together are $\frac{2}{3}$ (two thirds) of the whole segment. And three of them would be $\frac{3}{3}$ (three thirds), or the whole segment.

Let's do the same thing again with six equal divisions of the segment from 0 to 1 (see Figure 3). In this case we say each of the smaller segments has a length of $\frac{1}{6}$ (one sixth).

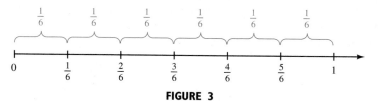

FIGURE 3

The same point we labeled with $\frac{1}{3}$ in Figure 2 is now labeled with $\frac{2}{6}$. Likewise, the point we labeled earlier with $\frac{2}{3}$ is now labeled $\frac{4}{6}$. It must be true then that

$$\frac{2}{6} = \frac{1}{3} \quad \text{and} \quad \frac{4}{6} = \frac{2}{3}$$

Actually, there are many fractions that name the same point as $\frac{1}{3}$. If we were to divide the segment between 0 and 1 into 12 equal parts, 4 of these 12 equal parts

$(\frac{4}{12})$ would be the same as $\frac{2}{6}$ or $\frac{1}{3}$. That is,

$$\frac{4}{12} = \frac{2}{6} = \frac{1}{3}$$

Even though these three fractions look different, each names the same point on the number line, as shown in Figure 4. All three fractions have the same *value* because they all represent the same number.

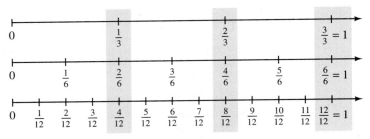

FIGURE 4

> **DEFINITION**   Fractions that represent the same number are said to be **equivalent**. Equivalent fractions may look different, but they must have the same value.

It is apparent that every fraction has many different representations, each of which is equivalent to the original fraction. The next two properties give us a way of changing the terms of a fraction without changing its value.

PROPERTY 1

Multiplying the numerator and denominator of a fraction by the same nonzero number never changes the value of the fraction.

PROPERTY 2

Dividing the numerator and denominator of a fraction by the same nonzero number never changes the value of the fraction.

EXAMPLE 2   Write $\frac{3}{4}$ as an equivalent fraction with denominator 20.

Solution   The denominator of the original fraction is 4. The fraction we are trying to find must have a denominator of 20. We know that if we multiply 4 by 5, we get 20. Property 1 indicates that we are free to multiply the denominator by 5 as long as we do the same to the numerator.

$$\frac{3}{4} = \frac{3 \cdot 5}{4 \cdot 5} = \frac{15}{20}$$

The fraction $\frac{15}{20}$ is equivalent to the fraction $\frac{3}{4}$.

ABSOLUTE VALUES AND OPPOSITES

Representing numbers on the number line lets us give each number two important properties: a direction from zero and a distance from zero. The direction from zero is represented by the sign in front of the number. (A number without a sign is understood to be positive.) The distance from zero is called the absolute value of the number, as the following definition indicates.

> **DEFINITION** The **absolute value** of a real number is its distance from zero on the number line. If x represents a real number, then the absolute value of x is written $|x|$.

EXAMPLES Write each expression without absolute value symbols.

3. $|5| = 5$ The number 5 is 5 units from zero.

4. $|-5| = 5$ The number -5 is 5 units from zero.

5. $\left|-\frac{1}{2}\right| = \frac{1}{2}$ The number $-\frac{1}{2}$ is $\frac{1}{2}$ unit from zero.

The absolute value of a number is *never* negative. It is the distance the number is from zero without regard to which direction it is from zero. When working with the absolute value of sums and differences, we must simplify the expression inside the absolute value symbols first, and then find the absolute value of the simplified expression.

EXAMPLES Simplify each expression.

6. $|8 - 3| = |5| = 5$

7. $|3 \cdot 2^3 + 2 \cdot 3^2| = |3 \cdot 8 + 2 \cdot 9| = |24 + 18| = |42| = 42$

8. $|9 - 2| - |8 - 6| = |7| - |2| = 7 - 2 = 5$

Another important concept associated with numbers on the number line is that of opposites. Here is the definition.

> **DEFINITION**   Numbers the same distance from zero but in opposite directions from zero are called **opposites.**

EXAMPLES   Give the opposite of each number.

	Number	Opposite	
9.	5	−5	5 and −5 are opposites.
10.	−3	3	−3 and 3 are opposites.
11.	$\dfrac{1}{4}$	$-\dfrac{1}{4}$	$\dfrac{1}{4}$ and $-\dfrac{1}{4}$ are opposites.
12.	−2.3	2.3	−2.3 and 2.3 are opposites.

Each negative number is the opposite of some positive number, and each positive number is the opposite of some negative number. The opposite of a negative number is a positive number. In symbols, if a represents a positive number, then

$$-(-a) = a$$

Opposites always have the same absolute value. When you add any two opposites, the result is always zero:

$$a + (-a) = 0$$

RECIPROCALS AND MULTIPLICATION WITH FRACTIONS

The last concept we want to cover in this section is the concept of reciprocals. Understanding reciprocals requires some knowledge of multiplication with fractions. To multiply two fractions, we simply multiply numerators and multiply denominators.

EXAMPLE 13   Multiply $\frac{3}{4} \cdot \frac{5}{7}$.

Solution   The product of the numerators is 15, and the product of the denominators is 28:

$$\frac{3}{4} \cdot \frac{5}{7} = \frac{3 \cdot 5}{4 \cdot 7} = \frac{15}{28}$$

EXAMPLE 14   Multiply $7(\frac{1}{3})$.

Solution   The number 7 can be thought of as the fraction $\frac{7}{1}$:

$$7\left(\frac{1}{3}\right) = \frac{7}{1}\left(\frac{1}{3}\right) = \frac{7 \cdot 1}{1 \cdot 3} = \frac{7}{3}$$

Note: In past math classes you may have written fractions like $\frac{7}{3}$ (improper fractions) as mixed numbers, such as $2\frac{1}{3}$. In algebra it is usually better to write them as improper fractions rather than mixed numbers.

E X A M P L E   1 5   Expand and multiply $(\frac{2}{3})^3$.

Solution   Using the definition of exponents from the previous section, we have

$$\left(\frac{2}{3}\right)^3 = \frac{2}{3} \cdot \frac{2}{3} \cdot \frac{2}{3} = \frac{8}{27}$$

We are now ready for the definition of reciprocals.

> **DEFINITION**   Two numbers whose product is 1 are called **reciprocals**.

E X A M P L E S   Give the reciprocal of each number.

	Number	*Reciprocal*	
16.	5	$\frac{1}{5}$	Because $5(\frac{1}{5}) = \frac{5}{1}(\frac{1}{5}) = \frac{5}{5} = 1$
17.	2	$\frac{1}{2}$	Because $2(\frac{1}{2}) = \frac{2}{1}(\frac{1}{2}) = \frac{2}{2} = 1$
18.	$\frac{1}{3}$	3	Because $\frac{1}{3}(3) = \frac{1}{3}(\frac{3}{1}) = \frac{3}{3} = 1$
19.	$\frac{3}{4}$	$\frac{4}{3}$	Because $\frac{3}{4}(\frac{4}{3}) = \frac{12}{12} = 1$
20.	$\frac{2}{5}$	$\frac{5}{2}$	Because $\frac{2}{5}(\frac{5}{2}) = \frac{10}{10} = 1$

Although we will not develop multiplication with negative numbers until later in the chapter, you should know that the reciprocal of a negative number is also a negative number. For example, the reciprocal of -4 is $-\frac{1}{4}$.

E X A M P L E   2 1   Find the next number in each sequence.

(a) $1, \frac{1}{2}, \frac{1}{3}, \frac{1}{4}, \ldots$ **(b)** $1, 1, \frac{1}{2}, \frac{1}{3}, \frac{1}{5}, \ldots$

Solution   As we did with the sequences in Section 1.1, we look for a pattern that defines the sequence.

(a) In the first case, the sequence is simply the reciprocals of the counting numbers. The next number in the sequence will be the reciprocal of 5, which is $\frac{1}{5}$.

(b) This sequence is formed by taking the reciprocals of the numbers in the Fibonacci sequence: 1, 1, 2, 3, 5, . . .

The next number in the Fibonacci sequence will be 8. Therefore, the next number in the sequence in question is $\frac{1}{8}$.

FACTS FROM
Geometry

Formulas for Area and Perimeter

A square, rectangle, and triangle are shown in the following figures. Note that we have labeled the dimensions of each with variables. The formulas for the perimeter and area of each object are given in terms of its dimensions.

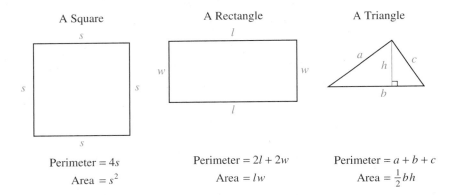

A Square	A Rectangle	A Triangle
Perimeter $= 4s$	Perimeter $= 2l + 2w$	Perimeter $= a + b + c$
Area $= s^2$	Area $= lw$	Area $= \frac{1}{2}bh$

The formula for perimeter gives us the distance around the outside of the object along its sides, while the formula for area gives us a measure of the amount of surface the object has.

Note: The vertical line labeled h in the triangle is its height, or altitude. It extends from the top of the triangle down to the base, meeting the base at an angle of 90°. The altitude of a triangle is always perpendicular to the base. The small square shown where the altitude meets the base is used to indicate that the angle formed is 90°.

EXAMPLE 22   Find the perimeter and area of each figure.

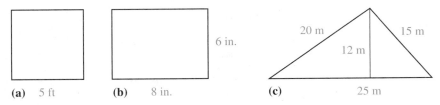

(a)   5 ft   **(b)**   8 in.   **(c)**   25 m

Solution   We use the preceding formulas to find the perimeter and the area. In each case, the units for perimeter are linear units, while the units for area are square units.

(a) Perimeter $= 4s = 4 \cdot 5$ feet $= 20$ feet
   Area $= s^2 = (5 \text{ feet})^2 = 25$ square feet

(b) Perimeter $= 2l + 2w = 2(8 \text{ inches}) + 2(6 \text{ inches}) = 28$ inches
   Area $= lw = (8 \text{ inches})(6 \text{ inches}) = 48$ square inches

(c) Perimeter $= a + b + c = (20 \text{ meters}) + (25 \text{ meters}) + (15 \text{ meters})$
$= 60$ meters

Area $= \dfrac{1}{2} bh = \dfrac{i}{2}(25 \text{ meters})(12 \text{ meters}) = 150$ square meters

Getting Ready for Class

After reading through the preceding section, respond in your own words and in complete sentences.

A. What is a real number?

B. Explain multiplication with fractions.

C. How do you find the opposite of a number?

D. Explain how you find the perimeter and the area of a rectangle.

PROBLEM SET 1.2

Draw a number line that extends from -5 to $+5$. Label the points with the following coordinates.

1. 5

2. -2

3. -4

4. -3

5. 1.5

6. -1.5

7. $\dfrac{9}{4}$

8. $\dfrac{8}{3}$

Write each of the following fractions as an equivalent fraction with denominator 24.

9. $\dfrac{3}{4}$

10. $\dfrac{5}{6}$

11. $\dfrac{1}{2}$

12. $\dfrac{1}{8}$

13. $\dfrac{5}{8}$

14. $\dfrac{7}{12}$

Write each fraction as an equivalent fraction with denominator 60.

15. $\dfrac{3}{5}$

16. $\dfrac{5}{12}$

17. $\dfrac{11}{30}$

18. $\dfrac{9}{10}$

For each of the following numbers, give the opposite, the reciprocal, and the absolute value. (Assume all variables are nonzero.)

19. 10

20. 8

21. $\dfrac{3}{4}$

22. $\dfrac{5}{7}$

23. $\dfrac{11}{2}$

24. $\dfrac{16}{3}$

25. -3

26. -5

27. $-\dfrac{2}{5}$

28. $-\dfrac{3}{8}$

29. x

30. a

Place one of the symbols $<$ or $>$ between each of the following to make the resulting statement true.

31. $-5 < -3$ **32.** $-8 < -1$ **33.** $-3 < -7$

34. $-6 < -5$ **35.** $|-4| < -|-4|$

36. $3 < -|-3|$ **37.** $7 < -|-7|$

38. $-7 < |-7|$ **39.** $-\dfrac{3}{4} < -\dfrac{1}{4}$ **40.** $-\dfrac{2}{3} < -\dfrac{1}{3}$

41. $-\dfrac{3}{2} < -\dfrac{3}{4}$ **42.** $-\dfrac{8}{3} < -\dfrac{17}{3}$

Simplify each expression.

43. $|8 - 2|$ **44.** $|6 - 1|$

45. $|5 \cdot 2^3 - 2 \cdot 3^2|$ **46.** $|2 \cdot 10^2 + 3 \cdot 10|$

47. $|7 - 2| - |4 - 2|$ **48.** $|10 - 3| - |4 - 1|$

49. $10 - |7 - 2(5 - 3)|$ **50.** $12 - |9 - 3(7 - 5)|$

51. $15 - |8 - 2(3 \cdot 4 - 9)| - 10$

52. $25 - |9 - 3(4 \cdot 5 - 18)| - 20$

Multiply the following.

53. $\dfrac{2}{3} \cdot \dfrac{4}{5}$ **54.** $\dfrac{1}{4} \cdot \dfrac{3}{5}$ **55.** $\dfrac{1}{2}\,(3)$

56. $\dfrac{1}{3}\,(2)$ **57.** $\dfrac{1}{4}\,(5)$ **58.** $\dfrac{1}{5}\,(4)$

59. $\dfrac{4}{3} \cdot \dfrac{3}{4}$ **60.** $\dfrac{5}{7} \cdot \dfrac{7}{5}$ **61.** $6\left(\dfrac{1}{6}\right)$

62. $8\left(\dfrac{1}{8}\right)$ **63.** $3 \cdot \dfrac{1}{3}$ **64.** $4 \cdot \dfrac{1}{4}$

Expand and multiply.

65. $\left(\dfrac{3}{4}\right)^2$ **66.** $\left(\dfrac{5}{6}\right)^2$ **67.** $\left(\dfrac{2}{3}\right)^3$

68. $\left(\dfrac{1}{2}\right)^3$ **69.** $\left(\dfrac{1}{10}\right)^4$ **70.** $\left(\dfrac{1}{10}\right)^5$

Find the next number in each sequence.

71. $1, \dfrac{1}{3}, \dfrac{1}{5}, \dfrac{1}{7}, \ldots$ (Reciprocals of odd numbers)

72. $\dfrac{1}{2}, \dfrac{1}{4}, \dfrac{1}{6}, \dfrac{1}{8}, \ldots$ (Reciprocals of even numbers)

73. $1, \dfrac{1}{4}, \dfrac{1}{9}, \dfrac{1}{16}, \ldots$ (Reciprocals of squares)

74. $1, \dfrac{1}{8}, \dfrac{1}{27}, \dfrac{1}{64}, \ldots$ (Reciprocals of cubes)

Find the perimeter and area of each figure.

75.

1 in.

1 in.

76.

15 mm

15 mm

77.

0.75 in.

1.5 in.

78.

1.5 cm

4.5 cm

79.

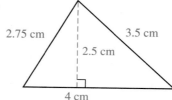

2.75 cm   3.5 cm

2.5 cm

4 cm

80.

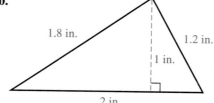

1.8 in.   1.2 in.

1 in.

2 in.

Applying the Concepts

81. Football Yardage A football team gains 6 yards on one play and then loses 8 yards on the next play. To what number on the number line does a loss of 8 yards correspond? The total yards gained or lost on the two plays corresponds to what negative number?

82. Checking Account Balance A woman has a balance of $20 in her checking account. If she writes a check for $30, what negative number can be used to represent the new balance in her checking account?

Temperature In the United States, temperature is measured on the Fahrenheit temperature scale. On this scale, water boils at 212 degrees and freezes at 32 degrees. To denote a temperature of 32 degrees on the Fahrenheit scale, we write

32°F, which is read "32 degrees Fahrenheit"

Use this information for Problems 83 and 84.

83. Temperature and Altitude Marilyn is flying from Seattle to San Francisco on a Boeing 737 jet. When the plane reaches an altitude of 35,000 feet, the temperature outside the plane is 64 degrees below zero Fahrenheit. Represent the temperature with a negative number. If the temperature outside the plane gets warmer by 10 degrees, what will the new temperature be?

84. Temperature Change At 10:00 in the morning in White Bear Lake, Minnesota, John notices the temperature outside is 10 degrees below zero Fahrenheit. Write the temperature as a negative number. An hour later it has warmed up by 6 degrees. What is the temperature at 11:00 that morning?

Wind Chill Table 2 is a table of wind chill temperatures. The top row gives the air temperature, while the first column is wind speed in miles per hour. The numbers within the table indicate how cold the weather will feel. For example, if the thermometer reads 30°F, and the wind is blowing at 15 miles per hour, the wind chill temperature is 9°F. Use Table 2 to answer Questions 85 and 86.

85. Reading Tables Find the wind chill temperature if the thermometer reads 20°F and the wind is blowing at 25 miles per hour.

86. Reading Tables Which will feel colder: a day with an air temperature of 10°F with a 25-mile-per-hour wind, or a day with an air temperature of −5°F and a 10-mile-per-hour wind?

87. Reading Tables and Charts The following bar chart gives the record low temperatures for each month of the year for Lake Placid, New York. The accompanying table on page 24 should display the same information. Use the bar chart to complete the table.

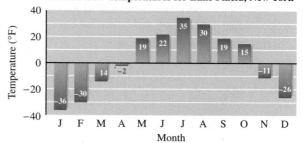

Record Low Temperatures for Lake Placid, New York

TABLE 2 Wind Chill Temperatures								
	Air Temperature (°F)							
Wind Speed	*30°*	*25°*	*20°*	*15°*	*10°*	*5°*	*0°*	*−5°*
10 mph	16°	10°	3°	−3°	−9°	−15°	−22°	−27°
15 mph	9°	2°	−5°	−11°	−18°	−25°	−31°	−38°
20 mph	4°	−3°	−10°	−17°	−24°	−31°	−39°	−46°
25 mph	1°	−7°	−15°	−22°	−29°	−36°	−44°	−51°
30 mph	−2°	−10°	−18°	−25°	−33°	−41°	−49°	−56°

Record Low Temperatures in Lake Placid, New York	
Month	**Temperature (°F)**
January	−36
February	
March	
April	
May	
June	
July	
August	
September	
October	
November	
December	

Daily Low Temperatures in Salt Lake City	
Day	**Temperature (°F)**
Monday	
Tuesday	
Wednesday	
Thursday	
Friday	
Saturday	
Sunday	

88. Reading Tables and Charts The following bar chart gives the low temperature for each day of one week in Salt Lake City, Utah. The accompanying table (top of next column) should display the same information. Use the bar chart to complete the table.

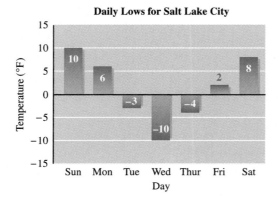

Daily Lows for Salt Lake City

89. Scuba Diving Steve is scuba diving near his home in Maui. At one point he is 100 feet below the surface. Represent this number with a negative number. If he descends another 5 feet, what negative number will represent his new position?

90. Checking Account Balance Pete and Heidi are married and have a joint checking account that allows them to withdraw money from the account at an automatic teller machine. With each withdrawal from the automatic teller, the bank charges an extra $2 to their account. Suppose the balance in the checkbook is $75. If Heidi writes a check for $55 and Pete withdraws $20 from the automatic teller machine, what will their new balance be?

91. Geometry Find the area and perimeter of an $8\frac{1}{2}$-by-11-inch piece of notebook paper.

92. Geometry Find the area and perimeter of an $8\frac{1}{2}$-by-$5\frac{1}{2}$-inch piece of paper.

Calories and Exercise Table 3 gives the amount of energy expended per hour for various activities, for a person weighing 120, 150, or 180 pounds. Use Table 3 to answer Questions 93–96.

TABLE 3 Energy Expended from Exercising			
	Calories per Hour		
Activity	**120 lb**	**150 lb**	**180 lb**
Bicycling	299	374	449
Bowling	212	265	318
Handball	544	680	816
Horseback trotting	278	347	416
Jazzercise	272	340	408
Jogging	544	680	816
Skiing (downhill)	435	544	653

93. Reading Tables Suppose you weigh 120 pounds. How many calories will you burn if you play handball for 2 hours and then ride your bicycle for an hour?

94. Reading Tables How many calories are burned by a person weighing 150 pounds who jogs for $\frac{1}{2}$ hour and then goes bicycling for 2 hours?

95. Reading Tables Two people go skiing. One weighs 180 pounds, and the other weighs 120 pounds. If they ski for 3 hours, how many more calories are burned by the person weighing 180 pounds?

96. Reading Tables Two people spend 3 hours bowling. If one weighs 120 pounds and the other weighs 150 pounds, how many more calories are burned during the evening by the person weighing 150 pounds?

1.3   Addition of Real Numbers

Suppose that you are playing a friendly game of poker with some friends, and you lose $3 on the first hand and $4 on the second hand. If you represent winning with positive numbers and losing with negative numbers, how can you translate this situation into symbols? Since you lost $3 and $4 for a total of $7, one way to represent this situation is with addition of negative numbers:

$$(-\$3) + (-\$4) = -\$7$$

From this equation, we see that the sum of two negative numbers is a negative number. To generalize addition with positive and negative numbers, we use the number line.

Since real numbers have both a distance from zero (absolute value) and a direction from zero (sign), we can think of addition of two numbers in terms of distance and direction from zero.

Let's look at a problem for which we know the answer. Suppose we want to add the numbers 3 and 4. The problem is written 3 + 4. To put it on the number line, we read the problem as follows:

1. The 3 tells us to "start at the origin and move 3 units in the positive direction."

2. The + sign is read "and then move."

3. The 4 means "4 units in the positive direction."

To summarize, 3 + 4 means to start at the origin, move 3 units in the *positive* direction and then 4 units in the *positive* direction.

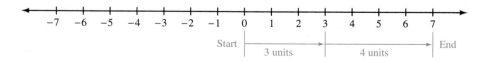

We end up at 7, which is the answer to our problem: 3 + 4 = 7.

Let's try other combinations of positive and negative 3 and 4 on the number line.

EXAMPLE 1   Add $3 + (-4)$.

Solution   Starting at the origin, move 3 units in the *positive* direction and then 4 units in the *negative* direction.

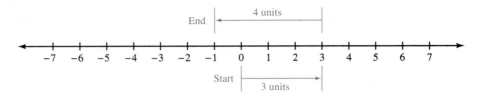

We end up at -1; therefore, $3 + (-4) = -1$.

EXAMPLE 2   Add $-3 + 4$.

Solution   Starting at the origin, move 3 units in the *negative* direction and then 4 units in the *positive* direction.

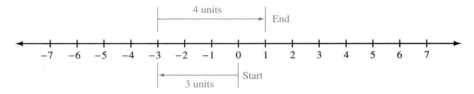

We end up at $+1$; therefore, $-3 + 4 = 1$.

EXAMPLE 3   Add $-3 + (-4)$.

Solution   Starting at the origin, move 3 units in the *negative* direction and then 4 units in the *negative* direction.

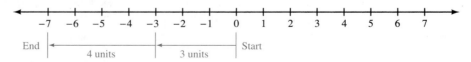

We end up at -7; therefore, $-3 + (-4) = -7$. Here is a summary of what we have just completed:

$$3 + 4 = 7$$

$$3 + (-4) = -1$$

$$-3 + 4 = 1$$

$$-3 + (-4) = -7$$

Let's do four more problems on the number line and then summarize our results in a rule we can use to add any two real numbers.

E X A M P L E   4   Add $5 + 7 = 12$.

Solution

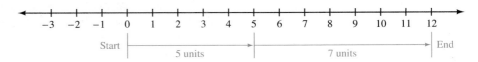

E X A M P L E   5   Add $5 + (-7) = -2$.

Solution

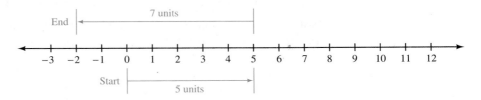

E X A M P L E   6   Add $-5 + 7 = 2$.

Solution

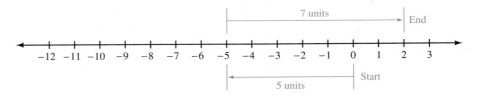

E X A M P L E   7   Add $-5 + (-7) = -12$.

Solution

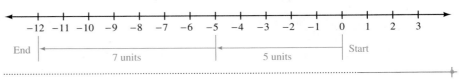

If we look closely at the results of the preceding addition problems, we can see that they support (or justify) the following rule.

> **RULE**
>
> To add two real numbers with
>
> 1. The *same* sign: Simply add their absolute values and use the common sign. (If both numbers are positive, the answer is positive. If both numbers are negative, the answer is negative.)
> 2. *Different* signs: Subtract the smaller absolute value from the larger. The answer will have the sign of the number with the larger absolute value.

This rule covers all possible combinations of addition with real numbers. You must memorize it. After you have worked a number of problems, it will seem almost automatic.

EXAMPLE 8   Add all combinations of positive and negative 10 and 13.

Solution   Rather than work these problems on the number line, we use the rule for adding positive and negative numbers to obtain our answers:

$$10 + 13 = 23$$
$$10 + (-13) = -3$$
$$-10 + 13 = 3$$
$$-10 + (-13) = -23$$

EXAMPLE 9   Add all possible combinations of positive and negative 12 and 17.

Solution   Applying the rule for adding positive and negative numbers, we have

$$12 + 17 = 29$$
$$12 + (-17) = -5$$
$$-12 + 17 = 5$$
$$-12 + (-17) = -29$$

EXAMPLE 10   Add $-3 + 2 + (-4)$.

Solution   Applying the rule for order of operations, we add left to right:

$$-3 + 2 + (-4) = -1 + (-4)$$
$$= -5$$

EXAMPLE 11   Add $-8 + [2 + (-5)] + (-1)$.

Solution Adding inside the brackets first, and then left to right, we have

$$-8 + [2 + (-5)] + (-1) = -8 + (-3) + (-1)$$
$$= -11 + (-1)$$
$$= -12$$

EXAMPLE 12 Simplify $-10 + 2(-8 + 11) + (-4)$.

Solution First, we simplify inside the parentheses. Then, we multiply. Finally, we add left to right:

$$-10 + 2(-8 + 11) + (-4) = -10 + 2(3) + (-4)$$
$$= -10 + 6 + (-4)$$
$$= -4 + (-4)$$
$$= -8$$

ARITHMETIC SEQUENCES

The pattern in a sequence of numbers is easy to identify when each number in the sequence comes from the preceding number by adding the same amount each time. This leads us to our next level of classification, in which we classify together groups of sequences with a common characteristic.

> **DEFINITION** An **arithmetic sequence** is a sequence of numbers in which each number (after the first number) comes from adding the same amount to the number before it.

Here is an example of an arithmetic sequence:

$$2, 5, 8, 11, \ldots$$

Each number is obtained by adding 3 to the number before it.

EXAMPLE 13 Each of the following is an arithmetic sequence. Find the next two numbers in each sequence.

(a) $7, 10, 13, \ldots$
(b) $9.5, 10, 10.5, \ldots$
(c) $5, 0, -5, \ldots$

Solution Since we know that each sequence is arithmetic, we know to look for the number that is added to each term to produce the next consecutive term.

(a) $7, 10, 13, \ldots$: Each term is found by adding 3 to the term before it. Therefore, the next two terms will be 16 and 19.
(b) $9.5, 10, 10.5, \ldots$: Each term comes from adding 0.5 to the term before it. Therefore, the next two terms will be 11 and 11.5.

(c) $5, 0, -5, \ldots$: Each term comes from adding -5 to the term before it. Therefore, the next two terms will be $-5 + (-5) = -10$ and $-10 + (-5) = -15$.

Getting Ready for Class

After reading through the preceding section, respond in your own words and in complete sentences.

A. Explain how you would add 3 and -5 on the number line.

B. How do you add two negative numbers?

C. What is an arithmetic sequence?

D. Why is the sum of a number and its opposite always 0?

PROBLEM SET 1.3

1. Add all positive and negative combinations of 3 and 5. (Look back to Examples 8 and 9.)

2. Add all positive and negative combinations of 6 and 4.

3. Add all positive and negative combinations of 15 and 20.

4. Add all positive and negative combinations of 18 and 12.

Work the following problems. You may want to begin by doing a few on the number line.

5. $6 + (-3)$	**6.** $7 + (-8)$
7. $13 + (-20)$	**8.** $15 + (-25)$
9. $18 + (-32)$	**10.** $6 + (-9)$
11. $-6 + 3$	**12.** $-8 + 7$
13. $-30 + 5$	**14.** $-18 + 6$
15. $-6 + (-6)$	**16.** $-5 + (-5)$
17. $-9 + (-10)$	**18.** $-8 + (-6)$
19. $-10 + (-15)$	**20.** $-18 + (-30)$

Work the following problems using the rule for addition of real numbers. You may want to refer back to the rule for order of operations.

21. $5 + (-6) + (-7)$	**22.** $6 + (-8) + (-10)$
23. $-7 + 8 + (-5)$	**24.** $-6 + 9 + (-3)$

25. $5 + [6 + (-2)] + (-3)$

26. $10 + [8 + (-5)] + (-20)$

27. $[6 + (-2)] + [3 + (-1)]$

28. $[18 + (-5)] + [9 + (-10)]$

29. $20 + (-6) + [3 + (-9)]$

30. $18 + (-2) + [9 + (-13)]$

31. $-3 + (-2) + [5 + (-4)]$

32. $-6 + (-5) + [-4 + (-1)]$

33. $(-9 + 2) + [5 + (-8)] + (-4)$

34. $(-7 + 3) + [9 + (-6)] + (-5)$

35. $[-6 + (-4)] + [7 + (-5)] + (-9)$

36. $[-8 + (-1)] + [8 + (-6)] + (-6)$

37. $(-6 + 9) + (-5) + (-4 + 3) + 7$

38. $(-10 + 4) + (-3) + (-3 + 8) + 6$

The problems that follow involve some multiplication. Be sure that you work inside the parentheses first, then multiply, and, finally, add left to right.

39. $-5 + 2(-3 + 7)$	**40.** $-3 + 4(-2 + 7)$
41. $9 + 3(-8 + 10)$	**42.** $4 + 5(-2 + 6)$

43. $-10 + 2(-6 + 8) + (-2)$

44. $-20 + 3(-7 + 10) + (-4)$

45. $2(-4 + 7) + 3(-6 + 8)$

46. $5(-2 + 5) + 7(-1 + 6)$

Each sequence below is an arithmetic sequence. In each case, find the next two numbers in the sequence.

47. 3, 8, 13, 18, . . .   **48.** 1, 5, 9, 13, . . .

49. 10, 15, 20, 25, . . .   **50.** 10, 16, 22, 28, . . .

51. 20, 15, 10, 5, . . .   **52.** 24, 20, 16, 12, . . .

53. 6, 0, −6, . . .   **54.** 1, 0, −1, . . .

55. 8, 4, 0, . . .   **56.** 5, 2, −1, . . .

57. Is the sequence of odd numbers an arithmetic sequence?

58. Is the sequence of squares an arithmetic sequence?

Recall that the word *sum* indicates addition. Write the numerical expression that is equivalent to each of the following phrases and then simplify.

59. The sum of 5 and 9   **60.** The sum of 6 and −3

61. Four added to the sum of −7 and −5

62. Six added to the sum of −9 and 1

63. The sum of −2 and −3 increased by 10

64. The sum of −4 and −12 increased by 2

Answer the following questions.

65. What number do you add to −8 to get −5?

66. What number do you add to 10 to get 4?

67. The sum of what number and −6 is −9?

68. The sum of what number and −12 is 8?

Applying the Concepts

69. Temperature Change   The temperature at noon is 12 degrees below zero Fahrenheit. By 1:00 it has risen 4 degrees. Write an expression using the numbers −12 and 4 to describe this situation.

70. Stock Value   On Monday a certain stock gains 2 points. On Tuesday it loses 3 points. Write an expression using positive and negative numbers with addition to describe this situation and then simplify.

71. Gambling   On three consecutive hands of draw poker, a gambler wins $10, loses $6, and then loses another $8. Write an expression using positive and negative numbers and addition to describe this situation and then simplify.

72. Number Problem   You know from your past experience with numbers that subtracting 5 from 8 results in 3: 8 − 5 = 3. What addition problem that starts with the number 8 gives the same result?

73. Checkbook Balance   Suppose that you balance your checkbook and find that you are overdrawn by $30. That is, your balance is −$30. Then you go to the bank and deposit $40. Translate this situation into an addition problem, the answer to which gives the new balance in your checkbook.

74. Checkbook Balance   The balance in your checkbook is −$25. If you make a deposit of $75 and then write a check for $18, what is the new balance?

75. Salary Increase   Colleen has a job for which her starting wage is $6.50 per hour. If she gets a raise of $0.25 every 6 months, write a sequence of numbers that gives her wage every 6 months for the first 3 years. Is the sequence an arithmetic sequence?

76. Coin Collection   Benja has a coin collection that she purchased for $650. Each year the collection increases $50 in value. Write a sequence of numbers that gives the value of the coin collection every year for the first 5 years. Is this sequence an arithmetic sequence?

77. Starting Salaries   The table and bar chart on p. 32 show the salaries for new teachers compared with starting salaries in other professions in 1998.
  (a) Use the information in the bar chart to fill in the missing entries in the table.
  (b) What is the average family income of a married couple graduating in 1998 if one majored in chemistry and the other in engineering?

Professions	Salaries
Engineering	
Computer Science	$40,920
	$40,523
	$36,036
Business Admin.	
Accounting	$33,702
Sales/Marketing	
Teaching	$25,735

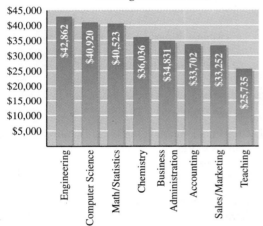

Starting Salaries in 1998

(b) Find the total digital camera sales from 1997 through 1999.

Year	Sales ($ millions)
1996	
1997	
1998	
	819

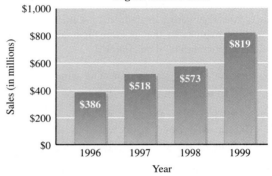

Digital Camera Sales

78. Digital Camera Sales The following bar chart shows the sales of digital cameras from 1996–1999.
(a) Use the information in the bar chart to fill in the missing entries in the table.

1600 × 1200 Resolution Digital Camera
Your Price: $599 ^{99}

1.4   Subtraction of Real Numbers

Suppose that the temperature at noon is 20° Fahrenheit and 12 hours later, at midnight, it has dropped to −15° Fahrenheit. What is the difference between the temperature at noon and the temperature at midnight? Intuitively, we know the difference in the two temperatures is 35°. We also know that the word *difference* indicates subtraction. The difference between 20 and −15 is written

$$20 - (-15)$$

It must be true that $20 - (-15) = 35$. In this section we will see how our definition for subtraction confirms that this last statement is in fact correct.

In the previous section we spent some time developing the rule for addition of real numbers. Since we want to make as few rules as possible, we can define subtraction in terms of addition. By doing so, we can then use the rule for addition to solve our subtraction problems.

RULE

To subtract one real number from another, simply add its opposite.

Algebraically, the rule is written like this: If a and b represent two real numbers, then it is always true that

$$\underbrace{a - b}_{\text{To subtract } b} = \underbrace{a + (-b)}_{\text{add the opposite of } b}$$

This is how subtraction is defined in algebra. This definition of subtraction will not conflict with what you already know about subtraction, but it will allow you to do subtraction using negative numbers.

E X A M P L E 1   Subtract all possible combinations of positive and negative 7 and 2.

Solution

$$\left. \begin{array}{l} 7 - 2 = \quad 7 + (-2) = 5 \\ -7 - 2 = -7 + (-2) = -9 \end{array} \right\} \quad \begin{array}{l} \text{Subtracting } +2 \text{ is the same} \\ \text{as adding } -2. \end{array}$$

$$\left. \begin{array}{l} 7 - (-2) = \quad 7 + 2 = 9 \\ -7 - (-2) = -7 + 2 = -5 \end{array} \right\} \quad \begin{array}{l} \text{Subtracting } -2 \text{ is the same} \\ \text{as adding } +2. \end{array}$$

Notice that each subtraction problem is first changed to an addition problem. The rule for addition is then used to arrive at the answer.

We have defined subtraction in terms of addition, and we still obtain answers consistent with the answers we are used to getting with subtraction. Moreover, we can now do subtraction problems involving both positive and negative numbers.

As you proceed through the following examples and the problem set, you will begin to notice shortcuts you can use in working the problems. You will not always have to change subtraction to addition of the opposite to be able to get answers quickly. Use all the shortcuts you wish as long as you consistently get the correct answers.

E X A M P L E 2   Subtract all combinations of positive and negative 8 and 13.

Solution

$$\left. \begin{array}{l} 8 - 13 = \quad 8 + (-13) = -5 \\ -8 - 13 = -8 + (-13) = -21 \end{array} \right\} \quad \begin{array}{l} \text{Subtracting } +13 \text{ is the} \\ \text{same as adding } -13. \end{array}$$

$$8 - (-13) = 8 + 13 = 21$$
$$-8 - (-13) = -8 + 13 = 5$$

Subtracting -13 is the same as adding $+13$.

EXAMPLES   Simplify each expression as much as possible.

3. $7 + (-3) - 5 = 7 + (-3) + (-5)$
$$= 4 + (-5)$$
$$= -1$$

Begin by changing all subtractions to additions.
Then add left to right.

4. $8 - (-2) - 6 = 8 + 2 + (-6)$
$$= 10 + (-6)$$
$$= 4$$

Begin by changing all subtractions to additions.
Then add left to right.

5. $-2 - (-3 + 1) - 5 = -2 - (-2) - 5$
$$= -2 + 2 + (-5)$$
$$= -5$$

Do what is in the parentheses first.

The next two examples involve multiplication and exponents as well as subtraction. Remember, according to the rule for order of operations, we evaluate the numbers containing exponents and multiply before we subtract.

EXAMPLE 6   Simplify $2 \cdot 5 - 3 \cdot 8 - 4 \cdot 9$.

Solution   First, we multiply left to right, and then we subtract.

$$2 \cdot 5 - 3 \cdot 8 - 4 \cdot 9 = 10 - 24 - 36$$
$$= -14 - 36$$
$$= -50$$

EXAMPLE 7   Simplify $3 \cdot 2^3 - 2 \cdot 4^2$.

Solution   We begin by evaluating each number that contains an exponent. Then we multiply before we subtract:

$$3 \cdot 2^3 - 2 \cdot 4^2 = 3 \cdot 8 - 2 \cdot 16$$
$$= 24 - 32$$
$$= -8$$

EXAMPLE 8   Subtract 7 from -3.

Solution   First, we write the problem in terms of subtraction. Then we change to addition of the opposite:

$$-3 - 7 = -3 + (-7)$$
$$= -10$$

E X A M P L E   9   Subtract -5 from 2.

Solution   Subtracting -5 is the same as adding $+5$:

$$2 - (-5) = 2 + 5$$
$$= 7$$

E X A M P L E   1 0   Find the difference of 9 and 2.

Solution   Written in symbols, the problem looks like this:

$$9 - 2 = 7$$

The difference of 9 and 2 is 7.

E X A M P L E   1 1   Find the difference of 3 and -5.

Solution   Subtracting -5 from 3 we have

$$3 - (-5) = 3 + 5$$
$$= 8$$

In the sport of drag racing, two cars at the starting line race to the finish line $\frac{1}{4}$ mile away. The car that crosses the finish line first wins the race.

Jim Rizzoli owns and races an alcohol dragster. On board the dragster is a computer that records data during each of Jim's races. Table 1 gives some of the data from a race Jim was in during the 1993 Winternationals. Figure 1 gives the same information visually.

TABLE 1	Speed of a Race Car
Time in Seconds	Speed in Miles/Hour
0	0
1	72.7
2	129.9
3	162.8
4	192.2
5	212.4
6	228.1

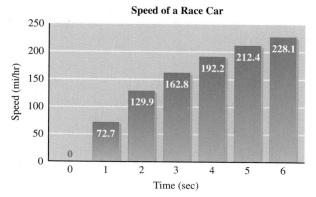

FIGURE 1

E X A M P L E   1 2   Use Table 1 to find the difference in speed after 5 seconds and after 2 seconds have elapsed during the race.

Solution We know the word *difference* implies subtraction. The speed at 2 seconds is 129.9 miles per hour, while the speed at 5 seconds is 212.4 miles per hour. Therefore, the expression that represents the solution to our problem looks like this:

$$212.4 - 129.9 = 82.5 \text{ miles per hour}$$

FACTS FROM

Geometry

Complementary and Supplementary Angles

If you have studied geometry at all, you know that there are 360° in a full rotation: the number of degrees swept out by the radius of a circle as it rotates once around the circle.

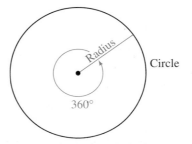

We can apply our knowledge of algebra to help solve some simple geometry problems. Before we do, however, we need to review some of the vocabulary associated with angles.

> **DEFINITION** In geometry, two angles that add to 90° are called **complementary angles.** In a similar manner, two angles that add to 180° are called **supplementary angles.** The following diagrams illustrate the relationships between angles that are complementary and between angles that are supplementary.

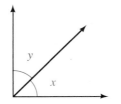

Complementary angles: $x + y = 90°$

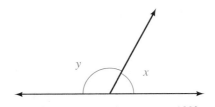

Supplementary angles: $x + y = 180°$

E X A M P L E   1 3   Find x in each of the following diagrams.

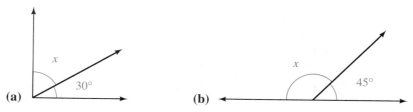

(a)   **(b)**

Solution   We use subtraction to find each angle.

(a) Since the two angles are complementary, we can find x by subtracting $30°$ from $90°$:

$$x = 90° - 30° = 60°$$

We say $30°$ and $60°$ are complementary angles. The complement of $30°$ is $60°$.

(b) The two angles in the diagram are supplementary. To find x, we subtract $45°$ from $180°$:

$$x = 180° - 45° = 135°$$

We say $45°$ and $135°$ are supplementary angles. The supplement of $45°$ is $135°$.

SUBTRACTION AND TAKING AWAY

For some people taking algebra for the first time, subtraction of positive and negative numbers can be a problem. These people may believe that $-5 - 9$ should be -4 or 4, not -14. If this is happening to you, you are probably thinking of subtraction in terms of taking one number away from another. Thinking of subtraction in this way works well with positive numbers if you always subtract the smaller number from the larger. In algebra, however, we encounter many situations other than this. The definition of subtraction, that $a - b = a + (-b)$, clearly indicates the correct way to use subtraction. That is, when working subtraction problems, you should think "addition of the opposite," not "take one number away from another." To be successful in algebra, you need to apply properties and definitions exactly as they are presented here.

 # Getting Ready for Class

After reading through the preceding section, respond in your own words and in complete sentences.

A. Why do we define subtraction in terms of addition?

B. Write the definition for $a - b$.

C. Explain in words how you would subtract 3 from -7.

D. What are complementary angles?

PROBLEM SET 1.4

The following problems are intended to give you practice with subtraction of positive and negative numbers. Remember, in algebra subtraction is not taking one number away from another. Instead, subtracting a number is equivalent to adding its opposite.

Subtract.

1. $5 - 8$ **2.** $6 - 7$ **3.** $3 - 9$
4. $2 - 7$ **5.** $5 - 5$ **6.** $8 - 8$
7. $-8 - 2$ **8.** $-6 - 3$ **9.** $-4 - 12$
10. $-3 - 15$ **11.** $-6 - 6$ **12.** $-3 - 3$
13. $-8 - (-1)$ **14.** $-6 - (-2)$ **15.** $15 - (-20)$
16. $20 - (-5)$ **17.** $-4 - (-4)$ **18.** $-5 - (-5)$

Simplify each expression by applying the rule for order of operations.

19. $3 - 2 - 5$ **20.** $4 - 8 - 6$
21. $9 - 2 - 3$ **22.** $8 - 7 - 12$
23. $-6 - 8 - 10$ **24.** $-5 - 7 - 9$
25. $-22 + 4 - 10$ **26.** $-13 + 6 - 5$
27. $10 - (-20) - 5$ **28.** $15 - (-3) - 20$
29. $8 - (2 - 3) - 5$ **30.** $10 - (4 - 6) - 8$
31. $7 - (3 - 9) - 6$ **32.** $4 - (3 - 7) - 8$
33. $5 - (-8 - 6) - 2$ **34.** $4 - (-3 - 2) - 1$
35. $-(5 - 7) - (2 - 8)$ **36.** $-(4 - 8) - (2 - 5)$
37. $-(3 - 10) - (6 - 3)$ **38.** $-(3 - 7) - (1 - 2)$
39. $16 - [(4 - 5) - 1]$ **40.** $15 - [(4 - 2) - 3]$
41. $5 - [(2 - 3) - 4]$ **42.** $6 - [(4 - 1) - 9]$
43. $21 - [-(3 - 4) - 2] - 5$
44. $30 - [-(10 - 5) - 15] - 25$

The following problems involve multiplication and exponents. Use the rule for order of operations to simplify each expression as much as possible.

45. $2 \cdot 8 - 3 \cdot 5$ **46.** $3 \cdot 4 - 6 \cdot 7$
47. $3 \cdot 5 - 2 \cdot 7$ **48.** $6 \cdot 10 - 5 \cdot 20$
49. $5 \cdot 9 - 2 \cdot 3 - 6 \cdot 2$ **50.** $4 \cdot 3 - 7 \cdot 1 - 9 \cdot 4$
51. $3 \cdot 8 - 2 \cdot 4 - 6 \cdot 7$ **52.** $5 \cdot 9 - 3 \cdot 8 - 4 \cdot 5$
53. $2 \cdot 3^2 - 5 \cdot 2^2$ **54.** $3 \cdot 7^2 - 2 \cdot 8^2$
55. $4 \cdot 3^3 - 5 \cdot 2^3$ **56.** $3 \cdot 6^2 - 2 \cdot 3^2 - 8 \cdot 6^2$

Rewrite each of the following phrases as an equivalent expression in symbols and then simplify.

57. Subtract 4 from -7. **58.** Subtract 5 from -19.
59. Subtract -8 from 12. **60.** Subtract -2 from 10.
61. Subtract -7 from -5. **62.** Subtract -9 from -3.
63. Subtract 17 from the sum of 4 and -5.
64. Subtract -6 from the sum of 6 and -3.

Recall that the word *difference* indicates subtraction. The difference of a and b is $a - b$, in that order. Write a numerical expression that is equivalent to each of the following phrases and then simplify.

65. The difference of 8 and 5.
66. The difference of 5 and 8.
67. The difference of -8 and 5.
68. The difference of -5 and 8.
69. The difference of 8 and -5.
70. The difference of 5 and -8.

Answer the following questions.

71. What number do you subtract from 8 to get -2?
72. What number do you subtract from 1 to get -5?
73. What number do you subtract from 8 to get 10?
74. What number do you subtract from 1 to get 5?

Applying the Concepts

75. Savings Account Balance A man with $1,500 in a savings account makes a withdrawal of $730. Write an expression using subtraction that describes this situation.

First Bank Account No. 12345			
Date	Withdrawals	Deposits	Balance
1/1/99			1,500
2/2/99	730		

76. Temperature Change The temperature inside a space shuttle is 73°F before reentry. During reentry, the temperature inside the craft increases 10°. Upon landing it drops 8°F. Write an expression using the numbers 73, 10, and 8 to describe this situation. What is the temperature inside the shuttle upon landing?

77. **Gambling** A man who has lost $35 playing roulette in Las Vegas wins $15 playing blackjack. He then loses $20 playing the wheel of fortune. Write an expression using the numbers −35, 15, and 20 to describe this situation and then simplify it.

78. **Altitude Change** An airplane flying at 10,000 feet lowers its altitude by 1,500 feet to avoid other air traffic. Then it increases its altitude by 3,000 feet to clear a mountain range. Write an expression that describes this situation and then simplify it.

79. **Checkbook Balance** Bob has $98 in his checking account when he writes a check for $65 and then another check for $53. Write a subtraction problem that gives the new balance in Bob's checkbook. What is his new balance?

80. **Temperature Change** The temperature at noon is 23°F. Six hours later it has dropped 19°F, and by midnight it has dropped another 10°F. Write a subtraction problem that gives the temperature at midnight. What is the temperature at midnight?

81. **Depreciation** Stacey buys a used car for $4,500. With each year that passes, the car drops $550 in value. Write a sequence of numbers that gives the value of the car at the beginning of each of the first 5 years she owns it. Can this sequence be considered an arithmetic sequence?

82. **Depreciation** Wade buys a computer system for $6,575. Each year after that he finds that the system is worth $1,250 less than it was the year before. Write a sequence of numbers that gives the value of the computer system at the beginning of each of the first 4 years he owns it. Can this sequence be considered an arithmetic sequence?

Drag Racing Table 2 (top of next column) extends the information given in Table 1 of this section. In addition to showing the time and speed of Jim Rizzoli's dragster during a race, it also shows the distance past the starting line that his dragster has traveled. Use the information in Table 2 to answer Problems 83–88.

83. Find the difference in the distance traveled by the dragster after 5 seconds and after 2 seconds.

84. How much faster is he traveling after 4 seconds than he is after 2 seconds?

85. How far from the starting line is he after 3 seconds?

TABLE 2 Speed and Distance for a Race Car

Time in Seconds	Speed in Miles/Hour	Distance Traveled in Feet
0	0	0
1	72.7	69
2	129.9	231
3	162.8	439
4	192.2	728
5	212.4	1,000
6	228.1	1,373

86. How far from the starting line is he when his speed is 192.2 miles per hour?

87. How many seconds have gone by between the time his speed is 162.8 miles per hour and the time at which he has traveled 1,000 feet?

88. How many seconds have gone by between the time at which he has traveled 231 feet and the time at which his speed is 228.1 miles per hour?

Look carefully at the relationship between the numbers in Table 1 on page 35 and the diagram in Figure 1 that appears next to it, then work the following problems.

89. **Garbage Production** The bar chart below shows the annual production of garbage in the United States for some specific years.

Year	Garbage (millions of tons)
1960	
1970	
1980	
	205
1997	

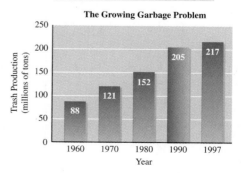

The Growing Garbage Problem

Trash Production (millions of tons)

1960: 88
1970: 121
1980: 152
1990: 205
1997: 217

Year

© *Tony Stone Images /Lonnie Duka*

Overall Plant Height

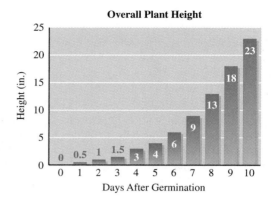

(a) Use the information in the bar chart on the previous page to fill in the missing entries in the table.

(b) How much more garbage was there in 1990 than in 1970?

90. Grass Growth The bar chart at the top of the next column shows the growth of a certain species of grass over a period of 10 days.

(a) Use the chart to fill in the missing entries in the table below.

(b) How much higher is the grass after 8 days than after 3 days?

© *Mark E. Gibson / Visuals Unlimited*

91. Wireless Phone Costs The bar chart on page 41 shows the projected cost of wireless phone use through 2003.

(a) Use the chart to fill in the missing entries in the table below.

(b) What is the difference in cost between 1998 and 1999?

Day	Plant Height (inches)
0	0
1	0.5
2	
	1.5
4	
5	4
6	
7	
	13
9	18
10	

Year	Cents/Minute
1998	33
1999	
2000	
	23
2002	
	20

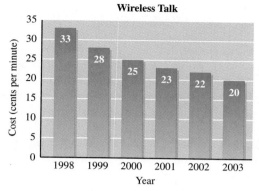

Wireless Talk

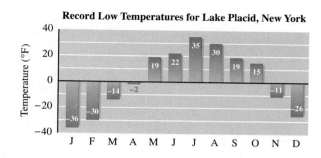

Record Low Temperatures for Lake Placid, New York

(a) Find the difference of the April record low and the May record low.
(b) Find the difference of the October record low and the November record low.

Find x in each of the following diagrams.

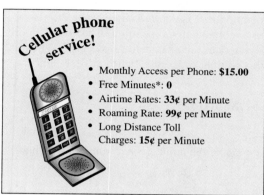

Cellular phone service!

- Monthly Access per Phone: **$15.00**
- Free Minutes*: **0**
- Airtime Rates: **33¢** per Minute
- Roaming Rate: **99¢** per Minute
- Long Distance Toll Charges: **15¢** per Minute

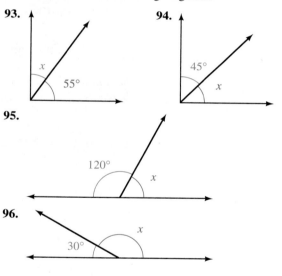

93.   **94.**   **95.**   **96.**

92. Temperature   The following table and bar chart both give the record low for each month of the year for Lake Placid, New York.

Month	Temperature (°F)
January	−36
February	−30
March	−14
April	−2
May	19
June	22
July	35
August	30
September	19
October	15
November	−11
December	−26

In this section we will list all the facts (properties) that you know from past experience are true about numbers in general. We will give each property a name so we can refer to it later in the book. Mathematics is very much like a game. The game involves numbers. The rules of the game are the properties and rules we are developing in this chapter. The goal of the game is to extend the basic rules to as many new situations as possible.

You know from past experience with numbers that it makes no difference in which order you add two numbers. That is, $3 + 5$ is the same as $5 + 3$. This fact about numbers is called the *commutative property of addition*. We say addition is a commutative operation. Changing the order of the numbers does not change the answer.

There is one other basic operation that is commutative. Since $3(5)$ is the same as $5(3)$, we say multiplication is a commutative operation. Changing the order of the two numbers you are multiplying does not change the answer.

For all properties listed in this section, a, b, and c represent real numbers.

COMMUTATIVE PROPERTY OF ADDITION

In symbols: $a + b = b + a$
In words: Changing the *order* of the numbers in a sum will not change the result.

COMMUTATIVE PROPERTY OF MULTIPLICATION

In symbols: $a \cdot b = b \cdot a$
In words: Changing the *order* of the numbers in a product will not change the result.

EXAMPLES

1. The statement $5 + 8 = 8 + 5$ is an example of the commutative property of addition.

2. The statement $2 \cdot y = y \cdot 2$ is an example of the commutative property of multiplication.

3. The expression $5 + x + 3$ can be simplified using the commutative property of addition:

$$5 + x + 3 = x + 5 + 3 \qquad \text{Commutative property of addition}$$
$$= x + 8 \qquad\qquad\;\; \text{Addition}$$

The other two basic operations, subtraction and division, are not commutative. The order in which we subtract or divide two numbers makes a difference in the answer.

Another property of numbers that you have used many times has to do with grouping. You know that when we add three numbers it makes no difference which two we add first. When adding $3 + 5 + 7$, we can add the 3 and 5 first and then the 7, or we can add the 5 and 7 first and then the 3. Mathematically, it looks like this: $(3 + 5) + 7 = 3 + (5 + 7)$. This property is true of multiplication as well. Operations that behave in this manner are called *associative* operations. The answer will not change when we change the association (or grouping) of the numbers.

ASSOCIATIVE PROPERTY OF ADDITION

In symbols: $a + (b + c) = (a + b) + c$
In words: Changing the *grouping* of the numbers in a sum will not change the result.

ASSOCIATIVE PROPERTY OF MULTIPLICATION

In symbols: $a(bc) = (ab)c$
In words: Changing the *grouping* of the numbers in a product will not change the result.

The following examples illustrate how the associative properties can be used to simplify expressions that involve both numbers and variables.

EXAMPLES   Simplify.

4. $4 + (5 + x) = (4 + 5) + x$ Associative property of addition
$ = 9 + x$ Addition

5. $5(2x) = (5 \cdot 2)x$ Associative property of multiplication
$ = 10x$ Multiplication

6. $\dfrac{1}{5}(5x) = \left(\dfrac{1}{5} \cdot 5\right)x$ Associative property of multiplication

$\phantom{\dfrac{1}{5}(5x)} = 1x$ Multiplication

$\phantom{\dfrac{1}{5}(5x)} = x$

The associative and commutative properties apply to problems that are either all multiplication or all addition. There is a third basic property that involves both addition and multiplication. It is called the *distributive property* and looks like this.

DISTRIBUTIVE PROPERTY

In symbols: $a(b + c) = ab + ac$
In words: Multiplication *distributes* over addition.

Note: Since subtraction is defined in terms of addition, it is also true that the distributive property applies to subtraction as well as addition. That is, $a(b - c) = ab - ac$ for any three real numbers a, b, and c.

You will see as we progress through the book that the distributive property is used very frequently in algebra. We can give a visual justification to the distributive property by finding the areas of rectangles. Figure 1 shows a large rectangle that is made up of two smaller rectangles.

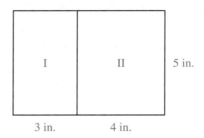

FIGURE I

We can find the area of the large rectangle two different ways.

Method 1 We can calculate the area of the large rectangle directly by finding its length and width. The width is 5 inches, and the length is (3 + 4) inches.

$$\text{Area of large rectangle} = 5(3 + 4)$$
$$= 5(7)$$
$$= 35 \text{ square inches}$$

Method 2 Since the area of the large rectangle is the sum of the areas of the two smaller rectangles, we find the area of each small rectangle and then add to find the area of the large rectangle.

$$\text{Area of large rectangle} = \text{Area of rectangle I} + \text{Area of rectangle II}$$
$$= \quad 5(3) \quad + \quad 5(4)$$
$$= \quad 15 \quad + \quad 20$$
$$= \quad 35 \text{ square inches}$$

In both cases the result is 35 square inches. Since the results are the same, the two original expressions must be equal. Stated mathematically, $5(3 + 4) = 5(3) + 5(4)$. We can either add the 3 and the 4 first and then multiply that sum by 5, or we can multiply the 3 and the 4 separately by 5 and then add the results. In either case we get the same answer.

Here are some examples that illustrate how we use the distributive property.

EXAMPLES Apply the distributive property to each expression and then simplify the result.

7. $2(x + 3) = 2(x) + 2(3)$ Distributive property
$$= 2x + 6 \qquad\qquad \text{Multiplication}$$

8. $5(2x - 8) = 5(2x) - 5(8)$ Distributive property

$\quad\quad\quad\quad\quad = 10x - 40$ Multiplication

Notice in this example that multiplication distributes over subtraction as well as addition.

9. $4(x + y) = 4x + 4y$ Distributive property

10. $5(2x + 4y) = 5(2x) + 5(4y)$ Distributive property

$\quad\quad\quad\quad\quad\quad = 10x + 20y$ Multiplication

11. $\dfrac{1}{2}(3x + 6) = \dfrac{1}{2}(3x) + \dfrac{1}{2}(6)$ Distributive property

$\quad\quad\quad\quad\quad\quad = \dfrac{3}{2}x + 3$ Multiplication

12. $4(2a + 3) + 8 = 4(2a) + 4(3) + 8$ Distributive property

$\quad\quad\quad\quad\quad\quad\quad = 8a + 12 + 8$ Multiplication

$\quad\quad\quad\quad\quad\quad\quad = 8a + 20$ Addition

SPECIAL NUMBERS

In addition to the three properties mentioned so far, we want to include in our list two special numbers that have unique properties. They are the numbers zero and one.

ADDITIVE IDENTITY PROPERTY

There exists a unique number 0 such that

In symbols: $a + 0 = a$ and $0 + a = a$

In words: Zero preserves identities under addition. (The identity of the number is unchanged after addition with 0.)

MULTIPLICATIVE IDENTITY PROPERTY

There exists a unique number 1 such that

In symbols: $a(1) = a$ and $1(a) = a$

In words: The number 1 preserves identities under multiplication. (The identity of the number is unchanged after multiplication by 1.)

ADDITIVE INVERSE PROPERTY

For each real number a, there exists a unique number $-a$ such that

In symbols: $a + (-a) = 0$

In words: Opposites add to 0.

MULTIPLICATIVE INVERSE PROPERTY

For every real number a, except 0, there exists a unique real number $\dfrac{1}{a}$ such that

In symbols: $a\left(\dfrac{1}{a}\right) = 1$

In words: Reciprocals multiply to 1.

Of all the basic properties listed, the commutative, associative, and distributive properties are the ones we will use most often. They are important because they will be used as justifications or reasons for many of the things we will do in the future.

The following examples illustrate how we use the preceding properties. Each one contains an algebraic expression that has been changed in some way. The property that justifies the change is written to the right.

EXAMPLES   State the property that justifies the given statement.

13. $x + 5 = 5 + x$ Commutative property of addition

14. $(2 + x) + y = 2 + (x + y)$ Associative property of addition

15. $6(x + 3) = 6x + 18$ Distributive property

16. $2 + (-2) = 0$ Additive inverse property

17. $3\left(\dfrac{1}{3}\right) = 1$ Multiplicative inverse property

18. $(2 + 0) + 3 = 2 + 3$ 0 is the identity element for addition.

19. $(2 + 3) + 4 = 3 + (2 + 4)$ Commutative and associative properties
                                              of addition

20. $(x + 2) + y = (x + y) + 2$ Commutative and associative properties
                                              of addition

As a final note on the properties of real numbers, we should mention that although some of the properties are stated for only two or three real numbers, they hold for as many numbers as needed. For example, the distributive property holds for expressions like $3(x + y + z + 5 + 2)$. That is,

$$3(x + y + z + 5 + 2) = 3x + 3y + 3z + 15 + 6$$

It is not important how many numbers are contained in the sum, only that it is a sum. Multiplication, you see, distributes over addition, whether there are 2 numbers in the sum or 200.

Getting Ready for Class

After reading through the preceding section, respond in your own words and in complete sentences.

A. What is the commutative property of addition?

B. Do you know from your experience with numbers that the commutative property of addition is true? Explain why.

C. Write the commutative property of multiplication in symbols and in words.

D. How do you rewrite expressions using the distributive property?

PROBLEM SET 1.5

State the property or properties that justify the following.

1. $3 + 2 = 2 + 3$

2. $5 + 0 = 5$

3. $4\left(\dfrac{1}{4}\right) = 1$

4. $10(0.1) = 1$

5. $4 + x = x + 4$

6. $3(x - 10) = 3x - 30$

7. $2(y + 8) = 2y + 16$

8. $3 + (4 + 5) = (3 + 4) + 5$

9. $(3 + 1) + 2 = 1 + (3 + 2)$

10. $(5 + 2) + 9 = (2 + 5) + 9$

11. $(8 + 9) + 10 = (8 + 10) + 9$

12. $(7 + 6) + 5 = (5 + 6) + 7$

13. $3(x + 2) = 3(2 + x)$

14. $2(7y) = (7 \cdot 2)y$

15. $x(3y) = 3(xy)$

16. $a(5b) = 5(ab)$

17. $4(xy) = 4(yx)$

18. $3[2 + (-2)] = 3(0)$

19. $8[7 + (-7)] = 8(0)$

20. $7(1) = 7$

Each of the following problems has a mistake in it. Correct the right-hand side.

21. $3(x + 2) = 3x + 2$

22. $5(4 + x) = 4 + 5x$

23. $9(a + b) = 9a + b$

24. $2(y + 1) = 2y + 1$

25. $3(0) = 3$

26. $5\left(\dfrac{1}{5}\right) = 5$

27. $3 + (-3) = 1$

28. $8(0) = 8$

29. $10(1) = 0$

30. $3 \cdot \dfrac{1}{3} = 0$

Use the associative property to rewrite each of the following expressions and then simplify the result. (See Examples 4, 5, and 6.)

31. $4 + (2 + x)$

32. $5 + (6 + x)$

33. $(x + 2) + 7$

34. $(x + 8) + 2$

35. $3(5x)$

36. $5(3x)$

37. $9(6y)$

38. $6(9y)$

39. $\dfrac{1}{2}(3a)$

40. $\dfrac{1}{3}(2a)$

41. $\dfrac{1}{3}(3x)$

42. $\dfrac{1}{4}(4x)$

43. $\dfrac{1}{2}(2y)$

44. $\dfrac{1}{7}(7y)$

45. $\dfrac{3}{4}\left(\dfrac{4}{3}x\right)$

46. $\dfrac{3}{2}\left(\dfrac{2}{3}x\right)$

47. $\dfrac{6}{5}\left(\dfrac{5}{6}a\right)$

48. $\dfrac{2}{5}\left(\dfrac{5}{2}a\right)$

Apply the distributive property to each of the following expressions. Simplify when possible.

49. $8(x + 2)$

50. $5(x + 3)$

51. $8(x - 2)$

52. $5(x - 3)$

53. $4(y + 1)$

54. $4(y - 1)$

55. $3(6x + 5)$

56. $3(5x + 6)$

57. $2(3a + 7)$

58. $5(3a + 2)$

59. $9(6y - 8)$

60. $2(7y - 4)$

61. $\dfrac{1}{2}(3x - 6)$ **62.** $\dfrac{1}{3}(2x - 6)$

63. $\dfrac{1}{3}(3x + 6)$ **64.** $\dfrac{1}{2}(2x + 4)$

65. $3(x + y)$ **66.** $2(x - y)$

67. $8(a - b)$ **68.** $7(a + b)$

69. $6(2x + 3y)$ **70.** $8(3x + 2y)$

71. $4(3a - 2b)$ **72.** $5(4a - 8b)$

73. $\dfrac{1}{2}(6x + 4y)$ **74.** $\dfrac{1}{3}(6x + 9y)$

75. $4(a + 4) + 9$ **76.** $6(a + 2) + 8$

77. $2(3x + 5) + 2$ **78.** $7(2x + 1) + 3$

79. $7(2x + 4) + 10$ **80.** $3(5x + 6) + 20$

Applying the Concepts

81. Getting Dressed While getting dressed for work, a man puts on his socks and puts on his shoes. Are the two statements "put on your socks" and "put on your shoes" commutative?

82. Getting Dressed Are the statements "put on your left shoe" and "put on your right shoe" commutative?

83. Skydiving A skydiver flying over the jump area is about to do two things: jump out of the plane and pull the rip cord. Are the two events "jump out of the plane" and "pull the rip cord" commutative? That is, will changing the order of the events always produce the same result?

84. Commutative Property Give an example of two events in your daily life that are commutative.

85. Division Give an example that shows that division is not a commutative operation. That is, find two numbers for which changing the order of division gives two different answers.

86. Subtraction Simplify the expression $10 - (5 - 2)$ and the expression $(10 - 5) - 2$ to show that subtraction is not an associative operation.

87. Take-Home Pay Jose works at a winery. His monthly salary is $2,400. To cover his taxes and retirement, the winery withholds $480 from each check. Calculate his yearly "take-home" pay using the numbers 2,400, 480, and 12. Do the calculation two different ways so that the results give further justification for the distributive property.

88. Hours Worked Carlo works as a waiter. He works 4 days a week. The lunch shift is 2 hours, and the dinner shift is 3 hours. Find the total number of hours he works per week, using the numbers 2, 3, and 4. Do the calculation two different ways so that the results give further justification for the distributive property.

89. Distributive Property Use the distributive property to rewrite the formula for the perimeter of a rectangle: $P = 2l + 2w$.

90. College Expenses Maria is estimating her expenses for attending college for a year. Tuition is $650 per academic quarter. She estimates she will spend $225 on books each quarter. If she plans on attending 3 academic quarters during the year, how much can she expect to spend? Do the calculation two different ways so that the results give further justification for the distributive property.

1.6   Multiplication of Real Numbers

Suppose that you own 5 shares of a stock and the price per share drops $3. How much money have you lost? Intuitively, we know the loss is $15. Since it is a loss, we can express it as $-\$15$. To describe this situation with numbers, we would write

5 shares   each loses $3   for a total of $15

$$5(-3) = -15$$

Reasoning in this manner, we conclude that the product of a positive number with a negative number is a negative number. Let's look at multiplication in more detail.

From our experience with counting numbers, we know that multiplication is simply repeated addition. That is, $3(5) = 5 + 5 + 5$. We will use this fact, along with our knowledge of negative numbers, to develop the rule for multiplication of any two real numbers. The following examples illustrate multiplication with three of the possible combinations of positive and negative numbers.

EXAMPLES   Multiply.

1. Two positives: $3(5) = 5 + 5 + 5$
$ = 15$ Positive answer

2. One positive: $3(-5) = -5 + (-5) + (-5)$
$ = -15$ Negative answer

3. One negative: $-3(5) = 5(-3)$ Commutative property
$ = -3 + (-3) + (-3) + (-3) + (-3)$
$ = -15$ Negative answer

4. Two negatives: $-3(-5) = ?$

With two negatives, $-3(-5)$, it is not possible to work the problem in terms of repeated addition. (It doesn't "make sense" to write -5 down a -3 number of times.) The answer is probably $+15$ (that's just a guess), but we need some justification for saying so. We will solve a different problem and, in so doing, get the answer to the problem $(-3)(-5)$.

Here is a problem to which we know the answer. We will work it two different ways.

$$-3[5 + (-5)] = -3(0) = 0$$

The answer is zero. We can also work the problem using the distributive property.

$$-3[5 + (-5)] = -3(5) + (-3)(-5)   \text{Distributive property}$$
$$ = -15   +   ?$$

Since the answer to the problem is 0, our ? must be $+15$. (What else could we add to -15 to get 0? Only $+15$.)

Note: You may have to read the explanation for Example 4 several times before you understand it completely. The purpose of the explanation in Example 4 is simply to justify the fact that the product of two negative numbers is a positive number. If you have no trouble believing that, then it is not so important that you understand everything in the explanation.

Here is a summary of the results we have obtained from the first four examples:

Original Numbers Have		The Answer Is
The same sign	$3(5) = 15$	Positive
Different signs	$3(-5) = -15$	Negative
Different signs	$-3(5) = -15$	Negative
The same sign	$-3(-5) = 15$	Positive

By examining Examples 1 through 4 and the preceding table, we can use the information there to write the following rule. This rule tells us how to multiply any two real numbers.

> **RULE**
>
> To multiply any two real numbers, simply multiply their absolute values. The sign of the answer is
>
> 1. *Positive* if both numbers have the same sign (both + or both −).
> 2. *Negative* if the numbers have opposite signs (one +, the other −).

The following examples illustrate how we use the preceding rule to multiply real numbers.

EXAMPLES   Multiply.

5. $-8(-3) = 24$ ⎫
6. $-10(-5) = 50$ ⎬ If the two numbers in the product have the same sign, the answer is positive.
7. $-4(-7) = 28$ ⎭

8. $5(-7) = -35$ ⎫
9. $-4(8) = -32$ ⎬ If the two numbers in the product have different signs, the answer is negative.
10. $-6(10) = -60$ ⎭

Note: Some students have trouble with the expression $-8(-3)$ because they want to subtract rather than multiply. Because we are very precise with the notation we use in algebra, the expression $-8(-3)$ has only one meaning—multiplication. A subtraction problem that uses the same numbers is $-8 - 3$. Compare the two following lists.

All Multiplication	*No Multiplication*
$5(4)$	$5 + 4$
$-5(4)$	$-5 + 4$
$5(-4)$	$5 - 4$
$-5(-4)$	$-5 - 4$

In the following examples, we combine the rule for order of operations with the rule for multiplication to simplify expressions. Remember, the rule for order of operations specifies that we are to work inside the parentheses first and then simplify numbers containing exponents. After this, we multiply and divide, left to right. The last step is to add and subtract, left to right.

EXAMPLES   Simplify as much as possible.

11. $-5(-3)(-4) = 15(-4)$
$$= -60$$

12. $4(-3) + 6(-5) - 10 = -12 + (-30) - 10$ Multiply.

$= -42 - 10$ Add.

$= -52$ Subtract.

13. $(-2)^3 = (-2)(-2)(-2)$ Definition of exponents

$= -8$ Multiply, left to right.

14. $-3(-2)^3 - 5(-4)^2 = -3(-8) - 5(16)$ Exponents first

$= 24 - 80$ Multiply.

$= -56$ Subtract.

15. $6 - 4(7 - 2) = 6 - 4(5)$ Inside parentheses first

$= 6 - 20$ Multiply.

$= -14$ Subtract.

MULTIPLYING FRACTIONS

Previously, we mentioned that to multiply two fractions we multiply numerators and multiply denominators. We can apply the rule for multiplication of positive and negative numbers to fractions in the same way we apply it to other numbers. We multiply absolute values: The product is positive if both fractions have the same sign and negative if they have different signs. Here are some examples.

EXAMPLES   Multiply.

16. $-\dfrac{3}{4}\left(\dfrac{5}{7}\right) = -\dfrac{3 \cdot 5}{4 \cdot 7}$ Different signs give a negative answer.

$= -\dfrac{15}{28}$

17. $-6\left(\dfrac{1}{2}\right) = -\dfrac{6}{1}\left(\dfrac{1}{2}\right)$ Different signs give a negative answer.

$= -\dfrac{6}{2}$

$= -3$

18. $-\dfrac{2}{3}\left(-\dfrac{3}{2}\right) = \dfrac{2 \cdot 3}{3 \cdot 2}$ Same signs give a positive answer.

$= \dfrac{6}{6}$

$= 1$

EXAMPLE 19   Figure 1 on page 52 gives the calories that are burned in 1 hour for a variety of forms of exercise by a person weighing 150 pounds. Figure 2 on page 52 gives the calories that are consumed by eating some popular fast foods. Find the net change in calories for a 150-pound person playing handball for 2 hours and then eating a Whopper.

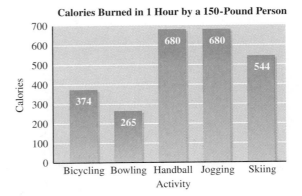

FIGURE 1

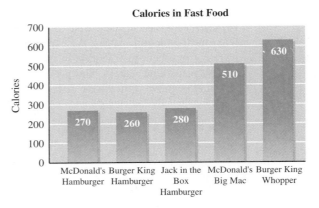

FIGURE 2

Solution The net change in calories will be the difference of the calories gained from eating and the calories lost from exercise.

$$\text{Net change in calories} = 630 - 2(680) = -730 \text{ calories}$$

We can use the rule for multiplication of real numbers, along with the associative property, to multiply expressions that contain numbers and variables.

EXAMPLES Apply the associative property and then multiply.

20. $-3(2x) = (-3 \cdot 2)x$ Associative property
$= -6x$ Multiplication

21. $6(-5y) = [6(-5)]y$ Associative property
$= -30y$ Multiplication

22. $-2\left(-\dfrac{1}{2}x\right) = \left[(-2)\left(-\dfrac{1}{2}\right)\right]x$ Associative property

$\qquad\qquad\quad = 1x$ Multiplication

$\qquad\qquad\quad = x$ Multiplication

The following examples show how we can use both the distributive property and multiplication with real numbers.

E X A M P L E S   Apply the distributive property to each expression.

23. $-2(a + 3) = -2a + (-2)(3)$ Distributive property

$\qquad\qquad\quad = -2a + (-6)$ Multiplication

$\qquad\qquad\quad = -2a - 6$

24. $-3(2x + 1) = -3(2x) + (-3)(1)$ Distributive property

$\qquad\qquad\quad = -6x + (-3)$ Multiplication

$\qquad\qquad\quad = -6x - 3$

25. $-\dfrac{1}{3}(2x - 6) = -\dfrac{1}{3}(2x) - \left(-\dfrac{1}{3}\right)(6)$ Distributive property

$\qquad\qquad\quad = -\dfrac{2}{3}x - (-2)$ Multiplication

$\qquad\qquad\quad = -\dfrac{2}{3}x + 2$

26. $-4(3x - 5) - 8 = -4(3x) - (-4)(5) - 8$ Distributive property

$\qquad\qquad\quad = -12x - (-20) - 8$ Multiplication

$\qquad\qquad\quad = -12x + 20 - 8$ Definition of subtraction

$\qquad\qquad\quad = -12x + 12$ Subtraction

GEOMETRIC SEQUENCES

A *geometric sequence* is a sequence of numbers in which each number (after the first number) comes from multiplying the number before it by the same amount each time. For example, the sequence

$$2, 6, 18, 54, \ldots$$

is a geometric sequence because each number is obtained by multiplying the number before it by 3.

E X A M P L E   2 7   Each of the following is a geometric sequence. Find the next number in each sequence.

(a) 5, 10, 20, . . .

(b) 3, −15, 75, . . .

(c) $\frac{1}{8}, \frac{1}{4}, \frac{1}{2}, \dots$

Solution   Since each sequence is a geometric sequence, we know that each term is obtained from the previous term by multiplying by the same number each time.

(a) 5, 10, 20, . . . : Starting with 5, each number is obtained from the previous number by multiplying by 2 each time, The next number will be $20 \cdot 2 = 40$.

(b) 3, −15, 75, . . . : The sequence starts with 3. After that, each number is obtained by multiplying by −5 each time. The next number will be $75(-5) = -375$.

(c) $\frac{1}{8}, \frac{1}{4}, \frac{1}{2}, \dots$: This sequence starts with $\frac{1}{8}$. Multiplying each number in the sequence by 2 produces the next number in the sequence. To extend the sequence, we multiply $\frac{1}{2}$ by 2:

$$\frac{1}{2} \cdot 2 = 1$$

The next number in the sequence is 1.

Getting Ready for Class

After reading through the preceding section, respond in your own words and in complete sentences.

A. How do you multiply two negative numbers?

B. How do you multiply two numbers with different signs?

C. Explain how some multiplication problems can be thought of as repeated addition.

D. What is a geometric sequence?

PROBLEM SET 1.6

Use the rule for multiplying two real numbers to find each of the following products.

1. $7(-6)$ **2.** $8(-4)$ **3.** $-7(3)$

4. $-5(4)$ **5.** $-8(2)$ **6.** $-16(3)$

7. $-3(-1)$ **8.** $-7(-1)$ **9.** $-11(-11)$

10. $-12(-12)$

Use the rule for order of operations to simplify each expression as much as possible.

11. $-3(2)(-1)$ **12.** $-2(3)(-4)$

13. $-3(-4)(-5)$ **14.** $-5(-6)(-7)$

15. $-2(-4)(-3)(-1)$ **16.** $-1(-3)(-2)(-1)$

17. $(-7)^2$ **18.** $(-8)^2$

19. $(-3)^3$ **20.** $(-2)^4$

21. $-2(2 - 5)$ **22.** $-3(3 - 7)$

23. $-5(8 - 10)$ **24.** $-4(6 - 12)$

25. $(4 - 7)(6 - 9)$ **26.** $(3 - 10)(2 - 6)$

27. $(-3 - 2)(-5 - 4)$ **28.** $(-3 - 6)(-2 - 8)$

29. $-3(-6) + 4(-1)$ **30.** $-4(-5) + 8(-2)$

31. $2(3) - 3(-4) + 4(-5)$

32. $5(4) - 2(-1) + 5(6)$　**33.** $4(-3)^2 + 5(-6)^2$

34. $2(-5)^2 + 4(-3)^2$　**35.** $7(-2)^3 - 2(-3)^3$

36. $10(-2)^3 - 5(-2)^4$　**37.** $6 - 4(8 - 2)$

38. $7 - 2(6 - 3)$　**39.** $9 - 4(3 - 8)$

40. $8 - 5(2 - 7)$

41. $-4(3 - 8) - 6(2 - 5)$

42. $-8(2 - 7) - 9(3 - 5)$　**43.** $7 - 2[-6 - 4(-3)]$

44. $6 - 3[-5 - 3(-1)]$

45. $7 - 3[2(-4 - 4) - 3(-1 - 1)]$

46. $5 - 3[7(-2 - 2) - 3(-3 + 1)]$

47. $8 - 6[-2(-3 - 1) + 4(-2 - 3)]$

48. $4 - 2[-3(-1 + 8) + 5(-5 + 7)]$

Multiply the following fractions. (See Examples 16–18.)

49. $-\dfrac{2}{3} \cdot \dfrac{5}{7}$

50. $-\dfrac{6}{5} \cdot \dfrac{2}{7}$

51. $-8\left(\dfrac{1}{2}\right)$

52. $-12\left(\dfrac{1}{3}\right)$

53. $-\dfrac{3}{4}\left(-\dfrac{4}{3}\right)$

54. $-\dfrac{5}{8}\left(-\dfrac{8}{5}\right)$

55. $\left(-\dfrac{3}{4}\right)^2$

56. $\left(-\dfrac{2}{5}\right)^2$

57. $\left(-\dfrac{2}{3}\right)^3$

58. $\left(-\dfrac{1}{2}\right)^3$

Find the following products. (See Examples 20, 21, and 22.)

59. $-2(4x)$　**60.** $-8(7x)$　**61.** $-7(-6x)$

62. $-8(-9x)$　**63.** $-\dfrac{1}{3}(-3x)$　**64.** $-\dfrac{1}{5}(-5x)$

65. $-4\left(-\dfrac{1}{4}x\right)$　**66.** $-2\left(-\dfrac{1}{2}x\right)$

Apply the distributive property to each expression; then simplify the result. (See Examples 23–26.)

67. $-4(a + 2)$　**68.** $-7(a + 6)$

69. $-\dfrac{1}{2}(3x - 6)$　**70.** $-\dfrac{1}{4}(2x - 4)$

71. $-3(2x - 5) - 7$　**72.** $-4(3x - 1) - 8$

73. $-5(3x + 4) - 10$　**74.** $-3(4x + 5) - 20$

75. Five added to the product of 3 and -10 is what number?

76. If the product of -8 and -2 is decreased by 4, what number results?

77. Write an expression for twice the product of -4 and x, and then simplify it.

78. Write an expression for twice the product of -2 and $3x$, and then simplify it.

79. What number results if 8 is subtracted from the product of -9 and 2?

80. What number results if -8 is subtracted from the product of -9 and 2?

Each of the following is a geometric sequence. In each case, find the next number in the sequence.

81. $1, 2, 4, \ldots$　**82.** $1, 5, 25, \ldots$

83. $10, -20, 40, \ldots$　**84.** $10, -30, 90, \ldots$

85. $1, \dfrac{1}{2}, \dfrac{1}{4}, \ldots$　**86.** $1, \dfrac{1}{3}, \dfrac{1}{9}, \ldots$

87. $8, 4, 2, \ldots$　**88.** $20, 10, 5, \ldots$

89. $3, -6, 12, \ldots$　**90.** $-3, 6, -12, \ldots$

Applying the Concepts

91. Stock Value　Suppose you own 20 shares of a stock. If the price per share drops \$3, how much money have you lost?

92. Stock Value　Imagine that you purchase 50 shares of a stock at a price of \$18 per share. If the stock is selling for \$11 a share a week after you purchased it, how much money have you lost?

93. Temperature Change　The temperature is 25°F at 5:00 in the afternoon. If the temperature drops 6°F every hour after that, what is the temperature at 9:00 in the evening?

94. Temperature Change　The temperature is -5°F at 6:00 in the evening. If the temperature drops 3°F every hour after that, what is the temperature at midnight?

95. Investment Value　Suppose you purchase \$500 worth of a mutual fund and find that the value of your purchase doubles every 2 years. Write a sequence of numbers that gives the value of your purchase every 2 years for the first 10 years you own it. Is this sequence a geometric sequence?

96. Number of Bacteria　A colony of 150 bacteria doubles in size every 8 hours. Starting with the original 150 bacteria, write a sequence of numbers that gives the number of bacteria in the colony every 8 hours during a 40-hour time span. Is this sequence a geometric sequence?

97. Reading Charts   Refer to Figures 1 and 2 in Example 19 in this section to find the net change in

calories for a 150-pound person who bowls for 3 hours and then eats 2 Whoppers.

98. Reading Charts   Refer to Figures 1 and 2 in Example 19 in this section to find the net change in calories for a 150-pound person who eats a Big Mac and then skis for 3 hours.

1.7   Division of Real Numbers

Suppose that you and four friends bought equal shares of an investment for a total of $15,000 and then sold it later for only $13,000. How much did each person lose? Since the total amount of money that was lost can be represented by $-\$2,000$ and there are 5 people with equal shares, we can represent each person's loss with division:

$$\frac{-\$2,000}{5} = -\$400$$

From this discussion it seems reasonable to say that a negative number divided by a positive number is a negative number. Here is a more detailed discussion of division with positive and negative numbers.

The last of the four basic operations is division. We will use the same approach to define division as we used for subtraction. That is, we will define division in terms of rules we already know.

Recall that we developed the rule for subtraction of real numbers by defining subtraction in terms of addition. We changed our subtraction problems to addition problems and then added to get our answers. Since we already have a rule for multiplication of real numbers and division is the inverse operation of multiplication, we will simply define division in terms of multiplication.

We know that division by the number 2 is the same as multiplication by $\frac{1}{2}$. That is, 6 divided by 2 is 3, which is the same as 6 times $\frac{1}{2}$. Similarly, dividing a number by 5 gives the same result as multiplying by $\frac{1}{5}$. We can extend this idea to all real numbers with the following rule.

> **RULE**
>
> If a and b represent any two real numbers (b cannot be 0), then it is always true that
>
> $$a \div b = \frac{a}{b} = a\left(\frac{1}{b}\right)$$

Division by a number is the same as multiplication by its reciprocal. Since every division problem can be written as a multiplication problem and since we already know the rule for multiplication of two real numbers, we do not have to write a new rule for division of real numbers. We will simply replace our division problem with multiplication and use the rule we already have.

E X A M P L E S   Write each division problem as an equivalent multiplication problem and then multiply.

1. $\dfrac{6}{2} = 6\left(\dfrac{1}{2}\right) = 3$ The product of two positives is positive.

2. $\dfrac{6}{-2} = 6\left(-\dfrac{1}{2}\right) = -3$

The product of a positive and a negative is a negative.

3. $\dfrac{-6}{2} = -6\left(\dfrac{1}{2}\right) = -3$

4. $\dfrac{-6}{-2} = -6\left(-\dfrac{1}{2}\right) = 3$ The product of two negatives is positive.

The second step in these examples is used only to show that we *can* write division in terms of multiplication. (In actual practice we wouldn't write $\frac{6}{2}$ as $6(\frac{1}{2})$.) The answers, therefore, follow from the rule for multiplication. That is, like signs produce a positive answer, and unlike signs produce a negative answer.

   Here are some examples. This time we will not show division as multiplication by the reciprocal. We will simply divide. If the original numbers have the same signs, the answer will be positive. If the original numbers have different signs, the answer will be negative.

E X A M P L E S   Divide.

5. $\dfrac{12}{6} = 2$ Like signs give a positive answer.

6. $\dfrac{12}{-6} = -2$ Unlike signs give a negative answer.

7. $\dfrac{-12}{6} = -2$ Unlike signs give a negative answer.

8. $\dfrac{-12}{-6} = 2$ Like signs give a positive answer.

9. $\dfrac{15}{-3} = -5$ Unlike signs give a negative answer.

10. $\dfrac{-40}{-5} = 8$ Like signs give a positive answer.

11. $\dfrac{-14}{2} = -7$ Unlike signs give a negative answer.

DIVISION WITH FRACTIONS

We can apply the definition of division to fractions. Since dividing by a fraction is equivalent to multiplying by its reciprocal, we can divide a number by the fraction $\frac{3}{4}$ by multiplying it by the reciprocal of $\frac{3}{4}$, which is $\frac{4}{3}$. For example,

$$\frac{2}{5} \div \frac{3}{4} = \frac{2}{5} \cdot \frac{4}{3} = \frac{8}{15}$$

You may have learned this rule in previous math classes. In some math classes, multiplication by the reciprocal is referred to as "inverting the divisor and multiplying." No matter how you say it, division by any number (except 0) is always equivalent to multiplication by its reciprocal. Here are additional examples that involve division by fractions.

EXAMPLES   Divide.

12. $\dfrac{2}{3} \div \dfrac{5}{7} = \dfrac{2}{3} \cdot \dfrac{7}{5}$ Rewrite as multiplication by the reciprocal.

$= \dfrac{14}{15}$ Multiply.

13. $-\dfrac{3}{4} \div \dfrac{7}{9} = -\dfrac{3}{4} \cdot \dfrac{9}{7}$ Rewrite as multiplication by the reciprocal.

$= -\dfrac{27}{28}$ Multiply.

14. $8 \div \left(-\dfrac{4}{5}\right) = \dfrac{8}{1}\left(-\dfrac{5}{4}\right)$ Rewrite as multiplication by the reciprocal.

$= -\dfrac{40}{4}$ Multiply.

$= -10$ Divide 40 by 4.

The last step in each of the following examples involves reducing a fraction to lowest terms. To reduce a fraction to lowest terms, we divide the numerator and denominator by the largest number that divides each of them exactly. For example, to reduce $\frac{15}{20}$ to lowest terms, we divide 15 and 20 by 5 to get $\frac{3}{4}$.

EXAMPLES   Simplify as much as possible.

15. $\dfrac{-4(5)}{6} = \dfrac{-20}{6}$ Simplify the numerator.

$= -\dfrac{10}{3}$ Reduce to lowest terms by dividing the numerator and denominator by 2.

16. $\dfrac{30}{-4-5} = \dfrac{30}{-9}$ Simplify the denominator.

$\phantom{\dfrac{30}{-4-5}} = -\dfrac{10}{3}$ Reduce to lowest terms by dividing the numerator and denominator by 3.

In the examples that follow, the numerators and denominators contain expressions that are somewhat more complicated than those we have seen thus far. To apply the rule for order of operations to these examples, we treat fraction bars the same way we treat grouping symbols. That is, fraction bars separate numerators and denominators so that each will be simplified separately.

EXAMPLES   Simplify.

17. $\dfrac{2(-3)+4}{12} = \dfrac{-6+4}{12}$ In the numerator, we multiply before we add.

$\phantom{\dfrac{2(-3)+4}{12}} = \dfrac{-2}{12}$ Addition

$\phantom{\dfrac{2(-3)+4}{12}} = -\dfrac{1}{6}$ Reduce to lowest terms by dividing the numerator and denominator by 2.

18. $\dfrac{5(-4)+6(-1)}{2(3)-4(1)} = \dfrac{-20+(-6)}{6-4}$ Multiplication before addition

$\phantom{\dfrac{5(-4)+6(-1)}{2(3)-4(1)}} = \dfrac{-26}{2}$ Simplify the numerator and denominator.

$\phantom{\dfrac{5(-4)+6(-1)}{2(3)-4(1)}} = -13$ Divide -26 by 2.

We must be careful when we are working with expressions such as $(-5)^2$ and -5^2, that we include the negative sign with the base only when parentheses indicate we are to do so.

Unless there are parentheses to indicate otherwise, we consider the base to be only the number directly below and to the left of the exponent. If we want to include a negative sign with the base, we must use parentheses.

To simplify a more complicated expression, we follow the same rule. For example,

$7^2 - 3^2 = 49 - 9$ The bases are 7 and 3; the sign between the two terms is a subtraction sign.

For another example,

$5^3 - 3^4 = 125 - 81$ We simplify exponents first, and then subtract.

EXAMPLES Simplify.

19. $\dfrac{5^2 - 3^2}{-5 + 3} = \dfrac{25 - 9}{-2}$ Simplify numerator and denominator
separately.

$= \dfrac{16}{-2}$

$= -8$

20. $\dfrac{(3 + 2)^2}{-3^2 - 2^2} = \dfrac{5^2}{-9 - 4}$ Simplify numerator and denominator
separately.

$= \dfrac{25}{-13}$

$= -\dfrac{25}{13}$

DIVISION WITH THE NUMBER 0

For every division problem, there is an associated multiplication problem involving the same numbers. For example, the following two problems say the same thing about the numbers 2, 3, and 6:

| *Division* | *Multiplication* |

$$\frac{6}{3} = 2 \qquad 6 = 2(3)$$

We can use this relationship between division and multiplication to clarify division involving the number 0.

First, dividing 0 by a number other than 0 is allowed and always results in 0. To see this, consider dividing 0 by 5. We know the answer is 0 because of the relationship between multiplication and division. This is how we write it:

$$\frac{0}{5} = 0 \qquad \text{because} \qquad 0 = 0(5)$$

On the other hand, dividing a nonzero number by 0 is not allowed in the real numbers. Suppose we were attempting to divide 5 by 0. We don't know if there is an answer to this problem, but if there is, let's say the answer is a number that we can represent with the letter n. If 5 divided by 0 is a number n, then

$$\frac{5}{0} = n \qquad \text{and} \qquad 5 = n(0)$$

This is impossible, however, because no matter what number n is, when we multiply it by 0 the answer must be 0. It can never be 5. In algebra, we say expressions like $\frac{5}{0}$ are undefined, because there is no answer to them. That is, division by 0 is not allowed in the real numbers.

The only other possibility for division involving the number 0 is 0 divided by 0. We will treat problems like $\frac{0}{0}$ as if they were undefined also.

Getting Ready for Class

After reading through the preceding section, respond in your own words and in complete sentences.

A. Why do we define division in terms of multiplication?

B. What is the reciprocal of a number?

C. How do we divide fractions?

D. Why is division by 0 not allowed with real numbers?

PROBLEM SET 1.7

Find the following quotients (divide).

1. $\dfrac{8}{-4}$

2. $\dfrac{10}{-5}$

3. $\dfrac{-48}{16}$

4. $\dfrac{-32}{4}$

5. $\dfrac{-7}{21}$

6. $\dfrac{-25}{100}$

7. $\dfrac{-39}{-13}$

8. $\dfrac{-18}{-6}$

9. $\dfrac{-6}{-42}$

10. $\dfrac{-4}{-28}$

11. $\dfrac{0}{-32}$

12. $\dfrac{0}{17}$

The following problems review all four operations with positive and negative numbers. Perform the indicated operations.

13. $-3 + 12$

14. $5 + (-10)$

15. $-3 - 12$

16. $5 - (-10)$

17. $-3(12)$

18. $5(-10)$

19. $-3 \div 12$

20. $5 \div (-10)$

Divide and reduce all answers to lowest terms.

21. $\dfrac{4}{5} \div \dfrac{3}{4}$

22. $\dfrac{6}{8} \div \dfrac{3}{4}$

23. $-\dfrac{5}{6} \div \left(-\dfrac{5}{8}\right)$

24. $-\dfrac{7}{9} \div \left(-\dfrac{1}{6}\right)$

25. $\dfrac{10}{13} \div \left(-\dfrac{5}{4}\right)$

26. $\dfrac{5}{12} \div \left(-\dfrac{10}{3}\right)$

27. $-\dfrac{5}{6} \div \dfrac{5}{6}$

28. $-\dfrac{8}{9} \div \dfrac{8}{9}$

29. $-\dfrac{3}{4} \div \left(-\dfrac{3}{4}\right)$

30. $-\dfrac{6}{7} \div \left(-\dfrac{6}{7}\right)$

The following problems involve more than one operation. Simplify as much as possible.

31. $\dfrac{3(-2)}{-10}$

32. $\dfrac{4(-3)}{24}$

33. $\dfrac{-5(-5)}{-15}$

34. $\dfrac{-7(-3)}{-35}$

35. $\dfrac{-8(-7)}{-28}$

36. $\dfrac{-3(-9)}{-6}$

37. $\dfrac{27}{4 - 13}$

38. $\dfrac{27}{13 - 4}$

39. $\dfrac{20 - 6}{5 - 5}$

40. $\dfrac{10 - 12}{3 - 3}$

41. $\dfrac{-3 + 9}{2 \cdot 5 - 10}$

42. $\dfrac{-4 + 8}{2 \cdot 4 - 8}$

43. $\dfrac{15(-5) - 25}{2(-10)}$

44. $\dfrac{10(-3) - 20}{5(-2)}$

45. $\dfrac{27 - 2(-4)}{-3(5)}$

46. $\dfrac{20 - 5(-3)}{10(-3)}$

47. $\dfrac{12 - 6(-2)}{12(-2)}$

48. $\dfrac{3(-4) + 5(-6)}{10 - 6}$

49. $\dfrac{5^2 - 2^2}{-5 + 2}$

50. $\dfrac{7^2 - 4^2}{-7 + 4}$

51. $\dfrac{8^2 - 2^2}{8^2 + 2^2}$

52. $\dfrac{4^2 - 6^2}{4^2 + 6^2}$

53. $\dfrac{(5 + 3)^2}{-5^2 - 3^2}$

54. $\dfrac{(7 + 2)^2}{-7^2 - 2^2}$

55. $\dfrac{(8 - 4)^2}{8^2 - 4^2}$

56. $\dfrac{(6 - 2)^2}{6^2 - 2^2}$

57. $\dfrac{-4 \cdot 3^2 - 5 \cdot 2^2}{-8(7)}$

58. $\dfrac{-2 \cdot 5^2 + 3 \cdot 2^3}{-3(13)}$

59. $\dfrac{3 \cdot 10^2 + 4 \cdot 10 + 5}{345}$

60. $\dfrac{5 \cdot 10^2 + 6 \cdot 10 + 7}{567}$

61. $\dfrac{7 - [(2 - 3) - 4]}{-1 - 2 - 3}$

62. $\dfrac{2 - [(3 - 5) - 8]}{-3 - 4 - 5}$

63. $\dfrac{6(-4) - 2(5 - 8)}{-6 - 3 - 5}$

64. $\dfrac{3(-4) - 5(9 - 11)}{-9 - 2 - 3}$

65. $\dfrac{3(-5 - 3) + 4(7 - 9)}{5(-2) + 3(-4)}$

66. $\dfrac{-2(6 - 10) - 3(8 - 5)}{6(-3) - 6(-2)}$

67. $\dfrac{|3 - 9|}{3 - 9}$

68. $\dfrac{|4 - 7|}{4 - 7}$

Answer the following questions.

69. What is the quotient of -12 and -4?

70. The quotient of -4 and -12 is what number?

71. What number do we divide by -5 to get 2?

72. What number do we divide by -3 to get 4?

73. Twenty-seven divided by what number is -9?

74. Fifteen divided by what number is -3?

75. If the quotient of -20 and 4 is decreased by 3, what number results?

76. If -4 is added to the quotient of 24 and -8, what number results?

Applying the Concepts

77. Investment Suppose that you and 3 friends bought equal shares of an investment for a total of $15,000 and then sold it later for only $13,600. How much did each person lose?

78. Investment If 8 people invest $500 each in a stamp collection and after a year the collection is worth $3,800, how much did each person lose?

79. Temperature Change Suppose that the temperature outside is dropping at a constant rate. If the temperature is 75°F at noon and drops to 61°F by 4:00 in the afternoon, by how much did the temperature change each hour?

80. Temperature Change In a chemistry class, a thermometer is placed in a beaker of hot water. The initial temperature of the water is 165°F. After 10 minutes, the water has cooled to 72°F. If the water temperature drops at a constant rate, by how much does the water temperature change each minute?

1.8   Subsets of the Real Numbers

In Section 1.2 we introduced the real numbers and defined them as the numbers associated with points on the real number line. At that time we said the real numbers include whole numbers, fractions, and decimals, as well as other numbers that are not as familiar to us as these numbers. In this section we take a more detailed look at the kinds of numbers that make up the set of real numbers.

Let's begin with the definition of a set, the starting point for all the branches of mathematics.

SETS

> **DEFINITION** A **set** is a collection of objects or things. The objects in the set are called *elements*, or *members*, of the set.

Sets are usually denoted by capital letters and elements of sets by lowercase letters. We use braces, { }, to enclose the elements of a set.

To show that an element is contained in a set we use the symbol $\in$. That is,

$x \in A$ is read "x is an element (member) of set A"

For example, if A is the set $\{1, 2, 3\}$, then $2 \in A$. On the other hand, $5 \notin A$ means 5 is not an element of set A.

> **DEFINITION**   Set A is a **subset** of set B, written $A \subset B$, if every element in A is also an element of B. That is,
>
> $A \subset B$ if and only if A is contained in B

EXAMPLES

1. The set of numbers used to count things is $\{1, 2, 3, \ldots\}$. The dots mean the set continues indefinitely in the same manner. This is an example of an *infinite* set.
2. The set of all numbers represented by the dots on the faces of a regular die is $\{1, 2, 3, 4, 5, 6\}$. This set is a subset of the set in Example 1. It is an example of a *finite* set, since it has a limited number of elements.

The diagrams shown here are called *Venn diagrams* after John Venn (1834–1923). They can be used to visualize operations with sets. The region inside the circle labeled A is set A; the region inside the circle labeled B is set B.

> **DEFINITION**   The set with no members is called the **empty,** or **null, set.** It is denoted by the symbol $\varnothing$.
>   The empty set is considered a subset of every set.

OPERATIONS WITH SETS

Two basic operations are used to combine sets: union and intersection.

> **DEFINITION**   The **union** of two sets A and B, written $A \cup B$, is the set of all elements that are either in A or in B, or in both A and B. The key word here is *or.* For an element to be in $A \cup B$ it must be in A or B. In symbols, the definition looks like this:
>
> $x \in A \cup B$ if and only if $x \in A$ or $x \in B$

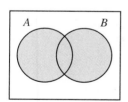

$A \cup B$

FIGURE 1 The *union* of two sets

> **DEFINITION**   The **intersection** of two sets A and B, written $A \cap B$, is the set of elements in both A and B. The key word in this definition is the word *and.* For an element to be in $A \cap B$ it must be in both A and B, or
>
> $x \in A \cap B$ if and only if $x \in A$ and $x \in B$

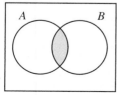

$A \cap B$

FIGURE 2 The *intersection* of two sets

EXAMPLES   Let $A = \{1, 3, 5\}$, $B = \{0, 2, 4\}$, and $C = \{1, 2, 3, \ldots\}$. Then

3. $A \cup B = \{0, 1, 2, 3, 4, 5\}$
4. $A \cap B = \varnothing$ (A and B have no elements in common)

5. $A \cap C = \{1, 3, 5\} = A$

6. $B \cup C = \{0, 1, 2, 3, \ldots\}$

Another notation we can use to describe sets is called *set builder* notation. Here is how we write our definition for the union of two sets A and B using set builder notation:

$$A \cup B = \{x \,|\, x \in A \text{ or } x \in B\}$$

The right side of this statement is read "the set of all x such that x is a member of A or x is a member of B." As you can see, the vertical line after the first x is read "such that."

EXAMPLE 7   If $A = \{1, 2, 3, 4, 5, 6\}$, find $C = \{x \,|\, x \in A \text{ and } x \geq 4\}$.

Solution   We are looking for all the elements of A that are also greater than or equal to 4. They are 4, 5, and 6. Using set notation, we have

$$C = \{4, 5, 6\}$$

SUBSETS OF THE REAL NUMBERS

The numbers that make up the set of real numbers can be classified as *counting numbers, whole numbers, integers, rational numbers,* and *irrational numbers;* each is said to be a *subset* of the real numbers.

Here is a detailed description of the major subsets of the real numbers.

The counting numbers are the numbers with which we count. They are the numbers 1, 2, 3, and so on.

$$\text{Counting numbers} = \{1, 2, 3, \ldots\}$$

EXAMPLE 8   Which of the numbers in the following set are not counting numbers?

$$\left\{-3, 0, \frac{1}{2}, 1, 1.5, 3\right\}$$

Solution   The numbers $-3, 0, \frac{1}{2}$, and 1.5 are not counting numbers.

The whole numbers include the counting numbers and the number 0.

$$\text{Whole numbers} = \{0, 1, 2, \ldots\}$$

The set of integers includes the whole numbers and the opposites of all the counting numbers.

$$\text{Integers} = \{\ldots, -3, -2, -1, 0, 1, 2, 3, \ldots\}$$

When we refer to positive integers, we are referring to the numbers 1, 2, 3, ... Likewise, the negative integers are $-1, -2, -3, . . .$. The number 0 is neither positive nor negative.

EXAMPLE 9   Which of the numbers in the following set are not integers?

$$\left\{-5, -1.75, 0, \frac{2}{3}, 1, \pi, 3\right\}$$

Solution   The only numbers in the set that are not integers are $-1.75, \frac{2}{3}$, and π.

The set of *rational numbers* is the set of numbers commonly called "fractions" together with the integers. The set of rational numbers is difficult to list in the same way we have listed the other sets, so we will use a different kind of notation:

$$\text{Rational numbers} = \left\{\frac{a}{b} \,\middle|\, a \text{ and } b \text{ are integers } (b \neq 0)\right\}$$

This notation is read "The set of elements $\frac{a}{b}$ such that a and b are integers (and b is not 0)." If a number can be put in the form $\frac{a}{b}$, where a and b are both from the set of integers, then it is called a rational number.

Rational numbers include any number that can be written as the ratio of two integers. That is, rational numbers are numbers that can be put in the form

$$\frac{\text{integer}}{\text{integer}}$$

EXAMPLE 10   Show why each of the numbers in the following set is a rational number.

$$\left\{-3, -\frac{2}{3}, 0, 0.333 \ . \ . \ . \ , 0.75\right\}$$

Solution   The number -3 is a rational number because it can be written as the ratio of -3 to 1. That is,

$$-3 = \frac{-3}{1}$$

Similarly, the number $-\frac{2}{3}$ can be thought of as the ratio of -2 to 3, while the number 0 can be thought of as the ratio of 0 to 1.

Any repeating decimal, such as 0.333 . . . (the dots indicate that the 3's repeat forever), can be written as the ratio of two integers. In this case 0.333 . . . is the same as the fraction $\frac{1}{3}$.

Finally, any decimal that terminates after a certain number of digits can be written as the ratio of two integers. The number 0.75 is equal to the fraction $\frac{3}{4}$ and is therefore a rational number.

Still other numbers exist, each of which is associated with a point on the real number line, that cannot be written as the ratio of two integers. In decimal form they never terminate and never repeat a sequence of digits indefinitely. They are called *irrational numbers* (because they are not rational):

Irrational numbers = {nonrational numbers; nonrepeating, nonterminating decimals}

We cannot write any irrational number in a form that is familiar to us, because they are all nonterminating, nonrepeating decimals. Since they are not rational, they cannot be written as the ratio of two integers. They have to be represented in other ways. One irrational number you have probably seen before is π. It is not 3.14. Rather, 3.14 is an approximation to π. It cannot be written as a decimal number. Other representations for irrational numbers are $\sqrt{2}$, $\sqrt{3}$, $\sqrt{5}$, $\sqrt{6}$, and, in general, the square root of any number that is not itself a perfect square. (If you are not familiar with square roots, you will be after Chapter 9.) Right now it is enough to know that some numbers on the number line cannot be written as the ratio of two integers or in decimal form. We call them irrational numbers.

The set of real numbers is the set of numbers that are either rational or irrational. That is, a real number is either rational or irrational.

Real numbers = {all rational numbers and all irrational numbers}

PRIME NUMBERS AND FACTORING

The following diagram shows the relationship between multiplication and factoring:

$$\text{Factors} \rightarrow 3 \cdot 4 = 12 \leftarrow \text{Product}$$

Multiplication (above), Factoring (below)

When we read the problem from left to right, we say the product of 3 and 4 is 12. Or we multiply 3 and 4 to get 12. When we read the problem in the other direction, from right to left, we say we have *factored* 12 into 3 times 4, or 3 and 4 are *factors* of 12.

The number 12 can be factored still further:

$$12 = 4 \cdot 3$$
$$= 2 \cdot 2 \cdot 3$$
$$= 2^2 \cdot 3$$

The numbers 2 and 3 are called *prime factors* of 12, because neither of them can be factored any further.

DEFINITION   If a and b represent integers, then a is said to be a **factor** (or divisor) of b if a divides b evenly—that is, if a divides b with no remainder.

> **DEFINITION**   A **prime number** is any positive integer larger than 1 whose only positive factors (divisors) are itself and 1.

Note: The number 15 is not a prime number since it has factors of 3 and 5. That is, $15 = 3 \cdot 5$. When a whole number larger than 1 is not prime, it is said to be *composite*.

Here is a list of the first few prime numbers.

$$\text{Prime numbers} = \{2, 3, 5, 7, 11, 13, 17, 19, 23, 29, 31, 37, 41, \ldots\}$$

When a number is not prime, we can factor it into the product of prime numbers. To factor a number into the product of primes, we simply factor it until it cannot be factored further.

EXAMPLE 11   Factor the number 60 into the product of prime numbers.

Solution   We begin by writing 60 as the product of any two positive integers whose product is 60, like 6 and 10:

$$60 = 6 \cdot 10$$

We then factor these numbers:

$$
\begin{aligned}
60 &= 6 \cdot 10 \\
&= (2 \cdot 3) \cdot (2 \cdot 5) \\
&= 2 \cdot 2 \cdot 3 \cdot 5 \\
&= 2^2 \cdot 3 \cdot 5
\end{aligned}
$$

Note: It is customary to write the prime factors in order from smallest to largest.

EXAMPLE 12   Factor the number 630 into the product of primes.

Solution   Let's begin by writing 630 as the product of 63 and 10:

$$
\begin{aligned}
630 &= 63 \cdot 10 \\
&= (7 \cdot 9) \cdot (2 \cdot 5) \\
&= 7 \cdot 3 \cdot 3 \cdot 2 \cdot 5 \\
&= 2 \cdot 3^2 \cdot 5 \cdot 7
\end{aligned}
$$

It makes no difference which two numbers we start with, as long as their product is 630. We will always get the same result because a number has only one set of prime factors.

$$630 = 18 \cdot 35$$
$$= 3 \cdot 6 \cdot 5 \cdot 7$$
$$= 3 \cdot 2 \cdot 3 \cdot 5 \cdot 7$$
$$= 2 \cdot 3^2 \cdot 5 \cdot 7$$

Note: There are some "tricks" to finding the divisors of a number. For instance, if a number ends in 0 or 5, then it is divisible by 5. If a number ends in an even number (0, 2, 4, 6, or 8), then it is divisible by 2. A number is divisible by 3 if the sum of its digits is divisible by 3. For example, 921 is divisible by 3 because the sum of its digits is $9 + 2 + 1 = 12$, which is divisible by 3.

When we have factored a number into the product of its prime factors, we not only know what prime numbers divide the original number, but we also know all of the other numbers that divide it as well. For instance, if we were to factor 210 into its prime factors, we would have $210 = 2 \cdot 3 \cdot 5 \cdot 7$, which means that 2, 3, 5, and 7 divide 210, as well as any combination of products of 2, 3, 5, and 7. That is, since 3 and 7 divide 210, then so does their product 21. Since 3, 5, and 7 each divide 210, then so does their product 105:

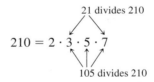

Although there are many ways in which factoring is used in arithmetic and algebra, one simple application is in reducing fractions to lowest terms.

Recall that we reduce fractions to lowest terms by dividing the numerator and denominator by the same number. We can use the prime factorization of numbers to help us reduce fractions with large numerators and denominators.

EXAMPLE 13　Reduce $\frac{210}{231}$ to lowest terms.

Solution　First, we factor 210 and 231 into the product of prime factors. Then we reduce to lowest terms by dividing the numerator and denominator by any factors they have in common.

$$\frac{210}{231} = \frac{2 \cdot 3 \cdot 5 \cdot 7}{3 \cdot 7 \cdot 11} \qquad \text{Factor the numerator and denominator completely.}$$

$$= \frac{2 \cdot \cancel{3} \cdot 5 \cdot \cancel{7}}{\cancel{3} \cdot \cancel{7} \cdot 11} \qquad \text{Divide the numerator and denominator by } 3 \cdot 7.$$

$$= \frac{2 \cdot 5}{11}$$

$$= \frac{10}{11}$$

Note: The small lines we have drawn through the factors that are common to the numerator and denominator are used to indicate that we have divided the numerator and denominator by those factors.

Getting Ready for Class

After reading through the preceding section, respond in your own words and in complete sentences.

A. What is a whole number?

B. How are factoring and multiplication related?

C. Is every integer also a rational number? Explain.

D. What is a prime number?

PROBLEM SET 1.8

For the following problems, let $A = \{0, 2, 4, 6\}$, $B = \{1, 2, 3, 4, 5\}$, and $C = \{1, 3, 5, 7\}$.

1. $A \cup B$

2. $A \cap B$

3. $A \cap C$

4. $B \cup C$

5. $A \cup (B \cap C)$

6. $C \cup (A \cap B)$

7. $\{x \mid x \in A \text{ and } x < 4\}$

8. $\{x \mid x \in B \text{ and } x > 3\}$

9. $\{x \mid x \in A \text{ and } x \notin B\}$

10. $\{x \mid x \in B \text{ and } x \notin C\}$

11. $\{x \mid x \in A \text{ or } x \in C\}$

12. $\{x \mid x \in A \text{ or } x \in B\}$

Given the numbers in the set $\{-3, -2.5, 0, 1, \frac{3}{2}, \sqrt{15}\}$:

13. List all the whole numbers.

14. List all the integers.

15. List all the rational numbers.

16. List all the irrational numbers.

17. List all the real numbers.

Given the numbers in the set
$\{-10, -8, -0.333 \ldots, -2, 9, \frac{25}{3}, \pi\}$:

18. List all the whole numbers.

19. List all the integers.

20. List all the rational numbers.

21. List all the irrational numbers.

22. List all the real numbers.

Identify the following statements as either true or false.

23. Every whole number is also an integer.

24. The set of whole numbers is a subset of the set of integers.

25. A number can be both rational and irrational.

26. The set of rational numbers and the set of irrational numbers have some elements in common.

27. Some whole numbers are also negative integers.

28. Every rational number is also a real number.

29. All integers are also rational numbers.

30. The set of integers is a subset of the set of rational numbers.

Label each of the following numbers as *prime* or *composite*. If a number is composite, then factor it completely.

31. 48

32. 72

33. 37

34. 23

35. 1,023

36. 543

Factor the following into the product of primes. When the number has been factored completely, write its prime factors from smallest to largest.

37. 144

38. 288

39. 38

40. 63   **41.** 105   **42.** 210

43. 180   **44.** 900   **45.** 385

46. 1,925   **47.** 121   **48.** 546

49. 420   **50.** 598   **51.** 620

52. 2,310

Reduce each fraction to lowest terms by first factoring the numerator and denominator into the product of prime factors and then dividing out any factors they have in common.

53. $\dfrac{105}{165}$ **54.** $\dfrac{165}{385}$ **55.** $\dfrac{525}{735}$

56. $\dfrac{550}{735}$ **57.** $\dfrac{385}{455}$ **58.** $\dfrac{385}{735}$

59. $\dfrac{322}{345}$ **60.** $\dfrac{266}{285}$ **61.** $\dfrac{205}{369}$

62. $\dfrac{111}{185}$ **63.** $\dfrac{215}{344}$ **64.** $\dfrac{279}{310}$

65. Factor 6^3 into the product of prime factors by first factoring 6 and then raising each of its factors to the third power.

66. Factor 12^2 into the product of prime factors by first factoring 12 and then raising each of its factors to the second power.

67. Factor $9^4 \cdot 16^2$ into the product of prime factors by first factoring 9 and 16 completely.

68. Factor $10^2 \cdot 12^3$ into the product of prime factors by first factoring 10 and 12 completely.

69. Simplify the expression $3 \cdot 8 + 3 \cdot 7 + 3 \cdot 5$ and then factor the result into the product of primes. (Notice one of the factors of the answer is 3.)

70. Simplify the expression $5 \cdot 4 + 5 \cdot 9 + 5 \cdot 3$ and then factor the result into the product of primes.

Recall the Fibonacci sequence we introduced earlier in this chapter.

Fibonacci sequence = 1, 1, 2, 3, 5, 8, . . .

Any number in the Fibonacci sequence is a *Fibonacci number.*

71. The Fibonacci numbers are not a subset of which of the following sets: real numbers, rational numbers, irrational numbers, whole numbers?

72. Name three Fibonacci numbers that are prime numbers.

73. Name three Fibonacci numbers that are composite numbers.

74. Is the sequence of odd numbers a subset of the Fibonacci numbers?

1.9   Addition and Subtraction With Fractions

You may recall from previous math classes that to add two fractions with the same denominator, you simply add their numerators and put the result over the common denominator:

$$\frac{3}{4} + \frac{2}{4} = \frac{3 + 2}{4} = \frac{5}{4}$$

The reason we add numerators but do not add denominators is that we must follow the distributive property. To see this, you first have to recall that $\frac{3}{4}$ can be written as $3 \cdot \frac{1}{4}$, and $\frac{2}{4}$ can be written as $2 \cdot \frac{1}{4}$ (dividing by 4 is equivalent to multiplying by $\frac{1}{4}$). Here is the addition problem again, this time showing the use of the distributive property:

$$\frac{3}{4} + \frac{2}{4} = 3 \cdot \frac{1}{4} + 2 \cdot \frac{1}{4}$$

$$= (3 + 2) \cdot \frac{1}{4} \qquad \text{Distributive property}$$

$$= 5 \cdot \frac{1}{4}$$

$$= \frac{5}{4}$$

What we have here is the sum of the numerators placed over the *common denominator*. In symbols we have the following.

ADDITION AND SUBTRACTION OF FRACTIONS

If a, b, and c are integers and c is not equal to 0, then

$$\frac{a}{c} + \frac{b}{c} = \frac{a + b}{c}$$

This rule holds for subtraction as well. That is,

$$\frac{a}{c} - \frac{b}{c} = \frac{a - b}{c}$$

In Examples 1–4, find the sum or difference. (Add or subtract as indicated.) Reduce all answers to lowest terms. (Assume all variables represent nonzero numbers.)

EXAMPLES

1. $\dfrac{3}{8} + \dfrac{1}{8} = \dfrac{3 + 1}{8}$ Add numerators; keep the same denominator.

$$= \frac{4}{8} \qquad \text{The sum of 3 and 1 is 4.}$$

$$= \frac{1}{2} \qquad \text{Reduce to lowest terms.}$$

2. $\dfrac{a + 5}{8} - \dfrac{3}{8} = \dfrac{a + 5 - 3}{8}$ Combine numerators; keep the same denominator.

$$= \frac{a + 2}{8}$$

3. $\dfrac{9}{x} - \dfrac{3}{x} = \dfrac{9 - 3}{x}$ Subtract numerators; keep the same denominator.

$$= \frac{6}{x} \qquad \text{The difference of 9 and 3 is 6.}$$

4. $\dfrac{3}{7} + \dfrac{2}{7} - \dfrac{9}{7} = \dfrac{3 + 2 - 9}{7}$

$\qquad\qquad = \dfrac{-4}{7}$

$\qquad\qquad = -\dfrac{4}{7} \qquad$ Unlike signs give a negative answer.

As Examples 1–4 indicate, addition and subtraction are simple, straightforward processes when all the fractions have the same denominator. We will now turn our attention to the process of adding fractions that have different denominators. In order to get started, we need the following definition.

> **DEFINITION** The **least common denominator (LCD)** for a set of denominators is the smallest number that is exactly divisible by each denominator. (Note that in some books the least common denominator is also called the *least common multiple*.)
>
> In other words, all the denominators of the fractions involved in a problem must divide into the least common denominator exactly. That is, they divide it without giving a remainder.

EXAMPLE 5 Find the LCD for the fractions $\dfrac{5}{12}$ and $\dfrac{7}{18}$.

Solution The least common denominator for the denominators 12 and 18 must be the smallest number divisible by both 12 and 18. We can factor 12 and 18 completely and then build the LCD from these factors. Factoring 12 and 18 completely gives us

$$12 = 2 \cdot 2 \cdot 3 \qquad 18 = 2 \cdot 3 \cdot 3$$

Now, if 12 is going to divide the LCD exactly, then the LCD must have factors of $2 \cdot 2 \cdot 3$. If 18 is to divide it exactly, it must have factors of $2 \cdot 3 \cdot 3$. We don't need to repeat the factors that 12 and 18 have in common:

12 divides the LCD

$\left. \begin{array}{l} 12 = 2 \cdot 2 \cdot 3 \\ 18 = 2 \cdot 3 \cdot 3 \end{array} \right\} \text{LCD} = 2 \cdot 2 \cdot 3 \cdot 3 = 36$

18 divides the LCD

In other words, first we write down the factors of 12, then we attach the factors of 18 that do not already appear as factors of 12. We start with $2 \cdot 2 \cdot 3$ because those are the factors of 12. Then we look at the first factor of 18. It is 2. Since 2 already appears in the expression $2 \cdot 2 \cdot 3$, we don't need to attach another one. Next, we look at the factors $3 \cdot 3$. The expression $2 \cdot 2 \cdot 3$ has one 3. In order for it to contain the expression $3 \cdot 3$, we attach another 3. The final expression, our LCD, is $2 \cdot 2 \cdot 3 \cdot 3$.

The LCD for 12 and 18 is 36. It is the smallest number that is divisible by both 12 and 18; 12 divides it exactly three times, and 18 divides it exactly two times.

We can use the results of Example 5 to find the sum of the fractions $\frac{5}{12}$ and $\frac{7}{18}$.

E X A M P L E   6   Add $\frac{5}{12} + \frac{7}{18}$.

Solution   We can add fractions only when they have the same denominators. In Example 5 we found the LCD for $\frac{5}{12}$ and $\frac{7}{18}$ to be 36. We change $\frac{5}{12}$ and $\frac{7}{18}$ to equivalent fractions that each have 36 for a denominator by applying Property 1 (see p. 16) for fractions:

$$\frac{5}{12} = \frac{5 \cdot 3}{12 \cdot 3} = \frac{15}{36}$$

$$\frac{7}{18} = \frac{7 \cdot 2}{18 \cdot 2} = \frac{14}{36}$$

The fraction $\frac{15}{36}$ is equivalent to $\frac{5}{12}$, since it was obtained by multiplying both the numerator and denominator by 3. Likewise, $\frac{14}{36}$ is equivalent to $\frac{7}{18}$, since it was obtained by multiplying the numerator and denominator by 2. All we have left to do is to add numerators:

$$\frac{15}{36} + \frac{14}{36} = \frac{29}{36}$$

The sum of $\frac{5}{12}$ and $\frac{7}{18}$ is the fraction $\frac{29}{36}$. Let's write the complete problem again step-by-step.

$$\frac{5}{12} + \frac{7}{18} = \frac{5 \cdot 3}{12 \cdot 3} + \frac{7 \cdot 2}{18 \cdot 2} \qquad \text{Rewrite each fraction as an equivalent}$$
$$\text{fraction with denominator 36.}$$

$$= \frac{15}{36} + \frac{14}{36}$$

$$= \frac{29}{36} \qquad \text{Add numerators; keep the common}$$
$$\text{denominator.}$$

E X A M P L E   7   Find the LCD for $\frac{3}{4}$ and $\frac{1}{6}$.

Solution   We factor 4 and 6 into products of prime factors and build the LCD from these factors:

$$\left. \begin{array}{l} 4 = 2 \cdot 2 \\ 6 = 2 \cdot 3 \end{array} \right\} \qquad \text{LCD} = 2 \cdot 2 \cdot 3 = 12$$

The LCD is 12. Both denominators divide it exactly; 4 divides 12 exactly three times, and 6 divides 12 exactly two times.

EXAMPLE 8 Add $\frac{3}{4} + \frac{1}{6}$.

Solution In Example 7 we found that the LCD for these two fractions is 12. We begin by changing $\frac{3}{4}$ and $\frac{1}{6}$ to equivalent fractions with denominator 12:

$$\frac{3}{4} = \frac{3 \cdot 3}{4 \cdot 3} = \frac{9}{12}$$

$$\frac{1}{6} = \frac{1 \cdot 2}{6 \cdot 2} = \frac{2}{12}$$

The fraction $\frac{9}{12}$ is equal to the fraction $\frac{3}{4}$, since it was obtained by multiplying the numerator and denominator of $\frac{3}{4}$ by 3. Likewise, $\frac{2}{12}$ is equivalent to $\frac{1}{6}$, since it was obtained by multiplying the numerator and denominator of $\frac{1}{6}$ by 2. To complete the problem, we add numerators:

$$\frac{9}{12} + \frac{2}{12} = \frac{11}{12}$$

The sum of $\frac{3}{4}$ and $\frac{1}{6}$ is $\frac{11}{12}$. Here is how the complete problem looks:

$$\frac{3}{4} + \frac{1}{6} = \frac{3 \cdot 3}{4 \cdot 3} + \frac{1 \cdot 2}{6 \cdot 2} \qquad \text{Rewrite each fraction as an equivalent fraction with denominator 12.}$$

$$= \frac{9}{12} + \frac{2}{12}$$

$$= \frac{11}{12} \qquad \text{Add numerators; keep the same denominator.}$$

EXAMPLE 9 Subtract $\frac{7}{15} - \frac{3}{10}$.

Solution Let's factor 15 and 10 completely and use these factors to build the LCD:

$$\left. \begin{array}{l} 15 = 3 \cdot 5 \\ 10 = 2 \cdot 5 \end{array} \right\} \text{LCD} = 2 \cdot 3 \cdot 5 = 30$$

15 divides the LCD

10 divides the LCD

Changing to equivalent fractions and subtracting, we have

$$\frac{7}{15} - \frac{3}{10} = \frac{7 \cdot 2}{15 \cdot 2} - \frac{3 \cdot 3}{10 \cdot 3} \qquad \text{Rewrite as equivalent fractions with the LCD for denominator.}$$

$$= \frac{14}{30} - \frac{9}{30}$$

$$= \frac{5}{30} \qquad\qquad \text{Subtract numerators; keep the LCD.}$$

$$= \frac{1}{6} \qquad\qquad \text{Reduce to lowest terms.}$$

As a summary of what we have done so far and as a guide to working other problems, we will now list the steps involved in adding and subtracting fractions with different denominators.

TO ADD OR SUBTRACT ANY TWO FRACTIONS

Step 1: Factor each denominator completely and use the factors to build the LCD. (Remember, the LCD is the smallest number divisible by each of the denominators in the problem.)

Step 2: Rewrite each fraction as an equivalent fraction that has the LCD for its denominator. This is done by multiplying both the numerator and denominator of the fraction in question by the appropriate whole number.

Step 3: Add or subtract the numerators of the fractions produced in step 2. This is the numerator of the sum or difference. The denominator of the sum or difference is the LCD.

Step 4: Reduce the fraction produced in step 3 to lowest terms if it is not already in lowest terms.

The idea behind adding or subtracting fractions is really very simple. We can add or subtract only fractions that have the same denominators. If the fractions we are trying to add or subtract do not have the same denominators, we rewrite each of them as an equivalent fraction with the LCD for a denominator.

Here are some further examples of sums and differences of fractions.

EXAMPLE 10   Add $\frac{1}{6} + \frac{1}{8} + \frac{1}{4}$.

Solution   We begin by factoring the denominators completely and building the LCD from the factors that result:

$$
\begin{aligned}
6 &= 2 \cdot 3 \\
8 &= 2 \cdot 2 \cdot 2 \\
4 &= 2 \cdot 2
\end{aligned}
\qquad
\text{LCD} = 2 \cdot 2 \cdot 2 \cdot 3 = 24
$$

8 divides the LCD

4 divides the LCD   6 divides the LCD

We then change to equivalent fractions and add as usual:

$$\frac{1}{6} + \frac{1}{8} + \frac{1}{4} = \frac{1 \cdot \mathbf{4}}{6 \cdot \mathbf{4}} + \frac{1 \cdot \mathbf{3}}{8 \cdot \mathbf{3}} + \frac{1 \cdot \mathbf{6}}{4 \cdot \mathbf{6}}$$

$$= \frac{4}{24} + \frac{3}{24} + \frac{6}{24}$$

$$= \frac{13}{24}$$

EXAMPLE 11 Subtract $3 - \frac{5}{6}$.

Solution The denominators are 1 (because $3 = \frac{3}{1}$) and 6. The smallest number divisible by both 1 and 6 is 6.

$$3 - \frac{5}{6} = \frac{3}{1} - \frac{5}{6}$$

$$= \frac{3 \cdot \mathbf{6}}{1 \cdot \mathbf{6}} - \frac{5}{6}$$

$$= \frac{18}{6} - \frac{5}{6}$$

$$= \frac{13}{6}$$

EXAMPLE 12 Find the next number in each sequence.

(a) $\frac{1}{2}, 0, -\frac{1}{2}, \ldots$ **(b)** $\frac{1}{2}, 1, \frac{3}{2}, \ldots$ **(c)** $\frac{1}{2}, \frac{1}{4}, \frac{1}{8}, \ldots$

Solution

(a) $\frac{1}{2}, 0, -\frac{1}{2}, \ldots$: Adding $-\frac{1}{2}$ to each term produces the next term. The fourth term will be $-\frac{1}{2} + (-\frac{1}{2}) = -1$. This is an arithmetic sequence.

(b) $\frac{1}{2}, 1, \frac{3}{2}, \ldots$: Each term comes from the term before it by adding $\frac{1}{2}$. The fourth term will be $\frac{3}{2} + \frac{1}{2} = 2$. This sequence is also an arithmetic sequence.

(c) $\frac{1}{2}, \frac{1}{4}, \frac{1}{8}, \ldots$: This is a geometric sequence in which each term comes from the term before it by multiplying by $\frac{1}{2}$ each time. The next term will be $\frac{1}{8} \cdot \frac{1}{2} = \frac{1}{16}$.

Getting Ready for Class

After reading through the preceding section, respond in your own words and in complete sentences.

A. How do we add two fractions that have the same denominators?

B. What is a least common denominator?

C. What is the first step in adding two fractions that have different denominators?

D. What is the last thing you do when adding two fractions?

PROBLEM SET 1.9

Find the following sums and differences and reduce to lowest terms. Add and subtract as indicated. Assume all variables represent nonzero numbers.

1. $\dfrac{3}{6} + \dfrac{1}{6}$ **2.** $\dfrac{2}{5} + \dfrac{3}{5}$ **3.** $\dfrac{3}{8} - \dfrac{5}{8}$

4. $\dfrac{1}{7} - \dfrac{6}{7}$ **5.** $-\dfrac{1}{4} + \dfrac{3}{4}$ **6.** $-\dfrac{4}{9} + \dfrac{7}{9}$

7. $\dfrac{x}{3} - \dfrac{1}{3}$ **8.** $\dfrac{x}{8} - \dfrac{1}{8}$ **9.** $\dfrac{1}{4} + \dfrac{2}{4} + \dfrac{3}{4}$

10. $\dfrac{2}{5} + \dfrac{3}{5} + \dfrac{4}{5}$ **11.** $\dfrac{x+7}{2} - \dfrac{1}{2}$ **12.** $\dfrac{x+5}{4} - \dfrac{3}{4}$

13. $\dfrac{1}{10} - \dfrac{3}{10} - \dfrac{4}{10}$ **14.** $\dfrac{3}{20} - \dfrac{1}{20} - \dfrac{4}{20}$

15. $\dfrac{1}{a} + \dfrac{4}{a} + \dfrac{5}{a}$ **16.** $\dfrac{5}{a} + \dfrac{4}{a} + \dfrac{3}{a}$

Find the LCD for each of the following; then use the methods developed in this section to add and subtract as indicated. Assume all variables represent nonzero numbers.

17. $\dfrac{1}{8} + \dfrac{3}{4}$ **18.** $\dfrac{1}{6} + \dfrac{2}{3}$ **19.** $\dfrac{3}{10} - \dfrac{1}{5}$

20. $\dfrac{5}{6} - \dfrac{1}{12}$ **21.** $\dfrac{4}{9} + \dfrac{1}{3}$ **22.** $\dfrac{1}{2} + \dfrac{1}{4}$

23. $2 + \dfrac{1}{3}$ **24.** $3 + \dfrac{1}{2}$ **25.** $-\dfrac{3}{4} + 1$

26. $-\dfrac{3}{4} + 2$ **27.** $\dfrac{1}{2} + \dfrac{2}{3}$ **28.** $\dfrac{2}{3} + \dfrac{1}{4}$

29. $\dfrac{5}{12} - \left(-\dfrac{3}{8}\right)$ **30.** $\dfrac{9}{16} - \left(-\dfrac{7}{12}\right)$

31. $-\dfrac{1}{20} + \dfrac{8}{30}$ **32.** $-\dfrac{1}{30} + \dfrac{9}{40}$

33. $\dfrac{17}{30} + \dfrac{11}{42}$ **34.** $\dfrac{19}{42} + \dfrac{13}{70}$

35. $\dfrac{25}{84} + \dfrac{41}{90}$ **36.** $\dfrac{23}{70} + \dfrac{29}{84}$

37. $\dfrac{13}{126} - \dfrac{13}{180}$ **38.** $\dfrac{17}{84} - \dfrac{17}{90}$

39. $\dfrac{3}{4} + \dfrac{1}{8} + \dfrac{5}{6}$ **40.** $\dfrac{3}{8} + \dfrac{2}{5} + \dfrac{1}{4}$

41. $\dfrac{1}{2} + \dfrac{1}{3} + \dfrac{1}{4} + \dfrac{1}{6}$ **42.** $\dfrac{1}{8} + \dfrac{1}{4} + \dfrac{1}{5} + \dfrac{1}{10}$

43. Find the sum of $\dfrac{3}{7}$, 2, and $\dfrac{1}{9}$.

44. Find the sum of 6, $\dfrac{6}{11}$, and 11.

45. Give the difference of $\dfrac{7}{8}$ and $\dfrac{1}{4}$.

46. Give the difference of $\dfrac{9}{10}$ and $\dfrac{1}{100}$.

Find the fourth term in each sequence.

47. $\dfrac{1}{3}$, 0, $-\dfrac{1}{3}$, . . .   **48.** $\dfrac{2}{3}$, 0, $-\dfrac{2}{3}$, . . .

49. $\dfrac{1}{3}$, 1, $\dfrac{5}{3}$, . . .   **50.** 1, $\dfrac{3}{2}$, 2, . . .

51. 1, $\dfrac{1}{5}$, $\dfrac{1}{25}$, . . .   **52.** 1, $-\dfrac{1}{2}$, $\dfrac{1}{4}$, . . .

CHAPTER 1 SUMMARY

Note: We will use the margins in the chapter summaries to give examples that correspond to the topic being reviewed whenever it is appropriate.

The number(s) in brackets next to each heading indicates the section(s) in which that topic is discussed.

Symbols [1.1]

$$a = b \qquad a \text{ is equal to } b$$

$$a \neq b \qquad a \text{ is not equal to } b$$

$$a < b \qquad a \text{ is less than } b$$

$$a \nless b \qquad a \text{ is not less than } b$$

$$a > b \qquad a \text{ is greater than } b$$

$$a \ngtr b \qquad a \text{ is not greater than } b$$

Examples

1. $2^5 = 2 \cdot 2 \cdot 2 \cdot 2 \cdot 2 = 32$

   $5^2 = 5 \cdot 5 = 25$

   $10^3 = 10 \cdot 10 \cdot 10 = 1{,}000$

   $1^4 = 1 \cdot 1 \cdot 1 \cdot 1 = 1$

2. $10 + (2 \cdot 3^2 - 4 \cdot 2)$

   $$= 10 + (2 \cdot 9 - 4 \cdot 2)$$
   $$= 10 + (18 - 8)$$
   $$= 10 + 10$$
   $$= 20$$

Exponents [1.1]

Exponents are notation used to indicate repeated multiplication. In the expression 3^4, 3 is the *base* and 4 is the *exponent*.

$$3^4 = 3 \cdot 3 \cdot 3 \cdot 3 = 81$$

Order of Operations [1.1]

When evaluating a mathematical expression, we will perform the operations in the following order, beginning with the expression in the innermost parentheses or brackets and working our way out.

1. Simplify all numbers with exponents, working from left to right if more than one of these numbers is present.
2. Then do all multiplications and divisions left to right.
3. Finally, perform all additions and subtractions left to right.

3. Add all combinations of positive and negative 10 and 13:

   $$10 + 13 = 23$$
   $$10 + (-13) = -3$$
   $$-10 + 13 = 3$$
   $$-10 + (-13) = -23$$

Addition of Real Numbers [1.3]

To add two real numbers with

1. The same sign: Simply add their absolute values and use the common sign.
2. Different signs: Subtract the smaller absolute value from the larger absolute value. The answer has the same sign as the number with the larger absolute value.

4. Subtracting 2 is the same as adding -2:

   $$7 - 2 = 7 + (-2) = 5$$

Subtraction of Real Numbers [1.4]

To subtract one number from another, simply add the opposite of the number you are subtracting. That is, if a and b represent real numbers, then

$$a - b = a + (-b)$$

5. $3(5) = 15$

$3(-5) = -15$

$-3(5) = -15$

$-3(-5) = 15$

6. $\frac{-6}{2} = -6(\frac{1}{2}) = -3$

$\frac{-6}{-2} = -6(-\frac{1}{2}) = 3$

7. $|5| = 5$

$|-5| = 5$

8. The numbers 3 and -3 are opposites; their sum is 0:

$3 + (-3) = 0$

9. The numbers 2 and $\frac{1}{2}$ are reciprocals; their product is 1:

$2\left(\frac{1}{2}\right) = 1$

Multiplication of Real Numbers [1.6]
To multiply two real numbers, simply multiply their absolute values. Like signs give a positive answer. Unlike signs give a negative answer.

Division of Real Numbers [1.7]
Division by a number is the same as multiplication by its reciprocal. Like signs give a positive answer. Unlike signs give a negative answer.

Absolute Value [1.2]
The absolute value of a real number is its distance from zero on the real number line. Absolute value is never negative.

Opposites [1.2, 1.5]
Any two real numbers the same distance from zero on the number line but in opposite directions from zero are called opposites. Opposites always add to zero.

Reciprocals [1.2, 1.5]
Any two real numbers whose product is 1 are called reciprocals. Every real number has a reciprocal except 0.

Properties of Real Numbers [1.5]

	For Addition	*For Multiplication*
Commutative:	$a + b = b + a$	$a \cdot b = b \cdot a$
Associative:	$a + (b + c) = (a + b) + c$	$a \cdot (b \cdot c) = (a \cdot b) \cdot c$
Identity:	$a + 0 = a$	$a \cdot 1 = a$
Inverse:	$a + (-a) = 0$	$a\left(\dfrac{1}{a}\right) = 1$
Distributive:	$a(b + c) = ab + ac$	

10. (a) 7 and 100 are counting numbers, but 0 and -2 are not.

(b) 0 and 241 are whole numbers, but -4 and $\frac{1}{2}$ are not.

(c) -15, 0, and 20 are integers.

(d) -4, $-\frac{1}{2}$, 0.75, and 0.666 . . . are rational numbers.

(e) $-\pi$, $\sqrt{3}$, and π are irrational numbers.

(f) All the numbers listed above are real numbers.

Subsets of the Real Numbers [1.8]

Counting numbers:	$\{1, 2, 3, \ . \ . \ .\}$
Whole numbers:	$\{0, 1, 2, 3, \ . \ . \ .\}$
Integers:	$\{ . \ . \ . \ , -3, -2, -1, 0, 1, 2, 3, \ . \ . \ .\}$
Rational numbers:	{all numbers that can be expressed as the ratio of two integers}
Irrational numbers:	{all numbers on the number line that cannot be expressed as the ratio of two integers}
Real numbers:	{all numbers that are either rational or irrational}

11. The number 150 can be factored into the product of prime numbers:

$$150 = 15 \cdot 10$$
$$= (3 \cdot 5)(2 \cdot 5)$$
$$= 2 \cdot 3 \cdot 5^2$$

12. The LCD for $\frac{5}{12}$ and $\frac{7}{18}$ is 36.

13. $\dfrac{5}{12} + \dfrac{7}{18} = \dfrac{5}{12} \cdot \dfrac{3}{3} + \dfrac{7}{18} \cdot \dfrac{2}{2}$

$\qquad = \dfrac{15}{36} + \dfrac{14}{36}$

$\qquad = \dfrac{29}{36}$

Factoring [1.8]
Factoring is the reverse of multiplication.

Multiplication

Factors $\longrightarrow 3 \cdot 5 = 15 \longleftarrow$ Product

Factoring

Least Common Denominator (LCD) [1.9]
The *least common denominator* (LCD) for a set of denominators is the smallest number that is exactly divisible by each denominator.

Addition and Subtraction of Fractions [1.9]
To add (or subtract) two fractions with a common denominator, add (or subtract) numerators and use the common denominator.

$$\frac{a}{c} + \frac{b}{c} = \frac{a + b}{c} \qquad \text{and} \qquad \frac{a}{c} - \frac{b}{c} = \frac{a - b}{c}$$

COMMON MISTAKES

1. Interpreting absolute value as changing the sign of the number inside the absolute value symbols. $|-5| = +5, |+5| = -5$. (The first expression is correct, the second one is not.) To avoid this mistake, remember: Absolute value is a distance, and distance is always measured in positive units.

2. Using the phrase "two negatives make a positive." This works only with multiplication and division. With addition, two negative numbers produce a negative answer. It is best not to use the phrase "two negatives make a positive" at all.

CHAPTER 1 REVIEW

The numbers in brackets refer to the sections of the text in which similar problems can be found.

Write the numerical expression that is equivalent to each phrase, and then simplify. [1.3, 1.4, 1.6, 1.7]

1. The sum of -7 and -10

2. Five added to the sum of -7 and 4

3. The sum of -3 and 12 increased by 5

4. The difference of 4 and 9

5. The difference of 9 and -3

6. The difference of -7 and -9

7. The product of -3 and -7 decreased by 6

8. Ten added to the product of 5 and -6

9. Twice the product of -8 and $3x$

10. The quotient of -25 and -5

11. The quotient of -40 and 8 decreased by 7

12. The quotient of -45 and 15 increased by 9

Locate and label the points on the number line with the following coordinates. [1.2]

13. 5 **14.** -4 **15.** $\dfrac{24}{8}$

16. $-\dfrac{11}{2}$ **17.** 1.25 **18.** -1.5

For each number give the absolute value. [1.2]

19. 12 **20.** -3 **21.** $-\dfrac{4}{5}$

22. $-\dfrac{7}{10}$

Simplify. [1.2]

23. $|-1.8|$ **24.** $-|-10|$

For each number, give the opposite and the reciprocal. [1.2]

25. 6 **26.** $\dfrac{3}{10}$ **27.** -9

28. $-\dfrac{12}{5}$

Multiply. [1.2, 1.6]

29. $\left(\dfrac{2}{5}\right)\left(\dfrac{3}{7}\right)$ **30.** $\dfrac{1}{2}(-10)$

31. $\left(-\dfrac{4}{5}\right)\left(\dfrac{25}{16}\right)$

Add. [1.3]

32. $-9 + 12$ **33.** $-18 + (-20)$

34. $-3 + 8 + (-5)$

35. $(-5) + (-10) + (-7)$

36. $-2 + (-8) + [-9 + (-6)]$

37. $(-21) + 40 + (-23) + 5$

Subtract. [1.4]

38. $6 - 9$ **39.** $14 - (-8)$

40. $-12 - (-8)$ **41.** $4 - 9 - 15$

Simplify. [1.4]

42. $-14 + 7 - 8$ **43.** $5 - (-10 - 2) - 3$

44. $6 - [(3 - 4) - 5]$

45. $20 - [-(10 - 3) - 8] - 7$

Find the products. [1.6]

46. $(-5)(6)$ **47.** $4(-3)$

48. $-2(3)(4)$ **49.** $(-1)(-3)(-1)(-4)$

Find the following quotients. [1.7]

50. $\dfrac{12}{-3}$ **51.** $\dfrac{-9}{36}$ **52.** $-\dfrac{8}{9} \div \dfrac{4}{3}$

Simplify. [1.1, 1.6, 1.7]

53. $4 \cdot 5 + 3$ **54.** $9 \cdot 3 + 4 \cdot 5$

55. $2^3 - 4 \cdot 3^2 + 5^2$

56. $12 - 3(2 \cdot 5 + 7) + 4$

57. $20 + 8 \div 4 + 2 \cdot 5$ **58.** $2(3 - 5) - (2 - 8)$

59. $-4(-5) + 10$

60. $(-2)(3) - (4)(-3) - 9$

61. $3(4 - 7)^2 - 5(3 - 8)^2$

62. $(-5 - 2)(-3 - 7)$

63. $\dfrac{4(-3)}{-6}$ **64.** $\dfrac{3^2 + 5^2}{(3 - 5)^2}$

65. $\dfrac{15 - 10}{6 - 6}$ **66.** $\dfrac{8(-5) - 24}{4(-2)}$

67. $\dfrac{2(-7) + (-11)(-4)}{7 - (-3)}$

State the property or properties that justify the following. [1.5]

68. $9(3y) = (9 \cdot 3)y$ **69.** $8(1) = 8$

70. $(4 + y) + 2 = (y + 4) + 2$

71. $5 + (-5) = 0$ **72.** $6\left(\dfrac{1}{6}\right) = 1$

73. $8 + 0 = 8$

74. $(4 + 2) + y = (4 + y) + 2$

75. $5(w - 6) = 5w - 30$

Use the associative property to rewrite each expression and then simplify the result. [1.5]

76. $7 + (5 + x)$ **77.** $4(7a)$

78. $\dfrac{1}{9}(9x)$ **79.** $\dfrac{4}{5}\left(\dfrac{5}{4}y\right)$

Apply the distributive property to each of the following expressions. Simplify when possible. [1.5, 1.6]

80. $7(2x + 3)$

81. $3(2a - 4)$

82. $\dfrac{1}{2}(5x - 6)$

83. $-\dfrac{1}{2}(3x - 6)$

For the set $\{\sqrt{7}, -\frac{1}{3}, 0, 5, -4.5, \frac{2}{5}, \pi, -3\}$ list all the [1.8]

84. Rational numbers

85. Whole numbers

86. Irrational numbers

87. Integers

Factor into the product of primes. [1.8]

88. 90

89. 840

Combine. [1.9]

90. $\dfrac{18}{35} + \dfrac{13}{42}$

91. $\dfrac{9}{70} + \dfrac{11}{84}$

Find the next number in each sequence. [1.1, 1.2, 1.3, 1.6, 1.9]

92. $10, 7, 4, 1, \ldots$

93. $10, -30, 90, -270, \ldots$

94. $1, 1, 2, 3, 5, \ldots$

95. $4, 6, 8, 10, \ldots$

96. $1, \dfrac{1}{2}, 0, -\dfrac{1}{2}, \ldots$

97. $1, -\dfrac{1}{2}, \dfrac{1}{4}, -\dfrac{1}{8}, \ldots$

CHAPTER 1 PROJECTS

THE BASICS

Students and Instructors: The end of each chapter in this book will contain a page like this one containing two projects. The group project is intended to be done in class. The research projects are to be completed outside of class. They can be done in groups or individually. In my classes, I use the research projects for extra credit. I require all research projects to be done on a word processor and to be free of spelling errors.

GROUP PROJECT

GAUSS'S METHOD FOR ADDING CONSECUTIVE INTEGERS

Number of People: 3

Time Needed: 15–20 minutes

Equipment: Paper and pencil

Background: There is a popular story about famous mathematician Karl Friedrich Gauss (1777–1855). As the story goes, a 9-year-old Gauss, and the rest of his class, was given what the teacher had hoped was busy work. They were told to find the sum of all the whole numbers from 1 to 100. While the rest of the class labored under the assignment, Gauss found the answer within a few

Karl Gauss, 1777–1855
(Corbis/Bettmann)

moments. He may have set up the problem like this:

$$\begin{array}{cccccccc} 1 & 2 & 3 & 4 & \cdots & 98 & 99 & 100 \\ \underline{100} & \underline{99} & \underline{98} & \underline{97} & \cdots & \underline{3} & \underline{2} & \underline{1} \end{array}$$

Procedure: Copy the above illustration to get started.

1. Add the numbers in the columns and write your answers below the line.
2. How many numbers do you have below the line?
3. Suppose you add all the numbers below the line; how will your result be related to the sum of all the whole numbers from 1 to 100?
4. Use multiplication to add all the numbers below the line.
5. What is the sum of all the whole numbers from 1 to 100?
6. Use Gauss's method to add all the whole numbers from 1 to 10. (The answer is 55.)
7. Explain, in your own words, Gauss's method for finding this sum so quickly.

R E S E A R C H   P R O J E C T

SOFIA KOVALEVSKAYA

Sofia Kovalevskaya (1850–1891) was born in Moscow, Russia, around the time that Gauss died. Both Gauss and Kovalevskaya showed talent in mathematics early in their lives, but unlike Gauss, Kovalevskaya had to struggle to receive an education, and have her work recognized. Recently, the Russian people honored her with the commemorative postage stamp shown here. Research the life of Sofia Kovalevskaya, noting the obstacles she faced in pursuing her dream of studying and producing research in mathematics. Then summarize your results into an essay.

C H A P T E R 1   **TEST**

Translate into symbols. [1.1]

1. The sum of x and 3 is 8.
2. The product of 5 and y is 15.

Simplify according to the rule for order of operations. [1.1]

3. $5^2 + 3(9 - 7) + 3^2$
4. $10 - 6 \div 3 + 2^3$

For each number, name the opposite, reciprocal, and absolute value. [1.2]

5. -4

6. $\dfrac{3}{4}$

Add. [1.3]

7. $3 + (-7)$

8. $|-9 + (-6)| + |-3 + 5|$

Subtract. [1.4]

9. $-4 - 8$

10. $9 - (7 - 2) - 4$

Match each expression below with the letter of the property that justifies it. [1.5]

11. $(x + y) + z = x + (y + z)$ *C*

12. $3(x + 5) = 3x + 15$ *e*

13. $5(3x) = (5 \cdot 3)x$ *d*

14. $(x + 5) + 7 = 7 + (x + 5)$ *a*

    a. Commutative property of addition

    b. Commutative property of multiplication

    c. Associative property of addition

    d. Associative property of multiplication

    e. Distributive property

Multiply. [1.6]

15. $-3(7)$

16. $-4(8)(-2)$

17. $8\left(-\dfrac{1}{4}\right)$

18. $\left(-\dfrac{2}{3}\right)^3$

Simplify using the rule for order of operations. [1.6]

19. $-3(-4) - 8$

20. $5(-6)^2 - 3(-2)^3$

Simplify as much as possible. [1.7]

21. $7 - 3(2 - 8)$

22. $4 - 2[-3(-1 + 5) + 4(-3)]$

23. $\dfrac{4(-5) - 2(7)}{-10 - 7}$

24. $\dfrac{2(-3 - 1) + 4(-5 + 2)}{-3(2) - 4}$

Apply the associative property and then simplify. [1.5, 1.6]

25. $3 + (5 + 2x)$

26. $-2(-5x)$

Multiply by applying the distributive property. [1.5, 1.6]

27. $2(3x + 5)$

28. $-\dfrac{1}{2}(4x - 2)$

From the set of numbers $\{1, 1.5, \sqrt{2}, \frac{3}{4}, -8\}$, list [1.8]

29. All the integers $1, -8$

30. All the rational numbers $1, 1.5, \frac{3}{4}, -8$

31. All the irrational numbers $\sqrt{2}$

32. All the real numbers   All

Factor into the product of primes. [1.8]

33. 592

34. 1,340

Combine. [1.9]

35. $\dfrac{5}{15} + \dfrac{11}{42}$

36. $\dfrac{5}{x} + \dfrac{3}{x}$

Write an expression in symbols that is equivalent to each written phrase and then simplify it.

37. The sum of 8 and -3 [1.1, 1.3]

38. The difference of -24 and 2 [1.1, 1.4]

39. The product of -5 and -4 [1.1, 1.6]

40. The quotient of -24 and -2 [1.1, 1.7]

Find the next number in each sequence. [1.1, 1.2, 1.3, 1.6, 1.9]

41. $-8, -3, 2, 7, \ldots$

42. $8, -4, 2, -1, \ldots$

Linear Equations and Inequalities

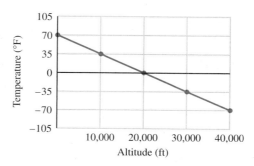

(Index Stock Imagery)

INTRODUCTION

Just before starting work on this edition of your text, I flew to Europe for vacation. From time to time the television screens on the plane displayed statistics about the flight. At one point during the flight the temperature outside the plane was −60°F. When I returned home, I did some research and found that the relationship between temperature T and altitude A can be described with the formula

$$T = -0.0035A + 70$$

when the temperature on the ground is 70°F. Table 1 and the line graph below also describe this relationship.

TABLE 1 Air Temperature and Altitude	
Altitude (feet)	**Temperature (°F)**
0	70
10,000	35
20,000	0
30,000	−35
40,000	−70

In this chapter we will start our work with formulas, and you will see how we use formulas to produce tables and line graphs like the ones above.

STUDY SKILLS

If you have successfully completed Chapter 1, then you have made a good start at developing the study skills necessary to succeed in all math classes. Here is the list of study skills for this chapter. Some are a continuation of the skills from Chapter 1, while others are new to this chapter.

1. **Continue to Set and Keep a Schedule** Sometimes I find students do well in Chapter 1 and then become overconfident. They begin to put in less time with their homework. Don't do it. Keep to the same schedule.

2. **Increase Effectiveness** You want to become more and more effective with the time you spend on your homework. You want to increase the amount of learning you obtain in the time you have set aside. Increase those activities that you feel are the most beneficial, and decrease those that have not given you the results you want.

3. **List Difficult Problems** Begin to make lists of problems that give you the most difficulty. These are the problems in which you are repeatedly making mistakes.

4. **Begin to Develop Confidence With Word Problems** It seems that the major difference between those people who are good at working word problems and those who are not is confidence. The people with confidence know that no matter how long it takes them, they will eventually be able to solve the problem. Those without confidence begin by saying to themselves, "I'll never be able to work this problem." Are you like that? If you are, what you need to do is put your old ideas about you and word problems aside for a while and make a decision to be successful. Sometimes that's all it takes. Instead of telling yourself that you can't do word problems, that you don't like them, or that they're not good for anything anyway, decide to do whatever it takes to master them.

Many of my students keep a notebook that contains everything that they need for the course: class notes, homework, quizzes, tests, and research projects. A three-ring binder with tabs is ideal. Organize your notebook so that you can easily get to any item you wish to look at.

Simplifying Expressions

If a cellular phone company charges \$35 per month plus \$0.25 for each minute, or fraction of a minute, that you use one of their cellular phones, then the amount of your monthly bill is given by the expression $35 + 0.25t$. To find the amount you will pay for using that phone 30 minutes in one month, you substitute 30 for t and simplify the resulting expression. This process is one of the topics we will study in this section.

As you will see in the next few sections, the first step in solving an equation is to simplify both sides as much as possible. In the first part of this section, we will practice simplifying expressions by combining what are called *similar* (or *like*) terms.

For our immediate purposes, a term is a number or a number and one or more variables multiplied together. For example, the number 5 is a term, as are the expressions $3x$, $-7y$, and $15xy$.

> **DEFINITION** Two or more terms with the same variable part are called **similar** (or **like**) terms.

The terms $3x$ and $4x$ are similar, since their variable parts are identical. Likewise, the terms $18y$, $-10y$, and $6y$ are similar terms.

To simplify an algebraic expression, we simply reduce the number of terms in the expression. We accomplish this by applying the distributive property along with our knowledge of addition and subtraction of positive and negative real numbers. The following examples illustrate the procedure.

EXAMPLES   Simplify by combining similar terms.

1. $3x + 4x = (3 + 4)x$ Distributive property
$\qquad\qquad = 7x$ Addition of 3 and 4

2. $7a - 10a = (7 - 10)a$ Distributive property
$\qquad\qquad = -3a$ Addition of 7 and -10

3. $18y - 10y + 6y = (18 - 10 + 6)y$ Distributive property
$\qquad\qquad\qquad = 14y$ Addition of 18, -10, and 6

When the expression we intend to simplify is more complicated, we use the commutative and associative properties first.

EXAMPLES   Simplify each expression.

4. $3x + 5 + 2x - 3 = 3x + 2x + 5 - 3$ Commutative property
$\qquad\qquad = (3x + 2x) + (5 - 3)$ Associative property
$\qquad\qquad = (3 + 2)x + (5 - 3)$ Distributive property
$\qquad\qquad = 5x + 2$ Addition

5. $4a - 7 - 2a + 3 = (4a - 2a) + (-7 + 3)$ Commutative and
                                            associative properties

$$= (4 - 2)a + (-7 + 3)$$ Distributive property
$$= 2a - 4$$ Addition

6. $5x + 8 - x - 6 = (5x - x) + (8 - 6)$ Commutative and
                                            associative properties

$$= (5 - 1)x + (8 - 6)$$ Distributive property
$$= 4x + 2$$ Addition

Notice that in each case the result has fewer terms than the original expression. Since there are fewer terms, the resulting expression is said to be simpler than the original expression.

SIMPLIFYING EXPRESSIONS CONTAINING PARENTHESES

If an expression contains parentheses, it is often necessary to apply the distributive property to remove the parentheses before combining similar terms.

EXAMPLE 7 Simplify the expression $5(2x - 8) - 3$.

Solution We begin by distributing the 5 across $2x - 8$. We then combine similar terms:

$$5(2x - 8) - 3 = 10x - 40 - 3$$ Distributive property
$$= 10x - 43$$

EXAMPLE 8 Simplify $7 - 3(2y + 1)$.

Solution By the rule for order of operations, we must multiply before we add or subtract. For that reason, it would be incorrect to subtract 3 from 7 first. Instead, we multiply -3 and $2y + 1$ to remove the parentheses and then combine similar terms:

$$7 - 3(2y + 1) = 7 - 6y - 3$$ Distributive property
$$= -6y + 4$$

EXAMPLE 9 Simplify $5(x - 2) - (3x + 4)$.

Solution We begin by applying the distributive property to remove the parentheses. The expression $-(3x + 4)$ can be thought of as $-1(3x + 4)$. Thinking of it in this way allows us to apply the distributive property:

$$-1(3x + 4) = -1(3x) + (-1)(4)$$
$$= -3x - 4$$

The complete solution looks like this:

$$5(x - 2) - (3x + 4) = 5x - 10 - 3x - 4 \qquad \text{Distributive property}$$
$$= 2x - 14 \qquad \text{Combine similar terms.}$$

As you can see from the explanation in Example 9, we use the distributive property to simplify expressions in which parentheses are preceded by a negative sign. In general we can write

$$-(a + b) = -1(a + b)$$
$$= -a + (-b)$$
$$= -a - b$$

The negative sign outside the parentheses ends up changing the sign of each term within the parentheses. In words, we say "the opposite of a sum is the sum of the opposites."

THE VALUE OF AN EXPRESSION

An expression like $3x + 2$ has a certain value depending on what number we assign to x. For instance, when x is 4, $3x + 2$ becomes $3(4) + 2$, or 14. When x is -8, $3x + 2$ becomes $3(-8) + 2$, or -22. The value of an expression is found by replacing the variable with a given number.

E X A M P L E S   Find the value of the following expressions by replacing the variable with the given number.

Expression	Value of the Variable	Value of the Expression
10. $3x - 1$	$x = 2$	$3(2) - 1 = 6 - 1 = 5$
11. $7a + 4$	$a = -3$	$7(-3) + 4 = -21 + 4 = -17$
12. $2x - 3 + 4x$	$x = -1$	$2(-1) - 3 + 4(-1)$
		$\quad = -2 - 3 + (-4) = -9$
13. $2x - 5 - 8x$	$x = 5$	$2(5) - 5 - 8(5)$
		$\quad = 10 - 5 - 40 = -35$
14. $y^2 - 6y + 9$	$y = 4$	$4^2 - 6(4) + 9 = 16 - 24 + 9 = 1$

Simplifying an expression should not change its value. That is, if an expression has a certain value when x is 5, then it will always have that value no matter how much it has been simplified, as long as x is 5. If we were to simplify the expression in Example 13 first, it would look like

$$2x - 5 - 8x = -6x - 5$$

When x is 5, the simplified expression $-6x - 5$ is

$$-6(5) - 5 = -30 - 5 = -35$$

It has the same value as the original expression when x is 5.

We can also find the value of an expression that contains two variables if we know the values for both variables.

E X A M P L E   1 5   Find the value of the expression $2x - 3y + 4$ when x is -5 and y is 6.

Solution   Substituting -5 for x and 6 for y, the expression becomes

$$2(-5) - 3(6) + 4 = -10 - 18 + 4$$
$$= -28 + 4$$
$$= -24$$

E X A M P L E   1 6   Find the value of the expression $x^2 - 2xy + y^2$ when x is 3 and y is -4.

Solution   Replacing each x in the expression with the number 3 and each y in the expression with the number -4 gives us

$$3^2 - 2(3)(-4) + (-4)^2 = 9 - 2(3)(-4) + 16$$
$$= 9 - (-24) + 16$$
$$= 33 + 16$$
$$= 49$$

MORE ABOUT SEQUENCES

As the next example indicates, when we substitute the counting numbers, in order, into algebraic expressions, we form some of the sequences of numbers that we studied in Chapter 1. To review, recall that the sequence of counting numbers (also called the sequence of positive integers) is

$$\text{Counting numbers} = 1, 2, 3, \ldots$$

E X A M P L E   1 7   Substitute 1, 2, 3, and 4 for n in the expression $2n - 1$.

Solution   Substituting as indicated, we have

When $n = 1, 2n - 1 = 2 \cdot 1 - 1 = 1.$

When $n = 2, 2n - 1 = 2 \cdot 2 - 1 = 3.$

When $n = 3, 2n - 1 = 2 \cdot 3 - 1 = 5.$

When $n = 4, 2n - 1 = 2 \cdot 4 - 1 = 7.$

As you can see, substituting the first four counting numbers into the formula $2n - 1$ produces the first four numbers in the sequence of odd numbers.

The next example is similar to Example 17 but uses tables to display the information.

EXAMPLE 18   Fill in the tables below to find the sequences formed by substituting the first four counting numbers into the expressions $2n$ and n^2.

(a)

n	1	2	3	4
$2n$				

(b)

n	1	2	3	4
n^2				

Solution   Proceeding as we did in the previous example, we substitute the numbers 1, 2, 3, and 4 into the given expressions.

(a) When $n = 1$, $2n = 2 \cdot 1 = 2$.
When $n = 2$, $2n = 2 \cdot 2 = 4$.
When $n = 3$, $2n = 2 \cdot 3 = 6$.
When $n = 4$, $2n = 2 \cdot 4 = 8$.

As you can see, the expression $2n$ produces the sequence of even numbers when n is replaced by the counting numbers. Placing these results into our first table gives us

n	1	2	3	4
$2n$	2	4	6	8

(b) The expression n^2 produces the sequence of squares when n is replaced by 1, 2, 3, and 4. In table form we have

n	1	2	3	4
n^2	1	4	9	16

Getting Ready for Class

After reading through the preceding section, respond in your own words and in complete sentences.

A. What are similar terms?

B. Explain how the distributive property is used to combine similar terms.

C. What is wrong with writing $3x + 4x = 7x^2$?

D. Explain how you would find the value of $5x + 3$ when x is 6.

92

PROBLEM SET 2.1

Simplify the following expressions.

1. $3x - 6x$ **2.** $7x - 5x$

3. $-2a + a$ **4.** $3a - a$

5. $7x + 3x + 2x$ **6.** $8x - 2x - x$

7. $3a - 2a + 5a$ **8.** $7a - a + 2a$

9. $4x - 3 + 2x$ **10.** $5x + 6 - 3x$

11. $3a + 4a + 5$ **12.** $6a + 7a + 8$

13. $2x - 3 + 3x - 2$ **14.** $6x + 5 - 2x + 3$

15. $3a - 1 + a + 3$ **16.** $-a + 2 + 8a - 7$

17. $-4x + 8 - 5x - 10$ **18.** $-9x - 1 + x - 4$

19. $7a + 3 + 2a + 3a$ **20.** $8a - 2 + a + 5a$

21. $5(2x - 1) + 4$ **22.** $2(4x - 3) + 2$

23. $7(3y + 2) - 8$ **24.** $6(4y + 2) - 7$

25. $-3(2x - 1) + 5$ **26.** $-4(3x - 2) - 6$

27. $5 - 2(a + 1)$ **28.** $7 - 8(2a + 3)$

Simplify the following expressions.

29. $6 - 4(x - 5)$ **30.** $12 - 3(4x - 2)$

31. $-9 - 4(2 - y) + 1$

32. $-10 - 3(2 - y) + 3$

33. $-6 + 2(2 - 3x) + 1$

34. $-7 - 4(3 - x) + 1$

35. $(4x - 7) - (2x + 5)$

36. $(7x - 3) - (4x + 2)$

37. $8(2a + 4) - (6a - 1)$

38. $9(3a + 5) - (8a - 7)$

39. $3(x - 2) + (x - 3)$

40. $2(2x + 1) - (x + 4)$

41. $4(2y - 8) - (y + 7)$

42. $5(y - 3) - (y - 4)$

43. $-9(2x + 1) - (x + 5)$

44. $-3(3x - 2) - (2x + 3)$

Evaluate the following expressions when x is 2. (Find the value of the expressions if x is 2.)

45. $3x - 1$ **46.** $4x + 3$

47. $-2x - 5$ **48.** $-3x + 6$

49. $x^2 - 8x + 16$ **50.** $x^2 - 10x + 25$

51. $(x - 4)^2$ **52.** $(x - 5)^2$

Find the value of the following expressions when x is -5. Then simplify the expression, and check to see that it has the same value for $x = -5$.

53. $7x - 4 - x - 3$ **54.** $3x + 4 + 7x - 6$

55. $5(2x + 1) + 4$ **56.** $2(3x - 10) + 5$

Find the value of each expression when x is -3 and y is 5.

57. $x^2 - 2xy + y^2$ **58.** $x^2 + 2xy + y^2$

59. $(x - y)^2$ **60.** $(x + y)^2$

61. $x^2 + 6xy + 9y^2$ **62.** $x^2 + 10xy + 25y^2$

63. $(x + 3y)^2$ **64.** $(x + 5y)^2$

Find the value of $12x - 3$ for each of the following values of x.

65. $\dfrac{1}{2}$ **66.** $\dfrac{1}{3}$ **67.** $\dfrac{1}{4}$

68. $\dfrac{1}{6}$ **69.** $\dfrac{3}{2}$ **70.** $\dfrac{2}{3}$

71. $\dfrac{3}{4}$ **72.** $\dfrac{5}{6}$

Substitute 1, 2, 3, and 4 for n in each of the following expressions.

73. $2n + 3$ **74.** $5n + 2$ **75.** $n^2 + 1$

76. $(n + 1)^2$

77. Fill in the tables below to find the sequences formed by substituting the first four counting numbers into the expressions $3n$ and n^3.

(a)

n	1	2	3	4
$3n$				

(b)

n	1	2	3	4
n^3				

78. Fill in the tables below to find the sequences formed by substituting the first four counting numbers into the expressions $2n - 1$ and $2n + 1$.

(a)

n	1	2	3	4
$2n - 1$				

(b)

n	1	2	3	4
$2n + 1$				

Applying the Concepts

79. Temperature and Altitude If the temperature on the ground is 70°F, then the temperature at A feet above the ground can be found from the expression $-0.0035A + 70$. Find the temperature at the following altitudes.

(a) 8,000 feet   (b) 12,000 feet   (c) 24,000 feet

(Index Stock Imagery)

80. Perimeter of a Rectangle The expression $2l + 2w$ gives the perimeter of a rectangle with length l

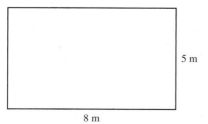

8 m

5 m

and width w. Find the perimeter of the rectangles with the following lengths and widths.

(a) Length = 8 meters, width = 5 meters
(b) Length = 10 feet, width = 3 feet

3 ft

10 ft

81. Cellular Phone Rates A cellular phone company charges $35 per month plus $0.25 for each minute, or fraction of a minute, that you use one of their cellular phones. The expression $35 + 0.25t$ gives the amount of money you will pay for using one of their phones for t minutes a month. Find the monthly bill for using one of their phones.

(a) 10 minutes in a month
(b) 20 minutes in a month
(c) 30 minutes in a month

82. Cost of Bottled Water A water bottling company charges $7.00 per month for their water dispenser and $1.10 for each gallon of water delivered. If you have g gallons of water delivered in a month, then the expression $7 + 1.1g$ gives the amount of your bill for that month. Find the monthly bill for each of the following deliveries.

(a) 10 gallons   (b) 20 gallons   (c) 30 gallons

Cool,
Refreshing
Spring Water

for only **$7.00** per month and
$1.10 per gallon!

83. Taxes We all have to pay taxes. Suppose that 21% of your monthly pay is withheld for federal income taxes and another 8% is withheld for Social Security, state income tax, and other miscellaneous items. If G is your monthly pay before any money is deducted (your gross pay), then the amount of money that you take home each month is given by the expression $G - 0.21G - 0.08G$. Simplify this expression, and then find your take-home pay if your gross pay is $1,250 per month.

84. Taxes If you work H hours a day and you pay 21% of your income for federal income tax and another 8% for miscellaneous deductions, then the expression $0.21H + 0.08H$ tells you how many hours a day you are working to pay for those taxes and miscellaneous deductions. Simplify this expression. If you work 8 hours a day under these conditions, how many hours do you work to pay for your taxes and miscellaneous deductions?

Translating Words Into Symbols

Translate each phrase into an algebraic expression. Then find the value of the expression when x is -2.

85. The difference of x and 5
86. The difference of 5 and x
87. Twice the sum of x and 10
88. The sum of twice x and 10
89. The quotient of 10 and x
90. The quotient of x and 10
91. The sum of $3x$ and -2, decreased by 5
92. The product of $3x$ and -2, decreased by 5

Review Problems

From here on, each problem set will end with some review problems. In mathematics it is very important to review work you have done previously. The more you review, the better you will understand the topics we cover and the longer you will remember them. Also, there are times when material that seemed confusing earlier becomes less confusing the second time around.

The following problems review material we covered previously in Section 1.9. Reviewing these problems will help you with the material in the next section.

Add and subtract as indicated.

93. $-3 - \dfrac{1}{2}$

94. $-5 - \dfrac{1}{3}$

95. $\dfrac{4}{5} + \dfrac{1}{10} + \dfrac{3}{8}$

96. $\dfrac{3}{10} + \dfrac{7}{25} + \dfrac{3}{4}$

2.2   Addition Property of Equality

When light comes into contact with any object, it is reflected, absorbed, and transmitted, as shown in Figure 1.

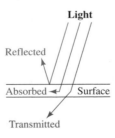

Light

Reflected

Absorbed ◄ Surface

Transmitted

FIGURE 1

For a certain type of glass, 88% of the light hitting the glass is transmitted through to the other side, while 6% of the light is absorbed into the glass. To find the percent of light that is reflected by the glass, we can solve the equation

$$88 + R + 6 = 100$$

Solving equations of this type is what we study in this section. To solve an equation we must find all replacements for the variable that make the equation a true statement.

> **DEFINITION**   The **solution set** for an equation is the set of all numbers that when used in place of the variable make the equation a true statement.

For example, the equation $x + 2 = 5$ has solution set $\{3\}$ because when x is 3 the equation becomes the true statement $3 + 2 = 5$, or $5 = 5$.

EXAMPLE 1   Is 5 a solution to $2x - 3 = 7$?

Solution   We substitute 5 for x in the equation and then simplify to see if a true statement results. A true statement means we have a solution; a false statement indicates the number we are using is not a solution.

$$\text{When} \qquad x = 5$$
$$\text{the equation} \qquad 2x - 3 = 7$$
$$\text{becomes} \qquad 2(5) - 3 \overset{?}{=} 7$$
$$10 - 3 \overset{?}{=} 7$$
$$7 = 7 \qquad \text{A true statement}$$

Since $x = 5$ turns the equation into the true statement $7 = 7$, we know 5 is a solution to the equation.

Note: We can use a question mark over the equal signs to show that we don't know yet whether the two sides of the equation are equal.

EXAMPLE 2   Is -2 a solution to $8 = 3x + 4$?

Solution   Substituting -2 for x in the equation, we have

$$8 \overset{?}{=} 3(-2) + 4$$
$$8 \overset{?}{=} -6 + 4$$
$$8 = -2 \qquad \text{A false statement}$$

Substituting -2 for x in the equation produces a false statement. Therefore, $x = -2$ is not a solution to the equation.

The important thing about an equation is its solution set. We therefore make the following definition in order to classify together all equations with the same solution set.

> **DEFINITION** Two or more equations with the same solution set are said to be **equivalent equations**.

Equivalent equations may look different but must have the same solution set.

EXAMPLE 3

(a) $x + 2 = 5$ and $x = 3$ are equivalent equations, since both have solution set $\{3\}$.

(b) $a - 4 = 3$, $a - 2 = 5$, and $a = 7$ are equivalent equations, since they all have solution set $\{7\}$.

(c) $y + 3 = 4$, $y - 8 = -7$, and $y = 1$ are equivalent equations, since they all have solution set $\{1\}$.

If two numbers are equal and we increase (or decrease) both of them by the same amount, the resulting quantities are also equal. We can apply this concept to equations. Adding the same amount to both sides of an equation always produces an equivalent equation—one with the same solution set. This fact about equations is called the *addition property of equality* and can be stated more formally as follows.

ADDITION PROPERTY OF EQUALITY

For any three algebraic expressions A, B, and C,

$$\text{if} \quad A = B$$

$$\text{then} \quad A + C = B + C$$

In words: Adding the same quantity to both sides of an equation will not change the solution set.

This property is just as simple as it seems. We can add any amount to both sides of an equation and always be sure we have not changed the solution set.

Consider the equation $x + 6 = 5$. We want to solve this equation for the value of x that makes it a true statement. We want to end up with x on one side of the equal sign and a number on the other side. Since we want x by itself, we will add -6 to both sides:

$$x + 6 + (-6) = 5 + (-6) \qquad \text{Addition property of equality}$$
$$x + 0 = -1 \qquad \text{Addition}$$
$$x = -1$$

All three equations say the same thing about x. They all say that x is -1. All three equations are equivalent. The last one is just easier to read.

Here are some further examples of how the addition property of equality can be used to solve equations.

E X A M P L E   4   Solve the equation $x - 5 = 12$ for x.

Solution   Since we want x alone on the left side, we choose to add $+5$ to both sides:

$$x - 5 + 5 = 12 + 5 \qquad \text{Addition property of equality}$$
$$x + 0 = 17$$
$$x = 17$$

To check our solution, we substitute 17 for x in the original equation:

$$\text{When} \qquad x = 17$$
$$\text{the equation} \qquad x - 5 = 12$$
$$\text{becomes} \qquad 17 - 5 \stackrel{?}{=} 12$$
$$12 = 12 \qquad \text{A true statement}$$

As you can see, our solution checks. The purpose for checking a solution to an equation is to catch any mistakes we may have made in the process of solving the equation.

E X A M P L E   5   Solve for a: $a + \frac{3}{4} = -\frac{1}{2}$

Solution   Since we want a by itself on the left side of the equal sign, we add the opposite of $\frac{3}{4}$ to each side of the equation.

$$a + \frac{3}{4} + \left(-\frac{3}{4}\right) = -\frac{1}{2} + \left(-\frac{3}{4}\right) \qquad \text{Addition property of equality}$$

$$a + 0 = -\frac{1}{2} \cdot \frac{2}{2} + \left(-\frac{3}{4}\right) \qquad \text{LCD on the right side is 4.}$$

$$a = -\frac{2}{4} + \left(-\frac{3}{4}\right) \qquad \tfrac{2}{4} \text{ is equivalent to } \tfrac{1}{2}.$$

$$a = -\frac{5}{4} \qquad \text{Add fractions.}$$

The solution is $a = -\frac{5}{4}$. To check our result, we replace a with $-\frac{5}{4}$ in the original equation. The left side then becomes $-\frac{5}{4} + \frac{3}{4}$, which reduces to $-\frac{1}{2}$, so our solution checks.

E X A M P L E   6   Solve for x: $7.3 + x = -2.4$

Solution   Again, we want to isolate x, so we add the opposite of 7.3 to both sides:

$$7.3 + (-\mathbf{7.3}) + x = -2.4 + (-\mathbf{7.3}) \qquad \text{Addition property of equality}$$

$$0 + x = -9.7$$

$$x = -9.7$$

Sometimes it is necessary to simplify each side of an equation before using the addition property of equality. The reason we simplify both sides first is that we want as few terms as possible on each side of the equation before we use the addition property of equality. The following examples illustrate this procedure.

EXAMPLE 7   Solve for x: $-x + 2 + 2x = 7 + 5$

Solution   We begin by combining similar terms on each side of the equation. Then we use the addition property to solve the simplified equation.

$$x + 2 = 12 \qquad\qquad\quad \text{Simplify both sides first.}$$

$$x + 2 + (-\mathbf{2}) = 12 + (-\mathbf{2}) \qquad \text{Addition property of equality}$$

$$x + 0 = 10$$

$$x = 10$$

EXAMPLE 8   Solve $4(2a - 3) - 7a = 2 - 5$.

Solution   We must begin by applying the distributive property to separate terms on the left side of the equation. Following that, we combine similar terms and then apply the addition property of equality.

$$4(2a - 3) - 7a = 2 - 5 \qquad\quad \text{Original equation}$$

$$8a - 12 - 7a = 2 - 5 \qquad\quad \text{Distributive property}$$

$$a - 12 = -3 \qquad\qquad\quad \text{Simplify each side.}$$

$$a - 12 + \mathbf{12} = -3 + \mathbf{12} \qquad \text{Add } \mathbf{12} \text{ to each side.}$$

$$a = 9 \qquad\qquad\qquad\quad \text{Addition}$$

To check our solution, we replace a with 9 in the original equation.

$$4(2 \cdot 9 - 3) - 7 \cdot 9 \stackrel{?}{=} 2 - 5$$

$$4(18 - 3) - 63 \stackrel{?}{=} -3$$

$$4(15) - 63 \stackrel{?}{=} -3$$

$$60 - 63 \stackrel{?}{=} -3$$

$$-3 = -3 \qquad \text{A true statement}$$

Our solution checks.

Note: Again, we place a question mark over the equal sign because we don't know yet whether the expressions on the left and right side of the equal sign will be equal.

So far in this section we have used the addition property of equality to add only numbers to both sides of an equation. It is often necessary to add a term involving a variable to both sides of an equation, as the following example indicates.

EXAMPLE 9   Solve $3x - 5 = 2x + 7$.

Solution   We can solve this equation in two steps. First, we add $-2x$ to both sides of the equation. When this has been done, x appears on the left side only. Second, we add 5 to both sides:

$$3x + (\mathbf{-2x}) - 5 = 2x + (\mathbf{-2x}) + 7 \qquad \text{Add } \mathbf{-2x} \text{ to both sides.}$$
$$x - 5 = 7 \qquad \text{Simplify each side.}$$
$$x - 5 + \mathbf{5} = 7 + \mathbf{5} \qquad \text{Add } \mathbf{5} \text{ to both sides.}$$
$$x = 12 \qquad \text{Simplify each side.}$$

A NOTE ON SUBTRACTION

Although the addition property of equality is stated for addition only, we can subtract the same number from both sides of an equation as well. Because subtraction is defined as addition of the opposite, subtracting the same quantity from both sides of an equation will not change the solution. If we were to solve the equation in Example 7 using subtraction instead of addition, the steps would look like this:

$$x + 2 = 12$$
$$x + 2 - \mathbf{2} = 12 - \mathbf{2} \qquad \text{Subtract } \mathbf{2} \text{ from each side.}$$
$$2 + \mathbf{2} - x = 10 - \mathbf{2} \qquad \text{Subtraction}$$

In my experience teaching algebra, I find that students make fewer mistakes if they think in terms of addition rather than subtraction. So, you are probably better off if you continue to use the addition property just the way we have used it in the examples in this section. But, if you are curious as to whether you can subtract the same number from both sides of an equation, the answer is yes.

Getting Ready for Class

After reading through the preceding section, respond in your own words and in complete sentences.

A. What is a solution to an equation?

B. What are equivalent equations?

C. Explain in words the addition property of equality.

D. How do you check a solution to an equation?

PROBLEM SET 2.2

Find the solution set for the following equations. Be sure to show when you have used the addition property of equality.

1. $x - 3 = 8$　　　　**2.** $x - 2 = 7$

3. $x + 2 = 6$　　　　**4.** $x + 5 = 4$

5. $a + \dfrac{1}{2} = -\dfrac{1}{4}$　　**6.** $a + \dfrac{1}{3} = -\dfrac{5}{6}$

7. $x + 2.3 = -3.5$　　**8.** $x + 7.9 = -3.4$

9. $y + 11 = -6$　　　**10.** $y - 3 = -1$

11. $x - \dfrac{5}{8} = -\dfrac{3}{4}$　　**12.** $x - \dfrac{2}{5} = -\dfrac{1}{10}$

13. $m - 6 = -10$　　**14.** $m - 10 = -6$

15. $6.9 + x = 3.3$　　**16.** $7.5 + x = 2.2$

17. $5 = a + 4$　　　**18.** $12 = a - 3$

19. $-\dfrac{5}{9} = x - \dfrac{2}{5}$　　**20.** $-\dfrac{7}{8} = x - \dfrac{4}{5}$

Simplify both sides of the following equations as much as possible, and then solve.

21. $4x + 2 - 3x = 4 + 1$

22. $5x + 2 - 4x = 7 - 3$

23. $8a - \dfrac{1}{2} - 7a = \dfrac{3}{4} + \dfrac{1}{8}$

24. $9a - \dfrac{4}{5} - 8a = \dfrac{3}{10} - \dfrac{1}{5}$

25. $-3 - 4x + 5x = 18$

26. $10 - 3x + 4x = 20$

27. $-11x + 2 + 10x + 2x = 9$

28. $-10x + 5 - 4x + 15x = 0$

29. $-2.5 + 4.8 = 8x - 1.2 - 7x$

30. $-4.8 + 6.3 = 7x - 2.7 - 6x$

31. $2y - 10 + 3y - 4y = 18 - 6$

32. $3y - 20 + 6y - 8y = 21$

33. $15 - 21 = 8x + 3x - 10x$

34. $23 - 17 = -7x - x + 9x$

35. $24 - 3 + 8a - 5a - 2a = 21$

36. $30 - 4 + 7a - 2 - 6a = 30$

The following equations contain parentheses. Apply the distributive property to remove the parentheses; then simplify each side before using the addition property of equality.

37. $2(x + 3) - x = 4$　　**38.** $5(x + 1) - 4x = 2$

39. $-3(x - 4) + 4x = 3 - 7$

40. $-2(x - 5) + 3x = 4 - 9$

41. $5(2a + 1) - 9a = 8 - 6$

42. $4(2a - 1) - 7a = 9 - 5$

43. $-(x + 3) + 2x - 1 = 6$

44. $-(x - 7) + 2x - 8 = 4$

45. $4y - 3(y - 6) + 2 = 8$

46. $7y - 6(y - 1) + 3 = 9$

47. $2(3x + 1) - 5(x + 2) = 1 - 10$

48. $4(2x + 1) - 7(x - 1) = 2 - 6$

49. $-3(2m - 9) + 7(m - 4) = 12 - 9$

50. $-5(m - 3) + 2(3m + 1) = 15 - 8$

Solve the following equations by the method used in Example 9 in this section. Check each solution in the original equation.

51. $4x = 3x + 2$　　　**52.** $6x = 5x - 4$

53. $8a = 7a - 5$　　　**54.** $9a = 8a - 3$

55. $2x = 3x + 1$　　　**56.** $4x = 3x + 5$

57. $3y + 4 = 2y + 1$　　**58.** $5y + 6 = 4y + 2$

59. $2m - 3 = m + 5$　　**60.** $8m - 1 = 7m - 3$

61. $4x - 7 = 5x + 1$　　**62.** $3x - 7 = 4x - 6$

63. $5x - \dfrac{2}{3} = 4x + \dfrac{4}{3}$　　**64.** $3x - \dfrac{5}{4} = 2x + \dfrac{1}{4}$

65. $8a - 7.1 = 7a + 3.9$

66. $10a - 4.3 = 9a + 4.7$

Applying the Concepts

67. Magic Square The sum of the numbers in each row, each column, and each diagonal of the following magic square is 6. Use this fact, along with the information in the first column of the magic square, to write an equation containing the variable x, and then solve the equation to find x. Next, write and solve equations that will give you y and z.

x	4	y
-2	2	6
5	z	1

68. Magic Square   The sum of the numbers in each row and each column of the square below is 0. Use this fact, along with the information in the second row of the magic square, to write an equation containing the variable a, and then solve the equation to find a. Next, write and solve an equation that will allow you to find the value of b. Next, write and solve equations that will give you c and d.

-3	d	b
a	-1	-6
-4	c	1

69. Light   When light comes into contact with any object, it is reflected, absorbed, and transmitted, as shown in the following figure. If T represents the percent of light transmitted, R the percent of light reflected, and A the percent of light absorbed by a surface, then the equation $T + R + A = 100$ shows one way these quantities are related.

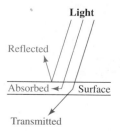

Light

Reflected

Absorbed   Surface

Transmitted

(a) For glass, $T = 88$ and $A = 6$, meaning that 88% of the light hitting the glass is transmitted

and 6% is absorbed. Substitute $T = 88$ and $A = 6$ into the equation $T + R + A = 100$ and solve for R to find the percent of light that is reflected.

(b) For flat black paint, $A = 95$ and no light is transmitted, meaning that $T = 0$. What percent of light is reflected by flat black paint?

(c) A pure white surface can reflect 98% of light, so $R = 98$. If no light is transmitted, what percent of light is absorbed by the pure white surface?

(d) Typically, shiny gray metals reflect 70–80% of light. Suppose a thick sheet of aluminum absorbs 25% of light. What percent of light is reflected by this shiny gray metal? (Assume no light is transmitted.)

70. Movie Tickets   A movie theater has a total of 300 seats. For a special Saturday night preview, they reserve 20 seats to give away to their VIP guests at no charge. If x represents the number of tickets they can sell for the preview, then x can be found by solving the equation $x + 20 = 300$.

(a) Solve the equation for x.

(b) If tickets for the preview are $7.50 each, what is the maximum amount of money they can make from ticket sales?

71. Geometry   The three angles shown in the triangle at the front of the tent in the following figure add up to 180°. Use this fact to write an equation containing x, and then solve the equation to find the number of degrees in the angle at the top of the triangle.

72. Geometry   The following figure shows part of a room. From a point on the floor, the angle of elevation to the top of the door is 47°, while the angle of elevation to the ceiling above the door is 59°. Use this diagram to write an equation involving x, and then solve the equation to find the number of degrees in the angle that extends from the top of the door to the ceiling.

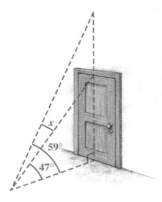

Translations   Translate each of the following sentences into an equation, and then solve the equation.

73. The difference of x and 3 is 8.

74. The difference of 3 and 8 is x.

75. The sum of $2y$ and 3 is equal to the sum of y and 5.

76. The sum of 3 and $2y$ is equal to the sum of 5 and y.

Review Problems

The following problems review material we covered in Section 1.5. Reviewing this material will help you in Section 2.3.

Apply the associative property to each expression and then simplify the result.

77. $3(6x)$ **78.** $5(4x)$ **79.** $\dfrac{1}{5}(5x)$

80. $\dfrac{1}{3}(3x)$ **81.** $8\left(\dfrac{1}{8}y\right)$ **82.** $6\left(\dfrac{1}{6}y\right)$

83. $-2\left(-\dfrac{1}{2}x\right)$ **84.** $-4\left(-\dfrac{1}{4}x\right)$

85. $-\dfrac{4}{3}\left(-\dfrac{3}{4}a\right)$ **86.** $-\dfrac{5}{2}\left(-\dfrac{2}{5}a\right)$

2.3   Multiplication Property of Equality

As we have mentioned before, we all have to pay taxes. According to Figure 1, people have been paying taxes for quite a long time.

FIGURE 1   Collection of taxes, ca. 3000 B.C. Clerks and scribes appear at the right, with pen and papyrus, and officials and taxpayers appear at the left.

If 21% of your monthly pay is withheld for federal income taxes and another 8% is withheld for Social Security, state income tax, and other miscellaneous items, leaving you with $987.50 a month in take-home pay, then the amount you earned before the deductions were removed from your check is given by the equation

$$G - 0.21G - 0.08G = 987.5$$

In this section we will learn how to solve equations of this type.

In the previous section, we found that adding the same number to both sides of an equation never changed the solution set. The same idea holds for multiplication by numbers other than zero. We can multiply both sides of an equation by the same nonzero number and always be sure we have not changed the solution set. (The reason we cannot multiply both sides by zero will become apparent later.) This fact about equations is called the *multiplication property of equality*, which can be stated formally as follows.

MULTIPLICATION PROPERTY OF EQUALITY

For any three algebraic expressions A, B, and C, where $C \neq 0$,

$$\text{if} \qquad A = B$$

$$\text{then} \qquad AC = BC$$

In words: Multiplying both sides of an equation by the same nonzero number will not change the solution set.

Suppose we want to solve the equation $5x = 30$. We have $5x$ on the left side but would like to have just x. We choose to multiply both sides by $\frac{1}{5}$ since $(\frac{1}{5})(5) = 1$. Here is the solution:

$$5x = 30$$

$$\frac{1}{5}(5x) = \frac{1}{5}(30) \qquad \text{Multiplication property of equality}$$

$$\left(\frac{1}{5} \cdot 5\right)x = \frac{1}{5}(30) \qquad \text{Associative property of multiplication}$$

$$1x = 6$$

$$x = 6$$

We chose to multiply by $\frac{1}{5}$ because it is the reciprocal of 5. We can see that multiplication by any number except zero will not change the solution set. If, however, we were to multiply both sides by zero, the result would always be $0 = 0$, since multiplication by zero always results in zero. Although the statement $0 = 0$ is true, we have lost our variable and cannot solve the equation. This is the only restriction of the multiplication property of equality. We are free to multiply both sides of an equation by any number except zero.

Here are some more examples that use the multiplication property of equality.

EXAMPLE 1   Solve for a: $-4a = 24$

Solution   Since we want a alone on the left side, we choose to multiply both sides by $-\frac{1}{4}$:

$$-\frac{1}{4}(-4a) = -\frac{1}{4}(24) \qquad \text{Multiplication property of equality}$$

$$\left[-\frac{1}{4}(-4)\right]a = -\frac{1}{4}(24) \qquad \text{Associative property}$$

$$a = -6$$

EXAMPLE 2   Solve for t: $-\frac{t}{3} = 5$

Solution   Since division by 3 is the same as multiplication by $\frac{1}{3}$, we can write $-\frac{t}{3}$ as $-\frac{1}{3}t$. To solve the equation, we multiply each side by the reciprocal of $-\frac{1}{3}$, which is -3.

$$-\frac{t}{3} = 5 \qquad \text{Original equation}$$

$$-\frac{1}{3}t = 5 \qquad \text{Dividing by 3 is equivalent to multiplying by } \frac{1}{3}.$$

$$-3\left(-\frac{1}{3}t\right) = -3(5) \qquad \text{Multiply each side by } -3.$$

$$\left[-3\left(-\frac{1}{3}\right)\right]t = -3(5) \qquad \text{Associative property}$$

$$t = -15 \qquad \text{Multiplication}$$

EXAMPLE 3   Solve $\frac{2}{3}y = 4$.

Solution   We can multiply both sides by $\frac{3}{2}$ and have $1y$ on the left side:

$$\frac{3}{2}\left(\frac{2}{3}y\right) = \frac{3}{2}(4) \qquad \text{Multiplication property of equality}$$

$$\left(\frac{3}{2} \cdot \frac{2}{3}\right)y = \frac{3}{2}(4) \qquad \text{Associative property}$$

$$y = 6 \qquad \text{Simplify } \frac{3}{2}(4) = \frac{3}{2}\left(\frac{4}{1}\right) = \frac{12}{2} = 6.$$

Notice in Examples 1 through 3 that if the variable is being multiplied by a number like -4 or $\frac{2}{3}$, we always multiply by the number's reciprocal, $-\frac{1}{4}$ or $\frac{3}{2}$, to end up with just the variable on one side of the equation.

E X A M P L E   4   Solve $5 + 8 = 10x + 20x - 4x$.

Solution   Our first step will be to simplify each side of the equation:

$$13 = 26x \qquad \text{Simplify both sides first.}$$

$$\frac{1}{26}(13) = \frac{1}{26}(26x) \qquad \text{Multiplication property of equality}$$

$$\frac{13}{26} = x \qquad \text{Multiplication}$$

$$\frac{1}{2} = x \qquad \text{Reduce to lowest terms.}$$

In the next three examples, we will use both the addition property of equality and the multiplication property of equality.

E X A M P L E   5   Solve for x: $6x + 5 = -13$

Solution   We begin by adding -5 to both sides of the equation:

$$6x + 5 + (-5) = -13 + (-5) \qquad \text{Add } -5 \text{ to both sides.}$$

$$6x = -18 \qquad \text{Simplify.}$$

$$\frac{1}{6}(6x) = \frac{1}{6}(-18) \qquad \text{Multiply both sides by } \frac{1}{6}.$$

$$x = -3$$

E X A M P L E   6   Solve for x: $5x = 2x + 12$

Solution   We begin by adding $-2x$ to both sides of the equation:

$$5x + (-2x) = 2x + (-2x) + 12 \qquad \text{Add } -2x \text{ to both sides.}$$

$$3x = 12 \qquad \text{Simplify.}$$

$$\frac{1}{3}(3x) = \frac{1}{3}(12) \qquad \text{Multiply both sides by } \frac{1}{3}.$$

$$x = 4 \qquad \text{Simplify.}$$

Notice that in Example 6 we used the addition property of equality first in order to combine all the terms containing x on the left side of the equation. Once this had been done, we used the multiplication property to isolate x on the left side.

EXAMPLE 7   Solve for x: $3x - 4 = -2x + 6$

Solution   We begin by adding $2x$ to both sides:

$$3x + 2x - 4 = -2x + 2x + 6 \qquad \text{Add } 2x \text{ to both sides.}$$

$$5x - 4 = 6 \qquad \text{Simplify.}$$

Now we add 4 to both sides:

$$5x - 4 + 4 = 6 + 4 \qquad \text{Add } 4 \text{ to both sides.}$$

$$5x = 10 \qquad \text{Simplify.}$$

$$\frac{1}{5}(5x) = \frac{1}{5}(10) \qquad \text{Multiply by } \frac{1}{5}.$$

$$x = 2 \qquad \text{Simplify.}$$

The next example involves fractions. You will see that the properties we use to solve equations containing fractions are the same as the properties we used to solve the previous equations. Also, the LCD that we used previously to add fractions can be used with the multiplication property of equality to simplify equations containing fractions.

EXAMPLE 8   Solve $\frac{2}{3}x + \frac{1}{2} = -\frac{3}{4}$.

Solution   We can solve this equation by applying our properties and working with the fractions, or we can begin by eliminating the fractions. Let's use both methods.

Method 1   Working with the fractions.

$$\frac{2}{3}x + \frac{1}{2} + \left(-\frac{1}{2}\right) = -\frac{3}{4} + \left(-\frac{1}{2}\right) \qquad \text{Add } -\frac{1}{2} \text{ to each side.}$$

$$\frac{2}{3}x = -\frac{5}{4} \qquad \text{Note that } -\frac{3}{4} + (-\frac{1}{2}) = -\frac{3}{4} + (-\frac{2}{4}).$$

$$\frac{3}{2}\left(\frac{2}{3}x\right) = \frac{3}{2}\left(-\frac{5}{4}\right) \qquad \text{Multiply each side by } \frac{3}{2}.$$

$$x = -\frac{15}{8}$$

Method 2   Eliminating the fractions in the beginning.

Our original equation has denominators of 3, 2, and 4. The LCD for these three denominators is 12, and it has the property that all three denominators will divide it evenly. Therefore, if we multiply both sides of our equation by 12, each denominator will divide into 12, and we will be left with an equation that does not contain any denominators other than 1.

$$12\left(\frac{2}{3}x + \frac{1}{2}\right) = 12\left(-\frac{3}{4}\right) \qquad \text{Multiply each side by the LCD } \mathbf{12.}$$

$$12\left(\frac{2}{3}x\right) + 12\left(\frac{1}{2}\right) = 12\left(-\frac{3}{4}\right) \qquad \text{Distributive property on the left side}$$

$$8x + 6 = -9 \qquad \text{Multiply.}$$

$$8x = -15 \qquad \text{Add } -6 \text{ to each side.}$$

$$x = -\frac{15}{8} \qquad \text{Multiply each side by } \tfrac{1}{8}.$$

As the third line in the previous expression indicates, multiplying each side of the equation by the LCD eliminates all the fractions from the equation.

As you can see, both methods yield the same solution.

A NOTE ON DIVISION

Since division is defined as multiplication by the reciprocal, multiplying both sides of an equation by the same number is equivalent to dividing both sides of the equation by the reciprocal of that number. That is, multiplying each side of an equation by $\frac{1}{3}$ and dividing each side of the equation by 3 are equivalent operations. If we were to solve the equation $3x = 18$ using division instead of multiplication, the steps would look like this:

$$3x = 18 \qquad \text{Original equation}$$

$$\frac{3x}{\mathbf{3}} = \frac{18}{\mathbf{3}} \qquad \text{Divide each side by } \mathbf{3.}$$

$$x = 6 \qquad \text{Division}$$

Using division instead of multiplication on a problem like this may save you some writing. On the other hand, with multiplication, it is easier to explain "why" we end up with just one x on the left side of the equation. (The "why" has to do with the associative property of multiplication.) My suggestion is that you continue to use multiplication to solve equations like this one until you understand the process completely. Then, if you find it more convenient, you can use division instead of multiplication.

Getting Ready for Class

After reading through the preceding section, respond in your own words and in complete sentences.

A. Explain in words the multiplication property of equality.

B. If an equation contains fractions, how do you use the multiplication property of equality to clear the equation of fractions?

C. Why is it okay to divide both sides of an equation by the same number?

D. Explain in words how you would solve the equation $3x = 7$, using the multiplication property of equality.

PROBLEM SET 2.3

Solve the following equations. Be sure to show your work.

1. $5x = 10$

2. $6x = 12$

3. $7a = 28$

4. $4a = 36$

5. $-8x = 4$

6. $-6x = 2$

7. $8m = -16$

8. $5m = -25$

9. $-3x = -9$

10. $-9x = -36$

11. $-7y = -28$

12. $-15y = -30$

13. $2x = 0$

14. $7x = 0$

15. $-5x = 0$

16. $-3x = 0$

17. $\dfrac{x}{3} = 2$

18. $\dfrac{x}{4} = 3$

19. $-\dfrac{m}{5} = 10$

20. $-\dfrac{m}{7} = 1$

21. $-\dfrac{x}{2} = -\dfrac{3}{4}$

22. $-\dfrac{x}{3} = \dfrac{5}{6}$

23. $\dfrac{2}{3}a = 8$

24. $\dfrac{3}{4}a = 6$

25. $-\dfrac{3}{5}x = \dfrac{9}{5}$

26. $-\dfrac{2}{5}x = \dfrac{6}{15}$

27. $-\dfrac{5}{8}y = -20$

28. $-\dfrac{7}{2}y = -14$

Simplify both sides as much as possible, and then solve.

29. $-4x - 2x + 3x = 24$

30. $7x - 5x + 8x = 20$

31. $4x + 8x - 2x = 15 - 10$

32. $5x + 4x + 3x = 4 + 8$

33. $-3 - 5 = 3x + 5x - 10x$

34. $10 - 16 = 12x - 6x - 3x$

35. $18 - 13 = \dfrac{1}{2}a + \dfrac{3}{4}a - \dfrac{5}{8}a$

36. $20 - 14 = \dfrac{1}{3}a + \dfrac{5}{6}a - \dfrac{2}{3}a$

Solve the following equations by multiplying both sides by -1.

37. $-x = 4$

38. $-x = -3$

39. $-x = -4$

40. $-x = 3$

41. $15 = -a$

42. $-15 = -a$

43. $-y = \dfrac{1}{2}$

44. $-y = -\dfrac{3}{4}$

Solve each of the following equations using the method shown in Examples 5–8 in this section.

45. $3x - 2 = 7$

46. $2x - 3 = 9$

47. $2a + 1 = 3$

48. $5a - 3 = 7$

49. $\dfrac{1}{8} + \dfrac{1}{2}x = \dfrac{1}{4}$

50. $\dfrac{1}{3} + \dfrac{1}{7}x = -\dfrac{8}{21}$

51. $6x = 2x - 12$

52. $8x = 3x - 10$

53. $2y = -4y + 18$

54. $3y = -2y - 15$

55. $-7x = -3x - 8$

56. $-5x = -2x - 12$

57. $8x + 4 = 2x - 5$

58. $5x + 6 = 3x - 6$

59. $x + \dfrac{1}{2} = \dfrac{1}{4}x - \dfrac{5}{8}$

60. $\dfrac{1}{3}x + \dfrac{2}{5} = \dfrac{1}{5}x - \dfrac{2}{5}$

Solve.

61. $6m - 3 = m + 2$

62. $6m - 5 = m + 5$

63. $\dfrac{1}{2}m - \dfrac{1}{4} = \dfrac{1}{12}m + \dfrac{1}{6}$

64. $\dfrac{1}{2}m - \dfrac{5}{12} = \dfrac{1}{12}m + \dfrac{5}{12}$

65. $9y + 2 = 6y - 4$

66. $6y + 14 = 2y - 2$

67. $\dfrac{3}{2}y + \dfrac{1}{3} = y - \dfrac{2}{3}$

68. $\dfrac{3}{2}y + \dfrac{7}{2} = \dfrac{1}{2}y - \dfrac{1}{2}$

Applying the Concepts

69. Break-Even Point   Movie theaters pay a certain price for the movies that you and I see. Suppose a theater pays $1,500 for each showing of a popular movie. If they charge $7.50 for each ticket they sell, then the equation $7.5x = 1,500$ gives the number of tickets they must sell to equal the $1,500 cost of showing the movie. This number is called the break-even point. Solve the equation for x to find the break-even point.

70. Stock Sales   Suppose you purchase x shares of a stock at $12 per share. After 6 months you decide to sell all your shares at $20 per share. Your broker charges you $15 for the trade, which gives you a profit of $3,985. In this situation, solving the equation

$$20x - 12x - 15 = 3,985$$

will tell you how many shares of the stock you purchased originally. How many shares did you purchase originally?

71. Basketball   Laura plays basketball for her community college. In one game she scored 13 points total, with a combination of free throws, field goals, and three-pointers. Each free throw is worth 1 point, each field goal is 2 points, and each 3-pointer is worth 3 points. If she made 1 free throw and 3 field goals, then solving the equation

$$1 + 3(2) + 3x = 13$$

will give us the number of three-pointers she made. Solve the equation to find the number of three-point shots Laura made.

72. Taxes   If 21% of your monthly pay is withheld for federal income taxes and another 8% is withheld for Social Security, state income tax, and other miscellaneous items, leaving you with $987.50 a month in take-home pay, then the amount you earned before the deductions were removed from your check is given by the equation

$$G - 0.21G - 0.08G = 987.5$$

Solve this equation to find your gross income.

73. Similar Triangles   If two triangles are similar, their corresponding sides are in proportion.

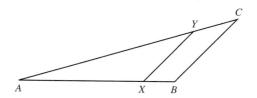

Triangle ABC is similar to triangle AXY means that

$$\frac{AX}{AB} = \frac{AY}{AC} = \frac{XY}{BC}$$

(a) If $AB = 12$, $AY = 15$, and $AC = 20$, find AX.
(b) If $XY = 8$, $BC = 10$, and $AX = 12$, find AB.
(c) If $YC = 6$, $XB = 4$, and $AX = 8$, find AC.

74. Rhind Papyrus   The *Rhind Papyrus* is an ancient Egyptian document that contains some mathematical riddles. One problem on the *Rhind Papyrus* asks the reader to find a quantity such that when it is added to one-fourth of itself the sum is 15. The equation that describes this situation is

$$x + \frac{1}{4}x = 15$$

Solve this equation.

(© The British Museum)

Translations   Translate each sentence below into an equation, and then solve the equation.

75. The sum of $3x$ and 2 is 19.

76. The difference of 4 and $6x$ is 20.

77. Twice the sum of x and 10 is 40.

78. The sum of twice x and 10 is 40.

Review Problems

The following problems review material we covered in Section 2.1. Reviewing this material will help you in the next section.

Simplify each expression.

79. $5(2x - 8) - 3$

80. $4(3x - 1) + 7$

81. $-2(3x + 5) + 3(x - 1)$

82. $6(x + 3) - 2(2x + 4)$

83. $7 - 3(2y + 1)$

84. $8 - 5(3y - 4)$

85. $4x - (9x - 3) + 4$

86. $x - (5x + 2) - 3$

2.4   Solving Linear Equations

We will now use the material we have developed in the first three sections of this chapter to build a method for solving any linear equation.

> **DEFINITION**   A **linear equation** in one variable is any equation that can be put in the form $ax + b = 0$, where a and b are real numbers and a is not zero.

Each of the equations we will solve in this section is a linear equation in one variable. The steps we use to solve a linear equation in one variable are listed here.

STRATEGY FOR SOLVING LINEAR EQUATIONS IN ONE VARIABLE

Step 1a: Use the distributive property to separate terms, if necessary.

1b: If fractions are present, consider multiplying both sides by the LCD to eliminate the fractions. If decimals are present, consider multiplying both sides by a power of 10 to clear the equation of decimals.

1c: Combine similar terms on each side of the equation.

Step 2: Use the addition property of equality to get all variable terms on one side of the equation and all constant terms on the other side. A variable term is a term that contains the variable (e.g., $5x$). A constant term is a term that does not contain the variable (e.g., the number 3).

Step 3: Use the multiplication property of equality to get x (that is, $1x$) by itself on one side of the equation.

Step 4: Check your solution in the original equation to be sure that you have not made a mistake in the solution process.

As you will see as you work through the examples in this section, it is not always necessary to use all four steps when solving equations. The number of steps used depends upon the equation. In Example 1 there are no fractions or decimals in the original equation, so step 1b will not be used. Likewise, after applying the distributive property to the left side of the equation in Example 1, there are no similar terms to combine on either side of the equation, making step 1c also unnecessary.

EXAMPLE I   Solve $2(x + 3) = 10$.

Solution   To begin, we apply the distributive property to the left side of the equation to separate terms:

Step 1a: $2x + 6 = 10$ Distributive property

Step 2: $2x + 6 + (-6) = 10 + (-6)$ Addition property of
                        $2x = 4$ equality

Step 3: $\dfrac{1}{2}(2x) = \dfrac{1}{2}(4)$ Multiply each side by $\dfrac{1}{2}$.

                  $x = 2$ The solution is 2.

The solution to our equation is 2. We check our work (to be sure we have not made either a mistake in applying the properties or an arithmetic mistake) by substituting 2 into our original equation and simplifying each side of the result separately.

Check: When $x = 2$

            the equation $2(x + 3) = 10$

Step 4:   becomes $2(2 + 3) \stackrel{?}{=} 10$

                        $2(5) \stackrel{?}{=} 10$

                        $10 = 10$ A true statement

Our solution checks.

The general method of solving linear equations is actually very simple. It is based on the properties we developed in Chapter 1 and on two very simple new properties. We can add any number to both sides of the equation and multiply both sides by any nonzero number. The equation may change in form, but the solution set will not. If we look back to Example 1, each equation looks a little different from each preceding equation. What is interesting and useful is that each equation says the same thing about x. They all say x is 2. The last equation, of course, is the easiest to read, and that is why our goal is to end up with x by itself.

EXAMPLE 2   Solve for x: $3(x - 5) + 4 = 13$

Solution   Our first step will be to apply the distributive property to the left side of the equation:

Step 1a: $3x - 15 + 4 = 13$ Distributive property

Step 1c: $3x - 11 = 13$ Simplify the left side.

Step 2: $\begin{cases} 3x - 11 + \mathbf{11} = 13 + \mathbf{11} \\ \\ 3x = 24 \end{cases}$ Add **11** to both sides.

Step 3: $\begin{cases} \dfrac{1}{3}(3x) = \dfrac{1}{3}(24) \\ \\ x = 8 \end{cases}$ Multiply both sides by $\frac{1}{3}$.

                   The solution is 8.

Check:

Step 4: $\begin{cases} \text{When} & x = 8 \\ \text{the equation} & 3(x - 5) + 4 = 13 \\ \text{becomes} & 3(8 - 5) + 4 \overset{?}{=} 13 \\ & 3(3) + 4 \overset{?}{=} 13 \\ & 9 + 4 \overset{?}{=} 13 \\ & 13 = 13 \quad \text{A true statement} \end{cases}$

E X A M P L E 3   Solve $5(x - 3) + 2 = 5(2x - 8) - 3$.

Solution   In this case we apply the distributive property on each side of the equation:

Step 1a: $5x - 15 + 2 = 10x - 40 - 3$ Distributive property

Step 1c: $5x - 13 = 10x - 43$ Simplify each side.

Step 2: $\begin{cases} 5x + (\mathbf{-5x}) - 13 = 10x + (\mathbf{-5x}) - 43 \\ -13 = 5x - 43 \\ -13 + \mathbf{43} = 5x - 43 + \mathbf{43} \\ 30 = 5x \end{cases}$

            Add **−5x** to both sides.

            Add **43** to both sides.

Step 3: $\begin{cases} \dfrac{1}{5}(30) = \dfrac{1}{5}(5x) \\ \\ 6 = x \end{cases}$ Multiply both sides by $\frac{1}{5}$.

                  The solution is 6.

Check:

Step 4: $\begin{cases} \text{Replacing } x \text{ with 6 in the original equation, we have} \\ 5(6 - 3) + 2 \overset{?}{=} 5(2 \cdot 6 - 8) - 3 \\ 5(3) + 2 \overset{?}{=} 5(12 - 8) - 3 \\ 5(3) + 2 \overset{?}{=} 5(4) - 3 \\ 15 + 2 \overset{?}{=} 20 - 3 \\ 17 = 17 \qquad\qquad \text{A true statement} \end{cases}$

Note: It makes no difference on which side of the equal sign x ends up. Most people prefer to have x on the left side because we read from left to right, and it seems to sound better to say x is 6 rather than 6 is x. Both expressions, however, have exactly the same meaning.

E X A M P L E   4   Solve the equation $0.08x + 0.09(x + 2{,}000) = 690$.

Solution   We can solve the equation in its original form by working with the decimals, or we can eliminate the decimals first by using the multiplication property of equality and solving the resulting equation. Both methods follow.

Method 1   Working with the decimals.

$0.08x + 0.09(x + 2{,}000) = 690$	Original equation
Step 1a: $0.08x + 0.09x + 0.09(2{,}000) = 690$	Distributive property
Step 1c: $\qquad\qquad 0.17x + 180 = 690$	Simplify the left side.
Step 2: $\begin{cases} 0.17x + 180 + (-\mathbf{180}) = 690 + (-\mathbf{180}) \\ \\ 0.17x = 510 \end{cases}$	Add $-\mathbf{180}$ to each side.
Step 3: $\begin{cases} \dfrac{0.17x}{\mathbf{0.17}} = \dfrac{510}{\mathbf{0.17}} \\ \\ x = 3{,}000 \end{cases}$	Divide each side by **0.17**.

Note that we divided each side of the equation by 0.17 to obtain our solution. This is still an application of the multiplication property of equality, since dividing by 0.17 is equivalent to multiplying by $\frac{1}{0.17}$.

Method 2   Eliminating the decimals in the beginning.

$0.08x + 0.09(x + 2{,}000) = 690$	Original equation
Step 1a: $\qquad 0.08x + 0.09x + 180 = 690$	Distributive property
Step 1b: $\begin{cases} \mathbf{100}(0.08x + 0.09x + 180) = \mathbf{100}(690) \\ \\ 8x + 9x + 18{,}000 = 69{,}000 \end{cases}$	Multiply both sides by **100**.
Step 1c: $\qquad\qquad 17x + 18{,}000 = 69{,}000$	Simplify the left side.
Step 2: $\qquad\qquad\qquad 17x = 51{,}000$	Add $-18{,}000$ to each side.
Step 3: $\begin{cases} \dfrac{17x}{\mathbf{17}} = \dfrac{51{,}000}{\mathbf{17}} \\ \\ x = 3{,}000 \end{cases}$	Divide each side by **17**.

Check: Substituting 3,000 for x in the original equation, we have

$$0.08(3,000) + 0.09(3,000 + 2,000) \overset{?}{=} 690$$

Step 4:

$$0.08(3,000) + 0.09(5,000) \overset{?}{=} 690$$

$$240 + 450 \overset{?}{=} 690$$

$$690 = 690 \qquad \text{A true statement}$$

EXAMPLE 5 Solve $7 - 3(2y + 1) = 16$.

Solution We begin by multiplying -3 times the sum of $2y$ and 1:

Step 1a:	$7 - 6y - 3 = 16$	Distributive property
Step 1c:	$-6y + 4 = 16$	Simplify the left side.

Step 2:
$$\begin{cases} -6y + 4 + (-4) = 16 + (-4) & \text{Add } -4 \text{ to both sides.} \\ -6y = 12 \end{cases}$$

Step 3:
$$\begin{cases} -\dfrac{1}{6}(-6y) = -\dfrac{1}{6}(12) & \text{Multiply both sides by } -\dfrac{1}{6}. \\ y = -2 \end{cases}$$

Step 4: Replacing y with -2 in the original equation yields a true statement.

There are two things to notice about the example that follows: first, the distributive property is used to remove parentheses that are preceded by a negative sign, and, second, the addition property and the multiplication property are not shown in as much detail as in the previous examples.

EXAMPLE 6 Solve $3(2x - 5) - (2x - 4) = 6 - (4x + 5)$.

Solution When we apply the distributive property to remove the grouping symbols and separate terms, we have to be careful with the signs. Remember, we can think of $-(2x - 4)$ as $-1(2x - 4)$, so that

$$-(2x - 4) = -1(2x - 4) = -2x + 4$$

It is not uncommon for students to make a mistake with this type of simplification and write the result as $-2x - 4$, which is incorrect. Here is the complete solution to our equation:

$3(2x - 5) - (2x - 4) = 6 - (4x + 5)$	Original equation
$6x - 15 - 2x + 4 = 6 - 4x - 5$	Distributive property
$4x - 11 = -4x + 1$	Simplify each side.
$8x - 11 = 1$	Add $4x$ to each side.

$$8x = 12 \qquad \text{Add 11 to each side.}$$

$$x = \frac{12}{8} \qquad \text{Multiply each side by } \tfrac{1}{8}.$$

$$x = \frac{3}{2} \qquad \text{Reduce to lowest terms.}$$

The solution, $\frac{3}{2}$, checks when replacing x in the original equation.

Getting Ready for Class

After reading through the preceding section, respond in your own words and in complete sentences.

A. What is the first step in solving a linear equation containing parentheses?

B. What is the last step in solving a linear equation?

C. Explain in words how you would solve the equation $2x - 3 = 8$.

D. If an equation contains decimals, what can you do to eliminate the decimals?

PROBLEM SET 2.4

Solve each of the following equations using the four steps shown in this section.

1. $2(x + 3) = 12$ **2.** $3(x - 2) = 6$

3. $6(x - 1) = -18$ **4.** $4(x + 5) = 16$

5. $2(4a + 1) = -6$ **6.** $3(2a - 4) = 12$

7. $14 = 2(5x - 3)$ **8.** $-25 = 5(3x + 4)$

9. $-2(3y + 5) = 14$ **10.** $-3(2y - 4) = -6$

11. $-5(2a + 4) = 0$ **12.** $-3(3a - 6) = 0$

13. $1 = \frac{1}{2}(4x + 2)$ **14.** $1 = \frac{1}{3}(6x + 3)$

15. $3(t - 4) + 5 = -4$ **16.** $5(t - 1) + 6 = -9$

Solve each equation.

17. $4(2y + 1) - 7 = 1$

18. $6(3y + 2) - 8 = -2$

19. $\frac{1}{2}(x - 3) = \frac{1}{4}(x + 1)$

20. $\frac{1}{3}(x - 4) = \frac{1}{2}(x - 6)$

21. $-0.7(2x - 7) = 0.3(11 - 4x)$

22. $-0.3(2x - 5) = 0.7(3 - x)$

23. $-2(3y + 1) = 3(1 - 6y) - 9$

24. $-5(4y - 3) = 2(1 - 8y) + 11$

25. $\frac{3}{4}(8x - 4) + 3 = \frac{2}{5}(5x + 10) - 1$

26. $\frac{5}{6}(6x + 12) + 1 = \frac{2}{3}(9x - 3) + 5$

27. $0.06x + 0.08(100 - x) = 6.5$

28. $0.05x + 0.07(100 - x) = 6.2$

29. $6 - 5(2a - 3) = 1$

30. $-8 - 2(3 - a) = 0$

31. $0.2x - 0.5 = 0.5 - 0.2(2x - 13)$

32. $0.4x - 0.1 = 0.7 - 0.3(6 - 2x)$

33. $2(t - 3) + 3(t - 2) = 28$

34. $-3(t - 5) - 2(2t + 1) = -8$

35. $5(x - 2) - (3x + 4) = 3(6x - 8) + 10$

36. $3(x - 1) - (4x - 5) = 2(5x - 1) - 7$

37. $2(5x - 3) - (2x - 4) = 5 - (6x + 1)$

38. $3(4x - 2) - (5x - 8) = 8 - (2x + 3)$

39. $-(3x + 1) - (4x - 7) = 4 - (3x + 2)$

40. $-(6x + 2) - (8x - 3) = 8 - (5x + 1)$

Review Problems

The problems that follow review material we covered in Sections 1.2, 1.5, and 1.6. Reviewing these problems will help you understand the next section.

Multiply.

41. $\dfrac{1}{2}(3)$ **42.** $\dfrac{1}{3}(2)$ **43.** $\dfrac{2}{3}(6)$

44. $\dfrac{3}{2}(4)$ **45.** $\dfrac{5}{9} \cdot \dfrac{9}{5}$ **46.** $\dfrac{3}{7} \cdot \dfrac{7}{3}$

Apply the distributive property, and then simplify each expression as much as possible.

47. $2(3x - 5)$ **48.** $4(2x - 6)$

49. $\dfrac{1}{2}(3x + 6)$ **50.** $\dfrac{1}{4}(2x + 8)$

51. $\dfrac{1}{3}(-3x + 6)$ **52.** $\dfrac{1}{2}(-2x + 6)$

2.5   Formulas

In this section we continue solving equations by working with formulas. To begin, here is the definition for a formula.

> **DEFINITION**   In mathematics, a **formula** is an equation that contains more than one variable.

The equation $P = 2l + 2w$, which tells us how to find the perimeter of a rectangle, is an example of a formula.

To begin our work with formulas, we will consider some examples in which we are given numerical replacements for all but one of the variables.

EXAMPLE 1   The perimeter P of a rectangular livestock pen is 40 feet. If the width w is 6 feet, find the length.

$w = 6$ ft

Solution   First, we substitute 40 for P and 6 for w in the formula $P = 2l + 2w$. Then we solve for l:

When	$P = 40$ and $w = 6$	
the formula	$P = 2l + 2w$	
becomes	$40 = 2l + 2(6)$	
or	$40 = 2l + 12$	Multiply 2 and 6.
	$28 = 2l$	Add -12 to each side.
	$14 = l$	Multiply each side by $\frac{1}{2}$.

To summarize our results, if a rectangular pen has a perimeter of 40 feet and a width of 6 feet, then the length must be 14 feet.

EXAMPLE 2   Find y when $x = 4$ in the formula $3x + 2y = 6$.

Solution   We substitute 4 for x in the formula and then solve for y:

When	$x = 4$	
the formula	$3x + 2y = 6$	
becomes	$3(4) + 2y = 6$	
or	$12 + 2y = 6$	Multiply 3 and 4.
	$2y = -6$	Add -12 to each side.
	$y = -3$	Multiply each side by $\frac{1}{2}$.

In the next examples we will solve a formula for one of its variables without being given numerical replacements for the other variables.

Consider the formula for the area of a triangle:

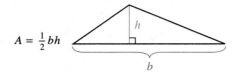

$$A = \tfrac{1}{2}bh$$

where A = area, b = length of the base, and h = the height of the triangle.

Suppose we want to solve this formula for h. What we must do is isolate the variable h on one side of the equal sign. We begin by multiplying both sides by 2:

$$2 \cdot A = 2 \cdot \frac{1}{2}bh$$

$$2A = bh$$

Then we divide both sides by b:

$$\frac{2A}{b} = \frac{bh}{b}$$

$$h = \frac{2A}{b}$$

The original formula $A = \tfrac{1}{2}bh$ and the final formula $h = \dfrac{2A}{b}$ both give the same relationship among A, b, and h. The first one has been solved for A, and the second one has been solved for h.

RULE

To solve a formula for one of its variables, we must isolate that variable on either side of the equal sign. All other variables and constants will appear on the other side.

EXAMPLE 3 Solve $3x + 2y = 6$ for y.

Solution To solve for y, we must isolate y on the left side of the equation. To begin, we use the addition property of equality to add $-3x$ to each side:

$3x + 2y = 6$	Original formula
$3x + (-3x) + 2y = (-3x) + 6$	Add $-3x$ to each side.
$2y = -3x + 6$	Simplify the left side.
$\dfrac{1}{2}(2y) = \dfrac{1}{2}(-3x + 6)$	Multiply each side by $\dfrac{1}{2}$.
$y = -\dfrac{3}{2}x + 3$	Multiplication

E X A M P L E   4   Solve $h = vt - 16t^2$ for v.

Solution   Let's begin by interchanging the left and right sides of the equation. That way, the variable we are solving for, v, will be on the left side.

$$vt - 16t^2 = h \qquad \text{Exchange sides.}$$

$$vt - 16t^2 + \mathbf{16t^2} = h + \mathbf{16t^2} \qquad \text{Add } \mathbf{16t^2} \text{ to each side.}$$

$$vt = h + 16t^2$$

$$\frac{vt}{t} = \frac{h + 16t^2}{t} \qquad \text{Divide each side by } \mathbf{t.}$$

$$v = \frac{h + 16t^2}{t}$$

We know we are finished because we have isolated the variable we are solving for on the left side of the equation and it does not appear on the other side.

FACTS FROM
Geometry

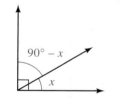

Complementary angles

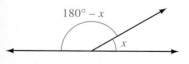

Supplementary angles

In Chapter 1 we defined complementary angles as angles that add to 90°. That is, if x and y are complementary angles, then

$$x + y = 90°$$

If we solve this formula for y, we obtain a formula equivalent to our original formula:

$$y = 90° - x$$

Since y is the complement of x, we can generalize by saying that the complement of angle x is the angle $90° - x$. By a similar reasoning process, we can say that the supplement of angle x is the angle $180° - x$. To summarize, if x is an angle, then

the complement of x is $90° - x$ and

the supplement of x is $180° - x$.

If you go on to take a trigonometry class, you will see this formula again.

E X A M P L E   5   Find the complement and the supplement of 25°.

Solution   We can use the formulas above with $x = 25°$.

The complement of 25° is $90° - 25° = 65°$.

The supplement of 25° is $180° - 25° = 155°$.

BASIC PERCENT PROBLEMS

The last examples in this section show how basic percent problems can be translated directly into equations. To understand these examples, you must recall that percent means "per hundred." That is, 75% is the same as $\frac{75}{100}$, 0.75, and, in reduced fraction form, $\frac{3}{4}$. Likewise, the decimal 0.25 is equivalent to 25%. To change a decimal to a percent, we move the decimal point two places to the right and write the % symbol. To change from a percent to a decimal, we drop the % symbol and move the decimal point two places to the left. The table that follows gives some of the most commonly used fractions and decimals and their equivalent percents.

Fraction	Decimal	Percent
$\frac{1}{2}$	0.5	50%
$\frac{1}{4}$	0.25	25%
$\frac{3}{4}$	0.75	75%
$\frac{1}{3}$	$0.33\frac{1}{3}$	$33\frac{1}{3}\%$
$\frac{2}{3}$	$0.66\frac{2}{3}$	$66\frac{2}{3}\%$
$\frac{1}{5}$	0.2	20%
$\frac{2}{5}$	0.4	40%
$\frac{3}{5}$	0.6	60%
$\frac{4}{5}$	0.8	80%

EXAMPLE 6   What number is 25% of 60?

Solution   To solve a problem like this, we let $x =$ the number in question (that is, the number we are looking for). Then, we translate the sentence directly into an equation by using an equal sign for the word *is* and multiplication for the word *of*. Here is how it is done:

$$\text{What number} \quad \text{is} \quad 25\% \quad \text{of} \quad 60?$$
$$x \qquad\qquad = 0.25 \ \cdot \ 60$$
$$x = 15$$

Notice that we must write 25% as a decimal in order to do the arithmetic in the problem.

The number 15 is 25% of 60.

EXAMPLE 7   What percent of 24 is 6?

Solution   Translating this sentence into an equation, as we did in Example 6, we have:

What percent of 24 is 6?

$$x \cdot 24 = 6$$

or $24x = 6$

Next, we multiply each side by $\frac{1}{24}$. (This is the same as dividing each side by 24).

$$\frac{1}{24}(24x) = \frac{1}{24}(6)$$

$$x = \frac{6}{24}$$

$$= \frac{1}{4}$$

$$= 0.25, \text{ or } 25\%$$

The number 6 is 25% of 24.

E X A M P L E   8   45 is 75% of what number?

Solution   Again, we translate the sentence directly:

45 is 75% of what number?

$$45 = 0.75 \cdot x$$

Next, we multiply each side by $\frac{1}{0.75}$ (which is the same as dividing each side by 0.75):

$$\frac{1}{0.75}(45) = \frac{1}{0.75}(0.75x)$$

$$\frac{45}{0.75} = x$$

$$60 = x$$

The number 45 is 75% of 60.

We can solve application problems involving percent by translating each problem into one of the three basic percent problems shown in Examples 6, 7, and 8.

E X A M P L E   9   The American Dietetics Association (ADA) recommends eating foods in which the calories from fat are less than 30% of the total calories. The nutrition labels from two kinds of granola bars are shown in Figure 1. For each bar, what percent of the total calories comes from fat?

Solution   The information needed to solve this problem is located toward the top of each label. Each serving of Bar I contains 210 calories, of which 70 calories come from fat. To find the percent of total calories that come from fat, we must answer this question:

70 is what percent of 210?

Bar I

Nutrition Facts
Serving Size 2 bars (47g)
Servings Per Container 6

Amount Per Serving

Calories	210
Calories from Fat	70

	% Daily Value*
Total Fat 8g	12%
Saturated Fat 1g	5%
Cholesterol 0mg	0%
Sodium 150mg	6%
Total Carbohydrate 32g	11%
Dietary Fiber 2g	10%
Sugars 12g	
Protein 4g	

* Percent Daily Values are based on a 2,000 calorie diet. Your daily values may be higher or lower depending on your calorie needs.

Bar II

Nutrition Facts
Serving Size 1 bar (21g)
Servings Per Container 8

Amount Per Serving

Calories	80
Calories from Fat	15

	% Daily Value*
Total Fat 1.5g	2%
Saturated Fat 0g	0%
Cholesterol 0mg	0%
Sodium 60mg	3%
Total Carbohydrate 16g	5%
Dietary Fiber 1g	4%
Sugars 5g	
Protein 2g	

* Percent Daily Values are based on a 2,000 calorie diet. Your daily values may be higher or lower depending on your calorie needs.

FIGURE 1

For Bar II, one serving contains 80 calories, of which 15 calories come from fat. To find the percent of total calories that come from fat, we must answer this question:

15 is what percent of 80?

Translating each question into symbols, we have

70 is what percent of 210?   15 is what percent of 80?

$70 = x \cdot 210$ $15 = x \cdot 80$

$x = \dfrac{70}{210}$ $x = \dfrac{15}{80}$

$x = 0.33$ to the nearest hundredth $x = 0.19$ to the nearest hundredth

$x = 33\%$ $x = 19\%$

Comparing the two bars, 33% of the calories in Bar I are fat calories, while 19% of the calories in Bar II are fat calories. According to the ADA, Bar II is the healthier choice.

Getting Ready for Class

After reading through the preceding section, respond in your own words and in complete sentences.

A. What is a formula?

B. How do you solve a formula for one of its variables?

C. What are complementary angles?

D. What does percent mean?

PROBLEM SET 2.5

Use the formula $P = 2l + 2w$ to find the length l of a rectangular lot if

1. The width w is 50 feet and the perimeter P is 300 feet.

2. The width w is 75 feet and the perimeter P is 300 feet.

Use the formula $2x + 3y = 6$ to find y if

3. x is 3   **4.** x is -2

5. x is 0   **6.** x is -3

Use the formula $2x - 5y = 20$ to find x if

7. y is 2   **8.** y is -4

9. y is 0   **10.** y is -6

Use the equation $y = 2x - 1$ to find x when

11. y is 7   **12.** y is 9

13. y is 3   **14.** y is -1

Solve each of the following for the indicated variable.

15. $A = lw$ for l **16.** $A = lw$ for w

17. $d = rt$ for r **18.** $d = rt$ for t

19. $V = lwh$ for h **20.** $V = lwh$ for l

21. $PV = nRT$ for P **22.** $PV = nRT$ for T

23. $P = a + b + c$ for a

24. $P = a + b + c$ for b **25.** $x - 3y = -1$ for x

26. $x + 3y = 2$ for x

27. $-3x + y = 6$ for y **28.** $2x + y = -17$ for y

29. $2x + 3y = 6$ for y **30.** $4x + 5y = 20$ for y

31. $6x + 3y = 12$ for y **32.** $3x + 6y = 12$ for y

33. $5x - 2y = 3$ for y **34.** $7x - 3y = 5$ for y

35. $P = 2l + 2w$ for w **36.** $P = 2l + 2w$ for l

37. $h = vt + 16t^2$ for v **38.** $h = vt - 16t^2$ for v

39. $A = \pi r^2 + 2\pi rh$ for h

40. $A = 2\pi r^2 + 2\pi rh$ for h

Solve each formula for y.

41. $\dfrac{x}{2} + \dfrac{y}{3} = 1$ **42.** $\dfrac{x}{5} + \dfrac{y}{4} = 1$

43. $\dfrac{x}{7} - \dfrac{y}{3} = 1$ **44.** $\dfrac{x}{4} - \dfrac{y}{9} = 1$

45. $-\dfrac{1}{4}x + \dfrac{1}{8}y = 1$ **46.** $-\dfrac{1}{9}x + \dfrac{1}{3}y = 1$

Find the complement and the supplement of each angle.

47. 30°   **48.** 60°

49. 45°   **50.** 15°

Translate each of the following into an equation, and then solve that equation.

51. What number is 25% of 40?

52. What number is 75% of 40?

53. What number is 12% of 2,000?

54. What number is 9% of 3,000?

55. What percent of 28 is 7?

56. What percent of 28 is 21?

57. What percent of 40 is 14?

58. What percent of 20 is 14?

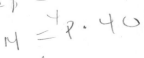

59. 32 is 50% of what number?

60. 16 is 50% of what number?

61. 240 is 12% of what number?

62. 360 is 12% of what number?

Applying the Concepts

63. Percent of People Voting In the 60 years between 1932 and 1992, the largest percentage of eligible voters who actually cast votes in a presidential election was 62.8%, when John Kennedy defeated Richard Nixon in 1960. Kennedy and Nixon split a total of nearly 68,840,000 votes.
(a) Let T = the total number of eligible voters and write an equation that relates T, 62.8%, and 68,840,000.
(b) Solve the equation you wrote in part (a) to find the number of people who were eligible to vote in the 1960 election.

64. Women in the Military The number of women in the military on active duty in 1981 was 171,418. This number was 76.8% of the number of women on active duty in 1984. How many women were on active duty in 1984?

65. Jewelry Content Silver jewelry is actually a mixture of silver and copper. Suppose you have a silver bracelet that weighs 20 grams and contains 12 grams of silver.
(a) What percent of the bracelet is silver?
(b) What percent of the bracelet is copper?

66. Treasury Bonds If you buy a $5,000 treasury bond that pays 6% interest each year, then in t years its value will be

$$V = 5,000(1.06)^t$$

Find the value of this bond after 2 years.

More About Temperatures As we mentioned in Chapter 1, in the U.S. system, temperature is measured on the Fahrenheit scale. In the metric system, temperature is measured on the Celsius scale. On the Celsius scale, water boils at 100 degrees and freezes at 0 degrees. To denote a temperature of 100 degrees on the Celsius scale, we write

100°C, which is read "100 degrees Celsius"

Table 1 is intended to give you an intuitive idea of the relationship between the two temperature scales. Table 2 gives the formulas, in both symbols and words, that are used to convert between the two scales.

TABLE 1

| | Temperature | |
Situation	Fahrenheit	Celsius
Water freezes	32°F	0°C
Room temperature	68°F	20°C
Normal body temperature	98.6°F	37°C
Water boils	212°F	100°C
Bake cookies	365°F	185°C

67. Let $F = 212$ in the formula $C = \frac{5}{9}(F - 32)$, and solve for C. Does the value of C agree with the information in Table 1?

68. Let $C = 100$ in the formula $F = \frac{9}{5}C + 32$, and solve for F. Does the value of F agree with the information in Table 1?

69. Let $F = 68$ in the formula $C = \frac{5}{9}(F - 32)$, and solve for C. Does the value of C agree with the information in Table 1?

70. Let $C = 37$ in the formula $F = \frac{9}{5}C + 32$, and solve for F. Does the value of F agree with the information in Table 1?

TABLE 2

To Convert From	Formula in Symbols	Formula in Words
Fahrenheit to Celsius	$C = \frac{5}{9}(F - 32)$	Subtract 32, multiply by 5, then divide by 9.
Celsius to Fahrenheit	$F = \frac{9}{5}C + 32$	Multiply by $\frac{9}{5}$, then add 32.

71. Solve the formula $F = \frac{9}{5}C + 32$ for C.

72. Solve the formula $C = \frac{5}{9}(F - 32)$ for F.

Nutrition Labels   The nutrition label on the top is from a quart of vanilla ice cream. The label on the bottom is from a pint of vanilla frozen yogurt. Use the information on these labels for Problems 73–76. Round your answers to the nearest tenth of a percent.

Vanilla Ice Cream

Nutrition Facts
Serving Size 1/2 cup (65g)
Servings 8

Amount/Serving		
Calories 150	Calories from Fat 90	
	% Daily Value*	
Total Fat 10g		**16%**
Saturated Fat 6g		**32%**
Cholesterol 35mg		**12%**
Sodium 30mg		**1%**
Total Carbohydrate 14g		**5%**
Dietary Fiber 0g		**0%**
Sugars 11g		
Protein 2g		

Vitamin A 6%	•	Vitamin C 0%
Calcium 6%	•	Iron 0%

* Percent Daily Values are based on a 2,000 calorie diet.

Vanilla Frozen Yogurt

Nutrition Facts
Serving Size 1/2 cup (98g)
Servings Per Container 4

Amount Per Serving		
Calories 160	Calories from Fat 25	
	% Daily Value*	
Total Fat 2.5g		**4%**
Saturated Fat 1.5g		**7%**
Cholesterol 45mg		**15%**
Sodium 55mg		**2%**
Total Carbohydrate 26g		**9%**
Dietary Fiber 0g		**0%**
Sugars 19g		
Protein 8g		

Vitamin A 0%	•	Vitamin C 0%
Calcium 25%	•	Iron 0%

* Percent Daily Values are based on a 2,000 calorie diet.

73. What percent of the calories in one serving of the vanilla ice cream are fat calories?

74. What percent of the calories in one serving of the frozen yogurt are fat calories?

75. One serving of frozen yogurt is 98 grams, of which 26 grams are carbohydrates. What percent of one serving are carbohydrates?

76. One serving of vanilla ice cream is 65 grams. Find the percent of one serving that is sugar.

77. Temperature and Altitude   As we mentioned in the introduction to this chapter, there is a relationship between air temperature and altitude. The higher above the Earth we are, the lower the temperature. If the temperature on the ground is 56°F, then the temperature T at altitude A is given by the formula

$$T = -0.0035A + 56$$

Use this formula to fill in the table below.

Altitude (feet)	Temperature (°F)
0	
9,000	
15,000	
21,000	
28,000	
40,000	

78. Complementary Angles   The diagram below shows sunlight hitting the ground. Angle α (alpha) is called the angle of inclination, and angle θ (theta) is called the angle of incidence. As the sun moves across the sky, the values of these angles change. Assume that $\alpha + \theta = 90°$, and fill in the table below.

α	θ
0°	
15°	
30°	
45°	
60°	
90°	

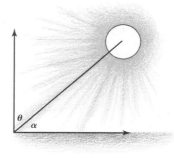

79. Circumference   The circumference of a circle is given by the formula $C = 2\pi r$. Find r if
(a) The circumference C is 44 meters and π is $\frac{22}{7}$.
(b) The circumference is 9.42 inches and π is 3.14.

80. Volume   The volume of a cylinder is given by the formula $V = \pi r^2 h$. Find the height h if

(a) The volume V is 42 cubic feet, the radius is $\frac{7}{22}$ feet, and π is $\frac{22}{7}$.

(b) The volume is 6.28 cubic centimeters, the radius is 3 centimeters, and π is 3.14.

Review Problems

The following problems review material we covered in Chapter 1. Reviewing these problems will help you in the next section.

Write an equivalent expression in words. Include the words *sum* and *difference* when possible.

81. $4 + 1 = 5$

82. $7 + 3 = 10$

83. $6 - 2 = 4$

84. $8 - 1 = 7$

For each of the following expressions, write an equivalent expression with numbers.

85. Twice the sum of 6 and 3

86. 4 added to the product of 5 and -1

87. The sum of twice 5 and 3 is 13.

88. Twice the difference of 8 and 2 is 12.

2.6   Applications

As you begin reading through the examples in this section, you may find yourself asking why some of these problems seem so contrived. The title of the section is "Applications," but many of the problems here don't seem to have much to do with "real life." You are right about that. Example 3 is what we refer to as an "age problem." But imagine a conversation in which you ask someone how old her children are and she replies, "Bill is 6 years older than Tom. Three years ago Bill's age was 4 times Tom's age. You figure it out." Although many of the "application" problems in this section are contrived, they are also good for practicing the strategy we will use to solve all application problems.

To begin this section, we list the steps used in solving application problems. We call this strategy the *Blueprint for Problem Solving*. It is an outline that will overlay the solution process we use on all application problems.

Blueprint for Problem Solving

Step 1: ***Read*** the problem, and then mentally ***list*** the items that are known and the items that are unknown.

Step 2: ***Assign a variable*** to one of the unknown items. (In most cases this will amount to letting $x =$ the item that is asked for in the problem.) Then ***translate*** the other ***information*** in the problem to expressions involving the variable.

Step 3: ***Reread*** the problem, and then ***write an equation,*** using the items and variables listed in steps 1 and 2, that describes the situation.

Step 4: ***Solve the equation*** found in step 3.

Step 5: ***Write*** your ***answer*** using a complete sentence.

Step 6: ***Reread*** the problem, and ***check*** your solution with the original words in the problem.

There are a number of substeps within each of the steps in our blueprint. For instance, with steps 1 and 2 it is always a good idea to draw a diagram or picture, if it helps visualize the relationship between the items in the problem. In other cases a table helps organize the information. As you gain more experience using the blueprint to solve application problems, you will find additional techniques that expand the blueprint.

To help with problems of the type shown next in Example 1, here are some common words and phrases and their mathematical translations.

Words	Algebra
The sum of a and b	$a + b$
The difference of a and b	$a - b$
The product of a and b	$a \cdot b$
The quotient of a and b	$\dfrac{a}{b}$
of	$\cdot$ (multiply)
is	$=$ (equals)
A number	x
4 more than x	$x + 4$
4 times x	$4x$
4 less than x	$x - 4$

NUMBER PROBLEMS

EXAMPLE   1   The sum of twice a number and 3 is 7. Find the number.

Solution   Using the Blueprint for Problem Solving as an outline, we solve the problem as follows:

Step 1: **Read** the problem, and then mentally **list** the items that are known and the items that are unknown.

> *Known items:*   The numbers 3 and 7
>
> *Unknown item:*   The number in question

Step 2: **Assign a variable** to one of the unknown items. Then **translate** the other **information** in the problem to expressions involving the variable.

> Let x = the number asked for in the problem,
>
> then "the sum of twice a number and three" translates to $2x + 3$.

Step 3: **Reread** the problem, and then **write an equation,** using the items and variables listed in steps 1 and 2, that describes the situation.

With all word problems, the word *is* translates to =.

$$\underbrace{\text{The sum of twice } x \text{ and } 3}_{2x + 3} \quad \underset{=\ 7}{\text{is } 7}$$

Step 4: *Solve the equation* found in step 3.

$$2x + 3 = 7$$
$$2x + 3 + (-3) = 7 + (-3)$$
$$2x = 4$$
$$\frac{1}{2}(2x) = \frac{1}{2}(4)$$
$$x = 2$$

Step 5: *Write* your **answer** using a complete sentence.

The number is 2.

Step 6: *Reread* the problem, and **check** your solution with the original words in the problem.

The sum of twice 2 and 3 is 7; a true statement.

You may find some examples and problems in this section, and the problem set that follows, that you can solve without using algebra or our blueprint. It is very important that you solve those problems using the methods we are showing here. The purpose behind these problems is to give you experience using the blueprint as a guide to solving problems written in words. Your answers are much less important than the work that you show to obtain your answer. You will be able to condense the steps in the blueprint later in the course. For now, though, you need to show your work in the same detail that we are showing in the examples in this section.

EXAMPLE 2   One number is 3 more than twice another; their sum is 18. Find the numbers.

Solution

Step 1: *Read and list.*

Known items:	Two numbers that add to 18. One is 3 more than twice the other.
Unknown items:	The numbers in question

Step 2: *Assign a variable and translate information.*
Let x = the first number. The other is $2x + 3$.

Step 3: *Reread and write an equation.*

$$\underbrace{\text{Their sum}}_{x + (2x + 3)} \text{ is } 18 \atop = 18$$

Step 4: Solve the equation.

$$x + (2x + 3) = 18$$
$$3x + 3 = 18$$
$$3x + 3 + (\mathbf{-3}) = 18 + (\mathbf{-3})$$
$$3x = 15$$
$$x = 5$$

Step 5: Write the answer.
The first number is 5. The other is $2 \cdot 5 + 3 = 13$.

Step 6: Reread and check.
The sum of 5 and 13 is 18, and 13 is 3 more than twice 5.

AGE PROBLEM

Remember as you read through the steps in the solutions to the examples in this section that step 1 is done mentally. Read the problem, and then mentally list the items that you know and the items that you don't know. The purpose of step 1 is to give you direction as you begin to work application problems. Finding the solution to an application problem is a process; it doesn't happen all at once. The first step is to read the problem with a purpose in mind. That purpose is to mentally note the items that are known and the items that are unknown.

E X A M P L E   3   Bill is 6 years older than Tom. Three years ago Bill's age was 4 times Tom's age. Find the age of each boy now.

Solution   Applying the Blueprint for Problem Solving, we have

Step 1: Read and list.

Known items:	Bill is 6 years older than Tom. Three years ago Bill's age was 4 times Tom's age.
Unknown items:	Bill's age and Tom's age

Step 2: Assign a variable and translate information.
Let $x = $ Tom's age now. That makes Bill $x + 6$ years old now. A table like the one shown here can help organize the information in an age problem. Notice how we placed the x in the box that corresponds to Tom's age now.

	Three Years Ago	Now
Bill		$x + 6$
Tom		x

If Tom is x years old now, three years ago he was $x - 3$ years old. If Bill is $x + 6$ years old now, three years ago he was $x + 6 - 3 =$

$x + 3$ years old. We use this information to fill in the remaining squares in the table.

	Three Years Ago	Now
Bill	$x + 3$	$x + 6$
Tom	$x - 3$	x

Step 3: *Reread and write an equation.*

Reading the problem again, we see that three years ago Bill's age was four times Tom's age. Writing this as an equation, we have Bill's age three years ago = 4 (Tom's age three years ago):

$$x + 3 = 4(x - 3)$$

Step 4: *Solve the equation.*

$$x + 3 = 4(x - 3)$$
$$x + 3 = 4x - 12$$
$$x + (-x) + 3 = 4x + (-x) - 12$$
$$3 = 3x - 12$$
$$3 + \mathbf{12} = 3x - 12 + \mathbf{12}$$
$$15 = 3x$$
$$x = 5$$

Step 5: *Write the answer.*

Tom is 5 years old. Bill is 11 years old.

Step 6: *Reread and check.*

If Tom is 5 and Bill is 11, then Bill is 6 years older than Tom. Three years ago Tom was 2 and Bill was 8. At that time, Bill's age was four times Tom's age. As you can see, the answers check with the original problem.

GEOMETRY PROBLEM

To understand Example 4 completely, you need to recall from Chapter 1 that the perimeter of a rectangle is the sum of the lengths of the sides. The formula for the perimeter is $P = 2l + 2w$.

EXAMPLE 4 The length of a rectangle is 5 inches more than twice the width. The perimeter is 34 inches. Find the length and width.

Solution When working problems that involve geometric figures, a sketch of the figure helps organize and visualize the problem.

Step 1: *Read and list.*

Known items: The figure is a rectangle. The length is 5 inches more than twice the width. The perimeter is 34 inches.

Unknown items: The length and the width

Step 2: *Assign a variable and translate information.*

Since the length is given in terms of the width (the length is 5 more than twice the width), we let x = the width of the rectangle. The length is 5 more than twice the width, so it must be $2x + 5$. The diagram below is a visual description of the relationships we have listed so far.

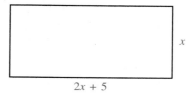

x

$2x + 5$

Step 3: *Reread and write an equation.*

The equation that describes the situation is

Twice the length + twice the width is the perimeter

$$2(2x + 5) \quad + \quad 2x \quad = \quad 34$$

Step 4: *Solve the equation.*

$$2(2x + 5) + 2x = 34$$
$$4x + 10 + 2x = 34 \qquad \text{Distributive property}$$
$$6x + 10 = 34 \qquad \text{Add } 4x \text{ and } 2x.$$
$$6x = 24 \qquad \text{Add } -10 \text{ to each side.}$$
$$x = 4 \qquad \text{Divide each side by 6.}$$

Step 5: *Write the answer.*

The width x is 4 inches. The length is $2x + 5 = 2(4) + 5 = 13$ inches.

Step 6: *Reread and check.*

If the length is 13 and the width is 4, then the perimeter must be $2(13) + 2(4) = 26 + 8 = 34$, which checks with the original problem.

COIN PROBLEM

EXAMPLE 5   Jennifer has $2.45 in dimes and nickels. If she has 8 more dimes than nickels, how many of each coin does she have?

Solution

Step 1: Read and list.

Known items: The type of coins, the total value of the coins, and that there are 8 more dimes than nickels

Unknown items: The number of nickels and the number of dimes

Step 2: Assign a variable and translate information.

If we let x = the number of nickels, then $x + 8$ = the number of dimes. Since the value of each nickel is 5 cents, the amount of money in nickels is $5x$. Similarly, since each dime is worth 10 cents, the amount of money in dimes is $10(x + 8)$. Here is a table that summarizes the information we have so far:

	Nickels	Dimes
Number	x	$x + 8$
Value (in cents)	$5x$	$10(x + 8)$

Step 3: Reread and write an equation.

Since the total value of all the coins is 245 cents, the equation that describes this situation is

Amount of money in nickels	+	Amount of money in dimes	=	Total amount of money
$5x$	+	$10(x + 8)$	=	245

Step 4: Solve the equation.

To solve the equation, we apply the distributive property first.

$$5x + 10x + 80 = 245 \qquad \text{Distributive property}$$

$$15x + 80 = 245 \qquad \text{Add } 5x \text{ and } 10x.$$

$$15x = 165 \qquad \text{Add } -80 \text{ to each side.}$$

$$x = 11 \qquad \text{Divide each side by 15.}$$

Step 5: Write the answer.

The number of nickels is $x = 11$.

Step 6: Reread and check.

The number of dimes is $x + 8 = 11 + 8 = 19$.

To check our results

$$
\begin{array}{ll}
\text{11 nickels are worth } 5(11) = & 55 \text{ cents} \\
\text{19 dimes are worth } 10(19) = & 190 \text{ cents} \\
\hline
\text{The total value is} & 245 \text{ cents} = \$2.45
\end{array}
$$

When you begin working the problems in the problem set that follows, there are a couple of things to remember. The first is that you may have to read the problems a number of times before you begin to see how to solve them. The second thing to remember is that word problems are not always solved correctly the first time you try them. Sometimes it takes a couple of attempts and some wrong answers before you can set up and solve these problems correctly.

 Getting Ready for Class

After reading through the preceding section, respond in your own words and in complete sentences.

A. What is the first step in the Blueprint for Problem Solving?

B. What is the last thing you do when solving an application problem?

C. What good does it do you to solve application problems even when they don't have much to do with real life?

D. Write an application problem whose solution depends on solving the equation $2x + 3 = 7$.

PROBLEM SET 2.6

Solve the following word problems. Follow the steps given in the Blueprint for Problem Solving.

Number Problems

1. The sum of a number and 5 is 13. Find the number.

2. The difference of 10 and a number is -8. Find the number.

3. The sum of twice a number and 4 is 14. Find the number.

4. The difference of 4 times a number and 8 is 16. Find the number.

5. 5 times the sum of a number and 7 is 30. Find the number.

6. 5 times the difference of twice a number and 6 is -20. Find the number.

7. One number is 2 more than another. Their sum is 8. Find both numbers.

8. One number is 3 less than another. Their sum is 15. Find the numbers.

9. One number is 4 less than 3 times another. If their sum is increased by 5, the result is 25. Find the numbers.

10. One number is 5 more than twice another. If their

sum is decreased by 10, the result is 22. Find the numbers.

Age Problems

11. Fred is 4 years older than Barney. Five years ago the sum of their ages was 48. How old are they now? (Begin by filling in the table.)

	Five Years Ago	Now
Fred		
Barney		x

12. Tim is 5 years older than JoAnn. Six years from now the sum of their ages will be 79. How old are they now?

	Now	Six Years From Now
Tim		
JoAnn	x	

13. Jack is twice as old as Lacy. In 3 years the sum of their ages will be 54. How old are they now?

14. John is 4 times as old as Martha. Five years ago the sum of their ages was 50. How old are they now?

15. Pat is 20 years older than his son Patrick. In 2 years Pat will be twice as old as Patrick. How old are they now?

16. Diane is 23 years older than her daughter Amy. In 6 years Diane will be twice as old as Amy. How old are they now?

Geometry Problems

17. The length of a rectangle is 5 inches more than the width. The perimeter is 34 inches. Find the length and width.

x

$x + 5$

18. The width of a rectangle is 3 feet less than the length. The perimeter is 10 feet. Find the length and width.

19. The perimeter of a square is 48 meters. Find the length of one side.

20. One side of a triangle is twice the shortest side. The third side is 3 feet more than the shortest side. The perimeter is 19 feet. Find all three sides.

21. The length of a rectangle is 3 inches less than twice the width. The perimeter is 54 inches. Find the length and width.

22. The length of a rectangle is 4 feet less than 3 times the width. The sum of the length and width is 14 more than the width. Find the width.

Coin Problems

23. Sue has $2.10 in dimes and nickels. If she has 9 more dimes than nickels, how many of each coin does she have? (Completing the table may help you get started.)

	Nickels	Dimes
Number	x	
Value (in cents)		

24. Mike has $1.55 in dimes and nickels. If he has 7 more nickels than dimes, how many of each coin does he have?

	Nickels	Dimes
Number		x
Value (in cents)		

25. Suppose you have $9.00 in dimes and quarters. How many of each coin do you have if you have twice as many quarters as dimes?

26. A collection of dimes and quarters has a total value of $2.20. If there are 3 times as many dimes as quarters, how many of each coin is in the collection?

27. Katie has a collection of nickels, dimes, and quarters with a total value of $4.35. There are 3 more dimes than nickels and 5 more quarters than nickels. How many of each coin is in her collection? (*Hint:* Let x = the number of nickels.)

28. Mary Jo has $3.90 worth of nickels, dimes, and quarters. The number of nickels is 3 more than the number of dimes. The number of quarters is 7 more than the number of dimes. How many of each coin does she have? (*Hint:* Let x = the number of dimes.)

Review Problems

The following problems review material we covered in Sections 1.1 and 1.2.

Write an equivalent statement in words. [1.1]

29. $4 < 10$

30. $4 \le 10$

31. $9 \ge -5$

32. $x - 2 > 4$

Place the symbol $<$ or the symbol $>$ between the quantities in each expression. [1.2]

33. 12   20

34. -12 20

35. -8 -6

36. -10 -20

Simplify. [1.2]

37. $|8 - 3| - |5 - 2|$

38. $|9 - 2| - |10 - 8|$

39. $15 - |9 - 3(7 - 5)|$

40. $10 - |7 - 2(5 - 3)|$

Now that you have worked through a number of application problems using our blueprint, you have probably noticed that step 3, in which we write an equation that describes the situation, is the key step. Anyone with experience solving application problems will tell you that there will be times when your first attempt at step 3 results in the wrong equation. Remember, mistakes are part of the process of learning to do things correctly. Many times the correct equation will become obvious after you have written an equation that is partially wrong. In any case, it is better to write an equation that is partially wrong and be actively involved with the problem than to write nothing at all. Application problems, like other problems in algebra, are not always solved correctly the first time.

In this section we continue our work with application problems. There are two main differences between the problems in this section and the problems you worked with in the previous section. First of all, not all of the problems in the problem set that follows this section fall into specific categories. Second, many of the problems in this section and problem set are more realistic in nature.

E X A M P L E   1   The total price for one of the cameras I used to produce the videotapes that accompany this book was $3,200. The total price is the retail price of the camera plus the sales tax. At that time, the sales tax rate in California was 7.5%. What was the retail price of the camera?

Solution   We solve this problem using our six-step Blueprint for Problem Solving:

Step 1: *Read and list.*
Known items: The total bill is $3,200. The sales tax rate is 7.5%, which is 0.075 in decimal form.
Unknown item: The retail price of the camera

Step 2: *Assign a variable and translate information.*
If we let x = the retail price of the camera, then to calculate the sales tax, we multiply the retail price x by the sales tax rate:

$$\text{Sales tax} = (\text{retail price})(\text{sales tax rate})$$
$$= 0.075x$$

Step 3: *Reread and write an equation.*

$$\text{Retail price} + \text{sales tax} = \text{total price}$$
$$x \quad + \quad 0.075x \quad = \quad 3{,}200$$

Step 4: *Solve the equation.*

$$x + 0.075x = 3{,}200$$
$$1.075x = 3{,}200$$
$$x = \frac{3{,}200}{1.075}$$
$$x = 2{,}976.74 \qquad \text{to the nearest hundredth}$$

135

Step 5: Write the answer.
The retail price of the camera was $2,976.74. The sales tax is
$3,200 − $2,976.74 = $223.26.

Step 6: Reread and check.
The retail price of the camera is $2,976.74. The tax on this is
0.075(2,976.74) = $223.26. Adding the retail price and the sales tax
we have the total bill, $3,200.00.

EXAMPLE 2   Suppose you invest a certain amount of money in an account
that earns 8% in annual interest. At the same time, you invest $2,000 more than that
in an account that pays 9% in annual interest. If the total interest from both accounts at the end of the year is $690, how much is invested in each account?

Solution

Step 1: Read and list.
Known items: The interest rates, the total interest earned, and how
much more is invested at 9%
Unknown items: The amounts invested in each account

Step 2: Assign a variable and translate information.
Let x = the amount of money invested at 8%. From this,
$x + 2,000$ = the amount of money invested at 9%. The interest earned
on x dollars invested at 8% is $0.08x$. The interest earned on $x + 2,000$
dollars invested at 9% is $0.09(x + 2,000)$.

Here is a table that summarizes this information:

	Dollars Invested at 8%	Dollars Invested at 9%
Number of	x	$x + 2,000$
Interest on	$0.08x$	$0.09(x + 2,000)$

Step 3: Reread and write an equation.
Since the total amount of interest earned from both accounts is $690,
the equation that describes the situation is

Interest earned at 8%	+	interest earned at 9%	=	total interest earned
$0.08x$	+ $0.09(x + 2,000)$ =			690

Step 4: Solve the equation.

$$0.08x + 0.09(x + 2,000) = 690$$

$$0.08x + 0.09x + 180 = 690 \qquad \text{Distributive property}$$

$$0.17x + 180 = 690 \qquad \text{Add } 0.08x \text{ and } 0.09x.$$

$$0.17x = 510 \qquad \text{Add } -180 \text{ to each side.}$$

$$x = 3,000 \qquad \text{Divide each side by } 0.17.$$

Step 5: ***Write the answer.***

The amount of money invested at 8% is $3,000, while the amount of money invested at 9% is $x + 2,000 = 3,000 + 2,000 = \$5,000$.

Step 6: ***Reread and check.***

The interest at 8% is 8% of 3,000 = 0.08(3,000) = \$240
The interest at 9% is 9% of 5,000 = 0.09(5,000) = \$450

The total interest is   \$690

FACTS FROM
Geometry

Labeling Triangles and the Sum of the Angles in a Triangle

One way to label the important parts of a triangle is to label the vertices with capital letters and the sides with small letters, as shown in Figure 1.

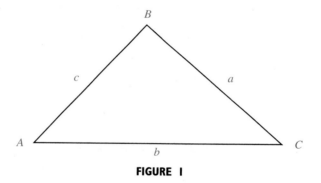

FIGURE 1

In Figure 1, notice that side a is opposite vertex A, side b is opposite vertex B, and side c is opposite vertex C. Also, since each vertex is the vertex of one of the angles of the triangle, we refer to the three interior angles as A, B, and C.

In any triangle, the sum of the interior angles is 180°. For the triangle shown in Figure 1, the relationship is written

$$A + B + C = 180°$$

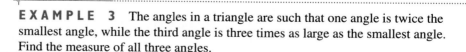

E X A M P L E   3   The angles in a triangle are such that one angle is twice the smallest angle, while the third angle is three times as large as the smallest angle. Find the measure of all three angles.

Solution

Step 1: ***Read and list.***

Known items: The sum of all three angles is 180°, one angle is twice the smallest angle, the largest angle is three times the smallest angle.

Unknown items: The measure of each angle

Step 2: ***Assign a variable and translate information.***

Let x be the smallest angle; then $2x$ will be the measure of another angle, and $3x$ will be the measure of the largest angle.

Step 3: ***Reread and write an equation.***

When working with geometric objects, drawing a generic diagram will sometimes help us visualize what it is that we are asked to find. In Figure 2, we draw a triangle with angles A, B, and C.

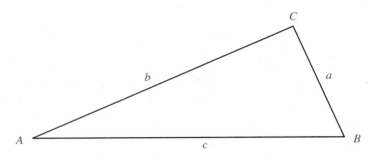

FIGURE 2

We can let the value of $A = x$, the value of $B = 2x$, and the value of $C = 3x$. We know that the sum of angles A, B, and C will be $180°$, so our equation becomes

$$x + 2x + 3x = 180°$$

Step 4: ***Solve the equation.***

$$x + 2x + 3x = 180°$$
$$6x = 180°$$
$$x = 30°$$

Step 5: ***Write the answer.***

The smallest angle A measures $30°$.
Angle B measures $2x$, or $2(30°) = 60°$.
Angle C measures $3x$, or $3(30°) = 90°$.

Step 6: ***Reread and check.***

The angles must add to $180°$:

$$A + B + C = 180°$$
$$30° + 60° + 90° = 180°$$
$$180° = 180° \qquad \text{Our answers check.}$$

Getting Ready for Class

After reading through the preceding section, respond in your own words and in complete sentences.

A. How do we label triangles?

B. What rule is always true about the three angles in a triangle?

C. Write an application problem whose solution depends on solving the equation $x + 0.075x = 500$.

D. Write an application problem whose solution depends on solving the equation $0.05x + 0.06(x + 200) = 67$.

PROBLEM SET 2.7

Interest Problems

1. Suppose you invest money in two accounts. One of the accounts pays 8% annual interest, while the other pays 9% annual interest. If you have $2,000 more invested at 9% than you have invested at 8%, how much do you have invested in each account if the total amount of interest you earn in a year is $860? (Begin by completing the following table.)

	Dollars Invested at 8%	Dollars Invested at 9%
Number of	x	
Interest on		

2. Suppose you invest a certain amount of money in an account that pays 11% interest annually and $4,000 more than that in an account that pays 12% annually. How much money do you have in each account if the total interest for a year is $940?

	Dollars Invested at 11%	Dollars Invested at 12%
Number of	x	
Interest on		

3. Tyler has two savings accounts that his grandparents opened for him. The two accounts pay 10% and 12% in annual interest; there is $500 more in the account that pays 12% than there is in the other

account. If the total interest for a year is $214, how much money does he have in each account?

4. Travis has a savings account that his parents opened for him. It pays 6% annual interest. His uncle also opened an account for him, but it pays 8% annual interest. If there is $800 more in the account that pays 6% and the total interest from both accounts is $104, how much money is in each of the accounts?

5. A stockbroker has money in three accounts. The interest rates on the three accounts are 8%, 9%, and 10%. If she has twice as much money invested at 9% as she has invested at 8%, and three times as much at 10% as she has at 8%, and the total interest for the year is $280, how much is invested at each rate? (*Hint:* Let $x = $ the amount invested at 8%.)

6. An accountant has money in three accounts that pay 9%, 10%, and 11% in annual interest. He has twice as much invested at 9% as he does at 10% and three times as much invested at 11% as he does at 10%. If the total interest from the three accounts is $610 for the year, how much is invested at each rate? (*Hint:* Let $x = $ the amount invested at 10%.)

Geometry Problems

7. Two angles in a triangle are equal, and their sum is equal to the third angle in the triangle. What are the measures of each of the three interior angles?

8. One angle in a triangle measures twice the smallest angle, while the largest angle is six times the smallest angle. Find the measures of all three angles.

9. The smallest angle in a triangle is $\frac{1}{5}$ as large as the largest angle. The third angle is twice the smallest angle. Find the three angles.

10. One angle in a triangle is half the largest angle, but three times the smallest. Find all three angles.

11. A right triangle has one 37° angle. Find the other two angles.

12. In a right triangle, one of the acute angles is twice as large as the other acute angle. Find the measure of the two acute angles.

Miscellaneous Problems

13. **Phone Bill** The cost of a long-distance phone call is $0.41 for the first minute and $0.32 for each additional minute. If the total charge for a long-distance call is $5.21, how many minutes was the call?

14. **Phone Bill** Danny, who is 1 year old, is playing with the telephone when he accidentally presses one of the buttons his mother has programmed to dial her friend Sue's number. Sue answers the phone and realizes Danny is on the other end. She talks to Danny, trying to get him to hang up. The cost for a call is $0.23 for the first minute and $0.14 for every minute after that. If the total charge for the call is $3.73, how long did it take Sue to convince Danny to hang up the phone?

15. **Hourly Wages** JoAnn works in the publicity office at the state university. She is paid $12 an hour for the first 35 hours she works each week and $18 an hour for every hour after that. If she makes $492 one week, how many hours did she work?

16. **Hourly Wages** Diane has a part-time job that pays her $6.50 an hour. During one week she works 26 hours and is paid $178.10. She realizes when she sees her check that she has been given a raise. How much per hour is that raise?

17. **Ticket Sales** Stacey is selling tickets to the school play. The tickets are $6.00 for adults and $4.50 for children. She sells twice as many adult tickets as children's tickets and brings in a total of $115.50. How many of each kind of ticket did she sell?

18. **Piano Lessons** Tyler is taking piano lessons. Since he doesn't practice as often as his parents would like him to, he has to pay for part of the lessons himself. His parents pay him $0.50 to do the laundry and $1.25 to mow the lawn. In one month, he does the laundry 6 more times than he mows the lawn. If his parents pay him $13.50 that month, how many times did he mow the lawn?

19. **Arrival Time** Jeff and Carla Cole are driving separately from San Luis Obispo, California, to the north shore of Lake Tahoe, a distance of 425 miles. Jeff leaves San Luis Obispo at 11:00 A.M. and averages 55 miles per hour on the drive. Carla leaves later, at 1:00 P.M., but averages 65 miles per hour. Which person arrives in Lake Tahoe first?

20. **Arrival Time** Referring to the information in Problem 19, how many minutes after the first person arrives does the second person arrive?

Cost of a Taxi Ride Recently, the Texas Junior College Teachers Association annual conference was held in Austin. At that time a taxi ride in Austin was $1.25 for the first $\frac{1}{5}$ of a mile and $0.25 for each additional $\frac{1}{5}$ of a mile. The charge for a taxi to wait is $12.00 per hour. Use this information for Problems 21 through 24.

21. If the distance from one of the convention hotels to the airport is 7.5 miles, how much will it cost to take a taxi from that hotel to the airport?

22. If you were to tip the driver of the taxi in Problem 21 15%, how much would it cost to take a taxi from the hotel to the airport?

23. Suppose the distance from one of the hotels to one of the western dance clubs in Austin is 12.4 miles. If the fare meter in the taxi gives the charge for that trip as $16.50, is the meter working correctly?

24. Suppose that the distance from a hotel to the airport is 8.2 miles and the ride takes 20 minutes. Is it more expensive to take a taxi to the airport or just to sit in the taxi?

Dance Lessons Ike and Nancy Lara give western dance lessons at the Elks Lodge on Sunday nights. The lessons cost $3.00 for members of the lodge and $5.00 for nonmembers. Half of the money collected for the lessons is paid to Ike and Nancy. The Elks Lodge keeps the other half. One Sunday night Ike counts 36 people in the dance lesson. Use this information to work Problems 25 through 28.

25. What is the least amount of money Ike and Nancy will make?

26. What is the largest amount of money Ike and Nancy will make?

27. At the end of the evening, the Elks Lodge gives Ike and Nancy a check for $80 to cover half of the receipts. Can this amount be correct?

28. Besides the number of people in the dance lesson, what additional information does Ike need to know in order to always be sure he is being paid the correct amount?

Applying the Concepts

29. Starting Salaries   *USA Today* reported in 1999 that the median starting salary of graduates of Stanford University's business school was $85,200, which was 7% more than it was the year before. What was the median starting salary of the 1998 graduates of Stanford University's business school? Round your answer to the nearest dollar.

30. Starting Salaries   In the same article mentioned in Problem 29, *USA Today* reported that the total compensation, salary, bonus, housing allowance, and so forth, for the 1999 graduates of Stanford's business school was $122,000, an increase of 8% over the previous year. What was the total compensation package for Stanford's business school graduates in 1998? Round your answer to the nearest dollar.

Review Problems

The problems that follow review material we covered in Chapter 1 on number sequences.

Find the next number in each sequence.

31. $8, 4, 0, -4, \ldots$

32. $8, -16, 32, -64, \ldots$

33. $12, -6, 3, -\dfrac{3}{2}, \ldots$

34. $-12, -6, 0, 6, \ldots$

35. $1, -4, 9, -16, \ldots$

36. $-1, 8, -27, 64, \ldots$

37. $2, \dfrac{3}{2}, 1, \dfrac{1}{2}, \ldots$

38. $1, \dfrac{1}{2}, 0, -\dfrac{1}{2}, \ldots$

2.8   Linear Inequalities

Linear inequalities are solved by a method similar to the one used in solving linear equations. The only real differences between the methods are in the multiplication property for inequalities and in graphing the solution set.

An inequality differs from an equation only with respect to the comparison symbol between the two quantities being compared. In place of the equal sign, we use $<$ (less than), $\leq$ (less than or equal to), $>$ (greater than), or $\geq$ (greater than or equal to). The addition property for inequalities is almost identical to the addition property for equality.

ADDITION PROPERTY FOR INEQUALITIES

For any three algebraic expressions A, B, and C,

$$\text{if} \qquad A < B$$

$$\text{then} \qquad A + C < B + C$$

In words: Adding the same quantity to both sides of an inequality will not change the solution set.

It makes no difference which inequality symbol we use to state the property. Adding the same amount to both sides always produces an inequality equivalent to the original inequality. Also, since subtraction can be thought of as addition of the opposite, this property holds for subtraction as well as addition.

EXAMPLE 1   Solve the inequality $x + 5 < 7$.

Solution   To isolate x, we add -5 to both sides of the inequality:

$$x + 5 < 7$$

$$x + 5 + (-5) < 7 + (-5) \qquad \text{Addition property for inequalities}$$

$$x < 2$$

We can go one step further here and graph the solution set. The solution set is all real numbers less than 2. To graph this set, we simply draw a straight line and label the center 0 (zero) for reference. Then we label the 2 on the right side of zero and extend an arrow beginning at 2 and pointing to the left. We use an open circle at 2, since it is not included in the solution set. Here is the graph.

EXAMPLE 2   Solve $x - 6 \leq -3$.

Solution   Adding 6 to each side will isolate x on the left side:

$$x - 6 \leq -3$$

$$x - 6 + \mathbf{6} \leq -3 + \mathbf{6} \qquad \text{Add } \mathbf{6} \text{ to both sides.}$$

$$x \leq 3$$

The graph of the solution set is

Notice that the dot at the 3 is darkened because 3 is included in the solution set. We will always use open circles on the graphs of solution sets with $<$ or $>$ and closed (darkened) circles on the graphs of solution sets with $\leq$ or $\geq$.

To see the idea behind the multiplication property for inequalities, we will consider three true inequality statements and explore what happens when we multiply both sides by a positive number and then what happens when we multiply by a negative number.

Consider the following three true statements:

$$3 < 5 \qquad -3 < 5 \qquad -5 < -3$$

Now multiply both sides by the positive number 4:

$$4(3) < 4(5) \qquad 4(-3) < 4(5) \qquad 4(-5) < 4(-3)$$

$$12 < 20 \qquad\qquad -12 < 20 \qquad\qquad -20 < -12$$

In each case, the inequality symbol in the result points in the same direction it did in the original inequality. We say the "sense" of the inequality doesn't change when we multiply both sides by a positive quantity.

Notice what happens when we go through the same process but multiply both sides by −4 instead of 4:

$$3 < 5 \qquad\qquad\qquad -3 < 5 \qquad\qquad\qquad -5 < -3$$

$$-4(3) > -4(5) \qquad -4(-3) > -4(5) \qquad -4(-5) > -4(-3)$$

$$-12 > -20 \qquad\qquad 12 > -20 \qquad\qquad 20 > 12$$

In each case, we have to change the direction in which the inequality symbol points to keep each statement true. Multiplying both sides of an inequality by a negative quantity *always* reverses the sense of the inequality. Our results are summarized in the multiplication property for inequalities.

MULTIPLICATION PROPERTY FOR INEQUALITIES

For any three algebraic expressions A, B, and C,

$$\text{if} \qquad A < B$$

$$\text{then} \qquad AC < BC \qquad \text{when } C \text{ is positive}$$

$$\text{and} \qquad AC > BC \qquad \text{when } C \text{ is negative}$$

In words: Multiplying both sides of an inequality by a positive number does not change the solution set. When multiplying both sides of an inequality by a negative number, it is necessary to reverse the inequality symbol in order to produce an equivalent inequality.

We can multiply both sides of an inequality by any nonzero number we choose. If that number happens to be negative, we must also reverse the sense of the inequality.

Note: Since division is defined in terms of multiplication, this property is also true for division. We can divide both sides of an inequality by any number we choose. If that number happens to be negative, we must also reverse the direction of the inequality symbol.

E X A M P L E   3   Solve $3a < 15$ and graph the solution.

Solution   We begin by multiplying each side by $\frac{1}{3}$. Since $\frac{1}{3}$ is a positive number, we do not reverse the direction of the inequality symbol:

$$3a < 15$$

$$\frac{1}{3}(3a) < \frac{1}{3}(15) \qquad \text{Multiply each side by } \frac{1}{3}.$$

$$a < 5$$

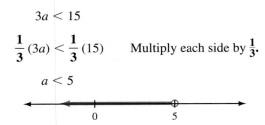

EXAMPLE 4   Solve $-3a \leq 18$ and graph the solution.

Solution   We begin by multiplying both sides by $-\frac{1}{3}$. Since $-\frac{1}{3}$ is a negative number, we must reverse the direction of the inequality symbol at the same time that we multiply by $-\frac{1}{3}$.

$$-3a \leq 18$$

$$-\frac{1}{3}(-3a) \geq -\frac{1}{3}(18) \qquad \begin{array}{l}\text{Multiply both sides by } -\frac{1}{3} \text{ and reverse}\\ \text{the direction of the inequality symbol.}\end{array}$$

$$a \geq -6$$

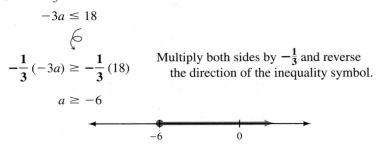

EXAMPLE 5   Solve $-\dfrac{x}{4} > 2$ and graph the solution.

Solution   To isolate x, we multiply each side by -4. Since -4 is a negative number, we must also reverse the direction of the inequality symbol:

$$-\frac{x}{4} > 2$$

$$-4\left(-\frac{x}{4}\right) < -4(2) \qquad \begin{array}{l}\text{Multiply each side by } -4 \text{ and reverse the}\\ \text{direction of the inequality symbol.}\end{array}$$

$$x < -8$$

To solve more complicated inequalities, we use the following steps.

> **STRATEGY FOR SOLVING LINEAR INEQUALITIES IN ONE VARIABLE**
>
> **Step 1a:** Use the distributive property to separate terms, if necessary.
>   **1b:** If fractions are present, consider multiplying both sides by the LCD to eliminate the fractions. If decimals are present, consider multiplying both sides by a power of 10 to clear the inequality of decimals.
>   **1c:** Combine similar terms on each side of the inequality.
> **Step 2:** Use the addition property for inequalities to get all variable terms on one side of the inequality and all constant terms on the other side.
> **Step 3:** Use the multiplication property for inequalities to get x by itself on one side of the inequality.
> **Step 4:** Graph the solution set.

EXAMPLE 6   Solve $2.5x - 3.48 < -4.9x + 2.07$.

Solution   We have two methods we can use to solve this inequality. We can simply apply our properties to the inequality the way it is currently written and work with the decimal numbers, or we can eliminate the decimals to begin with and solve the resulting inequality.

Method 1   Working with the decimals.

$2.5x - 3.48 < -4.9x + 2.07$	Original inequality
$2.5x + \mathbf{4.9x} - 3.48 < -4.9x + \mathbf{4.9x} + 2.07$	Add **4.9x** to each side.
$7.4x - 3.48 < 2.07$	
$7.4x - 3.48 + \mathbf{3.48} < 2.07 + \mathbf{3.48}$	Add **3.48** to each side.
$7.4x < 5.55$	
$\dfrac{7.4x}{\mathbf{7.4}} < \dfrac{5.55}{\mathbf{7.4}}$	Divide each side by **7.4.**
$x < 0.75$	

Method 2   Eliminating the decimals in the beginning.
Since the greatest number of places to the right of the decimal point in any of the numbers is 2, we can multiply each side of the inequality by 100 and we will be left with an equivalent inequality that contains only whole numbers.

$2.5x - 3.48 < -4.9x + 2.07$	Original inequality
$\mathbf{100}(2.5x - 3.48) < \mathbf{100}(-4.9x + 2.07)$	Multiply each side by 100.
$\mathbf{100}(2.5x) - \mathbf{100}(3.48) < \mathbf{100}(-4.9x) + \mathbf{100}(2.07)$	Distributive property
$250x - 348 < -490x + 207$	Multiplication
$740x - 348 < 207$	Add 490x to each side.

$$740x < 555 \qquad \text{Add 348 to each side.}$$

$$\frac{740x}{740} < \frac{555}{740} \qquad \begin{array}{l}\text{Divide each side}\\\text{by } \mathbf{740.}\end{array}$$

$$x < 0.75$$

The solution by either method is $x < 0.75$. Here is the graph:

E X A M P L E   7   Solve $3(x - 4) \geq -2$.

Solution

$$3x - 12 \geq -2 \qquad \text{Distributive property}$$

$$3x - 12 + \mathbf{12} \geq -2 + \mathbf{12} \qquad \text{Add } \mathbf{12} \text{ to both sides.}$$

$$3x \geq 10$$

$$\frac{\mathbf{1}}{\mathbf{3}}(3x) \geq \frac{\mathbf{1}}{\mathbf{3}}(10) \qquad \text{Multiply both sides by } \tfrac{1}{3}.$$

$$x \geq \frac{10}{3}$$

E X A M P L E   8   Solve and graph $2(1 - 3x) + 4 < 4x - 14$.

Solution

$$2 - 6x + 4 < 4x - 14 \qquad \text{Distributive property}$$

$$-6x + 6 < 4x - 14 \qquad \text{Simplify.}$$

$$-6x + 6 + (\mathbf{-6}) < 4x - 14 + (\mathbf{-6}) \qquad \text{Add } \mathbf{-6} \text{ to both sides.}$$

$$-6x < 4x - 20$$

$$-6x + (\mathbf{-4x}) < 4x + (\mathbf{-4x}) - 20 \qquad \text{Add } \mathbf{-4x} \text{ to both sides.}$$

$$-10x < -20$$

$$\left(-\frac{\mathbf{1}}{\mathbf{10}}\right)(-10x) > \left(-\frac{\mathbf{1}}{\mathbf{10}}\right)(-20) \qquad \begin{array}{l}\text{Multiply by } -\tfrac{1}{10} \text{ and reverse}\\\text{the sense of the inequality.}\end{array}$$

$$x > 2$$

E X A M P L E   9   Solve $2x - 3y < 6$ for y.

Solution   We can solve this formula for y by first adding $-2x$ to each side and then multiplying each side by $-\frac{1}{3}$. When we multiply by $-\frac{1}{3}$ we must reverse the direction of the inequality symbol. Since this is a formula, we will not graph the solution.

$$2x - 3y < 6 \qquad\qquad \text{Original formula}$$

$$2x + (\mathbf{-2x}) - 3y < (\mathbf{-2x}) + 6 \qquad\qquad \text{Add } \mathbf{-2x} \text{ to each side.}$$

$$-3y < -2x + 6$$

$$-\frac{1}{3}(-3y) > -\frac{1}{3}(-2x + 6) \qquad\qquad \text{Multiply each side by } -\frac{1}{3}.$$

$$y > \frac{2}{3}x - 2 \qquad\qquad \text{Distributive property}$$

When working application problems that involve inequalities, the phrases "at least" and "at most" translate as follows:

In Words	In Symbols
x is at least 30.	$x \geq 30$
x is at most 20.	$x \leq 20$

APPLYING THE CONCEPTS

Our next example involves consecutive integers. When we ask for consecutive integers, we mean integers that are next to each other on the number line, like 5 and 6, or 13 and 14, or -4 and -3. In the dictionary, consecutive is defined as following one another in uninterrupted order. If we ask for consecutive *odd* integers, then we mean odd integers that follow one another on the number line. For example, 3 and 5, 11 and 13, and -9 and -7 are consecutive odd integers. As you can see, to get from one odd integer to the next consecutive odd integer we add 2.

We can modify our Blueprint for Problem Solving to solve application problems whose solutions depend on writing and then solving inequalities.

E X A M P L E   1 0   The sum of two consecutive odd integers is at most 28. What are the possibilities for the first of the two integers?

Solution   When we use the phrase "their sum is at most 28," we mean that their sum is less than or equal to 28.

> **Step 1: Read and list.**
>   *Known items:*   Two consecutive odd integers. Their sum is less than or equal to 28.
>   *Unknown items:*   The numbers in question

Step 2: Assign a variable and translate information.
If we let $x =$ the first of the two consecutive integers, then $x + 2$ is the next consecutive one.

Step 3: Reread and write an inequality.
Their sum is at most 28.

$$x + (x + 2) \leq 28$$

Step 4: Solve the inequality.

$$2x + 2 \leq 28 \qquad \text{Simplify the left side.}$$

$$2x \leq 26 \qquad \text{Add } -2 \text{ to each side.}$$

$$x \leq 13 \qquad \text{Multiply each side by } \tfrac{1}{2}.$$

Step 5: Write the answer.
The first of the two integers must be an odd integer that is less than or equal to 13. The second of the two integers will be 2 more than whatever the first one is.

Step 6: Reread and check.
Suppose the first integer is 13. The next consecutive odd integer is 15. The sum of 15 and 13 is 28. If the first odd integer is less than 13, the sum of it and the next consecutive odd integer will be less than 28.

Getting Ready for Class

After reading through the preceding section, respond in your own words and in complete sentences.

A. State the addition property for inequalities.

B. How is the multiplication property for inequalities different from the multiplication property of equality?

C. When do we reverse the direction of an inequality symbol?

D. Under what conditions do we not change the direction of the inequality symbol when we multiply both sides of an inequality by a number?

PROBLEM SET 2.8

Solve the following inequalities using the addition property of inequalities. Graph each solution set.

1. $x - 5 < 7$

2. $x + 3 < -5$

3. $a - 4 \leq 8$

4. $a + 3 \leq 10$

5. $x - 4.3 > 8.7$

6. $x - 2.6 > 10.4$

7. $y + 6 \geq 10$

8. $y + 3 \geq 12$

9. $2 < x - 7$

10. $3 < x + 8$

Solve the following inequalities using the multiplication property of inequalities. If you multiply both sides by a negative number, be sure to reverse the direction of the inequality symbol. Graph the solution set.

11. $3x < 6$

12. $2x < 14$

13. $5a \le 25$ **14.** $4a \le 16$

15. $\dfrac{x}{3} > 5$ **16.** $\dfrac{x}{7} > 1$

17. $-2x > 6$ **18.** $-3x \ge 9$

19. $-3x \ge -18$ **20.** $-8x \ge -24$

21. $-\dfrac{x}{5} \le 10$ **22.** $-\dfrac{x}{9} \ge -1$

23. $-\dfrac{2}{3}y > 4$ **24.** $-\dfrac{3}{4}y > 6$

Solve the following inequalities. Graph the solution set in each case.

25. $2x - 3 < 9$ **26.** $3x - 4 < 17$

27. $-\dfrac{1}{5}y - \dfrac{1}{3} \le \dfrac{2}{3}$ **28.** $-\dfrac{1}{6}y - \dfrac{1}{2} \le \dfrac{2}{3}$

29. $-7.2x + 1.8 > -19.8$

30. $-7.8x - 1.3 > 22.1$

31. $\dfrac{2}{3}x - 5 \le 7$ **32.** $\dfrac{3}{4}x - 8 \le 1$

33. $-\dfrac{2}{5}a - 3 > 5$ **34.** $-\dfrac{4}{5}a - 2 > 10$

35. $5 - \dfrac{3}{5}y > -10$ **36.** $4 - \dfrac{5}{6}y > -11$

37. $0.3(a + 1) \le 1.2$ **38.** $0.4(a - 2) \le 0.4$

39. $2(5 - 2x) \le -20$ **40.** $7(8 - 2x) > 28$

41. $3x - 5 > 8x$ **42.** $8x - 4 > 6x$

43. $\dfrac{1}{3}y - \dfrac{1}{2} \le \dfrac{5}{6}y + \dfrac{1}{2}$

44. $\dfrac{7}{6}y + \dfrac{4}{3} \le \dfrac{11}{6}y - \dfrac{7}{6}$

45. $-2.8x + 8.4 < -14x - 2.8$

46. $-7.2x - 2.4 < -2.4x + 12$

47. $3(m - 2) - 4 \ge 7m + 14$

48. $2(3m - 1) + 5 \ge 8m - 7$

49. $3 - 4(x - 2) \le -5x + 6$

50. $8 - 6(x - 3) \le -4x + 12$

Solve each of the following formulas for y.

51. $3x + 2y < 6$ **52.** $-3x + 2y < 6$

53. $2x - 5y > 10$ **54.** $-2x - 5y > 5$

55. $-3x + 7y \le 21$ **56.** $-7x + 3y \le 21$

57. $2x - 4y \ge -4$ **58.** $4x - 2y \ge -8$

For each graph below, write an inequality whose solution is the graph.

59.

60.

61.

62.

Applying the Concepts

63. Consecutive Integers   The sum of two consecutive integers is at least 583. What are the possibilities for the first of the two integers?

64. Consecutive Integers   The sum of two consecutive integers is at most 583. What are the possibilities for the first of the two integers?

65. Number Problem   The sum of twice a number and six is less than ten. Find all solutions.

66. Number Problem   Twice the difference of a number and three is greater than or equal to the number increased by five. Find all solutions.

67. Number Problem   The product of a number and four is greater than the number minus eight. Find the solution set.

68. Number Problem   The quotient of a number and five is less than the sum of seven and two. Find the solution set.

69. Geometry Problem   The length of a rectangle is 3 times the width. If the perimeter is to be at least 48 meters, what are the possible values for the width? (If the perimeter is at least 48 meters, then it is greater than or equal to 48 meters.)

70. Geometry Problem   The length of a rectangle is 3 more than twice the width. If the perimeter is to be at least 51 meters, what are the possible values for the width? (If the perimeter is at least 51 meters, then it is greater than or equal to 51 meters.)

71. Geometry Problem   The numerical values of the three sides of a triangle are given by three consecutive even integers. If the perimeter is greater than 24 inches, what are the possibilities for the shortest side?

72. Geometry Problem The numerical values of the three sides of a triangle are given by three consecutive odd integers. If the perimeter is greater than 27 inches, what are the possibilities for the shortest side?

73. Car Heaters If you have ever gotten in a cold car early in the morning you know that the heater does not work until the engine warms up. This is because the heater relies on the heat coming off the engine. Write an inequality sign to express when the heater will work if the heater works only after the engine is 100°F.

74. Exercise When Kate exercises, she either swims or runs. She wants to spend a minimum of 8 hours a week exercising, and she wants to swim 3 times the amount she runs. What is the minimum amount of time she must spend doing each exercise?

75. Profit and Loss Movie theaters pay a certain price for the movies that you and I see. Suppose a theater pays $1,500 for each showing of a popular movie. If they charge $7.50 for each ticket they sell, then they will lose money if ticket sales are less than $1,500. On the other hand, they will make a profit if ticket sales are greater than $1,500. What is the range of tickets they can sell and still lose money? What is the range of tickets they can sell and make a profit?

76. Stock Sales Suppose you purchase x shares of a stock at $12 per share. After 6 months you decide to sell all your shares at $20 per share. Your broker charges you $15 for the trade. If your profit is at least $3,985, how many shares did you purchase in the first place?

Review Problems

Problems 77–82 review material we covered in Sections 1.5.

Match each numbered expression with one or more of the properties in (a)–(e).

77. $x + 4 = 4 + x$

78. $2(3x) = (2 \cdot 3)x$

79. $5(x - 3) = 5x - 15$

80. $x + (y + 4) = (x + y) + 4$

81. $x + (y + 4) = (x + 4) + y$

82. $7 \cdot 5 = 5 \cdot 7$

(a) Distributive property

(b) Commutative property of addition

(c) Associative property of addition

(d) Commutative property of multiplication

(e) Associative property of multiplication

CHAPTER 2   SUMMARY

Examples

1. The terms $2x$, $5x$, and $-7x$ are all similar since their variable parts are the same.

Similar Terms [2.1]

A *term* is a number or a number and one or more variables multiplied together. *Similar terms* are terms with the same variable part.

2. Simplify $3x + 4x$.
$$3x + 4x = (3 + 4)x$$
$$= 7x$$

Simplifying Expressions [2.1]

In this chapter we simplified expressions that contained variables by using the distributive property to combine similar terms.

3. The solution set for the equation $x + 2 = 5$ is $\{3\}$ because when x is 3 the equation is $3 + 2 = 5$, or $5 = 5$.

Solution Set [2.2]

The *solution set* for an equation (or inequality) is all the numbers that, when used in place of the variable, make the equation (or inequality) a true statement.

4. The equations $a - 4 = 3$ and $a - 2 = 5$ are equivalent since both have solution set $\{7\}$.

Equivalent Equations [2.2]

Two equations are called *equivalent* if they have the same solution set.

5. Solve $x - 5 = 12$.
$$x - 5 + 5 = 12 + 5$$
$$x + 0 = 17$$
$$x = 17$$

Addition Property of Equality [2.2]

When the same quantity is added to both sides of an equation, the solution set for the equation is unchanged. Adding the same amount to both sides of an equation produces an equivalent equation.

6. Solve $3x = 18$.
$$\frac{1}{3}(3x) = \frac{1}{3}(18)$$
$$x = 6$$

Multiplication Property of Equality [2.3]

If both sides of an equation are multiplied by the same nonzero number, the solution set is unchanged. Multiplying both sides of an equation by a nonzero quantity produces an equivalent equation.

7. Solve $2(x + 3) = 10$.
$$2x + 6 = 10$$
$$2x + 6 + (-6) = 10 + (-6)$$
$$2x = 4$$
$$\frac{1}{2}(2x) = \frac{1}{2}(4)$$
$$x = 2$$

Strategy for Solving Linear Equations in One Variable [2.4]

> ***Step 1a:*** Use the distributive property to separate terms, if necessary.
>
> ***1b:*** If fractions are present, consider multiplying both sides by the LCD to eliminate the fractions. If decimals are present, consider multiplying both sides by a power of 10 to clear the equation of decimals.
>
> ***1c:*** Combine similar terms on each side of the equation.
>
> ***Step 2:*** Use the addition property of equality to get all variable terms on one side of the equation and all constant terms on

the other side. A variable term is a term that contains the variable (for example, $5x$). A constant term is a term that does not contain the variable (the number 3, for example).

Step 3: Use the multiplication property of equality to get x (that is, $1x$) by itself on one side of the equation.

Step 4: Check your solution in the original equation to be sure that you have not made a mistake in the solution process.

8. Solving $P = 2l + 2w$ for l, we have

$$P - 2w = 2l$$

$$\frac{P - 2w}{2} = l$$

Formulas [2.5]

A *formula* is an equation with more than one variable. To solve a formula for one of its variables, we use the addition and multiplication properties of equality to move everything except the variable in question to one side of the equal sign so that the variable in question is alone on the other side.

Blueprint for Problem Solving [2.6, 2.7]

Step 1: **Read** the problem, and then mentally **list** the items that are known and the items that are unknown.

Step 2: **Assign a variable** to one of the unknown items. (In most cases this will amount to letting $x =$ the item that is asked for in the problem.) Then **translate** the other **information** in the problem to expressions involving the variable.

Step 3: **Reread** the problem, and then **write an equation,** using the items and variables listed in steps 1 and 2, that describes the situation.

Step 4: **Solve the equation** found in step 3.

Step 5: **Write** your **answer** using a complete sentence.

Step 6: **Reread** the problem, and **check** your solution with the original words in the problem.

---

9. Solve $x + 5 < 7$.

$$x + 5 + (-5) < 7 + (-5)$$

$$x < 2$$

Addition Property for Inequalities [2.8]

Adding the same quantity to both sides of an inequality produces an equivalent inequality, one with the same solution set.

10. Solve $-3a \leq 18$.

$$-\frac{1}{3}(-3a) \geq -\frac{1}{3}(18)$$

$$a \geq -6$$

Multiplication Property for Inequalities [2.8]

Multiplying both sides of an inequality by a positive number never changes the solution set. If both sides are multiplied by a negative number, the sense of the inequality must be reversed to produce an equivalent inequality.

11. Solve $3(x - 4) \geq -2$.

$$3x - 12 \geq -2$$

$$3x - 12 + \mathbf{12} \geq -2 + \mathbf{12}$$

$$3x \geq 10$$

$$\frac{\mathbf{1}}{\mathbf{3}}(3x) \geq \frac{\mathbf{1}}{\mathbf{3}}(10)$$

$$x \geq \frac{10}{3}$$

Strategy for Solving Linear Inequalities in One Variable [2.8]

Step 1a: Use the distributive property to separate terms, if necessary.

1b: If fractions are present, consider multiplying both sides by the LCD to eliminate the fractions. If decimals are present, consider multiplying both sides by a power of 10 to clear the inequality of decimals.

1c: Combine similar terms on each side of the inequality.

Step 2: Use the addition property for inequalities to get all variable terms on one side of the inequality and all constant terms on the other side.

Step 3: Use the multiplication property for inequalities to get x by itself on one side of the inequality.

Step 4: Graph the solution set.

COMMON MISTAKES

1. Trying to subtract away coefficients (the number in front of variables) when solving equations. For example:

$$4x = 12$$

$$4x - 4 = 12 - 4$$

$$x = 8 \leftarrow \text{Mistake}$$

It is not incorrect to add (-4) to both sides; it's just that $4x - 4$ is not equal to x. Both sides should be multiplied by $\frac{1}{4}$ to solve for x.

2. Forgetting to reverse the direction of the inequality symbol when multiplying both sides of an inequality by a negative number. For instance:

$$-3x < 12$$

$$-\frac{\mathbf{1}}{\mathbf{3}}(-3x) < -\frac{\mathbf{1}}{\mathbf{3}}(12) \leftarrow \text{Mistake}$$

$$x < -4$$

It is not incorrect to multiply both sides by $-\frac{1}{3}$. But if we do, we must also reverse the sense of the inequality.

CHAPTER 2 REVIEW

The numbers in brackets refer to the sections of the text in which similar problems can be found.

Simplify each expression as much as possible. [2.1]

1. $5x - 8x$ **2.** $6x - 3 - 8x$

3. $-a + 2 + 5a - 9$

4. $5(2a - 1) - 4(3a - 2)$

5. $6 - 2(3y + 1) - 4$ **6.** $4 - 2(3x - 1) - 5$

Find the value of each expression when x is 3. [2.1]

7. $7x - 2$ **8.** $-4x - 5 + 2x$

9. $-x - 2x - 3x$

Find the value of each expression when x is -2. [2.1]

10. $5x - 3$ **11.** $-3x + 2$ **12.** $7 - x - 3$

Solve each equation. [2.2, 2.3]

13. $x + 2 = -6$ **14.** $x - \dfrac{1}{2} = \dfrac{4}{7}$

15. $10 - 3y + 4y = 12$

16. $-3 - 4 = -y - 2 + 2y$

17. $2x = -10$ **18.** $3x = 0$

19. $\dfrac{x}{3} = 4$ **20.** $-\dfrac{x}{4} = 2$

21. $3a - 2 = 5a$ **22.** $\dfrac{7}{10} a = \dfrac{1}{5} a + \dfrac{1}{2}$

23. $3x + 2 = 5x - 8$ **24.** $6x - 3 = 2x + 7$

25. $0.7x - 0.1 = 0.5x - 0.1$

26. $0.2x - 0.3 = 0.8x - 0.3$

Solve each equation. Be sure to simplify each side first. [2.4]

27. $2(x - 5) = 10$ **28.** $12 = 2(5x - 4)$

29. $\dfrac{1}{2}(3t - 2) + \dfrac{1}{2} = \dfrac{5}{2}$

30. $\dfrac{3}{5}(5x - 10) = \dfrac{2}{3}(9x + 3)$

31. $2(3x + 7) = 4(5x - 1) + 18$

32. $7 - 3(y + 4) = 10$

Use the formula $4x - 5y = 20$ to find y if [2.5]

33. x is 5   **34.** x is 0

35. x is -5 **36.** x is 10

Solve each of the following formulas for the indicated variable. [2.5]

37. $2x - 5y = 10$ for y **38.** $5x - 2y = 10$ for y

39. $V = \pi r^2 h$ for h **40.** $P = 2l + 2w$ for w

41. What number is 86% of 240? [2.5]

42. What percent of 2,000 is 180? [2.5]

Solve each of the following word problems. In each case, be sure to show the equation that describes the situation. [2.6, 2.7]

43. Number Problem The sum of twice a number and 6 is 28. Find the number.

44. Geometry The length of a rectangle is 5 times as long as the width. If the perimeter is 60 meters, find the length and the width.

45. Investing A man invests a certain amount of money in an account that pays 9% annual interest. He invests $300 more than that in an account that pays 10% annual interest. If his total interest after a year is $125, how much does he have invested in each account?

46. Coin Problem A collection of 15 coins is worth $1.00. If the coins are dimes and nickels, how many of each coin are there?

Solve each inequality. [2.8]

47. $-2x < 4$ **48.** $-5x > -10$

49. $-\dfrac{a}{2} \le -3$ **50.** $-\dfrac{a}{3} > 5$

Solve each inequality, and graph the solution. [2.8]

51. $-4x + 5 > 37$ **52.** $2x + 10 < 5x - 11$

53. $2(3t + 1) + 6 \ge 5(2t + 4)$

CHAPTER 2 PROJECTS

LINEAR EQUATIONS AND INEQUALITIES

GROUP PROJECT

FINDING THE MAXIMUM HEIGHT OF A MODEL ROCKET

Number of People: 3

Time Needed: 20 minutes

Equipment: Paper and pencil

Background: In this chapter we used formulas to do some table building. Once we have a table, it is sometimes possible to use just the table information to extend what we know about the situation described by the table. In this project we take some basic information from a table and then look for patterns among the table entries. Once we have established the patterns we continue them and, in so doing, solve a realistic application problem.

Procedure: A model rocket is launched into the air. Table 1 gives the height of the rocket every second after takeoff, for the first 5 seconds. Figure 1 is a graphical representation of the information in Table 1.

TABLE 1	Height of a Model Rocket
Time (seconds)	Height (feet)
0	0
1	176
2	320
3	432
4	512
5	560

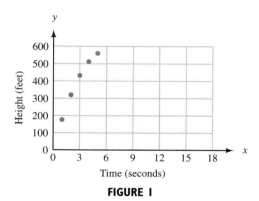

FIGURE 1

1. Table 1 is shown again on the next page. Fill in the first five entries in the First Differences column by finding the difference of consecutive heights. For example, the second entry in the First Differences column will be the difference of 320 and 176, which is 144.

TABLE 1	Height of a Model Rocket		
Time (seconds)	Height (feet)	First Differences	Second Differences
0	0		
1	176		
2	320		
3	432		
4	512		
5	560		

2. Start filling in the Second Differences column by finding the differences of the First Differences.

3. Once you see the pattern in the Second Differences table, fill in the rest of the entries.

4. Now, using the results in the Second Differences table, go back and complete the First Differences table.

5. Now, using the results in the First Differences table, go back and complete the Heights column in the original table.

6. Plot the rest of the points from Table 1 on the graph in Figure 1.

7. What is the maximum height of the rocket?

8. How long was the rocket in the air?

RESEARCH PROJECT

THE EQUAL SIGN

We have been using the equal sign, =, for some time now. It is interesting to note that the first published use of the symbol was in 1557, with the publication of *The Whetstone of Witte* by the English mathematician and physician Robert Recorde. Research the first use of the symbols we use for addition, subtraction, multiplication, and division and then write an essay on the subject from your results.

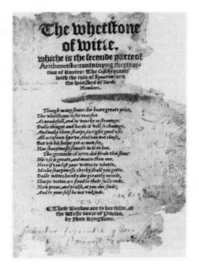

The Whetstone of Witte, 1557
(*Source: Smithsonian Institution Libraries, Photo No. 92-338*)

CHAPTER 2   TEST

Simplify each of the following expressions. [2.1]

1. $3x + 2 - 7x + 3$
2. $4a - 5 - a + 1$
3. $7 - 3(y + 5) - 4$
4. $8(2x + 1) - 5(x - 4)$
5. Find the value of $2x - 3 - 7x$ when $x = -5$. [2.1]
6. Find the value of $x^2 + 2xy + y^2$ when $x = 2$ and $y = 3$. [2.1]

Solve the following equations. [2.2, 2.3, 2.4]

7. $2x - 5 = 7$
8. $2y + 4 = 5y$
9. $\dfrac{1}{2}x - \dfrac{1}{10} = \dfrac{1}{5}x + \dfrac{1}{2}$
10. $\dfrac{2}{5}(5x - 10) = -5$
11. $-5(2x + 1) - 6 = 19$
12. $0.04x + 0.06(100 - x) = 4.6$
13. $2(t - 4) + 3(t + 5) = 2t - 2$
14. $2x - 4(5x + 1) = 3x + 17$
15. What number is 15% of 38? [2.5]
16. 240 is 12% of what number? [2.5]
17. If $2x - 3y = 12$, find x when $y = -2$. [2.5]
18. The formula for the volume of a cone is $V = \dfrac{1}{3}\pi r^2 h$. Find h if $V = 88$ cubic inches, $\pi = \dfrac{22}{7}$, and $r = 3$ inches. [2.5]
19. Solve $2x + 5y = 20$ for y. [2.5]

20. Solve $h = x + vt + 16t^2$ for v. [2.5]

Solve each word problem. [2.6, 2.7]

21. Age Problem Dave is twice as old as Rick. Ten years ago the sum of their ages was 40. How old are they now?

22. Geometry A rectangle is twice as long as it is wide. The perimeter is 60 inches. What are the length and width?

23. Coin Problem A man has a collection of dimes and quarters with a total value of $3.50. If he has 7 more dimes than quarters, how many of each coin does he have?

24. Investing A woman has money in two accounts. One account pays 7% annual interest, while the other pays 9% annual interest. If she has $600 more invested at 9% than she does at 7% and her total interest for a year is $182, how much does she have in each account?

Solve each inequality, and graph the solution. [2.8]

25. $2x + 3 < 5$

26. $-5a > 20$

27. $0.4 - 0.2x \geq 1$

28. $4 - 5(m + 1) \leq 9$

CHAPTERS 1–2 CUMULATIVE REVIEW

Simplify.

1. $6 + 3(6 + 2)$

2. $-11 + 17 + (-13)$

3. $7 - 9 - 12$

4. $8 - 4 \cdot 5$

5. $\dfrac{1}{5}(10x)$

6. $-6(5 - 11) - 4(13 - 6)$

7. $\left(-\dfrac{2}{3}\right)^3$

8. $-8(9x)$

9. $-\dfrac{3}{4} \div \dfrac{15}{16}$

10. $\dfrac{2}{5} \cdot \dfrac{4}{7}$

11. $\dfrac{-4(-6)}{-9}$

12. $\dfrac{2(-9) + 3(6)}{-7 - 8}$

13. $\dfrac{(5 - 3)^2}{5^2 - 3^2}$

14. $\dfrac{5}{9} + \dfrac{1}{3}$

15. $\dfrac{4}{21} - \dfrac{9}{35}$

16. $3(2x - 1) + 5(x + 2)$

Solve each equation.

17. $7x = 6x + 4$

18. $x + 12 = -4$

19. $-\dfrac{3}{5}x = 30$

20. $3x - 4 = 11$

21. $5x - 7 = x - 1$

22. $4(3x - 8) + 5(2x + 7) = 25$

23. $15 - 3(2t + 4) = 1$

24. $3(2a - 7) - 4(a - 3) = 15$

25. $\dfrac{1}{3}(x - 6) = \dfrac{1}{4}(x + 8)$

26. Solve $P = a + b + c$ for c.

27. Solve $3x + 4y = 12$ for y.

Solve each inequality.

28. $x - 3 \geq 2$

29. $-\dfrac{1}{2}x < 5$

Solve each inequality, and graph the solution.

30. $3(x - 4) \leq 6$

31. $-5x + 9 < -6$

32. Translate into symbols: The difference of x and 5 is 12.

33. Give the opposite, reciprocal, and absolute value of $-\dfrac{2}{3}$.

34. Find the perimeter and area:

3 in.

3 in.

35. Find the next number in the sequence 4, 1, −2, −5, ...

36. State the property or properties that justify the following: $5 + (-5) = 0$

37. Use the distributive property: $\frac{1}{4}(8x - 4)$

38. List all the rational numbers from the set $\{-5, -3, -1.7, 2.3, \frac{12}{7}, \pi\}$

39. Reduce to lowest terms: $\dfrac{234}{312}$

40. Evaluate $x^2 - 8x - 9$ when $x = -2$.

41. Evaluate $a^2 - 2ab + b^2$ when $a = 3, b = -2$.

42. What is 30% of 50?

43. Number Problem Twice a number increased by 7 is 31. Find the number.

44. Gambling A gambler won $5, then lost $8, won $1, and finally lost $7. How much did she win or lose?

45. Geometry A right triangle has one 42° angle. Find the other two angles.

46. Geometry The length of a rectangle is 4 inches more than the width. The perimeter is 28. Find the length and width.

47. Geometry Two angles are complementary. If one is 25°, find the other one.

48. Coin Problem Arbi has $2.20 in dimes and quarters. If he has 8 more dimes than quarters, how many coins of each does he have?

49. Investing I have invested money in two accounts. One account pays 5% annual interest, and the other pays 6%. I have $200 more in the 6% account than I have in the 5% account. If the total amount of interest was $56, how much do I have in each account?

50. Hourly Pay Carol tutors in the math lab. She gets paid $8 per hour for the first 15 hours and $10 per hour for each hour after that. She made $150 one week. How many hours did she work?

Graphing and Linear Systems

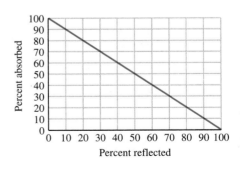

(© EyeWire, Inc.)

INTRODUCTION

When light comes into contact with a surface that does not transmit light, then all the light that contacts the surface is either reflected off the surface or absorbed into the surface. If we let R represent the percentage of light reflected and A the percentage of light absorbed, then the relationship between these two variables can be written as

$$R + A = 100$$

which is a linear equation in two variables. Table 1 and the following graph show the same relationship as that described by the equation. The table is a numerical description; the graph is a visual description.

TABLE I Reflected and Absorbed Light	
Percent Reflected	**Percent Absorbed**
0	100
20	80
40	60
60	40
80	20
100	0

In this chapter we learn how to build tables and draw graphs from linear equations in two variables.

STUDY SKILLS

1. **Getting Ready to Take an Exam**   Try to arrange your daily study habits so that you have very little studying to do the night before your next exam. The next two goals will help you achieve goal 1.

2. **Review With the Exam in Mind**   Each day you should review material that will be covered on the next exam. Your review should consist of working problems. Preferably, the problems you work should be problems from your list of difficult problems.

3. **Continue to List Difficult Problems**   This study skill was started in the previous chapter. You should continue to list and rework the problems that give you the most difficulty. It is this list that you will use to study for the next exam. Your goal is to go into the next exam knowing that you can successfully work any problem from your list of hard problems.

4. **Pay Attention to Instructions**   Taking a test is different from doing homework. When you take a test, the problems will be mixed up. When you do your homework, you usually work a number of similar problems. I sometimes have students who do very well on their homework, but become confused when they see the same problems on a test because they have not paid attention to the instructions on their homework. For example, suppose you see the equation $y = 3x - 2$ on your next test. By itself, the equation is simply a statement. There isn't anything to do unless the equation is accompanied by instructions. Each of the following is a valid instruction with respect to the equation $y = 3x - 2$, and the result of applying the instructions will be different in each case:

Find x when y is 10.	(Section 2.5)
Solve for x.	(Section 2.5)
Graph the equation.	(Section 3.3)
Find the intercepts.	(Section 3.4)

There are many things to do with the equation $y = 3x - 2$. If you train yourself to pay attention to the instructions that accompany a problem as you work through the assigned problems, you will not find yourself confused about what to do with a problem when you see it on a test.

In Chapter 1, we showed the relationship between the table of values for the speed of a race car and the corresponding bar chart. Table 1 and Figure 1 from the introduction are reproduced here for reference. In Figure 1, the horizontal line that shows the elapsed time in seconds is called the *horizontal axis,* and the vertical line that shows the speed in miles per hour is called the *vertical axis.*

TABLE 1	Speed of a Race Car
Time (seconds)	Speed (miles per hour)
0	0
1	72.7
2	129.9
3	162.8
4	192.2
5	212.4
6	228.1

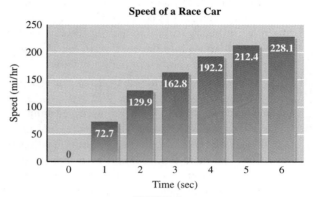

FIGURE 1

The data in Table 1 are called *paired data* because the information is organized so that each number in the first column is paired with a specific number in the second column. Each pair of numbers is associated with one of the solid bars in Figure 1. For example, the third bar in the bar chart is associated with the pair of numbers 3 seconds and 162.8 miles per hour. The first number, 3 seconds, is associated with the horizontal axis, and the second number, 162.8 miles per hour, is associated with the vertical axis.

SCATTER DIAGRAMS AND LINE GRAPHS

The information in Table 1 can be visualized with a *scatter diagram* and *line graph* as well. Figure 2 is a scatter diagram of the information in Table 1. We use dots instead of the bars shown in Figure 1 to show the speed of the race car at each second during the race. Figure 3 is called a *line graph.* It is constructed by taking the dots in Figure 2 and connecting each one to the next with a straight line. Notice that we have labeled the axes in these two figures a little differently than we did with the histogram (or bar chart), by making the axes intersect at the number 0.

The number sequences we have worked with in the past can also be written as paired data by associating each number in the sequence with its position in the sequence. For instance, in the sequence of odd numbers

$$1, 3, 5, 7, 9, \ldots$$

the number 7 is the fourth number in the sequence. Its position is 4, and its value is

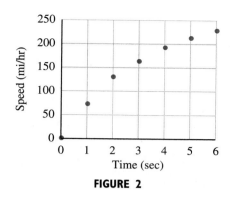

FIGURE 2

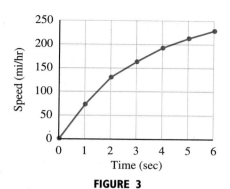

FIGURE 3

7. Here is the sequence of odd numbers written so that the position of each term is noted:

Position   1, 2, 3, 4, 5, . . .

Value   1, 3, 5, 7, 9, . . .

EXAMPLE 1 Tables 2 and 3 give the first five terms of the sequence of odd numbers and the sequence of squares as paired data. In each case construct a scatter diagram.

TABLE 2 Odd Numbers	
Position	**Value**
1	1
2	3
3	5
4	7
5	9

TABLE 3 Squares	
Position	**Value**
1	1
2	4
3	9
4	16
5	25

Solution The two scatter diagrams are based on the data from Tables 2 and 3 shown here. Notice how the dots in Figure 4 seem to line up in a straight line, while

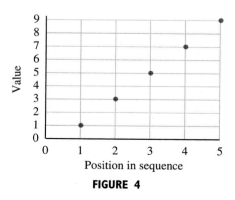

FIGURE 4

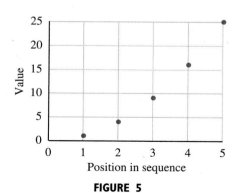

FIGURE 5

the dots in Figure 5 give the impression of a curve. We say the points in Figure 4 suggest a *linear* relationship between the two sets of data, while the points in Figure 5 suggest a *nonlinear* relationship.

As you know, each dot in Figures 4 and 5 corresponds to a pair of numbers, one of which is associated with the horizontal axis and the other with the vertical axis. Paired data play a very important role in the equations we will solve in the next section. To prepare ourselves for those equations, we need to expand the concept of paired data to include negative numbers. At the same time, we want to standardize the position of the axes in the diagrams that we use to visualize paired data.

> **DEFINITION**   A pair of numbers enclosed in parentheses and separated by a comma, such as $(-2, 1)$, is called an **ordered pair** of numbers. The first number in the pair is called the **x-coordinate** of the ordered pair; the second number is called the **y-coordinate**. For the ordered pair $(-2, 1)$, the x-coordinate is -2, and the y-coordinate is 1.

Ordered pairs of numbers are important in the study of mathematics because they give us a way to visualize solutions to equations. To see the visual component of ordered pairs, we need the diagram shown in Figure 6. It is called the *rectangular coordinate system.*

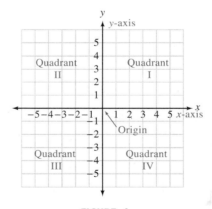

FIGURE 6

The rectangular coordinate system is built from two number lines oriented perpendicular to each other. The horizontal number line is exactly the same as our real number line and is called the *x*-axis. The vertical number line is also the same as our real number line, with the positive direction up and the negative direction down. It is called the *y*-axis. The point where the two axes intersect is called the origin. As you can see from Figure 6, the axes divide the plane into four quadrants, which are numbered I through IV in a counterclockwise direction.

GRAPHING ORDERED PAIRS

To graph the ordered pair (a, b), we start at the origin and move a units forward or back (forward if a is positive and back if a is negative). Then we move b units up or down (up if b is positive, down if b is negative). The point where we end up is the graph of the ordered pair (a, b).

E X A M P L E   2   Graph the ordered pairs $(3, 4)$, $(3, -4)$, $(-3, 4)$, and $(-3, -4)$.

Solution

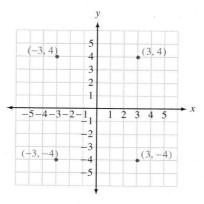

FIGURE 7

We can see in Figure 7 that when we graph ordered pairs, the x-coordinate corresponds to movement parallel to the x-axis (horizontal) and the y-coordinate corresponds to movement parallel to the y-axis (vertical).

E X A M P L E   3   Graph the ordered pairs $(-1, 3)$, $(2, 5)$, $(0, 0)$, $(0, -3)$, and $(4, 0)$.

Solution   See Figure 8.

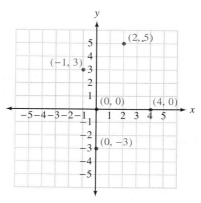

FIGURE 8

Note: If we do not label the axes of a coordinate system, we assume that each square is 1 unit long and 1 unit wide.

Getting Ready for Class

After reading through the preceding section, respond in your own words and in complete sentences.

A. What is an ordered pair of numbers?

B. Explain in words how you would graph the ordered pair (3, 4).

C. How do you construct a rectangular coordinate system?

D. Where is the origin on a rectangular coordinate system?

PROBLEM SET 3.1

Graph the following ordered pairs.

1. (3, 2)	**2.** (3, −2)	**3.** (−3, 2)
4. (−3, −2)	**5.** (5, 1)	**6.** (5, −1)
7. (1, 5)	**8.** (1, −5)	**9.** (−1, 5)
10. (−1, −5)	**11.** $(2, \frac{1}{2})$	**12.** $(3, \frac{3}{2})$
13. $(-4, -\frac{5}{2})$	**14.** $(-5, -\frac{3}{2})$	**15.** (3, 0)
16. (−2, 0)	**17.** (0, 5)	**18.** (0, 0)

19–28. Give the coordinates of each numbered point in the figure.

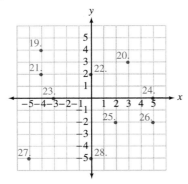

Graph the points (4, 3) and (−4, −1), and draw a straight line that passes through both of them. Then answer the following questions.

29. Does the point (2, 2) lie on the line?

30. Does the point (−2, 0) lie on the line?

31. Does the point (0, −2) lie on the line?

32. Does the point (−6, 2) lie on the line?

Graph the points (−2, 4) and (2, −4), and draw a straight line that passes through both of them. Then answer the following questions.

33. Does the point (0, 0) lie on the line?

34. Does the point (−1, 2) lie on the line?

35. Does the point (2, −1) lie on the line?

36. Does the point (1, −2) lie on the line?

Draw a straight line that passes through the points (3, 4) and (3, −4). Then answer the following questions.

37. Is the point (3, 0) on this line?

38. Is the point (0, 3) on this line?

39. Is there any point on this line with an *x*-coordinate other than 3?

40. If you extended the line, would it pass through a point with a *y*-coordinate of 10?

Draw a straight line that passes through the points (3, 4) and (−3, 4). Then answer the following questions.

41. Is the point (4, 0) on this line?

42. Is the point (0, 4) on this line?

43. Is there any point on this line with a *y*-coordinate other than 4?

44. If you extended the line, would it pass through a point with an *x*-coordinate of 10?

45. Draw a straight line that passes through the points (−2, −3) and (4, −3). What do all the points on this line have in common?

46. Draw a straight line that passes through the points (−2, −3) and (−2, 4). What do all the points on this line have in common?

47. For the points on a rectangular coordinate system, where will you find all the ordered pairs of the form (0, *y*)?

48. For the points on a rectangular coordinate system, where will you find all the ordered pairs of the form (*x*, 0)?

Applying the Concepts

49. Hourly Wages Jane is considering a job at Marcy's department store. The job pays $8.00 per hour. The following line graph shows how much Jane will make for working from 0 to 40 hours in a week. List three ordered pairs that lie on the line graph.

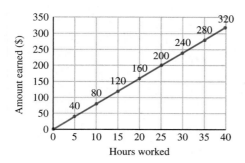

50. Hourly Wages Jane is also considering a job at Gigi's Boutique. This job pays $6.00 per hour plus $50 per week in commission. The following line

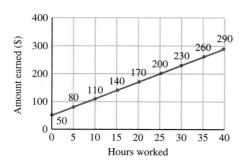

graph shows how much Jane will make for working from 0 to 40 hours in a week at Gigi's. List three ordered pairs that lie on the line graph.

51. Garbage Production The table and bar chart from Problem 89 in Problem Set 1.4 are shown here. Each gives the annual production of garbage in the United States for some specific years.

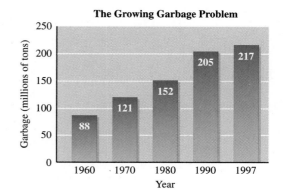

Year	Garbage (millions of tons)
1960	88
1970	121
1980	152
1990	205
1997	217

Use the information from the table and bar chart to construct a line graph using the template below.

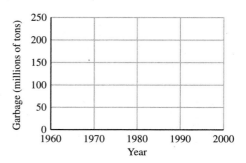

52. Grass Height The table and bar chart from Problem 90 in Problem Set 1.4 are shown on next page.

Each gives the growth of a certain species of grass over time.

Day	Plant Height (inches)
0	0
1	0.5
2	1
3	1.5
4	3
5	4
6	6
7	9
8	13
9	18
10	23

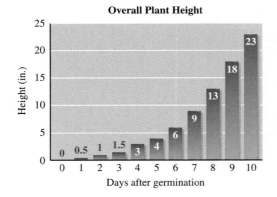

Overall Plant Height

Use the information from the table and chart to construct a line graph using the following template.

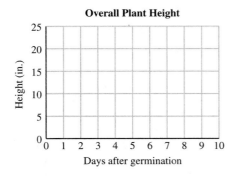

Overall Plant Height

53. Wireless Phone Costs   The table and bar chart from Problem 91 in Problem Set 1.4 are shown here. Each shows the projected cost of wireless phone use. Use the information from the table and chart to construct a line graph.

Year	Cents per Minute
1998	33
1999	28
2000	25
2001	23
2002	22
2003	20

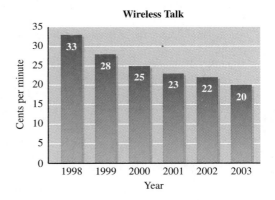

Wireless Talk

54. Digital Camera Sales   The table and bar chart from Problem 78 in Problem Set 1.3 are shown here. Each shows the sales of digital cameras from 1996 to 1999. Use the information from the table and chart to construct a line graph.

Year	Sales (in millions)
1996	$386
1997	$518
1998	$573
1999	$819

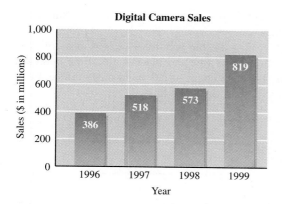

Digital Camera Sales

55. Unemployment Rates The following table shows adjusted unemployment rates in the United States for the years 1983 to 1993. Construct a scatter diagram of this information.

Unemployment Rates	
Year	**Unemployment Rate (%)**
1983	10.2
1984	7.8
1985	7.4
1986	7.0
1987	6.3
1988	5.4
1989	5.2
1990	5.4
1991	6.5
1992	7.3
1993	7.0

56. Tourist Attraction Attendance The annual number of visitors to some of the most popular tourist attractions in the mid-1990s is shown in the following table. Construct a line graph for the information in this table.

Visitors to Tourist Attractions	
Attraction	**Number of Visitors**
Walt Disney World	30,000,000
Disneyland	12,000,000
Great Smoky Mountains	9,000,000
Universal Studios, Orlando	6,000,000
Branson, Missouri	5,000,000
Grand Canyon	3,900,000
Sea World, Orlando	3,850,000
Sea World, San Diego	3,820,000
Knott's Berry Farm	3,800,000

Review Problems

The problems that follow review the material we covered on number sequences in Chapter 1.

Find the next number in each sequence.

57. 3, 10, 17, 24, . . .   **58.** 3, −4, −11, −18, . . .

59. 3, 1, $\frac{1}{3}, \frac{1}{9}$, . . .   **60.** 3, 21, 147, 1,029, . . .

61. 7, 4, 1, −2, . . .   **62.** 7, 10, 13, 16, . . .

63. 7, 21, 63, 189, . . .   **64.** 7, 1, $\frac{1}{7}, \frac{1}{49}$, . . .

65. 5, 6, 8, 11, . . .   **66.** 10, 11, 13, 16, . . .

3.2   **Solutions to Linear Equations in Two Variables**

In this section we will begin to investigate equations in two variables. As you will see, equations in two variables have pairs of numbers for solutions. Since we know how to use paired data to construct tables, histograms, and other charts, we can take our work with paired data further by using equations in two variables to construct tables of paired data. Let's begin this section by reviewing the relationship between equations in one variable and their solutions.

If we solve the equation $3x - 2 = 10$, the solution is $x = 4$. If we graph this solution, we simply draw the real number line and place a dot at the point whose coordinate is 4. The relationship between linear equations in one variable, their solutions, and the graphs of those solutions looks like this:

Equation	Solution	Graph of Solution Set

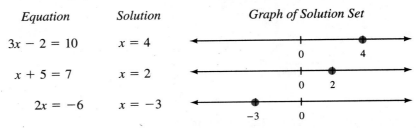

When the equation has one variable, the solution is a single number whose graph is a point on a line.

Now, consider the equation $2x + y = 3$. The first thing we notice is that there are two variables instead of one. Therefore, a solution to the equation $2x + y = 3$ will be not a single number but a pair of numbers, one for x and one for y, that makes the equation a true statement. One pair of numbers that works is $x = 2, y = -1$, because when we substitute them for x and y in the equation, we get a true statement. That is,

$$2(2) + (-1) \overset{?}{=} 3$$

$$4 - 1 = 3$$

$$3 = 3 \qquad \text{A true statement}$$

The pair of numbers $x = 2, y = -1$ is written as $(2, -1)$. As you know from Section 3.1, $(2, -1)$ is called an *ordered pair* because it is a pair of numbers written in a specific order. The first number is always associated with the variable x, and the second number is always associated with the variable y. We call the first number in the ordered pair the *x-coordinate* (or *x*-component) and the second number the *y-coordinate* (or *y*-component) of the ordered pair.

Let's look back to the equation $2x + y = 3$. The ordered pair $(2, -1)$ is not the only solution. Another solution is $(0, 3)$, because when we substitute 0 for x and 3 for y we get

$$2(0) + 3 \overset{?}{=} 3$$

$$0 + 3 = 3$$

$$3 = 3 \qquad \text{A true statement}$$

Still another solution is the ordered pair $(5, -7)$, because

$$2(5) + (-7) \overset{?}{=} 3$$

$$10 - 7 = 3$$

$$3 = 3 \qquad \text{A true statement}$$

As a matter of fact, for any number we want to use for x, there is another number we can use for y that will make the equation a true statement. An infinite number of ordered pairs satisfy (are solutions to) the equation $2x + y = 3$; we have listed just a few of them.

EXAMPLE 1   Given the equation $2x + 3y = 6$, complete the following ordered pairs so that they will be solutions to the equation: $(0, \quad)$, $(\quad , 1)$, $(3, \quad)$.

Solution   To complete the ordered pair $(0, \quad)$, we substitute 0 for x in the equation and then solve for y:

$$2(0) + 3y = 6$$
$$3y = 6$$
$$y = 2$$

The ordered pair is $(0, 2)$.

To complete the ordered pair $(\quad , 1)$, we substitute 1 for y in the equation and solve for x:

$$2x + 3(1) = 6$$
$$2x + 3 = 6$$
$$2x = 3$$
$$x = \frac{3}{2}$$

The ordered pair is $(\frac{3}{2}, 1)$.

To complete the ordered pair $(3, \quad)$, we substitute 3 for x in the equation and solve for y:

$$2(3) + 3y = 6$$
$$6 + 3y = 6$$
$$3y = 0$$
$$y = 0$$

The ordered pair is $(3, 0)$.

Notice in each case that once we have used a number in place of one of the variables, the equation becomes a linear equation in one variable. We then use the method explained in Chapter 2 to solve for that variable.

EXAMPLE 2   Complete the following table for the equation $2x - 5y = 20$.

x	y
0	
	2
	0
−5	

Solution   Filling in the table is equivalent to completing the following ordered pairs: (0,   ), (, 2), (, 0), (−5,   ). So we proceed as in Example 1.

When $x = 0$, we have

$$2(0) - 5y = 20$$
$$0 - 5y = 20$$
$$-5y = 20$$
$$y = -4$$

When $y = 2$, we have

$$2x - 5(2) = 20$$
$$2x - 10 = 20$$
$$2x = 30$$
$$x = 15$$

When $y = 0$, we have

$$2x - 5(0) = 20$$
$$2x - 0 = 20$$
$$2x = 20$$
$$x = 10$$

When $x = -5$, we have

$$2(-5) - 5y = 20$$
$$-10 - 5y = 20$$
$$-5y = 30$$
$$y = -6$$

The completed table looks like this:

x	y
0	−4
15	2
10	0
−5	−6

which is equivalent to the ordered pairs $(0, -4)$, $(15, 2)$, $(10, 0)$, and $(-5, -6)$.

EXAMPLE 3   Complete the following table for the equation $y = 2x - 1$.

x	y
0	
5	
	7
	3

Solution   When $x = 0$, we have   When $x = 5$, we have

$$y = 2(0) - 1 \qquad\qquad\qquad y = 2(5) - 1$$

$$y = 0 - 1 \qquad\qquad\qquad\quad y = 10 - 1$$

$$y = -1 \qquad\qquad\qquad\qquad y = 9$$

When $y = 7$, we have   When $y = 3$, we have

$$7 = 2x - 1 \qquad\qquad\qquad 3 = 2x - 1$$

$$8 = 2x \qquad\qquad\qquad\qquad 4 = 2x$$

$$4 = x \qquad\qquad\qquad\qquad\quad 2 = x$$

The completed table is

x	y
0	−1
5	9
4	7
2	3

which means the ordered pairs $(0, -1)$, $(5, 9)$, $(4, 7)$, and $(2, 3)$ are among the solutions to the equation $y = 2x - 1$.

EXAMPLE 4   Which of the ordered pairs $(2, 3)$, $(1, 5)$, and $(-2, -4)$ are solutions to the equation $y = 3x + 2$?

Solution   If an ordered pair is a solution to the equation, then it must satisfy the equation. That is, when the coordinates are used in place of the variables in the equation, the equation becomes a true statement.

$$\text{Try } (2, 3) \text{ in } y = 3x + 2:$$

$$3 \overset{?}{=} 3(2) + 2$$

$$3 = 6 + 2$$

$$3 = 8 \qquad\qquad \text{A false statement}$$

$$\text{Try } (1, 5) \text{ in } y = 3x + 2:$$

$$5 \overset{?}{=} 3(1) + 2$$

$$5 = 3 + 2$$

$$5 = 5 \qquad\qquad \text{A true statement}$$

$$\text{Try } (-2, -4) \text{ in } y = 3x + 2:$$

$$-4 \overset{?}{=} 3(-2) + 2$$

$$-4 = -6 + 2$$

$$-4 = -4 \qquad\qquad \text{A true statement}$$

The ordered pairs (1, 5) and (−2, −4) are solutions to the equation $y = 3x + 2$, and (2, 3) is not.

EXAMPLE 5 One of the rates listed in the GTE Mobilnet rate card in 1993 called for a flat rate of $15 per month plus $1.50 for each minute you used the phone. Write an equation in two variables that will let you calculate the monthly charge for talking x minutes. Then make a table that shows the cost for talking 10, 15, 20, 25, or 30 minutes in one month. Use the data in the table to construct a bar chart.

Solution If we let x = the number of minutes of phone use, then the charge for using the phone x minutes in one month will be $1.5x + 15$. If y is the charge for talking x minutes, then

$$y = 1.5x + 15$$

Now, GTE Mobilnet charges $1.50 for each minute or fraction of a minute, so the preceding equation will work only when x is a nonnegative integer. If we substitute $x = 10, 15, 20, 25$, and 30 into the equation for x, then the equation gives us the corresponding costs as shown in Table 1. A bar chart of the data in Table 1 is shown in Figure 1.

TABLE 1 Monthly Phone Costs	
Time (in minutes)	**Cost (in dollars)**
10	$30.00
15	$37.50
20	$45.00
25	$52.50
30	$60.00

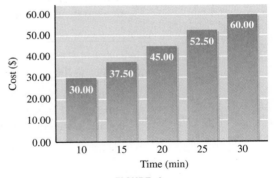

FIGURE 1

Getting Ready for Class

After reading through the preceding section, respond in your own words and in complete sentences.

A. How can you tell if an ordered pair is a solution to an equation?

B. How would you find a solution to $y = 3x − 5$?

C. Why is (3, 2) not a solution to $y = 3x − 5$?

D. How many solutions are there to an equation that contains two variables?

PROBLEM SET 3.2

For each equation, complete the given ordered pairs.

1. $2x + y = 6$ (0,   ), (3,   ), (, -6)
2. $3x - y = 5$ (0,   ), (1,   ), (, 5)
3. $3x + 4y = 12$ (0,   ), (, 0), (-4,   )
4. $5x - 5y = 20$ (0,   ), (, -2), (1,   )
5. $y = 4x - 3$ (1,   ), (, 0), (5,   )
6. $y = 3x - 5$ (, 13), (0,   ), (-2,   )
7. $y = 7x - 1$ (2,   ), (, 6), (0,   )
8. $y = 8x + 2$ (3,   ), (, 0), (, -6)
9. $x = -5$ (, 4), (, -3), (, 0)
10. $y = 2$ (5,   ), (-8,   ), ($\frac{1}{2}$,   )

For each of the following equations, complete the given table.

11. $y = 3x$

x	y
1	
-3	
	12
	18

12. $y = -2x$

x	y
-4	
0	
	10
	12

13. $y = 4x$

x	y
0	
	-2
-3	
	12

14. $y = -5x$

x	y
3	
	0
-2	
	-20

15. $x + y = 5$

x	y
2	
3	
	0
	-4

16. $x - y = 8$

x	y
0	
4	
	-3
	-2

17. $2x - y = 4$

x	y
	0
	2
1	
-3	

18. $3x - y = 9$

x	y
	0
	-9
5	
-4	

19. $y = 6x - 1$

x	y
0	
-1	
-3	
	8

20. $y = 5x + 7$

x	y
0	
-2	
-4	
	-8

For the following equations, tell which of the given ordered pairs are solutions.

21. $2x - 5y = 10$ (2, 3), (0, -2), ($\frac{5}{2}$, 1)
22. $3x + 7y = 21$ (0, 3), (7, 0), (1, 2)
23. $y = 7x - 2$ (1, 5), (0, -2), (-2, -16)
24. $y = 8x - 3$ (0, 3), (5, 16), (1, 5)
25. $y = 6x$ (1, 6), (-2, -12), (0, 0)
26. $y = -4x$ (0, 0), (2, 4), (-3, 12)
27. $x + y = 0$ (1, 1), (2, -2), (3, 3)
28. $x - y = 1$ (0, 1), (0, -1), (1, 2)
29. $x = 3$ (3, 0), (3, -3), (5, 3)
30. $y = -4$ (3, -4), (-4, 4), (0, -4)

Applying the Concepts

31. **Perimeter** If the perimeter of a rectangle is 30 inches, then the relationship between the length l and the width w is given by the equation

$$2l + 2w = 30$$

What is the length when the width is 3 inches?

32. **Perimeter** The relationship between the perimeter P of a square and the length of its side s is given

by the formula $P = 4s$. If each side of a square is 5 inches, what is the perimeter? If the perimeter of a square is 28 inches, how long is a side?

33. Phone Rates   MCIA, a long-distance phone company, charges $3.00 a month plus $0.10 per minute for long-distance calls. If you talk for x minutes during a month, then your total monthly cost y will be $y = 3 + 0.10x$. Use this equation to fill in the following table; then use the results to create a line graph on the template.

Minutes	Cost
0	
10	
20	
30	
40	
50	
60	
70	
80	
90	
100	

Minutes	Cost
0	
10	
20	
30	
40	
50	
60	
70	
80	
90	
100	

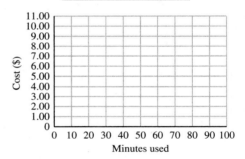

35. Internet Access   Billy is looking for an Internet service provider to get online. Computer Service

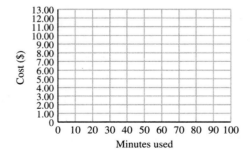

34. Phone Rates   One Bell, a long-distance phone company, charges $5.00 a month plus $0.05 per minute for long-distance calls. If you talk for x minutes during a month, then your total monthly cost y will be $y = 5 + 0.05x$. Use this equation to fill in the following table; then use the results to create a line graph on the template.

Hours	Cost
0	
1	
2	
3	
4	
5	
6	
7	
8	
9	
10	

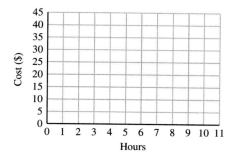

offers an Internet connection for $10 a month plus $3 for every hour you are connected to the Internet. If you are on the Internet for x hours, then your total cost y will be $y = 10 + 3x$. Use this equation to fill in the preceding table; then use the results to create a line graph on the template.

36. **Internet Access**   Joan wants to connect to the Internet. ICM World offers their Internet connection

Hours	Cost
0	
1	
2	
3	
4	
5	
6	
7	
8	
9	
10	

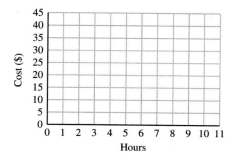

for $18 a month plus $1 for every hour you are connected. If you are on the Internet for x hours in 1 month, then your monthly cost y will be $y = 18 + x$. Use this equation to fill in the preceding table; then use the results to create a line graph on the template.

37. **Cost of Trash Service**   San Luis Garbage charges $13.00 per month for their basic service, plus $1.50 per bag for each trash bag left by the curb on the day the trash is picked up. Write an equation in two variables that will let you calculate the total monthly charge y for picking up x trash bags. Then use this equation to make a table that gives the cost for picking up 5, 7, 9, 11, and 13 trash bags in one month. Complete the table, and then use the information in the table to construct a bar chart.

Bags Collected	Cost (in dollars)
5	
7	
9	
11	
13	

38. **Cost of Trash Service**   Five Cities Garbage charges $21.85 per month for their basic service, plus $1.00 per bag for each trash bag left by the curb on the day the trash is picked up. Write an equation in two variables that will let you calculate the total monthly charge y for picking up x trash bags. Then use this equation to make a table that gives the cost for picking up 5, 7, 9, 11, and 13 trash bags in one month. Complete the table, and then use the information in the table to construct a bar chart.

Bags Collected	Cost (in dollars)
5	
7	
9	
11	
13	

39. Cost of Bottled Water   A water bottling company charges $7.00 per month for their water dispenser and $1.10 for each gallon of water delivered. Write an equation in two variables that will let you calculate the total monthly charge y for drinking x gallons of water in a month. Then use this equation to make a table that gives the cost for drinking 15, 20, 25, 30, and 35 gallons of water in 1 month. Use the information in the table to construct a bar chart.

40. Car Rental Cost   A car rental company charges $21.99 per day and $0.20 per mile to rent one of their cars. Write an equation in two variables that will let you calculate the total charge to rent one of these cars for a week and drive it x miles.

41. Find y when x is 4 in the formula $3x + 2y = 6$.

42. Find y when x is 0 in the formula $3x + 2y = 6$.

43. Find y when x is 0 in $y = -\dfrac{1}{3}x + 2$.

44. Find y when x is 3 in $y = -\dfrac{1}{3}x + 2$.

45. Find y when x is 2 in $y = \dfrac{3}{2}x - 3$.

46. Find y when x is 4 in $y = \dfrac{3}{2}x - 3$.

47. Solve $5x + y = 4$ for y.

48. Solve $-3x + y = 5$ for y.

49. Solve $3x - 2y = 6$ for y.

50. Solve $2x - 3y = 6$ for y.

Review Problems

The following problems review material we covered in Section 2.5.

3.3   Graphing Linear Equations in Two Variables

At the end of the previous section we used a bar chart to obtain a visual picture of *some* of the solutions to the equations $y = 1.5x + 15$. In this section we will use the rectangular coordinate system introduced in Section 3.1 to obtain a visual picture of *all* solutions to a linear equation in two variables. The process we use to obtain a visual picture of all solutions to an equation is called *graphing*. The picture itself is called the *graph* of the equation.

EXAMPLE 1   Graph the solution set for $x + y = 5$.

Solution   We know from the previous section that an infinite number of ordered pairs are solutions to the equation $x + y = 5$. We can't possibly list them all. What we can do is list a few of them and see if there is any pattern to their graphs.

Some ordered pairs that are solutions to $x + y = 5$ are (0, 5), (2, 3), (3, 2), (5, 0). The graph of each is shown in Figure 1.

Now, by passing a straight line through these points we can graph the solution set for the equation $x + y = 5$. Linear equations in two variables always have graphs that are straight lines. The graph of the solution set for $x + y = 5$ is shown in Figure 2.

Every ordered pair that satisfies $x + y = 5$ has its graph on the line, and any point on the line has coordinates that satisfy the equation. So, there is a one-to-one correspondence between points on the line and solutions to the equation.

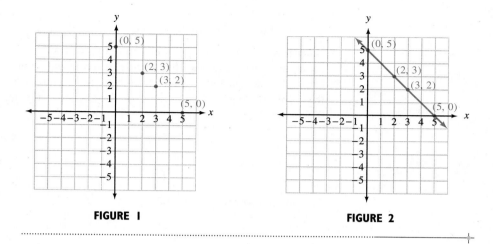

FIGURE 1   **FIGURE 2**

Our ability to graph an equation as we have done in Example 1 is due to the invention of the rectangular coordinate system. The French philosopher René Descartes (1595–1650) is the person usually credited with the invention of the rectangular coordinate system. As a philosopher, Descartes is responsible for the statement "I think, therefore I am." Until Descartes invented his coordinate system in 1637, algebra and geometry were treated as separate subjects. The rectangular coordinate system allows us to connect algebra and geometry by associating geometric shapes with algebraic equations.

Here is the precise definition for a linear equation in two variables.

> **DEFINITION**   Any equation that can be put in the form $ax + by = c$, where a, b, and c are real numbers and a and b are not both 0, is called a **linear equation in two variables.** The graph of any equation of this form is a straight line (that is why these equations are called "linear"). The form $ax + by = c$ is called **standard form.**

To graph a linear equation in two variables, we simply graph its solution set. That is, we draw a line through all the points whose coordinates satisfy the equation. Here are the steps to follow.

TO GRAPH A STRAIGHT LINE

Step 1: Find any three ordered pairs that satisfy the equation. This can be done by using a convenient number for one variable and solving for the other variable.

Step 2: Graph the three ordered pairs found in step 1. Actually, we need only two points to graph a straight line. The third point serves as a check. If all three points do not line up, there is a mistake in our work.

Step 3: Draw a straight line through the three points graphed in step 2.

EXAMPLE 2   Graph the equation $y = 3x - 1$.

Solution   Since $y = 3x - 1$ can be put in the form $ax + by = c$, it is a linear equation in two variables. Hence, the graph of its solution set is a straight line. We can find some specific solutions by substituting numbers for x and then solving for the corresponding values of y. We are free to choose any numbers for x, so let's use 0, 2, and -1.

Let $x = 0$: $y = 3(0) - 1$

$y = 0 - 1$

$y = -1$

The ordered pair $(0, -1)$ is one solution.

Let $x = 2$: $y = 3(2) - 1$

$y = 6 - 1$

$y = 5$

The ordered pair $(2, 5)$ is a second solution.

Let $x = -1$: $y = 3(-1) - 1$

$y = -3 - 1$

$y = -4$

The ordered pair $(-1, -4)$ is a third solution.

In table form

x	y
0	-1
2	5
-1	-4

Next, we graph the ordered pairs $(0, -1)$, $(2, 5)$, $(-1, -4)$ and draw a straight line through them.

The line we have drawn in Figure 3 is the graph of $y = 3x - 1$.

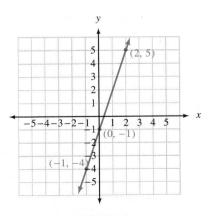

FIGURE 3

Example 2 again illustrates the connection between algebra and geometry that we mentioned previously. Descartes' rectangular coordinate system allows us to associate the equation $y = 3x - 1$ (an algebraic concept) with a specific straight line (a geometric concept). The study of the relationship between equations in algebra and their associated geometric figures is called *analytic geometry*. The rectangular coordinate system is often referred to as the *Cartesian coordinate system* in honor of Descartes.

EXAMPLE   3   Graph the equation $y = -\frac{1}{3}x + 2$.

Solution   We need to find three ordered pairs that satisfy the equation. To do so, we can let x equal any numbers we choose and find corresponding values of y. Since every value of x we substitute into the equation is going to be multiplied by $-\frac{1}{3}$, let's use numbers for x that are divisible by 3, like -3, 0, and 3. That way, when we multiply them by $-\frac{1}{3}$, the result will be an integer.

Let $x = -3$: $y = -\dfrac{1}{3}(-3) + 2$ In table form

$$y = 1 + 2$$

$$y = 3$$

x	y
-3	3
0	2
3	1

The ordered pair $(-3, 3)$ is one solution.

Let $x = 0$: $y = -\dfrac{1}{3}(0) + 2$

$$y = 0 + 2$$

$$y = 2$$

The ordered pair $(0, 2)$ is a second solution.

Let $x = 3$: $y = -\dfrac{1}{3}(3) + 2$

$$y = -1 + 2$$

$$y = 1$$

The ordered pair $(3, 1)$ is a third solution.

Graphing the ordered pairs $(-3, 3)$, $(0, 2)$, and $(3, 1)$ and drawing a straight line through their graphs, we have the graph of the equation $y = -\frac{1}{3}x + 2$, as shown in Figure 4.

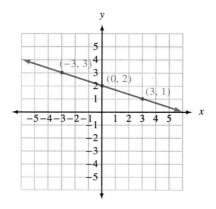

FIGURE 4

Note: In Example 3 the values of x we used, -3, 0, and 3, are referred to as convenient values of x because they are easier to work with than some other numbers. For instance, if we let $x = 2$ in our original equation, we would have to add $-\frac{2}{3}$ and 2 to find the corresponding value of y. Not only would the arithmetic be more difficult, the ordered pair we obtained would have a fraction for its y-coordinate, making it more difficult to graph accurately.

E X A M P L E   4   Graph the lines $x = 2$ and $y = -3$ on the same coordinate system.

Solution   The line $x = 2$ is the set of all points whose x-coordinate is 2. When the variable y does not appear in the equation, it means that y can be any number. The points $(2, 0)$, $(2, 3)$, $(2, -4)$, and $(2, 2)$ therefore all satisfy the equation $x = 2$, simply because their x-coordinates are 2. The graph of $x = 2$, along with the points we named, is given in Figure 5.

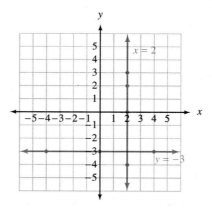

FIGURE 5

The line $y = -3$ is the set of all points whose y-coordinate is -3. The points $(0, -3)$, $(4, -3)$, and $(-4, -3)$ all satisfy the equation $y = -3$, because their y-coordinates are all -3. The graph of the line $y = -3$ is also shown in Figure 5.

EXAMPLE 5   Graph the solution set for $3x - 2y = 6$.

Solution   It will be easier to find convenient values of x to use in the equation if we first solve the equation for y. To do so, we add $-3x$ to each side, and then we multiply each side by $-\frac{1}{2}$.

$$3x - 2y = 6 \qquad\qquad \text{Original equation}$$

$$-2y = -3x + 6 \qquad\qquad \text{Add } -3x \text{ to each side.}$$

$$-\frac{1}{2}(-2y) = -\frac{1}{2}(-3x + 6) \qquad \text{Multiply each side by } -\frac{1}{2}.$$

$$y = \frac{3}{2}x - 3 \qquad\qquad \text{Simplify each side.}$$

Now, since each value of x will be multiplied by $\frac{3}{2}$, it will be to our advantage to choose values of x that are divisible by 2. That way, we will obtain values of y that do not contain fractions. This time, let's use 0, 2, and 4 for x.

When $x = 0$: $y = \frac{3}{2}(0) - 3$

$\qquad\qquad\qquad y = 0 - 3$

$\qquad\qquad\qquad y = -3 \qquad\qquad (0, -3)$ is one solution.

When $x = 2$: $y = \frac{3}{2}(2) - 3$

$\qquad\qquad\qquad y = 3 - 3$

$\qquad\qquad\qquad y = 0 \qquad\qquad (2, 0)$ is a second solution.

When $x = 4$: $y = \frac{3}{2}(4) - 3$

$\qquad\qquad\qquad y = 6 - 3$

$\qquad\qquad\qquad y = 3 \qquad\qquad (4, 3)$ is a third solution.

Graphing the ordered pairs $(0, -3)$, $(2, 0)$, and $(4, 3)$ and drawing a line through them, we have the graph shown in Figure 6.

Note: After reading through Example 5, many students ask why we didn't use -2 for x when we were finding ordered pairs that were solutions to the original equation. The answer is,

FIGURE 6

we could have. If we were to let $x = -2$, the corresponding value of y would have been -6. As you can see by looking at the graph in Figure 6, the ordered pair $(-2, -6)$ is on the graph.

Getting Ready for Class

After reading through the preceding section, respond in your own words and in complete sentences.

A. Explain how you would go about graphing the line $x + y = 5$.

B. When graphing straight lines, why is it a good idea to find three points, when every straight line is determined by only two points?

C. What kind of equations have vertical lines for graphs?

D. What kind of equations have horizontal lines for graphs?

PROBLEM SET 3.3

For the following equations, complete the given ordered pairs, and use the results to graph the solution set for the equation.

1. $x + y = 4$ $(0, \ ), (2, \ ), (\ , 0)$

2. $x - y = 3$ $(0, \ ), (2, \ ), (\ , 0)$

3. $x + y = 3$ $(0, \ ), (2, \ ), (\ , -1)$

4. $x - y = 4$ $(1, \ ), (-1, \ ), (\ , 0)$

5. $y = 2x$ $(0, \ ), (-2, \ ), (2, \ )$

6. $y = \dfrac{1}{2}x$ $(0, \ ), (-2, \ ), (2, \ )$

7. $y = \dfrac{1}{3}x$ $(-3, \ ), (0, \ ), (3, \ )$

8. $y = 3x$ $(-2, \ ), (0, \ ), (2, \ )$

9. $y = 2x + 1$ $(0, \ ), (-1, \ ), (1, \ )$

10. $y = -2x + 1$ $(0, \ ), (-1, \ ), (1, \ )$

11. $y = 4$ $(0, \ ), (-1, \ ), (2, \ )$

12. $x = 3$ $(\ , -2), (\ , 0), (\ , 5)$

13. $y = \dfrac{1}{2}x + 3$ $(-2, \ ), (0, \ ), (2, \ )$

14. $y = \dfrac{1}{2}x - 3$ $(-2,\ ), (0,\ ), (2,\ )$

15. $y = -\dfrac{2}{3}x + 1$ $(-3,\ ), (0,\ ), (3,\ )$

16. $y = -\dfrac{2}{3}x - 1$ $(-3,\ ), (0,\ ), (3,\ )$

Solve each equation for y. Then, complete the given ordered pairs, and use them to draw the graph.

17. $2x + y = 3$ $(-1,\ ), (0,\ ), (1,\ )$
18. $3x + y = 2$ $(-1,\ ), (0,\ ), (1,\ )$
19. $3x + 2y = 6$ $(0,\ ), (2,\ ), (4,\ )$
20. $2x + 3y = 6$ $(0,\ ), (3,\ ), (6,\ )$
21. $-x + 2y = 6$ $(-2,\ ), (0,\ ), (2,\ )$
22. $-x + 3y = 6$ $(-3,\ ), (0,\ ), (3,\ )$

Find three solutions to each of the following equations, and then graph the solution set.

23. $y = -\dfrac{1}{2}x$ **24.** $y = -2x$

25. $y = 3x - 1$ **26.** $y = -3x - 1$

27. $-2x + y = 1$ **28.** $-3x + y = 1$

29. $3x + 4y = 8$ **30.** $3x - 4y = 8$

31. $x = -2$ **32.** $y = 3$

33. $y = 2$ **34.** $x = -3$

35. Perimeter If the perimeter of a rectangle is 10 inches, then the equation that describes the relationship between the length l and width w is

$$2l + 2w = 10$$

Graph this equation using a coordinate system in which the horizontal axis is labeled l and the vertical axis is labeled w.

36. Perimeter The perimeter of a rectangle is 6 inches. Graph the equation that describes the relationship between the length l and width w.

37. The solutions to the equation $y = x$ are ordered pairs of the form (x, x). The x- and y-coordinates are equal. Graph the line $y = x$.

38. The solutions to a certain equation are ordered pairs of the form $(x, 2x)$. Find the equation and its graph.

39. Use the graph below to complete the table.

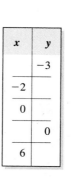
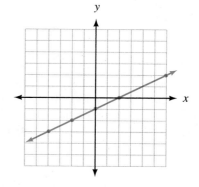

x	y
	-3
-2	
0	
	0
6	

40. Use the graph below to complete the table. (*Hint:* Some parts have two answers.)

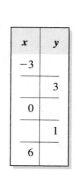
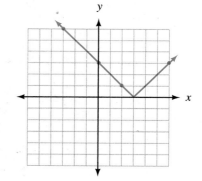

x	y
-3	
	3
0	
	1
6	

41. The y-coordinate of each ordered pair that satisfies the equation $y = |x|$ will never be negative. Each value of y is the absolute value of the corresponding value of x. Fill in the following ordered pairs so that they are solutions to the equation $y = |x|$. Then graph them and connect their graphs in a way that makes the most sense to you.

$(-3,\ )\,(-2,\ )\,(-1,\ )\,(0,\ )\,(1,\ )\,(2,\ )\,(3,\ )$

42. Fill in the following ordered pairs so that they are solutions to the equation $y = |x - 2|$. Graph each point, and then connect the graphs in a way that makes the most sense to you.

$(-1,\ )\,(0,\ )\,(1,\ )\,(2,\ )\,(3,\ )\,(4,\ )\,(5,\ )$

43. Graph the lines $y = x + 1$ and $y = x - 3$ on the same coordinate system. Can you tell from looking at these first two graphs where the graph of $y = x + 3$ would be?

44. Graph the lines $y = 2x + 2$ and $y = 2x - 1$ on the same coordinate system. Use the similarities between these two graphs to graph the line $y = 2x - 4$.

Applying the Concepts

45. Airlines Ocean Wide Airlines is leaving LAX going to New York. The airplane is on the ground at the airport. The plane takes off and maintains a constant climb rate. The airplane leaves from the airport, $(0, 0)$ on the coordinate system, and reaches its cruising altitude of 35,000 feet (approximately 6.6 miles) when it is 90 miles from the airport. The equation for the plane's altitude y, when it is x miles away from the airport, is $y = \dfrac{175}{2,376}x$.

Use the equation to fill in the following table, rounding your answers to the nearest tenth of a mile. Use the results to graph the equation on the template. (The graph will appear in the first quadrant only because x is greater than or equal to 0.)

Distance From Airport (miles)	Height (miles)
0	
15	
30	
45	
60	
75	
90	

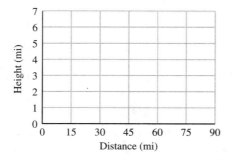

46. Airlines Ocean Wide Airlines is flying into LAX from Hawaii. The airplane has an altitude of 30,000

feet (approximately 5.7 miles) and is 65 miles from the airport when it begins its descent to the airport. The equation for the airplane's flight path during its descent is given by $y = -\dfrac{57}{650}x$. Use the equation to fill in the following table, rounding your answers to the nearest tenth of a mile. Then use your results to graph the equation on the template. (The graph will appear in the second quadrant only because we are giving its distance from the airport in negative numbers.)

Distance From Airport (miles)	Height (miles)
−65	
−55	
−45	
−35	
−25	
−15	
−5	
0	

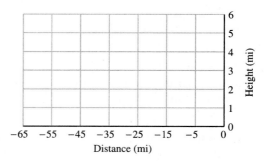

47. Taxes Suppose that 21% of your monthly pay is withheld for federal income taxes and another 8% is withheld for Social Security, state income tax, and other miscellaneous items. If G is your monthly pay before any money is deducted (your gross pay), the amount of money that you take home each month is given by $T = 0.71G$. Graph this equation on a coordinate system in which the horizontal axis is the G-axis and the vertical axis is the T-axis. Start your graph at $G = 0$ and end it at $G = 2,500$.

48. Temperature and Altitude As we mentioned previously, there is a relationship between air temperature and altitude. The higher above the Earth we are, the lower the temperature. If the temperature on the ground is 56°F, then the temperature T (in degrees Fahrenheit), at an altitude of A feet, is $T = -0.0035A + 56$. Graph this equation on a coordinate system in which the horizontal axis is the A-axis and the vertical axis is the T-axis. Start your graph at $A = 0$ and end it at $A = 40,000$.

Review Problems

The following problems review material we covered in Sections 2.3 and 2.4.

Solve each equation.

49. $3(x - 2) = 9$

50. $-4(x - 3) = -16$

51. $2(3x - 1) + 4 = -10$

52. $-5(2x + 3) - 10 = 15$

53. $6 - 2(4x - 7) = -4$

54. $5 - 3(2 - 3x) = 8$

55. $\dfrac{1}{2}x + 4 = \dfrac{2}{3}x + 5$

56. $\dfrac{1}{4}x - 3 = \dfrac{3}{2}x + 7$

3.4   More on Graphing: Intercepts

In this section we continue our work with graphing lines by finding the points where a line crosses the axes of our coordinate system. To do so, we use the fact that any point on the x-axis has a y-coordinate of 0, and any point on the y-axis has an x-coordinate of 0. We begin with the following definition.

> **DEFINITION** The **x-intercept** of a straight line is the x-coordinate of the point where the graph crosses the x-axis. The **y-intercept** is defined similarly. It is the y-coordinate of the point where the graph crosses the y-axis.

If the x-intercept is a, then the point $(a, 0)$ lies on the graph. (This is true because any point on the x-axis has a y-coordinate of 0.)

If the y-intercept is b, then the point $(0, b)$ lies on the graph. (This is true because any point on the y-axis has an x-coordinate of 0.)

Graphically, the relationship is shown in Figure 1.

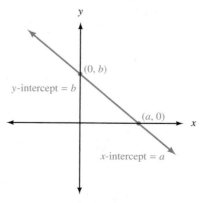

FIGURE I

EXAMPLE 1   Find the x- and y-intercepts for $3x - 2y = 6$, and then use them to draw the graph.

Solution   To find where the graph crosses the x-axis, we let $y = 0$. (The y-coordinate of any point on the x-axis is 0.)

x-intercept:

When $y = 0$

the equation $3x - 2y = 6$

becomes $3x - 2(0) = 6$

                            $3x - 0 = 6$

                            $x = 2$ Multiply each side by $\frac{1}{3}$.

The graph crosses the x-axis at $(2, 0)$, which means the x-intercept is 2.

y-intercept:

When $x = 0$

the equation $3x - 2y = 6$

becomes $3(0) - 2y = 6$

                            $0 - 2y = 6$

                            $-2y = 6$

                            $y = -3$ Multiply each side by $-\frac{1}{2}$.

The graph crosses the y-axis at $(0, -3)$, which means the y-intercept is -3.

Plotting the x- and y-intercepts and then drawing a line through them, we have the graph of $3x - 2y = 6$, as shown in Figure 2.

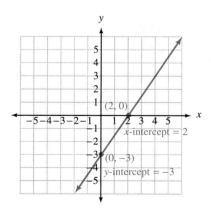

FIGURE 2

E X A M P L E   2   Graph $-x + 2y = 4$ by finding the intercepts and using them to draw the graph.

Solution   Again, we find the x-intercept by letting $y = 0$ in the equation and solving for x. Similarly, we find the y-intercept by letting $x = 0$ and solving for y.

x-intercept:

$$\text{When} \qquad\qquad\qquad y = 0$$
$$\text{the equation} \qquad -x + 2y = 4$$
$$\text{becomes} \qquad -x + 2(0) = 4$$
$$-x + 0 = 4$$
$$-x = 4$$
$$x = -4 \qquad \text{Multiply each side by } -1.$$

The x-intercept is -4, indicating that the point $(-4, 0)$ is on the graph of $-x + 2y = 4$.

y-intercept:

$$\text{When} \qquad\qquad\qquad x = 0$$
$$\text{the equation} \qquad -x + 2y = 4$$
$$\text{becomes} \qquad -0 + 2y = 4$$
$$2y = 4$$
$$y = 2 \qquad \text{Multiply each side by } \tfrac{1}{2}.$$

The y-intercept is 2, indicating that the point $(0, 2)$ is on the graph of $-x + 2y = 4$.

Plotting the intercepts and drawing a line through them, we have the graph of $-x + 2y = 4$, as shown in Figure 3.

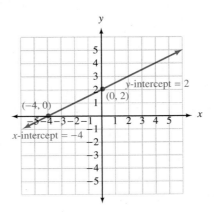

FIGURE 3

Graphing a line by finding the intercepts, as we have done in Examples 1 and 2, is an easy method of graphing if the equation has the form $ax + by = c$ and both the numbers a and b divide the number c evenly.

In our next example we use the intercepts to graph a line in which y is given in terms of x.

E X A M P L E   3   Use the intercepts for $y = -\frac{1}{3}x + 2$ to draw its graph.

Solution   We graphed this line previously in Example 3 of Section 3.3 by substituting three different values of x into the equation and solving for y. This time we will graph the line by finding the intercepts.

x-intercept:

When $y = 0$

the equation $y = -\frac{1}{3}x + 2$

becomes $0 = -\frac{1}{3}x + 2$

$-2 = -\frac{1}{3}x$ Add -2 to each side.

$6 = x$ Multiply each side by -3.

The x-intercept is 6, which means the graph passes through the point $(6, 0)$.

y-intercept:

When $x = 0$

the equation $y = -\frac{1}{3}x + 2$

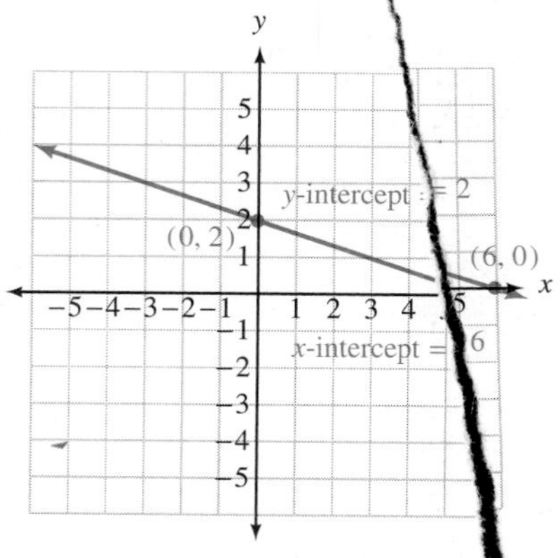

FIGURE 4

$$\text{becomes} \qquad y = -\frac{1}{3}(0) + 2$$

$$y = 2$$

The y-intercept is 2, which means the graph passes through the point $(0, 2)$.

The graph of $y = -\frac{1}{3}x + 2$ is shown in Figure 4. Compare this graph, and the method used to obtain it, with Example 3 in Section 3.3.

Getting Ready for Class

After reading through the preceding section, respond in your own words and in complete sentences.

A. What is the x-intercept for a graph?

B. What is the y-intercept for a graph?

C. How do we find the y-intercept for a line from the equation?

D. How do we graph a line using its intercepts?

PROBLEM SET 3.4

Find the x- and y-intercepts for the following equations. Then use the intercepts to graph each equation.

1. $2x + y = 4$

2. $2x + y = 2$

3. $-x + y = 3$

4. $-x + y = 4$

5. $-x + 2y = 2$

6. $-x + 2y = 4$

7. $5x + 2y = 10$

8. $2x + 5y = 10$

9. $4x - 2y = 8$

10. $2x - 4y = 8$

11. $-4x + 5y = 20$

12. $-5x + 4y = 20$

13. $y = 2x - 6$

14. $y = 2x + 6$

15. $y = 2x + 2$

16. $y = -2x + 2$

17. $y = 2x - 1$

18. $y = -2x - 1$

19. $y = \frac{1}{2}x + 3$

20. $y = \frac{1}{2}x - 3$

21. $y = -\frac{1}{3}x - 2$

22. $y = -\frac{1}{3}x + 2$

For each of the following lines the x-intercept and the y-intercept are both 0, which means the graph of each will go through the origin, $(0, 0)$. Graph each line by finding a point on each, other than the origin, and then drawing a line through that point and the origin.

23. $y = -2x$

24. $y = \frac{1}{2}x$

25. $y = 2x$

26. $y = -\frac{1}{2}x$

27. $y = \frac{1}{3}x$

28. $y = 3x$

29. $y = -\frac{1}{3}x$

30. $y = -3x$

31. $y = \frac{2}{3}x$

32. $y = \frac{3}{2}x$

33. Graph the line that passes through the point $(-4, 4)$ and has an x-intercept of -2. What is the y-intercept of this line?

34. Graph the line that passes through the point $(-3, 4)$ and has a y-intercept of 3. What is the x-intercept of this line?

35. A line passes through the point (1, 4) and has a y-intercept of 3. Graph the line and name its x-intercept.

36. A line passes through the point (3, 4) and has an x-intercept of 1. Graph the line and name its y-intercept.

37. Graph the line that passes through the points $(-2, 5)$ and $(5, -2)$. What are the x- and y-intercepts for this line?

38. Graph the line that passes through the points (5, 3) and $(-3, -5)$. What are the x- and y-intercepts for this line?

39. Use the graph below to complete the following table.

x	y
-2	
0	
	0
	-2

40. Use the graph below to complete the following table.

x	y
-2	
0	
	0
	6

41. The vertical line $x = 3$ has only one intercept. Graph $x = 3$, and name its intercept. (Remember, ordered pairs (x, y) that are solutions to the equation $x = 3$ are ordered pairs with an x-coordinate of 3 and any y-coordinate.)

42. Graph the vertical line $x = -2$. Then name its intercept.

43. The horizontal line $y = 4$ has only one intercept. Graph $y = 4$, and name its intercept. (Ordered pairs (x, y) that are solutions to the equation $y = 4$ are ordered pairs with a y-coordinate of 4 and any x-coordinate.)

44. Graph the horizontal line $y = -3$. Then name its intercept.

Applying the Concepts

45. Complementary Angles The following diagram shows sunlight hitting the ground. Angle α (alpha) is called the angle of inclination, and angle θ (theta) is called the angle of incidence. As the sun moves across the sky, the values of these angles change. Assume that $\alpha + \theta = 90$, where both α and θ are in degrees measure. Graph this equation on a coordinate system where the horizontal axis is the α-axis and the vertical axis is the θ-axis. Find the intercepts first, and limit your graph to the first quadrant only.

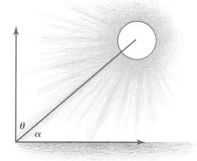

46. Light When light comes into contact with an impenetrable object, such as a thick piece of wood or metal, it is reflected or absorbed, but not transmitted, as shown in the following diagram. If we let R represent the percentage of light reflected and A the percentage of light absorbed by a surface, then the relationship between R and A is $R + A = 100$. Graph this equation on a coordinate system where the horizontal axis is the A-axis and the vertical axis is the *R*-axis. Find the intercepts first, and limit your graph to the first quadrant.

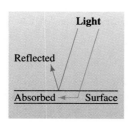

Light

Reflected

Absorbed ← Surface

The following problems review material we covered in Section 2.8.

Solve each inequality.

47. $-3x \geq 12$ **48.** $-2x > 10$

49. $-\dfrac{x}{3} \leq -1$ **50.** $-\dfrac{x}{5} < -2$

51. $-4x + 1 < 17$ **52.** $-3x + 2 \leq -7$

3.5   Solving Linear Systems by Graphing

Two linear equations considered at the same time make up what is called a *system of linear equations*. Both equations contain two variables and, of course, have graphs that are straight lines. The following are systems of linear equations:

$$x + \ y = 3 \qquad y = 2x + 1 \qquad 2x - \ y = 1$$
$$3x + 4y = 2 \qquad y = 3x + 2 \qquad 3x - 2y = 6$$

The solution set for a *system* of linear equations is all ordered pairs that are solutions to both equations. Since each linear equation has a graph that is a straight line, we can expect the intersection of the graphs to be a point whose coordinates are solutions to the system. That is, if we graph both equations on the same coordinate system, we can read the coordinates of the point of intersection and have the solution to our system. Here is an example.

EXAMPLE   I   Solve the following system by graphing.

$$x + y = 4$$
$$x - y = -2$$

Solution   On the same set of coordinate axes we graph each equation separately. Figure 1 shows both graphs, without showing the work necessary to get them. We

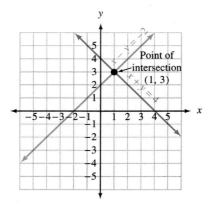

Point of intersection (1, 3)

FIGURE   I

can see from the graphs that they intersect at the point (1, 3). The point (1, 3) must therefore be the solution to our system, since it is the only ordered pair whose graph lies on both lines. Its coordinates satisfy both equations.

We can check our results by substituting the coordinates $x = 1$, $y = 3$ into both equations to see if they work.

When	$x = 1$	When	$x = 1$
and	$y = 3$	and	$y = 3$
the equation	$x + y = 4$	the equation	$x - y = -2$
becomes	$1 + 3 \overset{?}{=} 4$	becomes	$1 - 3 \overset{?}{=} -2$
or	$4 = 4$	or	$-2 = -2$

The point (1, 3) satisfies both equations.

Here are some steps to follow in solving linear systems by graphing.

STRATEGY FOR SOLVING A LINEAR SYSTEM BY GRAPHING

Step 1: Graph the first equation by the methods described in Section 3.3 or 3.4.
Step 2: Graph the second equation on the same set of axes used for the first equation.
Step 3: Read the coordinates of the point of intersection of the two graphs. The ordered pair is the solution to the system.
Step 4: Check the solution in both equations, if necessary.

EXAMPLE 2   Solve the following system by graphing.

$$x + 2y = 8$$
$$2x - 3y = 2$$

From Figure 2, we can see the solution for our system is (4, 2). We check this solution as follows.

When	$x = 4$	When	$x = 4$
and	$y = 2$	and	$y = 2$
the equation	$x + 2y = 8$	the equation	$2x - 3y = 2$

becomes $\quad\quad 4 + 2(2) \overset{?}{=} 8 \quad\quad$ becomes $\quad\quad 2(4) - 3(2) \overset{?}{=} 2$

$$4 + 4 = 8 \quad\quad\quad\quad\quad\quad 8 - 6 = 2$$

$$8 = 8 \quad\quad\quad\quad\quad\quad\quad 2 = 2$$

The point (4, 2) satisfies both equations and therefore must be the solution to our system.

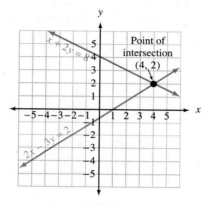

FIGURE 2

E X A M P L E   3   Solve this system by graphing.

$$y = 2x - 3$$

$$x = 3$$

Solution   Graphing both equations on the same set of axes, we have Figure 3.

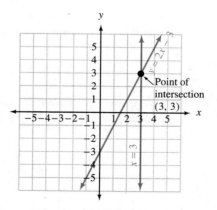

FIGURE 3

The solution to the system is the point (3, 3).

EXAMPLE 4   Solve by graphing.

$$y = x - 2$$
$$y = x + 1$$

Solution   Graphing both equations produces the lines shown in Figure 4. We can see in Figure 4 that the lines are parallel and therefore do not intersect. Our system has no ordered pair as a solution, since there is no ordered pair that satisfies both equations. We say the solution set is the empty set and write ∅.

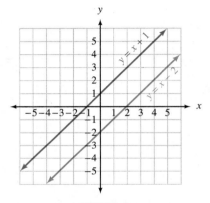

FIGURE 4

Example 4 is one example of two special cases associated with linear systems. The other special case happens when the two graphs coincide. Here is an example.

EXAMPLE 5   Graph the system.

$$2x + y = 4$$
$$4x + 2y = 8$$

Solution   Both graphs are shown in Figure 5. The two graphs coincide. The reason becomes apparent when we multiply both sides of the first equation by 2:

$$2x + y = 4$$
$$\mathbf{2}(2x + y) = \mathbf{2}(4) \qquad \text{Multiply both sides by } \mathbf{2}.$$
$$4x + 2y = 8$$

The equations have the same solution set. Any ordered pair that is a solution to one is a solution to the system. The system has an infinite number of solutions. (Any point on the line is a solution to the system.)

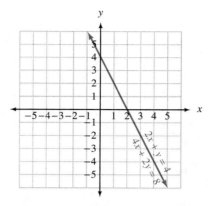

FIGURE 5

The two special cases illustrated in the previous two examples do not happen often. Usually, a system has a single ordered pair as a solution. Solving a system of linear equations by graphing is useful only when the ordered pair in the solution set has integers for coordinates. Two other solution methods work well in all cases. We will develop the other two methods in the next two sections.

Note: We sometimes use special vocabulary to describe the special cases shown in Examples 4 and 5. When a system of equations has no solution because the lines are parallel (as in Example 4), we say the system is *inconsistent.* When the lines coincide (as in Example 5), we say the system is *dependent.*

 ## Getting Ready for Class

After reading through the preceding section, respond in your own words and in complete sentences.

A. What is a system of two linear equations in two variables?

B. What is a solution to a system of linear equations?

C. How do we solve a system of linear equations by graphing?

D. Under what conditions will a system of linear equations not have a solution?

PROBLEM SET 3.5

Solve the following systems of linear equations by graphing.

1. $x + y = 3$
$x - y = 1$

2. $x + y = 2$
$x - y = 4$

3. $x + y = 1$
$-x + y = 3$

4. $x + y = 1$
$x - y = -5$

5. $x + y = 8$
$-x + y = 2$

6. $x + y = 6$
$-x + y = -2$

7. $3x - 2y = 6$
  $x - y = 1$

8. $5x - 2y = 10$
  $x - y = -1$

9. $6x - 2y = 12$
  $3x + y = -6$

10. $4x - 2y = 8$
  $2x + y = -4$

11. $4x + y = 4$
  $3x - y = 3$

12. $5x - y = 10$
  $2x + y = 4$

13. $x + 2y = 0$
  $2x - y = 0$

14. $3x + y = 0$
  $5x - y = 0$

15. $3x - 5y = 15$
  $-2x + y = 4$

16. $2x - 4y = 8$
  $2x - y = -1$

17. $y = 2x + 1$
  $y = -2x - 3$

18. $y = 3x - 4$
  $y = -2x + 1$

19. $x + 3y = 3$
  $y = x + 5$

20. $2x + y = -2$
  $y = x + 4$

21. $x + y = 2$
  $x = -3$

22. $x + y = 6$
  $y = 2$

23. $x = -4$
  $y = 6$

24. $x = 5$
  $y = -1$

25. $x + y = 4$
  $2x + 2y = -6$

26. $x - y = 3$
  $2x - 2y = 6$

27. $4x - 2y = 8$
  $2x - y = 4$

28. $3x - 6y = 6$
  $x - 2y = 4$

29. As you have probably guessed by now, it can be difficult to solve a system of equations by graphing if the solution to the system contains a fraction. The solution to the following system is $(\frac{1}{2}, 1)$. Solve the system by graphing.

$$y = -2x + 2$$

$$y = 4x - 1$$

30. The solution to the following system is $(\frac{1}{3}, -2)$. Solve the system by graphing.

$$y = 3x - 3$$

$$y = -3x - 1$$

31. A second difficulty can arise in solving a system of equations by graphing if one or both of the equations is difficult to graph. The solution to the following system is (2, 1). Solve the system by graphing.

$$3x - 8y = -2$$

$$x - y = 1$$

32. The solution to the following system is $(-3, 2)$. Solve the system by graphing.

$$2x + 5y = 4$$

$$x - y = -5$$

Applying the Concepts

33. Job Comparison   Jane is deciding between two sales positions. She can work for Marcy's and receive $8.00 per hour or for Gigi's, where she earns $6.00 per hour but also receives a $50 commission per week. The two lines in the following figure represent the money Jane will make for working at each of the jobs.

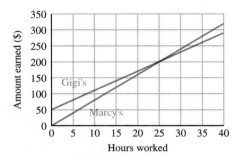

(a) From the figure, how many hours would Jane have to work in order to earn the same amount at each of the positions?

(b) If Jane expects to work less than 20 hours a week, which job should she choose?

(c) If Jane expects to work more than 30 hours a week, which job should she choose?

34. Truck Rental   You need to rent a moving truck for two days. Rider Moving Trucks charges $50 per day and $0.50 per mile. UMove Trucks charges $45 per day and $0.75 per mile. The following figure represents the cost of renting each of the trucks.

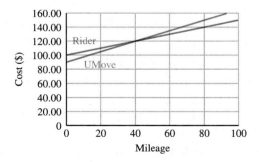

(a) From the figure, after how many miles would the trucks cost the same?

(b) Which company will give you a better deal if you drive less than 30 miles?

(c) Which company will give you a better deal if you drive more than 60 miles?

35. Internet Access Patrice has seen prices from two Internet companies, ICM World and Computer Service. She wants to make a decision about which one is cheaper. ICM World is $18 a month plus $1 for every hour on the Internet. Computer Service is $10 a month plus $3 for every hour on the Internet. The following graph represents the cost of the two Internet providers.

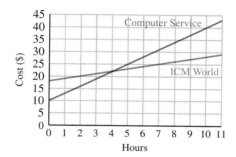

(a) From the graph, at how many hours would the cost of the two Internet providers be the same?
(b) If Patrice decides that she would be on the Internet for less than 3 hours a month, which plan is cheaper?
(c) If Patrice decides that she would be on the Internet for more than 6 hours a month, which plan is cheaper?

36. Phone Rates MaBell has a monthly charge of $5.00 and a rate of $0.05 per minute for long-distance calls. MCIA charges $3.00 per month and $0.10 per minute. The two lines in the following figure represent the total monthly cost for the two phone companies.

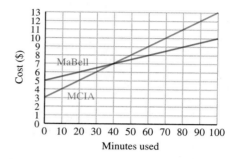

(a) From the figure, how many minutes would have to be used in order for the two plans to cost the same amount?
(b) If less than 30 minutes are used a month, which company has a better deal?
(c) If more than 60 minutes are used a month, which company has a better deal?

Review Problems

The following problems review material we covered in Section 2.1.

Find the value of each expression when x is -3.

37. $2x - 9$

38. $-4x + 3$

39. $9 - 6x$

40. $7 - 5x$

41. $4(3x + 2) + 1$

42. $3(6x + 5) + 10$

43. $2x^2 + 3x + 4$

44. $4x^2 + 3x + 2$

3.6   The Addition Method

The addition property states that if equal quantities are added to both sides of an equation, the solution set is unchanged. In the past we have used this property to help solve equations in one variable. We will now use it to solve systems of linear equations. Here is another way to state the addition property of equality.

Let A, B, C, and D represent algebraic expressions.

$$
\begin{array}{ll}
\text{If} & A = B \\
\text{and} & C = D \\
\hline
\text{then} & A + C = B + D
\end{array}
$$

Since C and D are equal (that is, they represent the same number), what we have done is to add the same amount to both sides of the equation $A = B$. Let's see how we can use this form of the addition property of equality to solve a system of linear equations.

EXAMPLE I   Solve the following system.

$$x + y = 4$$

$$x - y = 2$$

Solution   The system is written in the form of the addition property of equality as written in this section. It looks like this:

$$A = B$$

$$C = D$$

where A is $x + y$, B is 4, C is $x - y$, and D is 2.

We use the addition property of equality to add the left sides together and the right sides together.

$$x + y = 4$$
$$\underline{x - y = 2}$$
$$2x + 0 = 6$$

We now solve the resulting equation for x.

$$2x + 0 = 6$$

$$2x = 6$$

$$x = 3$$

The value we get for x is the value of the x-coordinate of the point of intersection of the two lines $x + y = 4$ and $x - y = 2$. To find the y-coordinate, we simply substitute $x = 3$ into either of the two original equations and get

$$3 + y = 4$$

$$y = 1$$

The solution to our system is the ordered pair (3, 1). It satisfies both equations.

When	$x = 3$	When	$x = 3$
and	$y = 1$	and	$y = 1$
the equation	$x + y = 4$	the equation	$x - y = 2$
becomes	$3 + 1 \overset{?}{=} 4$	becomes	$3 - 1 \overset{?}{=} 2$
or	$4 = 4$	or	$2 = 2$

Figure 1 is visual evidence that the solution to our system is (3, 1).

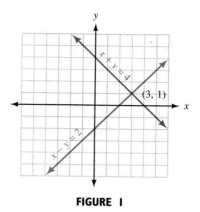

FIGURE 1

Note: The graphs shown for the first three examples are not part of the solution shown in each example. The graphs are there simply to show you that the results we obtain by the addition method are consistent with the results we would obtain by graphing.

The most important part of this method of solving linear systems is eliminating one of the variables when we add the left and right sides together. In our first example, the equations were written so that the y variable was eliminated when we added the left and right sides together. If the equations are not set up this way to begin with, we have to work on one or both of them separately before we can add them together to eliminate one variable.

E X A M P L E   2   Solve the following system.

$$x + 2y = \ \ 4$$
$$x - \ y = -5$$

Solution   Notice that if we were to add the equations together as they are, the resulting equation would have terms in both x and y. Let's eliminate the variable x by multiplying both sides of the second equation by -1 before we add the equations together. (As you will see, we can choose to eliminate either the x or the y variable.) Multiplying both sides of the second equation by -1 will not change its solution, so we do not need to be concerned that we have altered the system.

$$x + 2y = \ \ 4 \ \xrightarrow{\text{No change}} \ x + 2y = 4 \qquad \text{Add left and right sides to get}$$
$$x - \ y = -5 \ \xrightarrow[\text{Multiply by } -1]{} \ \underline{-x + \ y = 5}$$
$$0 + 3y = 9$$
$$3y = 9$$
$$y = 3 \quad \begin{cases} y\text{-coordinate of the} \\ \text{point of intersection} \end{cases}$$

Substituting $y = 3$ into either of the two original equations, we get $x = -2$. The solution to the system is $(-2, 3)$. It satisfies both equations. Figure 2 shows the solution to the system as the point where the two lines cross.

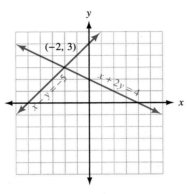

FIGURE 2

EXAMPLE 3   Solve the following system.

$$2x - y = 6$$
$$x + 3y = 3$$

Solution   Let's eliminate the y variable from the two equations. We can do this by multiplying the first equation by 3 and leaving the second equation unchanged.

$$2x - y = 6 \xrightarrow{\text{3 times both sides}} 6x - 3y = 18$$
$$x + 3y = 3 \xrightarrow[\text{No change}]{} x + 3y = 3$$

The important thing about our system now is that the coefficients (the numbers in front) of the y variables are opposites. When we add the terms on each side of the equal sign, then the terms in y will add to zero and be eliminated.

$$6x - 3y = 18$$
$$\underline{x + 3y = 3}$$
$$7x = 21 \qquad \text{Add corresponding terms.}$$

This gives us $x = 3$. Using this value of x in the second equation of our original system, we have

$$3 + 3y = 3$$
$$3y = 0$$
$$y = 0$$

We could substitute $x = 3$ into any of the equations with both x and y variables and also get $y = 0$. The solution to our system is the ordered pair $(3, 0)$. Figure 3 is a picture of the system of equations showing the solution $(3, 0)$.

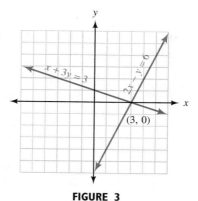

FIGURE 3

EXAMPLE 4   Solve the system.

$$2x + 3y = -1$$

$$3x + 5y = -2$$

Solution   Let's eliminate x from the two equations. If we multiply the first equation by 3 and the second by -2, the coefficients of x will be 6 and -6, respectively. The x terms in the two equations will then add to zero.

$$2x + 3y = -1 \xrightarrow{\text{Multiply by 3}} 6x + 9y = -3$$

$$3x + 5y = -2 \xrightarrow[\text{Multiply by } -2]{} -6x - 10y = 4$$

We now add the left and right sides of our new system together.

$$6x + 9y = -3$$
$$\underline{-6x - 10y = 4}$$
$$-y = 1$$
$$y = -1$$

Substituting $y = -1$ into the first equation in our original system, we have

$$2x + 3(-1) = -1$$

$$2x - 3 = -1$$

$$2x = 2$$

$$x = 1$$

The solution to our system is $(1, -1)$. It is the only ordered pair that satisfies both equations.

EXAMPLE 5 Solve the system.

$$3x + 5y = -7$$
$$5x + 4y = 10$$

Solution Let's eliminate y by multiplying the first equation by -4 and the second equation by 5.

$$
\begin{array}{lll}
3x + 5y = -7 & \xrightarrow{\text{Multiply by } -4} & -12x - 20y = 28 \\
5x + 4y = 10 & \xrightarrow[\text{Multiply by } 5]{} & \underline{25x + 20y = 50} \\
& & 13x \qquad\quad = 78 \\
& & x \qquad\qquad = 6
\end{array}
$$

Substitute $x = 6$ into either equation in our original system, and the result will be $y = -5$. The solution is therefore $(6, -5)$.

EXAMPLE 6 Solve the system.

$$\frac{1}{2}x - \frac{1}{3}y = 2$$
$$\frac{1}{4}x + \frac{2}{3}y = 6$$

Solution Although we could solve this system without clearing the equations of fractions, there is probably less chance for error if we have only integer coefficients to work with. So let's begin by multiplying both sides of the top equation by 6 and both sides of the bottom equation by 12, to clear each equation of fractions.

$$
\begin{array}{lll}
\frac{1}{2}x - \frac{1}{3}y = 2 & \xrightarrow{\text{Multiply by } 6} & 3x - 2y = 12 \\[2mm]
\frac{1}{4}x + \frac{2}{3}y = 6 & \xrightarrow[\text{Multiply by } 12]{} & 3x + 8y = 72
\end{array}
$$

Now we can eliminate x by multiplying the top equation by -1 and leaving the bottom equation unchanged.

$$
\begin{array}{lll}
3x - 2y = 12 & \xrightarrow{\text{Multiply by } -1} & -3x + 2y = -12 \\
3x + 8y = 72 & \xrightarrow[\text{No change}]{} & \underline{3x + 8y = \quad 72} \\
& & 10y = \quad 60 \\
& & y = \quad 6
\end{array}
$$

We can substitute $y = 6$ into any equation that contains both x and y. Let's use $3x - 2y = 12$.

$$3x - 2(6) = 12$$
$$3x - 12 = 12$$

$$3x = 24$$
$$x = 8$$

The solution to the system is $(8, 6)$.

Our next two examples will show what happens when we apply the addition method to a system of equations consisting of parallel lines and to a system in which the lines coincide.

E X A M P L E   7   Solve the system.

$$2x - y = 2$$
$$4x - 2y = 12$$

Solution   Let us choose to eliminate y from the system. We can do this by multiplying the first equation by -2 and leaving the second equation unchanged.

$$2x - y = 2 \xrightarrow{\text{Multiply by } -2} -4x + 2y = -4$$
$$4x - 2y = 12 \xrightarrow{\text{No change}} 4x - 2y = 12$$

If we add both sides of the resulting system, we have

$$-4x + 2y = -4$$
$$\underline{4x - 2y = 12}$$
$$0 + 0 = 8$$

or $\qquad\qquad 0 = 8 \qquad$ A false statement

Both variables have been eliminated, and we end up with the false statement $0 = 8$. We have tried to solve a system that consists of two parallel lines. There is no solution, and that is the reason we end up with a false statement. Figure 4 is a visual representation of the situation and is conclusive evidence that there is no solution to our system.

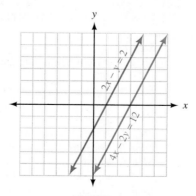

FIGURE 4

EXAMPLE 8   Solve the system.

$$4x - 3y = 2$$

$$8x - 6y = 4$$

Solution   Multiplying the top equation by -2 and adding, we can eliminate the variable x.

$$4x - 3y = 2 \xrightarrow{\text{Multiply by } -2} -8x + 6y = -4$$

$$8x - 6y = 4 \xrightarrow{\text{No change}} \underline{8x - 6y = 4}$$

$$0 = 0$$

Both variables have been eliminated, and the resulting statement $0 = 0$ is true. In this case the lines coincide because the equations are equivalent. The solution set consists of all ordered pairs that satisfy either equation.

The preceding two examples illustrate the two special cases in which the graphs of the equations in the system either coincide or are parallel.

Here is a summary of our results from these two examples:

Both variables are eliminated, and the resulting statement is false.	↔	The lines are parallel, and there is no solution to the system.
Both variables are eliminated, and the resulting statement is true.	↔	The lines coincide, and there are an infinite number of solutions to the system.

The main idea in solving a system of linear equations by the addition method is to use the multiplication property of equality on one or both of the original equations, if necessary, to make the coefficients of either variable opposites. The following box shows some steps to follow when solving a system of linear equations by the addition method.

STRATEGY FOR SOLVING A SYSTEM OF LINEAR EQUATIONS BY THE ADDITION METHOD

Step 1: Decide which variable to eliminate. (In some cases one variable will be easier to eliminate than the other. With some practice you will notice which one it is.)

Step 2: Use the multiplication property of equality on each equation separately to make the coefficients of the variable that is to be eliminated opposites.

Step 3: Add the respective left and right sides of the system together.

Step 4: Solve for the remaining variable.

Step 5: Substitute the value of the variable from step 4 into an equation containing both variables and solve for the other variable.

Step 6: Check your solution in both equations, if necessary.

Getting Ready for Class

After reading through the preceding section, respond in your own words and in complete sentences.

A. How is the addition property of equality used in the addition method of solving a system of linear equations?

B. What happens when we use the addition method to solve a system of linear equations consisting of two parallel lines?

C. What does it mean when we solve a system of linear equations by the addition method and we end up with the statement $0 = 8$?

D. What is the first step in solving a system of linear equations that contains fractions?

PROBLEM SET 3.6

Solve the following systems of linear equations by the addition method.

1. $x + y = 3$
$\quad x - y = 1$

2. $x + y = -2$
$\quad x - y = 6$

3. $\quad x + y = 10$
$\quad -x + y = 4$

4. $\quad x - y = 1$
$\quad -x - y = -7$

5. $\quad x - y = 7$
$\quad -x - y = 3$

6. $\quad x - y = 4$
$\quad 2x + y = 8$

7. $x + y = -1$
$\quad 3x - y = -3$

8. $\quad 2x - y = -2$
$\quad -2x - y = 2$

9. $\quad 3x + 2y = 1$
$\quad -3x - 2y = -1$

10. $-2x - 4y = 1$
$\quad 2x + 4y = -1$
$\qquad$ *une coincide*

Solve each of the following systems by eliminating the y variable.

11. $3x - y = 4$
$\quad 2x + 2y = 24$

12. $2x + y = 3$
$\quad 3x + 2y = 1$

13. $5x - 3y = -2$
$\quad 10x - y = 1$

14. $4x - y = -1$
$\quad 2x + 4y = 13$

15. $11x - 4y = 11$
$\quad 5x + y = 5$

16. $\quad 3x - y = 7$
$\quad 10x - 5y = 25$

Solve each of the following systems by eliminating the x variable.

17. $3x - 5y = 7$
$\quad -x + y = -1$

18. $4x + 2y = 32$
$\quad x + y = -2$

19. $-x - 8y = -1$
$\quad -2x + 4y = 13$

20. $-x + 10y = 1$
$\quad -5x + 15y = -9$

21. $-3x - y = 7$
$\quad 6x + 7y = 11$

22. $-5x + 2y = -6$
$\quad 10x + 7y = 34$

Solve each of the following systems of linear equations.

23. $6x - y = -8$
$\quad 2x + y = -16$

24. $5x - 3y = -3$
$\quad 3x + 3y = -21$

25. $\quad x + 3y = 9$
$\quad 2x - y = 4$

26. $\quad x + 2y = 0$
$\quad 2x - y = 0$

27. $\quad x - 6y = 3$
$\quad 4x + 3y = 21$

28. $8x + y = -1$
$\quad 4x - 5y = 16$

29. $2x + 9y = 2$
$\quad 5x + 3y = -8$

30. $5x + 2y = 11$
$\quad 7x + 8y = 7$

31. $\dfrac{1}{3}x + \dfrac{1}{4}y = \dfrac{7}{6}$
$\quad \dfrac{3}{2}x - \dfrac{1}{3}y = \dfrac{7}{3}$

32. $\dfrac{7}{12}x - \dfrac{1}{2}y = \dfrac{1}{6}$
$\quad \dfrac{2}{5}x - \dfrac{1}{3}y = \dfrac{11}{15}$

33. $3x + 2y = -1$
$\quad 6x + 4y = 0$

34. $8x - 2y = 2$
$\quad 4x - y = 2$

35. $11x + 6y = 17$
$\quad 5x - 4y = 1$

36. $\quad 3x - 8y = 7$
$\quad 10x - 5y = 45$

37. $\dfrac{1}{2}x + \dfrac{1}{6}y = \dfrac{1}{3}$
$\quad -x - \dfrac{1}{3}y = -\dfrac{1}{6}$

38. $-\dfrac{1}{3}x - \dfrac{1}{2}y = -\dfrac{2}{3}$
$\quad -\dfrac{2}{3}x - y = -\dfrac{4}{3}$

39. For some systems of equations it is necessary to apply the addition property of equality to each equation to line up the x-variables and y-variables before trying to eliminate a variable. Solve the following system by first writing each equation so that the variable terms with x in them come first, the variable terms with y in them come second, and the constant terms are on the right side of each equation.

$$4x - 5y = 17 - 2x$$

$$5y = 3x + 4$$

40. Solve the following system by first writing each equation so that the variable terms with x in them come first, the variable terms with y in them come second, and the constant terms are on the right side of each equation.

$$3x - 6y = -20 + 7x$$

$$4x = 3y - 34$$

41. Multiply both sides of the second equation in the following system by 100, and then solve as usual.

$$x + \quad y = 22$$

$$0.05x + 0.10y = 1.70$$

42. Multiply both sides of the second equation in the following system by 100, and then solve as usual.

$$x + \quad y = 15,000$$

$$0.06x + 0.07y = 980$$

Review Problems

The problems that follow review material we covered in Section 2.5.

Translate each of the following percent problems into an equation, and then solve the equation.

43. What number is 25% of 300?

44. 30 is what percent of 120?

45. 60 is 15% of what number?

46. 75 is 30% of what number?

3.7   The Substitution Method

There is a third method of solving systems of equations. It is the substitution method, and, like the addition method, it can be used on any system of linear equations. Some systems, however, lend themselves more to the substitution method than others do.

EXAMPLE 1   Solve the following system.

$$x + y = 2$$

$$y = 2x - 1$$

Solution   If we were to solve this system by the methods used in the previous section, we would have to rearrange the terms of the second equation so that similar terms would be in the same column. There is no need to do this, however, since the second equation tells us that y is $2x - 1$. We can replace the y variable in the first equation with the expression $2x - 1$ from the second equation. That is, we *substitute* $2x - 1$ from the second equation for y in the first equation. Here is what it looks like:

$$x + (2x - 1) = 2$$

The equation we end up with contains only the variable x. The y-variable has been eliminated by substitution.

Solving the resulting equation, we have

$$x + (2x - 1) = 2$$

$$3x - 1 = 2$$

$$3x = 3$$

$$x = 1$$

This is the x-coordinate of the solution to our system. To find the y-coordinate, we substitute $x = 1$ into the second equation of our system. (We could substitute $x = 1$ into the first equation also and have the same result.)

$$y = 2(1) - 1$$

$$y = 2 - 1$$

$$y = 1$$

The solution to our system is the ordered pair $(1, 1)$. It satisfies both of the original equations. Figure 1 provides visual evidence that the substitution method yields the correct solution.

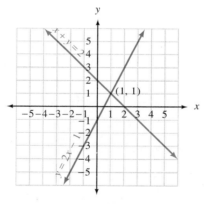

FIGURE 1

....

E X A M P L E   2   Solve the following system by the substitution method.

$$2x - 3y = 12$$

$$y = 2x - 8$$

Solution   Again, the second equation says y is $2x - 8$. Since we are looking for the ordered pair that satisfies both equations, the y in the first equation must also be $2x - 8$. Substituting $2x - 8$ from the second equation for y in the first equation, we have

$$2x - 3(2x - 8) = 12$$

This equation can still be read as $2x - 3y = 12$, because $2x - 8$ is the same as y. Solving the equation, we have

$$2x - 3(2x - 8) = 12$$
$$2x - 6x + 24 = 12$$
$$-4x + 24 = 12$$
$$-4x = -12$$
$$x = 3$$

To find the y-coordinate of our solution, we substitute $x = 3$ into the second equation in the original system.

$$\begin{aligned} \text{When} \qquad & x = 3 \\ \text{the equation} \qquad & y = 2x - 8 \\ \text{becomes} \qquad & y = 2(3) - 8 \\ & y = 6 - 8 = -2 \end{aligned}$$

The solution to our system is $(3, -2)$.

E X A M P L E   3   Solve the following system by solving the first equation for x and then using the substitution method:

$$x - 3y = -1$$
$$2x - 3y = 4$$

Solution   We solve the first equation for x by adding $3y$ to both sides to get

$$x = 3y - 1$$

Using this value of x in the second equation, we have

$$2(3y - 1) - 3y = 4$$
$$6y - 2 - 3y = 4$$
$$3y - 2 = 4$$
$$3y = 6$$
$$y = 2$$

Next, we find x.

$$\begin{aligned} \text{When} \qquad & y = 2 \\ \text{the equation} \qquad & x = 3y - 1 \\ \text{becomes} \qquad & x = 3(2) - 1 \\ & x = 6 - 1 \\ & x = 5 \end{aligned}$$

The solution to our system is $(5, 2)$.

Here are the steps to use in solving a system of equations by the substitution method.

STRATEGY FOR SOLVING A SYSTEM OF EQUATIONS BY THE SUBSTITUTION METHOD

Step 1: Solve either one of the equations for x or y. (This step is not necessary if one of the equations is already in the correct form, as in Examples 1 and 2.)

Step 2: Substitute the expression for the variable obtained in step 1 into the other equation and solve it.

Step 3: Substitute the solution from step 2 into any equation in the system that contains both variables and solve it.

Step 4: Check your results, if necessary.

EXAMPLE 4   Solve by substitution.

$$-2x + 4y = 14$$

$$-3x + \ y = 6$$

Solution   We can solve either equation for either variable. If we look at the system closely, it becomes apparent that solving the second equation for y is the easiest way to go. If we add $3x$ to both sides of the second equation, we have

$$y = 3x + 6$$

Substituting the expression $3x + 6$ back into the first equation in place of y yields the following result.

$$-2x + 4(3x + 6) = 14$$

$$-2x + 12x + 24 = 14$$

$$10x + 24 = 14$$

$$10x = -10$$

$$x = -1$$

Substituting $x = -1$ into the equation $y = 3x + 6$ leaves us with

$$y = 3(-1) + 6$$

$$y = -3 + 6$$

$$y = 3$$

The solution to our system is $(-1, 3)$.

EXAMPLE 5   Solve by substitution.

$$4x + 2y = 8$$
$$y = -2x + 4$$

Solution   Substituting the expression $-2x + 4$ for y from the second equation into the first equation, we have

$$4x + 2(-2x + 4) = 8$$
$$4x - 4x + 8 = 8$$
$$8 = 8 \qquad \text{A true statement}$$

Both variables have been eliminated, and we are left with a true statement. Recall from the last section that a true statement in this situation tells us the lines coincide. That is, the equations $4x + 2y = 8$ and $y = -2x + 4$ have exactly the same graph. Any point on that graph has coordinates that satisfy both equations and is a solution to the system.

EXAMPLE 6   The following table shows two contract rates charged by GTE Wireless for cellular phone use. At how many minutes will the two rates cost the same amount?

	Flat Rate	Plus	Per Minute Charge
Plan 1	$15		$1.50
Plan 2	$24.95		$0.75

Solution   If we let $y =$ the monthly charge for x minutes of phone use, then the equations for each plan are

$$\text{Plan 1:} \quad y = 1.5x + 15$$
$$\text{Plan 2:} \quad y = 0.75x + 24.95$$

We can solve this system by substitution by replacing the variable y in Plan 2 with the expression $1.5x + 15$ from Plan 1. If we do so, we have

$$1.5x + 15 = 0.75x + 24.95$$
$$0.75x + 15 = 24.95$$
$$0.75x = 9.95$$
$$x = 13.27 \qquad \text{to the nearest hundredth}$$

The monthly bill is based on the number of minutes you use the phone, with any fraction of a minute moving you up to the next minute. If you talk for a total of 13 minutes, you are billed for 13 minutes. If you talk for 13 minutes, 10 seconds, you are billed for 14 minutes. The number of minutes on your bill will always be a whole number. So, to calculate the cost for talking 13.27 minutes, we would replace x with 14 and find y. Let's compare the two plans at $x = 13$ minutes and at $x = 14$ minutes.

$$\text{Plan 1:} \qquad y = 1.5x + 15$$

$$\text{When } x = 13, y = \$34.50$$

$$\text{When } x = 14, y = \$36.00$$

$$\text{Plan 2:} \qquad y = 0.75x + 24.95$$

$$\text{When } x = 13, y = \$34.70$$

$$\text{When } x = 14, y = \$35.45$$

The two plans will never give the same cost for talking x minutes. If you talk 13 or fewer minutes, Plan 1 will cost less. If you talk for more than 13 minutes, you will be billed for 14 minutes, and Plan 2 will cost less than Plan 1.

Getting Ready for Class

After reading through the preceding section, respond in your own words and in complete sentences.

A. What is the first step in solving a system of linear equations by substitution?

B. When would substitution be more efficient than the addition method in solving a system of two linear equations?

C. What does it mean when we solve a system of linear equations by the substitution method and we end up with the statement $8 = 8$?

D. How would you begin solving the following system using the substitution method?

$$x + y = 2$$
$$y = 2x - 1$$

PROBLEM SET 3.7

Solve the following systems by substitution. Substitute the expression in the second equation into the first equation and solve.

1. $x + y = 11$
$\quad y = 2x - 1$

2. $x - y = -3$
$\quad y = 3x + 5$

3. $x + y = 20$
$\quad y = 5x + 2$

4. $3x - y = -1$
$\quad x = 2y - 7$

5. $-2x + y = -1$
$\quad y = -4x + 8$

6. $4x - y = 5$
$\quad y = -4x + 1$

7. $3x - 2y = -2$
$\quad x = -y + 6$

8. $2x - 3y = 17$
$\quad x = -y + 6$

9. $5x - 4y = -16$
$\quad y = 4$

10. $6x + 2y = 18$
$\quad x = 3$

11. $5x + 4y = 7$
$\quad y = -3x$

12. $10x + 2y = -6$
$\quad y = -5x$

Solve the following systems by solving one of the equations for x or y and then using the substitution method.

13. $x + 3y = 4$
  $x - 2y = -1$

14. $x - y = 5$
  $x + 2y = -1$

15. $2x + y = 1$
  $x - 5y = 17$

16. $2x - 2y = 2$
  $x - 3y = -7$

17. $3x + 5y = -3$
  $x - 5y = -5$

18. $2x - 4y = -4$
  $x + 2y = 8$

19. $5x + 3y = 0$
  $x - 3y = -18$

20. $x - 3y = -5$
  $x - 2y = 0$

21. $-3x - 9y = 7$
  $x + 3y = 12$

22. $2x + 6y = -18$
  $x + 3y = -9$

Solve the following systems using the substitution method.

23. $5x - 8y = 7$
  $y = 2x - 5$

24. $3x + 4y = 10$
  $y = 8x - 15$

25. $7x - 6y = -1$
  $x = 2y - 1$

26. $4x + 2y = 3$
  $x = 4y - 3$

27. $-3x + 2y = 6$
  $y = 3x$

28. $-2x - y = -3$
  $y = -3x$

29. $5x - 6y = -4$
  $x = y$

30. $2x - 4y = 0$
  $y = x$

31. $3x + 3y = 9$
  $y = 2x - 12$

32. $7x + 6y = -9$
  $y = -2x + 1$

33. $7x - 11y = 16$
  $y = 10$

34. $9x - 7y = -14$
  $x = 7$

35. $-4x + 4y = -8$
  $y = x - 2$

36. $-4x + 2y = -10$
  $y = 2x - 5$

Solve each system by substitution. You can eliminate the decimals if you like, but you don't have to. The solution will be the same in either case.

37. $0.05x + 0.10y = 1.70$
  $y = 22 - x$

38. $0.20x + 0.50y = 3.60$
  $y = 12 - x$

Applying the Concepts

39. **Gas Mileage** Daniel is trying to decide whether to buy a car or a truck. The truck he is considering will cost him $150 a month in loan payments, and it gets 20 miles per gallon in gas mileage. The car will cost $180 a month in loan payments, but it gets 35 miles per gallon in gas mileage. Daniel estimates that he will pay $1.40 per gallon for gas. This means that the monthly cost to drive the truck x miles will be $y = \frac{1.40}{20}x + 150$. The total monthly cost to drive the car x miles will be $y = \frac{1.40}{35}x + 180$. The following figure shows the graph of each equation.

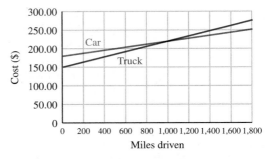

(a) At how many miles do the car and the truck cost the same to operate?
(b) If Daniel drives more than 1,200 miles, which will be cheaper?
(c) If Daniel drives fewer than 800 miles, which will be cheaper?
(d) Why do the graphs appear in the first quadrant only?

40. **Video Production** Pat runs a small company that duplicates videotapes. The daily cost and daily revenue for a company duplicating videos are shown in the following figure. The daily cost for duplicating x videos is $y = \frac{6}{5}x + 20$; the daily revenue (the amount of money he brings in each day) for duplicating x videos is $y = 1.7x$. The graphs of the two lines are shown in the following figure.

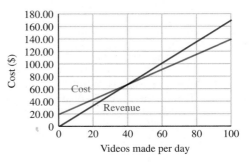

(a) Pat will "break even" when his cost and his revenue are equal. How many videos does he need to duplicate to break even?

(b) Pat will incur a loss when his revenue is less than his cost. If he duplicates 30 videos in one day, will he incur a loss?

(c) Pat will make a profit when his revenue is larger than his costs. For what values of x will Pat make a profit?

(d) Why does the graph appear in the first quadrant only?

41. **Cost of Bottled Water** The table shows the rates charged by two water-bottling companies for their monthly service, which includes rent on the water dispenser and the per gallon charge for water. If the water comes only in 5-gallon bottles, how many gallons must be used in a month for the two companies to charge the same amount?

	Flat Rate	Plus	Per Gallon Charge
Company 1	$7.00		$1.10
Company 2	$5.00		$1.15

42. **Cost of Renting a Car** The daily rates for renting a compact car from two rental companies are shown in the table. How many miles would a car have to be driven in a *week* for the two companies to charge the same amount?

	Daily Rate	Plus	Per Mile Charge
Company 1	$21.99		$0.20
Company 2	$19.99		$0.22

Review Problems

The problems that follow review material we covered in Sections 2.5, 2.6, and 2.7. Reviewing these problems will help you understand the next section.

43. **Geometry** A rectangle is 3 times as long as it is wide. If the perimeter is 24 meters, find the length and width.

44. **Geometry** The perimeter of a square is 48 inches. Find the length of its side.

45. **Coin Problem** A collection of coins consists of nickels and dimes and is worth $2.10. If there are 3 more dimes than nickels, how many of each coin are in the collection?

46. **Coin Problem** Mary has $5.75 in nickels and quarters. If she has 5 more quarters than nickels, how many of each coin does she have?

47. What number is 8% of 6,000?

48. 540 is 9% of what number?

49. **Investing** A man invests twice as much money at 10% annual interest as he does at 8% annual interest. If his total interest for the year is $224, how much does he have invested at each rate?

50. **Investing** A woman invests 3 times as much money at 8% annual interest as she does at 6% annual interest. If the total interest for the year is $66, how much does she have invested at each rate?

3.8 Applications

I have often heard students remark about the word problems in beginning algebra: "What does this have to do with real life?" Most of the word problems we will encounter don't have much to do with "real life." We are actually just practicing. Ultimately, all problems requiring the use of algebra are word problems. That is, they are stated in words first, then translated to symbols. The problem then is solved by some system of mathematics, like algebra. Most real applications involve calculus or higher levels of mathematics. So, if the problems we solve are upsetting or frustrating to you, then you are probably taking them too seriously.

The word problems in this section have two unknown quantities. We will write two equations in two variables (each of which represents one of the unknown quantities), which of course is a system of equations. We then solve the system by one of the methods developed in the previous sections of this chapter. Here are the steps to follow in solving these word problems.

> ### *Blueprint for Problem Solving*
> ### *Using a System of Equations*
>
> ***Step 1:*** ***Read*** the problem, and then mentally ***list*** the items that are known and the items that are unknown.
>
> ***Step 2:*** ***Assign variables*** to each of the unknown items. That is, let x = one of the unknown items and y = the other unknown item. Then ***translate*** the other ***information*** in the problem to expressions involving the two variables.
>
> ***Step 3:*** ***Reread*** the problem, and then ***write a system of equations,*** using the items and variables listed in steps 1 and 2, that describes the situation.
>
> ***Step 4:*** ***Solve the system*** found in step 3.
>
> ***Step 5:*** ***Write*** your ***answers*** using complete sentences.
>
> ***Step 6:*** ***Reread*** the problem, and ***check*** your solution with the original words in the problem.

Remember, the more problems you work, the more problems you will be able to work. If you have trouble getting started on the problem set, come back to the examples and work through them yourself. The examples are similar to the problems found in the problem set.

NUMBER PROBLEM

EXAMPLE 1   One number is 2 more than 5 times another number. Their sum is 20. Find the two numbers.

Solution   Applying the steps in our blueprint, we have the following:

Step 1: We know that the two numbers have a sum of 20 and that one of them is 2 more than 5 times the other. We don't know what the numbers themselves are.

Step 2: Let x represent one of the numbers and y represent the other. "One number is 2 more than 5 times another" translates to

$$y = 5x + 2$$

"Their sum is 20" translates to

$$x + y = 20$$

Step 3: The system that describes the situation must be

$$x + y = 20$$
$$y = 5x + 2$$

Step 4: We can solve this system by substituting the expression $5x + 2$ in the second equation for y in the first equation:

$$x + 5x + 2 = 20$$
$$6x + 2 = 20$$
$$6x = 18$$
$$x = 3$$

Using $x = 3$ in either of the first two equations and then solving for y, we get $y = 17$.

Step 5: So 17 and 3 are the numbers we are looking for.

Step 6: The number 17 is 2 more than 5 times 3, and the sum of 17 and 3 is 20.

INTEREST PROBLEM

E X A M P L E   2   Mr. Hicks had $15,000 to invest. He invested part at 6% and the rest at 7%. If he earned $980 in interest, how much did he invest at each rate?

Solution   Remember, step 1 is done mentally.

Step 1: We do not know the specific amounts invested in the two accounts. We do know that their sum is $15,000 and that the interest rates on the two accounts are 6% and 7%.

Step 2: Let $x =$ the amount invested at 6% and $y =$ the amount invested at 7%. Since Mr. Hicks invested a total of $15,000, we have

$$x + y = 15,000$$

The interest he earns comes from 6% of the amount invested at 6% and 7% of the amount invested at 7%. To find 6% of x, we multiply x by 0.06, which gives us $0.06x$. To find 7% of y, we multiply 0.07 times y and get $0.07y$.

$$\begin{matrix} \text{Interest} \\ \text{at } 6\% \end{matrix} + \begin{matrix} \text{interest} \\ \text{at } 7\% \end{matrix} = \begin{matrix} \text{total} \\ \text{interest} \end{matrix}$$

$$0.06x + 0.07y = 980$$

Step 3: The system is

$$x + y = 15,000$$
$$0.06x + 0.07y = 980$$

Step 4: We multiply the first equation by -6 and the second by 100 to eliminate x:

$$x + y = 15,000 \xrightarrow{\text{Multiply by } -6} -6x - 6y = -90,000$$
$$0.06x + 0.07y = 980 \xrightarrow[\text{Multiply by } 100]{} \underline{6x + 7y = 98,000}$$
$$y = 8,000$$

Substituting $y = 8{,}000$ into the first equation and solving for x, we get $x = 7{,}000$.

Step 5: He invested \$7,000 at 6% and \$8,000 at 7%.

Step 6: Checking our solutions in the original problem, we have: The sum of \$7,000 and \$8,000 is \$15,000, the total amount he invested. To complete our check, we find the total interest earned from the two accounts:

$$\text{The interest on \$7,000 at 6\% is } 0.06(7{,}000) = 420$$

$$\underline{\text{The interest on \$8,000 at 7\% is } 0.07(8{,}000) = 560}$$

$$\text{The total interest is} \hspace{5cm} \$980$$

COIN PROBLEM

E X A M P L E 3 John has \$1.70, all in dimes and nickels. He has a total of 22 coins. How many of each kind does he have?

Solution

Step 1: We know that John has 22 coins that are dimes and nickels. We know that a dime is worth 10 cents and a nickel is worth 5 cents. We do not know the specific number of dimes and nickels he has.

Step 2: Let $x =$ the number of nickels and $y =$ the number of dimes. The total number of coins is 22, so

$$x + y = 22$$

The total amount of money he has is \$1.70, which comes from nickels and dimes:

$$\begin{array}{ccccc} \text{Amount of money} & + & \text{amount of money} & = & \text{total amount} \\ \text{in nickels} & & \text{in dimes} & & \text{of money} \\ 0.05x & + & 0.10y & = & 1.70 \end{array}$$

Step 3: The system that represents the situation is

$$x + y = 22 \qquad \text{The number of coins}$$
$$0.05x + 0.10y = 1.70 \qquad \text{The value of the coins}$$

Step 4: We multiply the first equation by -5 and the second by 100 to eliminate the variable x:

$$\begin{array}{rcll} x + y = 22 & \xrightarrow{\text{Multiply by } -5} & -5x - 5y = -110 \\ 0.05x + 0.10y = 1.70 & \xrightarrow[\text{Multiply by 100}]{} & 5x + 10y = 170 \\ \hline & & 5y = 60 \\ & & y = 12 \end{array}$$

Substituting $y = 12$ into our first equation, we get $x = 10$.

Step 5: John has 12 dimes and 10 nickels.
Step 6: Twelve dimes and 10 nickels total 22 coins.

$$12 \text{ dimes are worth } 12(0.10) = 1.20$$

$$\underline{10 \text{ nickels are worth } 10(0.05) = 0.50}$$

The total value is $1.70

MIXTURE PROBLEM

EXAMPLE   4   How much 20% alcohol solution and 50% alcohol solution must be mixed to get 12 gallons of 30% alcohol solution?

Solution   To solve this problem we must first understand that a 20% alcohol solution is 20% alcohol and 80% water.

Step 1:   We know there are two solutions that together must total 12 gallons. We know 20% of one of the solutions is alcohol and the rest is water, while the other solution is 50% alcohol and 50% water. We do not know how many gallons of each individual solution we need.

Step 2:   Let $x =$ the number of gallons of 20% alcohol solution needed and $y =$ the number of gallons of 50% alcohol solution needed. Since the total number of gallons we will end up with is 12, and this 12 gallons must come from the two solutions we are mixing, our first equation is

$$x + y = 12$$

To obtain our second equation, we look at the amount of alcohol in our two original solutions and our final solution. The amount of alcohol in the x gallons of 20% solution is $0.20x$, while the amount of alcohol in y gallons of 50% solution is $0.50y$. The amount of alcohol in the 12 gallons of 30% solution is $0.30(12)$. Since the amount of alcohol we start with must equal the amount of alcohol we end up with, our second equation is

$$0.20x + 0.50y = 0.30(12)$$

The information we have so far can also be summarized with a table. Sometimes it is easier to see where the equations come from by looking at a table like the one that follows.

	20% Solution	50% Solution	Final Solution
Number of gallons	x	y	12
Gallons of alcohol	$0.20x$	$0.50y$	$0.30(12)$

Step 3: Our system of equations is

$$x + \quad y = 12$$

$$0.20x + 0.50y = 0.30(12)$$

Step 4: We can solve this system by substitution. Solving the first equation for y and substituting the result into the second equation, we have

$$0.20x + 0.50(12 - x) = 0.30(12)$$

Multiplying each side by 10 gives us an equivalent equation that is a little easier to work with.

$$2x + 5(12 - x) = 3(12)$$

$$2x + 60 - 5x = 36$$

$$-3x + 60 = 36$$

$$-3x = -24$$

$$x = 8$$

If x is 8, then y must be 4, because $x + y = 12$.

Step 5: It takes 8 gallons of 20% alcohol solution and 4 gallons of 50% alcohol solution to produce 12 gallons of 30% alcohol solution.

Step 6: Try it and see.

 # Getting Ready for Class

After reading through the preceding section, respond in your own words and in complete sentences.

A. What is the first step in the Blueprint for Problem Solving Using a System of Equations?

B. What is the last step in the Blueprint for Problem Solving Using a System of Equations?

C. How does step 3 in the Blueprint for Problem Solving Using a System of Equations differ from step 3 in the Blueprint for Problem Solving from Section 2.6?

D. Write an application problem for which the solution depends on solving a system of equations.

PROBLEM SET 3.8

Solve the following word problems. Be sure to show the equations used.

Number Problems

1. Two numbers have a sum of 25. One number is 5 more than the other. Find the numbers.
2. The difference of two numbers is 6. Their sum is 30. Find the two numbers.
3. The sum of two numbers is 15. One number is 4 times the other. Find the numbers.
4. The difference of two positive numbers is 28. One number is 3 times the other. Find the two numbers.
5. Two positive numbers have a difference of 5. The larger number is 1 more than twice the smaller. Find the two numbers.
6. One number is 2 more than 3 times another. Their sum is 26. Find the two numbers.
7. One number is 5 more than 4 times another. Their sum is 35. Find the two numbers.
8. The difference of two positive numbers is 8. The larger is twice the smaller decreased by 7. Find the two numbers.

Interest Problems

9. Mr. Wilson invested money in two accounts. His total investment was $20,000. If one account pays 6% in interest and the other pays 8% in interest, how much did he have in each account if he earned a total of $1,380 in interest in 1 year?
10. A total of $11,000 was invested. Part of the $11,000 was invested at 4%, and the rest was invested at 7%. If the investments earn $680 per year, how much was invested at each rate?
11. A woman invested 4 times as much at 5% as she did at 6%. The total amount of interest she earns in 1 year from both accounts is $520. How much did she invest at each rate?
12. Ms. Hagan invested twice as much money in an account that pays 7% interest as she did in an account that pays 6% in interest. Her total investment pays her $1,000 a year in interest. How much did she invest at each rate?

Coin Problems

13. Ron has 14 coins with a total value of $2.30. The coins are nickels and quarters. How many of each coin does he have?
14. Diane has $0.95 in dimes and nickels. She has a total of 11 coins. How many of each kind does she have?
15. Suppose Tom has 21 coins totaling $3.45. If he has only dimes and quarters, how many of each type does he have?
16. A coin collector has 31 dimes and nickels with a total face value of $2.40. (They are actually worth a lot more.) How many of each coin does she have?

Mixture Problems

17. How many liters of 50% alcohol solution and 20% alcohol solution must be mixed to obtain 18 liters of 30% alcohol solution?

	50% Solution	20% Solution	Final Solution
Number of liters	x	y	
Liters of alcohol			

18. How many liters of 10% alcohol solution and 5% alcohol solution must be mixed to obtain 40 liters of 8% alcohol solution?

	10% Solution	5% Solution	Final Solution
Number of liters			
Liters of alcohol			

19. A mixture of 8% disinfectant solution is to be made from 10% and 7% disinfectant solutions. How much of each solution should be used if 30 gallons of 8% solution are needed?

20. How much 50% antifreeze solution and 40% antifreeze solution should be combined to give 50 gallons of 46% antifreeze solution?

Miscellaneous Problems

21. For a Saturday matinee, adult tickets cost $5.50, while children under 12 pay only $4.00. If 70 tickets are sold for a total of $310, how many of the tickets were adult tickets and how many were sold to children under 12?

22. The Bishop's Peak 4-H club is having its annual fund-raising dinner. Adults pay $15 apiece, and children pay $10 apiece. If the number of adult tickets sold is twice the number of children's tickets sold and the total income for the dinner is $1,600, how many of each kind of ticket did the 4-H club sell?

23. A farmer has 96 feet of fence with which to make a corral. If he arranges it into a rectangle that is twice as long as it is wide, what are the dimensions?

24. If a 22-inch rope is to be cut into two pieces so that one piece is 3 inches longer than twice the other, how long is each piece?

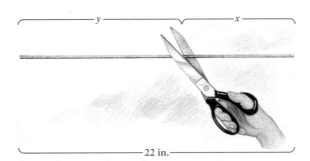

25. A gambler finishes a session of blackjack with $5 chips and $25 chips. If he has 45 chips in all, with a total value of $465, how many of each kind of chip does the gambler have?

26. Tyler has been saving his winning lottery tickets. He has 23 tickets that are worth a total of $175. If each ticket is worth either $5 or $10, how many of each does he have?

27. Mary Jo spends $2,550 to buy stock in two companies. She pays $11 a share to one of the companies and $20 a share to the other. If she ends up with a total of 150 shares, how many shares did she buy at $11 a share, and how many did she buy at $20 a share?

28. Kelly sells 62 shares of stock she owns for a total of $433. If the stock was in two different companies, one selling at $6.50 a share and the other at $7.25 a share, how many of each did she sell?

Review Problems

The following problems review material we covered in Chapter 2.

29. Simplify the expression $7 - 3(2x - 4) - 8$.

30. Find the value of $x^2 - 2xy + y^2$ when $x = 3$ and $y = -4$.

Solve each equation.

31. $-\dfrac{3}{2}x = 12$ **32.** $2x - 4 = 5x + 2$

33. $8 - 2(x + 7) = 2$

34. $3(2x - 5) - (2x - 4) = 6 - (4x + 5)$

35. Solve the formula $P = 2l + 2w$ for w.

Solve each inequality, and graph the solution.

36. $-4x < 20$ **37.** $3 - 2x > 5$

38. $3 - 4(x - 2) \geq -5x + 6$

39. Solve the formula $3x - 2y \leq 12$ for y.

40. What number is 12% of 2,000?

41. Geometry The length of a rectangle is 5 inches more than 3 times the width. If the perimeter is 26 inches, find the length and width.

CHAPTER 3 SUMMARY

Examples

1. The equation $3x + 2y = 6$ is an example of a linear equation in two variables.

Linear Equations in Two Variables [3.1, 3.2]

A linear equation in two variables is any equation that can be put in the form $ax + by = c$. The graph of every linear equation is a straight line.

2. The graph of $y = -\frac{2}{3}x - 1$ is shown below.

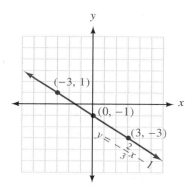

Strategy for Graphing a Straight Line [3.3]

Step 1: Find any three ordered pairs that satisfy the equation. This can be done by using a convenient number for one variable and solving for the other variable.

Step 2: Graph the three ordered pairs found in step 1. Actually, we need only two points to graph a straight line. The third point serves as a check. If all three points do not line up, there is a mistake in our work.

Step 3: Draw a straight line through the three points graphed in step 2.

3. To find the x-intercept for $3x + 2y = 6$, we let $y = 0$ and get

$$3x = 6$$

$$x = 2$$

In this case the x-intercept is 2, and the graph crosses the x-axis at $(2, 0)$.

Intercepts [3.4]

The x-intercept of an equation is the x-coordinate of the point where the graph crosses the x-axis. The y-intercept is the y-coordinate of the point where the graph crosses the y-axis. We find the y-intercept by substituting $x = 0$ into the equation and solving for y. The x-intercept is found by letting $y = 0$ and solving for x.

4. The solution to the system

$$x + 2y = 4$$
$$x - y = 1$$

is the ordered pair $(2, 1)$. It is the only ordered pair that satisfies both equations.

Definitions [3.5]

1. A *system of linear equations,* as the term is used in this book, is two linear equations that each contain the same two variables.

2. The *solution set* for a system of equations is the set of all ordered pairs that satisfy *both* equations. The solution set to a system of linear equations will contain the following:

Case I One ordered pair when the graphs of the two equations intersect at only one point (this is the most common situation)

Case II No ordered pairs when the graphs of the two equations are parallel lines

Case III An infinite number of ordered pairs when the graphs of the two equations coincide (are the same line)

5. Solving the system in Example 4 by graphing looks like

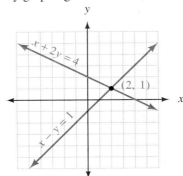

6. We can eliminate the y variable from the system in Example 4 by multiplying both sides of the second equation by 2 and adding the result to the first equation.

$$x + 2y = 4 \xrightarrow{\text{No change}} x + 2y = 4$$
$$x - y = 1 \xrightarrow[\text{Multiply by 2}]{} 2x - 2y = 2$$
$$3x \quad\quad = 6$$
$$x = 2$$

Substituting $x = 2$ into either of the original two equations gives $y = 1$. The solution is $(2, 1)$.

7. We can apply the substitution method to the system in Example 4 by first solving the second equation for x to get $x = y + 1$.

Substituting this expression for x into the first equation, we have

$$y + 1 + 2y = 4$$
$$3y + 1 = 4$$
$$3y = 3$$
$$y = 1$$

Using $y = 1$ in either of the original equations gives $x = 2$.

Strategy for Solving a System by Graphing [3.5]

Step 1: Graph the first equation.

Step 2: Graph the second equation on the same set of axes.

Step 3: The solution to the system consists of the coordinates of the point where the graphs cross each other (the coordinates of the point of intersection).

Step 4: Check the solution to see that it satisfies *both* equations, if necessary.

Strategy for Solving a System by the Addition Method [3.6]

Step 1: Look the system over to decide which variable will be easier to eliminate.

Step 2: Use the multiplication property of equality on each equation separately to ensure that the coefficients of the variable to be eliminated are opposites.

Step 3: Add the left and right sides of the system produced in step 2, and solve the resulting equation.

Step 4: Substitute the solution from step 3 back into any equation with both x and y variables, and solve.

Step 5: Check your solution in both equations, if necessary.

Strategy for Solving a System by the Substitution Method [3.7]

Step 1: Solve either of the equations for one of the variables (this step is not necessary if one of the equations has the correct form already).

Step 2: Substitute the results of step 1 into the other equation, and solve.

Step 3: Substitute the results of step 2 into an equation with both x and y variables, and solve. (The equation produced in step 1 is usually a good one to use.)

Step 4: Check your solution, if necessary.

Special Cases [3.5, 3.6, 3.7]

In some cases, using the addition or substitution method eliminates both variables. The situation is interpreted as follows.

1. If the resulting statement is *false,* then the lines are parallel, and there is no solution to the system.

2. If the resulting statement is *true,* then the equations represent the same line (the lines coincide). In this case any ordered pair that satisfies either equation is a solution to the system.

COMMON MISTAKES

The most common mistake encountered in solving linear systems is the failure to complete the problem. Here is an example.

$$x + y = 8$$
$$\underline{x - y = 4}$$
$$2x \quad\;\; = 12$$

$$x = 6$$

This is only half the solution. To find the other half, we must substitute the 6 back into one of the original equations and then solve for y.

Remember, solutions to systems of linear equations always consist of ordered pairs. We need an x-coordinate and a y-coordinate; $x = 6$ can never be a solution to a system of linear equations.

CHAPTER 3   REVIEW

The numbers in brackets refer to the sections of the text in which similar problems can be found.

For each equation, complete the given ordered pairs. [3.2]

1. $3x + y = 6$ $(4, \quad), (0, \quad), (\quad, 3), (\quad, 0)$

2. $2x - 5y = 20$ $(5, \quad), (0, \quad), (\quad, 2), (\quad, 0)$

3. $y = 2x - 6$ $(4, \quad), (\quad, -2), (\quad, 3)$

4. $y = 5x + 3$ $(2, \quad), (\quad, 0), (\quad, -3)$

5. $y = -3$ $(2, \quad), (-1, \quad), (-3, \quad)$

6. $x = 6$ $(\quad, 5), (\quad, 0), (\quad, -1)$

For the following equations, tell which of the given ordered pairs are solutions. [3.2]

7. $3x - 4y = 12$ $\left(-2, \dfrac{9}{2}\right), (0, 3), \left(2, -\dfrac{3}{2}\right)$

8. $y = 3x + 7$ $\left(-\dfrac{8}{3}, -1\right), \left(\dfrac{7}{3}, 0\right), (-3, -2)$

Graph the following ordered pairs. [3.1]

9. $(4, 2)$ **10.** $(-3, 1)$ **11.** $(0, 5)$

12. $(-2, -3)$ **13.** $(-3, 0)$ **14.** $\left(5, -\dfrac{3}{2}\right)$

For the following equations, complete the given ordered pairs, and use the results to graph the solution set for the equations. [3.3]

15. $x + y = -2$ **16.** $y = 3x$
$(\quad, 0), (0, \quad), (1, \quad)$ $(-1, \quad), (1, \quad), (\quad, 0)$

17. $y = 2x - 1$
$(1, \quad), (0, \quad), (\quad, -3)$

Graph the following equations. [3.3]

18. $3x - y = 3$ **19.** $y = -\dfrac{1}{3}x$ **20.** $y = 2x + 1$

21. $x = 5$ **22.** $y = -3$

Find the x- and y-intercepts for each equation. [3.4]

23. $3x - y = 6$ **24.** $2x - 6y = 24$

25. $y = x - 3$ **26.** $y = 3x - 6$

Solve the following systems by graphing. [3.5]

27. $x + y = 2$ **28.** $\quad x + y = -1$
$\;\;\; x - y = 6$ $\quad -x + y = 5$

29. $\quad 2x - 3y = 12$ **30.** $4x - 2y = 8$
$\quad -2x + y = -8$ $3x + y = 6$

31. $y = 2x - 3$
$y = -2x + 5$

32. $y = -x - 3$
$y = 3x + 1$

Solve the following systems by the addition method. [3.6]

33. $x - y = 4$
$x + y = -2$

34. $-x - y = -3$
$2x + y = 1$

35. $5x - 3y = 2$
$-10x + 6y = -4$

36. $2x + 3y = -2$
$3x - 2y = 10$

37. $-3x + 4y = 1$
$-4x + y = -3$

38. $-4x - 2y = 3$
$2x + y = 1$

39. $-2x + 5y = -11$
$7x - 3y = -5$

40. $-2x + 5y = -15$
$3x - 4y = 19$

Solve the following systems by substitution. [3.7]

41. $x + y = 5$
$y = -3x + 1$

42. $x - y = -2$
$y = -2x - 10$

43. $4x - 3y = -16$
$y = 3x + 7$

44. $5x + 2y = -2$
$y = -8x + 10$

45. $x - 4y = 2$
$-3x + 12y = -8$

46. $4x - 2y = 8$
$3x + y = -19$

47. $10x - 5y = 20$
$x + 6y = -11$

48. $3x - y = 2$
$-6x + 2y = -4$

Solve the following word problems. Be sure to show the equations used. [3.8]

49. Number Problem The sum of two numbers is 18. If twice the smaller number is 6 more than the larger, find the two numbers.

50. Number Problem The difference of two positive numbers is 16. One number is 3 times the other. Find the two numbers.

51. Investing A total of $12,000 was invested. Part of the $12,000 was invested at 4%, and the rest was invested at 5%. If the interest for one year is $560, how much was invested at each rate?

52. Investing A total of $14,000 was invested. Part of the $14,000 was invested at 6%, and the rest was invested at 8%. If the interest for one year is $1,060, how much was invested at each rate?

53. Coin Problem Barbara has $1.35 in dimes and nickels. She has a total of 17 coins. How many of each does she have?

54. Coin Problem Tom has $2.40 in dimes and quarters. He has a total of 15 coins. How many of each does he have?

55. Mixture Problem How many liters of 20% alcohol solution and 10% alcohol solution must be mixed to obtain 50 liters of a 12% alcohol solution?

56. Mixture Problem How many liters of 25% alcohol solution and 15% alcohol solution must be mixed to obtain 40 liters of a 20% alcohol solution?

CHAPTER 3   PROJECTS

LINEAR EQUATIONS AND INEQUALITIES IN TWO VARIABLES

GROUP PROJECT

READING GRAPHS

Number of People: 2–3

Time Needed: 5–10 minutes

Equipment: Pencil and Paper

Background: Although most of the graphs we have encountered in this chapter have been straight lines, many of the graphs that describe the world around us are not straight lines. In this group project we gain experience working with graphs that are not straight lines.

Procedure: Read the introduction to each problem below. Then use the graphs to answer the questions.

1. A patient is taking a prescribed dose of a medication every 4 hours during the day to relieve the symptoms of a cold. Figure 1 shows how the concentration of that medication in the patient's system changes over time. The 0 on the horizontal axis corresponds to the time the patient takes the first dose of the medication. (The units of concentration on the vertical axis are nanograms per milliliter.)

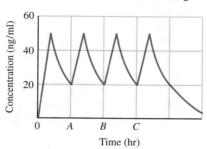

FIGURE I

(a) Explain what the steep vertical line segments show with regard to the patient and his medication.
(b) What has happened to make the graph fall off on the right?
(c) What is the maximum concentration of the medication in the patient's system during the time period shown in Figure 1?
(d) Find the values of A, B, and C.

2. **Reading Graphs** Figure 2 shows the number of people in line at a theater box office to buy tickets for a movie that starts at 7:30. The box office opens at 6:45.

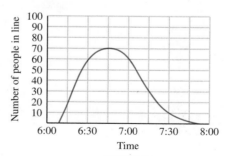

FIGURE 2

(a) How many people are in line at 6:30?
(b) How many people are in line when the box office opens?
(c) How many people are in line when the show starts?
(d) At what times are there 60 people in line?
(e) How long after the show starts is there no one left in line?

RESEARCH PROJECT

ZENO'S PARADOXES

Zeno of Elea was born at about the same time that Pythagoras died. He is responsible for three paradoxes that have come to be known as Zeno's paradoxes. One of the three has to do with a race between Achilles and a tortoise. Achilles is much faster than the tortoise, but the tortoise has a head start. According to Zeno's method of reasoning, Achilles can never pass the tortoise because each time he reaches the place where the tortoise was, the tortoise is gone. Research Zeno's paradox concerning Achilles and the tortoise. Put your findings into essay form that begins with a definition for the word "paradox." Then use Zeno's method of reasoning to describe a race between Achilles and the tortoise, if Achilles runs at 10 miles per hour, the tortoise at 1 mile per hour, and the tortoise has a 1-mile head start. Next, use the methods shown in this chapter to find the distance at which Achilles reaches the tortoise and the time at which Achilles reaches the tortoise. Conclude your essay by summarizing what you have done and showing how the two results you have obtained form a paradox.

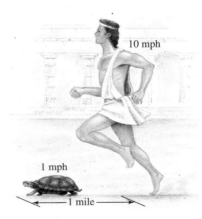

10 mph

1 mph

←——— 1 mile ———→

CHAPTER 3 TEST

1. Fill in the following ordered pairs for the equation $2x - 5y = 10$. [3.2]

   (0,)   (, 0)   (10,)   (, −3)

2. Which of the following ordered pairs are solutions to $y = 4x - 3$? [3.2]

   (2, 5)   (0, −3)   (3, 0)   (−2, 11)

Graph each line. [3.3]

3. $y = 3x - 2$

4. $x = -2$

Find the *x*- and *y*-intercepts. [3.4]

5. $3x - 5y = 15$

6. $y = \dfrac{3}{2}x + 1$

$y = mx + b$
$b = y$ int

Solve each system by graphing. [3.5]

7. $x + 2y = 5$
  $y = 2x$

8. $x - y = 5$
  $x = 3$

Solve each system by the addition method. [3.6]

9. $x - y = 1$
  $2x + y = -10$

10. $2x + y = 7$
  $3x + y = 12$

11. $7x + 8y = -2$
  $3x - 2y = 10$

12. $6x - 10y = 6$
  $9x - 15y = 9$

Solve each system by the substitution method. [3.7]

13. $3x + 2y = 20$
  $y = 2x + 3$

14. $3x - 6y = -6$
  $x = y + 1$

15. $7x - 2y = -4$
  $-3x + y = 3$

16. $2x - 3y = -7$
  $x + 3y = -8$

Solve the following word problems. In each case, be sure to show the system of equations that describes the situation. [3.8]

17. Number Problem The sum of two numbers is 12. Their difference is 2. Find the numbers.

18. Number Problem The sum of two numbers is 15. One number is 6 more than twice the other. Find the two numbers.

19. Investing Dr. Stork has $10,000 to invest. He would like to earn $980 per year in interest. How much should he invest at 9% if the rest is to be invested at 11%?

20. Coin Problem Diane has 12 coins that total $1.60. If the coins are all nickels and quarters, how many of each type does she have?

CHAPTERS 1–3 CUMULATIVE REVIEW

Simplify.

1. $3 \cdot 4 + 5$

2. $4 \cdot 3^2 + 4(6 - 3)$

3. $7[8 + (-5)] + 3(-7 + 12)$

4. $2(x - 5) + 8$

5. $8 - 6(5 - 9)$

6. $\dfrac{5(4 - 12) - 8(14 - 3)}{3 - 5 - 6}$

7. $\dfrac{2}{3} + \dfrac{3}{4} - \dfrac{1}{6}$

8. $7 - 5(2a - 3) + 7$

Solve each equation.

9. $-5 - 6 = -y - 3 + 2y$

10. $-2x = 0$

11. $3(x - 4) = 9$

12. $8 - 2(y + 4) = 12$

Solve each inequality, and graph the solution.

13. $0.3x + 0.7 \leq -2$

14. $5x + 10 \leq 7x - 14$

Graph on a rectangular coordinate system.

15. $y = -2x + 1$

16. $y = x$

17. $y = -\dfrac{2}{3}x$

18. If $A = \{1, 2, 3, 4\}$ and $B = \{-2, -1, 0, 1, 2\}$, find $A \cap B$.

19. Solve the following system by graphing.

$$2x + 3y = 3$$
$$4x + 6y = -4$$

20. Solve the following system by graphing.

$$3x - y = 3$$
$$2x + 3y = 2$$

Solve each system.

21. $x + y = 7$
  $2x + 2y = 14$

22. $2x + y = -1$
  $2x - 3y = 11$

23. $2x + 3y = 13$
  $x - y = -1$

24. $x + y = 13$
  $0.05x + 0.10y = 1$

25. $2x + 5y = 33$
  $x - 3y = 0$

26. $3x + 4y = 8$
  $x - y = 5$

27. $3x - 7y = 12$
  $2x + y = 8$

28. $5x + 6y = 9$
  $x - 2y = 5$

29. $2x - 3y = 7$
$y = 5x + 2$

30. $3x - 6y = 9$
$x = 2y + 3$

31. State the property or properties that justify the following: $a + 2 = 2 + a$

32. Find the next number in the geometric sequence: 2, 4, 8, 16, . . .

33. What is the quotient of -30 and 6?

34. Factor into primes: 180

35. What percent of 82 is 20.5?

36. What is 62% of 25?

Find the value of each expression when x is 3.

37. $-3x + 7 + 5x$

38. $-x - 4x - 2x$

39. Find the value of the expression $4x - 5$ when x is -2.

40. Given $4x - 5y = 12$, complete the ordered pair $(-2, \quad)$.

41. Write the expression, and then simplify: The difference of 5 and -8.

42. Complete the table: $y = \dfrac{3}{5}x - 2$

x	y
2	
	-1

43. Find the x- and y-intercepts: $3x - 4y = 12$

44. List all the integers from the set $\{-3, -0.2, 0, \frac{3}{4}, \sqrt{7}, 5\}$.

45. Use the formula $2x - 3y = 7$ to find y when $x = -1$.

46. Number Problem The sum of a number and 9 is 23. Find the number.

47. Geometry Problem The length of a rectangle is 5 centimeters more than twice the width. The perimeter is 44 centimeters. Find the width and length.

48. Cost of a Letter The cost of mailing a letter was 32¢ for the first ounce and 29¢ for each additional ounce. If the cost of mailing a letter was 90¢, how many ounces did the letter weigh?

49. Investing Barbara had money in two accounts. She had $900 more in an account paying 8% annual interest than she had in an account paying 6% annual interest. If she earned $240 in interest for the year, how much did she have invested in each account?

50. Coin Problem Joy has 15 coins with a value of $1.10. The coins are nickels and dimes. How many of each does she have?

Exponents and Polynomials

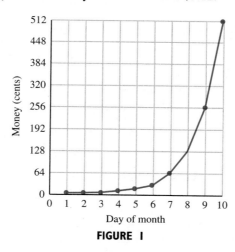

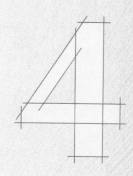

INTRODUCTION

If you were given a penny on the first day of September, and then each day after that you were given twice the amount of money you received the day before, how much money would you receive on September 30? To begin, Table 1 and Figure 1 show the amount of money you would receive on each of the first 10 days of the month. As you can see, on the tenth day of the month you would receive $5.12.

TABLE 1	Money That Doubles Each Day
Day	**Money (in cents)**
1	$1 = 2^0$
2	$2 = 2^1$
3	$4 = 2^2$
4	$8 = 2^3$
5	$16 = 2^4$
6	$32 = 2^5$
7	$64 = 2^6$
8	$128 = 2^7$
9	$256 = 2^8$
10	$512 = 2^9$

FIGURE 1

To find the amount of money on day 30, we could continue to double the amount on each of the next 20 days. Or, we could notice the pattern of exponents in the second column of the table and reason that the amount of money on day 30 would be 2^{29} cents, which is a very large number. In fact, 2^{29} cents is $5,368,709.12 — a little less than 5.4 million dollars. When you are finished with this chapter, you will have a good working knowledge of exponents.

STUDY SKILLS

The study skills for this chapter concern the way you approach new situations in mathematics. The first study skill applies to your natural instincts for what does and doesn't work in mathematics. The second study skill gives you a way of testing your instincts.

1. **Don't Let Your Intuition Fool You** As you become more experienced and more successful in mathematics, you will be able to trust your mathematical intuition. For now, though, it can get in the way of success. For example, if you ask a beginning algebra student to "subtract 3 from -5" many will answer -2 or 2. Both answers are incorrect, even though they may seem intuitively true.

2. **Test Properties About Which You Are Unsure** From time to time you will be in a situation in which you would like to apply a property or rule, but you are not sure it is true. You can always test a property or statement by substituting numbers for variables. For instance, I always have students who rewrite $(x + 3)^2$ as $x^2 + 9$, thinking that the two expressions are equivalent. The fact that the two expressions are not equivalent becomes obvious when we substitute 10 for x in each one.

   When $x = 10$, the expression $(x + 3)^2$ is

   $$(10 + 3)^2 = 13^2 = 169$$

   When $x = 10$, the expression $x^2 + 9$ is

   $$10^2 + 9 = 100 + 9 = 109$$

   Along the same lines, there may come a time when you are wondering if $\sqrt{x^2 + 25}$ is the same as $x + 5$. If you try $x = 10$ in each expression, you will find out quickly that the two expressions are not the same.

   When $x = 10$, the expression $\sqrt{x^2 + 25}$ is

   $$\sqrt{10^2 + 25} = \sqrt{125} \approx 11.2$$

   When $x = 10$, the expression $x + 5$ is

   $$10 + 5 = 15$$

   When you test the equivalence of expressions by substituting numbers for the variable, make it easy on yourself by choosing numbers that are easy to work with, such as 10. Don't try to verify the equivalence of expressions by substituting 0, 1, or 2 for the variable, as these numbers make expressions seem equivalent when in fact they are not.

It is not unusual, nor is it wrong, to try occasionally to apply a property that doesn't exist. If you have any doubt about generalizations you are making, test them by replacing variables with numbers and simplifying.

Multiplication With Exponents

Recall that an *exponent* is a number written just above and to the right of another number, which is called the *base*. In the expression 5^2, for example, the exponent is 2 and the base is 5. The expression 5^2 is read "5 to the second power" or "5 squared." The meaning of the expression is

$$5^2 = 5 \cdot 5 = 25$$

In the expression 5^3, the exponent is 3 and the base is 5. The expression 5^3 is read "5 to the third power" or "5 cubed." The meaning of the expression is

$$5^3 = 5 \cdot 5 \cdot 5 = 125$$

Here are some further examples.

EXAMPLES

1. $4^3 = 4 \cdot 4 \cdot 4 = 16 \cdot 4 = 64$ Exponent 3, base 4

2. $-3^4 = -3 \cdot 3 \cdot 3 \cdot 3 = -81$ Exponent 4, base 3

3. $(-2)^5 = (-2)(-2)(-2)(-2)(-2) = -32$ Exponent 5, base -2

4. $\left(-\dfrac{3}{4}\right)^2 = \left(-\dfrac{3}{4}\right)\left(-\dfrac{3}{4}\right) = \dfrac{9}{16}$ Exponent 2, base $-\dfrac{3}{4}$

QUESTION: In what way are $(-5)^2$ and -5^2 different?
ANSWER: In the first case, the base is -5. In the second case, the base is 5. The answer to the first is 25. The answer to the second is -25. Can you tell why? Would there be a difference in the answers if the exponent in each case were changed to 3?

We can simplify our work with exponents by developing some properties of exponents. We want to list the things we know are true about exponents and then use these properties to simplify expressions that contain exponents.

The first property of exponents applies to products with the same base. We can use the definition of exponents, as indicating repeated multiplication, to simplify expressions such as $7^4 \cdot 7^2$.

$$7^4 \cdot 7^2 = (7 \cdot 7 \cdot 7 \cdot 7)(7 \cdot 7)$$

$$= (7 \cdot 7 \cdot 7 \cdot 7 \cdot 7 \cdot 7)$$

$$= 7^6 \qquad \textit{Notice}: 4 + 2 = 6$$

As you can see, multiplication with the same base resulted in addition of exponents. We can summarize this result with the following property.

PROPERTY 1 FOR EXPONENTS

If a is any real number and r and s are integers, then

$$a^r \cdot a^s = a^{r+s}$$

In words: To multiply two expressions with the same base, add exponents and use the common base.

Here are some examples using Property 1.

EXAMPLES Use Property 1 to simplify the following expressions. Leave your answers in terms of exponents.

5. $5^3 \cdot 5^6 = 5^{3+6} = 5^9$

6. $x^7 \cdot x^8 = x^{7+8} = x^{15}$

7. $3^4 \cdot 3^8 \cdot 3^5 = 3^{4+8+5} = 3^{17}$

Note: In Examples 5, 6, and 7, notice that in each case the base in the original problem is the same as the base that appears in the answer and that the base is written only once in the answer. A very common mistake that people make when they first begin to use Property 1 is to write a 2 in front of the base in the answer. For example, people making this mistake would get $2x^{15}$ or $(2x)^{15}$ as the result in Example 6. To avoid this mistake, you must be sure you understand the meaning of Property 1 exactly as it is written.

Another common type of expression involving exponents is one in which an expression containing an exponent is raised to another power. The expression $(5^3)^2$ is an example:

$$(5^3)^2 = (5^3)(5^3)$$
$$= 5^{3+3}$$
$$= 5^6 \qquad \text{\textit{Notice}: } 3 \cdot 2 = 6.$$

This result offers justification for the second property of exponents.

PROPERTY 2 FOR EXPONENTS

If a is any real number and r and s are integers, then

$$(a^r)^s = a^{r \cdot s}$$

In words: A power raised to another power is the base raised to the product of the powers.

EXAMPLES Simplify the following expressions:

8. $(4^5)^6 = 4^{5 \cdot 6} = 4^{30}$

9. $(x^3)^5 = x^{3 \cdot 5} = x^{15}$

The third property of exponents applies to expressions in which the product of two or more numbers or variables is raised to a power. Let's look at how the expression $(2x)^3$ can be simplified:

$$(2x)^3 = (2x)(2x)(2x)$$
$$= (2 \cdot 2 \cdot 2)(x \cdot x \cdot x)$$
$$= 2^3 \cdot x^3 \qquad \textit{Notice: The exponent 3 distributes}$$
$$= 8x^3 \qquad \qquad \text{over the product } 2x.$$

We can generalize this result into a third property of exponents.

PROPERTY 3 FOR EXPONENTS

If a and b are any two real numbers and r is an integer, then

$$(ab)^r = a^r b^r$$

In words: The power of a product is the product of the powers.

Here are some examples using Property 3 to simplify expressions.

E X A M P L E S   Simplify the following expressions:

10. $(5x)^2 = 5^2 \cdot x^2$ Property 3
$\qquad\quad\;\; = 25x^2$

11. $(2xy)^3 = 2^3 \cdot x^3 \cdot y^3$ Property 3
$\qquad\quad\;\; = 8x^3 y^3$

12. $(3x^2)^3 = 3^3(x^2)^3$ Property 3
$\qquad\quad\;\; = 27x^6$ Property 2

13. $\left(-\dfrac{1}{4}x^2y^3\right)^2 = \left(-\dfrac{1}{4}\right)^2 (x^2)^2(y^3)^2$ Property 3

$$\qquad\qquad\qquad = \dfrac{1}{16}x^4y^6 \qquad\qquad\text{Property 2}$$

14. $(x^4)^3(x^2)^5 = x^{12} \cdot x^{10}$ Property 2
$\qquad\qquad\quad\;\; = x^{22}$ Property 1

15. $(2y)^3(3y^2) = 2^3y^3(3y^2)$ Property 3
$\qquad\qquad\quad\; = 8 \cdot 3(y^3 \cdot y^2)$ Commutative and associative
$\qquad\qquad\qquad\qquad\qquad\qquad\qquad\;$ properties

$\qquad\qquad\quad\; = 24y^5$ Property 1

16. $(2x^2y^5)^3(3x^4y)^2 = 2^3(x^2)^3(y^5)^3 \cdot 3^2(x^4)^2y^2$ Property 3
$\qquad\qquad\qquad\quad = 8x^6y^{15} \cdot 9x^8y^2$ Property 2
$\qquad\qquad\qquad\quad = (8 \cdot 9)(x^6x^8)(y^{15}y^2)$ Commutative and associative
$\qquad\qquad\qquad\qquad\qquad\qquad\qquad\qquad\qquad\;$ properties

$\qquad\qquad\qquad\quad = 72x^{14}y^{17}$ Property 1

FACTS FROM
Geometry

Volume of a Rectangular Solid

It is easy to see why the phrase "five squared" is associated with the expression 5^2. Simply find the area of the square shown in Figure 1 with a side of 5.

FIGURE 1

To see why the phrase "five cubed" is associated with the expression 5^3, we have to find the *volume* of a cube for which all three dimensions are 5 units long. The volume of a cube is a measure of the space occupied by the cube. To calculate the volume of the cube shown in Figure 2, we multiply the three dimensions together to get $5 \cdot 5 \cdot 5 = 5^3$.

FIGURE 2

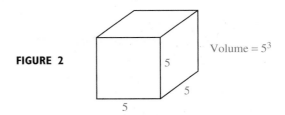

The cube shown in Figure 2 is a special case of a general category of three-dimensional geometric figures called *rectangular solids*. Rectangular solids have rectangles for sides, and all connecting sides meet at right angles. The three dimensions are length, width, and height. To find the volume of a rectangular solid, we find the product of the three dimensions, as shown in Figure 3.

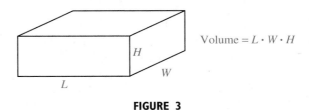

FIGURE 3

Note: If we include units with the dimensions of the diagrams, then the units for the area will be square units and the units for volume will be cubic units. More specifically, if a square has a side 5 inches long, then its area will be

$$A = (5 \text{ inches})^2 = 25 \text{ inches}^2$$

where the unit inches^2 stands for square inches.

If a cube has a side 5 inches long, then its volume will be

$$V = (5 \text{ inches})^3 = 125 \text{ inches}^3$$

where the unit inches^3 stands for cubic inches.

If a rectangular solid has a length of 5 inches, a width of 4 inches, and a height of 3 inches, then its volume is

$$V = (5 \text{ inches})(4 \text{ inches})(3 \text{ inches})$$

$$= 60 \text{ inches}^3$$

SCIENTIFIC NOTATION

Many branches of science require working with very large numbers. In astronomy, for example, distances commonly are given in light-years. A light-year is the distance light travels in a year. It is approximately

$$5,880,000,000,000 \text{ miles}$$

This number is difficult to use in calculations because of the number of zeros it contains. Scientific notation provides a way of writing very large numbers in a more manageable form.

DEFINITION   A number is in **scientific notation** when it is written as the product of a number between 1 and 10 and an integer power of 10. A number written in scientific notation has the form

$$n \times 10^r$$

where $1 \le n < 10$ and $r =$ an integer.

EXAMPLE 17   Write 376,000 in scientific notation.

Solution   We must rewrite 376,000 as the product of a number between 1 and 10 and a power of 10. To do so, we move the decimal point 5 places to the left so that it appears between the 3 and the 7. Then we multiply this number by 10^5. The number that results has the same value as our original number and is written in scientific notation:

$$376,000 = 3.76 \times 10^5$$

Moved 5 places.

Decimal point originally here.

Keeps track of the 5 places we moved the decimal point.

EXAMPLE 18 Write 4.52×10^3 in expanded form.

Solution Since 10^3 is 1,000, we can think of this as simply a multiplication problem. That is,

$$4.52 \times 10^3 = 4.52 \times 1,000 = 4,520$$

On the other hand, we can think of the exponent 3 as indicating the number of places we need to move the decimal point in order to write our number in expanded form. Since our exponent is positive 3, we move the decimal point three places to the right:

$$4.52 \times 10^3 = 4,520$$

 ## Getting Ready for Class

After reading through the preceding section, respond in your own words and in complete sentences.

A. Explain the difference between -5^2 and $(-5)^2$.

B. How do you multiply two expressions containing exponents when they each have the same base?

C. What is Property 2 for exponents?

D. When is a number written in scientific notation?

PROBLEM SET 4.1

Name the base and exponent in each of the following expressions. Then use the definition of exponents as repeated multiplication to simplify.

1. 4^2

2. 6^2

3. $(0.3)^2$

4. $(0.03)^2$

5. 4^3

6. 10^3

7. $(-5)^2$

8. -5^2

9. -2^3

10. $(-2)^3$

11. 3^4

12. $(-3)^4$

13. $\left(\dfrac{2}{3}\right)^2$

14. $\left(\dfrac{2}{3}\right)^3$

15. $\left(\dfrac{1}{2}\right)^4$

16. $\left(\dfrac{4}{5}\right)^2$

17. (a) Complete the following table.

Number x	Square x^2
1	
2	
3	
4	
5	
6	
7	

(b) Using the results of part (a), fill in the blank in the following statement: For numbers larger than 1, the square of the number is _____ than the number.

18. (a) Complete the following table.

Number x	Square x^2
$\dfrac{1}{2}$	
$\dfrac{1}{3}$	
$\dfrac{1}{4}$	
$\dfrac{1}{5}$	
$\dfrac{1}{6}$	
$\dfrac{1}{7}$	
$\dfrac{1}{8}$	

(b) Using the results of part (a), fill in the blank in the following statement: For numbers between 0 and 1, the square of the number is _____ than the number.

Use Property 1 for exponents to simplify each expression. Leave all answers in terms of exponents.

19. $x^4 \cdot x^5$

20. $x^7 \cdot x^3$

21. $y^{10} \cdot y^{20}$

22. $y^{30} \cdot y^{30}$

23. $2^5 \cdot 2^4 \cdot 2^3$

24. $4^2 \cdot 4^3 \cdot 4^4$

25. $x^4 \cdot x^6 \cdot x^8 \cdot x^{10}$

26. $x^{20} \cdot x^{18} \cdot x^{16} \cdot x^{14}$

Use Property 2 for exponents to write each of the following problems with a single exponent. (Assume all variables are positive numbers.)

27. $(x^2)^5$

28. $(x^5)^2$

29. $(5^4)^3$

30. $(5^3)^4$

31. $(y^3)^3$

32. $(y^2)^2$

33. $(2^5)^{10}$

34. $(10^5)^2$

35. $(a^3)^x$

36. $(a^5)^x$

37. $(b^x)^y$

38. $(b^r)^s$

Use Property 3 for exponents to simplify each of the following expressions.

39. $(4x)^2$

40. $(2x)^4$

41. $(2y)^5$

42. $(5y)^2$

43. $(-3x)^4$

44. $(-3x)^3$

45. $(0.5ab)^2$

46. $(0.4ab)^2$

47. $(4xyz)^3$

48. $(5xyz)^3$

Simplify the following expressions by using the properties of exponents.

49. $(2x^4)^3$

50. $(3x^5)^2$

51. $(4a^3)^2$

52. $(5a^2)^2$

53. $(x^2)^3(x^4)^2$

54. $(x^5)^2(x^3)^5$

55. $(a^3)^1(a^2)^4$

56. $(a^4)^1(a^1)^3$

57. $(2x)^3(2x)^4$

58. $(3x)^2(3x)^3$

59. $(3x^2)^3(2x)^4$

60. $(3x)^3(2x^3)^2$

61. $(4x^2y^3)^2$

62. $(9x^3y^5)^2$

63. $\left(\dfrac{2}{3}\, a^4b^5\right)^3$

64. $\left(\dfrac{3}{4}\, ab^7\right)^3$

65. Complete the following table, and then use the template to construct a line graph of the information in the table.

Number x	Square x^2
-3	
-2	
-1	
0	
1	
2	
3	

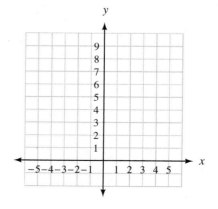

66. Complete the table, and then use the template to construct a line graph of the information in the table.

Number x	Cube x^3
−3	
−2	
−1	
0	
1	
2	
3	

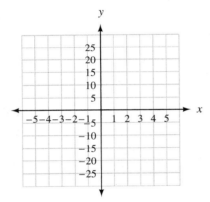

67. Complete the table. When you are finished, notice how the points in this table could be used to refine the line graph you created in Problem 65.

Number x	Square x^2
−2.5	
−1.5	
−0.5	
0	
0.5	
1.5	
2.5	

68. Complete the following table. When you are finished, notice that this table contains exactly the same entries as the table from Problem 67. This table uses fractions, while the table from Problem 67 uses decimals.

Number x	Square x^2
$-\frac{5}{2}$	
$-\frac{3}{2}$	
$-\frac{1}{2}$	
0	
$\frac{1}{2}$	
$\frac{3}{2}$	
$\frac{5}{2}$	

Write each number in scientific notation.

69. 43,200

70. 432,000

71. 570

72. 5,700

73. 238,000

74. 2,380,000

Write each number in expanded form.

75. 2.49×10^3

76. 2.49×10^4

77. 3.52×10^2

78. 3.52×10^5

79. 2.8×10^4

80. 2.8×10^3

Applying the Concepts

81. Volume of a Cube   Find the volume of a cube if each side is 3 inches long.

82. Volume of a Cube   Find the volume of a cube if each side is 3 feet long.

83. Volume of a Cube   A bottle of perfume is packaged in a box that is in the shape of a cube (see next page). Find the volume of the box if each side is 2.5 inches long. Round to the nearest tenth.

84. Volume of a Cube A television set is packaged in a box that is in the shape of a cube. Find the volume of the box if each side is 18 inches long.

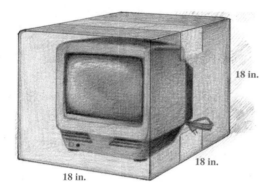

85. Volume of a Box A rented videotape is in a plastic container that has the shape of a rectangular solid. Find the volume of the container if the length is 8 inches, the width is 4.5 inches, and the height is 1 inch.

86. Volume of Your Book Your textbook is in the shape of a rectangular solid. Find the volume in cubic inches.

87. Volume of a Box If a box has a volume of 42 cubic feet, is it possible for you to fit inside the box? Explain your answer.

88. Volume of a Box A box has a volume of 45 cubic inches. Will a can of soup fit inside the box? Explain your answer.

89. Age in Seconds If you are 21 years old, you have been alive for over 650,000,000 seconds. Write this last number in scientific notation.

90. Distance Around the Earth The distance around the Earth at the equator is more than 130,000,000 feet. Write this number in scientific notation.

91. Lifetime Earnings If you earn at least $12 an hour and work full-time for 30 years, you will make at least 7.4×10^5 dollars. Write this last number in expanded form.

92. Heart Beats per Year If your pulse is 72, then in 1 year your heart will beat at least 3.78×10^7 times. Write this last number in expanded form.

93. Investing If you put $1,000 into a savings account every year from the time you are 25 years old until you are 55 years old, you will have more than 1.8×10^5 dollars in the account when you reach 55 years of age (assuming 10% annual interest). Write 1.8×10^5 in expanded form.

94. Investing If you put $20 into a savings account every month from the time you are 20 years old until you are 30 years old, you will have over 3.27×10^3 dollars in the account when you reach 30 years of age (assuming 6% annual interest compounded monthly). Write 3.27×10^3 in expanded form.

Displacement The displacement, in cubic inches, of a car engine is given by the formula

$$d = \pi \cdot s \cdot c \cdot \left(\frac{1}{2} \cdot b\right)^2$$

where s is the stroke and b is the bore, as shown in the figure, and c is the number of cylinders.

Calculate the engine displacement for each of the following cars. Use 3.14 to approximate π.

95. Ferrari Modena 8 cylinders, 3.35 inches of bore, 3.11 inches of stroke

96. Audi A8 8 cylinders, 3.32 inches of bore, 3.66 inches of stroke

97. Mitsubishi Eclipse 6 cylinders, 3.59 inches of bore, 2.99 inches of stroke

98. Porsche 911 GT3 6 cylinders, 3.94 inches of bore, 3.01 inches of stroke

99. Fractals Fractals are infinitely complex shapes of repeating patterns, which can be used to model the world around us. An example of a fractal is the Koch snowflake curve, which we construct as follows: Start with a straight line segment divided into three equal sections. This is Stage 0.

Stage 0

Next, cut out the middle section, and replace it with two pieces, each of which is the same length as the two outside sections. This becomes Stage 1.

Stage 1

To construct Stage 2, we remove the middle third of each of the four line segments in Stage 1 and re-

place them with two segments equal in length to the remaining segments. Here is Stage 2:

Stage 2

(a) If the original line (Stage 0) was 1 foot long, how long is the line in Stage 1?
(b) How long is the line in Stage 2?
(c) Construct Stage 3. How long is the line created in Stage 3?
(d) If you were to continue this process until you reached Stage 10, how long would the line in Stage 10 be?

100. Fractals Here is another fractal construction, similar to the Koch snowflake curve constructed in the previous problem. We start with a straight line segment divided into three equal sections. This is Stage 0.

Stage 0

Next, cut out the middle section, and replace it with three pieces, each of which is the same length as the two outside sections. Arrange the three new pieces so that they form three sides of a square. This becomes Stage 1.

Stage 1

To construct Stage 2, we remove the middle third of each of the five line segments in Stage 1 and replace them with three segments equal in length to the remaining segments. Arrange each new set of segments so that they form three sides of a square. Here is Stage 2:

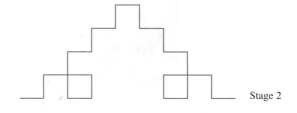

Stage 2

(a) If the original line (Stage 0) was 1 foot long, how long is the line in Stage 1?

(b) How long is the line in Stage 2?

(c) Construct Stage 3. How long is the line created in Stage 3?

(d) If you were to continue this process until you reached Stage 10, how long would the line in Stage 10 be?

Review Problems

The following problems review material on subtraction with negative numbers that we covered in Section 1.4.

Reviewing these problems will help you understand the material in the next section.

Subtract.

101. $4 - 7$ **102.** $-4 - 7$

103. $4 - (-7)$ **104.** $-4 - (-7)$

105. $15 - 20$ **106.** $15 - (-20)$

107. $-15 - (-20)$ **108.** $-15 - 20$

4.2   Division With Exponents

In Section 4.1 we found that multiplication with the same base results in addition of exponents, that is, $a^r \cdot a^s = a^{r+s}$. Since division is the inverse of multiplication, we can expect division with the same base to result in subtraction of exponents.

To develop the properties for exponents under division, we again apply the definition of exponents:

$$\frac{x^5}{x^3} = \frac{x \cdot x \cdot x \cdot x \cdot x}{x \cdot x \cdot x}$$

$$= \frac{x \cdot x \cdot x}{x \cdot x \cdot x}(x \cdot x)$$

$$= 1(x \cdot x)$$

$$= x^2 \quad \textit{Notice: } 5 - 3 = 2$$

$$\frac{2^4}{2^7} = \frac{2 \cdot 2 \cdot 2 \cdot 2}{2 \cdot 2 \cdot 2 \cdot 2 \cdot 2 \cdot 2 \cdot 2}$$

$$= \frac{2 \cdot 2 \cdot 2 \cdot 2}{2 \cdot 2 \cdot 2 \cdot 2} \cdot \frac{1}{2 \cdot 2 \cdot 2}$$

$$= \frac{1}{2 \cdot 2 \cdot 2}$$

$$= \frac{1}{2^3} \quad \textit{Notice: } 7 - 4 = 3$$

In both cases division with the same base resulted in subtraction of the smaller exponent from the larger. The problem is deciding whether the answer is a fraction or not. The problem is resolved quite easily by the following definition.

DEFINITION   If r is a positive integer, then $a^{-r} = \dfrac{1}{a^r} = \left(\dfrac{1}{a}\right)^r$ $(a \neq 0)$

The following examples illustrate how we use this definition to simplify expressions that contain negative exponents.

EXAMPLES Write each expression with a positive exponent and then simplify:

1. $2^{-3} = \dfrac{1}{2^3} = \dfrac{1}{8}$

2. $5^{-2} = \dfrac{1}{5^2} = \dfrac{1}{25}$

3. $3x^{-6} = 3 \cdot \dfrac{1}{x^6} = \dfrac{3}{x^6}$

Notice: Negative exponents do not indicate negative numbers. They indicate reciprocals.

Now let us look back to our original problem and try to work it again with the help of a negative exponent. We know that $2^4/2^7 = 1/2^3$. Let us decide now that with division of the same base, we will always subtract the exponent in the denominator from the exponent in the numerator and see if this conflicts with what we know is true.

$$\dfrac{2^4}{2^7} = 2^{4-7} \qquad \text{Subtracting the bottom exponent from the top exponent}$$

$$= 2^{-3} \qquad \text{Subtraction}$$

$$= \dfrac{1}{2^3} \qquad \text{Definition of negative exponents}$$

Subtracting the exponent in the denominator from the exponent in the numerator and then using the definition of negative exponents gives us the same result we obtained previously. We can now continue the list of properties of exponents we started in Section 4.1.

PROPERTY 4 FOR EXPONENTS

If a is any real number and r and s are integers, then

$$\dfrac{a^r}{a^s} = a^{r-s} \qquad (a \neq 0)$$

In words: To divide with the same base, subtract the exponent in the denominator from the exponent in the numerator and raise the base to the exponent that results.

The following examples show how we use Property 4 and the definition for negative exponents to simplify expressions involving division.

EXAMPLES Simplify the following expressions:

4. $\dfrac{x^9}{x^6} = x^{9-6} = x^3$

5. $\dfrac{x^4}{x^{10}} = x^{4-10} = x^{-6} = \dfrac{1}{x^6}$

6. $\dfrac{2^{15}}{2^{20}} = 2^{15-20} = 2^{-5} = \dfrac{1}{2^5} = \dfrac{1}{32}$

Our final property of exponents is similar to Property 3 from Section 4.1, but it involves division instead of multiplication. After we have stated the property, we will give a proof of it. The proof shows why this property is true.

> ### PROPERTY 5 FOR EXPONENTS
>
> If a and b are any two real numbers ($b \neq 0$) and r is an integer, then
> $$\left(\frac{a}{b}\right)^r = \frac{a^r}{b^r}$$
>
> *In words:*   A quotient raised to a power is the quotient of the powers.

Proof

$$\left(\frac{a}{b}\right)^r = \left(a \cdot \frac{1}{b}\right)^r \qquad \text{By the definition of division}$$

$$= a^r \cdot \left(\frac{1}{b}\right)^r \qquad \text{By Property 3}$$

$$= a^r \cdot b^{-r} \qquad \text{By the definition of negative exponents}$$

$$= a^r \cdot \frac{1}{b^r} \qquad \text{By the definition of negative exponents}$$

$$= \frac{a^r}{b^r} \qquad \text{By the definition of division}$$

EXAMPLES   Simplify the following expressions:

7. $\left(\dfrac{x}{2}\right)^3 = \dfrac{x^3}{2^3} = \dfrac{x^3}{8}$

8. $\left(\dfrac{5}{y}\right)^2 = \dfrac{5^2}{y^2} = \dfrac{25}{y^2}$

9. $\left(\dfrac{2}{3}\right)^4 = \dfrac{2^4}{3^4} = \dfrac{16}{81}$

ZERO AND ONE AS EXPONENTS

We have two special exponents left to deal with before our rules for exponents are complete: 0 and 1. To obtain an expression for x^1, we will solve a problem two different ways:

$$\left.\begin{array}{l} \dfrac{x^3}{x^2} = \dfrac{x \cdot x \cdot x}{x \cdot x} = x \\[2em] \dfrac{x^3}{x^2} = x^{3-2} = x^1 \end{array}\right\} \quad \text{Hence, } x^1 = x$$

Stated generally, this rule says that $a^1 = a$. This seems reasonable, and we will use it, since it is consistent with our property of division using the same base.

We use the same procedure to obtain an expression for x^0:

$$\left.\begin{array}{l} \dfrac{5^2}{5^2} = \dfrac{25}{25} = 1 \\[2em] \dfrac{5^2}{5^2} = 5^{2-2} = 5^0 \end{array}\right\} \quad \text{Hence, } 5^0 = 1$$

It seems, therefore, that the best definition of x^0 is 1 for all x except $x = 0$. In the case of $x = 0$, we have 0^0, which we will not define. This definition will probably seem awkward at first. Most people would like to define x^0 as 0 when they first encounter it. Remember, the zero in this expression is an exponent, so x^0 does not mean to multiply by zero. Thus, we can make the general statement that $a^0 = 1$ for all real numbers except $a = 0$.

Here are some examples involving the exponents 0 and 1.

EXAMPLES   Simplify the following expressions:

10. $8^0 = 1$

11. $8^1 = 8$

12. $4^0 + 4^1 = 1 + 4 = 5$

13. $(2x^2y)^0 = 1$

Here is a summary of the definitions and properties of exponents we have developed so far. For each definition or property in the list, a and b are real numbers, and r and s are integers.

DEFINITIONS	PROPERTIES
$a^{-r} = \dfrac{1}{a^r} = \left(\dfrac{1}{a}\right)^r \quad (a \neq 0)$	**1.** $a^r \cdot a^s = a^{r+s}$
$a^1 = a$	**2.** $(a^r)^s = a^{rs}$
$a^0 = 1 \quad (a \neq 0)$	**3.** $(ab)^r = a^r b^r$
	4. $\dfrac{a^r}{a^s} = a^{r-s} \quad (a \neq 0)$
	5. $\left(\dfrac{a}{b}\right)^r = \dfrac{a^r}{b^r} \quad (b \neq 0)$

Here are some additional examples. These examples use a combination of the preceding properties and definitions.

E X A M P L E S   Simplify each expression. Write all answers with positive exponents only:

14. $\dfrac{(5x^3)^2}{x^4} = \dfrac{25x^6}{x^4}$ Properties 2 and 3

$= 25x^2$ Property 4

15. $\dfrac{x^{-8}}{(x^2)^3} = \dfrac{x^{-8}}{x^6}$ Property 2

$= x^{-8-6}$ Property 4

$= x^{-14}$ Subtraction

$= \dfrac{1}{x^{14}}$ Definition of negative exponents

16. $\left(\dfrac{y^5}{y^3}\right)^2 = \dfrac{(y^5)^2}{(y^3)^2}$ Property 5

$= \dfrac{y^{10}}{y^6}$ Property 2

$= y^4$ Property 4

Notice in Example 16 that we could have simplified inside the parentheses first and then raised the result to the second power:

$$\left(\dfrac{y^5}{y^3}\right)^2 = (y^2)^2 = y^4$$

17. $(3x^5)^{-2} = \dfrac{1}{(3x^5)^2}$ Definition of negative exponents

$= \dfrac{1}{9x^{10}}$ Properties 2 and 3

18. $x^{-8} \cdot x^5 = x^{-8+5}$ Property 1

$= x^{-3}$ Addition

$= \dfrac{1}{x^3}$ Definition of negative exponents

19. $\dfrac{(a^3)^2 a^{-4}}{(a^{-4})^3} = \dfrac{a^6 a^{-4}}{a^{-12}}$ Property 2

$= \dfrac{a^2}{a^{-12}}$ Property 1

$= a^{14}$ Property 4

In the next two examples we use division to compare the area and volume of geometric figures.

EXAMPLE 20 Suppose you have two squares, one of which is larger than the other. If a side of the larger square is 3 times as long as a side of the smaller square, how many of the smaller squares will it take to cover up the larger square?

Solution If we let x represent the length of a side of the smaller square, then the length of a side of the larger square is $3x$. The area of each square, along with a diagram of the situation, is given in Figure 1.

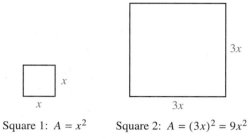

Square 1: $A = x^2$ Square 2: $A = (3x)^2 = 9x^2$

FIGURE 1

To find out how many smaller squares it will take to cover up the larger square, we divide the area of the larger square by the area of the smaller square.

$$\frac{\text{Area of square 2}}{\text{Area of square 1}} = \frac{9x^2}{x^2} = 9$$

It will take 9 of the smaller squares to cover the larger square.

EXAMPLE 21 Suppose you have two boxes, each of which is a cube. If the length of a side in the second box is 3 times as long as the length of a side of the first box, how many of the smaller boxes will fit inside the larger box?

Solution If we let x represent the length of a side of the smaller box, then the length of a side of the larger box is $3x$. The volume of each box, along with a diagram of the situation, is given in Figure 2.

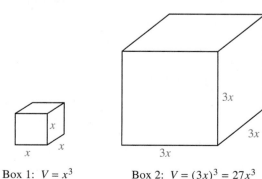

Box 1: $V = x^3$ Box 2: $V = (3x)^3 = 27x^3$

FIGURE 2

To find out how many smaller boxes will fit inside the larger box, we divide the volume of the larger box by the volume of the smaller box.

$$\frac{\text{Volume of box 2}}{\text{Volume of box 1}} = \frac{27x^3}{x^3} = 27$$

We can fit 27 of the smaller boxes inside the larger box.

MORE ON SCIENTIFIC NOTATION

Now that we have completed our list of definitions and properties of exponents, we can expand the work we did previously with scientific notation.

Recall that a number is in scientific notation when it is written in the form

$$n \times 10^r$$

where $1 \leq n < 10$ and r is an integer.

Since negative exponents give us reciprocals, we can use negative exponents to write very small numbers in scientific notation. For example, the number 0.00057, when written in scientific notation, is equivalent to 5.7×10^{-4}. Here's why:

$$5.7 \times 10^{-4} = 5.7 \times \frac{1}{10^4} = 5.7 \times \frac{1}{10,000} = \frac{5.7}{10,000} = 0.00057$$

Example 22 lists some other numbers in both scientific notation and expanded form.

E X A M P L E   2 2

Number Written the Long Way		Number Written Again in Scientific Notation
376,000	=	3.76×10^5
49,500	=	4.95×10^4
3,200	=	$3.2 \ \times 10^3$
591	=	5.91×10^2
46	=	$4.6 \ \times 10^1$
8	=	$8 \ \ \times 10^0$
0.47	=	$4.7 \ \times 10^{-1}$
0.093	=	$9.3 \ \times 10^{-2}$
0.00688	=	6.88×10^{-3}
0.0002	=	$2 \ \ \times 10^{-4}$
0.000098	=	$9.8 \ \times 10^{-5}$

Notice that in each case, when the number is written in scientific notation, the decimal point in the first number is placed so that the number is between 1 and 10.

The exponent on 10 in the second number keeps track of the number of places we moved the decimal point in the original number to get a number between 1 and 10:

$$376,000 = 3.76 \times 10^5$$

Moved 5 places.

Decimal point was originally here.

Keeps track of the 5 places we moved the decimal point.

$$0.00688 = 6.88 \times 10^{-3}$$

Moved 3 places.

Keeps track of the 3 places we moved the decimal point.

 ## Getting Ready for Class

After reading through the preceding section, respond in your own words and in complete sentences.

A. How do you divide two expressions containing exponents when they each have the same base?

B. Explain the difference between 3^2 and 3^{-2}.

C. If a positive base is raised to a negative exponent, can the result be a negative number?

D. Explain what happens when we use 0 as an exponent.

PROBLEM SET 4.2

Write each of the following with positive exponents, and then simplify, when possible.

1. 3^{-2} **2.** 3^{-3} **3.** 6^{-2}

4. 2^{-6} **5.** 8^{-2} **6.** 3^{-4}

7. 5^{-3} **8.** 9^{-2} **9.** $2x^{-3}$

10. $5x^{-1}$ **11.** $(2x)^{-3}$ **12.** $(5x)^{-1}$

13. $(5y)^{-2}$ **14.** $5y^{-2}$ **15.** 10^{-2}

16. 10^{-3}

17. Complete the following table.

Number x	Square x^2	Power of 2 2^x
−3		
−2		
−1		
0		
1		
2		
3		

18. Complete the following table.

Number x	Cube x^3	Power of 3 3^x
-3		
-2		
-1		
0		
1		
2		
3		

Use Property 4 to simplify each of the following expressions. Write all answers that contain exponents with positive exponents only.

19. $\dfrac{5^1}{5^3}$ **20.** $\dfrac{7^6}{7^8}$ **21.** $\dfrac{x^{10}}{x^4}$

22. $\dfrac{x^4}{x^{10}}$ **23.** $\dfrac{4^3}{4^0}$ **24.** $\dfrac{4^0}{4^3}$

25. $\dfrac{(2x)^7}{(2x)^4}$ **26.** $\dfrac{(2x)^4}{(2x)^7}$ **27.** $\dfrac{6^{11}}{6}$

28. $\dfrac{8^7}{8}$ **29.** $\dfrac{6}{6^{11}}$ **30.** $\dfrac{8}{8^7}$

31. $\dfrac{2^{-5}}{2^3}$ **32.** $\dfrac{2^{-5}}{2^{-3}}$ **33.** $\dfrac{2^5}{2^{-3}}$

34. $\dfrac{2^{-3}}{2^{-5}}$ **35.** $\dfrac{(3x)^{-5}}{(3x)^{-8}}$ **36.** $\dfrac{(2x)^{-10}}{(2x)^{-15}}$

Simplify the following expressions. Any answers that contain exponents should contain positive exponents only.

37. $(3xy)^4$ **38.** $(4xy)^3$ **39.** 10^0

40. 10^1 **41.** $(2a^2b)^1$ **42.** $(2a^2b)^0$

43. $(7y^3)^{-2}$ **44.** $(5y^4)^{-2}$ **45.** $x^{-3}x^{-5}$

46. $x^{-6} \cdot x^8$ **47.** $y^7 \cdot y^{-10}$ **48.** $y^{-4} \cdot y^{-6}$

49. $\dfrac{(x^2)^3}{x^4}$ **50.** $\dfrac{(x^5)^3}{x^{10}}$ **51.** $\dfrac{(a^4)^3}{(a^3)^2}$

52. $\dfrac{(a^5)^3}{(a^5)^2}$ **53.** $\dfrac{y^7}{(y^2)^8}$ **54.** $\dfrac{y^2}{(y^3)^4}$

55. $\left(\dfrac{y^7}{y^2}\right)^8$ **56.** $\left(\dfrac{y^2}{y^3}\right)^4$ **57.** $\dfrac{(x^{-2})^3}{x^{-5}}$

58. $\dfrac{(x^2)^{-3}}{x^{-5}}$ **59.** $\left(\dfrac{x^{-2}}{x^{-5}}\right)^3$ **60.** $\left(\dfrac{x^2}{x^{-5}}\right)^{-3}$

61. $\dfrac{(a^3)^2(a^4)^5}{(a^5)^2}$ **62.** $\dfrac{(a^4)^8(a^2)^5}{(a^3)^4}$ **63.** $\dfrac{(a^{-2})^3(a^4)^2}{(a^{-3})^{-2}}$

64. $\dfrac{(a^{-5})^{-3}(a^7)^{-1}}{(a^{-3})^5}$

65. Complete the following table, and then use the template to construct a line graph of the information in the table.

Number x	Power of 2 2^x
-3	
-2	
-1	
0	
1	
2	
3	

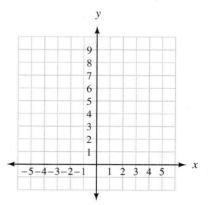

66. Complete the following table, and then use the template to construct a line graph of the information in the table.

Number x	Power of 3 3^x
−3	
−2	
−1	
0	
1	
2	
3	

Expanded Form	Scientific Notation $n \times 10^r$
0.000357	3.57×10^{-4}
0.00357	
0.0357	
0.357	
3.57	
35.7	
357	
3,570	
35,700	

74. Complete the following table.

Expanded Form	Scientific Notation $n \times 10^r$
0.000123	1.23×10^{-4}
	1.23×10^{-3}
	1.23×10^{-2}
	1.23×10^{-1}
	1.23×10^{0}
	1.23×10^{1}
	1.23×10^{2}
	1.23×10^{3}
	1.23×10^{4}

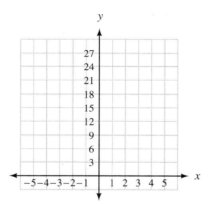

Write each of the following numbers in scientific notation.

67. 0.0048

68. 0.000048

69. 25

70. 35

71. 0.000009

72. 0.0009

73. Complete the following table.

Write each of the following numbers in expanded form.

75. 4.23×10^{-3}

76. 4.23×10^{3}

77. 8×10^{-5}

78. 8×10^{5}

79. 4.2×10^{0}

80. 4.2×10^{1}

Applying the Concepts

Scientific Notation Problems

81. Some home computers can do a calculation in 2×10^{-3} seconds. Write this number in expanded form.

82. Some of the cells in the human body have a radius of 3×10^{-5} inch. Write this number in expanded form.

83. One of the smallest ants in the world has a length of only 0.006 inch. Write this number in scientific notation.

84. Some cameras used in scientific research can take 1 picture every 0.000000167 second. Write this number in scientific notation.

85. The number 25×10^3 is not in scientific notation because 25 is larger than 10. Write 25×10^3 in scientific notation.

86. The number 0.25×10^3 is not in scientific notation because 0.25 is less than 1. Write 0.25×10^3 in scientific notation.

87. The number 23.5×10^4 is not in scientific notation because 23.5 is not between 1 and 10. Rewrite 23.5×10^4 in scientific notation.

88. The number 375×10^3 is not in scientific notation because 375 is not between 1 and 10. Rewrite 375×10^3 in scientific notation.

89. The number 0.82×10^{-3} is not in scientific notation because 0.82 is not between 1 and 10. Rewrite 0.82×10^{-3} in scientific notation.

90. The number 0.93×10^{-2} is not in scientific notation because 0.93 is not between 1 and 10. Rewrite 0.93×10^{-2} in scientific notation.

Comparing Areas   Suppose you have two squares, one of which is larger than the other. Suppose further that the side of the larger square is twice as long as the side of the smaller square.

91. If the length of the side of the smaller square is 10 inches, give the area of each square. Then find the number of smaller squares it will take to cover the larger square.

92. How many smaller squares will it take to cover the larger square if the length of the side of the smaller square is 1 foot?

93. If the length of the side of the smaller square is x, find the area of each square. Then find the number of smaller squares it will take to cover the larger square.

94. Suppose the length of the side of the larger square is 1 foot. How many smaller squares will it take to cover the larger square?

Comparing Volumes   Suppose you have two boxes, each of which is a cube. Suppose further that the length of a side of the second box is twice as long as the length of a side of the first box.

95. If the length of a side of the first box is 6 inches, give the volume of each box. Then find the number of smaller boxes that will fit inside the larger box.

96. How many smaller boxes can be placed inside the larger box if the length of a side of the second box is 1 foot?

97. If the length of a side of the first box is x, find the volume of each box. Then find the number of smaller boxes that will fit inside the larger box.

98. Suppose the length of a side of the larger box is 12 inches. How many smaller boxes will fit inside the larger box?

Review Problems

The following problems review material we covered in Section 2.1. They will help you understand some of the material in the next section.

Simplify the following expressions.

99. $4x + 3x$ 100. $9x + 7x$

101. $5a - 3a$ 102. $10a - 2a$

103. $4y + 5y + y$ 104. $6y - y + 2y$

4.3   **Operations With Monomials**

We have developed all the tools necessary to perform the four basic operations on the simplest of polynomials: monomials.

> **DEFINITION** A **monomial** is a one-term expression that is either a constant (number) or the product of a constant and one or more variables raised to whole-number exponents.

The following are examples of monomials:

$$-3 \qquad 15x \qquad -23x^2y \qquad 49x^4y^2z^4 \qquad \frac{3}{4}a^2b^3$$

The numerical part of each monomial is called the *numerical coefficient,* or just the *coefficient.* Monomials are also called *terms.*

MULTIPLICATION AND DIVISION OF MONOMIALS

There are two basic steps involved in the multiplication of monomials. First, we rewrite the products using the commutative and associative properties. Then, we simplify by multiplying coefficients and adding exponents of like bases.

EXAMPLES Multiply:

1. $(-3x^2)(4x^3) = (-3 \cdot 4)(x^2 \cdot x^3)$ Commutative and associative properties
$$= -12x^5 \qquad \text{Multiply coefficients; add exponents.}$$

2. $\left(\dfrac{4}{5}x^5 \cdot y^2\right)(10x^3 \cdot y) = \left(\dfrac{4}{5} \cdot 10\right)(x^5 \cdot x^3)(y^2 \cdot y)$ Commutative and associative properties
$$= 8x^8y^3 \qquad \text{Multiply coefficients; add exponents.}$$

You can see that in each case the work was the same—multiply coefficients and add exponents of the same base. We can expect division of monomials to proceed in a similar way. Since our properties are consistent, division of monomials will result in division of coefficients and subtraction of exponents of like bases.

EXAMPLES Divide:

3. $\dfrac{15x^3}{3x^2} = \dfrac{15}{3} \cdot \dfrac{x^3}{x^2}$ Write as separate fractions.
$$= 5x \qquad \text{Divide coefficients; subtract exponents.}$$

4. $\dfrac{39x^2y^3}{3xy^5} = \dfrac{39}{3} \cdot \dfrac{x^2}{x} \cdot \dfrac{y^3}{y^5}$ Write as separate fractions.
$$= 13x \cdot \dfrac{1}{y^2} \qquad \text{Divide coefficients; subtract exponents.}$$
$$= \dfrac{13x}{y^2} \qquad \text{Write answer as a single fraction.}$$

In Example 4, the expression y^3/y^5 simplifies to $1/y^2$ because of Property 4 for exponents and the definition of negative exponents. If we were to show all the work in this simplification process, it would look like this:

$$\frac{y^3}{y^5} = y^{3-5} \qquad \text{Property 4 for exponents}$$

$$= y^{-2} \qquad \text{Subtraction}$$

$$= \frac{1}{y^2} \qquad \text{Definition of negative exponents}$$

The point of this explanation is this: Even though we may not show all the steps when simplifying an expression involving exponents, the result we obtain can still be justified using the properties of exponents. We have not introduced any new properties in Example 4; we have just not shown the details of each simplification.

E X A M P L E   5 Divide:

$$\frac{25a^5b^3}{50a^2b^7} = \frac{25}{50} \cdot \frac{a^5}{a^2} \cdot \frac{b^3}{b^7} \qquad \text{Write as separate fractions.}$$

$$= \frac{1}{2} \cdot a^3 \cdot \frac{1}{b^4} \qquad \text{Divide coefficients; subtract exponents.}$$

$$= \frac{a^3}{2b^4} \qquad \text{Write answer as a single fraction.}$$

Notice in Example 5 that dividing 25 by 50 results in $\frac{1}{2}$. This is the same result we would obtain if we reduced the fraction $\frac{25}{50}$ to lowest terms, and there is no harm in thinking of it that way. Also, notice that the expression b^3/b^7 simplifies to $1/b^4$ by Property 4 for exponents and the definition of negative exponents, even though we have not shown the steps involved in doing so.

MULTIPLICATION AND DIVISION OF NUMBERS WRITTEN IN SCIENTIFIC NOTATION

We multiply and divide numbers written in scientific notation using the same steps we used to multiply and divide monomials.

E X A M P L E   6 Multiply $(4 \times 10^7)(2 \times 10^{-4})$.

Solution Since multiplication is commutative and associative, we can rearrange the order of these numbers and group them as follows:

$$(4 \times 10^7)(2 \times 10^{-4}) = (4 \times 2)(10^7 \times 10^{-4})$$

$$= 8 \times 10^3$$

Notice that we add exponents, $7 + (-4) = 3$, when we multiply with the same base.

EXAMPLE 7 Divide $\dfrac{9.6 \times 10^{12}}{3 \times 10^4}$.

Solution We group the numbers between 1 and 10 separately from the powers of 10 and proceed as we did in Example 6:

$$\frac{9.6 \times 10^{12}}{3 \times 10^4} = \frac{9.6}{3} \times \frac{10^{12}}{10^4}$$
$$= 3.2 \times 10^8$$

Notice that the procedure we used in both of these examples is very similar to multiplication and division of monomials, for which we multiplied or divided coefficients and added or subtracted exponents.

ADDITION AND SUBTRACTION OF MONOMIALS

Addition and subtraction of monomials will be almost identical, since subtraction is defined as addition of the opposite. With multiplication and division of monomials, the key was rearranging the numbers and variables using the commutative and associative properties. With addition, the key is application of the distributive property. We sometimes use the phrase *combine monomials* to describe addition and subtraction of monomials.

> **DEFINITION** Two terms (monomials) with the same variable part (same variables raised to the same powers) are called **similar** (or *like*) **terms**.

You can add only similar terms. This is because the distributive property (which is the key to addition of monomials) cannot be applied to terms that are not similar.

EXAMPLES Combine the following monomials.

8. $-3x^2 + 15x^2 = (-3 + 15)x^2$ Distributive property
$= 12x^2$ Add coefficients.

9. $9x^2y - 20x^2y = (9 - 20)x^2y$ Distributive property
$= -11x^2y$ Add coefficients.

10. $5x^2 + 8y^2$ In this case we cannot apply the distributive property, so we cannot add the monomials.

The next examples show how we simplify expressions containing monomials when more than one operation is involved.

E X A M P L E   I I   Simplify $\dfrac{(6x^4y)(3x^7y^5)}{9x^5y^2}$.

Solution   We begin by multiplying the two monomials in the numerator:

$$\frac{(6x^4y)(3x^7y^5)}{9x^5y^2} = \frac{18x^{11}y^6}{9x^5y^2} \qquad \text{Simplify numerator.}$$

$$= 2x^6y^4 \qquad \text{Divide.}$$

E X A M P L E   I 2   Simplify $\dfrac{(6.8 \times 10^5)(3.9 \times 10^{-7})}{7.8 \times 10^{-4}}$.

Solution   We group the numbers between 1 and 10 separately from the powers of 10:

$$\frac{(6.8)(3.9)}{7.8} \times \frac{(10^5)(10^{-7})}{10^{-4}} = 3.4 \times 10^{5+(-7)-(-4)}$$

$$= 3.4 \times 10^2$$

E X A M P L E   I 3   Simplify $\dfrac{14x^5}{2x^2} + \dfrac{15x^8}{3x^5}$.

Solution   Simplifying each expression separately and then combining similar terms gives

$$\frac{14x^5}{2x^2} + \frac{15x^8}{3x^5} = 7x^3 + 5x^3 \qquad \text{Divide.}$$

$$= 12x^3 \qquad \text{Add.}$$

E X A M P L E   I 4   A rectangular solid is twice as long as it is wide and half as high as it is wide. Write an expression for the volume.

Solution   We begin by making a diagram of the object (Figure 1) with the dimensions labeled as given in the problem.

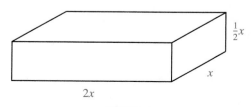

FIGURE   I

The volume is the product of the three dimensions:

$$V = 2x \cdot x \cdot \frac{1}{2}x = x^3$$

The box has the same volume as a cube with side x, as shown in Figure 2.

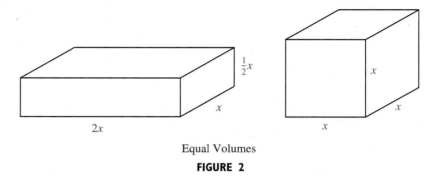

Equal Volumes

FIGURE 2

MORE ABOUT GRAPHING

Up to this point, the graphing we have done in this chapter has produced line graphs. On the next graph we draw, we connect the points with a smooth curve instead of line segments. This will give us an accurate picture of the graph of the equation $y = x^2$.

EXAMPLE 15   Graph the equation $y = x^2$.

Solution   We start by finding some ordered pairs that are solutions to the equation. We can choose any convenient numbers for x and then use the equation $y = x^2$ to find the corresponding values for y. Let's use the values $-3, -2, -1, 0, 1, 2,$ and 3 for x and find corresponding values for y. Here is how the table looks when we let x have these values:

x	$y = x^2$	y
-3	$y = (-3)^2 = 9$	9
-2	$y = (-2)^2 = 4$	4
-1	$y = (-1)^2 = 1$	1
0	$y = 0^2 = 0$	0
1	$y = 1^2 = 1$	1
2	$y = 2^2 = 4$	4
3	$y = 3^2 = 9$	9

The table gives us the solutions $(-3, 9)$, $(-2, 4)$, $(-1, 1)$, $(0, 0)$, $(1, 1)$, $(2, 4)$, and $(3, 9)$ for the equation $y = x^2$. We plot each of the points on a rectangular coordinate system and draw a smooth curve between them, as shown in Figure 3.

The graph is called a **parabola.** All equations of the form $y = ax^2$ ($a \neq 0$) produce parabolas when graphed. The point $(0, 0)$ is called the **vertex** of the parabola in Figure 3.

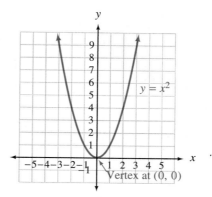

FIGURE 3

 ## Getting Ready for Class

After reading through the preceding section, respond in your own words and in complete sentences.

A. What is a monomial?

B. Describe how you would multiply $3x^2$ and $5x^2$.

C. Describe how you would add $3x^2$ and $5x^2$.

D. Describe how you would multiply two numbers written in scientific notation.

PROBLEM SET 4.3

Multiply.

1. $(3x^4)(4x^3)$

2. $(6x^5)(-2x^2)$

3. $(-2y^4)(8y^7)$

4. $(5y^{10})(2y^5)$

5. $(8x)(4x)$

6. $(7x)(5x)$

7. $(10a^3)(10a)(2a^2)$

8. $(5a^4)(10a)(10a^4)$

9. $(6ab^2)(-4a^2b)$

10. $(-5a^3b)(4ab^4)$

11. $(4x^2y)(3x^3y^3)(2xy^4)$

12. $(5x^6)(-10xy^4)(-2x^2y^6)$

Divide. Write all answers with positive exponents only.

13. $\dfrac{15x^3}{5x^2}$

14. $\dfrac{25x^5}{5x^4}$

15. $\dfrac{18y^9}{3y^{12}}$

16. $\dfrac{24y^4}{-8y^7}$

17. $\dfrac{32a^3}{64a^4}$

18. $\dfrac{25a^5}{75a^6}$

19. $\dfrac{21a^2b^3}{-7ab^5}$

20. $\dfrac{32a^5b^6}{8ab^5}$

21. $\dfrac{3x^3y^2z}{27xy^2z^3}$

22. $\dfrac{5x^5y^4z}{30x^3yz^2}$

23. Fill in the table.

a	b	ab	$\dfrac{a}{b}$	$\dfrac{b}{a}$
10	$5x$			
$20x^3$	$6x^2$			
$25x^5$	$5x^4$			
$3x^{-2}$	$3x^2$			
$-2y^4$	$8y^7$			

24. Fill in the table.

a	b	ab	$\dfrac{a}{b}$	$\dfrac{b}{a}$
$10y$	$2y^2$			
$10y^2$	$2y$			
$5y^3$	15			
5	$15y^3$			
$4y^{-3}$	$4y^3$			

Find each product. Write all answers in scientific notation.

25. $(3 \times 10^3)(2 \times 10^5)$ **26.** $(4 \times 10^8)(1 \times 10^6)$

27. $(3.5 \times 10^4)(5 \times 10^{-6})$

28. $(7.1 \times 10^5)(2 \times 10^{-8})$

29. $(5.5 \times 10^{-3})(2.2 \times 10^{-4})$

30. $(3.4 \times 10^{-2})(4.5 \times 10^{-6})$

Find each quotient. Write all answers in scientific notation.

31. $\dfrac{8.4 \times 10^5}{2 \times 10^2}$

32. $\dfrac{9.6 \times 10^{20}}{3 \times 10^6}$

33. $\dfrac{6 \times 10^8}{2 \times 10^{-2}}$

34. $\dfrac{8 \times 10^{12}}{4 \times 10^{-3}}$

35. $\dfrac{2.5 \times 10^{-6}}{5 \times 10^{-4}}$

36. $\dfrac{4.5 \times 10^{-8}}{9 \times 10^{-4}}$

Combine by adding or subtracting as indicated.

37. $3x^2 + 5x^2$ **38.** $4x^3 + 8x^3$

39. $8x^5 - 19x^5$ **40.** $75x^6 - 50x^6$

41. $2a + a - 3a$ **42.** $5a + a - 6a$

43. $10x^3 - 8x^3 + 2x^3$ **44.** $7x^5 + 8x^5 - 12x^5$

45. $20ab^2 - 19ab^2 + 30ab^2$

46. $18a^3b^2 - 20a^3b^2 + 10a^3b^2$

47. Fill in the table.

a	b	ab	$a + b$
$5x$	$3x$		
$4x^2$	$2x^2$		
$3x^3$	$6x^3$		
$2x^4$	$-3x^4$		
x^5	$7x^5$		

48. Fill in the table.

a	b	ab	$a - b$
$2y$	$3y$		
$-2y$	$3y$		
$4y^2$	$5y^2$		
y^3	$-3y^3$		
$5y^4$	$7y^4$		

Simplify. Write all answers with positive exponents only.

49. $\dfrac{(3x^2)(8x^5)}{6x^4}$

50. $\dfrac{(7x^3)(6x^8)}{14x^5}$

51. $\dfrac{(9a^2b)(2a^3b^4)}{18a^5b^7}$

52. $\dfrac{(21a^5b)(2a^8b^4)}{14ab}$

53. $\dfrac{(4x^3y^2)(9x^4y^{10})}{(3x^5y)(2x^6y)}$

54. $\dfrac{(5x^4y^4)(10x^3y^3)}{(25xy^5)(2xy^7)}$

Simplify each expression, and write all answers in scientific notation.

55. $\dfrac{(6 \times 10^8)(3 \times 10^5)}{9 \times 10^7}$

56. $\dfrac{(8 \times 10^4)(5 \times 10^{10})}{2 \times 10^7}$

57. $\dfrac{(5 \times 10^3)(4 \times 10^{-5})}{2 \times 10^{-2}}$

58. $\dfrac{(7 \times 10^6)(4 \times 10^{-4})}{1.4 \times 10^{-3}}$

59. $\dfrac{(2.8 \times 10^{-7})(3.6 \times 10^4)}{2.4 \times 10^3}$

60. $\dfrac{(5.4 \times 10^2)(3.5 \times 10^{-9})}{4.5 \times 10^6}$

Simplify.

61. $\dfrac{18x^4}{3x} + \dfrac{21x^7}{7x^4}$

62. $\dfrac{24x^{10}}{6x^4} + \dfrac{32x^7}{8x}$

63. $\dfrac{45a^6}{9a^4} - \dfrac{50a^8}{2a^6}$

64. $\dfrac{16a^9}{4a} - \dfrac{28a^{12}}{4a^4}$

65. $\dfrac{6x^7y^4}{3x^2y^2} + \dfrac{8x^5y^8}{2y^6}$

66. $\dfrac{40x^{10}y^{10}}{8x^2y^5} + \dfrac{10x^8y^8}{5y^3}$

Use your knowledge of the properties and definitions of exponents to find x in each of the following.

67. $4^x \cdot 4^5 = 4^7$

68. $\dfrac{5^x}{5^3} = 5^4$

69. $(7^3)^x = 7^{12}$

70. $\dfrac{3^x}{3^4} = 9$

Applying the Concepts

71. Comparing Expressions The statement

$$(a + b)^2 = a^2 + b^2$$

looks similar to Property 3 for exponents. However, it is not a property of exponents because almost every time we replace a and b with numbers, this expression becomes a false statement. Let $a = 4$ and $b = 5$ in the expressions $(a + b)^2$ and $a^2 + b^2$, and see what each simplifies to.

72. Comparing Expressions Show that the statement $(a - b)^2 = a^2 - b^2$ is not, in general, true by substituting 3 for a and 5 for b in each of the expressions and then simplifying each result.

73. Comparing Expressions Show that the expressions $(a + b)^2$ and $a^2 + 2ab + b^2$ are equal when $a = 3$ and $b = 4$.

74. Comparing Expressions Show that the expressions $(a - b)^2$ and $a^2 - 2ab + b^2$ are equal when $a = 7$ and $b = 5$.

75. Geometry A rectangle is twice as long as it is wide. Write an expression for the perimeter and an expression for the area of the rectangle.

76. Geometry A rectangle is 3 times as long as it is wide. Write an expression for the perimeter and an expression for the area of the rectangle.

77. Volume A rectangular solid is twice as long as it is wide. If the height is 4 inches, find an expression for the volume.

78. Volume A rectangular solid is 4 times as long as it is wide. If the height is 2 inches, find an expression for the volume.

79. Volume and Scientific Notation A delivery truck is being filled with boxes of prepackaged food. The cargo space of the truck is 8.5 feet wide, 55 feet long, and 10.1 feet high.
 (a) Find the volume of the truck, round your answer to the nearest hundred, then write the answer in scientific notation.
 (b) If each food box has a volume of 0.15 cubic foot, how many boxes can fit in the delivery truck? Round your answer to the nearest thousand and write your answer in scientific notation.

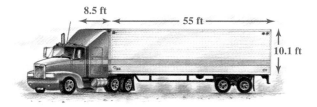

80. Pyramids and Scientific Notation The Great Pyramid at Giza (photo on next page) is one of the largest and oldest man-made structures in the world. It weighs over 10^{10} kilograms. If each stone making up the pyramid weighs approximately 4,000 kilograms, how many stones make up the structure? Write your answer in scientific notation.

© 1985 North Wind Pictures

For each equation, fill in the given ordered pairs and use them to draw the graph of the equation.

81. $y = x^2$ $\quad (-4, \quad), (-2, \quad), (-1, \quad), (0, \quad),$
$(1, \quad), (2, \quad), (4, \quad)$

82. $y = x^2$ $\quad (-2, \quad), (-1.5, \quad), (-\frac{1}{2}, \quad),$
$(0, \quad), (\frac{1}{2}, \quad), (1.5, \quad), (2, \quad)$

83. $y = 2x^2$ $\quad (-3, \quad), (-2, \quad), (-1, \quad), (0, \quad),$
$(1, \quad), (2, \quad), (3, \quad)$

84. $y = 3x^2$ $\quad (-3, \quad), (-2, \quad), (-1, \quad), (0, \quad),$
$(1, \quad), (2, \quad), (3, \quad)$

85. $y = \frac{1}{2}x^2$ $\quad (-4, \quad), (-2, \quad), (-1, \quad), (0, \quad),$
$(1, \quad), (2, \quad), (4, \quad)$

86. $y = \frac{1}{3}x^2$ $\quad (-3, \quad), (-1, \quad), (0, \quad), (1, \quad),$
$(3, \quad)$

87. $y = \frac{1}{4}x^2$ $\quad (-4, \quad), (-2, \quad), (-1, \quad), (0, \quad),$
$(1, \quad), (2, \quad), (4, \quad)$

88. $y = 4x^2$ $\quad (-2, \quad), (-1, \quad), (0, \quad), (1, \quad),$
$(2, \quad)$

Review Problems

The problems that follow review material we covered in Sections 2.1 and 3.3. Reviewing the problems from Section 2.1 will help you understand some of the material in the next section.

Find the value of each expression when $x = -2$. [2.1]

89. $-2x + 5$ $\qquad$ **90.** $-4x - 1$

91. $x^2 + 5x + 6$ $\qquad$ **92.** $x^2 - 5x + 6$

For each of the following equations complete the given ordered pairs so that each is a solution to the equation; then use the ordered pairs to graph the equation. [3.3]

93. $y = 2x + 2$ $\quad (-2, \quad), (0, \quad), (2, \quad)$

94. $y = 2x - 3$ $\quad (-1, \quad), (0, \quad), (2, \quad)$

95. $y = \frac{1}{3}x + 1$ $\quad (-3, \quad), (0, \quad), (3, \quad)$

96. $y = \frac{1}{2}x - 2$ $\quad (-2, \quad), (0, \quad), (2, \quad)$

4.4 **Addition and Subtraction of Polynomials**

In this section we will extend what we learned in Section 4.3 to expressions called polynomials. We begin this section with the definition of a polynomial.

DEFINITION A **polynomial** is a finite sum of monomials (terms).

Examples The following are polynomials:

$$3x^2 + 2x + 1 \qquad 15x^2y + 21xy^2 - y^2 \qquad 3a - 2b + 4c - 5d$$

Polynomials can be further classified by the number of terms they contain. A polynomial with two terms is called a binomial. If it has three terms, it is a trinomial. As stated before, a monomial has only one term.

DEFINITION   The **degree** of a polynomial in one variable is the highest power to which the variable is raised.

Examples

$3x^5 + 2x^3 + 1$	A trinomial of degree 5
$2x + 1$	A binomial of degree 1
$3x^2 + 2x + 1$	A trinomial of degree 2
$3x^5$	A monomial of degree 5
-9	A monomial of degree 0

There are no new rules for adding one or more polynomials. We rely only on our previous knowledge. Here are some examples.

EXAMPLE 1   Add $(2x^2 - 5x + 3) + (4x^2 + 7x - 8)$.

Solution   We use the commutative and associative properties to group similar terms together and then apply the distributive property to add

$(2x^2 - 5x + 3) + (4x^2 + 7x - 8)$

$$= (2x^2 + 4x^2) + (-5x + 7x) + (3 - 8) \qquad \text{Commutative and associative properties}$$

$$= (2 + 4)x^2 + (-5 + 7)x + (3 - 8) \qquad \text{Distributive property}$$

$$= 6x^2 + 2x - 5 \qquad \text{Addition}$$

The results here indicate that to add two polynomials, we add coefficients of similar terms.

EXAMPLE 2   Add $x^2 + 3x + 2x + 6$.

Solution   The only similar terms here are the two middle terms. We combine them as usual to get

$$x^2 + 3x + 2x + 6 = x^2 + 5x + 6$$

You will recall from Chapter 1 the definition of subtraction: $a - b = a + (-b)$. To subtract one expression from another, we simply add its opposite.

The letters a and b in the definition can each represent polynomials. The opposite of a polynomial is the opposite of each of its terms. When you subtract one polynomial from another, you subtract each of its terms.

EXAMPLE 3 Subtract $(3x^2 + x + 4) - (x^2 + 2x + 3)$.

Solution To subtract $x^2 + 2x + 3$, we change the sign of each of its terms and add. If you are having trouble remembering why we do this, remember that we can think of $-(x^2 + 2x + 3)$ as $-1(x^2 + 2x + 3)$. If we distribute the -1 across $x^2 + 2x + 3$, we get $-x^2 - 2x - 3$:

$(3x^2 + x + 4) - (x^2 + 2x + 3)$

$\qquad = 3x^2 + x + 4 - x^2 - 2x - 3 \qquad$ Take the opposite of each term in the second polynomial.

$\qquad = (3x^2 - x^2) + (x - 2x) + (4 - 3)$

$\qquad = 2x^2 - x + 1$

EXAMPLE 4 Subtract $-4x^2 + 5x - 7$ from $x^2 - x - 1$.

Solution The polynomial $x^2 - x - 1$ comes first, then the subtraction sign, and finally the polynomial $-4x^2 + 5x - 7$ in parentheses.

$(x^2 - x - 1) - (-4x^2 + 5x - 7)$

$\qquad = x^2 - x - 1 + 4x^2 - 5x + 7 \qquad$ We take the opposite of each term in the second polynomial.

$\qquad = (x^2 + 4x^2) + (-x - 5x) + (-1 + 7)$

$\qquad = 5x^2 - 6x + 6$

There are two important points to remember when adding or subtracting polynomials. First, to add or subtract two polynomials, you always add or subtract *coefficients* of similar terms. Second, the exponents never increase in value when you are adding or subtracting similar terms.

The last topic we want to consider in this section is finding the value of a polynomial for a given value of the variable.

To find the value of the polynomial $3x^2 + 1$ when x is 5, we replace x with 5 and simplify the result:

When	$x = 5$
the polynomial	$3x^2 + 1$
becomes	$3(5)^2 + 1 = 3(25) + 1$
	$= 75 + 1$
	$= 76$

E X A M P L E   5   Find the value of $3x^2 - 5x + 4$ when $x = -2$.

Solution

When $\qquad\qquad\qquad\qquad\qquad\qquad\qquad\qquad x = -2$

the polynomial $\qquad\qquad\qquad\qquad\qquad 3x^2 - 5x + 4$

becomes $\qquad\qquad\qquad 3(-2)^2 - 5(-2) + 4 = 3(4) + 10 + 4$

$$= 12 + 10 + 4$$

$$= 26$$

Getting Ready for Class

After reading through the preceding section, respond in your own words and in complete sentences.

A. What are similar terms?

B. What is the degree of a polynomial?

C. Describe how you would subtract one polynomial from another.

D. Describe how you would find the value of the polynomial $3x^2 - 5x + 4$ when x is -2.

PROBLEM SET 4.4

Identify each of the following polynomials as a trinomial, binomial, or monomial, and give the degree in each case.

1. $2x^3 - 3x^2 + 1$ $\qquad$ **2.** $4x^2 - 4x + 1$

3. $5 + 8a - 9a^3$ $\qquad$ **4.** $6 + 12x^3 + x^4$

5. $2x - 1$ $\qquad\qquad$ **6.** $4 + 7x$

7. $45x^2 - 1$ $\qquad\quad$ **8.** $3a^3 + 8$

9. $7a^2$ $\qquad\qquad\quad$ **10.** $90x$

11. -4 $\qquad\qquad\quad$ **12.** 56

Perform the following additions and subtractions.

13. $(2x^2 + 3x + 4) + (3x^2 + 2x + 5)$

14. $(x^2 + 5x + 6) + (x^2 + 3x + 4)$

15. $(3a^2 - 4a + 1) + (2a^2 - 5a + 6)$

16. $(5a^2 - 2a + 7) + (4a^2 - 3a + 2)$

17. $x^2 + 4x + 2x + 8$ $\qquad$ **18.** $x^2 + 5x - 3x - 15$

19. $6x^2 - 3x - 10x + 5$

20. $10x^2 + 30x - 2x - 6$

21. $x^2 - 3x + 3x - 9$ $\qquad$ **22.** $x^2 - 5x + 5x - 25$

23. $3y^2 - 5y - 6y + 10$

24. $y^2 - 18y + 2y - 12$

25. $(6x^3 - 4x^2 + 2x) + (9x^2 - 6x + 3)$

26. $(5x^3 + 2x^2 + 3x) + (2x^2 + 5x + 1)$

27. $\left(\dfrac{2}{3}x^2 - \dfrac{1}{5}x - \dfrac{3}{4}\right) + \left(\dfrac{4}{3}x^2 - \dfrac{4}{5}x + \dfrac{7}{4}\right)$

28. $\left(\dfrac{3}{8}x^3 - \dfrac{5}{7}x^2 - \dfrac{2}{5}\right) + \left(\dfrac{5}{8}x^3 - \dfrac{2}{7}x^2 + \dfrac{7}{5}\right)$

29. $(a^2 - a - 1) - (-a^2 + a + 1)$

30. $(5a^2 - a - 6) - (-3a^2 - 2a + 4)$

31. $\left(\dfrac{5}{9}x^3 + \dfrac{1}{3}x^2 - 2x + 1\right) - \left(\dfrac{2}{3}x^3 + x^2 + \dfrac{1}{2}x - \dfrac{3}{4}\right)$

32. $\left(4x^3 - \dfrac{2}{5}x^2 + \dfrac{3}{8}x - 1\right) - \left(\dfrac{9}{2}x^3 + \dfrac{1}{4}x^2 - x + \dfrac{5}{6}\right)$

33. $(4y^2 - 3y + 2) + (5y^2 + 12y - 4) - (13y^2 - 6y + 20)$

34. $(2y^2 - 7y - 8) - (6y^2 + 6y - 8) + (4y^2 - 2y + 3)$

35. Subtract $10x^2 + 23x - 50$ from $11x^2 - 10x + 13$.

36. Subtract $2x^2 - 3x + 5$ from $4x^2 - 5x + 10$.

37. Subtract $3y^2 + 7y - 15$ from $11y^2 + 11y + 11$.

38. Subtract $15y^2 - 8y - 2$ from $3y^2 - 3y + 2$.

39. Add $50x^2 - 100x - 150$ to $25x^2 - 50x + 75$.

40. Add $7x^2 - 8x + 10$ to $-8x^2 + 2x - 12$.

41. Subtract $2x + 1$ from the sum of $3x - 2$ and $11x + 5$.

42. Subtract $3x - 5$ from the sum of $5x + 2$ and $9x - 1$.

43. Find the value of the polynomial $x^2 - 2x + 1$ when x is 3.

44. Find the value of the polynomial $(x - 1)^2$ when x is 3.

45. Find the value of the polynomial $(y - 5)^2$ when y is 10.

46. Find the value of the polynomial $y^2 - 10y + 25$ when y is 10.

47. Find the value of $a^2 + 4a + 4$ when a is 2.

48. Find the value of $(a + 2)^2$ when a is 2.

Applying the Concepts

49. Packaging A crystal ball with a diameter of 6 inches is being packaged for shipment. If the crys-

tal ball is placed inside a circular cylinder with radius 3 inches and height 6 inches, how much volume will need to be filled with padding? (The volume of a sphere with radius r is $\dfrac{4}{3}\pi r^3$, and the volume of a right circular cylinder with radius r and height h is $\pi r^2 h$.)

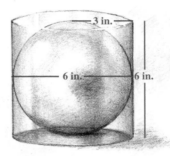

50. Packaging Suppose the circular cylinder of Problem 49 has a radius of 4 inches and a height of 7 inches. How much volume will need to be filled with padding?

Review Problems

The following problems review material we covered in Section 4.3. Reviewing these problems will help you understand the next section.

Multiply.

51. $3x(-5x)$

52. $-3x(-7x)$

53. $2x(3x^2)$

54. $x^2(3x)$

55. $3x^2(2x^2)$

56. $4x^2(2x^2)$

4.5   **Multiplication With Polynomials**

We begin our discussion of multiplication of polynomials by finding the product of a monomial and a trinomial.

E X A M P L E   1   Multiply $3x^2(2x^2 + 4x + 5)$.

Solution   Applying the distributive property gives us

$$3x^2(2x^2 + 4x + 5) = 3x^2(2x^2) + 3x^2(4x) + 3x^2(5) \qquad \text{Distributive property}$$
$$= 6x^4 + 12x^3 + 15x^2 \qquad \text{Multiplication}$$

The distributive property is the key to multiplication of polynomials. We can use it to find the product of any two polynomials. There are some shortcuts we can use in certain situations, however. Let's look at an example that involves the product of two binomials.

E X A M P L E   2   Multiply $(3x - 5)(2x - 1)$.

Solution

$$(3x - 5)(2x - 1) = 3x(2x - 1) - 5(2x - 1)$$
$$= 3x(2x) + 3x(-1) + (-5)(2x) + (-5)(-1)$$
$$= 6x^2 - 3x - 10x + 5$$
$$= 6x^2 - 13x + 5$$

If we look closely at the second and third lines of work in this example, we can see that the terms in the answer come from all possible products of terms in the first binomial with terms in the second binomial. This result is generalized as follows.

RULE

To multiply any two polynomials, multiply each term in the first with each term in the second.

There are two ways we can put this rule to work.

FOIL METHOD

If we look at the original problem in Example 2 and then at the answer, we see that the first term in the answer came from multiplying the first terms in each binomial:

$$3x \cdot 2x = 6x^2 \qquad \text{FIRST}$$

The middle term of the answer came from adding the product of the two outside terms to the product of the two inside terms in each binomial:

$$3x(-1) = -3x \qquad \text{OUTSIDE}$$
$$-5(2x) = \underline{-10x} \qquad \text{INSIDE}$$
$$-13x$$

The last term in the answer came from multiplying the two last terms:

$$-5(-1) = 5 \quad \text{LAST}$$

To summarize the FOIL method, we will multiply another two binomials.

E X A M P L E   3   Multiply $(2x + 3)(5x - 4)$.

Solution

$$(2x + 3)(5x - 4) = \underbrace{2x(5x)}_{} + \underbrace{2x(-4)}_{} + \underbrace{3(5x)}_{} + \underbrace{3(-4)}_{}$$

FIRST

OUTSIDE

INSIDE

LAST

$$= 10x^2 - 8x + 15x - 12$$
$$= 10x^2 + 7x - 12$$

With practice $-8x + 15x = 7x$ can be done mentally.

COLUMN METHOD

The FOIL method can be applied only when multiplying two binomials. To find products of polynomials with more than two terms, we use what is called the COLUMN method.

The COLUMN method of multiplying two polynomials is very similar to long multiplication with whole numbers. It is just another way of finding all possible products of terms in one polynomial with terms in another polynomial.

E X A M P L E   4   Multiply $(2x + 3)(3x^2 - 2x + 1)$.

Solution

$$
\begin{array}{r}
3x^2 - 2x + 1 \\
2x + 3 \\
\hline
6x^3 - 4x^2 + 2x \quad\quad \leftarrow 2x(3x^2 - 2x + 1) \\
9x^2 - 6x + 3 \leftarrow 3(3x^2 - 2x + 1) \\
\hline
6x^3 + 5x^2 - 4x + 3 \leftarrow \text{Add similar terms.}
\end{array}
$$

It will be to your advantage to become very fast and accurate at multiplying polynomials. You should be comfortable using either method. The following examples illustrate the three types of multiplication.

EXAMPLES Multiply:

5. $4a^2(2a^2 - 3a + 5) = 4a^2(2a^2) + 4a^2(-3a) + 4a^2(5)$

$$= 8a^4 - 12a^3 + 20a^2$$

6. $(x - 2)(y + 3) = x(y) + x(3) + (-2)(y) + (-2)(3)$

$$\qquad\qquad\quad\;\; \text{F}\qquad \text{O}\qquad \text{I}\qquad\;\; \text{L}$$

$$= xy + 3x - 2y - 6$$

7. $(x + y)(a - b) = x(a) + x(-b) + y(a) + y(-b)$

$$\qquad\qquad\quad\;\; \text{F}\qquad \text{O}\qquad \text{I}\qquad \text{L}$$

$$= xa - xb + ya - yb$$

8. $(5x - 1)(2x + 6) = 5x(2x) + 5x(6) + (-1)(2x) + (-1)(6)$

$$\qquad\qquad\qquad\quad \text{F}\qquad \text{O}\qquad\quad \text{I}\qquad\quad \text{L}$$

$$= 10x^2 + 30x + (-2x) + (-6)$$

$$= 10x^2 + 28x - 6$$

9. $(3x + 2)(x^2 - 5x + 6)$

$$
\begin{array}{r}
x^2 - \;\; 5x + \;\; 6 \\
3x + \;\; 2 \\
\hline
3x^3 - 15x^2 + 18x \qquad\;\; \\
2x^2 - 10x + 12 \\
\hline
3x^3 - 13x^2 + \;\; 8x + 12 \\
\end{array}
$$

EXAMPLE 10 The length of a rectangle is 3 units more than twice the width. Write an expression for the area of the rectangle.

Solution We begin by drawing a rectangle and labeling the width x. Since the length is 3 more than twice the width, we label the length $2x + 3$.

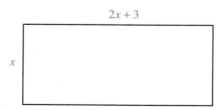

Since the area A of a rectangle is the product of the length and width, we write our formula for the area of this rectangle as

$$A = x(2x + 3)$$

$$A = 2x^2 + 3x \qquad \text{Multiply.}$$

REVENUE

Suppose that a store sells x items at p dollars per item. The total amount of money obtained by selling the items is called the *revenue*. It can be found by multiplying the number of items sold, x, by the price per item, p. For example, if 100 items are sold for $6 each, the revenue is $100(6) = \$600$. Similarly, if 500 items are sold for $8 each, the total revenue is $500(8) = \$4,000$. If we denote the revenue with the letter R, then the formula that relates R, x, and p is

$$\text{Revenue} = (\text{number of items sold})(\text{price of each item})$$

In symbols: $R = xp$.

EXAMPLE 11   A store selling diskettes for home computers knows from past experience that it can sell x diskettes each day at a price of p dollars per diskette, according to the equation $x = 800 - 100p$. Write a formula for the daily revenue that involves only the variables R and p.

Solution   From our previous discussion we know that the revenue R is given by the formula

$$R = xp$$

But, since $x = 800 - 100p$, we can substitute $800 - 100p$ for x in the revenue equation to obtain

$$R = (800 - 100p)p$$

$$R = 800p - 100p^2$$

This last formula gives the revenue, R, in terms of the price, p.

Getting Ready for Class

After reading through the preceding section, respond in your own words and in complete sentences.

A. How do we multiply two polynomials?

B. Describe how the distributive property is used to multiply a monomial and a polynomial.

C. Describe how you would use the FOIL method to multiply two binomials.

D. Give an example showing how the product of two binomials can be a trinomial.

PROBLEM SET 4.5

Multiply the following by applying the distributive property.

1. $2x(3x + 1)$

2. $4x(2x - 3)$

3. $2x^2(3x^2 - 2x + 1)$

4. $5x(4x^3 - 5x^2 + x)$

5. $2ab(a^2 - ab + 1)$

6. $3a^2b(a^3 + a^2b^2 + b^3)$

7. $y^2(3y^2 + 9y + 12)$

8. $5y(2y^2 - 3y + 5)$

9. $4x^2y(2x^3y + 3x^2y^2 + 8y^3)$

10. $6xy^3(2x^2 + 5xy + 12y^2)$

Multiply the following binomials. You should do about half the problems using the FOIL method and the other half using the COLUMN method. Remember, you want to be comfortable using both methods.

11. $(x + 3)(x + 4)$

12. $(x + 2)(x + 5)$

13. $(x + 6)(x + 1)$

14. $(x + 1)(x + 4)$

15. $\left(x + \dfrac{1}{2}\right)\left(x + \dfrac{3}{2}\right)$

16. $\left(x + \dfrac{3}{5}\right)\left(x + \dfrac{2}{5}\right)$

17. $(a + 5)(a - 3)$

18. $(a - 8)(a + 2)$

19. $(x - a)(y + b)$

20. $(x + a)(y - b)$

21. $(x + 6)(x - 6)$

22. $(x + 3)(x - 3)$

23. $\left(y + \dfrac{5}{6}\right)\left(y - \dfrac{5}{6}\right)$

24. $\left(y - \dfrac{4}{7}\right)\left(y + \dfrac{4}{7}\right)$

25. $(2x - 3)(x - 4)$

26. $(3x - 5)(x - 2)$

27. $(a + 2)(2a - 1)$

28. $(a - 6)(3a + 2)$

29. $(2x - 5)(3x - 2)$

30. $(3x + 6)(2x - 1)$

31. $(2x + 3)(a + 4)$

32. $(2x - 3)(a - 4)$

33. $(5x - 4)(5x + 4)$

34. $(6x + 5)(6x - 5)$

35. $\left(2x - \dfrac{1}{2}\right)\left(x + \dfrac{3}{2}\right)$

36. $\left(4x - \dfrac{3}{2}\right)\left(x + \dfrac{1}{2}\right)$

37. $(1 - 2a)(3 - 4a)$

38. $(1 - 3a)(3 + 2a)$

For each of the following problems, fill in the area of each small rectangle and square, and then add the results together to find the indicated product.

39. $(x + 2)(x + 3)$

40. $(x + 4)(x + 5)$

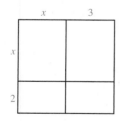

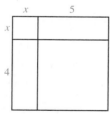

41. $(x + 1)(2x + 2)$

42. $(2x + 1)(2x + 2)$

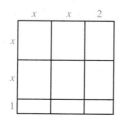

Multiply the following.

43. $(a - 3)(a^2 - 3a + 2)$

44. $(a + 5)(a^2 + 2a + 3)$

45. $(x + 2)(x^2 - 2x + 4)$

46. $(x + 3)(x^2 - 3x + 9)$

47. $(2x + 1)(x^2 + 8x + 9)$

48. $(3x - 2)(x^2 - 7x + 8)$

49. $(5x^2 + 2x + 1)(x^2 - 3x + 5)$

50. $(2x^2 + x + 1)(x^2 - 4x + 3)$

Multiply.

51. $(x^2 + 3)(2x^2 - 5)$

52. $(4x^3 - 8)(5x^3 + 4)$

53. $(3a^4 + 2)(2a^2 + 5)$

54. $(7a^4 - 8)(4a^3 - 6)$

55. $(x + 3)(x + 4)(x + 5)$

56. $(x - 3)(x - 4)(x - 5)$

Applying the Concepts

57. Area   The length of a rectangle is 5 units more than twice the width. Write an expression for the area of the rectangle.

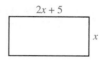

58. Area   The length of a rectangle is 2 units more than 3 times the width. Write an expression for the area of the rectangle.

59. Area   The width and length of a rectangle are given by two consecutive integers. Write an expression for the area of the rectangle.

60. Area   The width and length of a rectangle are given by two consecutive even integers. Write an expression for the area of the rectangle.

61. Revenue   A store selling typewriter ribbons knows that the number of ribbons it can sell each week, x, is related to the price per ribbon, p, by the equation $x = 1,200 - 100p$. Write an expression for the weekly revenue that involves only the variables R and p. (*Remember:* The equation for revenue is $R = xp$.)

62. Revenue   A store selling small portable radios knows from past experience that the number of radios it can sell each week, x, is related to the price per radio, p, by the equation $x = 1,300 - 100p$.

Write an expression for the weekly revenue that involves only the variables R and p.

63. Revenue The relationship between the number of calculators a company sells per day, x, and the price of each calculator, p, is given by the equation $x = 1,700 - 100p$. Write an expression for the daily revenue that involves only the variables R and p.

64. Revenue The relationship between the number of pencil sharpeners a company can sell each day, x, and the price of each sharpener, p, is given by the equation $x = 1,800 - 100p$. Write an expression for the daily revenue that involves only the variables R and p.

Review Problems

The following problems review material we covered in Sections 3.5, 3.6, 3.7, and 3.8.

65. Solve this system by graphing:
$$x + y = 4$$
$$x - y = 2$$

Solve each system by the addition method.

66. $3x + 2y = 1$
$\quad\ 2x + \ y = 3$

67. $2x + 3y = -1$
$\quad\ 3x + 5y = -2$

Solve each system by the substitution method.

68. $x + y = 20$
$\quad\ \ y = 5x + 2$

69. $2x - 6y = 2$
$\quad\ \ \ y = 3x + 1$

70. Investing A total of $1,200 is invested in two accounts. One of the accounts pays 8% interest annually, and the other pays 10% interest annually. If the total amount of interest earned from both accounts for the year is $104, how much is invested in each account?

71. Coin Problem Amy has $1.85 in dimes and quarters. If she has a total of 11 coins, how many of each coin does she have?

4.6 **Binomial Squares and Other Special Products**

In this section we will combine the results of the last section with our definition of exponents to find some special products.

EXAMPLE 1 Find the square of $(3x - 2)$.

Solution To square $(3x - 2)$, we multiply it by itself:

$$(3x - 2)^2 = (3x - 2)(3x - 2) \qquad \text{Definition of exponents}$$
$$= 9x^2 - 6x - 6x + 4 \qquad \text{FOIL method}$$
$$= 9x^2 - 12x + 4 \qquad \text{Combine similar terms.}$$

Notice that the first and last terms in the answer are the square of the first and last terms in the original problem and that the middle term is twice the product of the two terms in the original binomial.

EXAMPLES

2. $(a + b)^2 = (a + b)(a + b)$
$$= a^2 + 2ab + b^2$$

3. $(a - b)^2 = (a - b)(a - b)$
$$= a^2 - 2ab + b^2$$

Binomial squares having the form of Examples 2 and 3 occur very frequently in algebra. It will be to your advantage to memorize the following rule for squaring a binomial.

> **RULE**
>
> The square of a binomial is the sum of the square of the first term, the square of the last term, and twice the product of the two original terms. In symbols this rule is written as follows:
>
> $(x + y)^2 =$ x^2 $+$ $2xy$ $+$ y^2
> Square of first term   Twice product of the two terms   Square of last term

Note: A very common mistake when squaring binomials is to write

$$(a + b)^2 = a^2 + b^2$$

which just isn't true. The mistake becomes obvious when we substitute 2 for a and 3 for b:

$$(2 + 3)^2 \neq 2^2 + 3^2$$

$$25 \neq 13$$

Exponents do not distribute over addition or subtraction.

EXAMPLES   Multiply using the preceding rule:

	First term squared	*Twice their product*	*Last term squared*	*Answer*
4. $(x - 5)^2 =$	x^2	$+\ 2(x)(-5)\ +$	25	$= x^2 - 10x + 25$
5. $(x + 2)^2 =$	x^2	$+\ 2(x)(2)\ +$	4	$= x^2 + 4x + 4$
6. $(2x - 3)^2 =$	$4x^2$	$+\ 2(2x)(-3)\ +$	9	$= 4x^2 - 12x + 9$
7. $(5x - 4)^2 =$	$25x^2$	$+\ 2(5x)(-4)\ +$	16	$= 25x^2 - 40x + 16$

Another special product that occurs frequently is $(a + b)(a - b)$. The only difference in the two binomials is the sign between the two terms. The interesting thing about this type of product is that the middle term is always zero. Here are some examples.

EXAMPLES   Multiply using the FOIL method:

8. $(2x - 3)(2x + 3) = 4x^2 + 6x - 6x - 9$ FOIL method
$$= 4x^2 - 9$$

9. $(x - 5)(x + 5) = x^2 + 5x - 5x - 25$ FOIL method

$\qquad\qquad\qquad\quad = x^2 - 25$

10. $(3x - 1)(3x + 1) = 9x^2 + 3x - 3x - 1$ FOIL method

$\qquad\qquad\qquad\qquad = 9x^2 - 1$

Notice that in each case the middle term is zero and therefore doesn't appear in the answer. The answers all turn out to be the difference of two squares. Here is a rule to help you memorize the result.

RULE

When multiplying two binomials that differ only in the sign between their terms, subtract the square of the last term from the square of the first term.

$$(a - b)(a + b) = a^2 - b^2$$

Here are some problems that result in the difference of two squares.

EXAMPLES   Multiply using the preceding rule:

11. $(x + 3)(x - 3) = x^2 - 9$

12. $(a + 2)(a - 2) = a^2 - 4$

13. $(9a + 1)(9a - 1) = 81a^2 - 1$

14. $(2x - 5y)(2x + 5y) = 4x^2 - 25y^2$

15. $(3a - 7b)(3a + 7b) = 9a^2 - 49b^2$

Although all the problems in this section can be worked correctly using the methods in the previous section, they can be done much faster if the new rules are *memorized.* Here is a summary of the new rules:

$$(a + b)^2 = (a + b)(a + b) = a^2 + 2ab + b^2$$

$$(a - b)^2 = (a - b)(a - b) = a^2 - 2ab + b^2$$

$$(a - b)(a + b) = a^2 - b^2$$

EXAMPLE 16   Write an expression in symbols for the sum of the squares of three consecutive even integers. Then, simplify that expression.

Solution   If we let $x =$ the first of the even integers, then $x + 2$ is the next consecutive even integer and $x + 4$ is the one after that. An expression for the sum of their squares is

$x^2 + (x + 2)^2 + (x + 4)^2$ Sum of squares

$\qquad = x^2 + (x^2 + 4x + 4) + (x^2 + 8x + 16)$ Expand squares.

$\qquad = 3x^2 + 12x + 20$ Add similar terms.

Getting Ready for Class

After reading through the preceding section, respond in your own words and in complete sentences.

A. Describe how you would square the binomial $a + b$.

B. Explain why $(x + 3)^2$ cannot be $x^2 + 9$.

C. What kind of product results in the difference of two squares?

D. When multiplied out, how will $(x + 3)^2$ and $(x - 3)^2$ differ?

PROBLEM SET 4.6

Perform the indicated operations.

1. $(x - 2)^2$ **2.** $(x + 2)^2$

3. $(a + 3)^2$ **4.** $(a - 3)^2$

5. $(x - 5)^2$ **6.** $(x - 4)^2$

7. $\left(a - \dfrac{1}{2}\right)^2$ **8.** $\left(a + \dfrac{1}{2}\right)^2$

9. $(x + 10)^2$ **10.** $(x - 10)^2$

11. $(a + 0.8)^2$ **12.** $(a - 0.4)^2$

13. $(2x - 1)^2$ **14.** $(3x + 2)^2$

15. $(4a + 5)^2$ **16.** $(4a - 5)^2$

17. $(3x - 2)^2$ **18.** $(2x - 3)^2$

19. $(3a + 5b)^2$ **20.** $(5a - 3b)^2$

21. $(4x - 5y)^2$ **22.** $(5x + 4y)^2$

23. $(7m + 2n)^2$ **24.** $(2m - 7n)^2$

25. $(6x - 10y)^2$ **26.** $(10x + 6y)^2$

27. $(x^2 + 5)^2$ **28.** $(x^2 + 3)^2$

29. $(a^2 + 1)^2$ **30.** $(a^2 - 2)^2$

Comparing Expressions Fill in each table.

31.

x	$(x + 3)^2$	$x^2 + 9$	$x^2 + 6x + 9$
1			
2			
3			
4			

32.

x	$(x - 5)^2$	$x^2 + 25$	$x^2 - 10x + 25$
1			
2			
3			
4			

33.

a	b	$(a + b)^2$	$a^2 + b^2$	$a^2 + ab + b^2$	$a^2 + 2ab + b^2$
1	1				
3	5				
3	4				
4	5				

34.

a	b	$(a - b)^2$	$a^2 - b^2$	$a^2 - 2ab + b^2$
2	1			
5	2			
2	5			
4	3			

Multiply.

35. $(a + 5)(a - 5)$ **36.** $(a - 6)(a + 6)$

37. $(y - 1)(y + 1)$ **38.** $(y - 2)(y + 2)$

39. $(9 + x)(9 - x)$ **40.** $(10 - x)(10 + x)$

41. $(2x + 5)(2x - 5)$ **42.** $(3x + 5)(3x - 5)$

43. $\left(4x + \dfrac{1}{3}\right)\left(4x - \dfrac{1}{3}\right)$ **44.** $\left(6x + \dfrac{1}{4}\right)\left(6x - \dfrac{1}{4}\right)$

45. $(2a + 7)(2a - 7)$

46. $(3a + 10)(3a - 10)$

47. $(6 - 7x)(6 + 7x)$

48. $(7 - 6x)(7 + 6x)$

49. $(x^2 + 3)(x^2 - 3)$

50. $(x^2 + 2)(x^2 - 2)$

51. $(a^2 + 4)(a^2 - 4)$

52. $(a^2 + 9)(a^2 - 9)$

53. $(5y^4 - 8)(5y^4 + 8)$

54. $(7y^5 + 6)(7y^5 - 6)$

Multiply and simplify.

55. $(x + 3)(x - 3) + (x - 5)(x + 5)$

56. $(x - 7)(x + 7) + (x - 4)(x + 4)$

57. $(2x + 3)^2 - (4x - 1)^2$

58. $(3x - 5)^2 - (2x + 3)^2$

59. $(a + 1)^2 - (a + 2)^2 + (a + 3)^2$

60. $(a - 1)^2 + (a - 2)^2 - (a - 3)^2$

61. $(2x + 3)^3$ 62. $(3x - 2)^3$

63. **Shortcut** The formula for the difference of two squares can be used as a shortcut to multiplying certain whole numbers if they have the correct form. Use the difference of two squares formula to multiply 49(51) by first writing 49 as $(50 - 1)$ and 51 as $(50 + 1)$.

64. **Shortcut** Use the difference of two squares formula to multiply 101(99) by first writing 101 as $(100 + 1)$ and 99 as $(100 - 1)$.

65. **Comparing Expressions** Evaluate the expression $(x + 3)^2$ and the expression $x^2 + 6x + 9$ for $x = 2$.

66. **Comparing Expressions** Evaluate the expression $x^2 - 25$ and the expression $(x - 5)(x + 5)$ for $x = 6$.

67. **Number Problem** Write an expression for the sum of the squares of two consecutive integers. Then, simplify that expression.

68. **Number Problem** Write an expression for the sum of the squares of two consecutive odd integers. Then, simplify that expression.

69. **Number Problem** Write an expression for the sum of the squares of three consecutive integers. Then, simplify that expression.

70. **Number Problem** Write an expression for the sum of the squares of three consecutive odd integers. Then, simplify that expression.

71. **Area** We can use the concept of area to further justify our rule for squaring a binomial. The length of each side of the square shown in the figure is $a + b$. (The longer line segment has length a, and the shorter line segment has length b.) The area of the whole square is $(a + b)^2$. On the other hand, the whole area is the sum of the areas of the two smaller squares and the two small rectangles that make it up. Write the area of the two smaller squares and the two small rectangles and then add them together to verify the formula $(a + b)^2 = a^2 + 2ab + b^2$.

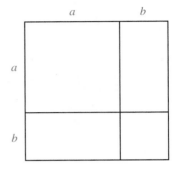

72. **Area** The length of each side of the large square shown in the figure is $x + 5$. Therefore, its area is $(x + 5)^2$. Find the area of the two smaller squares and the two smaller rectangles that make up the large square, and then add them together to verify the formula $(x + 5)^2 = x^2 + 10x + 25$.

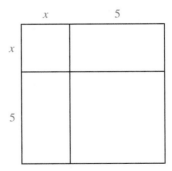

For each problem, fill in the area of each small rectangle and square, and then add the result together to find the indicated product.

73. $(2x + 1)^2$

74. $(2x + 3)^2$

Review Problems

The following problems review material we covered in Sections 4.3 and 3.5.

Simplify each expression (divide). [4.3]

75. $\dfrac{15x^2y}{3xy}$ **76.** $\dfrac{21xy^2}{3xy}$

77. $\dfrac{35a^6b^8}{70a^2b^{10}}$ **78.** $\dfrac{75a^2b^6}{25a^4b^3}$

Solve each system by graphing. [3.5]

79. $x + y = 2$ **80.** $x + y = 1$
$x - y = 4$ $x - y = -3$

81. $y = 2x + 3$ **82.** $y = 3x - 2$
$y = -2x - 1$ $y = -2x + 3$

4.7   Dividing a Polynomial by a Monomial

To divide a polynomial by a monomial, we will use the definition of division and apply the distributive property. Follow the steps in this example closely.

EXAMPLE 1   Divide $10x^3 - 15x^2$ by $5x$.

Solution

$$\frac{10x^3 - 15x^2}{5x} = (10x^3 - 15x^2)\frac{1}{5x}$$ Division by $5x$ is the same as multiplication by $\dfrac{1}{5x}$.

$$= 10x^3\left(\frac{1}{5x}\right) - 15x^2\left(\frac{1}{5x}\right)$$ Distribute $\dfrac{1}{5x}$ to both terms.

$$= \frac{10x^3}{5x} - \frac{15x^2}{5x}$$ Multiplication by $\dfrac{1}{5x}$ is the same as division by $5x$.

$$= 2x^2 - 3x$$ Division of monomials as done in Section 4.3

If we were to leave out the first steps, the problem would look like this:

$$\frac{10x^3 - 15x^2}{5x} = \frac{10x^3}{5x} - \frac{15x^2}{5x}$$

$$= 2x^2 - 3x$$

The problem is much shorter and clearer this way. You may leave out the first two steps from Example 1 when working problems in this section. They are part of Example 1 only to help show you why the following rule is true.

> **RULE**
>
> To divide a polynomial by a monomial, simply divide each term in the polynomial by the monomial.

Here are some further examples using our rule for division of a polynomial by a monomial.

E X A M P L E   2   Divide $\dfrac{3x^2 - 6}{3}$.

Solution   We begin by writing the 3 in the denominator under each term in the numerator. Then we simplify the result:

$$\frac{3x^2 - 6}{3} = \frac{3x^2}{3} - \frac{6}{3} \qquad \text{Divide each term in the numerator by 3.}$$

$$= x^2 - 2 \qquad \text{Simplify.}$$

E X A M P L E   3   Divide $\dfrac{4x^2 - 2}{2}$.

Solution   Dividing each term in the numerator by 2, we have

$$\frac{4x^2 - 2}{2} = \frac{4x^2}{2} - \frac{2}{2} \qquad \text{Divide each term in the numerator by 2.}$$

$$= 2x^2 - 1 \qquad \text{Simplify.}$$

E X A M P L E   4   Find the quotient of $27x^3 - 9x^2$ and $3x$.

Solution   We are again asked to divide the first polynomial by the second one:

$$\frac{27x^3 - 9x^2}{3x} = \frac{27x^3}{3x} - \frac{9x^2}{3x} \qquad \text{Divide each term by } 3x.$$

$$= 9x^2 - 3x \qquad \text{Simplify.}$$

E X A M P L E   5   Divide $(15x^2y - 21xy^2)$ by $(-3xy)$.

Solution   This is the same type of problem we have shown in the first four examples; it is just worded a little differently. Note that when we divide each term in the first polynomial by $-3xy$, the negative sign must be taken into account:

$$\frac{15x^2y - 21xy^2}{-3xy} = \frac{15x^2y}{-3xy} - \frac{21xy^2}{-3xy}$$ Divide each term by $-3xy$.

$$= -5x - (-7y)$$ Simplify.

$$= -5x + 7y$$ Simplify.

EXAMPLE 6 Divide $\dfrac{10x^3 - 5x^2 + 20x}{10x^3}$.

Solution Proceeding as we have in the first five examples, we distribute $10x^3$ under each term in the numerator:

$$\frac{10x^3 - 5x^2 + 20x}{10x^3} = \frac{10x^3}{10x^3} - \frac{5x^2}{10x^3} + \frac{20x}{10x^3}$$ Divide each term by $10x^3$.

$$= 1 - \frac{1}{2x} + \frac{2}{x^2}$$ Simplify.

Note that our answer is not a polynomial.

EXAMPLE 7 Divide $\dfrac{24x^3y^2 + 16x^2y^2 - 4x^2y^3}{8x^2y}$.

Solution Writing $8x^2y$ under each term in the numerator and then simplifying, we have

$$\frac{24x^3y^2 + 16x^2y^2 - 4x^2y^3}{8x^2y} = \frac{24x^3y^2}{8x^2y} + \frac{16x^2y^2}{8x^2y} - \frac{4x^2y^3}{8x^2y}$$

$$= 3xy + 2y - \frac{y^2}{2}$$

From the examples in this section, it is clear that to divide a polynomial by a monomial, we must divide each term in the polynomial by the monomial. Often, students taking algebra for the first time will make the following mistake:

$$\frac{x + 2}{2} = x + 1$$ Mistake

The mistake here is in not dividing both terms in the numerator by 2. The correct way to divide $x + 2$ by 2 looks like this:

$$\frac{x + 2}{2} = \frac{x}{2} + \frac{2}{2} = \frac{x}{2} + 1$$ Correct

Getting Ready for Class

After reading through the preceding section, respond in your own words and in complete sentences.

A. What property of real numbers is the key to dividing a polynomial by a monomial?

B. Describe how you would divide a polynomial by $5x$.

C. What part do the properties of exponents play in dividing a polynomial by a monomial?

D. Explain the mistake in the problem $\dfrac{x + 2}{2} = x + 1$.

PROBLEM SET 4.7

Divide the following polynomials by $5x$.

1. $5x^2 - 10x$

2. $10x^3 - 15x$

3. $15x - 10x^3$

4. $50x^3 - 20x^2$

5. $25x^2y - 10xy$

6. $15xy^2 + 20x^2y$

7. $35x^5 - 30x^4 + 25x^3$

8. $40x^4 - 30x^3 + 20x^2$

9. $50x^5 - 25x^3 + 5x$

10. $75x^6 + 50x^3 - 25x$

Divide the following by $-2a$.

11. $8a^2 - 4a$

12. $a^3 - 6a^2$

13. $16a^5 + 24a^4$

14. $30a^6 + 20a^3$

15. $8ab + 10a^2$

16. $6a^2b - 10ab^2$

17. $12a^3b - 6a^2b^2 + 14ab^3$

18. $4ab^3 - 16a^2b^2 - 22a^3b$

19. $a^2 + 2ab + b^2$

20. $a^2b - 2ab^2 + b^3$

Perform the following divisions (find the following quotients).

21. $\dfrac{6x + 8y}{2}$

22. $\dfrac{9x - 3y}{3}$

23. $\dfrac{7y - 21}{-7}$

24. $\dfrac{14y - 12}{2}$

25. $\dfrac{10xy - 8x}{2x}$

26. $\dfrac{12xy^2 - 18x}{-6x}$

27. $\dfrac{x^2y - x^3y^2}{x}$

28. $\dfrac{x^2y - x^3y^2}{x^2}$

29. $\dfrac{x^2y - x^3y^2}{-x^2y}$

30. $\dfrac{ab + a^2b^2}{ab}$

31. $\dfrac{a^2b^2 - ab^2}{-ab^2}$

32. $\dfrac{a^2b^2c + ab^2c^2}{abc}$

33. $\dfrac{x^3 - 3x^2y + xy^2}{x}$

34. $\dfrac{x^2 - 3xy^2 + xy^3}{x}$

35. $\dfrac{10a^2 - 15a^2b + 25a^2b^2}{5a^2}$

36. $\dfrac{11a^2b^2 - 33ab}{-11ab}$

37. $\dfrac{26x^2y^2 - 13xy}{-13xy}$

38. $\dfrac{6x^2y^2 - 3xy}{6xy}$

39. $\dfrac{4x^2y^2 - 2xy}{4xy}$

40. $\dfrac{6x^2a + 12x^2b - 6x^2c}{36x^2}$

41. $\dfrac{5a^2x - 10ax^2 + 15a^2x^2}{20a^2x^2}$

42. $\dfrac{12ax - 9bx + 18cx}{6x^2}$

43. $\dfrac{16x^5 + 8x^2 + 12x}{12x^3}$

44. $\dfrac{27x^2 - 9x^3 - 18x^4}{-18x^3}$

Divide. Assume all variables represent positive numbers.

45. $\dfrac{9a^{5m} - 27a^{3m}}{3a^{2m}}$

46. $\dfrac{26a^{3m} - 39a^{5m}}{13a^{3m}}$

47. $\dfrac{10x^{5m} - 25x^{3m} + 35x^m}{5x^m}$

48. $\dfrac{18x^{2m} + 24x^{4m} - 30x^{6m}}{6x^{2m}}$

Simplify each numerator, and then divide.

49. $\dfrac{2x^3(3x + 2) - 3x^2(2x - 4)}{2x^2}$

50. $\dfrac{5x^2(6x - 3) + 6x^3(3x - 1)}{3x}$

51. $\dfrac{(x + 2)^2 - (x - 2)^2}{2x}$

52. $\dfrac{(x - 3)^2 - (x + 3)^2}{3x}$

53. $\dfrac{(x + 5)^2 + (x + 5)(x - 5)}{2x}$

54. $\dfrac{(x - 4)^2 + (x + 4)(x - 4)}{2x}$

55. Comparing Expressions   Evaluate the expression $\dfrac{10x + 15}{5}$ and the expression $2x + 3$ when $x = 2$.

56. Comparing Expressions   Evaluate the expression $\dfrac{6x^2 + 4x}{2x}$ and the expression $3x + 2$ when $x = 5$.

57. Comparing Expressions   Show that the expression $\dfrac{3x + 8}{2}$ is not the same as the expression $3x + 4$ by replacing x with 10 in both expressions and simplifying the results.

58. Comparing Expressions   Show that the expression $\dfrac{x + 10}{x}$ is not equal to 10 by replacing x with 5 and simplifying.

Review Problems

The following problems review material we covered in Sections 3.6 and 3.7.

Solve each system of equations by the addition method. [3.6]

59. $x + y = 6$
$\quad\;\, x - y = 8$

60. $2x + y = 5$
$\quad\;\; -x + y = -4$

61. $2x - 3y = -5$
$\quad\;\; x + \;\; y = 5$

62. $2x - 4y = 10$
$\quad\;\; 3x - 2y = -1$

Solve each system by the substitution method. [3.7]

63. $x + y = 2$
$\quad\;\;\;\;\; y = 2x - 1$

64. $2x - 3y = 4$
$\quad\;\;\;\;\;\;\; x = 3y - 1$

65. $4x + 2y = 8$
$\quad\;\;\;\;\; y = -2x + 4$

66. $4x + 2y = 8$
$\quad\;\;\;\;\; y = -2x + 5$

4.8   ## Dividing a Polynomial by a Polynomial

Since long division for polynomials is very similar to long division with whole numbers, we will begin by reviewing a division problem with whole numbers. You may realize when looking at Example 1 that you don't have a very good idea why you proceed as you do with long division. What you do know is that the process always works. We are going to approach the explanations in this section in much the same manner. That is, we won't always be sure why the steps we will use are important, only that they always produce the correct result.

E X A M P L E   1   Divide $27\overline{)3{,}962}$.

Solution

$$
\begin{array}{r}
1 \quad \leftarrow \text{Estimate 27 into 39.} \\
27\overline{)3{,}962} \\
\underline{2\,7} \quad \leftarrow \text{Multiply } 1 \times 27 = 27. \\
1\,2 \quad \leftarrow \text{Subtract } 39 - 27 = 12.
\end{array}
$$

$$
\begin{array}{r}
1 \\
27\overline{)3{,}962} \\
\underline{2\,7\downarrow} \\
1\,26 \quad \leftarrow \text{Bring down the 6.}
\end{array}
$$

These are the four basic steps in long division. Estimate, multiply, subtract, and bring down the next term. To finish the problem, we simply perform the same four steps again:

$$
\begin{array}{r}
14 \quad \leftarrow \text{4 is the estimate.} \\
27\overline{)3{,}962} \\
\underline{2\,7} \\
1\,26 \\
\underline{1\,08}\downarrow \quad \leftarrow \text{Multiply to get 108.} \\
182 \quad \leftarrow \text{Subtract to get 18, and then bring down the 2.}
\end{array}
$$

One more time.

$$
\begin{array}{r}
146 \quad \leftarrow \text{6 is the estimate.} \\
27\overline{)3{,}962} \\
\underline{2\,7} \\
1\,26 \\
\underline{1\,08} \\
182 \\
\underline{162} \quad \leftarrow \text{Multiply to get 162.} \\
20 \quad \leftarrow \text{Subtract to get 20.}
\end{array}
$$

Since there is nothing left to bring down, we have our answer.

$$
\frac{3{,}962}{27} = 146 + \frac{20}{27}, \text{ or } 146\frac{20}{27}
$$

Here is how it works with polynomials.

EXAMPLE 2 Divide $\dfrac{x^2 - 5x + 8}{x - 3}$.

Solution

$$
\begin{array}{r}
x \\
x - 3 \overline{)\; x^2 - 5x + 8}
\end{array}
$$
← Estimate $x^2 \div x = x$.

$$
\begin{array}{r}
\not{-}\; x^2 \not{+}\; 3x \\
\hline
-\,2x
\end{array}
$$
← Multiply $x(x - 3) = x^2 - 3x$.
← Subtract $(x^2 - 5x) - (x^2 - 3x) = -2x$.

$$
\begin{array}{r}
x \\
x - 3 \overline{)\; x^2 - 5x + 8}
\end{array}
$$

$$
\begin{array}{r}
\not{-}\; x^2 \not{+}\; 3x \downarrow \\
\hline
-\,2x + 8
\end{array}
$$
← Bring down the 8.

Notice that to subtract one polynomial from another, we add its opposite. That is why we change the signs on $x^2 - 3x$ and add what we get to $x^2 - 5x$. (To subtract the second polynomial, simply change the signs and add.)

We perform the same four steps again:

$$
\begin{array}{r}
x - 2 \\
x - 3 \overline{)\; x^2 - 5x + 8}
\end{array}
$$
← -2 is the estimate $(-2x \div x = -2)$.

$$
\begin{array}{r}
\not{-}\; x^2 \not{+}\; 3x \downarrow \\
\hline
-\,2x + 8 \\
\not{+}\; 2x \not{-}\; 6 \\
\hline
2
\end{array}
$$
← Multiply $-2(x - 3) = -2x + 6$.
← Subtract $(-2x + 8) - (-2x + 6) = 2$.

Since there is nothing left to bring down, we have our answer:

$$
\frac{x^2 - 5x + 8}{x - 3} = x - 2 + \frac{2}{x - 3}
$$

To check our answer, we multiply $(x - 3)(x - 2)$ to get $x^2 - 5x + 6$. Then, adding on the remainder, 2, we have $x^2 - 5x + 8$.

EXAMPLE 3   Divide $\dfrac{6x^2 - 11x - 14}{2x - 5}$.

Solution

$$
\begin{array}{r}
3x + 2 \\
2x - 5 \overline{)\; 6x^2 - 11x - 14} \\
\not{-}\; 6x^2 \not{+}\; 15x \downarrow \\
\hline
+\; 4x - 14 \\
\not{-}\; 4x \not{+}\; 10 \\
\hline
-\; 4
\end{array}
$$

$$\frac{6x^2 - 11x - 14}{2x - 5} = 3x + 2 + \frac{-4}{2x - 5}$$

One other step is sometimes necessary. The two polynomials in a division problem must both be in descending powers of the variable and cannot skip any powers from the highest power down to the constant term.

EXAMPLE 4 Divide $\dfrac{2x^3 - 3x + 2}{x - 5}$.

Solution The problem will be much less confusing if we write $2x^3 - 3x + 2$ as $2x^3 + 0x^2 - 3x + 2$. Adding $0x^2$ does not change our original problem.

$$
\begin{array}{r}
2x^2 \\
x - 5 \overline{)\ 2x^3 + 0x^2 - 3x + 2}
\end{array}
$$

$\leftarrow$ Estimate $2x^3 \div x = 2x^2$.

$\leftarrow$ Multiply $2x^2(x - 5) = 2x^3 - 10x^2$.
$\leftarrow$ Subtract:
$$(2x^3 + 0x^2) - (2x^3 - 10x^2) = 10x^2$$
Bring down the next term.

Adding the term $0x^2$ gives us a column in which to write $10x^2$. (Remember, you can add and subtract only similar terms.)

Here is the completed problem:

$$
\begin{array}{r}
2x^2 + 10x + 47 \\
x - 5 \overline{)\ 2x^3 + 0x^2 - 3x + 2} \\
2x^3 - 10x^2 \\
\hline
10x^2 - 3x \\
10x^2 - 50x \\
\hline
47x + 2 \\
47x - 235 \\
\hline
237
\end{array}
$$

Our answer is $\dfrac{2x^3 - 3x + 2}{x - 5} = 2x^2 + 10x + 47 + \dfrac{237}{x - 5}$.

As you can see, long division with polynomials is a mechanical process. Once you have done it correctly a couple of times, it becomes very easy to produce the correct answer.

Getting Ready for Class

After reading through the preceding section, respond in your own words and in complete sentences.

A. What are the four steps used in long division with whole numbers?

B. How is division of two polynomials similar to long division with whole numbers?

C. What are the four steps used in long division with polynomials?

D. How do we use 0 when dividing the polynomial $2x^3 - 3x + 2$ by $x - 5$?

PROBLEM SET 4.8

Divide.

1. $\dfrac{x^2 - 5x + 6}{x - 3}$

2. $\dfrac{x^2 - 5x + 6}{x - 2}$

3. $\dfrac{a^2 + 9a + 20}{a + 5}$

4. $\dfrac{a^2 + 9a + 20}{a + 4}$

5. $\dfrac{x^2 - 6x + 9}{x - 3}$

6. $\dfrac{x^2 + 10x + 25}{x + 5}$

7. $\dfrac{2x^2 + 5x - 3}{2x - 1}$

8. $\dfrac{4x^2 + 4x - 3}{2x - 1}$

9. $\dfrac{2a^2 - 9a - 5}{2a + 1}$

10. $\dfrac{4a^2 - 8a - 5}{2a + 1}$

11. $\dfrac{x^2 + 5x + 8}{x + 3}$

12. $\dfrac{x^2 + 5x + 4}{x + 3}$

13. $\dfrac{a^2 + 3a + 2}{a + 5}$

14. $\dfrac{a^2 + 4a + 3}{a + 5}$

15. $\dfrac{x^2 + 2x + 1}{x - 2}$

16. $\dfrac{x^2 + 6x + 9}{x - 3}$

17. $\dfrac{x^2 + 5x - 6}{x + 1}$

18. $\dfrac{x^2 - x - 6}{x + 1}$

19. $\dfrac{a^2 + 3a + 1}{a + 2}$

20. $\dfrac{a^2 - a + 3}{a + 1}$

21. $\dfrac{2x^2 - 2x + 5}{2x + 4}$

22. $\dfrac{15x^2 + 19x - 4}{3x + 8}$

23. $\dfrac{6a^2 + 5a + 1}{2a + 3}$

24. $\dfrac{4a^2 + 4a + 3}{2a + 1}$

25. $\dfrac{6a^3 - 13a^2 - 4a + 15}{3a - 5}$

26. $\dfrac{2a^3 - a^2 + 3a + 2}{2a + 1}$

Fill in the missing terms in the numerator, and then use long division to find the quotients (see Example 4).

27. $\dfrac{x^3 + 4x + 5}{x + 1}$

28. $\dfrac{x^3 + 4x^2 - 8}{x + 2}$

29. $\dfrac{x^3 - 1}{x - 1}$

30. $\dfrac{x^3 + 1}{x + 1}$

31. $\dfrac{x^3 - 8}{x - 2}$

32. $\dfrac{x^3 + 27}{x + 3}$

Review Problems

The problems that follow review material we covered in Section 3.8.

Use systems of equations to solve the following word problems.

33. Number Problem The sum of two numbers is 25. One of the numbers is 4 times the other. Find the numbers.

34. Number Problem The sum of two numbers is 24. One of the numbers is 3 more than twice the other. Find the numbers.

35. Investing Suppose you have a total of $1,200 invested in two accounts. One of the accounts pays 8% annual interest, and the other pays 9% annual interest. If your total interest for the year is $100, how much money did you invest in each of the accounts?

36. **Investing** If you invest twice as much money in an account that pays 12% annual interest as you do in an account that pays 11% annual interest, how much do you have in each account if your total interest for a year is $210?

37. **Money Problem** If you have a total of $160 in $5 bills and $10 bills, how many of each type of bill do you have if you have 4 more $10 bills than $5 bills?

38. **Coin Problem** Suppose you have 20 coins worth a total of $2.80. If the coins are all nickels and quarters, how many of each type do you have?

39. **Mixture Problem** How many gallons of 20% antifreeze solution and 60% antifreeze solution must be mixed to get 16 gallons of 35% antifreeze solution?

40. **Mixture Problem** A chemist wants to obtain 80 liters of a solution that is 12% hydrochloric acid. How many liters of 10% hydrochloric acid solution and 20% hydrochloric acid solution should he mix to do so?

CHAPTER 4 SUMMARY

Examples

1. (a) $2^3 = 2 \cdot 2 \cdot 2 = 8$

(b) $x^5 \cdot x^3 = x^{5+3} = x^8$

(c) $\dfrac{x^5}{x^3} = x^{5-3} = x^2$

(d) $(3x)^2 = 3^2 \cdot x^2 = 9x^2$

(e) $\left(\dfrac{2}{3}\right)^3 = \dfrac{2^3}{3^3} = \dfrac{8}{27}$

(f) $(x^5)^3 = x^{5 \cdot 3} = x^{15}$

(g) $3^{-2} = \dfrac{1}{3^2} = \dfrac{1}{9}$

Exponents: Definition and Properties [4.1, 4.2]

Integer exponents indicate repeated multiplications.

$a^r \cdot a^s = a^{r+s}$ To multiply with the same base, you add exponents.

$\dfrac{a^r}{a^s} = a^{r-s}$ To divide with the same base, you subtract exponents.

$(ab)^r = a^r \cdot b^r$ Exponents distribute over multiplication.

$\left(\dfrac{a}{b}\right)^r = \dfrac{a^r}{b^r}$ Exponents distribute over division.

$(a^r)^s = a^{r \cdot s}$ A power of a power is the base raised to the product of the powers.

$a^{-r} = \dfrac{1}{a^r}$ Negative exponents imply reciprocals.

2. $(5x^2)(3x^4) = 15x^6$

Multiplication of Monomials [4.3]

To multiply two monomials, multiply coefficients and add exponents.

3. $\dfrac{12x^9}{4x^5} = 3x^4$

Division of Monomials [4.3]

To divide two monomials, divide coefficients and subtract exponents.

4. $768,000 = 7.68 \times 10^5$

$0.00039 = 3.9 \times 10^{-4}$

Scientific Notation [4.1, 4.2, 4.3]

A number is in scientific notation when it is written as the product of a number between 1 and 10 and an integer power of 10.

5. $(3x^2 - 2x + 1) + (2x^2 + 7x - 3)$
$= 5x^2 + 5x - 2$

Addition of Polynomials [4.4]

To add two polynomials, add coefficients of similar terms.

6. $(3x + 5) - (4x - 3)$
$= 3x + 5 - 4x + 3$
$= -x + 8$

Subtraction of Polynomials [4.4]

To subtract one polynomial from another, add the opposite of the second to the first.

7. (a) $2a^2(5a^2 + 3a - 2)$
$= 10a^4 + 6a^3 - 4a^2$

(b) $(x + 2)(3x - 1)$
$= 3x^2 - x + 6x - 2$
$= 3x^2 + 5x - 2$

(c) $2x^2 - 3x + 4$

$ 3x - 2$

$\overline{6x^3 - 9x^2 + 12x}$

$ - 4x^2 + 6x - 8$

$\overline{6x^3 - 13x^2 + 18x - 8}$

Multiplication of Polynomials [4.5]

To multiply a polynomial by a monomial, we apply the distributive property. To multiply two binomials we use the FOIL method. In other situations we use the COLUMN method. Each method achieves the same result: To multiply any two polynomials, we multiply each term in the first polynomial by each term in the second polynomial.

8. $(x + 3)^2 = x^2 + 6x + 9$

$(x - 3)^2 = x^2 - 6x + 9$

$(x + 3)(x - 3) = x^2 - 9$

Special Products [4.6]

$$\left. \begin{array}{l} (a + b)^2 = a^2 + 2ab + b^2 \\ (a - b)^2 = a^2 - 2ab + b^2 \end{array} \right\} \quad \text{Binomial squares}$$

$(a + b)(a - b) = a^2 - b^2 \qquad$ Difference of two squares

9. $\dfrac{12x^3 - 18x^2}{6x} = 2x^2 - 3x$

Dividing a Polynomial by a Monomial [4.7]

To divide a polynomial by a monomial, divide each term in the polynomial by the monomial.

10.

$$
\begin{array}{r}
x - 2 \\
x - 3 \overline{)\; x^2 - 5x + 8} \\
\underline{- + } \\
\cancel{+}\, x^2 \cancel{-}\, 3x \downarrow \\
\hline
- 2x + 8 \\
\underline{+ - } \\
\cancel{-}\, 2x \cancel{+}\, 6 \\
\hline
2
\end{array}
$$

Long Division With Polynomials [4.8]

Division with polynomials is similar to long division with whole numbers. The steps in the process are estimate, multiply, subtract, and bring down the next term. The divisors in all the long-division problems in this chapter were binomials.

COMMON MISTAKES

1. If a term contains a variable that is raised to a power, then the exponent on the variable is associated only with that variable, unless there are parentheses. That is, the expression $3x^2$ means $3 \cdot x \cdot x$, not $3x \cdot 3x$. It is a mistake to write $3x^2$ as $9x^2$. The only way to end up with $9x^2$ is to start with $(3x)^2$.

2. It is a mistake to add nonsimilar terms. For example, $2x$ and $3x^2$ are nonsimilar terms and therefore cannot be combined. That is, $2x + 3x^2 \neq 5x^3$. If you were to substitute 10 for x in the preceding expression, you would see that the two sides are not equal.

3. It is a mistake to distribute exponents over sums and differences. That is, $(a + b)^2 \neq a^2 + b^2$. Convince yourself of this by letting $a = 2$ and $b = 3$ and then simplifying both sides.

4. Another common mistake can occur when dividing a polynomial by a monomial. Here is an example:

$$\frac{x + 2}{2} = x + 1 \qquad \text{Mistake}$$

The mistake here is in not dividing both terms in the numerator by 2. The correct way to divide $x + 2$ by 2 looks like this:

$$\frac{x + 2}{2} = \frac{x}{2} + \frac{2}{2} \qquad \text{Correct}$$

$$= \frac{x}{2} + 1$$

CHAPTER 4 REVIEW

The numbers in brackets refer to the sections of the text in which similar problems can be found.

Simplify. [4.1]

1. $(-1)^3$

2. -8^2

3. $\left(\dfrac{3}{7}\right)^2$

4. $y^3 \cdot y^9$

5. $x^{15} \cdot x^7 \cdot x^5 \cdot x^3$

6. $(x^7)^5$

7. $(2^6)^4$

8. $(3y)^3$

9. $(-2xyz)^3$

Simplify each expression. Any answers that contain exponents should contain positive exponents only. [4.2]

10. 7^{-2}

11. $4x^{-5}$

12. $(3y)^{-3}$

13. $\dfrac{a^9}{a^3}$

14. $\left(\dfrac{x^3}{x^5}\right)^2$

15. $\dfrac{x^9}{x^{-6}}$

16. $\dfrac{x^{-7}}{x^{-2}}$

17. $(-3xy)^0$

18. $3^0 - 5^1 + 5^0$

Simplify. Any answers that contain exponents should contain positive exponents only. [4.1, 4.2]

19. $(3x^3y^2)^2$

20. $(2a^3b^2)^4(2a^5b^6)^2$

21. $(-3xy^2)^{-3}$

22. $\dfrac{(b^3)^4(b^2)^5}{(b^7)^3}$

23. $\dfrac{(x^{-3})^3(x^6)^{-1}}{(x^{-5})^{-4}}$

Simplify. Write all answers with positive exponents only. [4.3]

24. $\dfrac{(2x^4)(15x^9)}{6x^6}$

25. $\dfrac{(10x^3y^5)(21x^2y^6)}{(7xy^3)(5x^9y)}$

26. $\dfrac{21a^{10}}{3a^4} - \dfrac{18a^{17}}{6a^{11}}$

27. $\dfrac{8x^8y^3}{2x^3y} - \dfrac{10x^6y^9}{5xy^7}$

Simplify, and write all answers in scientific notation. [4.3]

28. $(3.2 \times 10^3)(2 \times 10^4)$

29. $\dfrac{4.6 \times 10^5}{2 \times 10^{-3}}$

30. $\dfrac{(4 \times 10^6)(6 \times 10^5)}{3 \times 10^8}$

Perform the following additions and subtractions. [4.4]

31. $(3a^2 - 5a + 5) + (5a^2 - 7a - 8)$

32. $(-7x^2 + 3x - 6) - (8x^2 - 4x + 7) + (3x^2 - 2x - 1)$

33. Subtract $8x^2 + 3x - 2$ from $4x^2 - 3x - 2$.

34. Find the value of $2x^2 - 3x + 5$ when $x = 3$.

Multiply. [4.5]

35. $3x(4x - 7)$

36. $8x^3y(3x^2y - 5xy^2 + 4y^3)$

37. $(a + 1)(a^2 + 5a - 4)$

38. $(x + 5)(x^2 - 5x + 25)$

39. $(3x - 7)(2x - 5)$

40. $\left(5y + \dfrac{1}{5}\right)\left(5y - \dfrac{1}{5}\right)$

41. $(a^2 - 3)(a^2 + 3)$

Perform the indicated operations. [4.6]

42. $(a - 5)^2$

43. $(3x + 4)^2$

44. $(y^2 + 3)^2$

45. Divide $10ab + 20a^2$ by $-5a$. [4.7]

46. Divide $40x^5y^4 - 32x^3y^3 - 16x^2y$ by $-8xy$.

Divide using long division. [4.8]

47. $\dfrac{x^2 + 15x + 54}{x + 6}$

48. $\dfrac{6x^2 + 13x - 5}{3x - 1}$

49. $\dfrac{x^3 + 64}{x + 4}$

50. $\dfrac{3x^2 - 7x + 10}{3x + 2}$

51. $\dfrac{2x^3 - 7x^2 + 6x + 10}{2x + 1}$

Volume A box is in the shape of a rectangular solid and is 3 times as long as it is wide. The height and the width are equal. [4.1, 4.2, 4.3]

52. Write an expression for the volume of this box.

53. If the width of this box is 2 feet, will it hold as much food as your refrigerator?

54. Area and Volume of a Sphere   A discotheque wants to install a new disco ball in the middle of the ceiling over the dance floor. The owners have determined that the radius r of the new ball must be 2 feet. [4.1]

(a) If the surface area of a sphere is $A = 4\pi r^2$, what is the surface area of the new disco ball?

(b) If the volume of a sphere is $V = \dfrac{4}{3}\pi r^3$, what is the volume of the new disco ball?

55. Area of a Circle   The Jacobs family is getting a round trampoline. Their square backyard has an area of 225 square feet. The smallest trampoline has a radius of 4 feet. The next largest has a radius of 6 feet, then 8 feet, and the largest has a radius of 10 feet. What is the largest trampoline that will fit in their backyard? [4.1]

CHAPTER 4   PROJECTS

EXPONENTS AND POLYNOMIALS

GROUP PROJECT

DISCOVERING PASCAL'S TRIANGLE

Number of People: 3

Time Needed: 20 minutes

Equipment: Paper and pencils

Background: The triangular array of numbers shown here is known as Pascal's triangle, after the French philosopher Blaise Pascal (1623 – 1662).

$$
\begin{array}{ccccccccccc}
 & & & & & 1 & & & & & \\
 & & & & 1 & & 1 & & & & \\
 & & & 1 & & 2 & & 1 & & & \\
 & & 1 & & 3 & & 3 & & 1 & & \\
 & 1 & & 4 & & 6 & & 4 & & 1 & \\
1 & & 5 & & 10 & & 10 & & 5 & & 1
\end{array}
$$

Procedure: Look at Pascal's triangle and discover how the numbers in each row of the triangle are obtained from the numbers in the row above it.

1. Once you have discovered how to extend the triangle, write the next two rows.

2. Pascal's triangle can be linked to the Fibonacci sequence by rewriting Pascal's triangle so that the 1's on the left side of the triangle line up under one another, and the other columns are equally spaced to the right of the first column. Rewrite Pascal's triangle as indicated and then look along the diagonals of the new array until you discover how the Fibonacci sequence can be obtained from it.

3. The diagram at the left shows Pascal's triangle as written in Japan in 1781. Use your knowledge of Pascal's triangle to translate the numbers written in Japanese into our number system. Then write down the Japanese numbers from 1 to 20.

Pascal's triangle in Japan

R E S E A R C H   P R O J E C T

BINOMIAL EXPANSIONS

The title on the following diagram is *Binomial Expansions* because each line gives the expansion of the binomial $x + y$ raised to a whole-number power.

Binomial Expansions

$$(x + y)^0 = \qquad\qquad 1$$
$$(x + y)^1 = \qquad\qquad x + y$$
$$(x + y)^2 = \qquad x^2 + 2xy + y^2$$
$$(x + y)^3 = x^3 + 3x^2y + 3xy^2 + y^3$$
$$(x + y)^4 =$$
$$(x + y)^5 =$$

The fourth row in the diagram was completed by expanding $(x + y)^3$ using the methods developed in this chapter. Next, complete the diagram by expanding the binomials $(x + y)^4$ and $(x + y)^5$ using the multiplication procedures you have learned in this chapter. Finally, study the completed diagram until you see patterns that will allow you to continue the diagram one more row without using multiplication. (One pattern that you will see is Pascal's triangle, which we mentioned in the preceding group project.) When you are finished, write an essay in which you describe what you have done and the results you have obtained.

CHAPTER 4   TEST

Simplify each of the following expressions. [4.1]

1. $(-3)^4$

2. $\left(\dfrac{3}{4}\right)^2$

3. $(3x^3)^2(2x^4)^3$

Simplify each expression. Write all answers with positive exponents only. [4.2]

4. 3^{-2}

5. $(3a^4b^2)^0$

6. $\dfrac{a^{-3}}{a^{-5}}$

7. $\dfrac{(x^{-2})^3(x^{-3})^{-5}}{(x^{-4})^{-2}}$

8. Write 0.0278 in scientific notation. [4.2]

9. Write 2.43×10^5 in expanded form. [4.2]

Simplify. Write all answers with positive exponents only. [4.3]

10. $\dfrac{35x^2y^4z}{70x^6y^2z}$

11. $\dfrac{(6a^2b)(9a^3b^2)}{18a^4b^3}$

12. $\dfrac{24x^7}{3x^2} + \dfrac{14x^9}{7x^4}$

13. $\dfrac{(2.4 \times 10^5)(4.5 \times 10^{-2})}{1.2 \times 10^{-6}}$

Add and subtract as indicated. [4.4]

14. $8x^2 - 4x + 6x + 2$

15. $(5x^2 - 3x + 4) - (2x^2 - 7x - 2)$

16. Subtract $3x - 4$ from $6x - 8$. [4.4]

17. Find the value of $2y^2 - 3y - 4$ when y is -2. [4.4]

Multiply. [4.5]

18. $2a^2(3a^2 - 5a + 4)$

19. $\left(x + \dfrac{1}{2}\right)\left(x + \dfrac{1}{3}\right)$

20. $(4x - 5)(2x + 3)$

21. $(x - 3)(x^2 + 3x + 9)$

Multiply. [4.6]

22. $(x + 5)^2$

23. $(3a - 2b)^2$

24. $(3x - 4y)(3x + 4y)$

25. $(a^2 - 3)(a^2 + 3)$

26. Divide $10x^3 + 15x^2 - 5x$ by $5x$. [4.7]

Divide. [4.8]

27. $\dfrac{8x^2 - 6x - 5}{2x - 3}$

28. $\dfrac{3x^3 - 2x + 1}{x - 3}$

29. Volume   Find the volume of a cube if the length of a side is 2.5 centimeters. [4.1]

30. Volume   Find the volume of a rectangular solid if the length is 5 times the width and the height is $\dfrac{1}{5}$ the width. [4.3]

CHAPTERS   1 – 4   CUMULATIVE REVIEW

Simplify.

1. $-\left(-\dfrac{3}{4}\right)$

2. $2 \cdot 7 + 10$

3. $6 \cdot 7 + 7 \cdot 9$

4. $\dfrac{12 - 3}{8 - 8}$

5. $6(4a + 2) - 3(5a - 1)$

6. $-7\left(-\dfrac{1}{7}\right)$

7. $-15 - (-3)$

8. $(-9)(-7)$

9. $(-9)(-5)$

10. $y^{10} \cdot y^6$

11. $(2y)^4$

12. $(3a^2b^5)^3(2a^6b^7)^2$

13. $\dfrac{(12xy^5)^3(16x^2y^2)}{(8x^3y^3)(3x^5y)}$

14. $\dfrac{3.5 \times 10^{-7}}{7 \times 10^{-3}}$

15. $(5x - 1)^2$

16. $(-4x^2 - 5x + 2) + (3x^2 - 6x + 1)$
$$- (-x^2 + 2x - 7)$$

17. $(x - 1)(x^2 + x + 1)$

Solve each equation.

18. $8 - 2y + 3y = 12$ **19.** $6a - 5 = 4a$

20. $18 = 3(2x - 2)$

21. $2(3x + 5) + 8 = 2x + 10$

Solve the inequality, and graph the solution set.

22. $-\dfrac{a}{6} > 4$ **23.** $-4x > 28$

24. $5 - 7x \geq 19$

25. $3(2t - 5) - 7 \leq 5(3t + 1) + 5$

Divide.

26. $\dfrac{4x^2 + 8x - 10}{2x - 3}$ **27.** $\dfrac{15x^5 - 10x^2 + 20x}{5x^5}$

Graph on a rectangular coordinate system.

28. $x + y = 3$ **29.** $x = 2$

30. $3x + 2y = 6$

Solve each system by graphing.

31. $x + y = 1$ **32.** $x - y = 3$
$\quad\ x + 2y = 2$ $\qquad 2x - 2y = 2$

33. $x + 2y = 4$
$\quad\ 3x + 6y = 12$

Solve each system.

34. $x + 2y = 5$ **35.** $x + 2y = 5$
$\quad\ x - y = 2$ $\qquad 3x + 6y = 14$

36. $2x + y = -3$
$\quad\ x - 3y = -5$

37. $\dfrac{1}{6}x + \dfrac{1}{4}y = 1$ **38.** $x + y = 9$
$\quad \dfrac{6}{5}x - y = \dfrac{8}{5}$ $\qquad y = x + 1$

39. $4x + 5y = 25$
$\quad\ 2y = x - 3$

40. Subtract 8 from -9.

41. Find the value of $8x - 3$ when $x = 3$.

42. Solve $A = \frac{1}{2}bh$ for h.

43. For the set $\{-3, -\sqrt{2}, 0, \frac{3}{4}, 1.5, \pi\}$ list all the irrational numbers.

44. Find x when y is 8 in the equation $y = 3x - 1$.

45. Which of the ordered pairs $(0, 3)$, $(4, 0)$, and $(\frac{16}{3}, 1)$ are solutions to the equation $3x - 4y = 12$?

46. Find the x- and y-intercepts for the equation $y = 2x + 4$.

47. Write 186,000 in scientific notation.

48. Write 9.87×10^{-4} in expanded form.

49. Multiply $(2 \times 10^5)(3 \times 10^{-8})$.

50. Mixture How many liters of a 40% solution and a 16% solution must be mixed to obtain 20 liters of a 22% solution?

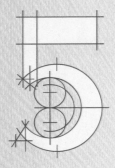

Factoring

INTRODUCTION

If you watch professional football on television, you will hear the announcers refer to "hang time" when the punter punts the ball. Hang time is the amount of time the ball is in the air, and it depends on only one thing, the initial vertical velocity imparted to the ball by the kicker's foot. We can find the hang time of a football by solving equations. Table 1 shows the equations to solve for hang time, given various initial vertical velocities. Figure 1 is a visual representation of some equations associated with the ones in Table 1. In Figure 1, you can find hang time on the horizontal axis.

TABLE 1 Hang Time for a Football		
Initial Vertical Velocity (feet/second)	**Equation in Factored Form**	**Hang Time (seconds)**
16	$16t(1 - t) = 0$	1
32	$16t(2 - t) = 0$	2
48	$16t(3 - t) = 0$	3
64	$16t(4 - t) = 0$	4
80	$16t(5 - t) = 0$	5

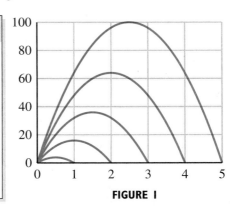

FIGURE 1

The equations in the second column of the table are in what is called "factored form." Once the equation is in factored form, hang time can be read from the second factor. In this chapter we develop techniques that allow us to factor a variety of polynomials. Factoring is the key to solving equations like the ones in Table 1.

STUDY SKILLS

The study skills for this chapter are about attitude. They are points of view that point toward success.

1. **Be Focused, Not Distracted** I have students who begin their assignments by asking themselves, "Why am I taking this class?" Or, "When am I ever going to use this stuff?" If you are asking yourself similar questions, you may be distracting yourself away from doing the things that will produce the results you want in this course. Don't dwell on questions and evaluations of the class that can be used as excuses for not doing well. If you want to succeed in this course, focus your energy and efforts toward success, rather than distracting yourself away from your goals.

2. **Be Resilient** Don't let setbacks keep you from your goals. You want to put yourself on the road to becoming a person who can succeed in this class or any class in college. Failing a test or quiz, or having a difficult time on some topics, is normal. No one goes through college without some setbacks. Don't let a temporary disappointment keep you from succeeding in this course. A low grade on a test or quiz is simply a signal that some reevaluation of your study habits needs to take place.

3. **Intend to Succeed** I always have a few students who simply go through the motions of studying without intending to master the material. It is more important to them to look like they are studying than to actually study. You need to study with the intention of being successful in the course. Intend to master the material, no matter what it takes.

The Greatest Common Factor and Factoring by Grouping

In Chapter 1 we used the following diagram to illustrate the relationship between multiplication and factoring.

$$\text{Multiplication}$$
$$\text{Factors} \longrightarrow 3 \cdot 5 = 15 \longleftarrow \text{Product}$$
$$\text{Factoring}$$

A similar relationship holds for multiplication of polynomials. Reading the following diagram from left to right, we say the product of the binomials $x + 2$ and $x + 3$ is the trinomial $x^2 + 5x + 6$. On the other hand, if we read in the other direction, we can say that $x^2 + 5x + 6$ factors into the product of $x + 2$ and $x + 3$.

$(x + 2)(x + 3) = x^2 + 3x + 2x + 6$

$x^2 + 5x + 6$

$$\text{Multiplication}$$
$$\text{Factors} \longrightarrow (x + 2)(x + 3) = x^2 + 5x + 6 \longleftarrow \text{Product}$$
$$\text{Factoring}$$

In this chapter we develop a systematic method of factoring polynomials.

In this section we will apply the distributive property to polynomials to factor from them what is called the greatest common factor.

> **DEFINITION** The **greatest common factor** for a polynomial is the largest monomial that divides (is a factor of) each term of the polynomial.

We use the term *largest monomial* to mean the monomial with the greatest coefficient and highest power of the variable.

EXAMPLE 1 Find the greatest common factor for the polynomial:

$$3x^5 + 12x^2$$

Solution The terms of the polynomial are $3x^5$ and $12x^2$. The largest number that divides the coefficients is 3, and the highest power of x that is a factor of x^5 and x^2 is x^2. Therefore, the greatest common factor for $3x^5 + 12x^2$ is $3x^2$. That is, $3x^2$ is the largest monomial that divides each term of $3x^5 + 12x^2$.

EXAMPLE 2 Find the greatest common factor for:

$$8a^3b^2 + 16a^2b^3 + 20a^3b^3$$

$4a^2b^2$

Solution   The largest number that divides each of the coefficients is 4. The highest power of the variable that is a factor of a^3b^2, a^2b^3, and a^3b^3 is a^2b^2. The greatest common factor for $8a^3b^2 + 16a^2b^3 + 20a^3b^3$ is $4a^2b^2$. It is the largest monomial that is a factor of each term.

Once we have recognized the greatest common factor of a polynomial, we can apply the distributive property and factor it out of each term. We rewrite the polynomial as the product of its greatest common factor with the polynomial that remains after the greatest common factor has been factored from each term in the original polynomial.

EXAMPLE 3   Factor the greatest common factor from $3x - 15$.

Solution   The greatest common factor for the terms $3x$ and 15 is 3. We can rewrite both $3x$ and 15 so that the greatest common factor 3 is showing in each term. It is important to realize that $3x$ means $3 \cdot x$. The 3 and the x are not "stuck" together:

$$3x - 15 = 3 \cdot x - 3 \cdot 5$$

Now, applying the distributive property, we have:

$$3 \cdot x - 3 \cdot 5 = 3(x - 5)$$

To check a factoring problem like this, we can multiply 3 and $x - 5$ to get $3x - 15$, which is what we started with. Factoring is simply a procedure by which we change sums and differences into products. In this case we changed the difference $3x - 15$ into the product $3(x - 5)$. Note, however, that we have not changed the meaning or value of the expression. The expression we end up with is equal to the expression we started with.

EXAMPLE 4   Factor the greatest common factor from:

$$5x^3 - 15x^2$$

Solution   The greatest common factor is $5x^2$. We rewrite the polynomial as:

$$5x^3 - 15x^2 = 5x^2 \cdot x - 5x^2 \cdot 3$$

Then we apply the distributive property to get:

$$5x^2 \cdot x - 5x^2 \cdot 3 = 5x^2(x - 3)$$

To check our work, we simply multiply $5x^2$ and $(x - 3)$ to get $5x^3 - 15x^2$, which is our original polynomial.

EXAMPLE 5   Factor the greatest common factor from:

$$16x^5 - 20x^4 + 8x^3$$

Solution   The greatest common factor is $4x^3$. We rewrite the polynomial so that we can see the greatest common factor $4x^3$ in each term; then we apply the distributive property to factor it out.

$$16x^5 - 20x^4 + 8x^3 = 4x^3 \cdot 4x^2 - 4x^3 \cdot 5x + 4x^3 \cdot 2$$
$$= 4x^3(4x^2 - 5x + 2)$$

EXAMPLE 6   Factor the greatest common factor from:

$$6x^3y - 18x^2y^2 + 12xy^3$$

Solution   The greatest common factor is $6xy$. We rewrite the polynomial in terms of $6xy$ and then apply the distributive property as follows:

$$6x^3y - 18x^2y^2 + 12xy^3 = 6xy \cdot x^2 - 6xy \cdot 3xy + 6xy \cdot 2y^2$$
$$= 6xy(x^2 - 3xy + 2y^2)$$

EXAMPLE 7   Factor the greatest common factor from:

$$3a^2b - 6a^3b^2 + 9a^3b^3$$

Solution   The greatest common factor is $3a^2b$:

$$3a^2b - 6a^3b^2 + 9a^3b^3 = 3a^2b(1) - 3a^2b(2ab) + 3a^2b(3ab^2)$$
$$= 3a^2b(1 - 2ab + 3ab^2)$$

FACTORING BY GROUPING

To develop our next method of factoring, called *factoring by grouping,* we start by examining the polynomial $xc + yc$. The greatest common factor for the two terms is c. Factoring c from each term we have:

$$xc + yc = c(x + y)$$

But suppose that c itself was a more complicated expression, such as $a + b$, so that the expression we were trying to factor was $x(a + b) + y(a + b)$, instead of $xc + yc$. The greatest common factor for $x(a + b) + y(a + b)$ is $(a + b)$. Factoring this common factor from each term looks like this:

$$x(a + b) + y(a + b) = (a + b)(x + y)$$

To see how all of this applies to factoring polynomials, consider the polynomial

$$xy + 3x + 2y + 6$$

There is no greatest common factor other than the number 1. However, if we group the terms together two at a time, we can factor an x from the first two terms and a 2 from the last two terms:

$$xy + 3x + 2y + 6 = x(y + 3) + 2(y + 3)$$

The expression on the right can be thought of as having two terms: $x(y + 3)$ and $2(y + 3)$. Each of these expressions contains the common factor $y + 3$, which can be factored out using the distributive property:

$$x(y + 3) + 2(y + 3) = (y + 3)(x + 2)$$

This last expression is in factored form. The process we used to obtain it is called factoring by grouping. Here are some additional examples.

EXAMPLE 8   Factor $ax + bx + ay + by$.

Solution   We begin by factoring x from the first two terms and y from the last two terms:

$$ax + bx + ay + by = x(a + b) + y(a + b)$$
$$= (a + b)(x + y)$$

To convince yourself that this is factored correctly, multiply the two factors $(a + b)$ and $(x + y)$.

EXAMPLE 9   Factor by grouping: $3ax - 2a + 15x - 10$.

Solution   First, we factor a from the first two terms and 5 from the last two terms. Then, we factor $3x - 2$ from the remaining two expressions:

$$3ax - 2a + 15x - 10 = a(3x - 2) + 5(3x - 2)$$
$$= (3x - 2)(a + 5)$$

Again, multiplying $(3x - 2)$ and $(a + 5)$ will convince you that these are the correct factors.

EXAMPLE 10   Factor $2x^2 + 5ax - 2xy - 5ay$.

Solution   From the first two terms we factor x. From the second two terms we must factor $-y$ so that the binomial that remains after we do so matches the binomial produced by the first two terms:

$$2x^2 + 5ax - 2xy - 5ay = x(2x + 5a) - y(2x + 5a)$$
$$= (2x + 5a)(x - y)$$

Another way to accomplish the same result is to use the commutative property to interchange the middle two terms and then factor by grouping:

$$2x^2 + 5ax - 2xy - 5ay = 2x^2 - 2xy + 5ax - 5ay \qquad \text{Commutative property}$$
$$= 2x(x - y) + 5a(x - y)$$
$$= (x - y)(2x + 5a)$$

This is the same as the result we obtained previously.

EXAMPLE 11 Factor $6x^2 - 3x - 4x + 2$ by grouping.

Solution The first two terms have $3x$ in common, and the last two terms have either a 2 or a -2 in common. Suppose we factor $3x$ from the first two terms and 2 from the last two terms. We get:

$$6x^2 - 3x - 4x + 2 = 3x(2x - 1) + 2(-2x + 1)$$

We can't go any further because there is no common factor that will allow us to factor further. However, if we factor -2, instead of 2, from the last two terms, our problem is solved:

$$6x^2 - 3x - 4x + 2 = 3x(2x - 1) - 2(2x - 1)$$
$$= (2x - 1)(3x - 2)$$

In this case, factoring -2 from the last two terms gives us an expression that can be factored further.

Getting Ready for Class

After reading through the preceding section, respond in your own words and in complete sentences.

A. What is the greatest common factor for a polynomial?

B. After factoring a polynomial, how can you check your result?

C. When would you try to factor by grouping?

D. What is the relationship between multiplication and factoring?

PROBLEM SET 5.1

Factor the following by taking out the greatest common factor.

1. $15x + 25$

2. $14x + 21$

3. $6a + 9$

4. $8a + 10$

5. $4x - 8y$

6. $9x - 12y$

7. $3x^2 - 6x - 9$

8. $2x^2 + 6x + 4$

9. $3a^2 - 3a - 60$

10. $2a^2 - 18a + 28$

11. $24y^2 - 52y + 24$

12. $18y^2 + 48y + 32$

13. $9x^2 - 8x^3$

14. $7x^3 - 4x^2$

15. $13a^2 - 26a^3$

16. $5a^2 - 10a^3$

17. $21x^2y - 28xy^2$

18. $30xy^2 - 25x^2y$

19. $22a^2b^2 - 11ab^2$

20. $15x^3 - 25x^2 + 30x$

21. $7x^3 + 21x^2 - 28x$

22. $16x^4 - 20x^2 - 16x$

23. $121y^4 - 11x^4$

24. $25a^4 - 5b^4$

25. $100x^4 - 50x^3 + 25x^2$

26. $36x^5 + 72x^3 - 81x^2$

27. $8a^2 + 16b^2 + 32c^2$

28. $9a^2 - 18b^2 - 27c^2$

29. $4a^2b - 16ab^2 + 32a^2b^2$

30. $5ab^2 + 10a^2b^2 + 15a^2b$

31. $121a^3b^2 - 22a^2b^3 + 33a^3b^3$

32. $20a^4b^3 - 18a^3b^4 + 22a^4b^4$

33. $12x^2y^3 - 72x^5y^3 - 36x^4y^4$

34. $49xy - 21x^2y^2 + 35x^3y^3$

Factor by grouping.

35. $xy + 5x + 3y + 15$ **36.** $xy + 2x + 4y + 8$

37. $xy + 6x + 2y + 12$ **38.** $xy + 2y + 6x + 12$

39. $ab + 7a - 3b - 21$ **40.** $ab + 3b - 7a - 21$

41. $ax - bx + ay - by$ **42.** $ax - ay + bx - by$

43. $2ax + 6x - 5a - 15$ **44.** $3ax + 21x - a - 7$

45. $3xb - 4b - 6x + 8$

46. $3xb - 4b - 15x + 20$ **47.** $x^2 + ax + 2x + 2a$

48. $x^2 + ax + 3x + 3a$ **49.** $x^2 - ax - bx + ab$

50. $x^2 + ax - bx - ab$

Factor by grouping. You can group the terms together two at a time or three at a time. Either way will produce the same result.

51. $ax + ay + bx + by + cx + cy$

52. $ax + bx + cx + ay + by + cy$

Factor the following polynomials by grouping the terms together two at a time.

53. $6x^2 + 9x + 4x + 6$ **54.** $6x^2 - 9x - 4x + 6$

55. $20x^2 - 2x + 50x - 5$

56. $20x^2 + 25x + 4x + 5$

57. $20x^2 + 4x + 25x + 5$

58. $20x^2 + 4x - 25x - 5$

59. $x^3 + 2x^2 + 3x + 6$

60. $x^3 - 5x^2 - 4x + 20$

61. $6x^3 - 4x^2 + 15x - 10$

62. $8x^3 - 12x^2 + 14x - 21$

63. The greatest common factor of the binomial $3x + 6$ is 3. The greatest common factor of the binomial $2x + 4$ is 2. What is the greatest common factor of their product $(3x + 6)(2x + 4)$, when it has been multiplied out?

64. The greatest common factors of the binomials $4x + 2$ and $5x + 10$ are 2 and 5, respectively. What is the greatest common factor of their product $(4x + 2)(5x + 10)$, when it has been multiplied out?

65. The following factorization is incorrect. Find the mistake, and correct the right-hand side:

$$12x^2 + 6x + 3 = 3(4x^2 + 2x)$$

66. Find the mistake in the following factorization, and then rewrite the right-hand side correctly:

$$10x^2 + 2x + 6 = 2(5x^2 + 3)$$

Applying the Concepts

67. Investing If you invest $1,000 in an account with an annual interest rate of r compounded annually, the amount of money you have in the account after one year is:

$$A = 1,000 + 1,000r$$

Write this formula again with the right side in factored form. Then, find the amount of money in this account at the end of one year if the interest rate is 12%.

68. Investing If you invest P dollars in an account with an annual interest rate of 8% compounded annually, then the amount of money in that account after one year is given by the formula:

$$A = P + 0.08P$$

Rewrite this formula with the right side in factored form, and then find the amount of money in the account at the end of one year if $500 was the initial investment.

69. Biological Growth If 1,000,000 bacteria are placed in a Petri dish and the bacteria have a growth rate of r (a percent expressed as a decimal) per hour, then one hour later the amount of bacteria will be $A = 1,000,000 + 1,000,000r$ bacteria.

(a) Factor the right side of the equation.

(b) If $r = 30\%$, find the number of bacteria present after one hour.

70. Biological Growth If there are B bacteria present initially in a Petri dish and their growth rate is r (a percent expressed as a decimal) per hour, then after one hour there will be $A = B + Br$ bacteria present.

(a) Factor the right side of this equation.

(b) The following bar graph shows the number of bacteria present initially and the number of bacteria present one hour later. Use the bar chart to find B and A in the preceding equation.

Bacterial growth

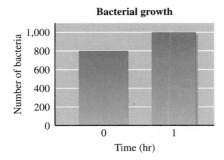

Time (hr)

C is the original amount and A is the amount after 1,000 years, then $A = C - Cr$, and r is called the decay rate (per thousand years).
(a) Factor and simplify the right side.
(b) From the following chart, find C, A, and r:

	1,000 Years Ago	**Now**
Amount of ^{14}C	5.00 kg	4.45 kg

71. Negative Growth Rate (Decline) Old family names in English disappear at a steady rate (and are replaced by new names that come into existence). If there were 7,000 English family names in the year 1500, and if they disappear at the rate of r (a percent expressed as a decimal) every 500 years, then $A = 7,000 - 7,000r$ of them will still be in use in the year 2000.
(a) Factor the right side of the equation.
(b) If $r = 23\%$, how many will still be in use in the year 2000?

72. Radioactive Decay Radioactive atoms such as ^{14}C (Carbon-14) decay—that is, after a certain amount of time, a certain percentage of the element will be lost by changing into a different element. If

Review Problems

The following problems review material we covered in Section 4.5. Reviewing these problems will help you with the next section.

Multiply using the FOIL method.

73. $(x - 7)(x + 2)$ **74.** $(x - 7)(x - 2)$

75. $(x - 3)(x + 2)$ **76.** $(x + 3)(x - 2)$

77. $(x + 3)(x^2 - 3x + 9)$

78. $(x - 2)(x^2 + 2x + 4)$

79. $(2x + 1)(x^2 + 4x - 3)$

80. $(3x + 2)(x^2 - 2x - 4)$

5.2   Factoring Trinomials

In this section we will factor trinomials in which the coefficient of the squared term is 1. The more familiar we are with multiplication of binomials, the easier factoring trinomials will be.

Recall multiplication of binomials from Chapter 4:

$$(x + 3)(x + 4) = x^2 + 7x + 12$$

$$(x - 5)(x + 2) = x^2 - 3x - 10$$

The first term in the answer is the product of the first terms in each binomial. The last term in the answer is the product of the last terms in each binomial. The middle term in the answer comes from adding the product of the outside terms to the product of the inside terms.

Let's have a and b represent real numbers and look at the product of $(x + a)$ and $(x + b)$:

$$(x + a)(x + b) = x^2 + ax + bx + ab$$
$$= x^2 + (a + b)x + ab$$

The coefficient of the middle term is the sum of a and b. The last term is the product of a and b. Writing this as a factoring problem, we have:

$$x^2 + \underset{\text{Sum}}{(a + b)x} + \underset{\text{Product}}{ab} = (x + a)(x + b)$$

To factor a trinomial in which the coefficient of x^2 is 1, we need only find the numbers a and b whose sum is the coefficient of the middle term and whose product is the constant term (last term).

EXAMPLE 1   Factor $x^2 + 8x + 12$.

Solution   The coefficient of x^2 is 1. We need two numbers whose sum is 8 and whose product is 12. The numbers are 6 and 2:

$$x^2 + 8x + 12 = (x + 6)(x + 2).$$

We can easily check our work by multiplying $(x + 6)$ and $(x + 2)$.

$$\text{Check:}\qquad (x + 6)(x + 2) = x^2 + 6x + 2x + 12$$
$$= x^2 + 8x + 12$$

EXAMPLE 2   Factor $x^2 - 2x - 15$.

Solution   The coefficient of x^2 is again 1. We need to find a pair of numbers whose sum is -2 and whose product is -15. Here are all the possibilities for products that are -15.

Products	Sums
$-1(15) = -15$	$-1 + 15 = 14$
$1(-15) = -15$	$1 + (-15) = -14$
$-5(3) = -15$	$-5 + 3 = -2$
$5(-3) = -15$	$5 + (-3) = 2$

The third line gives us what we want. The factors of $x^2 - 2x - 15$ are $(x - 5)$ and $(x + 3)$:

$$x^2 - 2x - 15 = (x - 5)(x + 3)$$

EXAMPLE 3   Factor $2x^2 + 10x - 28$.

Solution   The coefficient of x^2 is 2. We begin by factoring out the greatest common factor, which is 2:

$$2x^2 + 10x - 28 = 2(x^2 + 5x - 14)$$

Now, we factor the remaining trinomial by finding a pair of numbers whose sum is 5 and whose product is -14. Here are the possibilities:

Products	Sums
$-1(14) = -14$	$-1 + 14 = 13$
$1(-14) = -14$	$1 + (-14) = -13$
$-7(2) = -14$	$-7 + 2 = -5$
$7(-2) = -14$	$7 + (-2) = 5$

From the last line we see that the factors of $x^2 + 5x - 14$ are $(x + 7)$ and $(x - 2)$. Here is the complete problem:

$$2x^2 + 10x - 28 = 2(x^2 + 5x - 14)$$
$$= 2(x + 7)(x - 2)$$

Note: In Example 3 we began by factoring out the greatest common factor. The first step in factoring any trinomial is to look for the greatest common factor. If the trinomial in question has a greatest common factor other than 1, we factor it out first and then try to factor the trinomial that remains.

EXAMPLE 4   Factor $3x^3 - 3x^2 - 18x$.

Solution   We begin by factoring out the greatest common factor, which is $3x$. Then we factor the remaining trinomial. Without showing the table of products and sums as we did in Examples 2 and 3, here is the complete problem:

$$3x^3 - 3x^2 - 18x = 3x(x^2 - x - 6)$$
$$= 3x(x - 3)(x + 2)$$

EXAMPLE 5   Factor $x^2 + 8xy + 12y^2$.

Solution   This time we need two expressions whose product is $12y^2$ and whose sum is $8y$. The two expressions are $6y$ and $2y$ (see Example 1 in this section):

$$x^2 + 8xy + 12y^2 = (x + 6y)(x + 2y)$$

You should convince yourself that these factors are correct by finding their product.

 # Getting Ready for Class

After reading through the preceding section, respond in your own words and in complete sentences.

A. When the leading coefficient of a trinomial is 1, what is the relationship between the other two coefficients and the factors of the trinomial?

B. When factoring polynomials, what should you look for first?

C. How can you check to see that you have factored a trinomial correctly?

D. Describe how you would find the factors of $x^2 + 8x + 12$.

PROBLEM SET 5.2

Factor the following trinomials.

1. $x^2 + 7x + 12$ **2.** $x^2 + 7x + 10$

3. $x^2 + 3x + 2$ **4.** $x^2 + 7x + 6$

5. $a^2 + 10a + 21$ **6.** $a^2 - 7a + 12$

7. $x^2 - 7x + 10$ **8.** $x^2 - 3x + 2$

9. $y^2 - 10y + 21$ **10.** $y^2 - 7y + 6$

11. $x^2 - x - 12$ **12.** $x^2 - 4x - 5$

13. $y^2 + y - 12$ **14.** $y^2 + 3y - 18$

15. $x^2 + 5x - 14$ **16.** $x^2 - 5x - 24$

17. $r^2 - 8r - 9$ **18.** $r^2 - r - 2$

19. $x^2 - x - 30$ **20.** $x^2 + 8x + 12$

21. $a^2 + 15a + 56$ **22.** $a^2 - 9a + 20$

23. $y^2 - y - 42$ **24.** $y^2 + y - 42$

25. $x^2 + 13x + 42$ **26.** $x^2 - 13x + 42$

Factor the following problems completely. First, factor out the greatest common factor, and then factor the remaining trinomial.

27. $2x^2 + 6x + 4$ **28.** $3x^2 - 6x - 9$

29. $3a^2 - 3a - 60$ **30.** $2a^2 - 18a + 28$

31. $100x^2 - 500x + 600$

32. $100x^2 - 900x + 2,000$

33. $100p^2 - 1,300p + 4,000$

34. $100p^2 - 1,200p + 3,200$

35. $x^4 - x^3 - 12x^2$ **36.** $x^4 - 11x^3 + 24x^2$

37. $2r^3 + 4r^2 - 30r$ **38.** $5r^3 + 45r^2 + 100r$

39. $2y^4 - 6y^3 - 8y^2$ **40.** $3r^3 - 3r^2 - 6r$

41. $x^5 + 4x^4 + 4x^3$ **42.** $x^5 + 13x^4 + 42x^3$

43. $3y^4 - 12y^3 - 15y^2$ **44.** $5y^4 - 10y^3 + 5y^2$

45. $4x^4 - 52x^3 + 144x^2$ **46.** $3x^3 - 3x^2 - 18x$

Factor the following trinomials.

47. $x^2 + 5xy + 6y^2$ **48.** $x^2 - 5xy + 6y^2$

49. $x^2 - 9xy + 20y^2$ **50.** $x^2 + 9xy + 20y^2$

51. $a^2 + 2ab - 8b^2$ **52.** $a^2 - 2ab - 8b^2$

53. $a^2 - 10ab + 25b^2$ **54.** $a^2 + 6ab + 9b^2$

55. $a^2 + 10ab + 25b^2$ **56.** $a^2 - 6ab + 9b^2$

57. $x^2 + 2xa - 48a^2$ **58.** $x^2 - 3xa - 10a^2$

59. $x^2 - 5xb - 36b^2$ **60.** $x^2 - 13xb + 36b^2$

Factor completely.

61. $x^4 - 5x^2 + 6$ **62.** $x^6 - 2x^3 - 15$

63. $x^2 - 80x - 2,000$ **64.** $x^2 - 190x - 2,000$

65. $x^2 - x + \dfrac{1}{4}$ **66.** $x^2 - \dfrac{2}{3}x + \dfrac{1}{9}$

67. $x^2 + 0.6x + 0.08$ **68.** $x^2 + 0.8x + 0.15$

69. If one of the factors of $x^2 + 24x + 128$ is $x + 8$, what is the other factor?

70. If one factor of $x^2 + 260x + 2,500$ is $x + 10$, what is the other factor?

71. What polynomial, when factored, gives $(4x + 3)(x - 1)$?

72. What polynomial factors to $(4x - 3)(x + 1)$?

Review Problems

The problems that follow review material we covered in Sections 4.4 and 4.5. Reviewing the problems from Section 4.5 will help you with the next section.

Multiply using the FOIL method. [4.5]

73. $(6a + 1)(a + 2)$ **74.** $(6a - 1)(a - 2)$

75. $(3a + 2)(2a + 1)$ **76.** $(3a - 2)(2a - 1)$

77. $(6a + 2)(a + 1)$ **78.** $(3a + 1)(2a + 2)$

Subtract. [4.4]

79. $(5x^2 + 5x - 4) - (3x^2 - 2x + 7)$

80. $(7x^4 - 4x^2 - 5) - (2x^4 - 4x^2 + 5)$

81. Subtract $4x - 5$ from $7x + 3$.

82. Subtract $3x + 2$ from $-6x + 1$.

83. Subtract $2x^2 - 4x$ from $5x^2 - 5$.

84. Subtract $6x^2 + 3$ from $2x^2 - 4x$.

We will now consider trinomials whose greatest common factor is 1 and whose leading coefficient (the coefficient of the squared term) is a number other than 1. We present two methods for factoring trinomials of this type. The first method involves listing possible factors until the correct pair of factors is found. This requires a certain amount of trial and error. The second method is based on the factoring by grouping process, which we covered in Section 5.1. Either method can be used to factor trinomials whose leading coefficient is a number other than 1.

METHOD 1: FACTORING $ax^2 + bx + c$ BY TRIAL AND ERROR

Suppose we want to factor the trinomial $2x^2 - 5x - 3$. We know the factors (if they exist) will be a pair of binomials. The product of their first terms is $2x^2$, and the product of the last terms is -3. Let us list all the possible factors along with the trinomial that would result if we were to multiply them together. Remember, the middle term comes from the product of the inside terms plus the product of the outside terms.

Binomial Factors	First Term	Middle Term	Last Term
$(2x - 3)(x + 1)$	$2x^2$	$-x$	-3
$(2x + 3)(x - 1)$	$2x^2$	$+x$	-3
$(2x - 1)(x + 3)$	$2x^2$	$+5x$	-3
$(2x + 1)(x - 3)$	$2x^2$	$-5x$	-3

We can see from the last line that the factors of $2x^2 - 5x - 3$ are $(2x + 1)$ and $(x - 3)$. We call this the trial and error method of factoring trinomials. We look for possible factors that, when multiplied, will give the correct first and last terms, and then we see if we can adjust them to give the correct middle term.

EXAMPLE 1 Factor $6a^2 + 7a + 2$.

Solution We list all the possible pairs of factors that, when multiplied together, give a trinomial whose first term is $6a^2$ and whose last term is $+2$.

Binomial Factors	First Term	Middle Term	Last Term
$(6a + 1)(a + 2)$	$6a^2$	$+13a$	$+2$
$(6a - 1)(a - 2)$	$6a^2$	$-13a$	$+2$
$(3a + 2)(2a + 1)$	$6a^2$	$+7a$	$+2$
$(3a - 2)(2a - 1)$	$6a^2$	$-7a$	$+2$

The factors of $6a^2 + 7a + 2$ are $(3a + 2)$ and $(2a + 1)$.

Check: $(3a + 2)(2a + 1) = 6a^2 + 7a + 2$

Notice that in the preceding list we did not include the factors $(6a + 2)$ and $(a + 1)$. We do not need to try these, because the first factor has a 2 common to each term and so could be factored again, giving $2(3a + 1)(a + 1)$. Since our original trinomial, $6a^2 + 7a + 2$, did *not* have a greatest common factor of 2, neither of its factors will.

E X A M P L E   2   Factor $4x^2 - x - 3$.

Solution   We list all the possible factors that, when multiplied, give a trinomial whose first term is $4x^2$ and whose last term is -3.

Binomial Factors	First Term	Middle Term	Last Term
$(4x + 1)(x - 3)$	$4x^2$	$-11x$	-3
$(4x - 1)(x + 3)$	$4x^2$	$+11x$	-3
$(4x + 3)(x - 1)$	$4x^2$	$-x$	-3
$(4x - 3)(x + 1)$	$4x^2$	$+x$	-3
$(2x + 1)(2x - 3)$	$4x^2$	$-4x$	-3
$(2x - 1)(2x + 3)$	$4x^2$	$+4x$	-3

The third line shows that the factors are $(4x + 3)$ and $(x - 1)$.

Check: $(4x + 3)(x - 1) = 4x^2 - x - 3$

You will find that the more practice you have at factoring this type of trinomial, the faster you will get the correct factors. You will pick up some shortcuts along the way or may come across a system of eliminating some factors as possibilities. Whatever works best for you is the method you should use. Factoring is a very important tool, and you must be good at it.

E X A M P L E   3   Factor $12y^3 + 10y^2 - 12y$.

Solution   We begin by factoring out the greatest common factor, $2y$:

$$12y^3 + 10y^2 - 12y = 2y(6y^2 + 5y - 6)$$

We now list all possible factors of a trinomial with the first term $6y^2$ and last term -6, along with the associated middle terms.

Possible Factors	Middle Term When Multiplied
$(3y + 2)(2y - 3)$	$-5y$
$(3y - 2)(2y + 3)$	$+5y$
$(6y + 1)(y - 6)$	$-35y$
$(6y - 1)(y + 6)$	$+35y$

The second line gives the correct factors. The complete problem is:

$$12y^3 + 10y^2 - 12y = 2y(6y^2 + 5y - 6)$$
$$= 2y(3y - 2)(2y + 3)$$

EXAMPLE 4   Factor $30x^2y - 5xy^2 - 10y^3$.

Solution   The greatest common factor is $5y$:

$$30x^2y - 5xy^2 - 10y^3 = 5y(6x^2 - xy - 2y^2)$$
$$= 5y(2x + y)(3x - 2y)$$

Method 2: Factoring $ax^2 + bx + c$ by Grouping

Recall from Section 5.1 that we can use factoring by grouping to factor the polynomial $6x^2 - 3x - 4x + 2$. We begin by factoring $3x$ from the first two terms and -2 from the last two terms. For review, here is the complete problem:

$$6x^2 - 3x - 4x + 2 = 3x(2x - 1) - 2(2x - 1)$$
$$= (2x - 1)(3x - 2)$$

Now, let's back up a little and notice that our original polynomial $6x^2 - 3x - 4x + 2$ can be simplified to $6x^2 - 7x + 2$ by adding $-3x$ and $-4x$. This means that $6x^2 - 7x + 2$ can be factored to $(2x - 1)(3x - 2)$ by the grouping method shown in Section 5.1. The key to using this process is to rewrite the middle term $-7x$ as $-3x - 4x$.

To generalize this discussion, here are the steps we use to factor trinomials by grouping.

Strategy for Factoring $ax^2 + bx + c$ by Grouping

Step 1: Form the product ac.

Step 2: Find a pair of numbers whose product is ac and whose sum is b.

Step 3: Rewrite the polynomial to be factored so that the middle term bx is written as the sum of two terms whose coefficients are the two numbers found in step 2.

Step 4: Factor by grouping.

E X A M P L E   5   Factor $3x^2 - 10x - 8$ using these steps.

Solution   The trinomial $3x^2 - 10x - 8$ has the form $ax^2 + bx + c$, where $a = 3$, $b = -10$, and $c = -8$.

Step 1: The product ac is $3(-8) = -24$.

Step 2: We need to find two numbers whose product is -24 and whose sum is -10. Let's systematically begin to list all the pairs of numbers whose product is -24 to find the pair whose sum is -10.

Product	Sum
$-24(1) = -24$	$-24 + 1 = -23$
$-12(2) = -24$	$-12 + 2 = -10$

We stop here because we have found the pair of numbers whose product is -24 and whose sum is -10. The numbers are -12 and 2.

Step 3: We now rewrite our original trinomial so that the middle term, $-10x$, is written as the sum of $-12x$ and $2x$:

$$3x^2 - 10x - 8 = 3x^2 - 12x + 2x - 8$$

Step 4: Factoring by grouping, we have:

$$3x^2 - 12x + 2x - 8 = 3x(x - 4) + 2(x - 4)$$
$$= (x - 4)(3x + 2)$$

We can check our work by multiplying $x - 4$ and $3x + 2$ to get $3x^2 - 10x - 8$.

E X A M P L E   6   Factor $4x^2 - x - 3$.

Solution   In this case, $a = 4$, $b = -1$, and $c = -3$. The product ac is $4(-3) = -12$. We need a pair of numbers whose product is -12 and whose sum is -1. We begin by listing pairs of numbers whose product is -12 in order to find the pair whose sum is -1.

Product	Sum
$-12(1) = -12$	$-12 + 1 = -11$
$-6(2) = -12$	$-6 + 2 = -4$
$-4(3) = -12$	$-4 + 3 = -1$

We stop here because we have found the pair of numbers for which we are looking. They are -4 and 3. Next, we rewrite the middle term $-x$ as the sum $-4x + 3x$ and proceed to factor by grouping.

$$4x^2 - x - 3 = 4x^2 - 4x + 3x - 3$$
$$= 4x(x - 1) + 3(x - 1)$$
$$= (x - 1)(4x + 3)$$

Compare this procedure and the result with those shown in Example 2 of this section.

EXAMPLE 7   Factor $8x^2 - 2x - 15$.

Solution   The product ac is $8(-15) = -120$. There are many pairs of numbers whose product is -120. We are looking for the pair whose sum is also -2. The numbers are -12 and 10. Writing $-2x$ as $-12x + 10x$ and then factoring by grouping, we have:

$$8x^2 - 2x - 15 = 8x^2 - 12x + 10x - 15$$
$$= 4x(2x - 3) + 5(2x - 3)$$
$$= (2x - 3)(4x + 5)$$

EXAMPLE 8   A ball is tossed into the air with an upward velocity of 16 feet per second from the top of a building 32 feet high. The equation that gives the height of the ball above the ground at any time t is

$$h = 32 + 16t - 16t^2$$

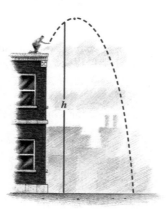

Factor the right side of this equation and then find h when t is 2.

Solution   We begin by factoring out the greatest common factor, 16. Then, we factor the trinomial that remains:

$$h = 32 + 16t - 16t^2$$
$$h = 16(2 + t - t^2)$$
$$h = 16(2 - t)(1 + t)$$

Letting $t = 2$ in the equation, we have

$$h = 16(2 - 2)(1 + 2)$$
$$= 16(0)(3)$$
$$= 0$$

When t is 2, h is 0.

Getting Ready for Class

After reading through the preceding section, respond in your own words and in complete sentences.

A. What is the first step in factoring a trinomial?

B. Describe the criteria you would use to set up a table of possible factors of a trinomial.

C. What does it mean if you factor a trinomial and one of your factors has a greatest common factor of 3?

D. Describe how you would look for possible factors of $6a^2 + 7a + 2$.

PROBLEM SET 5.3

Factor the following trinomials.

1. $2x^2 + 7x + 3$

2. $2x^2 + 5x + 3$

3. $2a^2 - a - 3$

4. $2a^2 + a - 3$

5. $3x^2 + 2x - 5$

6. $3x^2 - 2x - 5$

7. $3y^2 - 14y - 5$

8. $3y^2 + 14y - 5$

9. $6x^2 + 13x + 6$

10. $6x^2 - 13x + 6$

11. $4x^2 - 12xy + 9y^2$

12. $4x^2 + 12xy + 9y^2$

13. $4y^2 - 11y - 3$

14. $4y^2 + y - 3$

15. $20x^2 - 41x + 20$

16. $20x^2 + 9x - 20$

17. $20a^2 + 48ab - 5b^2$

18. $20a^2 + 29ab + 5b^2$

19. $20x^2 - 21x - 5$

20. $20x^2 - 48x - 5$

21. $12m^2 + 16m - 3$

22. $12m^2 + 20m + 3$

23. $20x^2 + 37x + 15$

24. $20x^2 + 13x - 15$

25. $12a^2 - 25ab + 12b^2$

26. $12a^2 + 7ab - 12b^2$

27. $3x^2 - xy - 14y^2$

28. $3x^2 + 19xy - 14y^2$

29. $14x^2 + 29x - 15$

30. $14x^2 + 11x - 15$

31. $6x^2 - 43x + 55$

32. $6x^2 - 7x - 55$

33. $15t^2 - 67t + 38$

34. $15t^2 - 79t - 34$

Factor each of the following completely. Look first for the greatest common factor.

35. $4x^2 + 2x - 6$

36. $6x^2 - 51x + 63$

37. $24a^2 - 50a + 24$

38. $18a^2 + 48a + 32$

39. $10x^3 - 23x^2 + 12x$

40. $10x^4 + 7x^3 - 12x^2$

41. $6x^4 - 11x^3 - 10x^2$

42. $6x^3 + 19x^2 + 10x$

43. $10a^3 - 6a^2 - 4a$

44. $6a^3 + 15a^2 + 9a$

45. $15x^3 - 102x^2 - 21x$

46. $2x^4 - 24x^3 + 64x^2$

47. $35y^3 - 60y^2 - 20y$

48. $14y^4 - 32y^3 + 8y^2$

49. $15a^4 - 2a^3 - a^2$

50. $10a^5 - 17a^4 + 3a^3$

51. $24x^2y - 6xy - 45y$

52. $8x^2y^2 + 26xy^2 + 15y^2$

53. $12x^2y - 34xy^2 + 14y^3$

54. $12x^2y - 46xy^2 + 14y^3$

55. Evaluate the expression $2x^2 + 7x + 3$ and the expression $(2x + 1)(x + 3)$ for $x = 2$.

56. Evaluate the expression $2a^2 - a - 3$ and the expression $(2a - 3)(a + 1)$ for $a = 5$.

57. What polynomial factors to $(2x + 3)(2x - 3)$?

58. What polynomial factors to $(5x + 4)(5x - 4)$?

59. What polynomial factors to $(x + 3)(x - 3)(x^2 + 9)$?

60. What polynomial factors to $(x + 2)(x - 2)(x^2 + 4)$?

Applying the Concepts

61. **Archery**   Margaret shoots an arrow into the air. The equation for the height (in feet) of the tip of the arrow is:

$$h = 8 + 62t - 16t^2$$

Factor the right side of this equation. Then fill in

the table for various heights of the arrow, using the factored form of the equation.

Time t (seconds)	Height h (feet)
0	
1	
2	
3	
4	

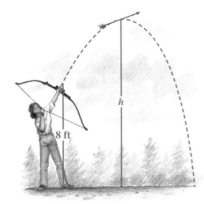

62. Coin Toss At the beginning of every football game, the referee flips a coin to see who will kick off. The equation that gives the height (in feet) of

the coin tossed in the air is:

$$h = 6 + 29t - 16t^2$$

(a) Factor this equation.

(b) Use the factored form of the equation to find the height of the quarter after 0 second, 1 second, and 2 seconds.

63. Constructing a Box Yesterday I was experimenting with how to cut and fold a certain piece of cardboard to make a box with different volumes. Unfortunately, today I have lost both the cardboard and most of my notes. I remember that I made the box by cutting equal squares from the corners, then folding up the side flaps:

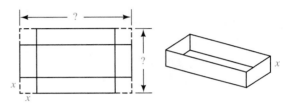

I don't remember how big the cardboard was, and I can only find the last page of notes, which says that, if x is the length of a side of a small square (in inches), then the volume is $V = 99x - 40x^2 + 4x^3$.

(a) Factor the right side of this expression completely.

(b) What were the dimensions of the original piece of cardboard?

64. Constructing a Box Repeat Problem 63 using the formula $V = 15x - 16x^2 + 4x^3$.

Review Problems

The following problems review material we covered in Section 4.6. Reviewing these problems will help you understand the next section.

Multiply.

65. $(x + 3)(x - 3)$

66. $(x + 5)(x - 5)$

67. $(6a + 1)(6a - 1)$

68. $(4a + 5)(4a - 5)$

69. $(x + 4)^2$

70. $(x - 5)^2$

71. $(2x + 3)^2$

72. $(2x - 3)^2$

In Chapter 4 we listed the following three special products:

$$(a + b)^2 = (a + b)(a + b) = a^2 + 2ab + b^2$$
$$(a - b)^2 = (a - b)(a - b) = a^2 - 2ab + b^2$$
$$(a + b)(a - b) = a^2 - b^2$$

Since factoring is the reverse of multiplication, we can also consider the three special products as three special factorings:

$$a^2 + 2ab + b^2 = (a + b)^2$$
$$a^2 - 2ab + b^2 = (a - b)^2$$
$$a^2 - b^2 = (a + b)(a - b)$$

Any trinomial of the form $a^2 + 2ab + b^2$ or $a^2 - 2ab + b^2$ can be factored by the methods of Section 5.3. The last line is the factorization of the difference of two squares. The difference of two squares always factors in this way. Again, these are patterns you must be able to recognize on sight.

E X A M P L E I   Factor $16x^2 - 25$.

Solution   We can see that the first term is a perfect square, and the last term is also. This fact becomes even more obvious if we rewrite the problem as:

$$16x^2 - 25 = (4x)^2 - (5)^2$$

The first term is the square of the quantity $4x$, and the last term is the square of 5. The completed problem looks like this:

$$16x^2 - 25 = (4x)^2 - (5)^2$$
$$= (4x + 5)(4x - 5)$$

To check our results, we multiply:

$$(4x + 5)(4x - 5) = 16x^2 + 20x - 20x - 25$$
$$= 16x^2 - 25$$

E X A M P L E 2   Factor $36a^2 - 1$.

Solution   We rewrite the two terms to show they are perfect squares and then factor. Remember, 1 is its own square, $1^2 = 1$.

$$36a^2 - 1 = (6a)^2 - (1)^2$$
$$= (6a + 1)(6a - 1)$$

To check our results, we multiply:

$$(6a + 1)(6a - 1) = 36a^2 + 6a - 6a - 1$$
$$= 36a^2 - 1$$

EXAMPLE 3   Factor $x^4 - y^4$.

Solution x^4 is the perfect square $(x^2)^2$, and y^4 is $(y^2)^2$:

$$x^4 - y^4 = (x^2)^2 - (y^2)^2$$
$$= (x^2 - y^2)(x^2 + y^2)$$

The factor $(x^2 - y^2)$ is itself the difference of two squares and therefore can be factored again. The factor $(x^2 + y^2)$ is the *sum* of two squares and cannot be factored again. The complete problem is this:

$$x^4 - y^4 = (x^2)^2 - (y^2)^2$$
$$= (x^2 - y^2)(x^2 + y^2)$$
$$= (x + y)(x - y)(x^2 + y^2)$$

Note: If you think the sum of two squares $x^2 + y^2$ factors, you should try it. Write down the factors you think it has, and then multiply them using the FOIL method. You won't get $x^2 + y^2$.

EXAMPLE 4   Factor $25x^2 - 60x + 36$.

Solution   Although this trinomial can be factored by the method we used in Section 5.3, we notice that the first and last terms are the perfect squares $(5x)^2$ and $(6)^2$. Before going through the method for factoring trinomials by listing all possible factors, we can check to see if $25x^2 - 60x + 36$ factors to $(5x - 6)^2$. We need only multiply to check:

$$(5x - 6)^2 = (5x - 6)(5x - 6)$$
$$= 25x^2 - 30x - 30x + 36$$
$$= 25x^2 - 60x + 36$$

The trinomial $25x^2 - 60x + 36$ factors to $(5x - 6)(5x - 6) = (5x - 6)^2$.

EXAMPLE 5   Factor $m^2 + 14m + 49$.

Solution   Since the first and last terms are perfect squares, we can try the factors $(m + 7)(m + 7)$:

$$(m + 7)^2 = (m + 7)(m + 7)$$
$$= m^2 + 7m + 7m + 49$$
$$= m^2 + 14m + 49$$

The factors of $m^2 + 14m + 49$ are $(m + 7)(m + 7) = (m + 7)^2$.

Note: As we have indicated before, perfect square trinomials like the ones in Examples 4 and 5 can be factored by the methods developed in previous sections. Recognizing that they factor to binomial squares simply saves time in factoring.

E X A M P L E   6   Factor $5x^2 + 30x + 45$.

Solution   We begin by factoring out the greatest common factor, which is 5. Then we notice that the trinomial that remains is a perfect square trinomial:

$$5x^2 + 30x + 45 = 5(x^2 + 6x + 9)$$
$$= 5(x + 3)^2$$

E X A M P L E   7   Factor $(x - 3)^2 - 25$.

Solution   This example has the form $a^2 - b^2$, where a is $x - 3$ and b is 5. We factor it according to the formula for the difference of two squares:

$$
\begin{aligned}
(x - 3)^2 - 25 &= (x - 3)^2 - 5^2 && \text{Write 25 as } 5^2. \\
&= [(x - 3) - 5][(x - 3) + 5] && \text{Factor.} \\
&= (x - 8)(x + 2) && \text{Simplify.}
\end{aligned}
$$

Notice in this example we could have expanded $(x - 3)^2$, subtracted 25, and then factored to obtain the same result:

$$
\begin{aligned}
(x - 3)^2 - 25 &= x^2 - 6x + 9 - 25 && \text{Expand } (x - 3)^2. \\
&= x^2 - 6x - 16 && \text{Simplify.} \\
&= (x - 8)(x + 2) && \text{Factor.}
\end{aligned}
$$

 Getting Ready for Class

After reading through the preceding section, respond in your own words and in complete sentences.

A. Describe how you factor the difference of two squares.
B. What is a perfect square trinomial?
C. How do you know when you've factored completely?
D. Describe how you would factor $25x^2 - 60x + 36$.

PROBLEM SET 5.4

Factor the following.

1. $x^2 - 9$

2. $x^2 - 25$

3. $a^2 - 36$

4. $a^2 - 64$

5. $x^2 - 49$

6. $x^2 - 121$

7. $4a^2 - 16$

8. $4a^2 + 16$

9. $9x^2 + 25$

10. $16x^2 - 36$

11. $25x^2 - 169$

12. $x^2 - y^2$

13. $9a^2 - 16b^2$

14. $49a^2 - 25b^2$

15. $9 - m^2$

16. $16 - m^2$

17. $25 - 4x^2$

18. $36 - 49y^2$

19. $2x^2 - 18$

20. $3x^2 - 27$

21. $32a^2 - 128$

22. $3a^3 - 48a$

23. $8x^2y - 18y$

24. $50a^2b - 72b$

25. $a^4 - b^4$

26. $a^4 - 16$

27. $16m^4 - 81$

28. $81 - m^4$

29. $3x^3y - 75xy^3$

30. $2xy^3 - 8x^3y$

Factor the following.

31. $x^2 - 2x + 1$

32. $x^2 - 6x + 9$

33. $x^2 + 2x + 1$

34. $x^2 + 6x + 9$

35. $a^2 - 10a + 25$

36. $a^2 + 10a + 25$

37. $y^2 + 4y + 4$

38. $y^2 - 8y + 16$

39. $x^2 - 4x + 4$

40. $x^2 + 8x + 16$

41. $m^2 - 12m + 36$

42. $m^2 + 12m + 36$

43. $4a^2 + 12a + 9$

44. $9a^2 - 12a + 4$

45. $49x^2 - 14x + 1$

46. $64x^2 - 16x + 1$

47. $9y^2 - 30y + 25$

48. $25y^2 + 30y + 9$

49. $x^2 + 10xy + 25y^2$

50. $25x^2 + 10xy + y^2$

51. $9a^2 + 6ab + b^2$

52. $9a^2 - 6ab + b^2$

Factor the following by first factoring out the greatest common factor.

53. $3a^2 + 18a + 27$

54. $4a^2 - 16a + 16$

55. $2x^2 + 20xy + 50y^2$

56. $3x^2 + 30xy + 75y^2$

57. $5x^3 + 30x^2y + 45xy^2$

58. $12x^2y - 36xy^2 + 27y^3$

Factor by grouping the first three terms together.

59. $x^2 + 6x + 9 - y^2$

60. $x^2 + 10x + 25 - y^2$

61. $x^2 + 2xy + y^2 - 9$

62. $a^2 + 2ab + b^2 - 25$

63. Find a value for b so that the polynomial $x^2 + bx + 49$ factors to $(x + 7)^2$.

64. Find a value of b so that the polynomial $x^2 + bx + 81$ factors to $(x + 9)^2$.

65. Find the value of c for which the polynomial $x^2 + 10x + c$ factors to $(x + 5)^2$.

66. Find the value of a for which the polynomial $ax^2 + 12x + 9$ factors to $(2x + 3)^2$.

Applying the Concepts

67. **Area**

(a) What is the area of the following figure?

(b) Factor the answer from part (a).

(c) Find a way to cut the figure into two pieces and put them back together to show that the factorization in part (b) is correct.

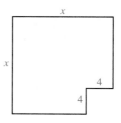

68. **Area**

(a) What is the area of the following figure?

(b) Factor the expression from part (a).

(c) Cut and rearrange the figure to show that the factorization is correct.

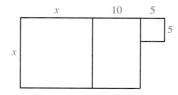

Find the area for the shaded regions; then write your result in factored form.

69.

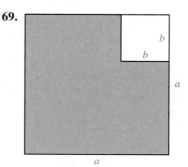

70.

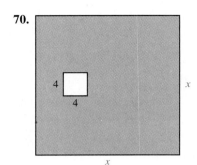

Review Problems

The following problems review material we covered in Section 4.8.

Use long division to divide.

71. $\dfrac{x^2 - 5x + 8}{x - 3}$

72. $\dfrac{x^2 + 7x + 12}{x + 4}$

73. $\dfrac{6x^2 + 5x + 3}{2x + 3}$

74. $\dfrac{x^3 + 27}{x + 3}$

5.5 Factoring: A General Review

In this section we will review the different methods of factoring that we presented in the previous sections of the chapter. This section is important because it will give you an opportunity to factor a variety of polynomials. Prior to this section, the polynomials you worked with were grouped together according to the method used to factor them. That is, in Section 5.4 all the polynomials you factored were either the difference of two squares or perfect square trinomials. What usually happens in a situation like this is that you become proficient at factoring the kind of polynomial you are working with at the time, but have trouble when given a variety of polynomials to factor.

We begin this section with a checklist that can be used in factoring polynomials of any type. When you have finished this section and the problem set that follows, you want to be proficient enough at factoring so that the checklist is second nature to you.

STRATEGY FOR FACTORING A POLYNOMIAL

Step 1: If the polynomial has a greatest common factor other than 1, then factor out the greatest common factor.

Step 2: If the polynomial has two terms (it is a binomial), then see if it is the difference of two squares. Remember, if it is the sum of two squares, it will not factor.

Step 3: If the polynomial has three terms (a trinomial), then either it is a perfect square trinomial, which will factor into the square of a binomial, or it is not a perfect square trinomial, in which case you use one of the methods developed in Section 5.3.

Step 4: If the polynomial has more than three terms, try to factor it by grouping.

Step 5: As a final check, see if any of the factors you have written can be factored further. If you have overlooked a common factor, you can catch it here.

Here are some examples illustrating how we use the checklist.

EXAMPLE 1   Factor $2x^5 - 8x^3$.

Solution   First, we check to see if the greatest common factor is other than 1. Since the greatest common factor is $2x^3$, we begin by factoring it out. Once we have done so, we notice that the binomial that remains is the difference of two squares:

$$2x^5 - 8x^3 = 2x^3(x^2 - 4) \qquad \text{Factor out the greatest}$$
$$\text{common factor, } 2x^3.$$

$$= 2x^3(x + 2)(x - 2) \qquad \text{Factor the difference}$$
$$\text{of two squares.}$$

Note that the greatest common factor $2x^3$ that we factored from each term in the first step of Example 1 remains as part of the answer to the problem. That is because it is one of the factors of the original binomial. Remember, the expression we end up with when factoring must be equal to the expression we start with. We can't just drop a factor and expect the resulting expression to equal the original expression.

EXAMPLE 2   Factor $3x^4 - 18x^3 + 27x^2$.

Solution   Step 1 is to factor out the greatest common factor, $3x^2$. After we have done so, we notice that the trinomial that remains is a perfect square trinomial, which will factor as the square of a binomial:

$$3x^4 - 18x^3 + 27x^2 = 3x^2(x^2 - 6x + 9) \qquad \text{Factor out } 3x^2.$$
$$= 3x^2(x - 3)^2 \qquad x^2 - 6x + 9 \text{ is the}$$
$$\text{square of } x - 3.$$

EXAMPLE 3   Factor $y^3 + 25y$.

Solution   We begin by factoring out the y that is common to both terms. The binomial that remains after we have done so is the sum of two squares, which does not factor, so after the first step we are finished:

$$y^3 + 25y = y(y^2 + 25) \qquad \text{Factor out the greatest common factor, } y;$$
$$\text{then notice that } y^2 + 25 \text{ cannot be}$$
$$\text{factored further.}$$

EXAMPLE 4   Factor $6a^2 - 11a + 4$.

Solution   Here we have a trinomial that does not have a greatest common factor other than 1. Since it is not a perfect square trinomial, we factor it by trial and error. That is, we look for binomial factors, the product of whose first terms is $6a^2$ and the product of whose last terms is 4. Then we look for the combination of these types of binomials whose product gives us a middle term of $-11a$. Without showing all the different possibilities, here is the answer:

$$6a^2 - 11a + 4 = (3a - 4)(2a - 1)$$

EXAMPLE 5   Factor $6x^3 - 12x^2 - 48x$.

Solution   This trinomial has a greatest common factor of $6x$. The trinomial that remains after the $6x$ has been factored from each term can be factored by trial and error:

$$6x^3 - 12x^2 - 48x = 6x(x^2 - 2x - 8)$$
$$= 6x(x - 4)(x + 2)$$

EXAMPLE 6   Factor $2ab^5 + 8ab^4 + 2ab^3$.

Solution   The greatest common factor is $2ab^3$. We begin by factoring it from each term. After that we find that the trinomial that remains cannot be factored further:

$$2ab^5 + 8ab^4 + 2ab^3 = 2ab^3(b^2 + 4b + 1)$$

EXAMPLE 7   Factor $xy + 8x + 3y + 24$.

Solution   Since our polynomial has four terms, we try factoring by grouping:

$$xy + 8x + 3y + 24 = x(y + 8) + 3(y + 8)$$
$$= (y + 8)(x + 3)$$

 Getting Ready for Class

After reading through the preceding section, respond in your own words and in complete sentences.

A. What is the first step in factoring any polynomial?

B. If a polynomial has four terms, what method of factoring should you try?

C. If a polynomial has two terms, what method of factoring should you try?

D. What is the last step in factoring any polynomial?

PROBLEM SET 5.5

Factor each of the following polynomials completely. That is, once you are finished factoring, none of the factors you obtain should be factorable. Also, note that the even-numbered problems are not necessarily similar to the odd-numbered problems that precede them in this problem set.

1. $x^2 - 81$

2. $x^2 - 18x + 81$

3. $x^2 + 2x - 15$

4. $15x^2 + 11x - 6$

5. $x^2 + 6x + 9$

6. $12x^2 - 11x + 2$

7. $y^2 - 10y + 25$

8. $21y^2 - 25y - 4$

9. $2a^3b + 6a^2b + 2ab$

10. $6a^2 - ab - 15b^2$

11. $x^2 + x + 1$

12. $2x^2 - 4x + 2$

13. $12a^2 - 75$

14. $18a^2 - 50$

15. $9x^2 - 12xy + 4y^2$

16. $x^3 - x^2$

17. $4x^3 + 16xy^2$

18. $16x^2 + 49y^2$

19. $2y^3 + 20y^2 + 50y$

20. $3y^2 - 9y - 30$

21. $a^6 + 4a^4b^2$

22. $5a^2 - 45b^2$

23. $xy + 3x + 4y + 12$

24. $xy + 7x + 6y + 42$

25. $x^4 - 16$

26. $x^4 - 81$

27. $xy - 5x + 2y - 10$

28. $xy - 7x + 3y - 21$

29. $5a^2 + 10ab + 5b^2$

30. $3a^3b^2 + 15a^2b^2 + 3ab^2$

31. $x^2 + 49$

32. $16 - x^4$

33. $3x^2 + 15xy + 18y^2$

34. $3x^2 + 27xy + 54y^2$

35. $2x^2 + 15x - 38$

36. $2x^2 + 7x - 85$

37. $100x^2 - 300x + 200$

38. $100x^2 - 400x + 300$

39. $x^2 - 64$

40. $9x^2 - 4$

41. $x^2 + 3x + ax + 3a$

42. $x^2 + 4x + bx + 4b$

43. $49a^7 - 9a^5$

44. $a^4 - 1$

45. $49x^2 + 9y^2$

46. $12x^4 - 62x^3 + 70x^2$

47. $25a^3 + 20a^2 + 3a$

48. $36a^4 - 100a^2$

49. $xa - xb + ay - by$

50. $xy - bx + ay - ab$

51. $48a^4b - 3a^2b$

52. $18a^4b^2 - 12a^3b^3 + 8a^2b^4$

53. $20x^4 - 45x^2$

54. $16x^3 + 16x^2 + 3x$

55. $3x^2 + 35xy - 82y^2$

56. $3x^2 + 37xy - 86y^2$

57. $16x^5 - 44x^4 + 30x^3$

58. $16x^2 + 16x - 1$

59. $2x^2 + 2ax + 3x + 3a$

60. $2x^2 + 2ax + 5x + 5a$

61. $y^4 - 1$

62. $25y^7 - 16y^5$

63. $12x^4y^2 + 36x^3y^3 + 27x^2y^4$

64. $16x^3y^2 - 4xy^2$

Review Problems

The problems that follow review material we covered in Sections 2.3 and 4.1. Reviewing the problems from Section 2.3 will help you understand the next section.

Solve each equation. [2.3]

65. $3x - 6 = 9$

66. $5x - 1 = 14$

67. $2x + 3 = 0$

68. $4x - 5 = 0$

69. $4x + 3 = 0$

70. $3x - 1 = 0$

Simplify, using the properties of exponents. [4.1]

71. $x^8 \cdot x^7$

72. $(x^5)^2$

73. $(3x^3)^2(2x^4)^3$

74. $(5x^2y)^2(4xy^3)^2$

75. Write the number 57,600 in scientific notation.

76. Write the number 4.3×10^5 in expanded form.

5.6 Solving Equations by Factoring

In this section we will use the methods of factoring developed in previous sections, along with a special property of 0, to solve quadratic equations.

> **DEFINITION**   Any equation that can be put in the form $ax^2 + bx + c = 0$, where a, b, and c are real numbers ($a \neq 0$), is called a **quadratic equation**. The equation $ax^2 + bx + c = 0$ is called **standard form** for a quadratic equation:
>
> an x^2 term   an x term   and a constant term
> $$a(\text{variable})^2 + b(\text{variable}) + (\text{absence of the variable}) = 0$$

The number 0 has a special property. If we multiply two numbers and the product is 0, then one or both of the original two numbers must be 0. In symbols, this property looks like this.

ZERO-FACTOR PROPERTY

Let a and b represent real numbers. If $a \cdot b = 0$, then $a = 0$ or $b = 0$.

Suppose we want to solve the quadratic equation $x^2 + 5x + 6 = 0$. We can factor the left side into $(x + 2)(x + 3)$. Then we have:

$$x^2 + 5x + 6 = 0$$

$$(x + 2)(x + 3) = 0$$

Now, $(x + 2)$ and $(x + 3)$ both represent real numbers. Their product is 0; therefore, either $(x + 3)$ is 0 or $(x + 2)$ is 0. Either way we have a solution to our equation. We use the property of 0 stated to finish the problem:

$$x^2 + 5x + 6 = 0$$

$$(x + 2)(x + 3) = 0$$

$$x + 2 = 0 \quad \text{or} \quad x + 3 = 0$$

$$x = -2 \quad \text{or} \quad x = -3$$

Our solution set is $\{-2, -3\}$. Our equation has two solutions. To check our solutions we have to check each one separately to see that they both produce a true statement when used in place of the variable:

When $x = -3$

the equation $x^2 + 5x + 6 = 0$

becomes $(-3)^2 + 5(-3) + 6 \stackrel{?}{=} 0$

$$9 + (-15) + 6 = 0$$

$$0 = 0$$

When $x = -2$

the equation $x^2 + 5x + 6 = 0$

becomes

$$(-2)^2 + 5(-2) + 6 \stackrel{?}{=} 0$$

$$4 + (-10) + 6 = 0$$

$$0 = 0$$

We have solved a quadratic equation by replacing it with two linear equations in one variable.

STRATEGY FOR SOLVING A QUADRATIC EQUATION BY FACTORING

Step 1: Put the equation in standard form, that is, 0 on one side and decreasing powers of the variable on the other.

Step 2: Factor completely.

Step 3: Use the zero-factor property to set each variable factor from step 2 to 0.

Step 4: Solve each equation produced in step 3.

Step 5: Check each solution, if necessary.

EXAMPLE   I   Solve the equation $2x^2 - 5x = 12$.

Solution

Step 1: We begin by adding -12 to both sides, so the equation is in standard form:

$$2x^2 - 5x = 12$$

$$2x^2 - 5x - 12 = 0$$

Step 2: We factor the left side completely:

$$(2x + 3)(x - 4) = 0$$

Step 3: We set each factor to 0:

$$2x + 3 = 0 \qquad \text{or} \qquad x - 4 = 0$$

Step 4: Solve each of the equations from step 3:

$$2x + 3 = 0 \qquad x - 4 = 0$$

$$2x = -3 \qquad x = 4$$

$$x = -\frac{3}{2}$$

Step 5: Substitute each solution into $2x^2 - 5x = 12$ to check:

Check: $-\dfrac{3}{2}$ Check: 4

$$2\left(-\frac{3}{2}\right)^2 - 5\left(-\frac{3}{2}\right) \stackrel{?}{=} 12 \qquad 2(4)^2 - 5(4) \stackrel{?}{=} 12$$

$$2\left(\frac{9}{4}\right) + 5\left(\frac{3}{2}\right) = 12 \qquad 2(16) - 20 = 12$$

$$\frac{9}{2} + \frac{15}{2} = 12 \qquad 32 - 20 = 12$$

$$\frac{24}{2} = 12 \qquad 12 = 12$$

$$12 = 12$$

E X A M P L E   2   Solve for a: $16a^2 - 25 = 0$

Solution   The equation is already in standard form:

$$16a^2 - 25 = 0$$

$$(4a - 5)(4a + 5) = 0 \qquad \text{Factor left side.}$$

$$4a - 5 = 0 \quad \text{or} \quad 4a + 5 = 0 \qquad \text{Set each factor to 0.}$$

$$4a = 5 \qquad\qquad 4a = -5 \qquad \text{Solve the resulting equations.}$$

$$a = \frac{5}{4} \qquad\qquad a = -\frac{5}{4}$$

E X A M P L E   3   Solve $4x^2 = 8x$.

Solution   We begin by adding $-8x$ to each side of the equation to put it in standard form. Then we factor the left side of the equation by factoring out the greatest common factor.

$$4x^2 = 8x$$

$$4x^2 - 8x = 0 \qquad \text{Add } -8x \text{ to each side.}$$

$$4x(x - 2) = 0 \qquad \text{Factor the left side.}$$

$$4x = 0 \quad \text{or} \quad x - 2 = 0 \qquad \text{Set each factor to 0.}$$

$$x = 0 \quad \text{or} \quad x = 2 \qquad \text{Solve the resulting equations.}$$

The solutions are 0 and 2.

E X A M P L E   4   Solve $x(2x + 3) = 44$.

Solution   We must multiply out the left side first and then put the equation in standard form:

$$x(2x + 3) = 44$$

$$2x^2 + 3x = 44 \qquad \text{Multiply out the left side.}$$

$$2x^2 + 3x - 44 = 0 \qquad \text{Add } -44 \text{ to each side.}$$

$$(2x + 11)(x - 4) = 0 \qquad \text{Factor the left side.}$$

$$2x + 11 = 0 \qquad \text{or} \qquad x - 4 = 0 \qquad \text{Set each factor to 0.}$$

$$2x = -11 \qquad \text{or} \qquad x = 4 \qquad \text{Solve the resulting equations.}$$

$$x = -\frac{11}{2}$$

The two solutions are $-\frac{11}{2}$ and 4.

E X A M P L E   5   Solve for x: $5^2 = x^2 + (x + 1)^2$

Solution   Before we can put this equation in standard form we must square the binomial. Remember, to square a binomial, we use the formula $(a + b)^2 = a^2 + 2ab + b^2$:

$$5^2 = x^2 + (x + 1)^2$$

$$25 = x^2 + x^2 + 2x + 1 \qquad \text{Expand } 5^2 \text{ and } (x + 1)^2.$$

$$25 = 2x^2 + 2x + 1 \qquad \text{Simplify the right side.}$$

$$0 = 2x^2 + 2x - 24 \qquad \text{Add } -25 \text{ to each side.}$$

$$0 = 2(x^2 + x - 12) \qquad \text{Begin factoring.}$$

$$0 = 2(x + 4)(x - 3) \qquad \text{Factor completely.}$$

$$x + 4 = 0 \qquad \text{or} \qquad x - 3 = 0 \qquad \text{Set each variable factor to 0.}$$

$$x = -4 \qquad \text{or} \qquad x = 3$$

Note, in the second to the last line, that we do not set 2 equal to 0. That is because 2 can never be 0. It is always 2. We only use the zero-factor property to set variable factors to 0 because they are the only factors that can possibly be 0.

   Also notice that it makes no difference which side of the equation is 0 when we write the equation in standard form.

   Although the equation in the next example is not a quadratic equation, it can be solved by the method shown in the first five examples.

E X A M P L E   6   Solve $24x^3 = -10x^2 + 6x$ for x.

Solution   First, we write the equation in standard form:

$$24x^3 + 10x^2 - 6x = 0 \qquad \text{Standard form}$$

$$2x(12x^2 + 5x - 3) = 0 \qquad \text{Factor out } 2x.$$

$$2x(3x - 1)(4x + 3) = 0 \qquad \text{Factor remaining trinomial.}$$

$$2x = 0 \quad \text{or} \quad 3x - 1 = 0 \quad \text{or} \quad 4x + 3 = 0 \qquad \text{Set factors to 0.}$$

$$x = 0 \quad \text{or} \quad x = \frac{1}{3} \quad \text{or} \quad x = -\frac{3}{4} \qquad \text{Solutions}$$

Getting Ready for Class

After reading through the preceding section, respond in your own words and in complete sentences.

A. When is a quadratic equation in standard form?

B. What is the first step in solving an equation by factoring?

C. Describe the zero-factor property in your own words.

D. Describe how you would solve the equation $2x^2 - 5x = 12$.

PROBLEM SET 5.6

The following equations are already in factored form. Use the special zero-factor property to set the factors to 0 and solve.

1. $(x + 2)(x - 1) = 0$ **2.** $(x + 3)(x + 2) = 0$

3. $(a - 4)(a - 5) = 0$ **4.** $(a + 6)(a - 1) = 0$

5. $x(x + 1)(x - 3) = 0$

6. $x(2x + 1)(x - 5) = 0$

7. $(3x + 2)(2x + 3) = 0$

8. $(4x - 5)(x - 6) = 0$

9. $m(3m + 4)(3m - 4) = 0$

10. $m(2m - 5)(3m - 1) = 0$

11. $2y(3y + 1)(5y + 3) = 0$

12. $3y(2y - 3)(3y - 4) = 0$

Solve the following equations.

13. $x^2 + 3x + 2 = 0$ **14.** $x^2 - x - 6 = 0$

15. $x^2 - 9x + 20 = 0$ **16.** $x^2 + 2x - 3 = 0$

17. $a^2 - 2a - 24 = 0$ **18.** $a^2 - 11a + 30 = 0$

19. $100x^2 - 500x + 600 = 0$

20. $100x^2 - 300x + 200 = 0$

21. $x^2 = -6x - 9$ **22.** $x^2 = 10x - 25$

23. $a^2 - 16 = 0$ **24.** $a^2 - 36 = 0$

25. $2x^2 + 5x - 12 = 0$ **26.** $3x^2 + 14x - 5 = 0$

27. $9x^2 + 12x + 4 = 0$

28. $12x^2 - 24x + 9 = 0$

29. $a^2 + 25 = 10a$ **30.** $a^2 + 16 = 8a$

31. $2x^2 = 3x + 20$ **32.** $6x^2 = x + 2$

33. $3m^2 = 20 - 7m$ **34.** $2m^2 = -18 + 15m$

35. $4x^2 - 49 = 0$ **36.** $16x^2 - 25 = 0$

37. $x^2 + 6x = 0$ **38.** $x^2 - 8x = 0$

39. $x^2 - 3x = 0$ **40.** $x^2 + 5x = 0$

41. $2x^2 = 8x$ **42.** $2x^2 = 10x$

43. $3x^2 = 15x$ **44.** $5x^2 = 15x$

45. $1,400 = 400 + 700x - 100x^2$

46. $2,700 = 700 + 900x - 100x^2$

47. $6x^2 = -5x + 4$ **48.** $9x^2 = 12x - 4$

49. $x(2x - 3) = 20$ **50.** $x(3x - 5) = 12$

51. $t(t + 2) = 80$ **52.** $t(t + 2) = 99$

53. $4,000 = (1,300 - 100p)p$

54. $3,200 = (1,200 - 100p)p$

55. $x(14 - x) = 48$ **56.** $x(12 - x) = 32$

57. $(x + 5)^2 = 2x + 9$ **58.** $(x + 7)^2 = 2x + 13$

59. $(y - 6)^2 = y - 4$ **60.** $(y + 4)^2 = y + 6$

61. $10^2 = (x + 2)^2 + x^2$

62. $15^2 = (x + 3)^2 + x^2$

63. $2x^3 + 11x^2 + 12x = 0$

64. $3x^3 + 17x^2 + 10x = 0$

65. $4y^3 - 2y^2 - 30y = 0$

66. $9y^3 + 6y^2 - 24y = 0$

67. $8x^3 + 16x^2 = 10x$

68. $24x^3 - 22x^2 = -4x$

69. $20a^3 = -18a^2 + 18a$

70. $12a^3 = -2a^2 + 10a$

Use factoring by grouping to solve the following equations.

71. $x^3 + 3x^2 - 4x - 12 = 0$

72. $x^3 + 5x^2 - 9x - 45 = 0$

73. $x^3 + x^2 - 16x - 16 = 0$

74. $4x^3 + 12x^2 - 9x - 27 = 0$

Review Problems

The following problems review material we covered in Sections 3.8 and 4.2.

The following word problems are taken from the book *Academic Algebra,* written by William J. Milne and published by the American Book Company in 1901.

Solve each problem. [3.8]

75. Cost of a Bicycle and a Suit A bicycle and a suit cost \$90. How much did each cost, if the bicycle cost 5 times as much as the suit?

76. Cost of a Cow and a Calf A man bought a cow and a calf for \$36, paying 8 times as much for the cow as for the calf. What was the cost of each?

77. Cost of a House and a Lot A house and a lot cost \$3,000. If the house cost 4 times as much as the lot, what was the cost of each?

78. Daily Wages A plumber and two helpers together earned \$7.50 per day. How much did each earn per day, if the plumber earned 4 times as much as each helper?

Use the properties of exponents to simplify each expression. [4.2]

79. 2^{-3}

80. 5^{-2}

81. $\dfrac{x^5}{x^{-3}}$

82. $\dfrac{x^{-2}}{x^{-5}}$

83. $\dfrac{(x^2)^3}{(x^{-3})^4}$

84. $\dfrac{(x^2)^{-4}(x^{-2})^3}{(x^{-3})^{-5}}$

85. Write the number 0.0056 in scientific notation.

86. Write the number 2.34×10^{-4} in expanded form.

5.7   Applications

In this section we will look at some application problems, the solutions to which require solving a quadratic equation. We will also introduce the Pythagorean theorem, one of the oldest theorems in the history of mathematics. The person whose name we associate with the theorem, Pythagoras (of Samos), was a Greek philosopher and mathematician who lived from about 560 B.C. to 480 B.C. According to the British philosopher Bertrand Russell, Pythagoras was "intellectually one of the most important men that ever lived."

Also in this section, the solutions to the examples show only the essential steps from our Blueprint for Problem Solving. Recall that step 1 is done mentally. We read the problem and mentally list the items that are known and the items that are unknown. This is an essential part of problem solving. However, now that you have had experience with application problems, you are doing step 1 automatically.

NUMBER PROBLEMS

E X A M P L E   I   The product of two consecutive odd integers is 63. Find the integers.

Solution   Let x = the first odd integer; then $x + 2$ = the second odd integer. An equation that describes the situation is:

$$x(x + 2) = 63 \qquad \text{(Their product is 63.)}$$

We solve the equation:

$$x(x + 2) = 63$$
$$x^2 + 2x = 63$$
$$x^2 + 2x - 63 = 0$$
$$(x - 7)(x + 9) = 0$$
$$x - 7 = 0 \qquad \text{or} \qquad x + 9 = 0$$
$$x = 7 \qquad \text{or} \qquad x = -9$$

If the first odd integer is 7, the next odd integer is $7 + 2 = 9$. If the first odd integer is -9, the next consecutive odd integer is $-9 + 2 = -7$. We have two pairs of consecutive odd integers that are solutions. They are 7, 9 and $-9, -7$.

We check to see that their products are 63:

$$7(9) = 63$$
$$-7(-9) = 63$$

Suppose we know that the sum of two numbers is 50. We want to find a way to represent each number using only one variable. If we let x represent one of the two numbers, how can we represent the other? Let's suppose for a moment that x turns out to be 30. Then the other number will be 20, because their sum is 50. That is, if two numbers add up to 50 and one of them is 30, then the other must be $50 - 30 = 20$. Generalizing this to any number x, we see that if two numbers have a sum of 50 and one of the numbers is x, then the other must be $50 - x$. The table that follows shows some additional examples.

If Two Numbers Have a Sum of	And One of Them Is	Then the Other Must Be
50	x	$50 - x$
100	x	$100 - x$
10	y	$10 - y$
12	n	$12 - n$

Now, let's look at an example that uses this idea.

EXAMPLE 2   The sum of two numbers is 13. Their product is 40. Find the numbers.

Solution   If we let x represent one of the numbers, then $13 - x$ must be the other number, because their sum is 13. Since their product is 40, we can write:

$$x(13 - x) = 40 \qquad \text{The product of the two numbers is 40.}$$

$$13x - x^2 = 40 \qquad \text{Multiply left side.}$$

$$x^2 - 13x = -40 \qquad \text{Multiply both sides by } -1 \text{ and reverse the order of the terms of the left side.}$$

$$x^2 - 13x + 40 = 0 \qquad \text{Add 40 to each side.}$$

$$(x - 8)(x - 5) = 0 \qquad \text{Factor the left side.}$$

$$x - 8 = 0 \quad \text{or} \quad x - 5 = 0$$

$$x = 8 \quad \text{or} \quad x = 5$$

The two solutions are 8 and 5. If x is 8, then the other number is $13 - x = 13 - 8 = 5$. Likewise, if x is 5, the other number is $13 - x = 13 - 5 = 8$. Therefore, the two numbers we are looking for are 8 and 5. Their sum is 13 and their product is 40.

GEOMETRY PROBLEMS

Many word problems dealing with area can best be described algebraically by quadratic equations.

EXAMPLE 3   The length of a rectangle is 3 more than twice the width. The area is 44 square inches. Find the dimensions (find the length and width).

Solution   As shown in Figure 1, let $x =$ the width of the rectangle. Then $2x + 3 =$ the length of the rectangle, because the length is three more than twice the width.

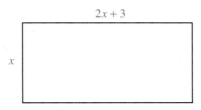

FIGURE 1

Since the area is 44 square inches, an equation that describes the situation is

$$x(2x + 3) = 44 \qquad \text{(Length} \cdot \text{width} = \text{area)}$$

We now solve the equation:

$$x(2x + 3) = 44$$
$$2x^2 + 3x = 44$$
$$2x^2 + 3x - 44 = 0$$
$$(2x + 11)(x - 4) = 0$$

$$2x + 11 = 0 \qquad \text{or} \qquad x - 4 = 0$$

$$x = -\frac{11}{2} \qquad \text{or} \qquad x = 4$$

The solution $x = -\frac{11}{2}$ cannot be used, since length and width are always given in positive units. The width is 4. The length is 3 more than twice the width or $2(4) + 3 = 11$.

$$\text{Width} = 4 \text{ inches}$$

$$\text{Length} = 11 \text{ inches}$$

The solutions check in the original problem, since $4(11) = 44$.

EXAMPLE 4   The numerical value of the area of a square is twice its perimeter. What is the length of its side?

Solution   As shown in Figure 2, let $x =$ the length of its side. Then $x^2 =$ the area of the square and $4x =$ the perimeter of the square:

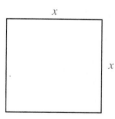

FIGURE 2

An equation that describes the situation is

$$x^2 = 2(4x) \qquad \text{The area is 2 times the perimeter.}$$

$$x^2 = 8x$$

$$x^2 - 8x = 0$$

$$x(x - 8) = 0$$

$$x = 0 \qquad \text{or} \qquad x = 8$$

Since $x = 0$ does not make sense in our original problem, we use $x = 8$. If the side has length 8, then the perimeter is $4(8) = 32$, and the area is $8^2 = 64$. Since 64 is twice 32, our solution is correct.

The Pythagorean Theorem

Next, we will work some problems involving the Pythagorean theorem, which we mentioned in the introduction to this section. It may interest you to know that Pythagoras formed a secret society around the year 540 B.C. Known as the Pythagoreans, members kept no written record of their work; everything was handed down by spoken word. They influenced not only mathematics, but religion, science, medicine, and music as well. Among other things, they discovered the correlation between musical notes and the reciprocals of counting numbers, $\frac{1}{2}, \frac{1}{3}, \frac{1}{4}$, and so on. In their daily lives, they followed strict dietary and moral rules to achieve a higher rank in future lives.

PYTHAGOREAN THEOREM

In any right triangle (Figure 3), the square of the longest side (called the hypotenuse) is equal to the sum of the squares of the other two sides (called legs).

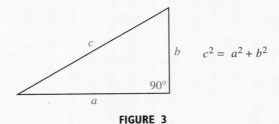

$$c^2 = a^2 + b^2$$

FIGURE 3

EXAMPLE 5 The three sides of a right triangle are three consecutive integers. Find the lengths of the three sides.

Solution Let x = the first integer (shortest side)

then $x + 1$ = the next consecutive integer

and $x + 2$ = the last consecutive integer (longest side)

A diagram of the triangle is shown in Figure 4.

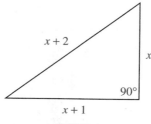

FIGURE 4

The Pythagorean theorem tells us that the square of the longest side $(x + 2)^2$ is equal to the sum of the squares of the two shorter sides, $(x + 1)^2 + x^2$. Here is the equation:

$$(x + 2)^2 = (x + 1)^2 + x^2$$

$$x^2 + 4x + 4 = x^2 + 2x + 1 + x^2 \qquad \text{Expand squares.}$$

$$x^2 - 2x - 3 = 0 \qquad \text{Standard form}$$

$$(x - 3)(x + 1) = 0 \qquad \text{Factor.}$$

$$x - 3 = 0 \quad \text{or} \quad x + 1 = 0 \qquad \text{Set factors to 0.}$$

$$x = 3 \quad \text{or} \quad x = -1$$

Since a triangle cannot have a side with a negative number for its length, we must not use -1 for a solution to our original problem; therefore, the shortest side is 3. The other two sides are the next two consecutive integers, 4 and 5.

EXAMPLE 6   The hypotenuse of a right triangle is 5 inches, and the lengths of the two legs (the other two sides) are given by two consecutive integers. Find the lengths of the two legs.

Solution   If we let $x =$ the length of the shortest side, then the other side must be $x + 1$. A diagram of the triangle is shown in Figure 5.

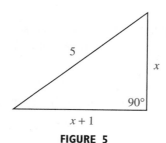

FIGURE 5

The Pythagorean theorem tells us that the square of the longest side, 5^2, is equal to the sum of the squares of the two shorter sides, $x^2 + (x + 1)^2$. Here is the equation:

$$5^2 = x^2 + (x + 1)^2 \qquad \text{Pythagorean theorem}$$

$$25 = x^2 + x^2 + 2x + 1 \qquad \text{Expand } 5^2 \text{ and } (x + 1)^2.$$

$$25 = 2x^2 + 2x + 1 \qquad \text{Simplify the right side.}$$

$$0 = 2x^2 + 2x - 24 \qquad \text{Add } -25 \text{ to each side.}$$

$$0 = 2(x^2 + x - 12) \qquad \text{Begin factoring.}$$

$$0 = 2(x + 4)(x - 3) \qquad \text{Factor completely.}$$

$$x + 4 = 0 \qquad \text{or} \qquad x - 3 = 0 \qquad \text{Set variable factors to 0.}$$

$$x = -4 \qquad \text{or} \qquad x = 3$$

Since a triangle cannot have a side with a negative number for its length, we cannot use -4; therefore, the shortest side must be 3 inches. The next side is $x + 1 = 3 + 1 = 4$ inches. Since the hypotenuse is 5, we can check our solutions with the Pythagorean theorem as shown in Figure 6.

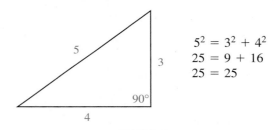

$$5^2 = 3^2 + 4^2$$
$$25 = 9 + 16$$
$$25 = 25$$

FIGURE 6

BUSINESS PROBLEMS

Our next two examples are from the world of business. If you are planning on taking finite mathematics, statistics, or business calculus in the future, these examples will give you a head start on some of the problems you will see in those classes.

EXAMPLE 7   A company can manufacture x hundred items for a total cost of $C = 300 + 500x - 100x^2$. How many items were manufactured if the total cost is $900?

Solution   We are looking for x when C is 900. We begin by substituting 900 for C in the cost equation. Then we solve for x:

When $\qquad\qquad\qquad C = 900$

the equation $\qquad\qquad C = 300 + 500x - 100x^2$

becomes $\qquad\qquad 900 = 300 + 500x - 100x^2$

We can write this equation in standard form by adding -300, $-500x$, and $100x^2$ to each side. The result looks like this:

$$100x^2 - 500x + 600 = 0$$

$$100(x^2 - 5x + 6) = 0 \qquad \text{Begin factoring.}$$

$$100(x - 2)(x - 3) = 0 \qquad \text{Factor completely.}$$

$$x - 2 = 0 \qquad \text{or} \qquad x - 3 = 0 \qquad \text{Set variable factors to 0.}$$

$$x = 2 \qquad \text{or} \qquad x = 3$$

Our solutions are 2 and 3, which means that the company can manufacture 2 hundred items or 3 hundred items for a total cost of $900.

EXAMPLE 8   A manufacturer of small portable radios knows that the number of radios she can sell each week is related to the price of the radios by the equation $x = 1{,}300 - 100p$ (x is the number of radios, and p is the price per radio). What price should she charge for the radios in order to have a weekly revenue of $4,000?

Solution   First, we must find the revenue equation. The equation for total revenue is $R = xp$, where x is the number of units sold and p is the price per unit. Since we want R in terms of p, we substitute $1{,}300 - 100p$ for x in the equation $R = xp$:

$$\text{If} \qquad R = xp$$

$$\text{and} \qquad x = 1{,}300 - 100p$$

$$\text{then} \qquad R = (1{,}300 - 100p)p$$

We want to find p when R is 4,000. Substituting 4,000 for R in the equation gives us:

$$4{,}000 = (1{,}300 - 100p)p$$

If we multiply out the right side, we have:

$$4{,}000 = 1{,}300p - 100p^2$$

To write this equation in standard form, we add $100p^2$ and $-1{,}300p$ to each side:

$$100p^2 - 1{,}300p + 4{,}000 = 0 \qquad \text{Add } 100p^2 \text{ and } -1{,}300p \text{ to each side.}$$

$$100(p^2 - 13p + 40) = 0 \qquad \text{Begin factoring.}$$

$$100(p - 5)(p - 8) = 0 \qquad \text{Factor completely.}$$

$$p - 5 = 0 \qquad \text{or} \qquad p - 8 = 0 \qquad \text{Set variable factors to 0.}$$

$$p = 5 \qquad \text{or} \qquad p = 8$$

If she sells the radios for $5 each or for $8 each, she will have a weekly revenue of $4,000.

Getting Ready for Class

After reading through the preceding section, respond in your own words and in complete sentences.

A. What are consecutive integers?

B. Explain the Pythagorean theorem in words.

C. Write an application problem for which the solution depends on solving the equation $x(x + 1) = 12$.

D. Write an application problem for which the solution depends on solving the equation $x(2x - 3) = 44$.

PROBLEM SET 5.7

Solve the following word problems. Be sure to show the equation used.

Number Problems

1. The product of two consecutive even integers is 80. Find the two integers.

2. The product of two consecutive integers is 72. Find the two integers.

3. The product of two consecutive odd integers is 99. Find the two integers.

4. The product of two consecutive integers is 132. Find the two integers.

5. The product of two consecutive even integers is 10 less than 5 times their sum. Find the two integers.

6. The product of two consecutive odd integers is 1 less than 4 times their sum. Find the two integers.

7. The sum of two numbers is 14. Their product is 48. Find the numbers.

8. The sum of two numbers is 12. Their product is 32. Find the numbers.

9. One number is 2 more than 5 times another. Their product is 24. Find the numbers.

10. One number is 1 more than twice another. Their product is 55. Find the numbers.

11. One number is 4 times another. Their product is 4 times their sum. Find the numbers.

12. One number is 2 more than twice another. Their product is 2 more than twice their sum. Find the numbers.

Geometry Problems

13. The length of a rectangle is 1 more than the width. The area is 12 square inches. Find the dimensions.

14. The length of a rectangle is 3 more than twice the width. The area is 44 square inches. Find the dimensions.

15. The height of a triangle is twice the base. The area is 9 square inches. Find the base.

16. The height of a triangle is 2 more than twice the base. The area is 20 square feet. Find the base.

17. The hypotenuse of a right triangle is 10 inches. The lengths of the two legs are given by two consecutive even integers. Find the lengths of the two legs.

18. The hypotenuse of a right triangle is 15 inches. One of the legs is 3 inches longer than the other. Find the lengths of the two legs.

19. The shorter leg of a right triangle is 5 meters. The hypotenuse is 1 meter longer than the longer leg. Find the length of the longer leg.

20. The shorter leg of a right triangle is 12 yards. If the hypotenuse is 20 yards, how long is the other leg?

Business Problems

21. A company can manufacture x hundred items for a total cost of $C = 400 + 700x - 100x^2$. Find x if the total cost is $1,400.

22. If the total cost C of manufacturing x hundred items is given by the equation $C = 700 + 900x - 100x^2$, find x when C is $2,700.

23. The total cost C of manufacturing x hundred video-tapes is given by the equation

$$C = 600 + 1,000x - 100x^2$$

Find x if the total cost is $2,200.

24. The total cost C of manufacturing x hundred pen and pencil sets is given by the equation $C = 500 + 800x - 100x^2$. Find x when C is $1,700.

25. A company that manufactures typewriter ribbons knows that the number of ribbons it can sell each week, x, is related to the price p per ribbon by the equation $x = 1,200 - 100p$. At what price should the company sell the ribbons if it wants the weekly revenue to be $3,200? (*Remember:* The equation for revenue is $R = xp$.)

26. A company manufactures diskettes for home computers. It knows from past experience that the number of diskettes it can sell each day, x, is related to the price p per diskette by the equation $x = 800 - 100p$. At what price should the company sell the diskettes if it wants the daily revenue to be $1,200?

27. The relationship between the number of calculators a company sells per week, x, and the price p of each calculator is given by the equation $x = 1,700 - 100p$. At what price should the calculators be sold if the weekly revenue is to be $7,000?

28. The relationship between the number of pencil sharpeners a company can sell each week, x, and the price p of each sharpener is given by the equation $x = 1,800 - 100p$. At what price should the sharpeners be sold if the weekly revenue is to be $7,200?

29. Pythagorean Theorem A 13-foot ladder is placed so that it reaches to a point on the wall that is 2 feet higher than twice the distance from the base of the wall to the base of the ladder.
   (a) How far from the wall is the base of the ladder?
   (b) How high does the ladder reach?

30. Height of a Projectile If a rocket is fired vertically into the air with a speed of 240 feet per second, its height at time t seconds is given by $h(t) = -16t^2 + 240t$. Here is a graph of its height at various times, with the details left out:

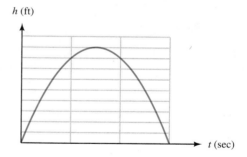

At what time(s) will the rocket be the following number of feet above the ground?
   (a) 704 feet
   (b) 896 feet
   (c) Why do parts (a) and (b) each have two answers?
   (d) How long will the rocket be in the air? (*Hint:* How high is it when it hits the ground?)
   (e) When the equation for part (d) is solved, one of the answers is $t = 0$ second. What does this represent?

31. Projectile Motion A gun fires a bullet almost straight up from the edge of a 100-foot cliff. If the bullet leaves the gun with a speed of 396 feet per second, its height at time t is given by $h(t) =$

$-16t^2 + 396t + 100$, measured from the ground below the cliff.

(a) When will the bullet land on the ground below the cliff? (*Hint:* What is its height when it lands? Remember that we are measuring from the ground below, not from the cliff.)

(b) Make a table showing the bullet's height every five seconds, from the time it is fired ($t = 0$) to the time it lands. (*Note:* It is faster to substitute into the factored form.)

32. Constructing a Box   I have a piece of cardboard that is twice as long as it is wide. If I cut a 2-inch by 2-inch square from each corner and fold up the resulting flaps, I get a box with a volume of 32 cubic inches. What are the dimensions of the cardboard?

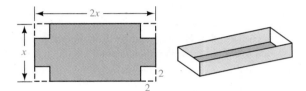

Review Problems

The problems that follow review material we covered in Chapter 4.

Simplify each expression. (Write all answers with positive exponents only.)

33. $(5x^3)^2(2x^6)^3$

34. 2^{-3}

35. $\dfrac{x^4}{x^{-3}}$

36. $\dfrac{(20x^2y^3)(5x^4y)}{(2xy^5)(10x^2y^3)}$

37. $(2 \times 10^{-4})(4 \times 10^5)$

38. $\dfrac{9 \times 10^{-3}}{3 \times 10^{-2}}$

39. $20ab^2 - 16ab^2 + 6ab^2$

40. Subtract $6x^2 - 5x - 7$ from $9x^2 + 3x - 2$.

Multiply.

41. $2x^2(3x^2 + 3x - 1)$

42. $(2x + 3)(5x - 2)$

43. $(3y - 5)^2$

44. $(a - 4)(a^2 + 4a + 16)$

45. $(2a^2 + 7)(2a^2 - 7)$

46. Divide $15x^{10} - 10x^8 + 25x^6$ by $5x^6$.

CHAPTER 5 SUMMARY

Examples

1. $8x^4 - 10x^3 + 6x^2$
$= 2x^2 \cdot 4x^2 - 2x^2 \cdot 5x + 2x^2 \cdot 3$
$= 2x^2(4x^2 - 5x + 3)$

Greatest Common Factor [5.1]

The largest monomial that divides each term of a polynomial is called the greatest common factor for that polynomial. We begin all factoring by factoring out the greatest common factor.

2. $x^2 + 5x + 6 = (x + 2)(x + 3)$
$x^2 - 5x + 6 = (x - 2)(x - 3)$
$6x^2 - x - 2 = (2x + 1)(3x - 2)$
$6x^2 + 7x + 2 = (2x + 1)(3x + 2)$

Factoring Trinomials [5.2, 5.3]

One method of factoring a trinomial is to list all pairs of binomials, the product of whose first terms gives the first term of the trinomial and the product of whose last terms gives the last term of the trinomial. We then choose the pair that gives the correct middle term for the original trinomial.

3. $x^2 + 10x + 25 = (x + 5)^2$
$x^2 - 10x + 25 = (x - 5)^2$
$x^2 - 25 = (x + 5)(x - 5)$

Special Factorings [5.4]

$$a^2 + 2ab + b^2 = (a + b)^2$$
$$a^2 - 2ab + b^2 = (a - b)^2$$
$$a^2 - b^2 = (a + b)(a - b)$$

4. (a) $2x^5 - 8x^3 = 2x^3(x^2 - 4)$
$= 2x^3(x + 2)(x - 2)$
(b) $3x^4 - 18x^3 + 27x^2$
$= 3x^2(x^2 - 6x + 9)$
$= 3x^2(x - 3)^2$
(c) $6x^3 - 12x^2 - 48x$
$= 6x(x^2 - 2x - 8)$
$= 6x(x - 4)(x + 2)$
(d) $x^2 + ax + bx + ab$
$= x(x + a) + b(x + a)$
$= (x + a)(x + b)$

Strategy for Factoring a Polynomial [5.5]

Step 1: If the polynomial has a greatest common factor other than 1, then factor out the greatest common factor.

Step 2: If the polynomial has two terms (it is a binomial), then see if it is the difference of two squares. Remember, if it is the sum of two squares, it will not factor.

Step 3: If the polynomial has three terms (a trinomial), then it is either a perfect square trinomial that will factor into the square of a binomial, or it is not a perfect square trinomial, in which case you use one of the methods developed in Section 5.3.

Step 4: If the polynomial has more than three terms, then try to factor it by grouping.

Step 5: As a final check, see if any of the factors you have written can be factored further. If you have overlooked a common factor, you can catch it here.

5. Solve $x^2 - 6x = -8$.

$$x^2 - 6x + 8 = 0$$
$$(x - 4)(x - 2) = 0$$

$x - 4 = 0$ or $x - 2 = 0$

$x = 4$ or $x = 2$

Both solutions check.

Strategy for Solving a Quadratic Equation [5.6]

Step 1: Write the equation in standard form:

$$ax^2 + bx + c = 0$$

Step 2: Factor completely.

Step 3: Set each variable factor equal to 0.

Step 4: Solve the equations found in step 3.

Step 5: Check solutions, if necessary.

6. The hypotenuse of a right triangle is 5 inches, and the lengths of the two legs (the other two sides) are given by two consecutive integers. Find the lengths of the two legs.

   If we let $x = $ the length of the shortest side, then the other side must be $x + 1$. The Pythagorean theorem tells us that

$$5^2 = x^2 + (x + 1)^2$$

$$25 = x^2 + x^2 + 2x + 1$$

$$25 = 2x^2 + 2x + 1$$

$$0 = 2x^2 + 2x - 24$$

$$0 = 2(x^2 + x - 12)$$

$$0 = 2(x + 4)(x - 3)$$

$x + 4 = 0$ or $x - 3 = 0$

$x = -4$ or $x = 3$

Since a triangle cannot have a side with a negative number for its length, we cannot use -4. The two sides are $x = 3$ and $x + 1 = 3 + 1 = 4$.

The Pythagorean Theorem [5.7]

In any right triangle, the square of the longest side (called the hypotenuse) is equal to the sum of the squares of the other two sides (called legs).

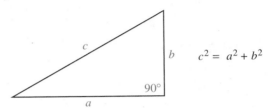

$$c^2 = a^2 + b^2$$

COMMON MISTAKE

It is a mistake to apply the zero-factor property to numbers other than zero. For example, consider the equation $(x - 3)(x + 4) = 18$. A fairly common mistake is to attempt to solve it with the following steps:

$$(x - 3)(x + 4) = 18$$

$x - 3 = 18$ or $x + 4 = 18 \leftarrow$ Mistake

$x = 21$ or $x = 14$

These are obviously not solutions, as a quick check will verify:

Check: $x = 21$

$(21 - 3)(21 + 4) \stackrel{?}{=} 18$

$18 \cdot 25 = 18$

$450 = 18$ $\xleftarrow{\text{False statements}}$

Check: $x = 14$

$(14 - 3)(14 + 4) \stackrel{?}{=} 18$

$11 \cdot 18 = 18$

$198 = 18$

The mistake is in setting each factor equal to 18. It is not necessarily true, when the product of two numbers is 18, that either one of them is itself 18. The correct solution looks like this:

$$(x - 3)(x + 4) = 18$$
$$x^2 + x - 12 = 18$$
$$x^2 + x - 30 = 0$$
$$(x + 6)(x - 5) = 0$$

$$x + 6 = 0 \qquad \text{or} \qquad x - 5 = 0$$
$$x = -6 \qquad \text{or} \qquad x = 5$$

To avoid this mistake, remember that before you factor a quadratic equation, you must write it in standard form. It is in standard form only when 0 is on one side and decreasing powers of the variable are on the other.

C H A P T E R   5   **R E V I E W**

The numbers in brackets refer to the sections of the text in which similar problems can be found.

Factor the following by factoring out the greatest common factor. [5.1]

1. $10x - 20$

2. $4x^3 - 9x^2$

3. $5x - 5y$

4. $7x^3 + 2x$

5. $8x + 4$

6. $2x^2 + 14x + 6$

7. $24y^2 - 40y + 48$

8. $30xy^3 - 45x^3y^2$

9. $49a^3 - 14b^3$

10. $6ab^2 + 18a^3b^3 - 24a^2b$

Factor by grouping. [5.1]

11. $xy + bx + ay + ab$

12. $xy + 4x - 5y - 20$

13. $2xy + 10x - 3y - 15$

14. $5x^2 - 4ax - 10bx + 8ab$

Factor the following trinomials. [5.2]

15. $y^2 + 9y + 14$

16. $w^2 + 15w + 50$

17. $a^2 - 14a + 48$

18. $r^2 - 18r + 72$

19. $y^2 + 20y + 99$

20. $y^2 + 8y + 12$

Factor the following trinomials. [5.3]

21. $2x^2 + 13x + 15$

22. $4y^2 - 12y + 5$

23. $5y^2 + 11y + 6$

24. $20a^2 - 27a + 9$

25. $6r^2 + 5rt - 6t^2$

26. $10x^2 - 29x - 21$

Factor the following if possible. [5.4]

27. $n^2 - 81$

28. $4y^2 - 9$

29. $x^2 + 49$

30. $36y^2 - 121x^2$

31. $64a^2 - 121b^2$

32. $64 - 9m^2$

Factor the following. [5.4]

33. $y^2 + 20y + 100$

34. $m^2 - 16m + 64$

35. $64t^2 + 16t + 1$

36. $16n^2 - 24n + 9$

37. $4r^2 - 12rt + 9t^2$

38. $9m^2 + 30mn + 25n^2$

Factor the following. [5.2]

39. $2x^2 + 20x + 48$

40. $a^3 - 10a^2 + 21a$

41. $3m^3 - 18m^2 - 21m$

42. $5y^4 + 10y^3 - 40y^2$

Factor the following trinomials. [5.3]

43. $8x^2 + 16x + 6$

44. $3a^3 - 14a^2 - 5a$

45. $20m^3 - 34m^2 + 6m$

46. $30x^2y - 55xy^2 + 15y^3$

Factor the following. [5.4]

47. $4x^2 + 40x + 100$

48. $4x^3 + 12x^2 + 9x$

49. $5x^2 - 45$

50. $12x^3 - 27xy^2$

Factor the following polynomials completely. [5.5]

51. $6a^3b + 33a^2b^2 + 15ab^3$

52. $x^5 - x^3$ **53.** $4y^6 + 9y^4$

54. $12x^5 + 20x^4y - 8x^3y^2$

55. $30a^4b + 35a^3b^2 - 15a^2b^3$

56. $18a^3b^2 + 3a^2b^3 - 6ab^4$

Solve. [5.6]

57. $(x - 5)(x + 2) = 0$

58. $3(2y + 5)(2y - 5) = 0$

59. $m^2 + 3m = 10$

60. $a^2 - 49 = 0$ **61.** $m^2 - 9m = 0$

62. $6y^2 = -13y - 6$ **63.** $9x^4 + 9x^3 = 10x^2$

Solve the following word problems. [5.7]

64. Number Problem The product of two consecutive even integers is 120. Find the two integers.

65. Number Problem The product of two consecutive integers is 110. Find the two integers.

66. Number Problem The product of two consecutive odd integers is 1 less than 3 times their sum. Find the integers.

67. Number Problem The sum of two numbers is 20. Their product is 75. Find the numbers.

68. Number Problem One number is 1 less than twice another. Their product is 66. Find the numbers.

69. Geometry The height of a triangle is 8 times the base. The area is 16 square inches. Find the base.

CHAPTER 5   PROJECTS

FACTORING

(GROUP PROJECT)

VISUAL FACTORING

Number of People: 2 or 3

Time Needed: 10–15 minutes

Equipment: Pencil, graph paper, and scissors

Background: When a geometric figure is divided into smaller figures, the area of the original figure and the area of any rearrangement of the smaller figures must be the same. We can use this fact to help visualize some factoring problems.

Procedure: Use the diagram below to work the following problems.

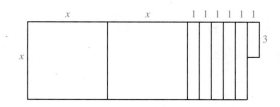

1. Write a polynomial involving x that gives the area of the diagram.
2. Factor the polynomial found in Part 1.
3. Copy the figure onto graph paper, then cut along the lines so that you end up with 2 squares and 6 rectangles.
4. Rearrange the pieces from Part 3 to show that the factorization you did in Part 2 is correct.

RESEARCH PROJECT

ARLIE O. PETTERS

It can seem at times as if all the mathematicians of note lived 100 or more years ago. However, that is not the case. There are mathematicians doing research today, who are discovering new mathematical ideas and extending what is known about mathematics. One of the current group of research mathematicians is Arlie O. Petters. Dr. Petters earned his Ph.D. from the Massachusetts Institute of Technology in 1991 and is currently working in the mathematics department at Duke University. Use the Internet to find out more about Dr. Petters' education, awards, and research interests. Then use your results to give a profile, in essay form, of a present-day working mathematician.

Arlie O. Petters

CHAPTER 5 TEST

Factor out the greatest common factor. [5.1]

1. $5x - 10$
2. $18x^2y - 9xy - 36xy^2$

Factor by grouping. [5.1]

3. $x^2 + 2ax - 3bx - 6ab$
4. $xy + 4x - 7y - 28$

Factor the following completely. [5.2–5.5]

5. $x^2 - 5x + 6$
6. $x^2 - x - 6$
7. $a^2 - 16$
8. $x^2 + 25$
9. $x^4 - 81$
10. $27x^2 - 75y^2$
11. $x^3 + 5x^2 - 9x - 45$
12. $x^2 - bx + 5x - 5b$

13. $4a^2 + 22a + 10$
14. $3m^2 - 3m - 18$
15. $6y^2 + 7y - 5$
16. $12x^3 - 14x^2 - 10x$

Solve the following quadratic equations. [5.6]

17. $x^2 + 7x + 12 = 0$
18. $x^2 - 4x + 4 = 0$
19. $x^2 - 36 = 0$
20. $x^2 = x + 20$
21. $x^2 - 11x = -30$
22. $y^3 = 16y$
23. $2a^2 = a + 15$
24. $30x^3 - 20x^2 = 10x$

Solve the following word problems. Be sure to show the equation used. [5.7]

25. **Number Problem** Two numbers have a sum of 20. Their product is 64. Find the numbers.

$3\ 5\ 7$ $a \cdot b = a + b + 7$

26. Consecutive Integers   The product of two consecutive odd integers is 7 more than their sum. Find the integers. $n,5$

27. Geometry   The length of a rectangle is 5 more than 3 times the width. The area is 42 square feet. Find the dimensions.

28. Geometry   One leg of a right triangle is 2 more than twice the other. The hypotenuse is 13 meters. Find the lengths of the two legs.

29. Production Cost   A company can manufacture x hundred items for a total cost C if $C = 200 + 500x - 100x^2$. How many items can be manufactured if the total cost is to be $800?

30. Price and Revenue   A manufacturer knows that the number of items he can sell each week, x, is related to the price p of each item by the equation $x = 900 - 100p$. What price should he charge for each item in order to have a weekly revenue of $1,800? (*Remember: R = xp.*)

CHAPTERS 1 – 5   CUMULATIVE REVIEW

Simplify.

1. $-|-9|$

2. $9 + (-7) + (-8)$

3. $20 - (-9)$

4. $\dfrac{-63}{-7}$

5. $\dfrac{9(-2)}{-2}$

6. $(-4)^3$

7. $\dfrac{-3(4 - 7) - 5(7 - 2)}{-5 - 2 - 1}$

8. $-a + 3 + 6a - 8$

9. $6 - 2(4a + 2) - 5$

10. $(x^4)^{10}$

11. $(9xy)^0$

12. $\dfrac{x^{-9}}{x^{-13}}$

13. $\dfrac{50x^8y^8}{25x^4y^2} + \dfrac{28x^7y^7}{14x^3y}$

14. $(5x - 2)(3x + 4)$

Solve each equation.

15. $3x = -18$

16. $\dfrac{x}{2} = 5$

17. $-\dfrac{x}{3} = 7$

18. $\dfrac{1}{2}(4t - 1) + \dfrac{1}{3} = -\dfrac{25}{6}$

19. $4m(m - 7)(2m - 7) = 0$

20. $16x^2 - 81 = 0$

21. $8x^2 = 10x + 3$

Solve each inequality.

22. $-2x > -8$

23. $7x + 1 \geq 2x - 5$

Graph on a rectangular coordinate system.

24. $y = -3x$

25. $y = 2$

26. $x + 2y = 3$

27. Find the x- and y-intercepts for the equation $10x - 3y = 30$.

28. Find the next number in the sequence: $\dfrac{1}{4}, \dfrac{1}{9}, \dfrac{1}{16}, \ldots$

29. Which ordered pairs are solutions to $2x + 5y = 10$: $(0, 2), (-5, 0), (4, \frac{2}{5})$?

30. Translate into symbols: the sum of $2x$ and 9 is 5.

31. Solve the following system by graphing:

$$y = x + 3$$
$$x + y = -1$$

Solve each system.

32. $-x + y = 3$
　　$x + y = 7$

33. $5x + 7y = -18$
　　$8x + 3y = 4$

34. $2x + y = 4$
　　$x = y - 1$

35. 　$x + y = 5,000$
　　$0.04x + 0.06y = 270$

36. $4x + 7y = -1$
　　$3x - 2y = -8$

Factor completely.

37. $n^2 - 5n - 36$

38. $14x^2 + 31xy - 10y^2$

39. $16 - a^2$

40. $49x^2 - 14x + 1$

41. $45x^2y - 30xy^2 + 5y^3$

42. $18x^3 - 3x^2y - 3xy^2$

43. $3xy + 15x - 2y - 10$

44. Give the opposite and reciprocal for -2.

45. What property justifies the statement $(2 + x) + 3 = (x + 2) + 3$?

46. Multiply and write your answer in scientific notation: $(5 \times 10^5)(2.1 \times 10^3)$.

47. Divide $28x^4y^4 - 14x^2y^3 + 21xy^2$ by $-7xy^2$.

48. Divide using long division $\dfrac{x^3 - 27}{x - 3}$.

49. Cutting a Board Into Two Pieces A 72-inch board is to be cut into two pieces. One piece is to be 4 inches longer than the other. How long is each piece?

50. Hamburgers and Fries Sheila bought burgers and fries for her children and some friends. The burgers cost $2.05 each, and the fries are $0.85 each. She bought a total of 14 items, for a total cost of $19.10. How many of each did she buy?

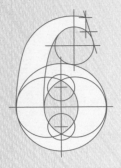

Rational Expressions

(Jon Riley/© Tony Stone Images)

INTRODUCTION

First Bank of San Luis Obispo charges $2.00 per month and $0.15 per check for a regular checking account. If we write x checks in one month, the total monthly cost of the checking account will be $C = 2.00 + 0.15x$. From this formula we see that the more checks we write in a month, the more we pay for the account. But, it is also true that the more checks we write in a month, the lower the cost per check. To find the average cost per check, we divide the total cost by the number of checks written:

$$\text{Average cost} = A = \frac{C}{x} = \frac{2.00 + 0.15x}{x}$$

We can use this formula to create Table 1 and Figure 1, giving us a visual interpretation of the relationship between the number of checks written and the average cost per check.

TABLE I Average Cost	
Number of Checks	**Average Cost Per Check**
1	2.15
2	1.15
5	0.55
10	0.35
15	0.28
20	0.25

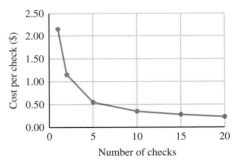

FIGURE I

As you can see, if we write one check per month, the cost per check is relatively high, $2.15. On the other hand, if we write 20 checks per month, each check costs us only $0.25. Using average cost per check is a good way to compare different checking accounts. The expression $\dfrac{2.00 + 0.15x}{x}$ in the average cost formula is a rational expression. When you have finished this chapter you will have a good working knowledge of rational expressions.

STUDY SKILLS

This is the last chapter in which we will mention study skills. You know by now what works best for you and what you have to do to achieve your goals for this course. From now on it is simply a matter of sticking with the things that work for you and avoiding the things that do not work. It seems simple, but as with anything that takes effort, it is up to you to see that you maintain the skills that get you where you want to be in the course.

If you intend to take more classes in mathematics and you want to ensure your success in those classes, then you can work toward this goal: *Become a student who can learn mathematics on your own.* Most people who have degrees in mathematics were students who could learn mathematics on their own. This doesn't mean that you have to learn it all on your own; it simply means that if you have to, you can learn it on your own. Attaining this goal gives you independence and puts you in control of your success in any math class you take.

In Chapter 1 we defined the set of rational numbers to be the set of all numbers that could be put in the form $\frac{a}{b}$, where a and b are integers ($b \neq 0$):

$$\text{Rational numbers} = \left\{ \frac{a}{b} \,\middle|\, a \text{ and } b \text{ are integers, } b \neq 0 \right\}$$

A *rational expression* is any expression that can be put in the form $\frac{P}{Q}$, where P and Q are polynomials and $Q \neq 0$:

$$\text{Rational expressions} = \left\{ \frac{P}{Q} \,\middle|\, P \text{ and } Q \text{ are polynomials, } Q \neq 0 \right\}$$

Each of the following is an example of a rational expression:

$$\frac{2x + 3}{x} \qquad \frac{x^2 - 6x + 9}{x^2 - 4} \qquad \frac{5}{x^2 + 6} \qquad \frac{2x^2 + 3x + 4}{2}$$

For the rational expression

$$\frac{x^2 - 6x + 9}{x^2 - 4}$$

the polynomial on top, $x^2 - 6x + 9$, is called the numerator, and the polynomial on the bottom, $x^2 - 4$, is called the denominator. The same is true of the other rational expressions.

We must be careful that we do not use a value of the variable that will give us a denominator of zero. Remember, division by zero is not defined.

E X A M P L E S State the restrictions on the variable in the following rational expressions:

1. $\dfrac{x + 2}{x - 3}$

Solution The variable x can be any real number except $x = 3$, since, when $x = 3$, the denominator is $3 - 3 = 0$. We state this restriction by writing $x \neq 3$.

2. $\dfrac{5}{x^2 - x - 6}$

Solution If we factor the denominator, we have $x^2 - x - 6 = (x - 3)(x + 2)$. If either of the factors is zero, the whole denominator is zero. Our restrictions are $x \neq 3$ and $x \neq -2$, since either one makes $x^2 - x - 6 = 0$.

We will not always list each restriction on a rational expression, but we should be aware of them and keep in mind that no rational expression can have a denominator of zero.

The two fundamental properties of rational expressions are listed next. We will use these two properties many times in this chapter.

PROPERTIES OF RATIONAL EXPRESSIONS

PROPERTY 1
Multiplying the numerator and denominator of a rational expression by the same nonzero quantity will not change the value of the rational expression.

PROPERTY 2
Dividing the numerator and denominator of a rational expression by the same nonzero quantity will not change the value of the rational expression.

We can use Property 2 to reduce rational expressions to lowest terms. Since this process is almost identical to the process of reducing fractions to lowest terms, let's recall how the fraction $\frac{6}{15}$ is reduced to lowest terms:

$$\frac{6}{15} = \frac{2 \cdot 3}{5 \cdot 3} \qquad \text{Factor numerator and denominator.}$$

$$= \frac{2 \cdot \cancel{3}}{5 \cdot \cancel{3}} \qquad \text{Divide out the common factor, 3.}$$

$$= \frac{2}{5} \qquad \text{Reduce to lowest terms.}$$

The same procedure applies to reducing rational expressions to lowest terms. The process is summarized in the following rule.

RULE

To reduce a rational expression to lowest terms, first factor the numerator and denominator completely and then divide both the numerator and denominator by any factors they have in common.

EXAMPLE 3   Reduce $\dfrac{x^2 - 9}{x^2 + 5x + 6}$ to lowest terms.

Solution   We begin by factoring:

$$\frac{x^2 - 9}{x^2 + 5x + 6} = \frac{(x - 3)(x + 3)}{(x + 2)(x + 3)}$$

Notice that both polynomials contain the factor $(x + 3)$. If we divide the numerator by $(x + 3)$, we are left with $(x - 3)$. If we divide the denominator by $(x + 3)$, we are left with $(x + 2)$. The complete problem looks like this:

$$\frac{x^2 - 9}{x^2 + 5x + 6} = \frac{(x - 3)\cancel{(x + 3)}}{(x + 2)\cancel{(x + 3)}}$$ Factor the numerator and
                          denominator completely.

$$= \frac{x - 3}{x + 2}$$ Divide out the common
                          factor, $x + 3$.

It is convenient to draw a line through the factors as we divide them out. It is especially helpful when the problems become longer.

E X A M P L E   4   Reduce to lowest terms $\dfrac{10a + 20}{5a^2 - 20}$.

Solution   We begin by factoring out the greatest common factor from the numerator and denominator:

$$\frac{10a + 20}{5a^2 - 20} = \frac{10(a + 2)}{5(a^2 - 4)}$$ Factor the greatest common factor
                          from the numerator and denominator.

$$= \frac{10\cancel{(a + 2)}}{5\cancel{(a + 2)}(a - 2)}$$ Factor the denominator as the
                          difference of two squares.

$$= \frac{2}{a - 2}$$ Divide out the common
                          factors 5 and $a + 2$.

E X A M P L E   5   Reduce $\dfrac{2x^3 + 2x^2 - 24x}{x^3 + 2x^2 - 8x}$ to lowest terms.

Solution   We begin by factoring the numerator and denominator completely. Then we divide out all factors common to the numerator and denominator. Here is what it looks like:

$$\frac{2x^3 + 2x^2 - 24x}{x^3 + 2x^2 - 8x} = \frac{2x(x^2 + x - 12)}{x(x^2 + 2x - 8)}$$ Factor out the greatest
                          common factor first.

$$= \frac{2\cancel{x}(x - 3)\cancel{(x + 4)}}{\cancel{x}(x - 2)\cancel{(x + 4)}}$$ Factor the remaining trinomials.

$$= \frac{2(x - 3)}{x - 2}$$ Divide out the factors common to
                          the numerator and denominator.

E X A M P L E   6   Reduce $\dfrac{x - 5}{x^2 - 25}$ to lowest terms.

Solution

$$\frac{x - 5}{x^2 - 25} = \frac{\cancel{x - 5}}{\cancel{(x - 5)}(x + 5)}$$ Factor numerator and denominator completely.

$$= \frac{1}{x + 5}$$ Divide out the common factor, $x - 5$.

RATIOS

For the rest of this section we will concern ourselves with ratios, a topic closely related to reducing fractions and rational expressions to lowest terms. Let's start with a definition.

DEFINITION   If a and b are any two numbers, $b \neq 0$, then the ratio of a and b is

$$\frac{a}{b}$$

As you can see, ratios are another name for fractions or rational numbers. They are a way of comparing quantities. Since we can also think of a/b as the quotient of a and b, ratios are also quotients. The following table gives some ratios in words and as fractions.

Ratio	As a Fraction	In Lowest Terms	
25 to 75	$\frac{25}{75}$	$\frac{1}{3}$	
8 to 2	$\frac{8}{2}$	$\frac{4}{1}$	With ratios it is common to leave the 1 in the denominator.
20 to 16	$\frac{20}{16}$	$\frac{5}{4}$	

EXAMPLE 7   A solution of hydrochloric acid (HCl) and water contains 49 milliliters of water and 21 milliliters of HCl. Find the ratio of HCl to water and of HCl to the total volume of the solution.

Solution   The ratio of HCl to water is 21 to 49, or

$$\frac{21}{49} = \frac{3}{7}$$

The amount of total solution volume is $49 + 21 = 70$ milliliters. Therefore, the ratio of HCl to total solution is 21 to 70, or

$$\frac{21}{70} = \frac{3}{10}$$

RATE EQUATION

Many of the problems in this chapter will use what is called the *rate equation*. You use this equation on an intuitive level when you are estimating how long it will take you to drive long distances. For example, if you drive at 50 miles per hour for 2 hours, you will travel 100 miles. Here is the rate equation:

$$\text{Distance} = \text{rate} \cdot \text{time}$$

$$d = r \cdot t$$

The rate equation has two equivalent forms, the more common of which is obtained by solving for *r*. Here it is:

$$r = \frac{d}{t}$$

The rate *r* in the rate equation is the ratio of distance to time and is also referred to as *average speed*. The units for rate are miles per hour, feet per second, kilometers per hour, and so on.

EXAMPLE 8   The Forest chair lift at the Northstar ski resort in Lake Tahoe is 5,603 feet long. If a ride on this chair lift takes 11 minutes, what is the average speed of the lift in feet per minute?

Solution   To find the speed of the lift, we find the ratio of distance covered to time. (Our answer is rounded to the nearest whole number.)

$$\text{Rate} = \frac{\text{distance}}{\text{time}} = \frac{5{,}603 \text{ feet}}{11 \text{ minutes}} = \frac{5{,}603}{11} \text{ feet/minute} = 509 \text{ feet/minute}$$

Note how we separate the numerical part of the problem from the units. In the next section, we will convert this rate to miles per hour.

L = 5,603 ft

Getting Ready for Class

After reading through the preceding section, respond in your own words and in complete sentences.

A. How do you reduce a rational expression to lowest terms?

B. What are the properties we use to manipulate rational expressions?

C. For what values of the variable is a rational expression undefined?

D. What is a ratio?

PROBLEM SET 6.1

Reduce the following rational expressions to lowest terms, if possible. Also, specify any restrictions on the variable in Problems 1 through 10.

1. $\dfrac{5}{5x - 10}$

2. $\dfrac{-4}{2x - 8}$

3. $\dfrac{a - 3}{a^2 - 9}$

4. $\dfrac{a + 4}{a^2 - 16}$

5. $\dfrac{x + 5}{x^2 - 25}$

6. $\dfrac{x - 2}{x^2 - 4}$

7. $\dfrac{2x^2 - 8}{4}$

8. $\dfrac{5x - 10}{x - 2}$

9. $\dfrac{2x - 10}{3x - 6}$

10. $\dfrac{4x - 8}{x - 2}$

11. $\dfrac{10a + 20}{5a + 10}$

12. $\dfrac{11a + 33}{6a + 18}$

13. $\dfrac{5x^2 - 5}{4x + 4}$

14. $\dfrac{7x^2 - 28}{2x + 4}$

15. $\dfrac{x - 3}{x^2 - 6x + 9}$

16. $\dfrac{x^2 - 10x + 25}{x - 5}$

17. $\dfrac{3x + 15}{3x^2 + 24x + 45}$

18. $\dfrac{5x + 15}{5x^2 + 40x + 75}$

19. $\dfrac{a^2 - 3a}{a^3 - 8a^2 + 15a}$

20. $\dfrac{a^2 + 3a}{a^3 - 2a^2 - 15a}$

21. $\dfrac{3x - 2}{9x^2 - 4}$

22. $\dfrac{2x - 3}{4x^2 - 9}$

23. $\dfrac{x^2 + 8x + 15}{x^2 + 5x + 6}$

24. $\dfrac{x^2 - 8x + 15}{x^2 - x - 6}$

25. $\dfrac{2m^3 - 2m^2 - 12m}{m^2 - 5m + 6}$

26. $\dfrac{2m^3 + 4m^2 - 6m}{m^2 - m - 12}$

27. $\dfrac{x^3 + 3x^2 - 4x}{x^3 - 16x}$

28. $\dfrac{3a^2 - 8a + 4}{9a^3 - 4a}$

29. $\dfrac{4x^3 - 10x^2 + 6x}{2x^3 + x^2 - 3x}$

30. $\dfrac{3a^3 - 8a^2 + 5a}{4a^3 - 5a^2 + 1a}$

31. $\dfrac{4x^2 - 12x + 9}{4x^2 - 9}$

32. $\dfrac{5x^2 + 18x - 8}{5x^2 + 13x - 6}$

33. $\dfrac{x + 3}{x^4 - 81}$

34. $\dfrac{x^2 + 9}{x^4 - 81}$

35. $\dfrac{3x^2 + x - 10}{x^4 - 16}$

36. $\dfrac{5x^2 - 26x + 24}{x^4 - 64}$

37. $\dfrac{42x^3 - 20x^2 - 48x}{6x^2 - 5x - 4}$

38. $\dfrac{36x^3 + 132x^2 - 135x}{6x^2 + 25x - 9}$

To reduce each of the following rational expressions to lowest terms, you will have to use factoring by grouping. Be sure to factor each numerator and denominator completely before dividing out any common factors. (Remember, factoring by grouping takes two steps.)

39. $\dfrac{xy + 3x + 2y + 6}{xy + 3x + 5y + 15}$

40. $\dfrac{xy + 7x + 4y + 28}{xy + 3x + 4y + 12}$

41. $\dfrac{x^2 - 3x + ax - 3a}{x^2 - 3x + bx - 3b}$

42. $\dfrac{x^2 - 6x + ax - 6a}{x^2 - 7x + ax - 7a}$

43. $\dfrac{xy + bx + ay + ab}{xy + bx + 3y + 3b}$

44. $\dfrac{x^2 + 5x + ax + 5a}{x^2 + 5x + bx + 5b}$

Write each ratio as a fraction in lowest terms.

45. 8 to 6   **46.** 6 to 8   **47.** 200 to 250

48. 250 to 200   **49.** 32 to 4   **50.** 4 to 32

Applying the Concepts

51. Cost and Average Cost As we mentioned in the introduction to this chapter, if a bank charges

$2.00 per month and $0.15 per check for one of its checking accounts, then the total monthly cost to write x checks is $C = 2.00 + 0.15x$, and the average cost of each of the x checks written is $A = \dfrac{2.00 + 0.15x}{x}$. Compare these two formulas by filling in the following table. Round to the nearest cent.

Checks Written x	Total Cost $2.00 + 0.15x$	Cost per Check $\dfrac{2.00 + 0.15x}{x}$
0		
5		
10		
15		
20		

52. Cost and Average Cost A rewritable CD-ROM drive for a computer costs $250. An individual CD for the drive costs $5.00 and can store 640 megabytes of information. The total cost of filling x CDs with information is $C = 250 + 5x$ dollars. The average cost per megabyte of information is given by $A = \dfrac{5x + 250}{640x}$. Compare the total cost and average cost per megabyte of storage by completing the following table. Round all answers to the nearest cent.

CDs Purchased x	Total Cost $250 + 5x$	Cost per Megabyte $\dfrac{5x + 250}{640x}$
0		
5		
10		
15		
20		

53. Speed of a Car A car travels 122 miles in 3 hours. Find the average speed of the car in miles per hour. Round to the nearest tenth.

54. Speed of a Bullet A bullet fired from a gun travels 4,500 feet in 3 seconds. Find the average speed of the bullet in feet per second.

55. Ferris Wheel The first Ferris wheel was designed and built by George Ferris in 1893. It was a large wheel with a circumference of 785 feet. If one trip around the circumference of the wheel took 20 minutes, find the average speed of a rider in feet per minute.

56. Ferris Wheel In 1897 a large Ferris wheel was built in Vienna; it is still in operation today. Known as *The Great Wheel,* it has a circumference of 618 feet. If one trip around the wheel takes 15 minutes, find the average speed of a rider on this wheel in feet per minute.

57. Ferris Wheel A Ferris wheel called *Colossus* was built in St. Louis in 1986. It has a circumference of 518 feet. If a trip around the circumference of *Colossus* takes 40 seconds, find the average speed of a rider in feet per second.

58. Ferris Wheel A person riding a Ferris wheel travels once around the wheel, a distance of 188 feet, in 30 seconds. What is the average speed of the rider in feet per second? Round to the nearest tenth.

59. Average Speed Tina is training for a biathlon. As part of her training, she runs an 8-mile course, 2 miles of which is on level ground and 6 miles of which is downhill. It takes her 20 minutes to run the level part of the course and 40 minutes to run the downhill part of the course. Find her average speed in minutes per mile and in miles per minute for each part of the course. Round to the nearest hundredth, if rounding is necessary.

60. Jogging A jogger covers a distance of 3 miles in 24 minutes. Find the average speed of the jogger in miles per minute.

61. Fuel Consumption   An economy car travels 168 miles on 3.5 gallons of gas. Give the average fuel consumption of the car in miles per gallon.

62. Fuel Consumption   A luxury car travels 100 miles on 8 gallons of gas. Give the average fuel consumption of the car in miles per gallon.

63. Comparing Expressions   Replace x with 5 and y with 4 in the expression

$$\frac{x^2 - y^2}{x - y}$$

and simplify the result. Is the result equal to $5 - 4$ or $5 + 4$?

64. Comparing Expressions   Replace x with 2 in the expression

$$\frac{x^3 - 1}{x - 1}$$

and simplify the result. Your answer should be equal to what you would get if you replaced x with 2 in $x^2 + x + 1$.

65. Comparing Expressions   Complete the following table; then show why the table turns out as it does.

x	$\dfrac{x - 3}{3 - x}$
-2	
-1	
0	
1	
2	

66. Comparing Expressions   Complete the following table; then show why the table turns out as it does.

x	$\dfrac{25 - x^2}{x^2 - 25}$
-4	
-2	
0	
2	
4	

67. Comparing Expressions   You know from reading through Example 6 in this section that $\dfrac{x - 5}{x^2 - 25} = \dfrac{1}{x + 5}$. Compare these expressions by completing the following table. (Be careful, not all the rows have equal entries.)

x	$\dfrac{x - 5}{x^2 - 25}$	$\dfrac{1}{x + 5}$
0		
-2		
2		
-5		
5		

68. Comparing Expressions   You know from your work in this section that $\dfrac{x^2 - 6x + 9}{x^2 - 9} = \dfrac{x - 3}{x + 3}$. Compare these expressions by completing the following table. (Be careful, not all the rows have equal entries.)

x	$\dfrac{x^2 - 6x + 9}{x^2 - 9}$	$\dfrac{x - 3}{x + 3}$
-3		
-2		
-1		
0		
1		
2		
3		

Review Problems

The following problems review material we covered in Sections 4.3 and 4.7.

Simplify. [4.3]

69. $\dfrac{27x^5}{9x^2} - \dfrac{45x^8}{15x^5}$

70. $\dfrac{36x^9}{4x} - \dfrac{45x^3}{5x^{-5}}$

71. $\dfrac{72a^3b^7}{9ab^5} + \dfrac{64a^5b^3}{8a^3b}$

72. $\dfrac{80a^5b^{11}}{10a^2b} + \dfrac{33a^6b^{12}}{11a^3b^2}$

Divide. [4.7]

73. $\dfrac{38x^7 + 42x^5 - 84x^3}{2x^3}$

74. $\dfrac{49x^6 - 63x^4 - 35x^2}{7x^2}$

75. $\dfrac{28a^5b^5 + 36ab^4 - 44a^4b}{4ab}$

76. $\dfrac{30a^3b - 12a^2b^2 + 6ab^3}{6ab}$

6.2   ## Multiplication and Division of Rational Expressions

Recall that to multiply two fractions we simply multiply numerators and multiply denominators and then reduce to lowest terms, if possible:

$$\frac{3}{4} \cdot \frac{10}{21} = \frac{30}{84} \leftarrow \text{Multiply numerators.}$$
$$\phantom{\frac{3}{4} \cdot \frac{10}{21} = \frac{30}{84}} \leftarrow \text{Multiply denominators.}$$
$$= \frac{5}{14} \leftarrow \text{Reduce to lowest terms.}$$

Recall also that the same result can be achieved by factoring numerators and denominators first and then dividing out the factors they have in common:

$$\frac{3}{4} \cdot \frac{10}{21} = \frac{3}{2 \cdot 2} \cdot \frac{2 \cdot 5}{3 \cdot 7} \qquad \text{Factor.}$$

$$= \frac{\cancel{3} \cdot \cancel{2} \cdot 5}{2 \cdot \cancel{2} \cdot \cancel{3} \cdot 7} \qquad \begin{array}{l}\text{Multiply numerators.}\\ \text{Multiply denominators.}\end{array}$$

$$= \frac{5}{14} \qquad \text{Divide out common factors.}$$

We can apply the second process to the product of two rational expressions, as the following example illustrates.

E X A M P L E   I   Multiply $\dfrac{x - 2}{x + 3} \cdot \dfrac{x^2 - 9}{2x - 4}$.

Solution   We begin by factoring numerators and denominators as much as possible. Then we multiply the numerators and denominators. The last step consists of dividing out all factors common to the numerator and denominator:

$$\frac{x-2}{x+3} \cdot \frac{x^2-9}{2x-4} = \frac{x-2}{x+3} \cdot \frac{(x-3)(x+3)}{2(x-2)}$$ Factor completely.

$$= \frac{(x-2)(x-3)(x+3)}{(x+3)(2)(x-2)}$$ Multiply numerators and denominators.

$$= \frac{x-3}{2}$$ Divide out common factors.

In Chapter 1 we defined division as the equivalent of multiplication by the reciprocal. This is how it looks with fractions:

$$\frac{4}{5} \div \frac{8}{9} = \frac{4}{5} \cdot \frac{9}{8}$$ Division as multiplication by the reciprocal

$$= \frac{2 \cdot 2 \cdot 3 \cdot 3}{5 \cdot 2 \cdot 2 \cdot 2}$$

$$= \frac{9}{10}$$

Factor and divide out common factors.

The same idea holds for division with rational expressions. The rational expression that follows the division symbol is called the *divisor*; to divide, we multiply by the reciprocal of the divisor.

E X A M P L E 2 Divide $\dfrac{3x-9}{x^2-x-20} \div \dfrac{x^2+2x-15}{x^2-25}$.

Solution We begin by taking the reciprocal of the divisor and writing the problem again in terms of multiplication. We then factor, multiply, and, finally, divide out all factors common to the numerator and denominator of the resulting expression. The complete problem looks like this:

$$\frac{3x-9}{x^2-x-20} \div \frac{x^2+2x-15}{x^2-25}$$

$$= \frac{3x-9}{x^2-x-20} \cdot \frac{x^2-25}{x^2+2x-15}$$ Multiply by the reciprocal of the divisor.

$$= \frac{3(x-3)}{(x+4)(x-5)} \cdot \frac{(x-5)(x+5)}{(x+5)(x-3)}$$ Factor.

$$= \frac{3(x-3)(x-5)(x+5)}{(x+4)(x-5)(x+5)(x-3)}$$ Divide out common factors.

$$= \frac{3}{x+4}$$

As you can see, factoring is the single most important tool we use in working with rational expressions. Most of the work we have done or will do with rational expressions is most easily accomplished if the rational expressions are in factored form. Here are some more examples of multiplication and division with rational expressions.

EXAMPLES

3. Multiply $\dfrac{3a + 6}{a^2} \cdot \dfrac{a}{2a + 4}$.

Solution

$$\dfrac{3a + 6}{a^2} \cdot \dfrac{a}{2a + 4}$$

$$= \dfrac{3(a + 2)}{a^2} \cdot \dfrac{a}{2(a + 2)} \qquad \text{Factor completely.}$$

$$= \dfrac{3(a + 2)a}{a^2(2)(a + 2)} \qquad \text{Multiply.}$$

$$= \dfrac{3}{2a} \qquad \begin{array}{l}\text{Divide numerator and denominator} \\ \text{by common factors } a(a + 2).\end{array}$$

4. Divide $\dfrac{x^2 + 7x + 12}{x^2 - 16} \div \dfrac{x^2 + 6x + 9}{2x - 8}$.

Solution

$$\dfrac{x^2 + 7x + 12}{x^2 - 16} \div \dfrac{x^2 + 6x + 9}{2x - 8}$$

$$= \dfrac{x^2 + 7x + 12}{x^2 - 16} \cdot \dfrac{2x - 8}{x^2 + 6x + 9} \qquad \begin{array}{l}\text{Division is multiplication} \\ \text{by the reciprocal.}\end{array}$$

$$= \dfrac{(x + 3)(x + 4)(2)(x - 4)}{(x - 4)(x + 4)(x + 3)(x + 3)} \qquad \text{Factor and multiply.}$$

$$= \dfrac{2}{x + 3} \qquad \text{Divide out common factors.}$$

In Example 4 we factored and multiplied the two expressions in a single step. This saves writing the problem one extra time.

EXAMPLE 5   Multiply $(x^2 - 49)\left(\dfrac{x + 4}{x + 7}\right)$.

Solution   We can think of the polynomial $x^2 - 49$ as having a denominator of 1. Thinking of $x^2 - 49$ in this way allows us to proceed as we did in previous examples:

$$(x^2 - 49)\left(\frac{x + 4}{x + 7}\right) = \frac{x^2 - 49}{1} \cdot \frac{x + 4}{x + 7} \qquad \text{Write } x^2 - 49 \text{ with denominator 1.}$$

$$= \frac{(x + 7)(x - 7)(x + 4)}{x + 7} \qquad \text{Factor and multiply.}$$

$$= (x - 7)(x + 4) \qquad \text{Divide out common factors.}$$

We can leave the answer in this form or multiply to get $x^2 - 3x - 28$. In this section let's agree to leave our answers in factored form.

E X A M P L E   6   Multiply $a(a + 5)(a - 5)\left(\dfrac{a + 4}{a^2 + 5a}\right)$.

Solution   We can think of the expression $a(a + 5)(a - 5)$ as having a denominator of 1:

$$a(a + 5)(a - 5)\left(\frac{a + 4}{a^2 + 5a}\right)$$

$$= \frac{a(a + 5)(a - 5)}{1} \cdot \frac{a + 4}{a^2 + 5a}$$

$$= \frac{a(a + 5)(a - 5)(a + 4)}{a(a + 5)} \qquad \text{Factor and multiply.}$$

$$= (a - 5)(a + 4) \qquad \text{Divide out common factors.}$$

UNIT ANALYSIS

Unit analysis is a method of converting between units of measure by multiplying by the number 1. Here is our first illustration: Suppose you are flying in a commercial airliner and the pilot tells you the plane has reached its cruising altitude of 35,000 feet. How many miles is the plane above the ground?

If you know that 1 mile is 5,280 feet, then it is simply a matter of deciding what to do with the two numbers, 5,280 and 35,000. By using unit analysis, this decision is unnecessary:

$$35{,}000 \text{ feet} = \frac{35{,}000 \text{ feet}}{1} \cdot \frac{1 \text{ mile}}{5{,}280 \text{ feet}}$$

We treat the units common to the numerator and denominator in the same way we treat factors common to the numerator and denominator; common units can be divided out, just as common factors are. In the previous expression, we have feet common to the numerator and denominator. Dividing them out leaves us with miles only. Here is the complete problem:

$$35,000 \text{ feet} = \frac{35,000 \text{ \cancel{feet}}}{1} \cdot \frac{1 \text{ mile}}{5,280 \text{ \cancel{feet}}}$$

$$= \frac{35,000}{5,280} \text{ miles}$$

$$= 6.6 \text{ miles to the nearest tenth of a mile}$$

The expression $\dfrac{1 \text{ mile}}{5,280 \text{ feet}}$ is called a *conversion factor.* It is simply the number 1 written in a convenient form. Because it is the number 1, we can multiply any other number by it and always be sure we have not changed that number. The key to unit analysis is choosing the right conversion factors.

EXAMPLE 7   The Mall of America in the Twin Cities covers 78 acres of land. If 1 square mile = 640 acres, how many square miles does the Mall of America cover? Round your answer to the nearest hundredth of a square mile.

Solution   We are starting with acres and want to end up with square miles. We need to multiply by a conversion factor that will allow acres to divide out and leave us with square miles:

$$78 \text{ acres} = \frac{78 \text{ \cancel{acres}}}{1} \cdot \frac{1 \text{ square mile}}{640 \text{ \cancel{acres}}}$$

$$= \frac{78}{640} \text{ square mile}$$

$$= 0.12 \text{ square mile to the nearest hundredth}$$

The next example is a continuation of Example 8 from Section 6.1.

EXAMPLE 8   The Forest chair lift at the Northstar ski resort in Lake Tahoe is 5,603 feet long. If a ride on this chair lift takes 11 minutes, what is the average speed of the lift in miles per hour?

Solution   First, we find the speed of the lift in feet per second, as we did in Example 8 of Section 6.1, by taking the ratio of distance to time.

$$\text{Rate} = \frac{\text{distance}}{\text{time}} = \frac{5,603 \text{ feet}}{11 \text{ minutes}} = \frac{5,603}{11} \text{ feet per minute}$$

$$= 509 \text{ feet per minute}$$

Next, we convert feet per minute to miles per hour. To do this, we need to know that

$$1 \text{ mile} = 5{,}280 \text{ feet}$$
$$1 \text{ hour} = 60 \text{ minutes}$$

$$\text{Speed} = 509 \text{ feet per minute} = \frac{509 \, \cancel{\text{feet}}}{1 \, \cancel{\text{minute}}} \cdot \frac{1 \text{ mile}}{5{,}280 \, \cancel{\text{feet}}} \cdot \frac{60 \, \cancel{\text{minutes}}}{1 \text{ hour}}$$

$$= \frac{509 \cdot 60}{5{,}280} \text{ miles per hour}$$

$$= 5.8 \text{ miles per hour to the nearest tenth}$$

Getting Ready for Class

After reading through the preceding section, respond in your own words and in complete sentences.

A. How do we multiply rational expressions?

B. Explain the steps used to divide rational expressions.

C. What part does factoring play in multiplying and dividing rational expressions?

D. Why are all conversion factors the same as the number 1?

PROBLEM SET 6.2

Multiply or divide as indicated. Be sure to reduce all answers to lowest terms. (That is, the numerator and denominator of the answer should not have any factors in common.)

1. $\dfrac{x + y}{3} \cdot \dfrac{6}{x + y}$

2. $\dfrac{x - 1}{x + 1} \cdot \dfrac{5}{x - 1}$

3. $\dfrac{2x + 10}{x^2} \cdot \dfrac{x^3}{4x + 20}$

4. $\dfrac{3x^4}{3x - 6} \cdot \dfrac{x - 2}{x^2}$

5. $\dfrac{9}{2a - 8} \div \dfrac{3}{a - 4}$

6. $\dfrac{8}{a^2 - 25} \div \dfrac{16}{a + 5}$

7. $\dfrac{x + 1}{x^2 - 9} \div \dfrac{2x + 2}{x + 3}$

8. $\dfrac{11}{x - 2} \div \dfrac{22}{2x^2 - 8}$

9. $\dfrac{a^2 + 5a}{7a} \cdot \dfrac{4a^2}{a^2 + 4a}$

10. $\dfrac{4a^2 + 4a}{a^2 - 25} \cdot \dfrac{a^2 - 5a}{8a}$

11. $\dfrac{y^2 - 5y + 6}{2y + 4} \div \dfrac{2y - 6}{y + 2}$

12. $\dfrac{y^2 - 7y}{3y^2 - 48} \div \dfrac{y^2 - 9}{y^2 - 7y + 12}$

13. $\dfrac{2x - 8}{x^2 - 4} \cdot \dfrac{x^2 + 6x + 8}{x - 4}$

14. $\dfrac{x^2 + 5x + 1}{7x - 7} \cdot \dfrac{x - 1}{x^2 + 5x + 1}$

15. $\dfrac{x - 1}{x^2 - x - 6} \cdot \dfrac{x^2 + 5x + 6}{x^2 - 1}$

16. $\dfrac{x^2 - 3x - 10}{x^2 - 4x + 3} \cdot \dfrac{x^2 - 5x + 6}{x^2 - 3x - 10}$

17. $\dfrac{a^2 + 10a + 25}{a + 5} \div \dfrac{a^2 - 25}{a - 5}$

18. $\dfrac{a^2 + a - 2}{a^2 + 5a + 6} \div \dfrac{a - 1}{a}$

19. $\dfrac{y^3 - 5y^2}{y^4 + 3y^3 + 2y^2} \div \dfrac{y^2 - 5y + 6}{y^2 - 2y - 3}$

20. $\dfrac{y^2 - 5y}{y^2 + 7y + 12} \div \dfrac{y^3 - 7y^2 + 10y}{y^2 + 9y + 18}$

21. $\dfrac{2x^2 + 17x + 21}{x^2 + 2x - 35} \cdot \dfrac{x^2 - 25}{2x^2 - 7x - 15}$

22. $\dfrac{x^2 - 13x + 42}{4x^2 + 31x + 21} \cdot \dfrac{4x^2 - 5x - 6}{x^2 - 4}$

23. $\dfrac{2x^2 + 10x + 12}{4x^2 + 24x + 32} \cdot \dfrac{2x^2 + 18x + 40}{x^2 + 8x + 15}$

24. $\dfrac{3x^2 - 3}{6x^2 + 18x + 12} \cdot \dfrac{2x^2 - 8}{x^2 - 3x + 2}$

25. $\dfrac{2a^2 + 7a + 3}{a^2 - 16} \div \dfrac{4a^2 + 8a + 3}{2a^2 - 5a - 12}$

26. $\dfrac{3a^2 + 7a - 20}{a^2 + 3a - 4} \div \dfrac{3a^2 - 2a - 5}{a^2 - 2a + 1}$

27. $\dfrac{4y^2 - 12y + 9}{y^2 - 36} \div \dfrac{2y^2 - 5y + 3}{y^2 + 5y - 6}$

28. $\dfrac{5y^2 - 6y + 1}{y^2 - 1} \div \dfrac{16y^2 - 9}{4y^2 + 7y + 3}$

29.
$$\dfrac{x^2 - 1}{6x^2 + 42x + 60} \cdot \dfrac{7x^2 + 17x + 6}{x + 1} \cdot \dfrac{6x + 30}{7x^2 - 11x - 6}$$

30. $\dfrac{4x^2 - 1}{3x - 15} \cdot \dfrac{4x^2 - 17x - 15}{4x^2 - 9x - 9} \cdot \dfrac{3x - 3}{x^2 - 9}$

31. $\dfrac{18x^3 + 21x^2 - 60x}{21x^2 - 25x - 4} \cdot \dfrac{28x^2 - 17x - 3}{16x^3 + 28x^2 - 30x}$

32. $\dfrac{56x^3 + 54x^2 - 20x}{8x^2 - 2x - 15} \cdot \dfrac{6x^2 + 5x - 21}{63x^3 + 129x^2 - 42x}$

Multiply the following expressions using the method shown in Examples 5 and 6 in this section.

33. $(x^2 - 9)\left(\dfrac{2}{x + 3}\right)$

34. $(x^2 - 9)\left(\dfrac{-3}{x - 3}\right)$

35. $a(a + 5)(a - 5)\left(\dfrac{2}{a^2 - 25}\right)$

36. $a(a^2 - 4)\left(\dfrac{a}{a + 2}\right)$

37. $(x^2 - x - 6)\left(\dfrac{x + 1}{x - 3}\right)$

38. $(x^2 - 2x - 8)\left(\dfrac{x + 3}{x - 4}\right)$

39. $(x^2 - 4x - 5)\left(\dfrac{-2x}{x + 1}\right)$

40. $(x^2 - 6x + 8)\left(\dfrac{4x}{x - 2}\right)$

Each of the following problems involves some factoring by grouping. Remember, before you can divide out factors common to the numerators and denominators of a product, you must factor completely.

41. $\dfrac{x^2 - 9}{x^2 - 3x} \cdot \dfrac{2x + 10}{xy + 5x + 3y + 15}$

42. $\dfrac{x^2 - 16}{x^2 - 4x} \cdot \dfrac{3x + 18}{xy + 6x + 4y + 24}$

43. $\dfrac{2x^2 + 4x}{x^2 - y^2} \cdot \dfrac{x^2 + 3x + xy + 3y}{x^2 + 5x + 6}$

44. $\dfrac{x^2 - 25}{3x^2 + 3xy} \cdot \dfrac{x^2 + 4x + xy + 4y}{x^2 + 9x + 20}$

45. $\dfrac{x^3 - 3x^2 + 4x - 12}{x^4 - 16} \cdot \dfrac{3x^2 + 5x - 2}{3x^2 - 10x + 3}$

46. $\dfrac{x^3 - 5x^2 + 9x - 45}{x^4 - 81} \cdot \dfrac{5x^2 + 18x + 9}{5x^2 - 22x - 15}$

Simplify each expression. Work inside parentheses first, and then divide out common factors.

47. $\left(1 - \dfrac{1}{2}\right)\left(1 - \dfrac{1}{3}\right)\left(1 - \dfrac{1}{4}\right)\left(1 - \dfrac{1}{5}\right)$

48. $\left(1 + \dfrac{1}{2}\right)\left(1 + \dfrac{1}{3}\right)\left(1 + \dfrac{1}{4}\right)\left(1 + \dfrac{1}{5}\right)$

The dots in the following problems represent factors not written that are in the same pattern as the surrounding factors. Simplify.

49.

$\left(1 - \dfrac{1}{2}\right)\left(1 - \dfrac{1}{3}\right)\left(1 - \dfrac{1}{4}\right) \cdots \left(1 - \dfrac{1}{99}\right)\left(1 - \dfrac{1}{100}\right)$

50.

$\left(1 - \dfrac{1}{3}\right)\left(1 - \dfrac{1}{4}\right)\left(1 - \dfrac{1}{5}\right) \cdots \left(1 - \dfrac{1}{98}\right)\left(1 - \dfrac{1}{99}\right)$

Applying the Concepts

51. Mount Whitney The top of Mount Whitney, the highest point in California, is 14,494 feet above sea level. Give this height in miles to the nearest tenth of a mile.

52. Motor Displacement The relationship between liters and cubic inches, both of which are measures of volume, is 0.0164 liter = 1 cubic inch. If a Ford Mustang has a motor with a displacement of 4.9

liters, what is the displacement in cubic inches? Round your answer to the nearest cubic inch.

53. Speed of Sound The speed of sound is 1,088 feet per second. Convert the speed of sound to miles per hour. Round your answer to the nearest whole number.

54. Average Speed A car travels 122 miles in 3 hours. Find the average speed of the car in feet per second. Round to the nearest whole number.

55. Ferris Wheel As we mentioned in Problem Set 6.1, the first Ferris wheel was built in 1893. It was a large wheel with a circumference of 785 feet. If one trip around the circumference of the wheel took 20 minutes, find the average speed of a rider in miles per hour. Round to the nearest hundredth.

56. Ferris Wheel In 1897 a large Ferris wheel was built in Vienna. Known as *The Great Wheel*, it has a circumference of 618 feet. If one trip around the wheel takes 15 minutes, find the average speed of a rider on this wheel in miles per hour. Round to the nearest hundredth.

57. Ferris Wheel A Ferris wheel called *Colossus* has a circumference of 518 feet. If a trip around the circumference of *Colossus* takes 40 seconds, find the average speed of a rider in miles per hour. Round to the nearest tenth.

58. Ferris Wheel A person riding a Ferris wheel travels once around the wheel, a distance of 188 feet, in 30 seconds. What is the average speed of the rider in miles per hour?

59. Average Speed Tina is training for a biathlon. As part of her training, she runs an 8-mile course, 2 miles of which is on level ground and 6 miles of which is downhill. It takes her 20 minutes to run the level part of the course and 40 minutes to run the downhill part of the course. Find her average speed in miles per hour for each part of the course.

60. Jogging   A jogger covers a distance of 3 miles in 24 minutes. Find the average speed of the jogger in miles per hour.

Review Problems

The following problems review material we covered in Sections 1.9 and 4.3. Reviewing these problems will help you in the next section.

Add the following fractions. [1.9]

61. $\dfrac{1}{2} + \dfrac{5}{2}$ **62.** $\dfrac{2}{3} + \dfrac{8}{3}$ **63.** $2 + \dfrac{3}{4}$

64. $1 + \dfrac{4}{7}$ **65.** $\dfrac{1}{10} + \dfrac{3}{14}$ **66.** $\dfrac{1}{12} + \dfrac{11}{30}$

Simplify each term, then add. [4.3]

67. $\dfrac{10x^4}{2x^2} + \dfrac{12x^6}{3x^4}$ **68.** $\dfrac{32x^8}{8x^3} + \dfrac{27x^7}{3x^2}$

69. $\dfrac{12a^2b^5}{3ab^3} + \dfrac{14a^4b^7}{7a^3b^5}$ **70.** $\dfrac{16a^3b^2}{4ab} + \dfrac{25a^6b^5}{5a^4b^4}$

6.3   ## Addition and Subtraction of Rational Expressions

In Chapter 1 we combined fractions having the same denominator by combining their numerators and putting the result over the common denominator. We use the same process to add two rational expressions with the same denominator.

EXAMPLES

1. Add $\dfrac{5}{x} + \dfrac{3}{x}$.

Solution   Adding numerators, we have:

$$\frac{5}{x} + \frac{3}{x} = \frac{8}{x}$$

2. Add $\dfrac{x}{x^2 - 9} + \dfrac{3}{x^2 - 9}$.

Solution   Since both expressions have the same denominator, we add numerators and reduce to lowest terms:

$$\frac{x}{x^2 - 9} + \frac{3}{x^2 - 9} = \frac{x + 3}{x^2 - 9}$$

$$= \frac{\cancel{x + 3}}{\cancel{(x + 3)}(x - 3)} \left.\begin{array}{c} \\ \\ \end{array}\right\} \begin{array}{l}\text{Reduce to lowest terms by} \\ \text{factoring the denominator} \\ \text{and then dividing out com-}\end{array}$$

$$= \frac{1}{x - 3} \qquad \text{mon factor } x + 3.$$

Remember, it is the distributive property that allows us to add rational expressions by simply adding numerators. Because of this, we must begin all addition problems involving rational expressions by first making sure all the expressions have the same denominator.

> **DEFINITION** The **least common denominator** (LCD) for a set of denominators is the simplest quantity that is exactly divisible by all the denominators.

EXAMPLE 3   Add $\frac{1}{10} + \frac{3}{14}$.

Solution

Step 1: Find the LCD for 10 and 14. To do so, we factor each denominator and build the LCD from the factors:

$$\left. \begin{array}{l} 10 = 2 \cdot 5 \\ 14 = 2 \cdot 7 \end{array} \right\} \quad LCD = 2 \cdot 5 \cdot 7 = 70$$

We know the LCD is divisible by 10 because it contains the factors 2 and 5. It is also divisible by 14 because it contains the factors 2 and 7.

Step 2: Change to equivalent fractions that each have denominator 70. To accomplish this task, we multiply the numerator and denominator of each fraction by the factor of the LCD that is not also a factor of its denominator:

Original Fractions		Denominators in Factored Form		Multiply by Factor Needed to Obtain LCD		These Have the Same Value as the Original Fractions
$\frac{1}{10}$	$=$	$\frac{1}{2 \cdot 5}$	$=$	$\frac{1}{2 \cdot 5} \cdot \frac{7}{7}$	$=$	$\frac{7}{70}$
$\frac{3}{14}$	$=$	$\frac{3}{2 \cdot 7}$	$=$	$\frac{3}{2 \cdot 7} \cdot \frac{5}{5}$	$=$	$\frac{15}{70}$

The fraction $\frac{7}{70}$ has the same value as the fraction $\frac{1}{10}$. Likewise, the fractions $\frac{15}{70}$ and $\frac{3}{14}$ are equivalent; they have the same value.

Step 3: Add numerators and put the result over the LCD:

$$\frac{7}{70} + \frac{15}{70} = \frac{7 + 15}{70} = \frac{22}{70}$$

Step 4: Reduce to lowest terms:

$$\frac{22}{70} = \frac{11}{35} \qquad \text{Divide numerator and denominator by 2.}$$

The main idea in adding fractions is to write each fraction again with the LCD for a denominator. Once we have done that, we simply add numerators. The same process can be used to combine rational expressions, as the next example illustrates.

EXAMPLE 4   Subtract $\dfrac{3}{x} - \dfrac{1}{2}$.

Solution

Step 1: The LCD for x and 2 is $2x$. It is the smallest expression divisible by x and by 2.

Step 2: To change to equivalent expressions with the denominator $2x$, we multiply the first fraction by $\frac{2}{2}$ and the second by $\dfrac{x}{x}$:

$$\frac{3}{x} \cdot \frac{2}{2} = \frac{6}{2x}$$

$$\frac{1}{2} \cdot \frac{x}{x} = \frac{x}{2x}$$

Step 3: Subtracting numerators of the rational expressions in step 2, we have:

$$\frac{6}{2x} - \frac{x}{2x} = \frac{6-x}{2x}$$

Step 4: Since $6 - x$ and $2x$ do not have any factors in common, we cannot reduce any further. Here is the complete problem:

$$\frac{3}{x} - \frac{1}{2} = \frac{3}{x} \cdot \frac{2}{2} - \frac{1}{2} \cdot \frac{x}{x}$$

$$= \frac{6}{2x} - \frac{x}{2x}$$

$$= \frac{6-x}{2x}$$

EXAMPLE 5   Add $\dfrac{5}{2x-6} + \dfrac{x}{x-3}$.

Solution   If we factor $2x - 6$, we have $2x - 6 = 2(x - 3)$. We need only multiply the second rational expression in our problem by $\frac{2}{2}$ to have two expressions with the same denominator:

$$\frac{5}{2x-6} + \frac{x}{x-3} = \frac{5}{2(x-3)} + \frac{x}{x-3}$$

$$= \frac{5}{2(x-3)} + \frac{2}{2}\left(\frac{x}{x-3}\right)$$

$$= \frac{5}{2(x-3)} + \frac{2x}{2(x-3)}$$

$$= \frac{2x+5}{2(x-3)}$$

EXAMPLE 6   Add $\dfrac{1}{x+4} + \dfrac{8}{x^2-16}$.

Solution   After writing each denominator in factored form, we find that the least common denominator is $(x+4)(x-4)$. To change the first rational expression to an equivalent rational expression with the common denominator, we multiply its numerator and denominator by $x-4$:

$$\frac{1}{x+4} + \frac{8}{x^2-16}$$

$$= \frac{1}{x+4} + \frac{8}{(x+4)(x-4)} \qquad \text{Factor each denominator.}$$

$$= \frac{1}{x+4} \cdot \frac{x-4}{x-4} + \frac{8}{(x+4)(x-4)} \qquad \begin{array}{l}\text{Change to equivalent} \\ \text{rational expressions.}\end{array}$$

$$= \frac{x-4}{(x+4)(x-4)} + \frac{8}{(x+4)(x-4)} \qquad \text{Simplify.}$$

$$= \frac{x+4}{(x+4)(x-4)} \qquad \text{Add numerators.}$$

$$= \frac{1}{x-4} \qquad \begin{array}{l}\text{Divide out common} \\ \text{factor } x+4.\end{array}$$

Note: In the last step we reduced the rational expression to lowest terms by dividing out the common factor of $x+4$.

EXAMPLE 7   Add $\dfrac{2}{x^2+5x+6} + \dfrac{x}{x^2-9}$.

Solution

Step 1: We factor each denominator and build the LCD from the factors:

$$\left. \begin{array}{l} x^2+5x+6 = (x+2)(x+3) \\ x^2-9 = (x+3)(x-3) \end{array} \right\} \qquad \text{LCD} = (x+2)(x+3)(x-3)$$

Step 2: Change to equivalent rational expressions:

$$\frac{2}{x^2+5x+6} = \frac{2}{(x+2)(x+3)} \cdot \frac{(x-3)}{(x-3)} = \frac{2x-6}{(x+2)(x+3)(x-3)}$$

$$\frac{x}{x^2-9} = \frac{x}{(x+3)(x-3)} \cdot \frac{(x+2)}{(x+2)} = \frac{x^2+2x}{(x+2)(x+3)(x-3)}$$

Step 3: Add numerators of the rational expressions produced in step 2:

$$\frac{2x - 6}{(x + 2)(x + 3)(x - 3)} + \frac{x^2 + 2x}{(x + 2)(x + 3)(x - 3)}$$

$$= \frac{x^2 + 4x - 6}{(x + 2)(x + 3)(x - 3)}$$

The numerator and denominator do not have any factors in common.

EXAMPLE 8   Subtract $\dfrac{x + 4}{2x + 10} - \dfrac{5}{x^2 - 25}$.

Solution   We begin by factoring each denominator:

$$\frac{x + 4}{2x + 10} - \frac{5}{x^2 - 25} = \frac{x + 4}{2(x + 5)} - \frac{5}{(x + 5)(x - 5)}$$

The LCD is $2(x + 5)(x - 5)$. Completing the problem, we have:

$$= \frac{x + 4}{2(x + 5)} \cdot \frac{(x - 5)}{(x - 5)} + \frac{-5}{(x + 5)(x - 5)} \cdot \frac{2}{2}$$

$$= \frac{x^2 - x - 20}{2(x + 5)(x - 5)} + \frac{-10}{2(x + 5)(x - 5)}$$

$$= \frac{x^2 - x - 30}{2(x + 5)(x - 5)}$$

To see if this expression will reduce, we factor the numerator into $(x - 6)(x + 5)$:

$$= \frac{(x - 6)\cancel{(x + 5)}}{2\cancel{(x + 5)}(x - 5)}$$

$$= \frac{x - 6}{2(x - 5)}$$

Note: Note that we replaced subtraction by addition of the opposite. There seems to be less chance for error when this is done on longer problems.

EXAMPLE 9   Write an expression for the sum of a number and its reciprocal, and then simplify that expression.

Solution   If we let $x =$ the number, then its reciprocal is $\dfrac{1}{x}$. To find the sum of the number and its reciprocal, we add them:

$$x + \frac{1}{x}$$

The first term x can be thought of as having a denominator of 1. Since the denominators are 1 and x, the least common denominator is x.

$$x + \frac{1}{x} = \frac{x}{1} + \frac{1}{x} \qquad \text{Write } x \text{ as } \frac{x}{1}.$$

$$= \frac{x}{1} \cdot \frac{x}{x} + \frac{1}{x} \qquad \text{The LCD is } x.$$

$$= \frac{x^2}{x} + \frac{1}{x}$$

$$= \frac{x^2 + 1}{x} \qquad \text{Add numerators.}$$

Getting Ready for Class

After reading through the preceding section, respond in your own words and in complete sentences.

A. How do we add two rational expressions that have the same denominator?

B. What is the least common denominator for two fractions?

C. What role does factoring play in finding a least common denominator?

D. Explain how to find a common denominator for two rational expressions.

PROBLEM SET 6.3

Find the following sums and differences.

1. $\dfrac{3}{x} + \dfrac{4}{x}$

2. $\dfrac{5}{x} + \dfrac{3}{x}$

3. $\dfrac{9}{a} - \dfrac{5}{a}$

4. $\dfrac{8}{a} - \dfrac{7}{a}$

5. $\dfrac{1}{x+1} + \dfrac{x}{x+1}$

6. $\dfrac{x}{x-3} - \dfrac{3}{x-3}$

7. $\dfrac{y^2}{y-1} - \dfrac{1}{y-1}$

8. $\dfrac{y^2}{y+3} - \dfrac{9}{y+3}$

9. $\dfrac{x^2}{x+2} + \dfrac{4x+4}{x+2}$

10. $\dfrac{x^2-6x}{x-3} + \dfrac{9}{x-3}$

11. $\dfrac{x^2}{x-2} - \dfrac{4x-4}{x-2}$

12. $\dfrac{x^2}{x-5} - \dfrac{10x-25}{x-5}$

13. $\dfrac{x+2}{x+6} - \dfrac{x-4}{x+6}$

14. $\dfrac{x+5}{x+2} - \dfrac{x+3}{x+2}$

15. $\dfrac{y}{2} - \dfrac{2}{y}$

16. $\dfrac{3}{y} + \dfrac{y}{3}$

17. $\dfrac{1}{2} + \dfrac{a}{3}$

18. $\dfrac{2}{3} + \dfrac{2a}{5}$

19. $\dfrac{x}{x+1} + \dfrac{3}{4}$

20. $\dfrac{x}{x-3} + \dfrac{1}{3}$

21. $\dfrac{x+1}{x-2} - \dfrac{4x+7}{5x-10}$

22. $\dfrac{3x+1}{2x-6} - \dfrac{x+2}{x-3}$

23. $\dfrac{4x-2}{3x+12} - \dfrac{x-2}{x+4}$

24. $\dfrac{6x+5}{5x-25} - \dfrac{x+2}{x-5}$

25. $\dfrac{6}{x(x-2)} + \dfrac{3}{x}$

26. $\dfrac{10}{x(x+5)} - \dfrac{2}{x}$

27. $\dfrac{4}{a} - \dfrac{12}{a^2+3a}$

28. $\dfrac{5}{a} + \dfrac{20}{a^2-4a}$

29. $\dfrac{2}{x+5} - \dfrac{10}{x^2-25}$

30. $\dfrac{6}{x^2-1} + \dfrac{3}{x+1}$

31. $\dfrac{x-4}{x-3} + \dfrac{6}{x^2-9}$ **32.** $\dfrac{x+1}{x-1} - \dfrac{4}{x^2-1}$

33. $\dfrac{a-4}{a-3} + \dfrac{5}{a^2-a-6}$

34. $\dfrac{a+2}{a+1} + \dfrac{7}{a^2-5a-6}$

35. $\dfrac{8}{x^2-16} - \dfrac{7}{x^2-x-12}$

36. $\dfrac{6}{x^2-9} - \dfrac{5}{x^2-x-6}$

37. $\dfrac{4y}{y^2+6y+5} - \dfrac{3y}{y^2+5y+4}$

38. $\dfrac{3y}{y^2+7y+10} - \dfrac{2y}{y^2+6y+8}$

39. $\dfrac{4x+1}{x^2+5x+4} - \dfrac{x+3}{x^2+4x+3}$

40. $\dfrac{2x-1}{x^2+x-6} - \dfrac{x+2}{x^2+5x+6}$

41. $\dfrac{1}{x} + \dfrac{x}{3x+9} - \dfrac{3}{x^2+3x}$

42. $\dfrac{1}{x} + \dfrac{x}{2x+4} - \dfrac{2}{x^2+2x}$

Complete the following tables.

43.

Number x	Reciprocal $\dfrac{1}{x}$	Sum $1+\dfrac{1}{x}$	Sum $\dfrac{x+1}{x}$
1			
2			
3			
4			

44.

Number x	Reciprocal $\dfrac{1}{x}$	Difference $1-\dfrac{1}{x}$	Difference $\dfrac{x-1}{x}$
1			
2			
3			
4			

45.

x	$x+\dfrac{4}{x}$	$\dfrac{x^2+4}{x}$	$x+4$
1			
2			
3			
4			

46.

x	$2x+\dfrac{6}{x}$	$\dfrac{2x^2+6}{x}$	$2x+6$
1			
2			
3			
4			

Add and subtract as indicated.

47. $1 + \dfrac{1}{x+2}$ **48.** $1 - \dfrac{1}{x+2}$

49. $1 - \dfrac{1}{x+3}$ **50.** $1 + \dfrac{1}{x+3}$

Applying the Concepts

51. Number Problem Write an expression for the sum of a number and twice its reciprocal. Then, simplify that expression. (If the reciprocal of a number is $\dfrac{1}{x}$, then twice that is $\dfrac{2}{x}$, not $\dfrac{1}{2x}$.)

52. Number Problem Write an expression for the sum of a number and 3 times its reciprocal. Then, simplify that expression.

53. Number Problem One number is twice another. Write an expression for the sum of their reciprocals. Then, simplify that expression. (*Hint:* The numbers are x and $2x$. Their reciprocals are, respectively, $\dfrac{1}{x}$ and $\dfrac{1}{2x}$.)

54. Number Problem One number is three times another. Write an expression for the sum of their reciprocals. Then, simplify that expression.

Review Problems

The following problems review material we covered in Sections 2.4 and 5.6. Reviewing these problems will help you understand the next section.

Solve each equation. [2.4]

55. $2x + 3(x - 3) = 6$ **56.** $4x - 2(x - 5) = 6$

57. $x - 3(x + 3) = x - 3$

58. $x - 4(x + 4) = x - 4$

59. $7 - 2(3x + 1) = 4x + 3$

60. $8 - 5(2x - 1) = 2x + 4$

Solve each quadratic equation. [5.6]

61. $x^2 + 5x + 6 = 0$ **62.** $x^2 - 5x + 6 = 0$

63. $x^2 - x = 6$ **64.** $x^2 + x = 6$

65. $x^2 - 5x = 0$ **66.** $x^2 - 6x = 0$

6.4 Equations Involving Rational Expressions

The first step in solving an equation that contains one or more rational expressions is to find the LCD for all denominators in the equation. Once the LCD has been found, we multiply both sides of the equation by it. The resulting equation should be equivalent to the original one (unless we inadvertently multiplied by zero) and free from any denominators except the number 1.

EXAMPLE 1 Solve $\dfrac{x}{3} + \dfrac{5}{2} = \dfrac{1}{2}$ for x.

Solution The LCD for 3 and 2 is 6. If we multiply both sides by 6, we have:

$$6\left(\frac{x}{3} + \frac{5}{2}\right) = 6\left(\frac{1}{2}\right) \qquad \text{Multiply both sides by 6.}$$

$$6\left(\frac{x}{3}\right) + 6\left(\frac{5}{2}\right) = 6\left(\frac{1}{2}\right) \qquad \text{Distributive property}$$

$$2x + 15 = 3$$

$$2x = -12$$

$$x = -6$$

We can check our solution by replacing x with -6 in the original equation:

$$-\frac{6}{3} + \frac{5}{2} \overset{?}{=} \frac{1}{2}$$

$$\frac{1}{2} = \frac{1}{2}$$

Multiplying both sides of an equation containing fractions by the LCD clears the equation of all denominators, because the LCD has the property that all denominators will divide it evenly.

EXAMPLE 2 Solve $\dfrac{3}{x-1} = \dfrac{3}{5}$ for x.

Solution The LCD for $(x-1)$ and 5 is $5(x-1)$. Multiplying both sides by $5(x-1)$, we have:

$$5(x-1) \cdot \frac{3}{x-1} = 5(x-1) \cdot \frac{3}{5}$$

$$5 \cdot 3 = (x-1) \cdot 3$$

$$15 = 3x - 3$$

$$18 = 3x$$

$$6 = x$$

If we substitute $x = 6$ into the original equation, we have:

$$\frac{3}{6-1} \overset{?}{=} \frac{3}{5}$$

$$\frac{3}{5} = \frac{3}{5}$$

The solution set is $\{6\}$.

EXAMPLE 3 Solve $1 - \dfrac{5}{x} = \dfrac{-6}{x^2}$.

Solution The LCD is x^2. Multiplying both sides by x^2, we have:

$$x^2\left(1 - \frac{5}{x}\right) = x^2\left(\frac{-6}{x^2}\right) \qquad \text{Multiply both sides by } x^2.$$

$$x^2(1) - x^2\left(\frac{5}{x}\right) = x^2\left(\frac{-6}{x^2}\right) \qquad \begin{array}{l}\text{Apply distributive property} \\ \text{to the left side.}\end{array}$$

$$x^2 - 5x = -6 \qquad \text{Simplify each side.}$$

We have a quadratic equation, which we write in standard form, factor, and solve as we did in Section 5.6:

$$x^2 - 5x + 6 = 0 \qquad \text{Standard form}$$

$$(x-2)(x-3) = 0 \qquad \text{Factor.}$$

$$x - 2 = 0 \quad \text{or} \quad x - 3 = 0 \qquad \text{Set factors equal to 0.}$$

$$x = 2 \quad \text{or} \quad x = 3$$

The two possible solutions are 2 and 3. Checking each in the original equation, we find they both give true statements. They are both solutions to the original equation:

Check: $x = 2$ Check: $x = 3$

$$1 - \frac{5}{2} \overset{?}{=} \frac{-6}{4} \qquad\qquad 1 - \frac{5}{3} \overset{?}{=} \frac{-6}{9}$$

$$\frac{2}{2} - \frac{5}{2} = -\frac{3}{2} \qquad\qquad \frac{3}{3} - \frac{5}{3} = -\frac{2}{3}$$

$$-\frac{3}{2} = -\frac{3}{2} \qquad\qquad -\frac{2}{3} = -\frac{2}{3}$$

E X A M P L E   4   Solve $\dfrac{x}{x^2 - 9} - \dfrac{3}{x - 3} = \dfrac{1}{x + 3}$.

Solution   The factors of $x^2 - 9$ are $(x + 3)(x - 3)$. The LCD, then, is $(x + 3)(x - 3)$:

$$\cancel{(x + 3)(x - 3)} \cdot \frac{x}{\cancel{(x + 3)(x - 3)}} + (x + 3)\cancel{(x - 3)} \cdot \frac{-3}{\cancel{x - 3}}$$

$$= \cancel{(x + 3)}(x - 3) \cdot \frac{1}{\cancel{x + 3}}$$

$$x + (x + 3)(-3) = (x - 3)1$$

$$x + (-3x) + (-9) = x - 3$$

$$-2x - 9 = x - 3$$

$$-3x = 6$$

$$x = -2$$

The solution is $x = -2$. It checks when replaced for x in the original equation.

E X A M P L E   5   Solve $\dfrac{x}{x - 3} + \dfrac{3}{2} = \dfrac{3}{x - 3}$.

Solution   We begin by multiplying each term on both sides of the equation by $2(x - 3)$:

$$2\cancel{(x - 3)} \cdot \frac{x}{\cancel{x - 3}} + 2(x - 3) \cdot \frac{3}{2} = 2\cancel{(x - 3)} \cdot \frac{3}{\cancel{x - 3}}$$

$$2x + (x - 3) \cdot 3 = 2 \cdot 3$$

$$2x + 3x - 9 = 6$$

$$5x - 9 = 6$$

$$5x = 15$$

$$x = 3$$

Our only possible solution is $x = 3$. If we substitute $x = 3$ into our original equation, we get:

$$\frac{3}{3-3} + \frac{3}{2} \overset{?}{=} \frac{3}{3-3}$$

$$\frac{3}{0} + \frac{3}{2} = \frac{3}{0}$$

Two of the terms are undefined, so the equation is meaningless. What has happened is that we have multiplied both sides of the original equation by zero. The equation produced by doing this is not equivalent to our original equation. We must always check our solution when we multiply both sides of an equation by an expression containing the variable in order to make sure we have not multiplied both sides by zero.

Our original equation has no solutions. That is, there is no real number x such that:

$$\frac{x}{x-3} + \frac{3}{2} = \frac{3}{x-3}$$

The solution set is $\varnothing$.

EXAMPLE 6   Solve $\dfrac{a+4}{a^2+5a} = \dfrac{-2}{a^2-25}$ for a.

Solution   Factoring each denominator, we have:

$$a^2 + 5a = a(a+5)$$

$$a^2 - 25 = (a+5)(a-5)$$

The LCD is $a(a+5)(a-5)$. Multiplying both sides of the equation by the LCD gives us:

$$a(a+5)(a-5) \cdot \frac{a+4}{a(a+5)} = \frac{-2}{(a+5)(a-5)} \cdot a(a+5)(a-5)$$

$$(a-5)(a+4) = -2a$$

$$a^2 - a - 20 = -2a$$

The result is a quadratic equation, which we write in standard form, factor, and solve:

$$a^2 + a - 20 = 0 \qquad \text{Add } 2a \text{ to both sides.}$$

$$(a+5)(a-4) = 0 \qquad \text{Factor.}$$

$$a+5 = 0 \qquad \text{or} \qquad a-4 = 0 \qquad \text{Set each factor to 0.}$$

$$a = -5 \qquad \text{or} \qquad a = 4$$

The two possible solutions are -5 and 4. There is no problem with the 4. It checks when substituted for a in the original equation. However, -5 is not a solution. Substituting -5 into the original equation gives:

$$\frac{-5 + 4}{(-5)^2 + 5(-5)} \overset{?}{=} \frac{-2}{(-5)^2 - 25}$$

$$\frac{-1}{0} = \frac{-2}{0}$$

This indicates -5 is not a solution. The solution is 4.

Getting Ready for Class

After reading through the preceding section, respond in your own words and in complete sentences.

A. What is the first step in solving an equation that contains rational expressions?

B. Explain how to find the LCD used to clear an equation of fractions.

C. When will a possible solution to an equation containing rational expressions not be an actual solution?

D. When must we check solutions to an equation containing rational expressions?

PROBLEM SET 6.4

Solve the following equations. Be sure to check each answer in the original equation if you multiply both sides by an expression that contains the variable.

1. $\dfrac{x}{3} + \dfrac{1}{2} = -\dfrac{1}{2}$

2. $\dfrac{x}{2} + \dfrac{4}{3} = -\dfrac{2}{3}$

3. $\dfrac{4}{a} = \dfrac{1}{5}$

4. $\dfrac{2}{3} = \dfrac{6}{a}$

5. $\dfrac{3}{x} + 1 = \dfrac{2}{x}$

6. $\dfrac{4}{x} + 3 = \dfrac{1}{x}$

7. $\dfrac{3}{a} - \dfrac{2}{a} = \dfrac{1}{5}$

8. $\dfrac{7}{a} + \dfrac{1}{a} = 2$

9. $\dfrac{3}{x} + 2 = \dfrac{1}{2}$

10. $\dfrac{5}{x} + 3 = \dfrac{4}{3}$

11. $\dfrac{1}{y} - \dfrac{1}{2} = -\dfrac{1}{4}$

12. $\dfrac{3}{y} - \dfrac{4}{5} = -\dfrac{1}{5}$

13. $1 - \dfrac{8}{x} = \dfrac{-15}{x^2}$

14. $1 - \dfrac{3}{x} = \dfrac{-2}{x^2}$

15. $\dfrac{x}{2} - \dfrac{4}{x} = -\dfrac{7}{2}$

16. $\dfrac{x}{2} - \dfrac{5}{x} = -\dfrac{3}{2}$

17. $\dfrac{x - 3}{2} + \dfrac{2x}{3} = \dfrac{5}{6}$

18. $\dfrac{x - 2}{3} + \dfrac{5x}{2} = 5$

19. $\dfrac{x + 1}{3} + \dfrac{x - 3}{4} = \dfrac{1}{6}$

20. $\dfrac{x + 2}{3} + \dfrac{x - 1}{5} = -\dfrac{3}{5}$

21. $\dfrac{6}{x + 2} = \dfrac{3}{5}$

22. $\dfrac{4}{x + 3} = \dfrac{1}{2}$

23. $\dfrac{3}{y - 2} = \dfrac{2}{y - 3}$

24. $\dfrac{5}{y + 1} = \dfrac{4}{y + 2}$

25. $\dfrac{x}{x-2} + \dfrac{2}{3} = \dfrac{2}{x-2}$

26. $\dfrac{x}{x-5} + \dfrac{1}{5} = \dfrac{5}{x-5}$

27. $\dfrac{x}{x-2} + \dfrac{3}{2} = \dfrac{9}{2(x-2)}$

28. $\dfrac{x}{x+1} + \dfrac{4}{5} = \dfrac{-14}{5(x+1)}$

29. $\dfrac{5}{x+2} + \dfrac{1}{x+3} = \dfrac{-1}{x^2+5x+6}$

30. $\dfrac{3}{x-1} + \dfrac{2}{x+3} = \dfrac{-3}{x^2+2x-3}$

31. $\dfrac{8}{x^2-4} + \dfrac{3}{x+2} = \dfrac{1}{x-2}$

32. $\dfrac{10}{x^2-25} - \dfrac{1}{x-5} = \dfrac{3}{x+5}$

33. $\dfrac{a}{2} + \dfrac{3}{a-3} = \dfrac{a}{a-3}$

34. $\dfrac{a}{2} + \dfrac{4}{a-4} = \dfrac{a}{a-4}$

35. $\dfrac{6}{y^2-4} = \dfrac{4}{y^2+2y}$ **36.** $\dfrac{2}{y^2-9} = \dfrac{5}{y^2-3y}$

37. $\dfrac{2}{a^2-9} = \dfrac{3}{a^2+a-12}$

38. $\dfrac{2}{a^2-1} = \dfrac{6}{a^2-2a-3}$

39. $\dfrac{3x}{x-5} - \dfrac{2x}{x+1} = \dfrac{-42}{x^2-4x-5}$

40. $\dfrac{4x}{x-4} - \dfrac{3x}{x-2} = \dfrac{-3}{x^2-6x+8}$

41. $\dfrac{2x}{x+2} = \dfrac{x}{x+3} - \dfrac{3}{x^2+5x+6}$

42. $\dfrac{3x}{x-4} = \dfrac{2x}{x-3} + \dfrac{6}{x^2-7x+12}$

43. The following table is the answer to Problem 45 in Problem Set 6.3. Solve the equation

$$x + \dfrac{4}{x} = 5$$

and compare your solutions with the entries in the table.

x	$x+\dfrac{4}{x}$	$\dfrac{x^2+4}{x}$	$x+4$
1	5	5	5
2	4	4	6
3	$\dfrac{13}{3}$	$\dfrac{13}{3}$	7
4	5	5	8

44. The following table is the answer to Problem 46 in Problem Set 6.3. Solve the equation

$$2x + \dfrac{6}{x} = 8$$

and compare your solutions with the entries in the table.

x	$2x+\dfrac{6}{x}$	$\dfrac{2x^2+6}{x}$	$2x+6$
1	8	8	8
2	7	7	10
3	8	8	12
4	$\dfrac{19}{2}$	$\dfrac{19}{2}$	14

Review Problems

The problems that follow review material we covered in Sections 2.6 and 5.7. Reviewing these problems will help you with the next section.

Solve each word problem. [2.6]

45. Number Problem   If twice the difference of a number and 3 were decreased by 5, the result would be 3. Find the number.

46. Number Problem   If 3 times the sum of a number and 2 were increased by 6, the result would be 27. Find the number.

47. Geometry   The length of a rectangle is 5 more than twice the width. The perimeter is 34 inches. Find the length and width.

48. Geometry   The length of a rectangle is 2 more than 3 times the width. The perimeter is 44 feet. Find the length and width.

Solve each problem. Be sure to show the equation that describes the situation. [5.7]

49. Number Problem The product of two consecutive even integers is 48. Find the two integers.

50. Number Problem The product of two consecutive odd integers is 35. Find the two integers.

51. Geometry The hypotenuse (the longest side) of a

right triangle is 10 inches, and the lengths of the two legs (the other two sides) are given by two consecutive even integers. Find the lengths of the two legs.

52. Geometry One leg of a right triangle is 2 more than twice the other. If the hypotenuse is 13 feet, find the lengths of the two legs.

6.5 Applications

In this section we will solve some word problems whose equations involve rational expressions. Like the other word problems we have encountered, the more you work with them, the easier they become.

EXAMPLE 1 One number is twice another. The sum of their reciprocals is $\frac{9}{2}$. Find the two numbers.

Solution Let $x =$ the smaller number. The larger then must be $2x$. Their reciprocals are $\frac{1}{x}$ and $\frac{1}{2x}$, respectively. An equation that describes the situation is:

$$\frac{1}{x} + \frac{1}{2x} = \frac{9}{2}$$

We can multiply both sides by the LCD $2x$ and then solve the resulting equation:

$$2x\left(\frac{1}{x}\right) + 2x\left(\frac{1}{2x}\right) = 2x\left(\frac{9}{2}\right)$$

$$2 + 1 = 9x$$

$$3 = 9x$$

$$x = \frac{3}{9} = \frac{1}{3}$$

The smaller number is $\frac{1}{3}$. The other number is twice as large, or $\frac{2}{3}$. If we add their reciprocals, we have:

$$\frac{3}{1} + \frac{3}{2} = \frac{6}{2} + \frac{3}{2} = \frac{9}{2}$$

The solutions check with the original problem.

EXAMPLE 2 A boat travels 30 miles up a river in the same amount of time it takes to travel 50 miles down the same river. If the current is 5 miles per hour, what is the speed of the boat in still water?

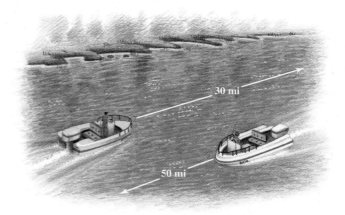

Solution   The easiest way to work a problem like this is with a table. The top row of the table is labeled with d for distance, r for rate, and t for time. The left column of the table is labeled with the two trips: upstream and downstream. Here is what the table looks like:

	d	r	t
Upstream			
Downstream			

The next step is to read the problem over again and fill in as much of the table as we can with the information in the problem. The distance the boat travels upstream is 30 and the distance downstream is 50. Since we are asked for the speed of the boat in still water, we will let that be x. If the speed of the boat in still water is x, then its speed upstream (against the current) must be $x - 5$, and its speed downstream (with the current) must be $x + 5$. Putting these four quantities into the appropriate positions in the table, we have

	d	r	t
Upstream	30	$x - 5$	
Downstream	50	$x + 5$	

The last positions in the table are filled in by using the equation $t = \dfrac{d}{r}$.

	d	r	t
Upstream	30	$x - 5$	$\dfrac{30}{x - 5}$
Downstream	50	$x + 5$	$\dfrac{50}{x + 5}$

Reading the problem again, we find that the time for the trip upstream is equal to the time for the trip downstream. Setting these two quantities equal to each other, we have our equation:

$$\text{Time (downstream)} = \text{time (upstream)}$$

$$\frac{50}{x + 5} = \frac{30}{x - 5}$$

The LCD is $(x + 5)(x - 5)$. We multiply both sides of the equation by the LCD to clear it of all denominators. Here is the solution:

$$\cancel{(x + 5)}(x - 5) \cdot \frac{50}{\cancel{x + 5}} = (x + 5)\cancel{(x - 5)} \cdot \frac{30}{\cancel{x - 5}}$$

$$50x - 250 = 30x + 150$$

$$20x = 400$$

$$x = 20$$

The speed of the boat in still water is 20 miles per hour.

EXAMPLE   3   Tina is training for a biathlon. To train for the bicycle portion, she rides her bike 15 miles up a hill and then 15 miles back down the same hill. The complete trip takes her 2 hours. If her downhill speed is 20 miles per hour faster than her uphill speed, how fast does she ride uphill?

**Up and back
total time = 2 hours**

Solution   Again, we make a table. As in the previous example, we label the top row with distance, rate, and time. We label the left column with the two trips, uphill and downhill.

	d	r	t
Uphill			
Downhill			

Next, we fill in the table with as much information as we can from the problem. We know the distance traveled is 15 miles uphill and 15 miles downhill, which allows us to fill in the distance column. To fill in the rate column, we first note that she rides 20 miles per hour faster downhill than uphill. Therefore, if we let x equal her rate uphill, then her rate downhill is $x + 20$. Filling in the table with this information gives us

	d	r	t
Uphill	15	x	
Downhill	15	$x + 20$	

Since time is distance divided by rate, $t = \dfrac{d}{r}$, we can fill in the last column in the table.

	d	r	t
Uphill	15	x	$\dfrac{15}{x}$
Downhill	15	$x + 20$	$\dfrac{15}{x + 20}$

Rereading the problem, we find that the total time (the time riding uphill plus the time riding downhill) is 2 hours. We write our equation as follows:

$$\text{Time (uphill)} + \text{time (downhill)} = 2$$

$$\frac{15}{x} + \frac{15}{x + 20} = 2$$

We solve this equation for x by first finding the LCD and then multiplying each term in the equation by it in order to clear the equation of all denominators. Our LCD is $x(x + 20)$. Here is our solution:

$$x(x + 20)\,\frac{15}{x} + x(x + 20)\,\frac{15}{x + 20} = 2 \cdot [x(x + 20)]$$

$$15(x + 20) + 15x = 2x(x + 20)$$

$$15x + 300 + 15x = 2x^2 + 40x$$

$$0 = 2x^2 + 10x - 300$$

$$0 = x^2 + 5x - 150 \qquad \text{Divide both}$$

$$0 = (x + 15)(x - 10) \qquad \text{sides by 2.}$$

$$x = -15 \quad \text{or} \quad x = 10$$

Since we cannot have a negative speed, our only solution is $x = 10$. Tina rides her bike at a rate of 10 miles per hour when going uphill. (Her downhill speed is $x + 20 = 30$ miles per hour.)

EXAMPLE 4   An inlet pipe can fill a water tank in 10 hours, and an outlet pipe can empty the same tank in 15 hours. By mistake, both pipes are left open. How long will it take to fill the water tank with both pipes open?

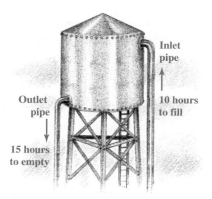

Inlet
pipe

10 hours
to fill

Outlet
pipe

15 hours
to empty

Solution   Let x = amount of time to fill the tank with both pipes open.

One method of solving this type of problem is to think in terms of how much of the job is done by a pipe in 1 hour.

1. If the inlet pipe fills the tank in 10 hours, then in 1 hour the inlet pipe fills $\frac{1}{10}$ of the tank.

2. If the outlet pipe empties the tank in 15 hours, then in 1 hour the outlet pipe empties $\frac{1}{15}$ of the tank.

3. If it takes x hours to fill the tank with both pipes open, then in 1 hour the tank is $\frac{1}{x}$ full.

Here is how we set up the equation. In one hour,

$$\frac{1}{10} \quad - \quad \frac{1}{15} \quad = \quad \frac{1}{x}$$

Amount of water let   Amount of water let   Total amount of
in by inlet pipe   out by outlet pipe   water in tank

The LCD for our equation is $30x$. We multiply both sides by the LCD and solve:

$$30x\left(\frac{1}{10}\right) - 30x\left(\frac{1}{15}\right) = 30x\left(\frac{1}{x}\right)$$

$$3x - 2x = 30$$

$$x = 30$$

It takes 30 hours with both pipes open to fill the tank.

Note: In solving a problem of this type, we have to assume that the thing doing the work (whether it is a pipe, a person, or a machine) is working at a constant rate. That is, as much work gets done in the first hour as is done in the last hour and any other hour in between.

EXAMPLE 5   Graph the equation $y = \dfrac{1}{x}$.

Solution   Since this is the first time we have graphed an equation of this form, we will make a table of values for x and y that satisfy the equation. Before we do, let's make some generalizations about the graph (Figure 1).

First, notice that, since y is equal to 1 divided by x, y will be positive when x is positive. (The quotient of two positive numbers is a positive number.) Likewise, when x is negative, y will be negative. In other words, x and y will always have the same sign. Thus, our graph will appear in quadrants 1 and 3 only, because in those quadrants x and y have the same sign.

Next, notice that the expression $\dfrac{1}{x}$ will be undefined when x is 0, meaning that there is no value of y corresponding to $x = 0$. Because of this, the graph will not cross the y-axis. Further, the graph will not cross the x-axis either. If we try to find the x-intercept by letting $y = 0$, we have

$$0 = \frac{1}{x}$$

But there is no value of x to divide into 1 to obtain 0. Therefore, since there is no solution to this equation, our graph will not cross the x-axis.

To summarize, we can expect to find the graph in quadrants I and III only, and the graph will cross neither axis.

x	y
-3	$-\frac{1}{3}$
-2	$-\frac{1}{2}$
-1	-1
$-\frac{1}{2}$	-2
$-\frac{1}{3}$	-3
0	Undefined
$\frac{1}{3}$	3
$\frac{1}{2}$	2
1	1
2	$\frac{1}{2}$
3	$\frac{1}{3}$

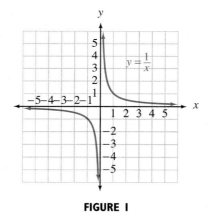

FIGURE 1

EXAMPLE 6   Graph the equation $y = \dfrac{-6}{x}$.

Solution   Since y is -6 divided by x, when x is positive, y will be negative (a negative divided by a positive is negative), and when x is negative, y will be positive (a negative divided by a negative). Thus, the graph (Figure 2) will appear in quadrants II and IV only. As was the case in Example 5, the graph will not cross either axis.

x	y
−6	1
−3	2
−2	3
−1	6
0	Undefined
1	−6
2	−3
3	−2
6	−1

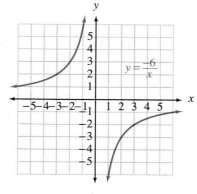

FIGURE 2

Getting Ready for Class

After reading through the preceding section, respond in your own words and in complete sentences.

A. Write an application problem for which the solution depends on solving the equation $\dfrac{1}{2} + \dfrac{1}{3} = \dfrac{1}{x}$.

B. How does the current of a river affect the speed of a motor boat traveling against the current?

C. How does the current of a river affect the speed of a motor boat traveling in the same direction as the current?

D. What is the relationship between the total number of minutes it takes for a drain to empty a sink and the amount of water that drains out of the sink in 1 minute?

PROBLEM SET 6.5

Number Problems

1. One number is 3 times as large as another. The sum of their reciprocals is $\dfrac{16}{3}$. Find the two numbers.

2. If $\dfrac{3}{5}$ is added to twice the reciprocal of a number, the result is 1. Find the number.

3. The sum of a number and its reciprocal is $\dfrac{13}{6}$. Find the number.

4. The sum of a number and 10 times its reciprocal is 7. Find the number.

5. If a certain number is added to both the numerator and denominator of the fraction $\dfrac{7}{9}$, the result is $\dfrac{5}{7}$. Find the number.

6. The numerator of a certain fraction is 2 more than the denominator. If $\dfrac{1}{3}$ is added to the fraction, the result is 2. Find the fraction.

7. The sum of the reciprocals of two consecutive even integers is $\dfrac{5}{12}$. Find the integers.

8. The sum of the reciprocals of two consecutive integers is $\dfrac{7}{12}$. Find the two integers.

Motion Problems

9. A boat travels 26 miles up a river in the same amount of time it takes to travel 38 miles down the same river. If the current is 3 miles per hour, what is the speed of the boat in still water?

	d	*r*	*t*
Upstream			
Downstream			

10. A boat can travel 9 miles up a river in the same amount of time it takes to travel 11 miles down the same river. If the current is 2 miles per hour, what is the speed of the boat in still water?

	d	*r*	*t*
Upstream			
Downstream			

11. An airplane flying against the wind travels 140 miles in the amount of time it would take the same plane to travel 160 miles with the wind. If the wind speed is a constant 20 miles per hour, how fast would the plane travel in still air?

12. An airplane flying against the wind travels 500 miles in the amount of time that it would take to travel 600 miles with the wind. If the speed of the wind is 50 miles per hour, what is the speed of the plane in still air?

13. One plane can travel 20 miles per hour faster than another. One of them goes 285 miles in the time it takes the other to go 255 miles. What are their speeds?

14. One car travels 300 miles in the amount of time it takes a second car traveling 5 miles per hour slower than the first to go 275 miles. What are the speeds of the cars?

15. Tina, whom we mentioned in Example 3 of this section, is training for a biathlon. To train for the running portion of the race, she runs 8 miles each day over the same course. The first 2 miles of the course are on level ground; the last 6 miles are downhill. She runs 3 miles per hour slower on level ground than she runs downhill. If the complete

course takes 1 hour, how fast does she run on the downhill part of the course?

16. Jerri is training for the same biathlon as Tina (Example 3 and Problem 15). To train for the bicycle portion of the race, she rides 24 miles out a straight road, and then turns around and rides 24 miles back. The trip out is against the wind; the trip back is with the wind. If she rides 10 miles per hour faster with the wind than she does against the wind and the complete trip out and back takes 2 hours, how fast does she ride when she rides against the wind?

17. To train for the running of a triathlon, Jerri jogs 1 hour each day over the same 9-mile course. Five miles of the course are downhill, while the other 4 miles are on level ground. Jerri figures that she runs 2 miles per hour faster downhill than she runs on level ground. Find the rate at which Jerri runs on level ground.

18. Travis paddles his kayak in the harbor at Morro Bay, California, where the incoming tide has caused a current in the water. From the point where he enters the water, he paddles 1 mile against the current, and then turns around and paddles 1 mile back to where he started. His average speed when paddling with the current is 4 miles per hour faster than his speed against the current. If the complete trip (out and back) takes him 1.2 hours, find his average speed when he paddles against the current.

Work Problems

19. An inlet pipe can fill a pool in 12 hours, and an outlet pipe can empty it in 15 hours. If both pipes are left open, how long will it take to fill the pool?

20. A water tank can be filled in 20 hours by an inlet pipe and emptied in 25 hours by an outlet pipe. How long will it take to fill the tank if both pipes are left open?

21. A bathtub can be filled by the cold-water faucet in 10 minutes and by the hot-water faucet in 12 minutes. How long does it take to fill the tub if both faucets are open?

22. A water faucet can fill a sink in 6 minutes, and the drain can empty it in 4 minutes. If the sink is full, how long will it take to empty if both the faucet and the drain are open?

23. A sink can be filled by the cold-water faucet in 3 minutes. The drain can empty a full sink in 4 minutes. If the sink is empty and both the cold-water faucet and the drain are open, how long will it take the sink to overflow?

24. A bathtub can be filled by the cold-water faucet in 9 minutes and by the hot-water faucet in 10 minutes. The drain can empty the tub in 5 minutes. Can the tub be filled if both faucets and the drain are open?

25. Average Cost As we mentioned in the introduction to this chapter, if a bank charges $2.00 per month and $0.15 per check for one of its checking accounts, then the average cost of each of the x checks written is $A = \dfrac{2.00 + 0.15x}{x}$. The following table was constructed using this formula. Use the template to construct a line graph using the information in the table.

Checks Written x	Average Cost per Check $A = \dfrac{2.00 + 0.15x}{x}$
1	2.15
5	0.55
10	0.35
15	0.28
20	0.25
25	0.23

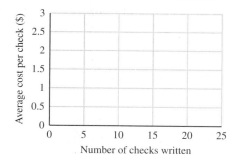

26. Average Cost A rewritable CD-ROM drive for a computer costs $250. An individual CD for the drive costs $5.00 and can store 640 megabytes of information. The average cost per megabyte to fill x CDs with information is $A = \dfrac{5x + 250}{640x}$. The following table was constructed using this formula.

Use the template to construct a line graph using the information in the table.

CDs Purchased x	Cost per Megabyte $A = \dfrac{5x + 250}{640x}$
1	0.40
5	0.09
10	0.05
15	0.03
20	0.03
25	0.02

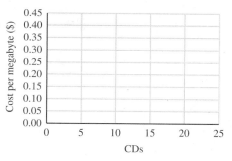

Graph each of the following equations.

27. $y = \dfrac{-4}{x}$

28. $y = \dfrac{4}{x}$

29. $y = \dfrac{8}{x}$

30. $y = \dfrac{-8}{x}$

31. Graph $y = \dfrac{3}{x}$ and $x + y = 4$ on the same coordinate system. At what points do the two graphs intersect?

32. Graph $y = \dfrac{4}{x}$ and $x - y = 3$ on the same coordinate system. At what points do the two graphs intersect?

Review Problems

The following problems review material we covered in Chapter 5.

33. Factor out the greatest common factor for
$$15a^3b^3 - 20a^2b - 35ab^2$$

34. Factor by grouping: $3ax - 2a + 15x - 10$

Factor completely.

35. $x^2 - 4x - 12$ **36.** $4x^2 - 20xy + 25y^2$

37. $x^4 - 16$ **38.** $2x^2 + xy - 21y^2$

39. $5x^3 - 25x^2 - 30x$

Solve each equation.

40. $x^2 - 9x + 18 = 0$ **41.** $x^2 - 6x = 0$

42. $8x^2 = -2x + 15$ **43.** $x(x + 2) = 80$

44. Number Problem The product of two consecutive even integers is 4 more than twice their sum. Find the two integers.

45. Geometry The hypotenuse of a right triangle is 15 inches. One of the legs is 3 inches more than the other. Find the lengths of the two legs.

6.6 **Complex Fractions**

A complex fraction is a fraction or rational expression that contains other fractions in its numerator or denominator. Each of the following is a complex fraction:

$$\frac{\dfrac{1}{2}}{\dfrac{2}{3}} \qquad \frac{x + \dfrac{1}{y}}{y + \dfrac{1}{x}} \qquad \frac{\dfrac{a + 1}{a^2 - 9}}{\dfrac{2}{a + 3}}$$

We will begin this section by simplifying the first of these complex fractions. Before we do, though, let's agree on some vocabulary. So that we won't have to use phrases such as the numerator of the denominator, let's call the numerator of a complex fraction the *top* and the denominator of a complex fraction the *bottom*.

EXAMPLE 1 Simplify $\dfrac{\dfrac{1}{2}}{\dfrac{2}{3}}$.

Solution There are two methods we can use to solve this problem.

Method 1 We can multiply the top and bottom of this complex fraction by the LCD for both fractions. In this case the LCD is 6:

$$\frac{\dfrac{1}{2}}{\dfrac{2}{3}} = \frac{6 \cdot \dfrac{1}{2}}{6 \cdot \dfrac{2}{3}} = \frac{3}{4}$$

Method 2   We can treat this as a division problem. Instead of dividing by $\frac{2}{3}$, we can multiply by its reciprocal $\frac{3}{2}$:

$$\frac{\dfrac{1}{2}}{\dfrac{2}{3}} = \frac{1}{2} \cdot \frac{3}{2} = \frac{3}{4}$$

Using either method, we obtain the same result.

E X A M P L E   2   Simplify:

$$\frac{\dfrac{2x^3}{y^2}}{\dfrac{4x}{y^5}}$$

Solution

Method 1   The LCD for each rational expression is y^5. Multiplying the top and bottom of the complex fraction by y^5, we have:

$$\frac{\dfrac{2x^3}{y^2}}{\dfrac{4x}{y^5}} = \frac{y^5 \cdot \dfrac{2x^3}{y^2}}{y^5 \cdot \dfrac{4x}{y^5}} = \frac{2x^3 y^3}{4x} = \frac{x^2 y^3}{2}$$

Method 2   Instead of dividing by $\dfrac{4x}{y^5}$, we can multiply by its reciprocal, $\dfrac{y^5}{4x}$:

$$\frac{\dfrac{2x^3}{y^2}}{\dfrac{4x}{y^5}} = \frac{2x^3}{y^2} \cdot \frac{y^5}{4x} = \frac{x^2 y^3}{2}$$

Again the result is the same, whether we use Method 1 or Method 2.

E X A M P L E   3   Simplify:

$$\frac{x + \dfrac{1}{y}}{y + \dfrac{1}{x}},$$

Solution   To apply Method 2 as we did in the first two examples, we would have to simplify the top and bottom separately to obtain a single rational expression for both before we could multiply by the reciprocal. It is much easier, in this case, to multiply the top and bottom by the LCD xy:

$$\frac{x + \dfrac{1}{y}}{y + \dfrac{1}{x}} = \frac{xy\left(x + \dfrac{1}{y}\right)}{xy\left(y + \dfrac{1}{x}\right)} \qquad \text{Multiply top and bottom by } xy.$$

$$= \frac{xy \cdot x + xy \cdot \dfrac{1}{y}}{xy \cdot y + xy \cdot \dfrac{1}{x}} \qquad \text{Distributive property}$$

$$= \frac{x^2 y + x}{xy^2 + y} \qquad \text{Simplify.}$$

We can factor an x from $x^2 y + x$ and a y from $xy^2 + y$ and then reduce to lowest terms:

$$= \frac{x\cancel{(xy + 1)}}{y\cancel{(xy + 1)}}$$

$$= \frac{x}{y}$$

EXAMPLE 4   Simplify:

$$\frac{1 - \dfrac{4}{x^2}}{1 - \dfrac{1}{x} - \dfrac{6}{x^2}}$$

Solution   The simplest way to simplify this complex fraction is to multiply the top and bottom by the LCD, x^2:

$$\frac{1 - \dfrac{4}{x^2}}{1 - \dfrac{1}{x} - \dfrac{6}{x^2}} = \frac{x^2\left(1 - \dfrac{4}{x^2}\right)}{x^2\left(1 - \dfrac{1}{x} - \dfrac{6}{x^2}\right)} \qquad \begin{array}{l}\text{Multiply top and}\\ \text{bottom by } x^2.\end{array}$$

$$= \frac{x^2 \cdot 1 - x^2 \cdot \dfrac{4}{x^2}}{x^2 \cdot 1 - x^2 \cdot \dfrac{1}{x} - x^2 \cdot \dfrac{6}{x^2}} \qquad \text{Distributive property}$$

$$= \frac{x^2 - 4}{x^2 - x - 6} \qquad \text{Simplify.}$$

$$= \frac{(x-2)\cancel{(x+2)}}{(x-3)\cancel{(x+2)}} \qquad \text{Factor.}$$

$$= \frac{x-2}{x-3} \qquad \text{Reduce.}$$

In our next example, we examine the relationship between the numbers in the Fibonacci sequence involving complex fractions.

EXAMPLE 5   Simplify each term in the following sequence, and then explain how this sequence is related to the Fibonacci sequence:

$$1 + \frac{1}{1+1}, \; 1 + \cfrac{1}{1 + \cfrac{1}{1+1}}, \; 1 + \cfrac{1}{1 + \cfrac{1}{1 + \cfrac{1}{1+1}}}, \; \dots$$

Solution   We can simplify our work somewhat if we notice that the first term $1 + \dfrac{1}{1+1}$ is the larger denominator in the second term, and that the second term is the largest denominator in the third term:

First term: $1 + \dfrac{1}{1+1} = 1 + \dfrac{1}{2} = \dfrac{2}{2} + \dfrac{1}{2} = \dfrac{3}{2}$

Second term: $1 + \cfrac{1}{1 + \cfrac{1}{1+1}} = 1 + \cfrac{1}{\cfrac{3}{2}} = 1 + \dfrac{2}{3} = \dfrac{3}{3} + \dfrac{2}{3} = \dfrac{5}{3}$

Third term: $1 + \cfrac{1}{1 + \cfrac{1}{1 + \cfrac{1}{1+1}}} = 1 + \cfrac{1}{\cfrac{5}{3}} = 1 + \dfrac{3}{5} = \dfrac{5}{5} + \dfrac{3}{5} = \dfrac{8}{5}$

Here are the simplified numbers for the first three terms in our sequence:

$$\frac{3}{2}, \frac{5}{3}, \frac{8}{5}, \dots$$

Recall the Fibonacci sequence:

$$1, 1, 2, 3, 5, 8, 13, 21, \dots$$

As you can see, each term in the sequence we have simplified is the ratio of two consecutive numbers in the Fibonacci sequence. If the pattern continues in this manner, the next number in our sequence will be $\dfrac{13}{8}$.

Getting Ready for Class

After reading through the preceding section, respond in your own words and in complete sentences.

A. What is a complex fraction?

B. Explain one method of simplifying complex fractions.

C. How is a least common denominator used to simplify a complex fraction?

D. What types of complex fractions can be rewritten as division problems?

PROBLEM SET 6.6

Simplify each complex fraction.

1. $\dfrac{\frac{3}{4}}{\frac{1}{8}}$

2. $\dfrac{\frac{1}{3}}{\frac{5}{6}}$

3. $\dfrac{\frac{2}{3}}{\frac{3}{4}}$

4. $\dfrac{\frac{5}{1}}{\frac{1}{2}}$

5. $\dfrac{\frac{x^2}{y}}{\frac{x}{y^3}}$

6. $\dfrac{\frac{x^5}{y^3}}{\frac{x^2}{y^8}}$

7. $\dfrac{\frac{4x^3}{y^6}}{\frac{8x^2}{y^7}}$

8. $\dfrac{\frac{6x^4}{y}}{\frac{2x}{y^5}}$

9. $\dfrac{y + \frac{1}{x}}{x + \frac{1}{y}}$

10. $\dfrac{y - \frac{1}{x}}{x - \frac{1}{y}}$

11. $\dfrac{1 + \frac{1}{a}}{1 - \frac{1}{a}}$

12. $\dfrac{\frac{1}{a} - 1}{\frac{1}{a} + 1}$

13. $\dfrac{\frac{x+1}{x^2-9}}{\frac{2}{x+3}}$

14. $\dfrac{\frac{3}{x-5}}{\frac{x+1}{x^2-25}}$

15. $\dfrac{\frac{1}{a+2}}{\frac{1}{a^2-a-6}}$

16. $\dfrac{\frac{1}{a^2+5a+6}}{\frac{1}{a+3}}$

17. $\dfrac{1 - \frac{9}{y^2}}{1 - \frac{1}{y} - \frac{6}{y^2}}$

18. $\dfrac{1 - \frac{4}{y^2}}{1 - \frac{2}{y} - \frac{8}{y^2}}$

19. $\dfrac{\frac{1}{y} + \frac{1}{x}}{\frac{1}{xy}}$

20. $\dfrac{\frac{1}{xy}}{\frac{1}{y} - \frac{1}{x}}$

21. $\dfrac{1 - \frac{1}{a^2}}{1 - \frac{1}{a}}$

22. $\dfrac{1 + \frac{1}{a}}{1 - \frac{1}{a^2}}$

23. $\dfrac{\frac{1}{10x} - \frac{y}{10x^2}}{\frac{1}{10} - \frac{y}{10x}}$

24. $\dfrac{\frac{1}{2x} + \frac{y}{2x^2}}{\frac{1}{4} + \frac{y}{4x}}$

25. $\dfrac{\frac{1}{a+1} + 2}{\frac{1}{a+1} + 3}$

26. $\dfrac{\frac{2}{a+1} + 3}{\frac{3}{a+1} + 4}$

Although the following problems do not contain complex fractions, they do involve more than one operation. Simplify inside the parentheses first, and then multiply.

27. $\left(1 - \dfrac{1}{x}\right)\left(1 - \dfrac{1}{x+1}\right)\left(1 - \dfrac{1}{x+2}\right)$

28. $\left(1 + \dfrac{1}{x}\right)\left(1 + \dfrac{1}{x+1}\right)\left(1 + \dfrac{1}{x+2}\right)$

29. $\left(1 + \dfrac{1}{x+3}\right)\left(1 + \dfrac{1}{x+2}\right)\left(1 + \dfrac{1}{x+1}\right)$

30. $\left(1 - \dfrac{1}{x+3}\right)\left(1 - \dfrac{1}{x+2}\right)\left(1 - \dfrac{1}{x+1}\right)$

31. Simplify each term in the following sequence.

$$2 + \cfrac{1}{2+1}, \; 2 + \cfrac{1}{2+\cfrac{1}{2+1}}, \; 2 + \cfrac{1}{2+\cfrac{1}{2+\cfrac{1}{2+1}}}, \; \ldots$$

32. Simplify each term in the following sequence.

$$2 + \cfrac{3}{2+3}, \; 2 + \cfrac{3}{2+\cfrac{3}{2+3}}, \; 2 + \cfrac{3}{2+\cfrac{3}{2+\cfrac{3}{2+3}}}, \; \ldots$$

Complete the following tables.

33.

Number x	Reciprocal $\dfrac{1}{x}$	Quotient $\dfrac{x}{\dfrac{1}{x}}$	Square x^2
1			
2			
3			
4			

34.

Number x	Reciprocal $\dfrac{1}{x}$	Quotient $\dfrac{\dfrac{1}{x}}{x}$	Square x^2
1			
2			
3			
4			

35.

Number x	Reciprocal $\dfrac{1}{x}$	Sum $1+\dfrac{1}{x}$	Quotient $\dfrac{1+\dfrac{1}{x}}{\dfrac{1}{x}}$
1			
2			
3			
4			

36.

Number x	Reciprocal $\dfrac{1}{x}$	Difference $1-\dfrac{1}{x}$	Quotient $\dfrac{1-\dfrac{1}{x}}{\dfrac{1}{x}}$
1			
2			
3			
4			

Review Problems

The following problems review material we covered in Section 2.8.

Solve each inequality.

37. $2x + 3 < 5$ **38.** $3x - 2 > 7$

39. $-3x \leq 21$ **40.** $-5x \geq -10$

41. $-2x + 8 > -4$ **42.** $-4x - 1 < 11$

43. $4 - 2(x + 1) \geq -2$ **44.** $6 - 2(x + 3) \leq -8$

A proportion is two equal ratios. That is, if $\dfrac{a}{b}$ and $\dfrac{c}{d}$ are ratios, then:

$$\frac{a}{b} = \frac{c}{d}$$

is a proportion.

Each of the four numbers in a proportion is called a *term* of the proportion. We number the terms as follows:

First term $\longrightarrow \quad a \quad \quad c \leftarrow$ Third term

Second term $\longrightarrow \quad b \quad \quad d \leftarrow$ Fourth term

The first and fourth terms are called the *extremes,* and the second and third terms are called the *means:*

Means $\quad \dfrac{a}{b} = \dfrac{c}{d} \quad$ Extremes

For example, in the proportion:

$$\frac{3}{8} = \frac{12}{32}$$

the extremes are 3 and 32, and the means are 8 and 12.

MEANS-EXTREMES PROPERTY

If a, b, c, and d are real numbers with $b \neq 0$ and $d \neq 0$, then

$$\text{if} \quad \frac{a}{b} = \frac{c}{d}$$

$$\text{then} \quad ad = bc$$

In words: In any proportion, the product of the extremes is equal to the product of the means.

This property of proportions comes from the multiplication property of equality. We can use it to solve for a missing term in a proportion.

EXAMPLE I Solve the proportion $\dfrac{3}{x} = \dfrac{6}{7}$ for x.

Solution We could solve for x by using the method developed in Section 6.4, that is, multiplying both sides by the LCD $7x$. Instead, let's use our new means-extremes property:

$$\frac{3}{x} = \frac{6}{7} \qquad \text{Extremes are 3 and 7;}$$
$$\text{means are } x \text{ and 6.}$$

$$21 = 6x \qquad \text{Product of extremes} =$$
$$\text{product of means}$$

$$\frac{21}{6} = x \qquad \text{Divide both sides by 6.}$$

$$x = \frac{7}{2} \qquad \text{Reduce to lowest terms.}$$

E X A M P L E   2   Solve for x: $\dfrac{x + 1}{2} = \dfrac{3}{x}$

Solution   Again, we want to point out that we could solve for x by using the method we used in Section 6.4. Using the means-extremes property is simply an alternative to the method developed in Section 6.4:

$$\frac{x + 1}{2} = \frac{3}{x} \qquad \text{Extremes are } x + 1 \text{ and } x;$$
$$\text{means are 2 and 3.}$$

$$x^2 + x = 6 \qquad \text{Product of extremes} =$$
$$\text{product of means}$$

$$x^2 + x - 6 = 0 \qquad \text{Standard form for a}$$
$$\text{quadratic equation}$$

$$(x + 3)(x - 2) = 0 \qquad \text{Factor.}$$

$$x + 3 = 0 \qquad \text{or} \qquad x - 2 = 0 \qquad \text{Set factors equal to 0.}$$

$$x = -3 \qquad \text{or} \qquad x = 2$$

This time we have two solutions: -3 and 2.

E X A M P L E   3   A manufacturer knows that during a production run, 8 out of every 100 parts produced by a certain machine will be defective. If the machine produces 1,450 parts, how many can be expected to be defective?

Solution   The ratio of defective parts to total parts produced is $\frac{8}{100}$. If we let x represent the number of defective parts out of the total of 1,450 parts, then we can write this ratio again as $\dfrac{x}{1,450}$. This gives us a proportion to solve:

Defective parts
in numerator
$$\frac{x}{1,450} = \frac{8}{100} \qquad \text{Extremes are } x \text{ and 100;}$$
$$\text{means are 1,450 and 8.}$$

Total parts
in denominator
$$100x = 11,600 \qquad \text{Product of extremes} =$$
$$x = 116 \qquad \text{product of means}$$

The manufacturer can expect 116 defective parts out of the total of 1,450 parts if the machine usually produces 8 defective parts for every 100 parts it produces.

 Getting Ready for Class

After reading through the preceding section, respond in your own words and in complete sentences.

A. What is a proportion?

B. What are the means and extremes of a proportion?

C. What is the relationship between the means and the extremes in a proportion? (It is called the means-extremes property of proportions.)

D. How are ratios and proportions related?

PROBLEM SET 6.7

Solve each of the following proportions.

1. $\dfrac{x}{2} = \dfrac{6}{12}$

2. $\dfrac{x}{4} = \dfrac{6}{8}$

3. $\dfrac{2}{5} = \dfrac{4}{x}$

4. $\dfrac{3}{8} = \dfrac{9}{x}$

5. $\dfrac{10}{20} = \dfrac{20}{x}$

6. $\dfrac{15}{60} = \dfrac{60}{x}$

7. $\dfrac{a}{3} = \dfrac{5}{12}$

8. $\dfrac{a}{2} = \dfrac{7}{20}$

9. $\dfrac{2}{x} = \dfrac{6}{7}$

10. $\dfrac{4}{x} = \dfrac{6}{7}$

11. $\dfrac{x+1}{3} = \dfrac{4}{x}$

12. $\dfrac{x+1}{6} = \dfrac{7}{x}$

13. $\dfrac{x}{2} = \dfrac{8}{x}$

14. $\dfrac{x}{9} = \dfrac{4}{x}$

15. $\dfrac{4}{a+2} = \dfrac{a}{2}$

16. $\dfrac{3}{a+2} = \dfrac{a}{5}$

17. $\dfrac{1}{x} = \dfrac{x-5}{6}$

18. $\dfrac{1}{x} = \dfrac{x-6}{7}$

Applying the Concepts

19. Baseball A baseball player gets 6 hits in the first 18 games of the season. If he continues hitting at the same rate, how many hits will he get in the first 45 games?

20. Basketball A basketball player makes 8 of 12 free throws in the first game of the season. If she shoots with the same accuracy in the second game, how many of the 15 free throws she attempts will she make?

21. Mixture Problem A solution contains 12 milliliters of alcohol and 16 milliliters of water. If another solution is to have the same concentration of alcohol in water but is to contain 28 milliliters of water, how much alcohol must it contain?

22. Mixture Problem A solution contains 15 milliliters of HCl and 42 milliliters of water. If another solution is to have the same concentration of HCl in water but is to contain 140 milliliters of water, how much HCl must it contain?

23. Nutrition If 100 grams of ice cream contain 13 grams of fat, how much fat is in 350 grams of ice cream?

24. Nutrition A 6-ounce serving of grapefruit juice contains 159 grams of water. How many grams of water are in 20 ounces of grapefruit juice?

25. Map Reading A map is drawn so that every 3.5 inches on the map corresponds to an actual distance of 100 miles. If the actual distance between two cities is 420 miles, how far apart are they on the map?

26. Map Reading   The scale on a map indicates that 1 inch on the map corresponds to an actual distance of 105 miles. Two cities are 4.5 inches apart on the map. What is the actual distance between the two cities?

27. Distance   A man drives his car 245 miles in 5 hours. At this rate, how far will he travel in 7 hours?

28. Distance   An airplane flies 1,380 miles in 3 hours. How far will it fly in 5 hours?

Review Problems

The following problems review material we covered in Sections 6.1, 6.2, and 6.3.

Reduce to lowest terms. [6.1]

29. $\dfrac{x^2 - x - 6}{x^2 - 9}$
    30. $\dfrac{xy + 5x + 3y + 15}{x^2 + ax + 3x + 3a}$

Multiply and divide. [6.2]

31. $\dfrac{x^2 - 25}{x + 4} \cdot \dfrac{2x + 8}{x^2 - 9x + 20}$

32. $\dfrac{3x + 6}{x^2 + 4x + 3} \div \dfrac{x^2 + x - 2}{x^2 + 2x - 3}$

Add and subtract. [6.3]

33. $\dfrac{x}{x^2 - 16} + \dfrac{4}{x^2 - 16}$

34. $\dfrac{2}{x^2 - 1} - \dfrac{5}{x^2 + 3x - 4}$

CHAPTER 6   SUMMARY

Examples

1. We can reduce $\frac{6}{8}$ to lowest terms by dividing the numerator and denominator by their greatest common factor 2:

$$\frac{6}{8} = \frac{2 \cdot 3}{2 \cdot 4} = \frac{3}{4}$$

Rational Numbers [6.1]

Any number that can be put in the form $\frac{a}{b}$, where a and b are integers $(b \neq 0)$, is called a rational number.

Multiplying or dividing the numerator and denominator of a rational number by the same nonzero number never changes the value of the rational number.

2. We reduce rational expressions to lowest terms by factoring the numerator and denominator and then dividing out any factors they have in common:

$$\frac{x-3}{x^2-9} = \frac{x-3}{(x-3)(x+3)} = \frac{1}{x+3}$$

Rational Expressions [6.1]

Any expression of the form $\frac{P}{Q}$, where P and Q are polynomials $(Q \neq 0)$, is a rational expression.

Multiplying or dividing the numerator and denominator of a rational expression by the same nonzero quantity always produces a rational expression equivalent to the original one.

3. $\dfrac{x-1}{x^2+2x-3} \cdot \dfrac{x^2-9}{x-2}$

$$= \frac{x-1}{(x+3)(x-1)} \cdot \frac{(x-3)((x+3)}{x-2}$$

$$= \frac{x-3}{x-2}$$

Multiplication [6.2]

To multiply two rational numbers or two rational expressions, multiply numerators, multiply denominators, and divide out any factors common to the numerator and denominator:

For rational numbers $\dfrac{a}{b}$ and $\dfrac{c}{d}$, $\dfrac{a}{b} \cdot \dfrac{c}{d} = \dfrac{ac}{bd}$

For rational expressions $\dfrac{P}{Q}$ and $\dfrac{R}{S}$, $\dfrac{P}{Q} \cdot \dfrac{R}{S} = \dfrac{PR}{QS}$

4. $\dfrac{2x}{x^2-25} \div \dfrac{4}{x-5}$

$$= \frac{2x}{(x-5)(x+5)} \cdot \frac{(x-5)}{4}$$

$$= \frac{x}{2(x+5)}$$

Division [6.2]

To divide by a rational number or rational expression, simply multiply by its reciprocal:

For rational numbers $\dfrac{a}{b}$ and $\dfrac{c}{d}$, $\dfrac{a}{b} \div \dfrac{c}{d} = \dfrac{a}{b} \cdot \dfrac{d}{c}$

For rational expressions $\dfrac{P}{Q}$ and $\dfrac{R}{S}$, $\dfrac{P}{Q} \div \dfrac{R}{S} = \dfrac{P}{Q} \cdot \dfrac{S}{R}$

5. $\dfrac{3}{x-1} + \dfrac{x}{2}$

$= \dfrac{3}{x-1} \cdot \dfrac{2}{2} + \dfrac{x}{2} \cdot \dfrac{x-1}{x-1}$

$= \dfrac{6}{2(x-1)} + \dfrac{x^2-x}{2(x-1)}$

$= \dfrac{x^2-x+6}{2(x-1)}$

Addition [6.3]

To add two rational numbers or rational expressions, find a common denominator, change each expression to an equivalent expression having the common denominator, then add numerators, and reduce if possible:

For rational numbers $\dfrac{a}{c}$ and $\dfrac{b}{c}$, $\dfrac{a}{c} + \dfrac{b}{c} = \dfrac{a+b}{c}$

For rational expressions $\dfrac{P}{S}$ and $\dfrac{Q}{S}$, $\dfrac{P}{S} + \dfrac{Q}{S} = \dfrac{P+Q}{S}$

6. $\dfrac{x}{x^2-4} - \dfrac{2}{x^2-4}$

$= \dfrac{x-2}{x^2-4}$

$= \dfrac{x-2}{(x-2)(x+2)}$

$= \dfrac{1}{x+2}$

Subtraction [6.3]

To subtract a rational number or rational expression, simply add its opposite:

For rational numbers $\dfrac{a}{c}$ and $\dfrac{b}{c}$, $\dfrac{a}{c} - \dfrac{b}{c} = \dfrac{a}{c} + \left(\dfrac{-b}{c} \right)$

For rational expressions $\dfrac{P}{S}$ and $\dfrac{Q}{S}$, $\dfrac{P}{S} - \dfrac{Q}{S} = \dfrac{P}{S} + \left(\dfrac{-Q}{S} \right)$

7. Solve $\dfrac{1}{2} + \dfrac{3}{x} = 5$.

$2x\left(\dfrac{1}{2}\right) + 2x\left(\dfrac{3}{x}\right) = 2x(5)$

$x + 6 = 10x$

$6 = 9x$

$x = \dfrac{2}{3}$

Equations [6.4]

To solve equations involving rational expressions, first find the least common denominator (LCD) for all denominators. Then multiply both sides by the LCD and solve as usual. Check all solutions in the original equation to be sure there are no undefined terms.

8. $\dfrac{1 - \dfrac{4}{x}}{x - \dfrac{16}{x}} = \dfrac{x\left(1 - \dfrac{4}{x}\right)}{x\left(x - \dfrac{16}{x}\right)}$

$= \dfrac{x-4}{x^2-16}$

$= \dfrac{x-4}{(x-4)(x+4)}$

$= \dfrac{1}{x+4}$

Complex Fractions [6.6]

A rational expression that contains a fraction in its numerator or denominator is called a complex fraction. The most common method of simplifying a complex fraction is to multiply the top and bottom by the LCD for all denominators.

9. Solve for x: $\dfrac{3}{x} = \dfrac{5}{20}$

$$3 \cdot 20 = 5 \cdot x$$

$$60 = 5x$$

$$x = 12$$

Ratio and Proportion [6.1, 6.7]

The ratio of a to b is:

$$\frac{a}{b}$$

Two equal ratios form a proportion. In the proportion

$$\frac{a}{b} = \frac{c}{d}$$

a and d are the *extremes,* and b and c are the *means.* In any proportion the product of the extremes is equal to the product of the means.

C H A P T E R 6 **REVIEW**

The numbers in brackets refer to the sections of the text in which similar problems can be found.

Reduce to lowest terms. Also specify any restriction on the variable. [6.1]

1. $\dfrac{7}{14x - 28}$

2. $\dfrac{a + 6}{a^2 - 36}$

3. $\dfrac{8x - 4}{4x + 12}$

4. $\dfrac{x + 4}{x^2 + 8x + 16}$

5. $\dfrac{3x^3 + 16x^2 - 12x}{2x^3 + 9x^2 - 18x}$

6. $\dfrac{x + 2}{x^4 - 16}$

7. $\dfrac{x^2 + 5x - 14}{x + 7}$

8. $\dfrac{a^2 + 16a + 64}{a + 8}$

9. $\dfrac{xy + bx + ay + ab}{xy + 5x + ay + 5a}$

Multiply or divide as indicated. [6.2]

10. $\dfrac{3x + 9}{x^2} \cdot \dfrac{x^3}{6x + 18}$

11. $\dfrac{x^2 + 8x + 16}{x^2 + x - 12} \div \dfrac{x^2 - 16}{x^2 - x - 6}$

12. $(a^2 - 4a - 12)\left(\dfrac{a - 6}{a + 2}\right)$

13. $\dfrac{3x^2 - 2x - 1}{x^2 + 6x + 8} \div \dfrac{3x^2 + 13x + 4}{x^2 + 8x + 16}$

Find the following sums and differences. [6.3]

14. $\dfrac{2x}{2x + 3} + \dfrac{3}{2x + 3}$

15. $\dfrac{x^2}{x - 9} - \dfrac{18x - 81}{x - 9}$

16. $\dfrac{a + 4}{a + 8} - \dfrac{a - 9}{a + 8}$

17. $\dfrac{x}{x + 9} + \dfrac{5}{x}$

18. $\dfrac{5}{4x + 20} + \dfrac{x}{x + 5}$

19. $\dfrac{3}{x^2 - 36} - \dfrac{2}{x^2 - 4x - 12}$

20. $\dfrac{3a}{a^2 + 8a + 15} - \dfrac{2}{a + 5}$

Solve each equation. [6.4]

21. $\dfrac{3}{x} + \dfrac{1}{2} = \dfrac{5}{x}$

22. $\dfrac{a}{a - 3} = \dfrac{3}{2}$

23. $1 - \dfrac{7}{x} = \dfrac{-6}{x^2}$

24. $\dfrac{3}{x + 6} - \dfrac{1}{x - 2} = \dfrac{-8}{x^2 + 4x - 12}$

25. $\dfrac{2}{y^2 - 16} = \dfrac{10}{y^2 + 4y}$

26. **Number Problem**   The sum of a number and 7 times its reciprocal is $\frac{16}{3}$. Find the number. [6.5]

27. **Distance, Rate, and Time**   A boat travels 48 miles up a river in the same amount of time it takes to travel 72 miles down the same river. If the current is 3 miles per hour, what is the speed of the boat in still water? [6.5]

28. **Filling a Pool**   An inlet pipe can fill a pool in 21 hours, while an outlet pipe can empty it in 28 hours. If both pipes are left open, how long will it take to fill the pool? [6.5]

Simplify each complex fraction. [6.6]

29. $\dfrac{\dfrac{x+4}{x^2-16}}{\dfrac{2}{x-4}}$

30. $\dfrac{1-\dfrac{9}{y^2}}{1+\dfrac{4}{y}-\dfrac{21}{y^2}}$

31. $\dfrac{\dfrac{1}{a-2}+4}{\dfrac{1}{a-2}+1}$

32. Write the ratio of 40 to 100 as a fraction in lowest terms. [6.7]

33. If there are 60 seconds in 1 minute, what is the ratio of 40 seconds to 3 minutes? [6.7]

Solve each proportion. [6.7]

34. $\dfrac{x}{9}=\dfrac{4}{3}$

35. $\dfrac{a}{3}=\dfrac{12}{a}$

36. $\dfrac{8}{x-2}=\dfrac{x}{6}$

CHAPTER 6   **PROJECTS**

RATIONAL EXPRESSIONS

GROUP PROJECT

Number of People: 3

Time Needed: 10–15 minutes

Equipment: Pencil and paper

Procedure: The following four problems all involve the same rational expressions. Many times, students who have worked problems successfully on their homework have trouble when they take a test on rational expressions because the problems are mixed up and do not have similar instructions. Noticing similarities and differences among the types of problems involving rational expressions can help with this situation.

Problem 1: Add: $\dfrac{-2}{x^2-2x-3}+\dfrac{3}{x^2-9}$

Problem 2: Divide: $\dfrac{-2}{x^2-2x-3}\div\dfrac{3}{x^2-9}$

Problem 3: Solve: $\dfrac{-2}{x^2-2x-3}+\dfrac{3}{x^2-9}=-1$

Problem 4: Simplify:
$$\frac{\dfrac{-2}{x^2 - 2x - 3}}{\dfrac{3}{x^2 - 9}}$$

1. Which of the previous problems do not require the use of a least common denominator?
2. Which two problems may involve multiplying by the least common denominator?
3. Which of the problems will have an answer that is one or two numbers, but no variables?
4. Work each of the four problems.

RESEARCH PROJECT

FERRIS WHEELS AND THE THIRD MAN

(Hulton Getty/Tony Stone Images)

Among the large Ferris wheels built around the turn of the century was one built in Vienna in 1897. It is the only one of those large wheels that is still in operation today. Known as the Riesenrad, it has a diameter of 197 feet and can carry a total of 800 people. A brochure that gives some statistics associated with the Riesenrad indicates that passengers riding it travel at 2 feet 6 inches per second. You can check the accuracy of this number by watching the movie *The Third*

(Selznick Films/The Kobal Collection)

Man. In the movie, Orson Welles rides the Riesenrad through one complete revolution. Play *The Third Man* on a VCR, so you can view the Riesenrad in operation. Use the pause button and the timer on the VCR to time how long it takes Orson Welles to ride once around the wheel. Then calculate his average speed during the ride. Use your results to either prove or disprove the claim that passengers travel at 2 feet 6 inches per second on the Riesenrad. When you have finished, write your procedures and results in essay form.

CHAPTER 6   TEST

Reduce to lowest terms. [6.1]

1. $\dfrac{x^2 - 16}{x^2 - 8x + 16}$

2. $\dfrac{10a + 20}{5a^2 + 20a + 20}$

3. $\dfrac{xy + 7x + 5y + 35}{x^2 + ax + 5x + 5a}$

Multiply or divide as indicated. [6.2]

4. $\dfrac{3x - 12}{4} \cdot \dfrac{8}{2x - 8}$

5. $\dfrac{x^2 - 49}{x + 1} \div \dfrac{x + 7}{x^2 - 1}$

6. $\dfrac{x^2 - 3x - 10}{x^2 - 8x + 15} \div \dfrac{3x^2 + 2x - 8}{x^2 + x - 12}$

7. $(x^2 - 9)\left(\dfrac{x + 2}{x + 3}\right)$

Add or subtract as indicated. [6.3]

8. $\dfrac{3}{x - 2} - \dfrac{6}{x - 2}$

9. $\dfrac{x}{x^2 - 9} + \dfrac{4}{4x - 12}$

10. $\dfrac{2x}{x^2 - 1} + \dfrac{x}{x^2 - 3x + 2}$

Solve the following equations. [6.4]

11. $\dfrac{7}{5} = \dfrac{x + 2}{3}$

12. $\dfrac{10}{x + 4} = \dfrac{6}{x} - \dfrac{4}{x}$

13. $\dfrac{3}{x - 2} - \dfrac{4}{x + 1} = \dfrac{5}{x^2 - x - 2}$

Solve the following problems. [6.5]

14. Speed of a Boat   A boat travels 26 miles up a river in the same amount of time it takes to travel 34 miles down the same river. If the current is 2 miles per hour, what is the speed of the boat in still water?

15. Emptying a Pool   An inlet pipe can fill a pool in 15 hours, and an outlet pipe can empty it in 12 hours. If the pool is full and both pipes are open, how long will it take to empty?

Solve the following problems involving ratio and proportion. [6.7]

16. Ratio   A solution of alcohol and water contains 27 milliliters of alcohol and 54 milliliters of water. What is the ratio of alcohol to water and the ratio of alcohol to total volume?

17. Proportion   A manufacturer knows that during a production run 8 out of every 100 parts produced by a certain machine will be defective. If the machine produces 1,650 parts, how many can be expected to be defective?

Simplify each complex fraction. [6.6]

18. $\dfrac{1 + \dfrac{1}{x}}{1 - \dfrac{1}{x}}$

19. $\dfrac{1 - \dfrac{16}{x^2}}{1 - \dfrac{2}{x} - \dfrac{8}{x^2}}$

CHAPTERS 1–6   CUMULATIVE REVIEW

Simplify.

1. $8 - 11$

2. $-20 + 14$

3. $\dfrac{-48}{12}$

4. $\dfrac{1}{6}(-18)$

5. $5x - 4 - 9x$

6. $8 - x - 4$

7. 9^{-2}

8. $\left(\dfrac{x^5}{x^3}\right)^{-2}$

9. $4^1 + 9^0 + (-7)^0$

10. $\dfrac{(x^{-4})^{-3}(x^{-3})^4}{x^0}$

11. $(4a^3 - 10a^2 + 6) - (6a^3 + 5a - 7)$

12. $(a^2 + 7)(a^2 - 7)$

13. $\dfrac{x^2}{x - 7} - \dfrac{14x - 49}{x - 7}$

14. $\dfrac{6}{10x + 30} + \dfrac{x}{x + 3}$

15. $\dfrac{x - 2}{\dfrac{x^2 + 6x + 8}{4}}{\dfrac{}{x + 4}}$

Solve each equation.

16. $\dfrac{3}{4}(8x - 12) = \dfrac{1}{2}(4x + 4)$

17. $x - \dfrac{3}{4} = \dfrac{5}{6}$

18. $4x - 3 = 8x + 5$

19. $98r^2 - 18 = 0$

20. $6x^4 = 33x^3 - 42x^2$

21. $\dfrac{5}{x} - \dfrac{1}{3} = \dfrac{3}{x}$

22. $\dfrac{4}{x - 5} - \dfrac{3}{x + 2} = \dfrac{28}{x^2 - 3x - 10}$

23. $\dfrac{x}{3} = \dfrac{6}{x - 3}$

Solve each system.

24. $x + 2y = 1$
$x - y = 4$

25. $9x + 14y = -4$
$15x - 8y = 9$

26. $x = 4y - 3$
$2x - 5y = 9$

27. $\dfrac{1}{2}x + \dfrac{1}{3}y = -1$
$\dfrac{1}{3}x = \dfrac{1}{4}y + 5$

Graph each equation on a rectangular coordinate system.

28. $y = -3x + 2$

29. $y = -\dfrac{3}{2}x$

30. Solve the inequality $-\dfrac{a}{3} \le -2$.

Factor completely.

31. $xy + 5x + ay + 5a$

32. $a^2 + 2a - 35$

33. $20y^2 - 27y + 9$

34. $4r^2 - 9t^2$

35. $16x^2 + 72xy + 81y^2$

36. Solve the following system by graphing.
$2x + y = 4$
$3x - 2y = 6$

37. Which of the ordered pairs $(3, -1)$, $(1, -3)$, $(-2, 9)$ are solutions to the equation $y = 2x - 5$?

38. For the equation $2x + 5y = 10$, find the x- and y-intercepts.

39. What is 33% of 220?

40. Write 0.00135 in scientific notation.

41. Find the next number in the sequence $\dfrac{1}{2}, -\dfrac{1}{4}, \dfrac{1}{8}, \ldots$

42. Add -2 to the product of -3 and 4.

43. What property justifies the statement $(2 + x) + 3 = 2 + (x + 3)$?

44. For the set $\{-3, -\sqrt{2}, 0, \dfrac{3}{4}, 1.5, \pi\}$ list all the rational numbers.

45. Multiply $\dfrac{6x - 12}{6x + 12} \cdot \dfrac{3x + 3}{12x - 24}$.

46. Divide $\dfrac{x^2 - 3x - 28}{x + 4}$.

47. Reduce to lowest terms $\dfrac{2xy + 10x + 3y + 15}{3xy + 15x + 2y + 10}$.

48. Solve $P = 2l + 2w$ for l.

49. Number Problem   The sum of two numbers is 40. The difference of the two numbers is 18. Find the numbers.

50. Investing   Ms. Jones invested \$18,000 in two accounts. One account pays 6% simple interest, and the other pays 8%. Her total interest for the year was \$1,290. How much did she have in each account?

Transitions

INTRODUCTION

Suppose you decide to buy a cellular phone and are trying to decide between two rate plans. Plan A is $18.95 per month plus $.48 for each minute, or

(Brandtner & Staede/Tony Stone Images)

fraction of a minute, that you use the phone. Plan B is $34.95 per month plus $.36 for each minute, or fraction of a minute. The monthly cost C for each plan can be represented with a linear equation in two variables:

$$\text{Plan A:} \quad C = 0.48x + 18.95$$

$$\text{Plan B:} \quad C = 0.36x + 34.95$$

To compare the two plans, we use the table and graph shown below.

TABLE 1	Monthly Cellular Phone Charges	
Number of Minutes x	**Monthly Cost**	
	Plan A (dollars)	**Plan B** (dollars)
0	18.95	34.95
40	38.15	49.35
80	57.35	63.75
120	76.55	78.15
160	95.75	92.55
200	114.95	106.95
240	134.15	121.35

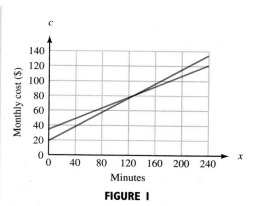

FIGURE 1

The point of intersection of the two lines in Figure 1 is the point at which the monthly costs of the two plans are equal. Among the topics in this chapter is a review of the methods of finding that point of intersection.

This chapter marks the transition from introductory algebra to intermediate algebra. Some of the material here is a review of material we covered earlier, and some of the material is new. If you cover all of the sections in this chapter, you will review all of the important points contained in the first six chapters of the book. So, it is a good idea to put some extra time and effort into this chapter to ensure that you get a good start with the rest of the course. Let's begin by reviewing the methods we use to solve equations.

A **linear equation in one variable** is any equation that can be put in the form

$$ax + b = c$$

where a, b, and c are constants and $a \neq 0$. For example, each of the equations

$$5x + 3 = 2 \qquad 2x = 7 \qquad 2x + 5 = 0$$

are linear because they can be put in the form $ax + b = c$. In the first equation, $5x$, 3, and 2 are called **terms** of the equation—$5x$ is a variable term; 3 and 2 are constant terms.

> **DEFINITION** The **solution set** for an equation is the set of all numbers that, when used in place of the variable, make the equation a true statement.

> **DEFINITION** Two or more equations with the same solution set are called **equivalent equations**.

EXAMPLE 1 The equations $2x - 5 = 9$, $x - 1 = 6$, and $x = 7$ are all equivalent equations since the solution set for each is $\{7\}$.

PROPERTIES OF EQUALITY

The first property states that adding the same quantity to both sides of an equation preserves equality. Or, more importantly, adding the same amount to both sides of an equation *never changes* the solution set. This property is called the **addition property of equality** and is stated in symbols as follows.

> **ADDITION PROPERTY OF EQUALITY**
>
> For any three algebraic expressions A, B, and C,
>
> $$\text{if} \qquad A = B$$
> $$\text{then} \qquad A + C = B + C$$
>
> *In words:* Adding the same quantity to both sides of an equation will not change the solution.

Our second new property is called the **multiplication property of equality** and is stated like this.

MULTIPLICATION PROPERTY OF EQUALITY

For any three algebraic expressions A, B, and C, where $C \neq 0$,

$$\text{if} \qquad A = B$$

$$\text{then} \qquad AC = BC$$

In words: Multiplying both sides of an equation by the same nonzero quantity will not change the solution.

Note: Since subtraction is defined in terms of addition, and division is defined in terms of multiplication, we do not need to introduce separate properties for subtraction and division. The solution set for an equation will never be changed by subtracting the same amount from both sides or by dividing both sides by the same nonzero quantity.

EXAMPLE 2   Find the solution set for $3a - 5 = -6a + 1$.

Solution   To solve for a we must isolate it on one side of the equation. Let's decide to isolate a on the left side by adding $6a$ to both sides of the equation.

$$3a - 5 = -6a + 1$$

$$3a + \mathbf{6a} - 5 = -6a + \mathbf{6a} + 1 \qquad \text{Add } \mathbf{6a} \text{ to both sides.}$$

$$9a - 5 = 1$$

$$9a - 5 + \mathbf{5} = 1 + \mathbf{5} \qquad \text{Add } \mathbf{5} \text{ to both sides.}$$

$$9a = 6$$

$$\frac{\mathbf{1}}{\mathbf{9}}(9a) = \frac{\mathbf{1}}{\mathbf{9}}(6) \qquad \text{Multiply both sides by } \frac{\mathbf{1}}{\mathbf{9}}.$$

$$a = \frac{2}{3} \qquad\qquad \frac{1}{9}(6) = \frac{6}{9} = \frac{2}{3}$$

Note: From Chapter 1 we know that multiplication by a number and division by its reciprocal always produce the same result. Because of this fact, instead of multiplying each side of our equation by $\frac{1}{9}$, we could just as easily divide each side by 9. If we did so, the last two lines in our solution would look like this:

$$\frac{9a}{9} = \frac{6}{9}$$

$$a = \frac{2}{3}$$

There will be times when we solve equations and end up with a negative sign in front of the variable. The next example shows how to handle this situation.

EXAMPLE 3 Solve each equation.

(a) $-x = 4$ **(b)** $-y = -8$

Solution Neither equation can be considered solved because of the negative sign in front of the variable. To eliminate the negative signs we simply multiply each side of the equations by -1.

(a) $\qquad -x = 4$ **(b)** $\qquad -y = -8$

$\qquad -1(-x) = -1(4) \qquad\qquad -1(-y) = -1(-8)$ Multiply each side

$\qquad\qquad x = -4 \qquad\qquad\qquad\qquad y = 8$ by -1.

The next example involves fractions. The least common denominator, which is the smallest expression that is divisible by each of the denominators, can be used with the multiplication property of equality to simplify equations containing fractions.

EXAMPLE 4 Solve $\dfrac{2}{3}x + \dfrac{1}{2} = -\dfrac{3}{8}$.

Solution We can solve this equation by applying our properties and working with fractions, or we can begin by eliminating the fractions. Let's use both methods.

Method 1 Working with the fractions.

$$\frac{2}{3}x + \frac{1}{2} + \left(-\frac{1}{2}\right) = -\frac{3}{8} + \left(-\frac{1}{2}\right) \qquad \text{Add } -\frac{1}{2} \text{ to each side.}$$

$$\frac{2}{3}x = -\frac{7}{8} \qquad\qquad -\frac{3}{8} + \left(-\frac{1}{2}\right) = -\frac{3}{8} + \left(-\frac{4}{8}\right)$$

$$\frac{3}{2}\left(\frac{2}{3}x\right) = \frac{3}{2}\left(-\frac{7}{8}\right) \qquad \text{Multiply each side by } \frac{3}{2}.$$

$$x = -\frac{21}{16}$$

Method 2 Eliminating the fractions in the beginning.

Our original equation has denominators of 3, 2, and 8. The least common denominator, abbreviated LCD, for these three denominators is 24, and it has the property that all three denominators will divide it evenly. Therefore, if we multiply both sides of our equation by 24, each denominator will divide into 24, and we will be left with an equation that does not contain any denominators other than 1.

$$24\left(\frac{2}{3}x + \frac{1}{2}\right) = 24\left(-\frac{3}{8}\right)$$ Multiply each side by the LCD **24.**

$$24\left(\frac{2}{3}x\right) + 24\left(\frac{1}{2}\right) = 24\left(-\frac{3}{8}\right)$$ Distributive property on the left side

$$16x + 12 = -9$$ Multiply.

$$16x = -21$$ Add -12 to each side.

$$x = -\frac{21}{16}$$ Multiply each side by $\frac{1}{16}$.

As the third line above indicated, multiplying each side of the equation by the LCD eliminates all the fractions from the equation. Both methods yield the same solution. To check our solution, we substitute $x = -21/16$ back into our original equation to obtain

$$\frac{2}{3}\left(-\frac{21}{16}\right) + \frac{1}{2} \overset{?}{=} -\frac{3}{8}$$

$$-\frac{7}{8} + \frac{1}{2} \overset{?}{=} -\frac{3}{8}$$

$$-\frac{7}{8} + \frac{4}{8} \overset{?}{=} -\frac{3}{8}$$

$$-\frac{3}{8} = -\frac{3}{8}$$

As we can see, our solution checks.

Note: We are placing question marks over the equal signs because we don't know yet if the expressions on the left will be equal to the expressions on the right.

EXAMPLE 5   Solve the equation $0.06x + 0.05(10,000 - x) = 560$.

Solution   We can solve the equation in its original form by working with the decimals, or we can eliminate the decimals first by using the multiplication property of equality and solve the resulting equation. Here are both methods.

Method 1   Working with the decimals.

$$0.06x + 0.05(10,000 - x) = 560$$ Original equation

$$0.06x + 0.05(10,000) - 0.05x = 560$$ Distributive property

$$0.01x + 500 = 560$$ Simplify the left side.

$$0.01x + 500 + (-500) = 560 + (-500)$$ Add -500 to each side.

$$0.01x = 60$$

$$\frac{0.01x}{0.01} = \frac{60}{0.01}$$ Divide each side by **0.01.**

$$x = 6,000$$

Method 2 Eliminating the decimals in the beginning.

To move the decimal point two places to the right in $0.06x$ and 0.05, we multiply each side of the equation by 100.

$0.06x + 0.05(10,000 - x) = 560$	Original equation
$0.06x + 500 - 0.05x = 560$	Distributive property
$\mathbf{100}(0.06x) + \mathbf{100}(500) - \mathbf{100}(0.05x) = \mathbf{100}(560)$	Multiply each side by **100**.
$6x + 50,000 - 5x = 56,000$	Multiply.
$x + 50,000 = 56,000$	Simplify the left side.
$x = 6,000$	Add $-50,000$ to each side.

Using either method, the solution to our equation is 6,000. We check our work (to be sure we have not made a mistake in applying the properties or an arithmetic mistake) by substituting 6,000 into our original equation and simplifying each side of the result separately.

Check: Substituting 6,000 for x in the original equation, we have

$$0.06(6,000) + 0.05(10,000 - 6,000) \stackrel{?}{=} 560$$

$$0.06(6,000) + 0.05(4,000) \stackrel{?}{=} 560$$

$$360 + 200 \stackrel{?}{=} 560$$

$$560 = 560 \qquad \text{A true statement}$$

Here is a list of steps to use as a guideline for solving linear equations in one variable.

STRATEGY FOR SOLVING LINEAR EQUATIONS IN ONE VARIABLE

Step 1a: Use the distributive property to separate terms, if necessary.

    1b: If fractions are present, consider multiplying both sides by the LCD to eliminate the fractions. If decimals are present, consider multiplying both sides by a power of 10 to clear the equation of decimals.

    1c: Combine similar terms on each side of the equation.

Step 2: Use the addition property of equality to get all variable terms on one side of the equation and all constant terms on the other side. A variable term is a term that contains the variable. A constant term is a term that does not contain the variable (the number 3, for example).

Step 3: Use the multiplication property of equality to get x by itself on one side of the equation.

Step 4: Check your solution in the original equation to be sure that you have not made a mistake in the solution process.

As you will see as you work through the problems in the problem set, it is not always necessary to use all four steps when solving equations. The number of steps used depends on the equation. In Example 6 there are no fractions or decimals in the original equation, so step 1b will not be used.

E X A M P L E   6   Solve the equation $8 - 3(4x - 2) + 5x = 35$.

Solution   We must begin distributing the -3 across the quantity $4x - 2$. (It would be a mistake to subtract 3 from 8 first, since the rule for order of operations indicates we are to do multiplication before subtraction.) After we have simplified the left side of our equation, we apply the addition property and the multiplication property. In this example, we will show only the result:

Step 1a: $\begin{cases} 8 - 3(4x - 2) + 5x = 35 \qquad \text{Original equation} \\ \quad\;\; \downarrow \quad\;\; \downarrow \\ 8 - 12x + 6 + 5x = 35 \qquad \text{Distributive property} \end{cases}$

Step 1c: $\qquad\qquad -7x + 14 = 35 \qquad$ Simplify.

Step 2: $\qquad\qquad\qquad\; -7x = 21 \qquad$ Add -14 to each side.

Step 3: $\qquad\qquad\qquad\quad x = -3 \qquad$ Multiply by $-\dfrac{1}{7}$.

Step 4: When x is replaced by -3 in the original equation, a true statement results. Therefore, -3 is the solution to our equation.

IDENTITIES AND EQUATIONS WITH NO SOLUTION

There are two special cases associated with solving linear equations in one variable, which are illustrated in the following examples.

E X A M P L E   7   Solve for x: $2(3x - 4) = 3 + 6x$

Solution   Applying the distributive property to the left side gives us

$$6x - 8 = 3 + 6x \qquad \text{Distributive property}$$

Now, if we add $-6x$ to each side, we are left with

$$-8 = 3$$

which is a false statement. This means that there is no solution to our equation. Any number we substitute for x in the original equation will lead to a similar false statement.

E X A M P L E   8   Solve for x: $-15 + 3x = 3(x - 5)$

Solution   We start by applying the distributive property to the right side.

$$-15 + 3x = 3x - 15 \qquad \text{Distributive property}$$

If we add $-3x$ to each side, we are left with the true statement

$$-15 = -15$$

In this case, our result tells us that any number we use in place of x in the original equation will lead to a true statement. Therefore, all real numbers are solutions to our equation. We say the original equation is an **identity,** because the left side is always identically equal to the right side.

SOLVING EQUATIONS BY FACTORING

Next we will use our knowledge of factoring to solve equations. Most of the equations we will see are *quadratic equations*. Here is the definition of a quadratic equation.

> **DEFINITION**   Any equation that can be written in the form
>
> $$ax^2 + bx + c = 0$$
>
> where $a, b,$ and c are constants and a is not 0 ($a \neq 0$), is called a *quadratic equation*. The form $ax^2 + bx + c = 0$ is called *standard form* for quadratic equations.

Each of the following is a quadratic equation:

$$2x^2 = 5x + 3 \qquad 5x^2 = 75 \qquad 4x^2 - 3x + 2 = 0$$

Note: The third equation is clearly a quadratic equation since it is in standard form. (Notice that a is 4, b is -3, and c is 2.) The first two equations are also quadratic because they could be put in the form $ax^2 + bx + c = 0$ by using the addition property of equality.

Notation For a quadratic equation written in standard form, the first term ax^2 is called the *quadratic term;* the second term bx is the *linear term;* and the last term c is called the *constant term.*

In the past we have noticed that the number 0 is a special number. There is another property of 0 that is the key to solving quadratic equations. It is called the *zero-factor property.*

> **ZERO-FACTOR PROPERTY**
>
> For all real numbers r and s,
>
> $$r \cdot s = 0 \qquad \text{if and only if} \qquad r = 0 \qquad \text{or} \qquad s = 0 \quad \text{(or both)}$$

EXAMPLE 9   Solve $x^2 - 2x - 24 = 0$.

Solution   We begin by factoring the left side as $(x - 6)(x + 4)$ and get

$$(x - 6)(x + 4) = 0$$

Now both $(x - 6)$ and $(x + 4)$ represent real numbers. We notice that their product is 0. By the zero-factor property, one or both of them must be 0:

$$x - 6 = 0 \qquad \text{or} \qquad x + 4 = 0$$

We have used factoring and the zero-factor property to rewrite our original second-degree equation as two first-degree equations connected by the word *or*. Completing the solution, we solve the two first-degree equations:

$$x - 6 = 0 \qquad \text{or} \qquad x + 4 = 0$$
$$x = 6 \qquad \text{or} \qquad x = -4$$

We check our solutions in the original equation as follows:

Check $x = 6$	Check $x = -4$
$6^2 - 2(6) - 24 \stackrel{?}{=} 0$	$(-4)^2 - 2(-4) - 24 \stackrel{?}{=} 0$
$36 - 12 - 24 \stackrel{?}{=} 0$	$16 + 8 - 24 \stackrel{?}{=} 0$
$0 = 0$	$0 = 0$

In both cases the result is a true statement, which means that both 6 and -4 are solutions to the original equation.

Although the next equation is not quadratic, the method we use is similar.

E X A M P L E   1 0   Solve $\frac{1}{3}x^3 = \frac{5}{6}x^2 + \frac{1}{2}x$.

Solution   We can simplify our work if we clear the equation of fractions. Multiplying both sides by the LCD, 6, we have

$$6 \cdot \tfrac{1}{3}x^3 = 6 \cdot \tfrac{5}{6}x^2 + 6 \cdot \tfrac{1}{2}x$$
$$2x^3 = 5x^2 + 3x$$

Next we add $-5x^2$ and $-3x$ to each side so that the right side will become 0.

$$2x^3 - 5x^2 - 3x = 0 \qquad \text{Standard form}$$

We factor the left side and then use the zero-factor property to set each factor to 0.

$$x(2x^2 - 5x - 3) = 0 \qquad \text{Factor out the greatest common factor.}$$

$$x(2x + 1)(x - 3) = 0 \qquad \text{Continue factoring.}$$

$$x = 0 \quad \text{or} \quad 2x + 1 = 0 \quad \text{or} \quad x - 3 = 0 \qquad \text{Zero-factor property}$$

Solving each of the resulting equations, we have

$$x = 0 \quad \text{or} \quad x = -\tfrac{1}{2} \quad \text{or} \quad x = 3$$

To generalize the preceding examples, here are the steps used in solving a quadratic equation by factoring.

TO SOLVE AN EQUATION BY FACTORING

Step 1: Write the equation in standard form.
Step 2: Factor the left side.
Step 3: Use the zero-factor property to set each factor equal to 0.
Step 4: Solve the resulting linear equations.
Step 5: Check the solutions in the original equation.

EXAMPLE 11   Solve $100x^2 = 300x$.

Solution   We begin by writing the equation in standard form and factoring:

$$100x^2 = 300x$$
$$100x^2 - 300x = 0 \qquad \text{Standard form}$$
$$100x(x - 3) = 0 \qquad \text{Factor.}$$

Using the zero-factor property to set each factor to 0, we have

$$100x = 0 \qquad \text{or} \qquad x - 3 = 0$$
$$x = 0 \qquad \text{or} \qquad x = 3$$

The two solutions are 0 and 3.

EXAMPLE 12   Solve $(x - 2)(x + 1) = 4$.

Solution   We begin by multiplying the two factors on the left side. (Notice that it would be incorrect to set each of the factors on the left side equal to 4. The fact that the product is 4 does not imply that either of the factors must be 4.)

$$(x - 2)(x + 1) = 4$$
$$x^2 - x - 2 = 4 \qquad \text{Multiply the left side.}$$
$$x^2 - x - 6 = 0 \qquad \text{Standard form}$$
$$(x - 3)(x + 2) = 0 \qquad \text{Factor.}$$
$$x - 3 = 0 \qquad \text{or} \qquad x + 2 = 0 \qquad \text{Zero-factor property}$$
$$x = 3 \qquad \text{or} \qquad x = -2$$

EXAMPLE 13   Solve for x: $x^3 + 2x^2 - 9x - 18 = 0$

Solution   We start with factoring by grouping.

$$x^3 + 2x^2 - 9x - 18 = 0$$

$$x^2(x + 2) - 9(x + 2) = 0$$
$$(x + 2)(x^2 - 9) = 0$$
$$(x + 2)(x - 3)(x + 3) = 0 \qquad \text{The difference of two squares}$$

$x + 2 = 0 \qquad$ or $\qquad x - 3 = 0 \qquad$ or $\qquad x + 3 = 0 \qquad$ Set factors to 0.

$\qquad x = -2 \qquad$ or $\qquad x = 3 \qquad$ or $\qquad x = -3$

We have three solutions: -2, 3, and -3.

APPLICATIONS

The method, or strategy, that we use to solve application problems is called the *Blueprint for Problem Solving.* It is an outline that will overlay the solution process we use on all application problems.

Blueprint for Problem Solving

Step 1: Read the problem, and then mentally **list** the items that are known and the items that are unknown.

Step 2: Assign a variable to one of the unknown items. (In most cases this will amount to letting $x =$ the item that is asked for in the problem.) Then **translate** the other **information** in the problem to expressions involving the variable.

Step 3: Reread the problem, and then **write an equation,** using the items and variables listed in steps 1 and 2, that describes the situation.

Step 4: Solve the equation found in step 3.

Step 5: Write your answer using a complete sentence.

Step 6: Reread the problem, and **check** your solution with the original words in the problem.

E X A M P L E   1 4   The sum of the squares of two consecutive integers is 25. Find the two integers.

Solution   We apply the Blueprint for Problem Solving to solve this application problem. Remember, step 1 in the blueprint is done mentally.

Step 1: Read and list.
   Known items: Two consecutive integers. If we add their squares, the result is 25.
   Unknown items: The two integers

Step 2: Assign a variable and translate information.
   Let $x =$ the first integer; then $x + 1 =$ the next consecutive integer.

Step 3: *Reread and write an equation.*

Since the sum of the squares of the two integers is 25, the equation that describes the situation is

$$x^2 + (x + 1)^2 = 25$$

Step 4: *Solve the equation.*

$$x^2 + (x + 1)^2 = 25$$
$$x^2 + (x^2 + 2x + 1) = 25$$
$$2x^2 + 2x - 24 = 0$$
$$x^2 + x - 12 = 0$$
$$(x + 4)(x - 3) = 0$$
$$x = -4 \quad \text{or} \quad x = 3$$

Step 5: *Write the answer.*

If $x = -4$, then $x + 1 = -3$. If $x = 3$, then $x + 1 = 4$. The two integers are -4 and -3, or the two integers are 3 and 4.

Step 6: *Reread and check.*

The two integers in each pair are consecutive integers, and the sum of the squares of either pair is 25.

Getting Ready for Class

After reading through the preceding section, respond in your own words and in complete sentences.

A. What are equivalent equations?

B. Describe how to eliminate fractions in an equation.

C. Suppose when solving an equation your result is the statement "$3 = -3$." What would you conclude about the solution to the equation?

D. What is the zero-factor property?

PROBLEM SET 7.1

Solve each of the following equations.

1. $2x - 4 = 6$

2. $3x - 5 = 4$

3. $-3 - 4x = 15$

4. $-8 - 5x = -6$

5. $-300y + 100 = 500$

6. $-20y + 80 = 30$

7. $-\frac{3}{5}a + 2 = 8$

8. $-\frac{5}{3}a + 3 = 23$

9. $-x = 2$

10. $-x = \frac{1}{2}$

11. $-a = -\frac{3}{4}$

12. $-a = -5$

13. $7y - 4 = 2y + 11$

14. $8y - 2 = 6y - 10$

15. $5(y + 2) - 4(y + 1) = 3$

16. $6(y - 3) - 5(y + 2) = 8$

17. $6 - 7(m - 3) = -1$

18. $3 - 5(2m - 5) = -2$

19. $5 = 7 - 2(3x - 1) + 4x$

20. $20 = 8 - 5(2x - 3) + 4x$

21. $\frac{1}{2}x + \frac{1}{4} = \frac{1}{3}x + \frac{5}{4}$ 22. $\frac{2}{3}x - \frac{3}{4} = \frac{1}{6}x + \frac{21}{4}$

23. $x^2 - 5x - 6 = 0$ 24. $x^2 + 5x - 6 = 0$

25. $x^3 - 5x^2 + 6x = 0$ 26. $x^3 + 5x^2 + 6x = 0$

27. $60x^2 - 130x + 60 = 0$

28. $90x^2 + 60x - 80 = 0$

29. $\frac{1}{5}y^2 - 2 = -\frac{3}{10}y$ 30. $\frac{1}{2}y^2 + \frac{5}{3} = \frac{17}{6}y$

31. $-100x = 10x^2$ 32. $800x = 100x^2$

33. $(x + 6)(x - 2) = -7$

34. $(x - 7)(x + 5) = -20$

35. $(x + 1)^2 = 3x + 7$ 36. $(x + 2)^2 = 9x$

37. $x^3 + 3x^2 - 4x - 12 = 0$

38. $x^3 + 5x^2 - 4x - 20 = 0$

39. $5 - 2x = 3x + 1$ 40. $7 - 3x = 8x - 4$

41. $\frac{1}{10}t^2 - \frac{5}{2} = 0$ 42. $\frac{2}{7}t^2 - \frac{7}{2} = 0$

43. $7 + 3(x + 2) = 4(x + 1)$

44. $5 + 2(4x - 4) = 3(2x - 1)$

45. $-\frac{2}{5}x + \frac{2}{15} = \frac{2}{3}$ 46. $-\frac{1}{6}x + \frac{2}{3} = \frac{1}{4}$

47. $\frac{1}{2}x + \frac{1}{3}x + \frac{1}{4}x = \frac{13}{12}$ 48. $\frac{1}{3}x + \frac{1}{4}x + \frac{1}{5}x = \frac{47}{60}$

49. $(2r + 3)(2r - 1) = -(3r + 1)$

50. $(3r + 2)(r - 1) = -(7r - 7)$

51. $9a^3 = 16a$ 52. $16a^3 = 25a$

53. $4x^3 + 12x^2 - 9x - 27 = 0$

54. $9x^3 + 18x^2 - 4x - 8 = 0$

Identities and Equations With No Solution

55. There is no solution to the equation

$$6x - 2(x - 5) = 4x + 3$$

That is, there is no real number to use in place of x that will turn the equation into a true statement. What happens when you try to solve the equation?

56. The equation $4x - 8 = 2(2x - 4)$ is called an **identity** because every real number is a solution. That is, replacing x with any real number will result in a true statement. Try to solve the equation.

Solve each equation, if possible.

57. $3x - 6 = 3(x + 4)$ 58. $7x - 14 = 7(x - 2)$

59. $4y + 2 - 3y + 5 = 3 + y + 4$

60. $7y + 5 - 2y - 3 = 6 + 5y - 4$

61. $2(4t - 1) + 3 = 5t + 4 + 3t$

62. $5(2t - 1) + 1 = 2t - 4 + 8t$

Applying the Concepts

63. **Perimeter**   A rectangle is twice as long as it is wide. The perimeter is 60 feet. Find the dimensions.

64. **Perimeter**   The length of a rectangle is 5 times the width. The perimeter is 48 inches. Find the dimensions.

65. **Livestock Pen**   A livestock pen is built in the shape of a rectangle that is twice as long as it is wide. The perimeter is 48 feet. If the material used to build the pen is $1.75 per foot for the longer sides and $2.25 per foot for the shorter sides (the shorter sides have gates, which increase the cost per foot), find the cost to build the pen.

66. **Garden**   A garden is in the shape of a square with a perimeter of 42 feet. The garden is surrounded by two fences. One fence is around the perimeter of the garden, while the second fence is 3 feet from the first fence, as the following figure indicates. If the material used to build the two fences is $1.28 per foot, what was the total cost of the fences?

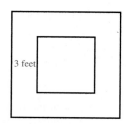

3 feet

67. **Sales Tax**   A woman owns a small, cash-only business in a state that requires her to charge a 6% sales tax on each item she sells. At the beginning of the day she has $250 in the cash register. At the end of the day she has $1,204 in the register. How much money should she send to the state government for the sales tax she collected?

68. **Sales Tax**   A store is located in a state that requires a 6% tax on all items sold. If the store brings

in a total of $3,392 in one day, how much of that total was sales tax?

Maximum Heart Rate In exercise physiology, a person's maximum heart rate, in beats per minute, is found by subtracting his age in years from 220. So if A represents your age in years, then your maximum heart rate is

$$M = 220 - A$$

Use this formula to complete the following tables.

69.

Age (years)	Maximum Heart Rate (beats per minute)
18	
19	
20	
21	
22	
23	

70.

Age (years)	Maximum Heart Rate (beats per minute)
15	
20	
25	
30	
35	
40	

Training Heart Rate A person's training heart rate, in beats per minute, is his resting heart rate plus 60% of the difference between his maximum heart rate and his resting heart rate. If resting heart rate is R and maximum heart rate is M, then the formula that gives training heart rate is

$$T = R + 0.6(M - R)$$

Use this formula along with the results of Problems 69 and 70 to fill in the following two tables.

71. For a 20-year-old person

Resting Heart Rate (beats per minute)	Training Heart Rate (beats per minute)
60	
62	
64	
68	
70	
72	

72. For a 40-year-old person

Resting Heart Rate (beats per minute)	Training Heart Rate (beats per minute)
60	
62	
64	
68	
70	
72	

73. Consecutive Integers The sum of the squares of two consecutive odd integers is 34. Find the two integers.

74. Consecutive Integers The sum of the squares of two consecutive even integers is 100. Find the two integers.

75. Pythagorean Theorem A 25-foot ladder is leaning against a building. The base of the ladder is 7 feet from the side of the building. How high does the ladder reach along the side of the building?

7 ft

76. Pythagorean Theorem Noreen wants to place her 13-foot ramp against the side of her house so that the top of the ramp rests on a ledge that is 5 feet

5 ft

above the ground. How far will the base of the ramp be from the house?

77. Geometry   The length of a rectangle is 2 feet more than 3 times the width. If the area is 16 square feet, find the width and the length.

78. Geometry   The length of a rectangle is 4 yards more than twice the width. If the area is 70 square yards, find the width and the length.

Review Problems

The problems that follow review material from Section 6.4.

Solve each equation.

79. $\dfrac{x}{3} - \dfrac{1}{2} = \dfrac{5}{2}$

80. $\dfrac{3}{x} + \dfrac{1}{5} = \dfrac{4}{5}$

81. $1 - \dfrac{5}{x} = \dfrac{-6}{x^2}$

82. $1 - \dfrac{1}{x} = \dfrac{6}{x^2}$

83. $\dfrac{a}{a-4} - \dfrac{a}{2} = \dfrac{4}{a-4}$

84. $\dfrac{a}{a-3} - \dfrac{a}{2} = \dfrac{3}{a-3}$

7.2   Equations With Absolute Value

In Chapter 1 we defined the absolute value of x, $|x|$, to be the distance between x and 0 on the number line. The absolute value of a number measures its distance from 0.

EXAMPLE 1   Solve for x: $|x| = 5$

Solution   Using the definition of absolute value, we can read the equation as, "The distance between x and 0 on the number line is 5." If x is 5 units from 0, then x can be 5 or -5:

$$\text{If } |x| = 5 \qquad \text{then } x = 5 \qquad \text{or} \qquad x = -5$$

In general, then, we can see that any equation of the form $|a| = b$ is equivalent to the equations $a = b$ or $a = -b$, as long as $b > 0$.

EXAMPLE 2   Solve $|2a - 1| = 7$.

Solution   We can read this question as "$2a - 1$ is 7 units from 0 on the number line." The quantity $2a - 1$ must be equal to 7 or -7:

$$|2a - 1| = 7$$

$$2a - 1 = 7 \qquad \text{or} \qquad 2a - 1 = -7$$

We have transformed our absolute value equation into two equations that do not involve absolute value. We can solve each equation using the method in

Section 2.4:

$$2a - 1 = 7 \quad \text{or} \quad 2a - 1 = -7$$

$$2a = 8 \quad \text{or} \quad \quad 2a = -6 \qquad \text{Add } +1 \text{ to both sides.}$$

$$a = 4 \quad \text{or} \quad \quad a = -3 \qquad \text{Multiply by } \tfrac{1}{2}.$$

Our solution set is $\{4, -3\}$.

To check our solutions, we put them into the original absolute value equation:

When	$a = 4$	When	$a = -3$
the equation	$|2a - 1| = 7$	the equation	$|2a - 1| = 7$
becomes	$|2(4) - 1| = 7$	becomes	$|2(-3) - 1| = 7$
	$|7| = 7$		$|-7| = 7$
	$7 = 7$		$7 = 7$

EXAMPLE 3   Solve $\left|\tfrac{2}{3}x - 3\right| + 5 = 12$.

Solution   In order to use the definition of absolute value to solve this equation, we must isolate the absolute value on the left side of the equal sign. To do so, we add -5 to both sides of the equation to obtain

$$\left|\frac{2}{3}x - 3\right| = 7$$

Now that the equation is in the correct form, we can write

$$\frac{2}{3}x - 3 = 7 \quad \text{or} \quad \frac{2}{3}x - 3 = -7$$

$$\frac{2}{3}x = 10 \quad \text{or} \quad \frac{2}{3}x = -4 \qquad \text{Add } +3 \text{ to both sides.}$$

$$x = 15 \quad \text{or} \quad x = -6 \qquad \text{Multiply by } \tfrac{3}{2}.$$

The solution set is $\{15, -6\}$.

EXAMPLE 4   Solve $|3a - 6| = -4$.

Solution   The solution set is $\varnothing$ because the left side cannot be negative and the right side is negative. No matter what we try to substitute for the variable a, the quantity $|3a - 6|$ will always be positive or zero. It can never be -4.

Consider the statement $|a| = |b|$. What can we say about a and b? We know they are equal in absolute value. By the definition of absolute value, they are the same distance from 0 on the number line. They must be equal to each other or opposites of each other. In symbols, we write

$$|a| = |b| \Leftrightarrow a = b \quad \text{or} \quad a = -b$$

$$\uparrow \qquad\qquad\quad \uparrow \qquad\qquad\quad \uparrow$$

Equal in   Equals   or   Opposites
absolute value

EXAMPLE 5   Solve $|3a + 2| = |2a + 3|$.

Solution   The quantities $3a + 2$ and $2a + 3$ have equal absolute values. They are, therefore, the same distance from 0 on the number line. They must be equals or opposites:

$$|3a + 2| = |2a + 3|$$

Equals		*Opposites*
$3a + 2 = 2a + 3$	or	$3a + 2 = -(2a + 3)$
$a + 2 = 3$		$3a + 2 = -2a - 3$
$a = 1$		$5a + 2 = -3$
		$5a = -5$
		$a = -1$

The solution set is $\{1, -1\}$.

It makes no difference in the outcome of the problem if we take the opposite of the first or second expression. It is very important, once we have decided which one to take the opposite of, that we take the opposite of both its terms and not just the first term. That is, the opposite of $2a + 3$ is $-(2a + 3)$, which we can think of as $-1(2a + 3)$. Distributing the -1 across *both* terms, we have

$$-1(2a + 3) = -2a - 3$$

EXAMPLE 6   Solve $|x - 5| = |x - 7|$.

Solution   As was the case in Example 5, the quantities $x - 5$ and $x - 7$ must be equal or they must be opposites, because their absolute values are equal:

Equals		*Opposites*
$x - 5 = x - 7$	or	$x - 5 = -(x - 7)$
$-5 = -7$		$x - 5 = -x + 7$
No solution here		$2x - 5 = 7$
		$2x = 12$
		$x = 6$

Since the first equation leads to a false statement, it will not give us a solution. (If either of the two equations were to reduce to a true statement, it would mean all real numbers would satisfy the original equation.) In this case, our only solution is $x = 6$.

Getting Ready for Class

After reading through the preceding section, respond in your own words and in complete sentences.

A. Why do some of the equations in this section have two solutions instead of one?

B. Translate $|x| = 6$ into words using the definition of absolute value.

C. Explain in words what the equation $|x - 3| = 4$ means with respect to distance on the number line.

D. When is the statement $|x| = x$ true?

PROBLEM SET 7.2

Use the definition of absolute value to solve each of the following problems.

1. $|x| = 4$ **2.** $|x| = 7$

3. $2 = |a|$ **4.** $5 = |a|$

5. $|x| = -3$ **6.** $|x| = -4$

7. $|a| + 2 = 3$ **8.** $|a| - 5 = 2$

9. $|y| + 4 = 3$ **10.** $|y| + 3 = 1$

11. $4 = |x| - 2$ **12.** $3 = |x| - 5$

13. $|x - 2| = 5$ **14.** $|x + 1| = 2$

15. $|a - 4| = \dfrac{5}{3}$ **16.** $|a + 2| = \dfrac{7}{5}$

17. $1 = |3 - x|$ **18.** $2 = |4 - x|$

19. $\left|\dfrac{3}{5}a + \dfrac{1}{2}\right| = 1$ **20.** $\left|\dfrac{2}{7}a + \dfrac{3}{4}\right| = 1$

21. $60 = |20x - 40|$ **22.** $800 = |400x - 200|$

23. $|2x + 1| = -3$ **24.** $|2x - 5| = -7$

25. $\left|\dfrac{3}{4}x - 6\right| = 9$ **26.** $\left|\dfrac{4}{5}x - 5\right| = 15$

27. $\left|1 - \dfrac{1}{2}a\right| = 3$ **28.** $\left|2 - \dfrac{1}{3}a\right| = 10$

Solve each equation.

29. $|3x + 4| + 1 = 7$ **30.** $|5x - 3| - 4 = 3$

31. $|3 - 2y| + 4 = 3$ **32.** $|8 - 7y| + 9 = 1$

33. $3 + |4t - 1| = 8$ **34.** $2 + |2t - 6| = 10$

35. $\left|9 - \dfrac{3}{5}x\right| + 6 = 12$ **36.** $\left|4 - \dfrac{2}{7}x\right| + 2 = 14$

37. $5 = \left|\dfrac{2x}{7} + \dfrac{4}{7}\right| - 3$ **38.** $7 = \left|\dfrac{3x}{5} + \dfrac{1}{5}\right| + 2$

39. $2 = -8 + \left|4 - \dfrac{1}{2}y\right|$ **40.** $1 = -3 + \left|2 - \dfrac{1}{4}y\right|$

Solve the following equations.

41. $|3a + 1| = |2a - 4|$ **42.** $|5a + 2| = |4a + 7|$

43. $\left|x - \dfrac{1}{3}\right| = \left|\dfrac{1}{2}x + \dfrac{1}{6}\right|$

44. $\left|\dfrac{1}{10}x - \dfrac{1}{2}\right| = \left|\dfrac{1}{5}x + \dfrac{1}{10}\right|$

45. $|y - 2| = |y + 3|$ **46.** $|y - 5| = |y - 4|$

47. $|3x - 1| = |3x + 1|$ **48.** $|5x - 8| = |5x + 8|$

Solve the following equations.

49. $|3 - m| = |m + 4|$ **50.** $|5 - m| = |m + 8|$

51. $|0.03 - 0.01x| = |0.04 + 0.05x|$

52. $|0.07 - 0.01x| = |0.08 - 0.02x|$

53. $|x - 2| = |2 - x|$ **54.** $|x - 4| = |4 - x|$

55. $\left|\dfrac{x}{5} - 1\right| = \left|1 - \dfrac{x}{5}\right|$ **56.** $\left|\dfrac{x}{3} - 1\right| = \left|1 - \dfrac{x}{3}\right|$

Review Problems

The problems below review material we covered in Sections 2.5 and 2.8. Reviewing these problems will help you with the next section.

57. Solve $d = rt$ for t. [2.5]

58. Solve $5x - 3y = 15$ for y. [2.5]

59. What percent of 60 is 15? [2.5]

60. Name the complement and supplement of 15°. [2.5]

Solve each inequality. [2.8]

61. $x - 5 > 8$ **62.** $-5x \leq 10$

63. $\frac{1}{4}x \geq 1$ **64.** $3x - 5 < 7$

65. $4 - 2x < 12$ **66.** $2(3t - 1) + 5 > -3$

7.3   Compound Inequalities and Interval Notation

The instrument panel on most cars includes a temperature gauge. The one shown below indicates that the normal operating temperature for the engine is from 50°F to 270°F.

We can represent the same situation with an inequality by writing $50 \leq F \leq 270$, where F is the temperature in degrees Fahrenheit. This inequality is a compound inequality. In this section we present the notation and definitions associated with compound inequalities.

Let's begin by reviewing. Here is an example showing how we solved inequalities in Chapter 2.

EXAMPLE 1   Solve $3x + 3 < 2x - 1$, and graph the solution.

Solution   We use the addition property for inequalities to write all the variable terms on one side and all constant terms on the other side:

$$3x + 3 < 2x - 1$$

$$3x + (-2x) + 3 < 2x + (-2x) - 1 \qquad \text{Add } -2x \text{ to each side.}$$

$$x + 3 < -1$$

$$x + 3 + (-3) < -1 + (-3) \qquad \text{Add } -3 \text{ to each side.}$$

$$x < -4$$

The solution set is all real numbers that are less than -4. To show this we can use set notation and write

$$\{x \mid x < -4\}$$

Or we can graph the solution set on the number line using an open circle at -4 to show that -4 is not part of the solution set:

This graph gives rise to the following notation, called *interval notation,* that is an alternative way to write the solution set.

$$(-\infty, -4)$$

The English mathematician John Wallis (1616–1703) was the first person to use the ∞ symbol to represent infinity. When we encounter the interval $(3, \infty)$ we read it as "the interval from 3 to infinity," and we mean the set of real numbers that are greater than 3. Likewise, the interval $(-\infty, -4)$ is read "the interval from negative infinity to -4," which is all real numbers less than -4.

The preceding expression indicates that the solution set is all real numbers from negative infinity up to, but not including, -4. (If the endpoint is included in the solution set, we use a bracket, [or], instead of a parenthesis.)

We have three equivalent representations for the solution set to our original inequality. Here are all three together.

Set Notation	Graph	Interval Notation

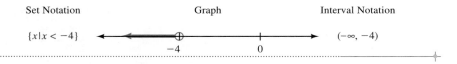

The *union* of two sets A and B is the set of all elements that are in A or in B. The word *or* is the key word in the definition. The *intersection* of two sets A and B is the set of elements contained in both A and B. The key word in this definition is *and*. We can put the words *and* and *or* together with our methods of graphing inequalities to find the solution sets for compound inequalities.

> **DEFINITION**   A **compound inequality** is two or more inequalities connected by the word *and* or *or.*

EXAMPLE 2   Graph the solution set for the compound inequality

$$x < -1 \qquad \text{or} \qquad x \geq 3$$

Solution   Graphing each inequality separately, we have

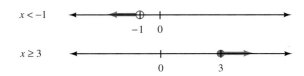

Since the two inequalities are connected by *or,* we want to graph their union: that is, we graph all points that are on either the first graph or the second graph. Essentially, we put the two graphs together on the same number line.

Set Notation	Graph	Interval Notation
$\{x \mid x < -1 \text{ or } x \geq 3\}$	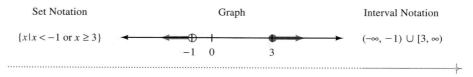	$(-\infty, -1) \cup [3, \infty)$

EXAMPLE 3   Graph the solution set for the compound inequality

$$x > -2 \quad \text{and} \quad x < 3$$

Solution   Graphing each inequality separately, we have

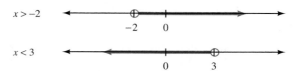

Since the two inequalities are connected by the word *and,* we will graph their intersection, which consists of all points that are common to both graphs; that is, we graph the region where the two graphs overlap.

Graph

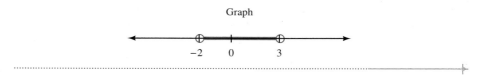

Notation   Sometimes compound inequalities that use the word *and* can be written in a shorter form. For example, the compound inequality $-2 < x$ and $x < 3$ can be written as $-2 < x < 3$. The word *and* does not appear when an inequality is written in this form. It is implied. The solution set for $-2 < x$ and $x < 3$ is

It is all the numbers between -2 and 3 on the number line. It seems reasonable, then, that this graph should be the graph of

$$-2 < x < 3$$

In both the graph and the inequality, x is said to be between -2 and 3. We call this last inequality a continued inequality.

EXAMPLE 4   Solve and graph the solution set for

$$2x - 1 \geq 3 \quad \text{and} \quad -3x > -12$$

Solution   Solving the two inequalities separately, we have

$$2x - 1 \geq 3 \qquad \text{and} \qquad -3x > -12$$

$$2x \geq 4$$

$$x \geq 2 \qquad \text{and} \qquad -\frac{1}{3}(-3x) < -\frac{1}{3}(-12)$$

$$x < 4$$

Since the word *and* connects the two graphs, we will graph their intersection—the points they have in common:

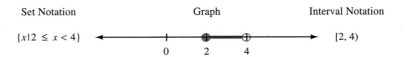

Set Notation	Graph	Interval Notation
$\{x \mid 2 \leq x < 4\}$		$[2, 4)$

EXAMPLE 5   Solve and graph $-3 \leq 2x - 1 \leq 9$.

Solution   To solve for *x,* we must add 1 to the center expression and then divide the result by 2. Whatever we do to the center expression, we must also do to the two expressions on the ends. In this way we can be sure we are producing equivalent inequalities. The solution set will not be affected.

$$-3 \leq 2x - 1 \leq 9$$

$$-2 \leq \quad 2x \quad \leq 10 \qquad \text{Add 1 to each expression.}$$

$$-1 \leq \quad x \quad \leq 5 \qquad \text{Multiply each expression by } \tfrac{1}{2}.$$

Set Notation	Graph	Interval Notation
$\{x \mid -1 \leq x \leq 5\}$		$[-1, 5]$

Getting Ready for Class

After reading through the preceding section, respond in your own words and in complete sentences.

A. What is a compound inequality?

B. Explain the shorthand notation that can be used to write two inequalities connected by the word *and*.

C. Write two inequalities connected by the word *and* that together are equivalent to $-1 < x < 2$.

D. When using interval notation to denote a section of the real number line, when do we use parentheses, (and), and when do we use brackets, [and]?

PROBLEM SET 7.3

Graph the following compound inequalities.

1. $x < -1$ or $x > 5$ **2.** $x \le -2$ or $x \ge -1$

3. $x < -3$ or $x \ge 0$ **4.** $x < 5$ and $x > 1$

5. $x \le 6$ and $x > -1$ **6.** $x \le 7$ and $x > 0$

7. $x > 2$ and $x < 4$ **8.** $x < 2$ or $x > 4$

9. $x \ge -2$ and $x \le 4$ **10.** $x \le 2$ or $x \ge 4$

11. $x < 5$ and $x > -1$ **12.** $x > 5$ or $x < -1$

13. $-1 < x < 3$ **14.** $-1 \le x \le 3$

15. $-3 < x \le -2$ **16.** $-5 \le x \le 0$

Solve the following. Graph the solution sets and write each answer using interval notation.

17. $3x - 1 < 5$ or $5x - 5 > 10$

18. $x + 1 < -3$ or $x - 2 > 6$

19. $x - 2 > -5$ and $x + 7 < 13$

20. $3x + 2 \le 11$ and $2x + 2 \ge 0$

21. $11x \le 22$ or $12x \ge 36$

22. $-5x < 25$ and $-2x \ge -12$

23. $3x - 5 \le 10$ and $2x + 1 \ge -5$

24. $5x + 8 < -7$ or $3x - 8 > 10$

25. $2x - 3 < 8$ and $3x + 1 > -10$

26. $11x - 8 > 3$ or $12x + 7 < -5$

27. $2x - 1 < 3$ and $3x - 2 > 1$

28. $3x + 9 < 7$ or $2x - 7 > 11$

Solve and graph each of the following.

29. $-1 \le x - 5 \le 2$ **30.** $0 \le x + 2 \le 3$

31. $-4 \le 2x \le 6$ **32.** $-5 < 5x < 10$

33. $-3 < 2x + 1 < 5$ **34.** $-7 \le 2x - 3 \le 7$

35. $0 \le 3x + 2 \le 7$ **36.** $2 \le 5x - 3 \le 12$

37. $-7 < 2x + 3 < 11$ **38.** $-5 < 6x - 2 < 8$

39. $-1 \le 4x + 5 \le 9$ **40.** $-8 \le 7x - 1 \le 13$

For each graph below, write an inequality whose solution is the graph.

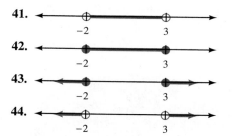

41.

42.

43.

44.

Applying the Concepts

45. Triangle Inequality   For any triangle, the triangle inequality states that the sum of any two sides must be greater than the third side. Given the following triangle *RST*

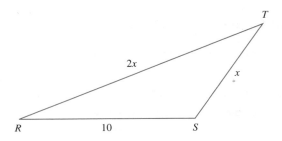

(a) Write three inequalities, based on the triangle inequality, that must be true.

(b) Use your results from part (a) to fill in the blanks in the following inequality:

$$\underline{\quad} < x < \underline{\quad}$$

46. Civil War Cost After the American Civil War was fully begun, its cost for the North was at least $30,000,000 per month and occasionally as great as $90,000,000 per month.

(a) Write an inequality for the total cost to the North during the 4 years of the Civil War.

(b) One textbook gave the total cost of the Civil War for the North as $2,845,907,626.56. Is this number a solution to the inequality you wrote in part (a)?

47. Engine Temperature The engine in a car gives off a lot of heat due to the combustion in the cylinders. The water used to cool the engine keeps the temperature within the range $50 \le F \le 270$ where F is in degrees Fahrenheit. Graph this inequality on a number line.

48. Engine Temperature To find the engine temperature range from Problem 47 in degrees Celsius, we use the fact that $F = \dfrac{9}{5}C + 32$ to rewrite the

inequality as

$$50 \le \frac{9}{5}C + 32 \le 270$$

Solve this inequality and graph the solution set.

49. Number Problem The sum of a number and 5 is between 10 and 20. Find the number.

50. Number Problem The difference of a number and 2 is between 6 and 14. Find the number.

51. Number Problem The difference of twice a number and 3 is between 5 and 7. Find the number.

52. Number Problem The sum of twice a number and 5 is between 7 and 13. Find the number.

53. Perimeter The length of a rectangle is 4 inches longer than the width. If the perimeter is between 20 inches and 30 inches, find all possible values for the width.

54. Perimeter The length of a rectangle is 6 feet longer than the width. If the perimeter is between 24 feet and 36 feet, find all possible values for the width.

Review Problems

The problems that follow review the material we covered in Chapter 1.

Simplify each expression.

55. $-|-5|$

56. $\left(-\dfrac{2}{3}\right)^3$

57. $-3 - 4(-2)$

58. $2^4 + 3^3 \div 9 - 4^2$

59. $5|3 - 8| - 6|2 - 5|$

60. $7 - 3(2 - 6)$

61. $5 - 2[-3(5 - 7) - 8]$

62. $\dfrac{5 + 3(7 - 2)}{2(-3) - 4}$

63. Find the difference of -3 and -9.

64. If you add -4 to the product of -3 and 5, what number results?

65. Apply the distributive property to $\dfrac{1}{2}(4x - 6)$.

66. Use the associative property to simplify $-6\left(\dfrac{1}{3}x\right)$.

For the set $\left\{-3, -\dfrac{4}{5}, 0, \dfrac{5}{8}, 2, \sqrt{5}\right\}$, which numbers are

67. Integers?

68. Rational numbers?

In this section we will again apply the definition of absolute value to solve inequalities involving absolute value. Again, the absolute value of x, which is $|x|$, represents the distance that x is from 0 on the number line. We will begin by considering three absolute value expressions and their verbal translations:

Expression	*In Words*
$|x| = 7$	x is exactly 7 units from 0 on the number line.
$|a| < 5$	a is less than 5 units from 0 on the number line.
$|y| \geq 4$	y is greater than or equal to 4 units from 0 on the number line.

Once we have translated the expression into words, we can use the translation to graph the original equation or inequality. The graph is then used to write a final equation or inequality that does not involve absolute value.

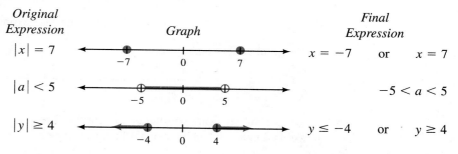

Original Expression	*Graph*	*Final Expression*
$|x| = 7$		$x = -7$ or $x = 7$
$|a| < 5$		$-5 < a < 5$
$|y| \geq 4$		$y \leq -4$ or $y \geq 4$

Although we will not always write out the verbal translation of an absolute value inequality, it is important that we understand the translation. Our second expression, $|a| < 5$, means a is within 5 units of 0 on the number line. The graph of this relationship is

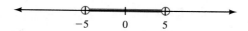

which can be written with the following continued inequality:

$$-5 < a < 5$$

We can follow this same kind of reasoning to solve more complicated absolute value inequalities.

EXAMPLE 1 Graph the solution set: $|2x - 5| < 3$

Solution The absolute value of $2x - 5$ is the distance that $2x - 5$ is from 0 on the number line. We can translate the inequality as, "$2x - 5$ is less than 3 units from 0 on the number line." That is, $2x - 5$ must appear between -3 and 3 on the number line.

A picture of this relationship is

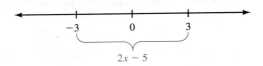

Using the picture, we can write an inequality without absolute value that describes the situation:

$$-3 < 2x - 5 < 3$$

Next, we solve the continued inequality by first adding $+5$ to all three members and then multiplying all three by $\frac{1}{2}$.

$$-3 < 2x - 5 < 3$$

$$2 < \quad 2x \quad < 8 \qquad \text{Add} + 5 \text{ to all three expressions.}$$

$$1 < \quad x \quad < 4 \qquad \text{Multiply each expression by } \frac{1}{2}.$$

The graph of the solution set is

We can see from the solution that for the absolute value of $2x - 5$ to be within 3 units of 0 on the number line, x must be between 1 and 4.

E X A M P L E   2   Solve and graph $|3a + 7| \leq 4$.

Solution   We can read the inequality as, "The distance between $3a + 7$ and 0 is less than or equal to 4." Or, "$3a + 7$ is within 4 units of 0 on the number line." This relationship can be written without absolute value as

$$-4 \leq 3a + 7 \leq 4$$

Solving as usual, we have

$$-4 \leq 3a + 7 \leq 4$$

$$-11 \leq \quad 3a \quad \leq -3 \qquad \text{Add} -7 \text{ to all three members.}$$

$$-\frac{11}{3} \leq \quad a \quad \leq -1 \qquad \text{Multiply each expression by } \frac{1}{3}.$$

We can see from Examples 1 and 2 that to solve an inequality involving absolute value, we must be able to write an equivalent expression that does not involve absolute value.

EXAMPLE 3   Solve $|x - 3| > 5$, and graph the solution.

Solution   We interpret the absolute value inequality to mean that $x - 3$ is more than 5 units from 0 on the number line. The quantity $x - 3$ must be either above $+5$ or below -5. Here is a picture of the relationship:

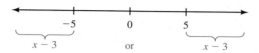

An inequality without absolute value that also describes this situation is

$$x - 3 < -5 \quad \text{or} \quad x - 3 > 5$$

Adding $+3$ to both sides of each inequality we have

$$x < -2 \quad \text{or} \quad x > 8$$

Here are three ways to write our result.

Set Notation	Graph	Interval Notation
$\{x \mid x < -2 \text{ or } x > 8\}$	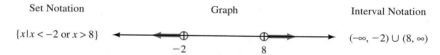	$(-\infty, -2) \cup (8, \infty)$

EXAMPLE 4   Graph the solution set: $|4t - 3| \geq 9$.

Solution   The quantity $4t - 3$ is greater than or equal to 9 units from 0. It must be either above $+9$ or below -9.

$$4t - 3 \leq -9 \quad \text{or} \quad 4t - 3 \geq 9$$
$$4t \leq -6 \quad \text{or} \quad 4t \geq 12 \quad \text{Add } +3.$$
$$t \leq -\frac{6}{4} \quad \text{or} \quad t \geq \frac{12}{4} \quad \text{Multiply by } \tfrac{1}{4}.$$
$$t \leq -\frac{3}{2} \quad \text{or} \quad t \geq 3$$

We can use the results of our first few examples and the material in the previous sections to summarize the information we have related to absolute value equations and inequalities.

REWRITING ABSOLUTE VALUE EQUATIONS AND INEQUALITIES

If c is a positive real number, then each of the following statements on the left is equivalent to the corresponding statement on the right.

With Absolute Value	*Without Absolute Value*		
$\|x\| = c$	$x = -c$	or	$x = c$
$\|x\| < c$		$-c < x < c$	
$\|x\| > c$	$x < -c$	or	$x > c$
$\|ax + b\| = c$	$ax + b = -c$	or	$ax + b = c$
$\|ax + b\| < c$		$-c < ax + b < c$	
$\|ax + b\| > c$	$ax + b < -c$	or	$ax + b > c$

EXAMPLE 5   Solve and graph $|2x + 3| + 4 < 9$.

Solution   Before we can apply the method of solution we used in the previous examples, we must isolate the absolute value on one side of the inequality. To do so, we add -4 to each side.

$$|2x + 3| + 4 < 9$$

$$|2x + 3| + 4 + (-4) < 9 + (-4)$$

$$|2x + 3| < 5$$

From this last line we know that $2x + 3$ must be between -5 and $+5$.

$$-5 < 2x + 3 < 5$$

$$-8 < \quad 2x \quad < 2 \qquad \text{Add} - 3 \text{ to each expression.}$$

$$-4 < \quad x \quad < 1 \qquad \text{Multiply each expression by } \tfrac{1}{2}.$$

Here are three equivalent ways to write our solution.

Set Notation	Graph	Interval Notation
$\{x \mid -4 < x < 1\}$		$(-4, 1)$

EXAMPLE 6   Solve and graph $|4 - 2t| > 2$.

Solution   The inequality indicates that $4 - 2t$ is less than -2 or greater than $+2$. Writing this without absolute value symbols, we have

$$4 - 2t < -2 \qquad \text{or} \qquad 4 - 2t > 2$$

To solve these inequalities we begin by adding -4 to each side.

$$4 + (-4) - 2t < -2 + (-4) \qquad \text{or} \qquad 4 + (-4) - 2t > 2 + (-4)$$

$$-2t < -6 \qquad\qquad \text{or} \qquad\qquad -2t > -2$$

Next we must multiply both sides of each inequality by $-\frac{1}{2}$. When we do so, we must also reverse the direction of each inequality symbol.

$$-2t < -6 \qquad \text{or} \qquad -2t > -2$$

$$-\frac{1}{2}(-2t) > -\frac{1}{2}(-6) \qquad \text{or} \qquad -\frac{1}{2}(-2t) < -\frac{1}{2}(-2)$$

$$t > 3 \qquad \text{or} \qquad t < 1$$

Although in situations like this we are used to seeing the "less than" symbol written first, the meaning of the solution is clear. We want to graph all real numbers that are either greater than 3 or less than 1. Here is the graph.

Since absolute value always results in a nonnegative quantity, we sometimes come across special solution sets when a negative number appears on the right side of an absolute value inequality.

EXAMPLE 7   Solve $|7y - 1| < -2$.

Solution   The *left* side is never negative because it is an absolute value. The *right* side is negative. We have a positive quantity less than a negative quantity, which is impossible. The solution set is the empty set, $\varnothing$. There is no real number to substitute for y to make this inequality a true statement.

EXAMPLE 8   Solve $|6x + 2| > -5$.

Solution   This is the opposite case from that in Example 7. No matter what real number we use for x on the *left* side, the result will always be positive, or zero. The *right* side is negative. We have a positive quantity greater than a negative quantity. Every real number we choose for x gives us a true statement. The solution set is the set of all real numbers.

Getting Ready for Class

After reading through the preceding section, respond in your own words and in complete sentences.

A. Write an inequality containing absolute value, the solution to which is all the numbers between -5 and 5 on the number line.

B. Translate $|x| \geq 3$ into words using the definition of absolute value.

C. Explain in words what the inequality $|x - 5| < 2$ means with respect to distance on the number line.

D. Why is there no solution to the inequality $|2x - 3| < 0$?

PROBLEM SET 7.4

Solve each of the following inequalities using the definition of absolute value. Graph the solution set in each case.

1. $|x| < 3$

2. $|x| \leq 7$

3. $|x| \geq 2$

4. $|x| > 4$

5. $|x| + 2 < 5$

6. $|x| - 3 < -1$

7. $|t| - 3 > 4$

8. $|t| + 5 > 8$

9. $|y| < -5$

10. $|y| > -3$

11. $|x| \geq -2$

12. $|x| \leq -4$

13. $|x - 3| < 7$

14. $|x + 4| < 2$

15. $|a + 5| \geq 4$

16. $|a - 6| \geq 3$

Solve each inequality and graph the solution set.

17. $|a - 1| < -3$

18. $|a + 2| \geq -5$

19. $|2x - 4| < 6$

20. $|2x + 6| < 2$

21. $|3y + 9| \geq 6$

22. $|5y - 1| \geq 4$

23. $|2k + 3| \geq 7$

24. $|2k - 5| \geq 3$

25. $|x - 3| + 2 < 6$

26. $|x + 4| - 3 < -1$

27. $|2a + 1| + 4 \geq 7$

28. $|2a - 6| - 1 \geq 2$

29. $|3x + 5| - 8 < 5$

30. $|6x - 1| - 4 \leq 2$

Solve each inequality and graph the solution set. Keep in mind that if you multiply or divide both sides of an inequality by a negative number you must reverse the sense of the inequality.

31. $|5 - x| > 3$

32. $|7 - x| > 2$

33. $\left| 3 - \dfrac{2}{3}x \right| \geq 5$

34. $\left| 3 - \dfrac{3}{4}x \right| \geq 9$

35. $\left| 2 - \dfrac{1}{2}x \right| > 1$

36. $\left| 3 - \dfrac{1}{3}x \right| > 1$

Solve each inequality.

37. $|x - 1| < 0.01$

38. $|x + 1| < 0.01$

39. $|2x + 1| \geq \dfrac{1}{5}$

40. $|2x - 1| \geq \dfrac{1}{8}$

41. $\left| \dfrac{3x - 2}{5} \right| \leq \dfrac{1}{2}$

42. $\left| \dfrac{4x - 3}{2} \right| \leq \dfrac{1}{3}$

43. $\left| 2x - \dfrac{1}{5} \right| < 0.3$

44. $\left| 3x - \dfrac{3}{5} \right| < 0.2$

45. Write the continued inequality $-4 \leq x \leq 4$ as a single inequality involving absolute value.

46. Write the continued inequality $-8 \leq x \leq 8$ as a single inequality involving absolute value.

47. Write $-1 \leq x - 5 \leq 1$ as a single inequality involving absolute value.

48. Write $-3 \leq x + 2 \leq 3$ as a single inequality involving absolute value.

Applying the Concepts

49. Television Channel Capacity   Channel capacity varies significantly on televisions throughout the United States. A recent study found that 30% of the televisions in the United States are equipped to receive c channels, where c satisfies $|c - 24| \leq 12$. What is the minimum and maximum number of channels that these television sets are capable of receiving?

50. Speed Limits   The interstate speed limit for cars is 75 miles per hour in Nebraska, Nevada, New Mexico, Oklahoma, South Dakota, Utah, and Wyoming, and is the highest in the nation. To discourage passing, minimum speeds are also posted, so that the difference between the fastest and slowest moving traffic is no more than 20 miles per hour. Write an absolute value inequality that describes the relationship between the minimum allowable speed and a maximum speed of 75 miles per hour.

51. Speed Limits   Suppose the speed limit on Highway 101 in California is 65 miles per hour, and slower cars cannot travel less than 20 miles per hour slower than the fastest cars. Write an absolute value inequality that describes the relationship between the minimum allowable speed and a maximum speed of 65 miles per hour.

52. Wavelengths of Light   When white light from the sun passes through a prism, it is broken down into bands of light that form colors. The range of wavelengths for each color is unique, and each may be expressed as an inequality. The wavelength, v, (in nanometers) of some common colors are:

Blue: $424 < v < 491$
Green: $491 < v < 575$
Yellow: $575 < v < 585$
Orange: $585 < v < 647$
Red: $647 < v < 700$

When a fireworks display made of copper is burned, it lets out light with wavelengths, v, that satisfy the relationship $|v - 455| < 23$. Write this inequality without absolute values, find the range of possible values for v, and then using the preceding list of wavelengths, determine the color of that copper fireworks display.

Review Problems

The problems that follow review material we covered in Chapter 5.

Factor out the greatest common factor.

53. $16x^4 - 20x^3 + 8x^2$

54. $4a^3b - 12a^2b^2 + 28ab^3$

Factor by grouping.

55. $2ax - 3a + 8x - 12$

56. $x^3 - 5x^2 + 4x - 20$

Factor completely.

57. $x^2 - 2x - 35$ **58.** $6a^2 + 19a - 7$

59. $x^2 - xy - 6y^2$ **60.** $x^2 - 6x + 9$

61. $x^2 - 9$ **62.** $9a^2 - 41ab + 49b^2$

Solve each equation.

63. $9x^2 = 12x - 4$ **64.** $(x + 7)^2 = 2x + 13$

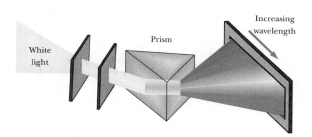

White light · Prism · Increasing wavelength

| 7.5 | **Factoring the Sum and Difference of Two Cubes** |

In Chapter 5 we factored a variety of polynomials. Among the polynomials we factored were polynomials that were the difference of two squares. The formula we used to factor the difference of two squares looks like this:

$$a^2 - b^2 = (a + b)(a - b)$$

If we ran across a binomial that had the form of the difference of two squares, we factored it by applying this formula. For example, to factor $x^2 - 25$, we simply notice that it can be written in the form $x^2 - 5^2$, which looks like the difference of two squares. According to the formula above, this binomial factors into $(x + 5)(x - 5)$.

In this section we want to use two new formulas that will allow us to factor the sum and difference of two cubes. For example, we want to factor the binomial $x^3 - 8$, which is the difference of two cubes. (To see that it is the difference of two cubes, notice that it can be written as $x^3 - 2^3$.) We also want to factor $y^3 + 27$, which is the sum of two cubes. (To see this, notice that $y^3 + 27$ can be written as $y^3 + 3^3$.)

The formulas that allow us to factor the sum of two cubes and the difference of two cubes are not as simple as the formula for factoring the difference of two squares. Here is what they look like:

$$a^3 + b^3 = (a + b)(a^2 - ab + b^2)$$

$$a^3 - b^3 = (a - b)(a^2 + ab + b^2)$$

Let's begin our work with these two formulas by showing that they are true. To do so, we multiply out the right side of each formula.

EXAMPLE 1   Verify the two formulas.

Solution   We verify the formulas by multiplying the right sides and comparing the results with the left sides:

$$
\begin{array}{r}
a^2 - ab + b^2 \\
a + b \\
\hline
a^3 - a^2b + ab^2 \\
a^2b - ab^2 + b^3 \\
\hline
a^3 \qquad\qquad\quad + b^3
\end{array}
$$

The first formula is correct.

$$
\begin{array}{r}
a^2 + ab + b^2 \\
a - b \\
\hline
a^3 + a^2b + ab^2 \\
- a^2b - ab^2 - b^3 \\
\hline
a^3 \qquad\qquad\quad - b^3
\end{array}
$$

The second formula is correct.

Here are some examples that use the formulas for factoring the sum and difference of two cubes.

EXAMPLE 2   Factor $x^3 - 8$.

Solution   Since the two terms are perfect cubes, we write them as such and apply the formula:

$$x^3 - 8 = x^3 - 2^3$$
$$= (x - 2)(x^2 + 2x + 2^2)$$
$$= (x - 2)(x^2 + 2x + 4)$$

EXAMPLE 3 Factor $y^3 + 27$.

Solution Proceeding as we did in Example 2, we first write 27 as 3^3. Then, we apply the formula for factoring the sum of two cubes, which is $a^3 + b^3 = (a + b)(a^2 - ab + b^2)$:

$$y^3 + 27 = y^3 + 3^3$$
$$= (y + 3)(y^2 - 3y + 3^2)$$
$$= (y + 3)(y^2 - 3y + 9)$$

EXAMPLE 4 Factor: $64 + t^3$

Solution The first term is the cube of 4 and the second term is the cube of t. Therefore,

$$64 + t^3 = 4^3 + t^3$$
$$= (4 + t)(16 - 4t + t^2)$$

EXAMPLE 5 Factor: $27x^3 + 125y^3$

Solution Writing both terms as perfect cubes, we have

$$27x^3 + 125y^3 = (3x)^3 + (5y)^3$$
$$= (3x + 5y)(9x^2 - 15xy + 25y^2)$$

EXAMPLE 6 Factor: $a^3 - \frac{1}{8}$

Solution The first term is the cube of a, while the second term is the cube of $\frac{1}{2}$:

$$a^3 - \frac{1}{8} = a^3 - \left(\frac{1}{2}\right)^3$$
$$= \left(a - \frac{1}{2}\right)\left(a^2 + \frac{1}{2}a + \frac{1}{4}\right)$$

EXAMPLE 7 Factor: $x^6 - y^6$

Solution We have a choice of how we initially want to write the two terms. We can write the expression as the difference of two squares, $(x^3)^2 - (y^3)^2$, or as the difference of two cubes, $(x^2)^3 - (y^2)^3$. It is better to start with the difference of two squares if we have a choice:

$$x^6 - y^6 = (x^3)^2 - (y^3)^2$$
$$= (x^3 - y^3)(x^3 + y^3)$$
$$= (x - y)(x^2 + xy + y^2)(x + y)(x^2 - xy + y^2)$$

Factor completely, using the sum and difference of two cubes.

Try this example again, writing the first line as the difference of two cubes instead of the difference of two squares. It will become apparent why it is better to use the difference of two squares.

FACTORING: A GENERAL REVIEW

We end this section by reviewing all the different methods of factoring we have covered. To begin, here is a list of the steps that can be used to factor polynomials of any type.

TO FACTOR A POLYNOMIAL

Step 1: If the polynomial has a greatest common factor other than 1, then factor out the greatest common factor.

Step 2: If the polynomial has two terms (a binomial), then see if it is the difference of two squares, or the sum or difference of two cubes, and then factor accordingly. (*Note:* If it is the *sum* of two squares, it will not factor.)

Step 3: If the polynomial has three terms (a trinomial), then either it is a perfect square trinomial, which will factor into the square of a binomial, or it is not a perfect square trinomial, in which case you can use one of the methods developed in Section 5.3.

Step 4: If the polynomial has more than three terms, try to factor it by grouping.

Step 5: As a final check, see if any of the factors you have written can be factored further. If you have overlooked a common factor, you can catch it here.

Here are some examples illustrating how we use these five steps. There are no new factoring problems here. The problems are all similar to the problems you have seen before, but they are not grouped according to type.

EXAMPLE 8   Factor: $2x^5 - 8x^3$

Solution   First we check to see if the greatest common factor is other than 1. Since the greatest common factor is $2x^3$, we begin by factoring it out. Once we have done this, we notice that the binomial that remains is the difference of two squares, which we factor according to the formula $a^2 - b^2 = (a + b)(a - b)$.

$$2x^5 - 8x^3 = 2x^3(x^2 - 4) \qquad \text{Factor out the greatest common factor, } 2x^3.$$

$$= 2x^3(x + 2)(x - 2) \qquad \text{Factor the difference of two squares.}$$

E X A M P L E   9   Factor: $3x^4 - 18x^3 + 27x^2$

Solution   Step 1 is to factor out the greatest common factor, $3x^2$. After we have done this, we notice that the trinomial that remains is a perfect square trinomial, which will factor as the square of a binomial:

$$3x^4 - 18x^3 + 27x^2 = 3x^2(x^2 - 6x + 9) \qquad \text{Factor out } 3x^2.$$
$$= 3x^2(x - 3)^2 \qquad \begin{array}{l} x^2 - 6x + 9 \text{ is the} \\ \text{square of } x - 3. \end{array}$$

E X A M P L E   1 0   Factor: $y^3 + 25y$

Solution   We begin by factoring out the y that is common to both terms. The binomial that remains after we have done this is the sum of two squares, which does not factor. So, after the first step, we are finished:

$$y^3 + 25y = y(y^2 + 25)$$

E X A M P L E   1 1   Factor: $6a^2 - 11a + 4$

Solution   Here we have a trinomial that does not have a greatest common factor other than 1. Since it is not a perfect square trinomial, we factor it by trial and error. Without showing all the different possibilities, here is the answer:

$$6a^2 - 11a + 4 = (3a - 4)(2a - 1)$$

E X A M P L E   1 2   Factor: $2x^4 + 16x$

Solution   This binomial has a greatest common factor of $2x$. The binomial that remains after the $2x$ has been factored from each term is the sum of two cubes, which we factor according to the formula $a^3 + b^3 = (a + b)(a^2 - ab + b^2)$.

$$2x^4 + 16x = 2x(x^3 + 8) \qquad \text{Factor } 2x \text{ from each term.}$$
$$= 2x(x + 2)(x^2 - 2x + 4) \qquad \text{Factor the sum of two cubes}$$

E X A M P L E   1 3   Factor: $2ab^5 + 8ab^4 + 2ab^3$

Solution   The greatest common factor is $2ab^3$. We begin by factoring it from each term. After that, we find that the trinomial that remains cannot be factored further:

$$2ab^5 + 8ab^4 + 2ab^3 = 2ab^3(b^2 + 4b + 1)$$

E X A M P L E   1 4   Factor: $4x^2 - 6x + 2ax - 3a$

Solution   This polynomial has four terms, so we factor by grouping:

$$4x^2 - 6x + 2ax - 3a = 2x(2x - 3) + a(2x - 3)$$
$$= (2x - 3)(2x + a)$$

E X A M P L E   I 5   Factor completely: $x^3 + 2x^2 - 9x - 18$

Solution   We use factoring by grouping to begin and then factor the difference of two squares:

$$x^3 + 2x^2 - 9x - 18 = x^2(x + 2) - 9(x + 2)$$
$$= (x + 2)(x^2 - 9)$$
$$= (x + 2)(x - 3)(x + 3)$$

 # Getting Ready for Class

After reading through the preceding section, respond in your own words and in complete sentences.

A. In what cases can you factor a binomial?

B. What is a perfect square trinomial?

C. How do you know when you've factored completely?

D. If a polynomial has four terms, what method of factoring should you try?

PROBLEM SET 7.5

Factor each perfect square trinomial.

1. $x^2 - 6x + 9$ **2.** $x^2 + 10x + 25$

3. $a^2 - 12a + 36$ **4.** $36 - 12a + a^2$

5. $25 - 10t + t^2$ **6.** $64 + 16t + t^2$

7. $4y^4 - 12y^2 + 9$ **8.** $9y^4 + 12y^2 + 4$

9. $16a^2 + 40ab + 25b^2$

10. $25a^2 - 40ab + 16b^2$

11. $\frac{1}{25} + \frac{1}{10}t^2 + \frac{1}{16}t^4$ **12.** $\frac{1}{9} - \frac{1}{3}t^3 + \frac{1}{4}t^6$

13. $(x + 2)^2 + 6(x + 2) + 9$

14. $(x + 5)^2 + 4(x + 5) + 4$

Factor completely.

15. $49x^2 - 64y^2$ **16.** $81x^2 - 49y^2$

17. $4a^2 - \frac{1}{4}$ **18.** $25a^2 - \frac{1}{25}$

19. $x^2 - \frac{9}{25}$ **20.** $x^2 - \frac{25}{36}$

21. $25 - t^2$ **22.** $64 - t^2$

23. $16a^4 - 81$ **24.** $81a^4 - 16b^4$

25. $x^2 - 10x + 25 - y^2$

26. $x^2 - 6x + 9 - y^2$

27. $a^2 + 8a + 16 - b^2$

28. $a^2 + 12a + 36 - b^2$

29. $x^3 + 2x^2 - 25x - 50$

30. $x^3 + 4x^2 - 9x - 36$

31. $2x^3 + 3x^2 - 8x - 12$

32. $3x^3 + 2x^2 - 27x - 18$

33. $4x^3 + 12x^2 - 9x - 27$

34. $9x^3 + 18x^2 - 4x - 8$

Factor each of the following as the sum or difference of two cubes.

35. $x^3 - y^3$ **36.** $x^3 + y^3$

37. $a^3 + 8$ **38.** $a^3 - 8$

39. $y^3 - 1$ **40.** $y^3 + 1$

41. $10r^3 - 1,250$ **42.** $10r^3 + 1,250$

43. $64 + 27a^3$ **44.** $27 - 64a^3$

45. $t^3 + \frac{1}{27}$ **46.** $t^3 - \frac{1}{27}$

Factor each of the following polynomials completely, if possible. That is, once you are finished factoring, none of the factors you obtain should be factorable.

47. $x^2 - 81$ **48.** $x^2 - 18x + 81$

49. $x^2 + 2x - 15$ **50.** $15x^2 + 13x - 6$

51. $x^2y^2 + 2y^2 + x^2 + 2$ **52.** $21y^2 - 25y - 4$

53. $2a^3b + 6a^2b + 2ab$ **54.** $6a^2 - ab - 15b^2$

55. $x^2 + x + 1$

56. $x^2y + 3y + 2x^2 + 6$

57. $12a^2 - 75$ **58.** $18a^2 - 50$

59. $25 - 10t + t^2$ **60.** $t^2 + 4t + 4 - y^2$

61. $4x^3 + 16xy^2$ **62.** $16x^2 + 49y^2$

63. $x^3 + 5x^2 - 9x - 45$

64. $x^3 + 5x^2 - 16x - 80$

65. $x^2 + 49$ **66.** $16 - x^4$

67. $x^2(x - 3) - 14x(x - 3) + 49(x - 3)$

68. $x^2 + 3ax - 2bx - 6ab$

69. $8 - 14x - 15x^2$ **70.** $5x^4 + 14x^2 - 3$

71. $r^2 - \frac{1}{25}$ **72.** $27 - r^3$

73. $49x^2 + 9y^2$

74. $12x^4 - 62x^3 + 70x^2$

75. $100x^2 - 100x - 600$

76. $100x^2 - 100x - 1,200$

77. $3x^4 - 14x^2 - 5$ **78.** $8 - 2x - 15x^2$

79. $24a^5b - 3a^2b$

80. $18a^4b^2 - 24a^3b^3 + 8a^2b^4$

81. $64 - r^3$ **82.** $r^2 - \frac{1}{9}$

83. $20x^4 - 45x^2$ **84.** $16x^3 + 16x^2 + 3x$

85. $16x^5 - 44x^4 + 30x^3$ **86.** $16x^2 + 16x - 1$

87. $y^6 - 1$ **88.** $25y^7 - 16y^5$

89. $50 - 2a^2$ **90.** $4a^2 + 2a + \frac{1}{4}$

91. $x^2 - 4x + 4 - y^2$

92. $x^2 - 12x + 36 - b^2$

93. Find two values of b that will make
$9x^2 + bx + 25$ a perfect square trinomial.

94. Find a value of c that will make $49x^2 - 42x + c$
a perfect square trinomial.

Applying the Concepts

Compound Interest   If \$100 is invested in an account with an annual interest rate of r compounded twice a year, then the amount of money in the account at the end of the year is given by the formula

$$A = 100\left(1 + r + \frac{r^2}{4}\right)$$

If the annual interest rate is 12%, find the amount of money in the account at the end of 1 year

95. . . . without factoring the right side of the formula.

96. . . . by first factoring the right side of the formula completely.

97. Volume Between Two Cubes   Recall that the volume of a cube is $V = (\text{side})^3$. A cube with side of length r inches is placed inside a cube of side p inches. Write an expression, in factored form, for the volume of the space between the two cubes.

98. Binomial Squares and Area   The area model you completed in Problem 71 on page 276 gives a visual interpretation of the factoring problem $a^2 + 2ab + b^2$. Using that diagram as a reference, draw a similar diagram for the factorization of $4x^2 + 4xy + y^2$.

99. Binomial Squares and Area   Refer to the preceding exercise. The following diagram is the start of an area model for $x^2 - 2xy + y^2 = (x - y)^2$. Finish labeling this area model.

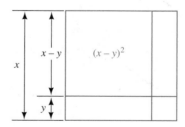

100. Genetics   The Austrian monk *Gregor Mendel* (1822–1884) showed that genes from both the mother and father determine what traits are inherited by a new generation. The Punnett square shown on the next page is used by biologists to study the probabilities associated with inherited traits. Explain how a Punnett square is modeled on the square of a binomial.

Male genes

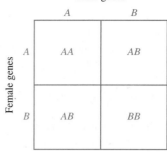

Female genes

	A	B
A	AA	AB
B	AB	BB

Review Problems

The following problems review material we covered in Sections 3.5, 3.6, 3.7, and 3.8.

Solve each system by graphing. [3.5]

101. $x - y = -2$
$x + y = 4$

102. $2x + 4y = 20$
$3x - 2y = 6$

Solve each system by the addition method. [3.6]

103. $x + y = 4$
$x - y = -2$

104. $5x + 2y = 11$
$7x + 8y = 18$

Solve each system by the substitution method. [3.7]

105. $x + y = 2$
$y = 2x - 1$

106. $4x + 2y = 3$
$x = 4y - 1$

107. Ticket Problem   A total of 925 tickets were sold for a game for a total of $1,150. If adult tickets sold for $2.00 and children's tickets sold for $1.00, how many of each kind of ticket were sold? [3.8]

108. Interest   Mr. Jones has $20,000 to invest. He invests part at 6% and the rest at 7%. If he earns $1,280 in interest after 1 year, how much did he invest at each rate? [3.8]

109. Mixture Problem   How many gallons of 20% alcohol solution and 50% alcohol solution must be mixed to get 9 gallons of 30% alcohol solution? [3.8]

110. Mixture Problem   How many ounces of 30% hydrochloric acid solution and 80% hydrochloric acid solution must be mixed to get 10 ounces of 50% hydrochloric acid solution? [3.8]

7.6   ## Review of Systems of Equations in Two Variables

In Chapter 3 we found that two linear equations considered together form a *linear system* of equations. For example,

$$3x - 2y = 6$$
$$2x + 4y = 20$$

is a linear system. The solution set to the system is the set of all ordered pairs that satisfy both equations. If we graph each equation on the same set of axes, we can see the solution set (see Figure 1).

The point (4, 3) lies on both lines and therefore must satisfy both equations. It is obvious from the graph that it is the only point that does so. The solution set for the system is {(4, 3)}.

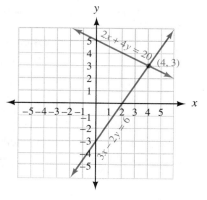

FIGURE I

More generally, if $a_1x + b_1y = c_1$ and $a_2x + b_2y = c_2$ are linear equations, then the solution set for the system

$$a_1x + b_1y = c_1$$
$$a_2x + b_2y = c_2$$

can be illustrated through one of the graphs in Figure 2.

Case I   The two lines intersect at one and only one point. The coordinates of the point give the solution to the system. This is what usually happens.

Case II   The lines are parallel and therefore have no points in common. The solution set to the system is the empty set, $\emptyset$. In this case, we say the equations are *inconsistent*.

Case III   The lines coincide. That is, their graphs represent the same line. The solution set consists of all ordered pairs that satisfy either equation. In this case, the equations are said to be *dependent*.

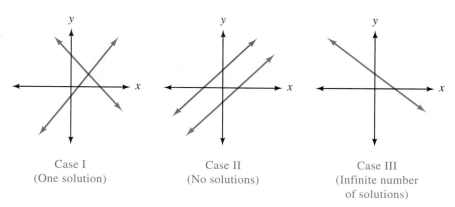

Case I	Case II	Case III
(One solution)	(No solutions)	(Infinite number of solutions)

FIGURE 2

THE ADDITION METHOD

In the beginning of this section we found the solution set for the system

$$3x - 2y = 6$$
$$2x + 4y = 20$$

by graphing each equation and then reading the solution set from the graph. Solving a system of linear equations by graphing is the least accurate method. If the coordinates of the point of intersection are not integers, it can be very difficult to read the solution set from the graph. As you know, there is another method of solving a linear system that does not depend on the graph. It is called the *addition method*.

EXAMPLE I   Solve the system.

$$4x + 3y = 10$$
$$2x + y = 4$$

Solution   If we multiply the bottom equation by -3, the coefficients of y in the resulting equation and the top equation will be opposites:

$$4x + 3y = 10 \xrightarrow{\text{No change}} 4x + 3y = 10$$
$$2x + y = 4 \xrightarrow{\text{Multiply by } -3} -6x - 3y = -12$$

Adding the left and right sides of the resulting equations, we have

$$\begin{array}{r} 4x + 3y = 10 \\ -6x - 3y = -12 \\ \hline -2x = -2 \end{array}$$

The result is a linear equation in one variable. We have eliminated the variable y from the equations by addition. (It is for this reason we call this method of solving a linear system the *addition method*.) Solving $-2x = -2$ for x, we have

$$x = 1$$

This is the x-coordinate of the solution to our system. To find the y-coordinate, we substitute $x = 1$ into any of the equations containing both the variables x and y. Let's try the second equation in our original system:

$$2(1) + y = 4$$
$$2 + y = 4$$
$$y = 2$$

This is the y-coordinate of the solution to our system. The ordered pair $(1, 2)$ is the solution to the system.

Note: If we had put $x = 1$ into the first equation in our system, we would have obtained $y = 2$ also:

$$4(1) + 3y = 10$$
$$3y = 6$$
$$y = 2$$

E X A M P L E   2   Solve the system.

$$3x - 5y = -2$$
$$2x - 3y = 1$$

Solution   We can eliminate either variable. Let's decide to eliminate the variable x. We can do so by multiplying the top equation by 2 and the bottom equation by -3, and then adding the left and right sides of the resulting equations:

$$3x - 5y = -2 \xrightarrow{\text{Multiply by 2}} 6x - 10y = -4$$
$$2x - 3y = 1 \xrightarrow[\text{Multiply by } -3]{} \underline{-6x + 9y = -3}$$
$$-y = -7$$
$$y = 7$$

The y-coordinate of the solution to the system is 7. Substituting this value of y into any of the equations with both x- and y-variables gives $x = 11$. The solution to the system is $(11, 7)$. It is the only ordered pair that satisfies both equations.

E X A M P L E   3   Solve the system.

$$2x - 3y = 4$$
$$4x + 5y = 3$$

Solution   We can eliminate x by multiplying the top equation by -2 and adding it to the bottom equation:

$$2x - 3y = 4 \xrightarrow{\text{Multiply by } -2} -4x + 6y = -8$$
$$4x + 5y = 3 \xrightarrow[\text{No change}]{} \underline{4x + 5y = 3}$$
$$11y = -5$$
$$y = -\frac{5}{11}$$

The y-coordinate of our solution is $-\frac{5}{11}$. If we were to substitute this value of y back into either of our original equations, we would find the arithmetic necessary to solve for x cumbersome. For this reason, it is probably best to go back to the original system and solve it a second time—for x instead of y. Here is how we do that:

$$2x - 3y = 4 \xrightarrow{\text{Multiply by 5}} 10x - 15y = 20$$

$$4x + 5y = 3 \xrightarrow[\text{Multiply by 3}]{} 12x + 15y = 9$$

$$22x = 29$$

$$x = \frac{29}{22}$$

The solution to our system is $(\frac{29}{22}, -\frac{5}{11})$.

EXAMPLE 4 Solve the system.

$$5x - 2y = 1$$

$$-10x + 4y = 3$$

Solution We can eliminate y by multiplying the first equation by 2 and adding the result to the second equation:

$$5x - 2y = 1 \xrightarrow{\text{Multiply by 2}} 10x - 4y = 2$$

$$-10x + 4y = 3 \xrightarrow[\text{No change}]{} -10x + 4y = 3$$

$$0 = 5$$

The result is the false statement $0 = 5$, which indicates there is no solution to the system. If we were to graph the two lines, we would find that they are parallel. In a case like this, we say the system is *inconsistent*. Whenever both variables have been eliminated and the resulting statement is false, the solution set for the system will be the empty set, $\varnothing$.

EXAMPLE 5 Solve the system.

$$4x - 3y = 2$$

$$8x - 6y = 4$$

Solution Multiplying the top equation by -2 and adding, we can eliminate the variable x:

$$4x - 3y = 2 \xrightarrow{\text{Multiply by } -2} -8x + 6y = -4$$

$$8x - 6y = 4 \xrightarrow[\text{No change}]{} 8x - 6y = 4$$

$$0 = 0$$

Both variables have been eliminated and the resulting statement $0 = 0$ is true. In this case the lines coincide and the system is said to be *dependent*. The solution set consists of all ordered pairs that satisfy either equation. We can write the solution set as $\{(x, y) | 4x - 3y = 2\}$ or $\{(x, y) | 8x - 6y = 4\}$.

The previous two examples illustrate the two special cases in which the graphs of the equations in the system either coincide or are parallel. In both cases the left-hand sides of the equations were multiples of each other. In the case of the dependent equations the right-hand sides were also multiples. We can generalize these observations as follows:

The equations in the system

$$a_1x + b_1y = c_1$$

$$a_2x + b_2y = c_2$$

will be inconsistent (their graphs are parallel lines) if

$$\frac{a_1}{a_2} = \frac{b_1}{b_2} \neq \frac{c_1}{c_2}$$

and will be dependent (their graphs will coincide) if

$$\frac{a_1}{a_2} = \frac{b_1}{b_2} = \frac{c_1}{c_2}$$

EXAMPLE 6   Solve the system.

$$\frac{1}{2}x - \frac{1}{3}y = 2$$

$$\frac{1}{4}x + \frac{2}{3}y = 6$$

Solution   Although we could solve this system without clearing the equations of fractions, there is probably less chance for error if we have only integer coefficients to work with. So let's begin by multiplying both sides of the top equation by 6, and both sides of the bottom equation by 12, to clear each equation of fractions:

$$\frac{1}{2}x - \frac{1}{3}y = 2 \xrightarrow{\text{Times 6}} 3x - 2y = 12$$

$$\frac{1}{4}x + \frac{2}{3}y = 6 \xrightarrow{\text{Times 12}} 3x + 8y = 72$$

Now we can eliminate x by multiplying the top equation by -1 and leaving the bottom equation unchanged:

$$3x - 2y = 12 \xrightarrow{\text{Times } -1} -3x + 2y = -12$$

$$3x + 8y = 72 \xrightarrow[\text{No change}]{} \underline{3x + 8y = 72}$$

$$10y = 60$$

$$y = 6$$

We can substitute $y = 6$ into any equation that contains both x and y. Let's use $3x - 2y = 12$.

$$3x - 2(6) = 12$$

$$3x - 12 = 12$$

$$3x = 24$$
$$x = 8$$

The solution to the system is $(8, 6)$.

THE SUBSTITUTION METHOD

We now review a third method of solving a linear system. The method is called the *substitution method* and is shown in the following examples.

EXAMPLE 7   Solve the system.

$$2x - 3y = -6$$
$$y = 3x - 5$$

Solution   The second equation tells us y is $3x - 5$. Substituting the expression $3x - 5$ for y in the first equation, we have

$$2x - 3(3x - 5) = -6$$

The result of the substitution is the elimination of the variable y. Solving the resulting linear equation in x as usual, we have

$$2x - 9x + 15 = -6$$
$$-7x + 15 = -6$$
$$-7x = -21$$
$$x = 3$$

Putting $x = 3$ into the second equation in the original system, we have

$$y = 3(3) - 5$$
$$= 9 - 5$$
$$= 4$$

The solution to the system is $(3, 4)$.

EXAMPLE 8   Solve by substitution.

$$2x + 3y = 5$$
$$x - 2y = 6$$

Solution   In order to use the substitution method, we must solve one of the two equations for x or y. We can solve for x in the second equation by adding $2y$ to both sides:

$$x - 2y = 6$$
$$x = 2y + 6 \qquad \text{Add } 2y \text{ to both sides.}$$

Substituting the expression $2y + 6$ for x in the first equation of our system, we have

$$2(2y + 6) + 3y = 5$$
$$4y + 12 + 3y = 5$$
$$7y + 12 = 5$$
$$7y = -7$$
$$y = -1$$

Using $y = -1$ in either equation in the original system, we find $x = 4$. The solution is $(4, -1)$.

Note: Both the substitution method and the addition method can be used to solve any system of linear equations in two variables. Systems like the one in Example 7, however, are easier to solve using the substitution method, since one of the variables is already written in terms of the other. A system like the one in Example 2 is easier to solve using the addition method, since solving for one of the variables would lead to an expression involving fractions. The system in Example 8 could be solved easily by either method, since solving the second equation for x is a one-step process.

APPLICATIONS

To review, here is the Blueprint for Problem Solving that we use to solve application problems that involve systems of equations.

Blueprint for Problem Solving Using a System of Equations

Step 1: **Read** the problem, and then mentally **list** the items that are known and the items that are unknown.

Step 2: **Assign variables** to each of the unknown items. That is, let $x =$ one of the unknown items and $y =$ the other unknown item. Then **translate** the other **information** in the problem to expressions involving the two variables.

Step 3: **Reread** the problem, and then **write a system of equations,** using the items and variables listed in steps 1 and 2, that describes the situation.

Step 4: **Solve the system** found in step 3.

Step 5: **Write your answers** using complete sentences.

Step 6: **Reread** the problem, and **check** your solution with the original words in the problem.

EXAMPLE 9

It takes 2 hours for a boat to travel 28 miles downstream (with the current). The same boat can travel 18 miles upstream (against the current) in 3 hours. What is the speed of the boat in still water, and what is the speed of the current of the river?

Solution

Step 1: *Read and list.*

A boat travels 18 miles upstream and 28 miles downstream. The trip upstream takes 3 hours. The trip downstream takes 2 hours. We don't know the speed of the boat or the speed of the current.

Step 2: *Assign variables and translate information.*

Let x = the speed of the boat in still water and let y = the speed of the current. The average speed (rate) of the boat upstream is $x - y$, since it is traveling against the current. The rate of the boat downstream is $x + y$, since the boat is traveling with the current.

Step 3: *Write a system of equations.*

Putting the information into a table, we have

	d (distance, miles)	r (rate, mph)	t (time, h)
Upstream	18	$x - y$	3
Downstream	28	$x + y$	2

The formula for the relationship between distance d, rate r, and time t is $d = rt$ (the rate equation). Since $d = r \cdot t$, the system we need to solve the problem is

$$18 = (x - y) \cdot 3$$

$$28 = (x + y) \cdot 2$$

which is equivalent to

$$6 = x - y$$

$$14 = x + y$$

Step 4: **Solve the system.**

Adding the two equations, we have

$$20 = 2x$$

$$x = 10$$

Substituting $x = 10$ into $14 = x + y$, we see that

$$y = 4$$

Step 5: **Write answers.**

The speed of the boat in still water is 10 miles per hour; the speed of the current is 4 miles per hour.

Step 6: **Reread and check.**

The boat travels at $10 + 4 = 14$ miles per hour downstream, so in 2 hours it will travel $14 \cdot 2 = 28$ miles. The boat travels at $10 - 4 = 6$ miles per hour upstream, so in 3 hours it will travel $6 \cdot 3 = 18$ miles.

Getting Ready for Class

After reading through the preceding section, respond in your own words and in complete sentences.

A. Two linear equations, each with the same two variables, form a system of equations. How do we define a solution to this system? That is, what form will a solution have, and what properties does a solution possess?

B. When would substitution be more efficient than the addition method in solving two linear equations?

C. Explain what an inconsistent system of linear equations looks like graphically and what would result algebraically when attempting to solve the system.

D. When might the graphing method of solving a system of equations be more desirable than the other techniques, and when might it be less desirable?

PROBLEM SET 7.6

Solve each system by graphing both equations on the same set of axes and then reading the solution from the graph.

1. $3x - 2y = 6$
$x - y = 1$

2. $5x - 2y = 10$
$x - y = -1$

3. $y = \dfrac{3}{5}x - 3$
$2x - y = -4$

4. $y = \dfrac{1}{2}x - 2$
$2x - y = -1$

5. $y = \dfrac{1}{2}x$
$y = -\dfrac{3}{4}x + 5$

6. $y = \dfrac{2}{3}x$
$y = -\dfrac{1}{3}x + 6$

7. $3x + 3y = -2$
$y = -x + 4$

8. $2x - 2y = 6$
$y = x - 3$

9. $2x - y = 5$
$y = 2x - 5$

10. $x + 2y = 5$
$y = -\dfrac{1}{2}x + 3$

Solve each of the following systems by the addition method.

11. $x + y = 5$
$3x - y = 3$

12. $x - y = 4$
$-x + 2y = -3$

13. $3x + y = 4$
$4x + y = 5$

14. $6x - 2y = -10$
$6x + 3y = -15$

15. $3x - 2y = 6$
$6x - 4y = 12$

16. $4x + 5y = -3$
$-8x - 10y = 3$

17. $x + 2y = 0$
$2x - 6y = 5$

18. $x + 3y = 3$
$2x - 9y = 1$

19. $2x - 5y = 16$
$4x - 3y = 11$

20. $5x - 3y = -11$
$7x + 6y = -12$

21. $6x + 3y = -1$
$9x + 5y = 1$

22. $5x + 4y = -1$
$7x + 6y = -2$

23. $4x + 3y = 14$
$9x - 2y = 14$

24. $7x - 6y = 13$
$6x - 5y = 11$

25. $2x - 5y = 3$
$-4x + 10y = 3$

26. $3x - 2y = 1$
$-6x + 4y = -2$

27. $\dfrac{1}{4}x - \dfrac{1}{6}y = -2$
$-\dfrac{1}{6}x + \dfrac{1}{5}y = 4$

28. $-\dfrac{1}{3}x + \dfrac{1}{4}y = 0$
$\dfrac{1}{5}x - \dfrac{1}{10}y = 1$

29. $\dfrac{1}{2}x + \dfrac{1}{3}y = 13$
$\dfrac{2}{5}x + \dfrac{1}{4}y = 10$

30. $\dfrac{1}{2}x + \dfrac{1}{3}y = \dfrac{2}{3}$
$\dfrac{2}{3}x + \dfrac{2}{5}y = \dfrac{14}{15}$

31. $\dfrac{2}{3}x + \dfrac{2}{5}y = 4$
$\dfrac{1}{3}x - \dfrac{1}{2}y = -\dfrac{1}{3}$

32. $\dfrac{1}{2}x - \dfrac{1}{3}y = \dfrac{5}{6}$
$-\dfrac{2}{5}x + \dfrac{1}{2}y = -\dfrac{9}{10}$

Solve each of the following systems by the substitution method.

33. $7x - y = 24$
$x = 2y + 9$

34. $3x - y = -8$
$y = 6x + 3$

35. $6x - y = 10$
$y = -\dfrac{3}{4}x - 1$

36. $2x - y = 6$
$y = -\dfrac{4}{3}x + 1$

37. $x - y = 4$
$2x - 3y = 6$

38. $x + y = 3$
$2x + 3y = -4$

39. $y = 3x - 2$
$y = 4x - 4$

40. $y = 5x - 2$
$y = -2x + 5$

41. $2x - y = 5$
$4x - 2y = 10$

42. $-10x + 8y = -6$
$y = \dfrac{5}{4}x$

43. $\dfrac{1}{3}x - \dfrac{1}{2}y = 0$
$x = \dfrac{3}{2}y$

44. $\dfrac{2}{5}x - \dfrac{2}{3}y = 0$
$y = \dfrac{3}{5}x$

You may want to read Example 3 again before solving the systems that follow.

45. $4x - 7y = 3$
$5x + 2y = -3$

46. $3x - 4y = 7$
$6x - 3y = 5$

47. $9x - 8y = 4$
$2x + 3y = 6$

48. $4x - 7y = 10$
$-3x + 2y = -9$

49. $3x - 5y = 2$
$7x + 2y = 1$

50. $4x - 3y = -1$
$5x + 8y = 2$

51. Multiply both sides of the second equation in the following system by 100, and then solve as usual.

$$x + y = 10{,}000$$
$$0.06x + 0.05y = 560$$

52. Multiply both sides of the second equation in the following system by 10, and then solve as usual.

$$x + y = 12$$
$$0.20x + 0.50y = 0.30(12)$$

53. What value of c will make the following system a dependent system (one in which the lines coincide)?

$$6x - 9y = 3$$
$$4x - 6y = c$$

54. What value of c will make the following system a dependent system?

$$5x - 7y = c$$
$$-15x + 21y = 9$$

Number Problems

55. One number is 3 more than twice another. The sum of the numbers is 18. Find the two numbers.

56. The sum of two numbers is 32. One of the numbers is 4 less than 5 times the other. Find the two numbers.

Ticket and Interest Problems

57. If tickets for a show cost $2.00 for adults and $1.50 for children, how many of each kind of ticket were sold if a total of 300 tickets were sold for $525?

58. A man invests $17,000 in two accounts. One account earns 5% interest per year and the other 6.5%. If his total interest after 1 year is $970, how much did he invest at each rate?

Mixture Problems

59. A mixture of 16% disinfectant solution is to be made from 20% and 14% disinfectant solutions. How much of each solution should be used if 15 gallons of the 16% solution are needed?

60. How much 25% antifreeze and 50% antifreeze should be combined to give 40 gallons of 30% antifreeze?

Rate Problems

61. It takes a boat 2 hours to travel 24 miles downstream and 3 hours to travel 18 miles upstream. What is the speed of the boat in still water? What is the speed of the current of the river?

62. A boat on a river travels 20 miles downstream in only 2 hours. It takes the same boat 6 hours to travel 12 miles upstream. What are the speed of the boat and the speed of the current?

63. An airplane flying with the wind can cover a certain distance in 2 hours. The return trip against the wind takes $2\frac{1}{2}$ hours. How fast is the plane and what is the speed of the wind, if the distance is 600 miles?

64. An airplane covers a distance of 1,500 miles in 3 hours when it flies with the wind and $3\frac{1}{3}$ hours when it flies against the wind. What is the speed of the plane in still air?

Review Problems

The problems that follow review material we covered in Chapter 4.

Simplify each expression. Write all answers with positive exponents only.

65. $(3x^2)^3(5y^4)^2$

66. 3^{-2}

67. $\dfrac{x^5}{x^{-2}}$

68. $\dfrac{(40a^3b^4)(7ab^2)}{(14a^{-2}b)(20a^2b^5)}$

69. $(5 \times 10^{-8})(4 \times 10^3)$

70. Add: $(7x^2 - 5x + 3) + (4x^2 + 3x - 2)$

71. Subtract $5a^2 - 2a + 1$ from $7a^2 + 8a + 9$.

Multiply.

72. $(3x - 7)(2x + 1)$

73. $(4a + 9)(4a - 9)$

74. $(3y + 1)^2$

75. $6xy^2(4x^2 - 3xy + 2y^2)$

76. Divide: $\dfrac{10x^4 - 5x^3 + 20x^2}{10x^3}$

77. Divide: $\dfrac{x^2 - 5x + 8}{x - 3}$

78. Divide $4x^2 + 20x + 25$ by $2x + 5$.

7.7 ## Systems of Linear Equations in Three Variables

A solution to an equation in three variables such as

$$2x + y - 3z = 6$$

is an ordered triple of numbers (x, y, z). For example, the ordered triples $(0, 0, -2)$, $(2, 2, 0)$, and $(0, 9, 1)$ are solutions to the equation $2x + y - 3z = 6$, since they produce a true statement when their coordinates are replaced for x, y, and z in the equation.

> **DEFINITION** The **solution set** for a system of three linear equations in three variables is the set of ordered triples that satisfy all three equations.

EXAMPLE I   Solve the system.

$$x + y + z = 6 \quad (1)$$
$$2x - y + z = 3 \quad (2)$$
$$x + 2y - 3z = -4 \quad (3)$$

Solution   We want to find the ordered triple (x, y, z) that satisfies all three equations. We have numbered the equations so it will be easier to keep track of where they are and what we are doing.

There are many ways to proceed. The main idea is to take two different pairs of equations and eliminate the same variable from each pair. We begin by adding equations (1) and (2) to eliminate the y-variable. The resulting equation is numbered (4):

$$\begin{array}{rl} x + y + z = 6 & (1) \\ 2x - y + z = 3 & (2) \\ \hline 3x \quad\quad + 2z = 9 & (4) \end{array}$$

Adding twice equation (2) to equation (3) will also eliminate the variable y. The resulting equation is numbered (5):

$$\begin{array}{rl} 4x - 2y + 2z = 6 & \text{Twice (2)} \\ x + 2y - 3z = -4 & (3) \\ \hline 5x \quad\quad - z = 2 & (5) \end{array}$$

Equations (4) and (5) form a linear system in two variables. By multiplying equation (5) by 2 and adding the result to equation (4), we succeed in eliminating the variable z from the new pair of equations:

$$\begin{array}{rl} 3x + 2z = 9 & (4) \\ 10x - 2z = 4 & \text{Twice (5)} \\ \hline 13x \quad\quad = 13 & \\ x = 1 & \end{array}$$

Substituting $x = 1$ into equation (4), we have

$$3(1) + 2z = 9$$
$$2z = 6$$
$$z = 3$$

Using $x = 1$ and $z = 3$ in equation (1) gives us

$$1 + y + 3 = 6$$
$$y + 4 = 6$$
$$y = 2$$

The solution is the ordered triple $(1, 2, 3)$.

E X A M P L E   2   Solve the system.

$$2x + y - z = 3 \quad (1)$$
$$3x + 4y + z = 6 \quad (2)$$
$$2x - 3y + z = 1 \quad (3)$$

Solution   It is easiest to eliminate z from the equations. The equation produced by adding (1) and (2) is

$$5x + 5y = 9 \quad (4)$$

The equation that results from adding (1) and (3) is

$$4x - 2y = 4 \quad (5)$$

Equations (4) and (5) form a linear system in two variables. We can eliminate the variable y from this system as follows:

$$5x + 5y = 9 \xrightarrow{\text{Multiply by 2}} 10x + 10y = 18$$
$$4x - 2y = 4 \xrightarrow[\text{Multiply by 5}]{} \underline{20x - 10y = 20}$$
$$30x \qquad = 38$$
$$x = \frac{38}{30}$$
$$= \frac{19}{15}$$

Substituting $x = \frac{19}{15}$ into equation (5) or equation (4) and solving for y gives

$$y = \frac{8}{15}$$

Using $x = \frac{19}{15}$ and $y = \frac{8}{15}$ in equation (1), (2), or (3) and solving for z results in

$$z = \frac{1}{15}$$

The ordered triple that satisfies all three equations is $(\frac{19}{15}, \frac{8}{15}, \frac{1}{15})$.

E X A M P L E   3   Solve the system.

$$2x + 3y - z = 5 \quad (1)$$
$$4x + 6y - 2z = 10 \quad (2)$$
$$x - 4y + 3z = 5 \quad (3)$$

Solution   Multiplying equation (1) by -2 and adding the result to equation (2) looks like this:

$$-4x - 6y + 2z = -10 \qquad -2 \text{ times (1)}$$
$$\underline{4x + 6y - 2z = \quad 10} \qquad (2)$$
$$0 = \quad 0$$

All three variables have been eliminated, and we are left with a true statement. As was the case in Section 7.6, this implies that the two equations are dependent. With a system of three equations in three variables, however, a dependent system can have no solution or an infinite number of solutions. After we have concluded the examples in this section, we will discuss the geometry behind these systems. Doing so will give you some additional insight into dependent systems.

E X A M P L E 4 Solve the system.

$$x - 5y + 4z = 8 \quad (1)$$
$$3x + y - 2z = 7 \quad (2)$$
$$-9x - 3y + 6z = 5 \quad (3)$$

Solution Multiplying equation (2) by 3 and adding the result to equation (3) produces

$$
\begin{array}{ll}
9x + 3y - 6z = 21 & \text{3 times (2)} \\
\underline{-9x - 3y + 6z = 5} & \text{(3)} \\
 0 = 26 &
\end{array}
$$

In this case all three variables have been eliminated, and we are left with a false statement. The two equations are inconsistent; there are no ordered triples that satisfy both equations. The solution set for the system is the empty set, $\varnothing$. If equations (2) and (3) have no ordered triples in common, then certainly (1), (2), and (3) do not either.

E X A M P L E 5 Solve the system.

$$x + 3y = 5 \quad (1)$$
$$6y + z = 12 \quad (2)$$
$$x - 2z = -10 \quad (3)$$

Solution It may be helpful to rewrite the system as

$$x + 3y = 5 \quad (1)$$
$$6y + z = 12 \quad (2)$$
$$x - 2z = -10 \quad (3)$$

Equation (2) does not contain the variable x. If we multiply equation (3) by -1 and add the result to equation (1), we will be left with another equation that does not contain the variable x:

$$
\begin{array}{ll}
x + 3y = 5 & \text{(1)} \\
\underline{-x + 2z = 10} & \text{-1 times (3)} \\
 3y + 2z = 15 & \text{(4)}
\end{array}
$$

Equations (2) and (4) form a linear system in two variables. Multiplying equation (2) by -2 and adding the result to equation (4) eliminates the variable z:

$$6y + z = 12 \xrightarrow{\text{Multiply by } -2} -12y - 2z = -24$$
$$3y + 2z = 15 \xrightarrow{\text{No change}} \underline{ 3y + 2z = 15}$$
$$-9y = -9$$
$$y = 1$$

Using $y = 1$ in equation (4) and solving for z, we have

$$z = 6$$

Substituting $y = 1$ into equation (1) gives

$$x = 2$$

The ordered triple that satisfies all three equations is $(2, 1, 6)$.

EXAMPLE 6 A coin collection consists of 14 coins with a total value of $1.35. If the coins are nickels, dimes, and quarters, and the number of nickels is 3 less than twice the number of dimes, how many of each coin is there in the collection?

Solution This problem will require three variables and three equations.

Step 1: Read and list.

We have 14 coins with a total value of $1.35. The coins are nickels, dimes, and quarters. The number of nickels is 3 less than twice the number of dimes. We do not know how many of each coin we have.

Step 2: Assign variables and translate information.

Since we have three types of coins, we will have to use three variables. Let's let $x = $ the number of nickels, $y = $ the number of dimes, and $z = $ the number of quarters.

Step 3: Write a system of equations.

Since the total number of coins is 14, our first equation is

$$x + y + z = 14$$

Since the number of nickels is 3 less than twice the number of dimes, a second equation is

$$x = 2y - 3 \qquad \text{which is equivalent to} \qquad x - 2y = -3$$

Our last equation is obtained by considering the value of each coin and the total value of the collection. Let's write the equation in terms of cents, so we won't have to clear it of decimals later.

$$5x + 10y + 25z = 135$$

Here is our system, with the equations numbered for reference:

$$x + y + z = 14 \qquad (1)$$
$$x - 2y = -3 \qquad (2)$$
$$5x + 10y + 25z = 135 \qquad (3)$$

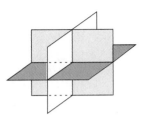

Case 1 The three planes have exactly one point in common. In this case we get one solution to our system, as in Examples 1, 2, and 5.

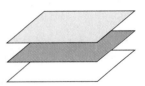

Case 2 The three planes have no points in common because they are all parallel to one another. The system they represent is an inconsistent system.

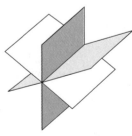

Case 3 The three planes intersect in a line. Any point on the line is a solution to the system of equations represented by the planes, so there is an infinite number of solutions to the system. This is an example of a dependent system.

Step 4: *Solve the system.*

Let's begin by eliminating x from the first and second equations, and the first and third equations. Adding -1 times the second equation to the first equation gives us an equation in only y and z. We call this equation (4).

$$3y + z = 17 \qquad (4)$$

Adding -5 times equation (1) to equation (3) gives us

$$5y + 20z = 65 \qquad (5)$$

We can eliminate z from equations (4) and (5) by adding -20 times (4) to (5). Here is the result:

$$-55y = -275$$
$$y = 5$$

Substituting $y = 5$ into equation (4) gives us $z = 2$. Substituting $y = 5$ and $z = 2$ into equation (1) gives us $x = 7$.

Step 5: *Write answers.*

The collection consists of 7 nickels, 5 dimes, and 2 quarters.

Step 6: *Reread and check.*

The total number of coins is $7 + 5 + 2 = 14$. The number of nickels, 7, is 3 less than twice the number of dimes, 5. To find the total value of the collection, we have

The value of the 7 nickels is $7(0.05) = \$0.35$
The value of the 5 dimes is $5(0.10) = \$0.50$
The value of the 2 quarters is $2(0.25) = \$0.50$

The total value of the collection is $\$1.35$

THE GEOMETRY BEHIND EQUATIONS IN THREE VARIABLES

We can graph an ordered triple on a coordinate system with three axes. The graph will be a point in space. The coordinate system is drawn in perspective; you have to imagine that the x-axis comes out of the paper and is perpendicular to both the y-axis and the z-axis. To graph the point $(3, 4, 5)$, we move 3 units in the x-direction, 4 units in the y-direction, and then 5 units in the z-direction, as shown in Figure 1.

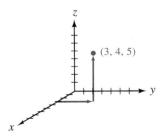

FIGURE 1

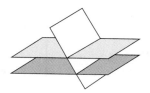

Case 4   Two of the planes are parallel; the third plane intersects each of the parallel planes. In this case the three planes have no points in common. There is no solution to the system; it is an inconsistent system.

Although in actual practice it is sometimes difficult to graph equations in three variables, if we were to graph a linear equation in three variables, we would find that the graph was a plane in space. A system of three equations in three variables is represented by three planes in space.

There are a number of possible ways in which these three planes can intersect, some of which are shown in the margin on this page. And there are still other possibilities that are not among those shown in the margin here and on page 454.

In Example 3 we found that equations (1) and (2) were dependent equations. They represent the same plane. That is, they have all their points in common. But the system of equations that they came from has either no solution or an infinite number of solutions. It all depends on the third plane. If the third plane coincides with the first two, then the solution to the system is a plane. If the third plane is parallel to the first two, then there is no solution to the system. And, finally, if the third plane intersects the first two, but does not coincide with them, then the solution to the system is that line of intersection.

In Example 4 we found that trying to eliminate a variable from the second and third equations resulted in a false statement. This means that the two planes represented by these equations are parallel. It makes no difference where the third plane is; there is no solution to the system in Example 4. (If we were to graph the three planes from Example 4, we would obtain a diagram similar to Case 2 or Case 4 in the margin.)

If, in the process of solving a system of linear equations in three variables, we eliminate all the variables from a pair of equations and are left with a false statement, we will say the system is inconsistent. If we eliminate all the variables and are left with a true statement, then we will say the system is a dependent one.

Getting Ready for Class

After reading through the preceding section, respond in your own words and in complete sentences.

A. What is an ordered triple of numbers?

B. Explain what it means for (1, 2, 3) to be a solution to a system of linear equations in three variables.

C. Explain in a general way the procedure you would use to solve a system of three linear equations in three variables.

D. How do you know when a system of linear equations in three variables has no solution?

PROBLEM SET 7.7

Solve the following systems.

1. $x + y + z = 4$
$x - y + 2z = 1$
$x - y - 3z = -4$

2. $x - y - 2z = -1$
$x + y + z = 6$
$x + y - z = 4$

3. $x + y + z = 6$
$x - y + 2z = 7$
$2x - y - 4z = -9$

4. $x + y + z = 0$
$x + y - z = 6$
$x - y + 2z = -7$

5. $x + 2y + z = 3$
$2x - y + 2z = 6$
$3x + y - z = 5$

6. $2x + y - 3z = -14$
$x - 3y + 4z = 22$
$3x + 2y + z = 0$

7. $2x + 3y - 2z = 4$
$x + 3y - 3z = 4$
$3x - 6y + z = -3$

8. $4x + y - 2z = 0$
$2x - 3y + 3z = 9$
$-6x - 2y + z = 0$

9. $-x + 4y - 3z = 2$
$2x - 8y + 6z = 1$
$3x - y + z = 0$

10. $4x + 6y - 8z = 1$
$-6x - 9y + 12z = 0$
$x - 2y - 2z = 3$

11. $\frac{1}{2}x - y + z = 0$

$2x + \frac{1}{3}y + z = 2$

$x + y + z = -4$

12. $\frac{1}{3}x + \frac{1}{2}y + z = -1$

$x - y + \frac{1}{5}z = 1$

$x + y + z = 5$

13. $2x - y - 3z = 1$
$x + 2y + 4z = 3$
$4x - 2y - 6z = 2$

14. $3x + 2y + z = 3$
$x - 3y + z = 4$
$-6x - 4y - 2z = 1$

15. $2x - y + 3z = 4$
$x + 2y - z = -3$
$4x + 3y + 2z = -5$

16. $6x - 2y + z = 5$
$3x + y + 3z = 7$
$x + 4y - z = 4$

17. $x + y = 9$
$y + z = 7$
$x - z = 2$

18. $x - y = -3$
$x + z = 2$
$y - z = 7$

19. $2x + y = 2$
$y + z = 3$
$4x - z = 0$

20. $2x + y = 6$
$3y - 2z = -8$
$x + z = 5$

21. $2x - 3y = 0$
$6y - 4z = 1$
$x + 2z = 1$

22. $3x + 2y = 3$
$y + 2z = 2$
$6x - 4z = 1$

23. $\frac{1}{2}x + \frac{2}{3}y = \frac{5}{2}$

$\frac{1}{5}x - \frac{1}{2}z = -\frac{3}{10}$

$\frac{1}{3}y - \frac{1}{4}z = \frac{3}{4}$

24. $\frac{1}{2}x - \frac{1}{3}y = \frac{1}{6}$

$\frac{1}{3}y - \frac{1}{3}z = 1$

$\frac{1}{5}x - \frac{1}{2}z = -\frac{4}{5}$

25. $\frac{1}{2}x - \frac{1}{4}y + \frac{1}{2}z = -2$

$\frac{1}{4}x - \frac{1}{12}y - \frac{1}{3}z = \frac{1}{4}$

$\frac{1}{6}x + \frac{1}{3}y - \frac{1}{2}z = \frac{3}{2}$

26. $\frac{1}{2}x + \frac{1}{2}y + z = \frac{1}{2}$

$\frac{1}{2}x - \frac{1}{4}y - \frac{1}{4}z = 0$

$\frac{1}{4}x + \frac{1}{12}y + \frac{1}{6}z = \frac{1}{6}$

Applying the Concepts

27. Electric Current In the following diagram of an electrical circuit, x, y, and z represent the amount of current (in amperes) flowing across the 5-ohm, 20-ohm, and 10-ohm resistors, respectively. (In circuit diagrams resistors are represented by $-W$ and potential differences by $\dashv\vdash$.)

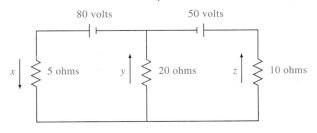

The system of equations used to find the three currents x, y, and z is

$$x - y - z = 0$$
$$5x + 20y = 80$$
$$20y - 10z = 50$$

Solve the system for all variables.

28. Cost of a Rental Car   If a car rental company charges $10 a day and 8¢ a mile to rent one of its cars, then the cost z, in dollars, to rent a car for x days and drive y miles can be found from the equation

$$z = 10x + 0.08y$$

(a) How much does it cost to rent a car for 2 days and drive it 200 miles under these conditions?

(b) A second company charges $12 a day and 6¢ a mile for the same car. Write an equation that gives the cost z, in dollars, to rent a car from this company for x days and drive it y miles.

(c) A car is rented from each of the companies mentioned in (a) and (b) for 2 days. To find the mileage at which the cost of renting the cars from each of the two companies will be equal, solve the following system for y:

$$z = 10x + 0.08y$$
$$z = 12x + 0.06y$$
$$x = 2$$

Interest Problems

29. A man invests $2,200 in three accounts that pay 6%, 8%, and 9% in annual interest, respectively. He has three times as much invested at 9% as he does at 6%. If his total interest for the year is $178, how much is invested at each rate?

30. A student has money in three accounts that pay 5%, 7%, and 8% in annual interest. She has three times as much invested at 8% as she does at 5%. If the total amount she has invested is $1,600 and her interest for the year comes to $115, how much money does she have in each account?

Coin Problems

31. A collection of nickels, dimes, and quarters consists of 9 coins with a total value of $1.20. If the number of dimes is equal to the number of nickels, find the number of each type of coin.

32. A coin collection consists of 12 coins with a total value of $1.20. If the collection consists only of nickels, dimes, and quarters, and the number of dimes is two more than twice the number of nickels, how many of each type of coin are in the collection?

33. Height of a Ball   A ball is tossed into the air so that the height after 1, 3, and 5 seconds is as given in the following table. If the relationship between the height of the ball h and the time t is quadratic, then the relationship can be written as

$$h = at^2 + bt + c$$

Use the information in the table to write a system of three equations in three variables a, b, and c. Solve the system to find the exact relationship between h and t.

t (sec)	h (ft)
1	128
3	128
5	0

34. Height of a Ball   A ball is tossed into the air and its height above the ground after 1, 3, and 4 seconds is recorded as shown in the following table. The relationship between the height of the ball h and the time t is quadratic and can be written as

$$h = at^2 + bt + c$$

Use the information in the table to write a system of three equations in three variables a, b, and c. Solve the system to find the exact relationship between the variables h and t.

t (sec)	h (ft)
1	96
3	64
4	0

Review Problems

The following problems review material we covered in Chapter 6.

35. Reduce to lowest terms $\dfrac{x^2 - x - 6}{x^2 - 9}$.

36. Divide using long division $\dfrac{x^2 - 2x + 6}{x - 4}$.

Perform the indicated operations.

37. $\dfrac{x^2 - 25}{x + 4} \cdot \dfrac{2x + 8}{x^2 - 9x + 20}$

38. $\dfrac{3x + 6}{x^2 + 4x + 3} \div \dfrac{x^2 + x - 2}{x^2 + 2x - 3}$

39. $\dfrac{x}{x^2 - 16} + \dfrac{4}{x^2 - 16}$

40. $\dfrac{2}{x^2 - 1} - \dfrac{5}{x^2 + 3x - 4}$

41. $\dfrac{1 - \dfrac{25}{x^2}}{1 - \dfrac{8}{x} + \dfrac{15}{x^2}}$

Solve each equation.

42. $\dfrac{x}{2} - \dfrac{5}{x} = -\dfrac{3}{2}$

43. $\dfrac{x}{x^2 - 9} - \dfrac{3}{x - 3} = \dfrac{1}{x + 3}$

44. Speed of a Boat A boat travels 30 miles up a river in the same amount of time it takes to travel 50 miles down the same river. If the current is 5 miles per hour, what is the speed of the boat in still water?

45. Filling a Pool A pool can be filled by an inlet pipe in 8 hours. The drain will empty the pool in 12 hours. How long will it take to fill the pool if both the inlet pipe and the drain are open?

46. Mixture Problem If 30 liters of a certain solution includes 2 liters of alcohol, how much alcohol is in 45 liters of the same solution?

CHAPTER 7 SUMMARY

Examples

1. Solve: $3(2x - 1) = 9$.

$$3(2x - 1) = 9$$

$$6x - 3 = 9$$

$$6x - 3 + 3 = 9 + 3$$

$$6x = 12$$

$$\frac{1}{6}(6x) = \frac{1}{6}(12)$$

$$x = 2$$

Strategy for Solving Linear Equations in One Variable [7.1]

Step 1: (a) Use the distributive property to separate terms, if necessary.

(b) If fractions are present, consider multiplying both sides by the LCD to eliminate the fractions. If decimals are present, consider multiplying both sides by a power of 10 to clear the equation of decimals.

(c) Combine similar terms on each side of the equation.

Step 2: Use the addition property of equality to get all variable terms on one side of the equation and all constant terms on the other side. A variable term is a term that contains the variable (for example, $5x$). A constant term is a term that does not contain the variable (the number 3, for example).

Step 3: Use the multiplication property of equality to get x (that is, $1x$) by itself on one side of the equation.

Step 4: Check your solution in the original equation to be sure that you have not made a mistake in the solution process.

2. Solve: $x^2 - 5x = -6$.

$$x^2 - 5x + 6 = 0$$

$$(x - 3)(x - 2) = 0$$

$$x - 3 = 0 \text{ or } x - 2 = 0$$

$$x = 3 \text{ or } \quad x = 2$$

To Solve a Quadratic Equation by Factoring [7.1]

Step 1: Write the equation in standard form.
Step 2: Factor the left side.
Step 3: Use the zero-factor property to set each factor equal to 0.
Step 4: Solve the resulting linear equations.
Step 5: Check the solutions in the original equation.

3. The perimeter of a rectangle is 32 inches. If the length is 3 times the width, find the dimensions.

Step 1: This step is done mentally.
Step 2: Let $x =$ the width. Then the length is $3x$.
Step 3: The perimeter is 32; therefore

$$2x + 2(3x) = 32$$

Step 4: $$8x = 32$$

$$x = 4$$

Step 5: The width is 4 inches. The length is $3(4) = 12$ inches.
Step 6: The perimeter is $2(4) + 2(12)$, which is 32. The length is 3 times the width.

Blueprint for Problem Solving [7.1, 7.6]

Step 1: **Read** the problem, and then mentally **list** the items that are known and the items that are unknown.

Step 2: **Assign a variable** to one of the unknown items. If you are using a system of equations, assign variables to each unknown item. Then **translate** the other **information** in the problem to expressions involving the variable, or variables.

Step 3: **Reread** the problem, and then **write an equation,** or system of equations, using the items and variables listed in steps 1 and 2, that describes the situation.

Step 4: **Solve the equation** or the system of equations found in step 3.

Step 5: **Write** your **answer** using a complete sentence.

Step 6: **Reread** the problem and **check** your solution with the original words in the problem.

4. To solve $|2x - 1| + 2 = 7$, we first isolate the absolute value on the left side by adding -2 to each side to obtain

$$|2x - 1| = 5$$

$$2x - 1 = 5 \text{ or } 2x - 1 = -5$$

$$2x = 6 \text{ or } \quad 2x = -4$$

$$x = 3 \text{ or } \quad\quad x = -2$$

5. Set notation

$$\{x \mid -2 \le x \le 3\}$$

Graph

$$-2 \qquad\qquad 3$$

Interval Notation

$$[-2, 3]$$

6. To solve $|x - 3| + 2 < 6$, we first add -2 to both sides to obtain

$$|x - 3| < 4$$

which is equivalent to

$$-4 < x - 3 < 4$$

$$-1 < \quad x \quad < 7$$

7. Factor completely.

(a) $3x^3 - 6x^2 = 3x^2(x - 2)$

(b) $x^2 - 9 = (x + 3)(x - 3)$

$x^3 - 8 = (x - 2)(x^2 + 2x + 4)$

$x^3 + 27 = (x + 3)(x^2 - 3x + 9)$

(c) $x^2 - 6x + 9 = (x - 3)^2$

$6x^2 - 7x - 5$

$$= (2x + 1)(3x - 5)$$

(d) $x^2 + ax + bx + ab$

$$= x(x + a) + b(x + a)$$

$$= (x + a)(x + b)$$

Absolute Value Equations [7.2]

To solve an equation that involves absolute value, we isolate the absolute value on one side of the equation and then rewrite the absolute value equation as two separate equations that do not involve absolute value. In general, if b is a positive real number, then

$$|a| = b \quad \text{is equivalent to} \quad a = b \quad \text{or} \quad a = -b$$

Compound Inequalities [7.3]

Two inequalities connected by the word *and* or *or* form a compound inequality. If the connecting word is *or*, we graph all points that are on either graph. If the connecting word is *and*, we graph only those points that are common to both graphs. The inequality $-2 \le x \le 3$ is equivalent to the compound inequality $-2 \le x$ and $x \le 3$.

Absolute Value Inequalities [7.4]

To solve an inequality that involves absolute value, we first isolate the absolute value on the left side of the inequality symbol. Then we rewrite the absolute value inequality as an equivalent continued or compound inequality that does not contain absolute value symbols. In general, if b is a positive real number, then

$$|a| < b \quad \text{is equivalent to} \quad -b < a < b$$

and $\quad |a| > b \quad \text{is equivalent to} \quad a < -b \quad \text{or} \quad a > b$

To Factor Polynomials in General [7.5]

Step 1: If the polynomial has a greatest common factor other than 1, then factor out the greatest common factor.

Step 2: If the polynomial has two terms (it is a binomial), then see if it is the difference of two squares, or the sum or difference of two cubes, and then factor accordingly. Remember, if it is the sum of two squares it will not factor.

Step 3: If the polynomial has three terms (a trinomial), then it is either a perfect square trinomial, which will factor into the square of a binomial, or it is not a perfect square trinomial, in which case you use one of the methods developed in Section 5.3.

Step 4: If the polynomial has more than three terms, then try to factor it by grouping.

Step 5: As a final check, see if any of the factors you have written can be factored further. If you have overlooked a common factor, you can catch it here.

8. We can eliminate the y-variable from the system below by multiplying both sides of the second equation by 2 and adding the result to the first equation:

$$x + 2y = 4 \xrightarrow{\text{No change}} x + 2y = 4$$

$$x - y = 1 \xrightarrow{\text{Times 2}} \underline{2x - 2y = 2}$$

$$3x \quad\quad = 6$$

$$x = 2$$

Substituting $x = 2$ into either of the original two equations gives $y = 1$. The solution is (2, 1).

9. We can apply the substitution method to the system in Example 8 by first solving the second equation for x to get

$$x = y + 1$$

Substituting this expression for x into the first equation we have

$$y + 1 + 2y = 4$$

$$3y + 1 = 4$$

$$3y = 3$$

$$y = 1$$

Using $y = 1$ in either of the original equations gives $x = 2$.

10. If the two lines are parallel, then the system will be inconsistent, and the solution is $\varnothing$. If the two lines coincide, then the system is dependent.

To Solve a System by the Addition Method [7.6]

Step 1: Look the system over to decide which variable will be easier to eliminate.

Step 2: Use the multiplication property of equality on each equation separately if necessary to ensure that the coefficients of the variable to be eliminated are opposites.

Step 3: Add the left and right sides of the system produced in step 2, and solve the resulting equation.

Step 4: Substitute the solution from step 3 back into any equation with both x- and y-variables, and solve.

Step 5: Check your solution in both equations if necessary.

To Solve a System by the Substitution Method [7.6]

Step 1: Solve either of the equations for one of the variables (this step is not necessary if one of the equations has the correct form already).

Step 2: Substitute the results of step 1 into the other equation, and solve.

Step 3: Substitute the results of step 2 into an equation with both x- and y-variables, and solve. (The equation produced in step 1 is usually a good one to use.)

Step 4: Check your solution if necessary.

Inconsistent and Dependent Equations [7.6, 7.7]

Two linear equations that have no solutions in common are said to be *inconsistent,* while two linear equations that have all their solutions in common are said to be *dependent.*

Solve each equation. [7.1]

1. $4x - 2 = 7x + 7$

2. $\dfrac{3y}{4} - \dfrac{1}{2} + \dfrac{3y}{2} = 2 - y$

3. $8 - 3(2t + 1) = 5(t + 2)$

4. $6 + 4(1 - 3t) = -3(t - 4) + 2$

5. $2x^2 - 5x = 12$ **6.** $81a^2 = 1$

7. $(x - 2)(x - 3) = 2$

8. $9x^3 + 18x^2 - 4x - 8 = 0$

9. Geometry The length of a rectangle is 3 times the width. The perimeter is 32 feet. Find the length and width. [7.1]

10. Geometry The lengths of the three sides of a right triangle are given by three consecutive integers. Find the three sides. [7.1]

Solve each equation. [7.2]

11. $|x - 3| = 1$ **12.** $|x - 2| = 3$

13. $|2y - 3| = 5$ **14.** $|3y - 2| = 7$

15. $|4x - 3| + 2 = 11$ **16.** $|6x - 2| + 4 = 16$

Solve each inequality. Write your answer using interval notation. [7.3]

17. $\frac{3}{4}x + 1 \le 10$ **18.** $600 - 300x < 900$

19. $\frac{1}{3} \le \frac{1}{6}x \le 1$ **20.** $-\frac{1}{2} \le \frac{1}{6}x \le \frac{1}{3}$

21. $5t + 1 \le 3t - 2$ or $-7t \le -21$

22. $6t - 3 \le t + 1$ or $-8t \le -16$

Solve each inequality and graph the solution set. [7.4]

23. $|y - 2| < 3$ **24.** $|5x - 1| > 3$

25. $|2t + 1| - 3 < 2$ **26.** $|2t + 1| - 1 < 5$

27. $|5 + 8t| + 4 \le 1$ **28.** $|5x - 8| \ge -2$

Factor completely. [7.5]

29. $x^4 - 16$ **30.** $3a^4 + 18a^2 + 27$

31. $a^3 - 8$

32. $5x^3 + 30x^2y + 45xy^2$

33. $3a^3b - 27ab^3$

34. $x^2 - 10x + 25 - y^2$

35. $36 - 25a^2$

36. $x^3 + 4x^2 - 9x - 36$

Solve each system using the addition method. [7.6]

37. $6x - 5y = -5$
    $3x + y = 1$

38. $6x + 4y = 8$
    $9x + 6y = 12$

39. $-7x + 4y = -1$
    $5x - 3y = 0$

40. $\dfrac{1}{2}x - \dfrac{3}{4}y = -4$
    $\dfrac{1}{4}x + \dfrac{3}{2}y = 13$

Solve each system by the substitution method. [7.6]

41. $x + y = 2$
    $y = x - 1$

42. $2x - 3y = 5$
    $y = 2x - 7$

43. $3x + 7y = 6$
    $x = -3y + 4$

44. $5x - y = 4$
    $y = 5x - 3$

Solve each system. [7.7]

45. $x + y + z = 6$
    $x - y - 3z = -8$
    $x + y - 2z = -6$

46. $3x + 2y + z = 4$
    $2x - 4y + z = -1$
    $x + 6y + 3z = -4$

47. $5x - 2y + z = 6$
    $-3x + 4y - z = 2$
    $6x - 8y + 2z = -4$

48. $4x - 6y + 8z = 4$
    $5x + y - 2z = 4$
    $6x - 9y + 12z = 6$

49. $2x - y = 5$
    $3x - 2z = -2$
    $5y + z = -1$

50. $x - y = 2$
    $y - z = -3$
    $x - z = -1$

CHAPTER 7 PROJECTS

TRANSITIONS

GROUP PROJECT

Number of People: 2 or 3

Time Needed: 10–15 minutes

Equipment: Pencil and paper

Background: The break-even point for a company occurs when the revenue from sales of a product equals the cost of producing the product. This group project is designed to give you more insight into revenue, cost, and break-even point.

Procedure: A company is planning to open a factory to manufacture calculators.

1. It costs them $120,000 to open the factory, and it will cost $10 for each calculator they make. What is the expression for C, the cost of making x calculators?

2. They can sell the calculators for $50 each. What is the expression for R, their revenue from selling x calculators? Remember that $R = px$, where p is the price per calculator.

3. Graph both the cost equation C and the revenue equation R on a coordinate system like the one below.

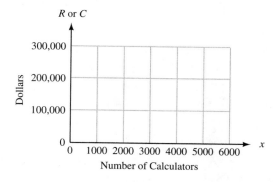

4. The break-even point is the value of x (the number of calculators) for which the revenue is equal to the cost. Where is the break-even point on the graph you produced in Part 3? Estimate the break-even point from the graph.

5. Set the cost equal to the revenue and solve to find x, to find the exact value of the break-even point. How many calculators do they need to make and sell to exactly break even? What will be their revenue and their cost for that many calculators?

6. Write an inequality that gives the values of x that will produce a profit for the company. (A profit occurs when the revenue is larger than the cost.)

7. Write an inequality that gives the values of x that will produce a loss for the company. (A loss occurs when the cost is larger than the revenue.)

8. Profit is the difference between revenue and cost, or $P = R - C$. Write the equation for profit and then graph it on a coordinate system like the one below.

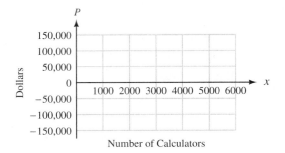

9. How do you recognize the break-even point and the regions of loss and profit on the graph you produced in Part 8?

RESEARCH PROJECT

MARIA GAETANA AGNESI (1718–1799)

January 9, 1999 was the 200th anniversary of the death of Maria Agnesi, the author of *Instituzioni analitiche ad uso della gioventu italiana* (1748), a calculus textbook considered to be the best book of its time, and the first surviving math-

Culver Pictures

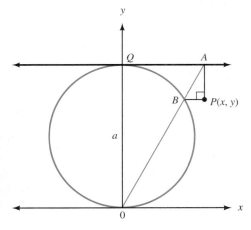

ematical work written by a woman. Maria Agnesi is also famous for a curve that is named for her. The curve is called the Witch of Agnesi in English, but its actual translation is the Locus of Agnesi. The foundation of the curve is shown in the figure on page 464. Research the Witch of Agnesi and then explain, in essay form, how the diagram in the figure is used to produce the Witch of Agnesi. Include a rough sketch of the curve, starting with a circle of diameter a as shown in the figure. Then, for comparison, sketch the curve again, starting with a circle of diameter $2a$.

C H A P T E R 7   **TEST**

Solve the following equations. [7.1]

1. $5 - \frac{4}{7}a = -11$ **2.** $3x^2 = 5x + 2$

3. $100x^3 = 500x^2$

4. $5(x - 1) - 2(2x + 3) = 5x - 4$

5. $(x + 1)(x + 2) = 12$

Solve each of the following. [7.1]

6. Geometry A rectangle is twice as long as it is wide. The perimeter is 36 inches. Find the dimensions.

7. Velocity If an object is thrown straight up into the air with an initial velocity of 32 feet per second, then its height h (in feet) above the ground at any time t (in seconds) is given by the formula $h = 32t - 16t^2$. Find the times at which the object is on the ground by letting $h = 0$ in the equation and solving for t.

Solve the following equations. [7.2]

8. $\left|\frac{1}{4}x - 1\right| = \frac{1}{2}$ **9.** $|3 - 2x| + 5 = 2$

Solve the following inequalities. Write the solution set using interval notation, then graph the solution set. [7.3]

10. $5 - \frac{3}{2}x > -1$ or $2x - 5 \geq 7$

11. $-3 \leq 5x - 1 \leq 9$

Solve the following inequalities and graph the solutions. [7.4]

12. $|6x - 1| > 7$ **13.** $|3x - 5| - 4 \leq 3$

Factor completely. [7.5]

14. $x^2 + x - 12$

15. $16a^4 - 81y^4$ **16.** $t^3 + \frac{1}{8}$

17. $4a^5b - 24a^4b^2 - 64a^3b^3$

Solve the following systems. [7.6]

18. $2x - 5y = -8$ **19.** $\frac{1}{3}x - \frac{1}{6}y = 3$
    $3x + y = 5$ $-\frac{1}{5}x + \frac{1}{4}y = 0$

20. $2x - 5y = 24$
     $y = 3x - 10$

21. Solve the system. [7.7]

$2x - y + z = 9$
$x + y - 3z = -2$
$3x + y - z = 6$

22. Number Problem A number is 1 less than twice another. Their sum is 14. Find the two numbers. [7.6]

23. Investing John invests twice as much money at 6% as he does at 5%. If his investments earn a total of $680 in 1 year, how much does he have invested at each rate? [7.6]

24. Coin Problem A collection of nickels, dimes, and quarters consists of 15 coins with a total value of $1.10. If the number of nickels is 1 less than 4 times the number of dimes, how many of each coin are contained in the collection? [7.7]

Simplify.

1. $15 - 12 \div 4 - 3 \cdot 2$

2. $6(11 - 13)^3 - 5(8 - 11)^2$

3. $\left(\dfrac{2}{5}\right)^{-2}$

4. $4(3x - 2) + 3(2x + 5)$

5. $5 - 3[2x - 4(x - 2)]$

6. $(3y + 2)^2 - (3y - 2)^2$

7. $(2x + 3)(x^2 - 4x + 2)$

Solve.

8. $-4y - 2 = 6y + 8$

9. $-6 + 2(2x + 3) = 0$

10. $3x^2 = 17x - 10$ **11.** $|2x - 3| + 7 = 1$

Solve each system.

12. $\begin{aligned} 2x - 5y &= -7 \\ -3x + 4y &= 0 \end{aligned}$ **13.** $\begin{aligned} 8x + 6y &= 4 \\ 12x + 9y &= 8 \end{aligned}$

14. $\begin{aligned} 2x + y &= 3 \\ y &= -2x + 3 \end{aligned}$ **15.** $\begin{aligned} 2x + \;\; y &= \;\; 8 \\ 4y - \;\; z &= -9 \\ 3x - 2z &= -6 \end{aligned}$

Solve each inequality, and graph the solution.

16. $3 < \dfrac{1}{4}x + 4 < 5$ **17.** $|2x - 7| \le 3$

18. $|2x - 7| - 5 \ge 6$

19. Solve $-3t \ge 12$. Write your answer with interval notation.

20. Solve for x: $ax - 4 = bx + 9$.

21. Solve the compound inequality
$$-2x > 6 \quad \text{or} \quad 3x - 7 > 2$$

22. Give the opposite and reciprocal of -7.

23. Add $-\dfrac{2}{9}$ to the product of -6 and $\dfrac{7}{54}$.

24. Translate into an inequality: x is between -5 and 5.

25. For the set $\{-1, 0, \sqrt{2}, 2.35, \sqrt{3}, 4\}$ list all elements belonging to the set of *rational numbers*.

26. Reduce $\dfrac{a^2 - b^2}{a^3 - b^3}$ to lowest terms.

27. Solve $3 + \dfrac{1}{x} = \dfrac{10}{x^2}$.

28. Write as positive exponents and simplify:
$$\left(\tfrac{2}{13}\right)^{-2} - \left(\tfrac{2}{5}\right)^{-2}$$

29. Simplify and write answer with positive exponents:
$$(5x^{-3}y^2z^{-2})(6x^{-5}y^4z^{-3})$$

30. Write 0.000469 in scientific notation.

31. Simplify and write in scientific notation:
$$\dfrac{(7 \times 10^{-5})(21 \times 10^{-6})}{3 \times 10^{-12}}$$

32. Write 1.23×10^{-4} in expanded form.

33. State any restrictions on the variable in the expression $\dfrac{x + 5}{x^2 - 25}$.

Graph on a rectangular coordinate system.

34. $0.03x + 0.04y = 0.04$ **35.** $y = \dfrac{2}{3}x - 4$

36. Solve the system
$$\begin{aligned} x - 2y - 2z &= -3 \\ x + 2y - z\; &= 4 \\ 2x - y - 3z &= -1 \end{aligned}$$

Factor completely.

37. $16y^2 + 2y + \dfrac{1}{16}$

38. $6a^2x + 2x + 3a^2y^2 + y^2$

39. $x^3 - 8$ **40.** $16a^4 - 81b^4$

41. Divide: $\dfrac{2x^2 - 7x + 9}{x - 2}$

42. Multiply: $\dfrac{2y^2 - 4y}{2y^2 - 2} \cdot \dfrac{y^2 - 2y - 3}{y^2 - 5y + 6}$

43. Add: $\dfrac{-2}{x^2 - 2x - 3} + \dfrac{3}{x^2 - 9}$

44. Simplify: $\dfrac{1 - \dfrac{4}{x^2}}{1 - \dfrac{1}{x} - \dfrac{6}{x^2}}$

45. Speed A boat travels 36 miles down a river in 3 hours. If it takes the boat 9 hours to travel the same distance going up the river, what is the speed of the boat? What is the speed of the current of the river?

46. Concentric Circles Recall that the area of a circle is $A = \pi r^2$. Two circles are concentric if they have the same center. Find the area between two concentric circles in which the larger circle has radius a and the smaller circle has radius b. Express your answer in a form in which π is a factor.

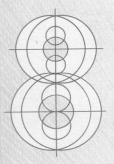

Equations and Inequalities in Two Variables

(© SuperStock)

INTRODUCTION

A student is heating water in a chemistry lab. As the water heats, she records the temperature readings from two thermometers, one giving temperature in degrees Fahrenheit and the other in degrees Celsius. Table 1 shows some of the data she collects. Figure 1 is a scatter diagram that gives a visual representation of the data in Table 1.

TABLE I	Corresponding Temperatures
In Degrees Fahrenheit	**In Degrees Celsius**
77	25
95	35
167	75
212	100

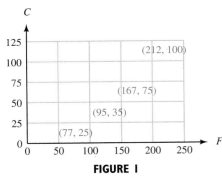

FIGURE I

The exact relationship between the Fahrenheit and Celsius temperature scales is given by the formula

$$C = \frac{5}{9}(F - 32)$$

We have three ways to describe the relationship between the two temperature scales: a table, a graph, and an equation. But, most important to us, we don't need to accept this formula on faith. In Problem Set 8.2, you will derive the formula from the data in Table 1.

Let's begin this section by reviewing the definition of a linear equation in two variables, and how we graph a linear equation using intercepts.

> **DEFINITION** Any equation that can be put in the form $ax + by = c$, where a, b, and c are real numbers and a and b are not both 0, is called a **linear equation in two variables**. The graph of any equation of this form is a straight line (that is why these equations are called "linear"). The form $ax + by = c$ is called **standard form**.

INTERCEPTS

> **DEFINITION** The *x-intercept* of the graph of an equation is the x-coordinate of the point where the graph crosses the x-axis. The **y-intercept** is defined similarly.

EXAMPLE 1 Find the x- and y-intercepts for $2x + 3y = 6$; then graph the solution set.

Solution To find the y-intercept we let $x = 0$.

$$\text{When} \qquad x = 0$$
$$\text{we have} \qquad 2(0) + 3y = 6$$
$$3y = 6$$
$$y = 2$$

The y-intercept is 2, and the graph crosses the y-axis at the point (0, 2).

$$\text{When} \qquad y = 0$$
$$\text{we have} \qquad 2x + 3(0) = 6$$
$$2x = 6$$
$$x = 3$$

The x-intercept is 3, so the graph crosses the x-axis at the point (3, 0). We use these results to graph the solution set for $2x + 3y = 6$. The graph is shown in Figure 1 on the next page.

469

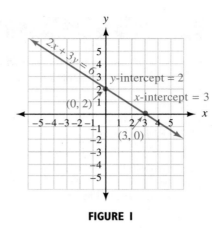

FIGURE 1

Note: Graphing straight lines by finding the intercepts works best when the coefficients of x and y are factors of the constant term.

 ## Using TECHNOLOGY

GRAPHING CALCULATORS AND COMPUTER GRAPHING PROGRAMS

A variety of computer programs and graphing calculators are currently available to help us graph equations and then obtain information from those graphs much faster than we could with paper and pencil. We will not give instructions for all the available calculators. Most of the instructions we give are generic in form. You will have to use the manual that came with your calculator to find the specific instructions for your calculator.

Graphing with Trace and Zoom

All graphing calculators have the ability to graph a function and then trace over the points on the graph, giving their coordinates. Furthermore, all graphing calculators can zoom in and out on a graph that has been drawn. To graph a linear equation on a graphing calculator, we first set the graph window. Most calculators call the smallest value of x Xmin and the largest value of x Xmax. The counterpart values of y are Ymin and Ymax. We will use the notation

Window: X from -5 to 4, Y from -3 to 2

to stand for a window in which

$$Xmin = -5 \qquad Ymin = -3$$
$$Xmax = 4 \qquad Ymax = 2$$

(continued)

Set your calculator with the following window:

Window: X from -10 to 10, Y from -10 to 10

Graph the equation $Y = -X + 8$. On the TI-82/83, you use the $\boxed{Y=}$ key to enter the equation; you enter a negative sign with the $\boxed{(-)}$ key. The graph will be similar to the one shown in Figure 2.

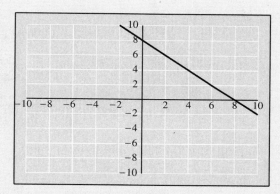

FIGURE 2

Use the Trace feature of your calculator to name three points on the graph. Next, use the Zoom feature of your calculator to zoom out so your window is twice as large.

Solving for *y* First

To graph the equation $2x + 3y = 6$ on a graphing calculator, you must first solve it for *y*. When you do so, you will get $y = -\frac{2}{3}x + 2$, which you enter into your calculator as $Y = -(2/3)X + 2$.

Hint on Tracing

If you are going to use the Trace feature and you want the *x*-coordinates to be exact numbers, set your window so that the range of X inputs is a multiple of the number of horizontal pixels on your calculator screen. On the TI-82/83, the screen has 94 pixels across. Here are a few convenient trace windows:

X from -4.7 to 4.7	To trace to the nearest tenth
X from -47 to 47	To trace to the nearest integer
X from 0 to 9.4	To trace to the nearest tenth
X from 0 to 94	To trace to the nearest integer
X from -94 to 94	To trace to the nearest even integer

THE SLOPE OF A LINE

A highway sign tells us we are approaching a 6% downgrade. As we drive down this hill, each 100 feet we travel horizontally is accompanied by a 6-foot drop in elevation.

In mathematics we say the slope of the highway is $-0.06 = -\frac{6}{100} = -\frac{3}{50}$. The *slope* is the ratio of the vertical change to the accompanying horizontal change.

Highway sign   Mathematical model

In defining the slope of a straight line, we are looking for a number to associate with a straight line that does two things. First, we want the slope of a line to measure the "steepness" of the line. That is, in comparing two lines, the slope of the steeper line should have the larger numerical value. Second, we want a line that *rises* going from left to right to have a *positive* slope. We want a line that *falls* going from left to right to have a *negative* slope. (A line that neither rises nor falls going from left to right must, therefore, have 0 slope.)

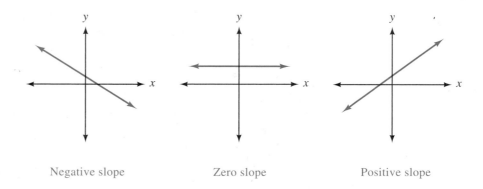

Negative slope   Zero slope   Positive slope

Geometrically, we can define the *slope* of a line as the ratio of the vertical change to the horizontal change encountered when moving from one point to another on the line. The vertical change is sometimes called the *rise*. The horizontal change is called the *run*.

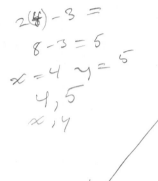

EXAMPLE 2   Find the slope of the line $y = 2x - 3$.

Solution   In order to use our geometric definition, we first graph $y = 2x - 3$ (Figure 3). We then pick any two convenient points and find the ratio of rise to run. By convenient points we mean points with integer coordinates. If we let $x = 2$ in the equation, then $y = 1$. Likewise if we let $x = 4$, then y is 5.

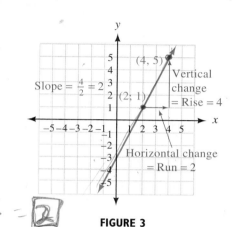

FIGURE 3

Our line has a slope of 2.

Notice that we can measure the vertical change (rise) by subtracting the y-coordinates of the two points shown in Figure 3: $5 - 1 = 4$. The horizontal change (run) is the difference of the x-coordinates: $4 - 2 = 2$. This gives us a second way of defining the slope of a line.

DEFINITION   The **slope** of the line between two points (x_1, y_1) and (x_2, y_2) is given by

$$\text{Slope} = m = \frac{\text{Rise}}{\text{Run}} = \frac{y_2 - y_1}{x_2 - x_1}$$

$\underset{\text{Geometric form}}{\uparrow} \qquad \underset{\text{Algebraic form}}{\uparrow}$

EXAMPLE 3   Find the slope of the line through $(-2, -3)$ and $(-5, 1)$.

Solution

$$m = \frac{y_2 - y_1}{x_2 - x_1} = \frac{1 - (-3)}{-5 - (-2)} = \frac{4}{-3} = -\frac{4}{3}$$

Looking at the graph of the line between the two points (Figure 4, next page), we can see our geometric approach does not conflict with our algebraic approach.

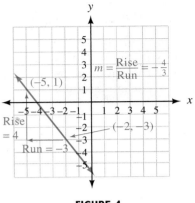

FIGURE 4

We should note here that it does not matter which ordered pair we call (x_1, y_1) and which we call (x_2, y_2). If we were to reverse the order of subtraction of both the x- and y-coordinates in the preceding example, we would have

$$m = \frac{-3 - 1}{-2 - (-5)} = \frac{-4}{3} = -\frac{4}{3}$$

which is the same as our previous result.

Note: The two most common mistakes students make when first working with the formula for the slope of a line are

1. Putting the difference of the x-coordinates over the difference of the y-coordinates.
2. Subtracting in one order in the numerator and then subtracting in the opposite order in the denominator. You would make this mistake in Example 3 if you wrote $1 - (-3)$ in the numerator and then $-2 - (-5)$ in the denominator.

E X A M P L E   4   Find the slope of the line containing $(3, -1)$ and $(3, 4)$.

Solution   Using the definition for slope, we have

$$m = \frac{-1 - 4}{3 - 3} = \frac{-5}{0}$$

The expression $\frac{-5}{0}$ is undefined. That is, there is no real number to associate with it. In this case, we say the line *has no slope*.

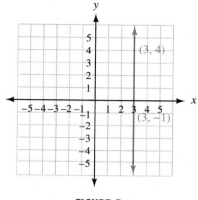

FIGURE 5

The graph of our line is shown in Figure 5. Our line with no slope is a vertical line. All vertical lines have no slope. (And all horizontal lines, as we mentioned earlier, have 0 slope.)

SLOPES OF PARALLEL AND PERPENDICULAR LINES

In geometry we call lines in the same plane that never intersect parallel. For two lines to be nonintersecting, they must rise or fall at the same rate. In other words, two lines are *parallel* if and only if they have the *same slope.*

Although it is not as obvious, it is also true that two nonvertical lines are *perpendicular* if and only if the *product of their slopes is* -1. This is the same as saying their slopes are negative reciprocals.

We can state these facts with symbols as follows: If line l_1 has slope m_1 and line l_2 has slope m_2, then

$$l_1 \text{ is parallel to } l_2 \Leftrightarrow m_1 = m_2$$

and

$$l_1 \text{ is perpendicular to } l_2 \Leftrightarrow m_1 \cdot m_2 = -1$$

$$\left(\text{or } m_1 = \frac{-1}{m_2} \right)$$

For example, if a line has a slope of $\frac{2}{3}$, then any line parallel to it has a slope of $\frac{2}{3}$. Any line perpendicular to it has a slope of $-\frac{3}{2}$ (the negative reciprocal of $\frac{2}{3}$).

Although we cannot give a formal proof of the relationship between the slopes of perpendicular lines at this level of mathematics, we can offer some justification for the relationship. Figure 6, next page, shows the graphs of two lines. One of the lines has a slope of $\frac{2}{3}$; the other has a slope of $-\frac{3}{2}$. As you can see, the lines are perpendicular.

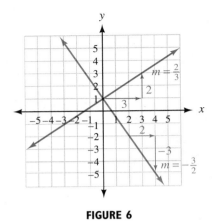

FIGURE 6

Using TECHNOLOGY

FAMILIES OF CURVES

We can use a graphing calculator to investigate the effects of the numbers a and b on the graph of $y = ax + b$. To see how the number b affects the graph, we can hold a constant and let b vary. Doing so will give us a *family* of curves. Suppose we set $a = 1$ and then let b take on integer values from -3 to 3. The equations we obtain are

$$y = x - 3$$

$$y = x - 2$$

$$y = x - 1$$

$$y = x$$

$$y = x + 1$$

$$y = x + 2$$

$$y = x + 3$$

We will give three methods of graphing this set of equations on a graphing calculator.

Method 1: Y-Variables List

To use the Y-variables list, enter each equation at one of the Y variables, set the graph window, then graph. The calculator will graph the equations in order,

(continued)

starting with Y_1 and ending with Y_7. Following is the Y-variables list, an appropriate window, and a sample of the type of graph obtained (Figure 7).

$Y_1 = X - 3$

$Y_2 = X - 2$

$Y_3 = X - 1$

$Y_4 = X$

$Y_5 = X + 1$

$Y_6 = X + 2$

$Y_7 = X + 3$

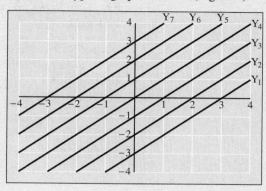

FIGURE 7

Window: X from -4 to 4, Y from -4 to 4

Method 2: Programming

The same result can be obtained by programming your calculator to graph $y = x + b$ for $b = -3, -2, -1, 0, 1, 2,$ and 3. Here is an outline of a program that will do this. Check the manual that came with your calculator to find the commands for your calculator.

Step 1:   Clear screen

Step 2:   Set window for X from -4 to 4 and Y from -4 to 4

Step 3: $-3 \rightarrow B$

Step 4:   Label 1

Step 5:   Graph $Y = X + B$

Step 6: $B + 1 \rightarrow B$

Step 7:   If $B < 4$, Goto 1

Step 8:   End

Method 3: Using Lists

On the TI-82/83 you can set Y_1 as follows:

$$Y_1 = X + \{-3, -2, -1, 0, 1, 2, 3\}$$

When you press $\boxed{\text{GRAPH}}$ the calculator will graph each line from $y = x + (-3)$ to $y = x + 3$.

Each of the three methods will produce graphs similar to those in Figure 7.

Getting Ready for Class

After reading through the preceding section, respond in your own words and in complete sentences.

A. If you were looking at a graph that described the performance of a stock you had purchased, why would it be better if the slope of the line were positive, rather than negative?

B. Describe the behavior of a line with a negative slope.

C. Would you rather climb a hill with a slope of $\frac{1}{2}$ or a slope of 3? Explain why.

D. Describe how to obtain the slope of a line if you know the coordinates of two points on the line.

PROBLEM SET 8.1

Find the slope of each of the following lines from the given graph.

1.

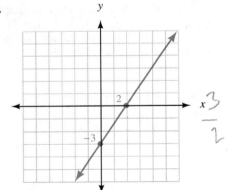

$x \dfrac{3}{2}$

3.

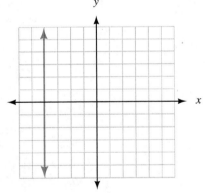

2.

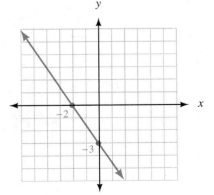

4.

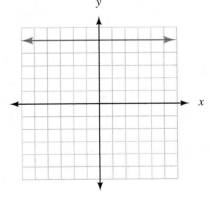

5.

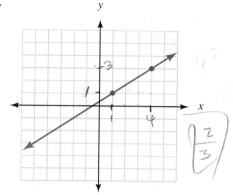

$\dfrac{2}{3}$

6.

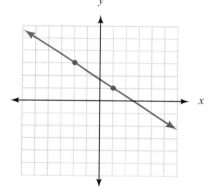

Find the slope of the line through the following pairs of points. Then, plot each pair of points, draw a line through them, and indicate the rise and run in the graph in the manner shown in Example 3.

7. (2, 1), (4, 4)

8. (3, 1), (5, 4)

9. (1, 4), (5, 2)

10. (1, 3), (5, 2)

11. (1, −3), (4, 2)

12. (2, −3), (5, 2)

13. (−3, −2), (1, 3)

14. (−3, −1), (1, 4)

15. (−3, 2), (3, −2)

16. (−3, 3), (3, −1)

17. (2, −5), (3, −2)

18. (2, −4), (3, −1)

For each of the equations in Problems 19–22, complete the table, and then use the results to find the slope of the graph of the equation.

19. $2x + 3y = 6$

x	y
0	
	0

20. $3x − 2y = 6$

x	y
0	
	0

21. $y = \frac{2}{3}x − 5$

x	y
0	
3	

22. $y = −\frac{3}{4}x + 2$

x	y
0	
4	

23. Finding Slope From Intercepts   Graph the line that has an x-intercept of 3 and a y-intercept of −2. What is the slope of this line?

24. Finding Slope From Intercepts   Graph the line that has an x-intercept of 2 and a y-intercept of −3. What is the slope of this line?

25. Finding Slope From Intercepts   Graph the line with x-intercept 4 and y-intercept 2. What is the slope of this line?

26. Finding Slope From Intercepts   Graph the line with x-intercept −4 and y-intercept −2. What is the slope of this line?

27. Parallel Lines   Find the slope of any line parallel to the line through (2, 3) and (−8, 1).

28. Parallel Lines   Find the slope of any line parallel to the line through (2, 5) and (5, −3).

29. Perpendicular Lines   Line l contains the points (5, −6) and (5, 2). Give the slope of any line perpendicular to l.

30. Perpendicular Lines   Line l contains the points (3, 4) and (−3, 1). Give the slope of any line perpendicular to l.

31. Determine if each of the following tables could represent ordered pairs from an equation of a line.

(a)

x	y
0	5
1	7
2	9
3	11

(b)

x	y
−2	−5
0	−2
2	0
4	1

32. The following lines have slope 2, $\frac{1}{2}$, 0, and −1. Match each line to its slope value.

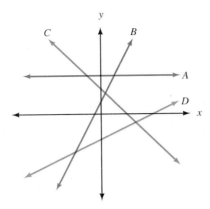

37. Find the slope of the line segment labeled A. What units would you attach to this number?

38. Find the slope of the line segment labeled C. Be sure to attach units to your answer.

39. Is the temperature changing faster during the 1st minute or the 16th minute?

40. Line segments B and D both have 0 slope. Explain what this means in terms of the melting ice.

Value of a Used Car The 1998 edition of a popular consumer car price book gives the values shown in the table for Volkswagen Jettas in good condition. The line graph was drawn from the information in the table. Use this information to solve Problems 41–44.

Applying the Concepts

33. Slope of a Sand Pile A pile of sand at a construction site is in the shape of a cone. If the slope of the side of the pile is $\frac{2}{3}$ and the pile is 8 feet high, how wide is the diameter of the base of the pile?

34. Slope of a Pyramid The slope of the sides of one of the ancient pyramids in Egypt is $\frac{13}{10}$. If the base of the pyramid is 750 feet, how tall is the pyramid?

Heating a Block of Ice A block of ice with an initial temperature of $-20°C$ is heated at a steady rate. The graph shows how the temperature changes as the ice melts to become water and the water boils to become steam and water. (Problems 35–40)

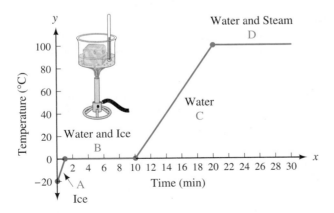

Year	Age in 1998	Value ($)
1994	4	8,525
1995	3	9,575
1996	2	11,950
1997	1	13,200
1998	0	15,250

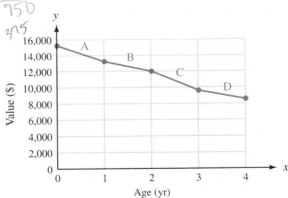

41. Find the slope of the line segment labeled B. What units should you attach to this number?

42. Find the slope of line segment C. Be sure to include units with your answer.

43. From the graph, does the value of the car decrease more from 1 to 2 years, or from 2 to 3 years?

44. From the graph, does the value of the car decrease more from 2 to 3 years, or from 3 to 4 years?

35. How long does it take all the ice to melt?

36. From the time the heat is applied to the block of ice, how long is it before the water boils?

45. Slope of a Highway A sign at the top of the Cuesta Grade, outside of San Luis Obispo, reads

"7% downgrade next 3 miles." The diagram in the figure below is a model of the Cuesta Grade that takes into account the information on that sign.

(a) At point B, the graph crosses the y-axis at 1,106 feet. How far is it from the origin to point A?

(b) What is the slope of the Cuesta Grade?

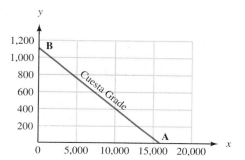

46. **Graph Reading** A fraternity is planning a fund-raising concert at the Veterans Memorial Building, which can accommodate a maximum of 300 ticket holders. Tickets are sold at $10 each. Their expenses are $455 for the band and $125 rent for the Veterans building. If 150 or more tickets are sold, the fraternity is required to hire a security guard for $200. The following diagram can be used to find the profit they will make by selling various numbers of tickets. As you can see, if no one buys a ticket, their profit will be −$580, the amount they will pay for the band and for rent.

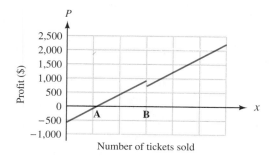

(a) Both line segments have the same slope. What is that slope?

(b) Point A is the break-even point. It corresponds to the number of tickets they must sell so that their income is equal to their expenses. How many tickets is this?

(c) How many tickets does point B represent?

(d) How many tickets must they sell to make a profit of $500?

47. **Baseball Salaries** The average salary of a major league baseball player in 1967 was $25,000, and in 1975 it was $40,000. What is the slope of the line that connects these two points; that is, what is the average yearly salary increase for this period?

    The average salary of a major league baseball player in 1990 was $600,000, and in 1994 it was $1,300,000. What is the slope of the line that connects these two points; that is, what is the average yearly salary increase for this period?

48. **Blood Pressure** The average systolic blood pressure for a 26-year-old is 113 and for a 40-year-old is 120.

(a) Using the age as the horizontal axis and the blood pressure as the vertical axis, plot this blood pressure data. Draw a line connecting them.

(b) Determine the slope of the line connecting the two points from part (a).

(c) What does the slope from part (b) represent in terms of blood pressure and age?

Review Problems

The following problems review material we covered in Section 2.5.

49. If $3x + 2y = 12$, find y when x is 4.

50. If $y = 3x - 1$, find x when y is 0.

51. Solve the formula $3x + 2y = 12$ for y.

52. Solve the formula $y = 3x - 1$ for x.

53. Solve the formula $A = P + Prt$ for t.

54. Solve the formula $S = \pi r^2 + 2\pi rh$ for h.

Extending the Concepts

The colored box around Problem 55 indicates you are to use a graphing calculator.

55. Use your Y-variables list or write a program to graph the family of curves $Y = 2X + B$ for $B = -3, -2, -1, 0, 1, 2,$ and 3.

56. Use your Y-variables list or write a program to graph the family of curves $Y = -2X + B$ for $B = -3, -2, -1, 0, 1, 2,$ and 3.

57. Use your Y-variables list or write a program to graph the family of curves $Y = AX$ for $A = -3$, -2, -1, 0, 1, 2, and 3.

58. Use your Y-variables list or write a program to graph the family of curves $Y = AX + 2$ for $A = -3$, -2, -1, 0, 1, 2, and 3.

59. Use your Y-variables list or write a program to graph the family of curves $Y = AX$ for $A = \frac{1}{4}, \frac{1}{3}, \frac{1}{2}$, 1, 2, and 3.

60. Use your Y-variables list or write a program to graph the family of curves $Y = AX - 2$ for $A = \frac{1}{4}, \frac{1}{3}, \frac{1}{2}$, 1, 2, and 3.

8.2   The Equation of a Line

The table and illustrations below show some corresponding temperatures on the Fahrenheit and Celsius temperature scales. For example, water freezes at 32°F and 0°C, and boils at 212°F and 100°C.

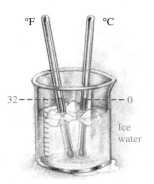

Degrees Celsius	Degrees Fahrenheit
0	32
25	77
50	122
75	167
100	212

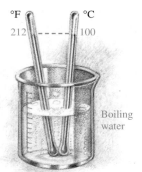

If we plot all the points in the table using the *x*-axis for temperatures on the Celsius scale and the *y*-axis for temperatures on the Fahrenheit scale, we see that they line up in a straight line (Figure 1). This means that a linear equation in two

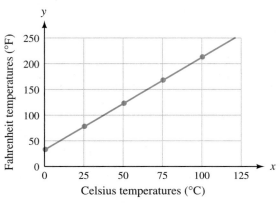

FIGURE 1

variables will give a perfect description of the relationship between the two scales. That equation is

$$F = \frac{9}{5}C + 32$$

The techniques we use to find the equation of a line from a set of points is what this section is all about.

Suppose line l has slope m and y-intercept b. What is the equation of l? Since the y-intercept is b, we know the point $(0, b)$ is on the line. If (x, y) is any other point on l, then using the definition for slope, we have

$$\frac{y - b}{x - 0} = m \qquad \text{Definition of slope}$$

$$y - b = mx \qquad \text{Multiply both sides by } x.$$

$$y = mx + b \qquad \text{Add } b \text{ to both sides.}$$

This last equation is known as the *slope-intercept form* of the equation of a straight line.

SLOPE-INTERCEPT FORM OF THE EQUATION OF A LINE

The equation of any line with slope m and y-intercept b is given by

$$y = mx + b$$

$$\nearrow \qquad \uparrow$$

Slope y-intercept

When the equation is in this form, the *slope* of the line is always the *coefficient of x*, and the *y-intercept* is always the *constant term*.

EXAMPLE 1   Find the equation of the line with slope $-\frac{4}{3}$ and y-intercept 5. Then graph the line.

Solution   Substituting $m = -\frac{4}{3}$ and $b = 5$ into the equation $y = mx + b$, we have

$$y = -\frac{4}{3}x + 5$$

Finding the equation from the slope and y-intercept is just that easy. If the slope is m and the y-intercept is b, then the equation is always $y = mx + b$. Now, let's graph the line.

Since the y-intercept is 5, the graph goes through the point $(0, 5)$. To find a second point on the graph, we start at $(0, 5)$ and move 4 units down (that's a rise of -4) and 3 units to the right (a run of 3). The point we end up at is $(3, 1)$. Drawing a line that passes through $(0, 5)$ and $(3, 1)$, we have the graph of our equation. (Note that we could also let the rise $= 4$ and the run $= -3$ and obtain the same graph.) The graph is shown in Figure 2, on the next page.

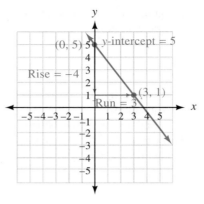

FIGURE 2

EXAMPLE 2   Give the slope and y-intercept for the line $2x - 3y = 5$.

Solution   To use the slope-intercept form, we must solve the equation for y in terms of x:

$$2x - 3y = 5$$

$$-3y = -2x + 5 \qquad \text{Add } -2x \text{ to both sides.}$$

$$y = \frac{2}{3}x - \frac{5}{3} \qquad \text{Divide by } -3.$$

The last equation has the form $y = mx + b$. The slope must be $m = \frac{2}{3}$, and the y-intercept is $b = -\frac{5}{3}$.

EXAMPLE 3   Graph the equation $2x + 3y = 6$ using the slope and y-intercept.

Solution   Although we could graph this equation using the methods developed in Section 3.3 (by finding ordered pairs that are solutions to the equation and drawing a line through their graphs), it is sometimes easier to graph a line using the slope-intercept form of the equation.

Solving the equation for y, we have

$$2x + 3y = 6$$

$$3y = -2x + 6 \qquad \text{Add } -2x \text{ to both sides.}$$

$$y = -\frac{2}{3}x + 2 \qquad \text{Divide by 3.}$$

The slope is $m = -\frac{2}{3}$ and the y-intercept is $b = 2$. Therefore, the point $(0, 2)$ is on the graph, and the ratio of rise to run going from $(0, 2)$ to any other point on the line is $-\frac{2}{3}$. If we start at $(0, 2)$ and move 2 units up (that's a rise of 2) and 3 units to the left (a run of -3), we will be at another point on the graph. (We could also go down 2 units and right 3 units and still be assured of ending up at another point on the line, since $\frac{2}{-3}$ is the same as $\frac{-2}{3}$.)

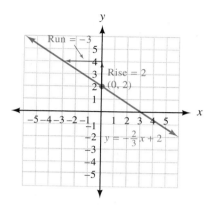

FIGURE 3

Note: As we mentioned previously in Chapter 3, the rectangular coordinate system is the tool we use to connect algebra and geometry. Example 3 illustrates this connection, as do the many other examples in this chapter. In Example 3, Descartes's rectangular coordinate system allows us to associate the equation $2x + 3y = 6$ (an algebraic concept) with the straight line (a geometric concept) shown in Figure 3.

A second useful form of the equation of a straight line is the point-slope form.

Let line l contain the point (x_1, y_1) and have slope m. If (x, y) is any other point on l, then by the definition of slope we have

$$\frac{y - y_1}{x - x_1} = m$$

Multiplying both sides by $(x - x_1)$ gives us

$$(x - x_1) \cdot \frac{y - y_1}{x - x_1} = m(x - x_1)$$

$$y - y_1 = m(x - x_1)$$

This last equation is known as the *point-slope form* of the equation of a straight line.

POINT-SLOPE FORM OF THE EQUATION OF A LINE

The equation of the line through (x_1, y_1) with slope m is given by

$$y - y_1 = m(x - x_1)$$

This form of the equation of a straight line is used to find the equation of a line, either given one point on the line and the slope, or given two points on the line.

EXAMPLE 4 Find the equation of the line with slope -2 that contains the point $(-4, 3)$. Write the answer in slope-intercept form.

Solution

Using $(x_1, y_1) = (-4, 3)$ and $m = -2$

in $y - y_1 = m(x - x_1)$ Point-slope form

gives us $y - 3 = -2(x + 4)$ *Note:* $x - (-4) = x + 4$

          $y - 3 = -2x - 8$ Multiply out right side.

          $y = -2x - 5$ Add 3 to each side.

Figure 4 is the graph of the line that contains $(-4, 3)$ and has a slope of -2. Notice that the y-intercept on the graph matches that of the equation we found.

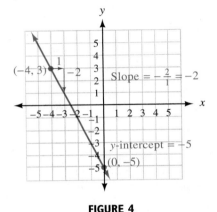

FIGURE 4

EXAMPLE 5 Find the equation of the line that passes through the points $(-3, 3)$ and $(3, -1)$.

Solution   We begin by finding the slope of the line:

$$m = \frac{3 - (-1)}{-3 - 3} = \frac{4}{-6} = -\frac{2}{3}$$

Using $(x_1, y_1) = (3, -1)$ and $m = -\frac{2}{3}$ in $y - y_1 = m(x - x_1)$ yields

$$y + 1 = -\frac{2}{3}(x - 3)$$

$$y + 1 = -\frac{2}{3}x + 2 \qquad \text{Multiply out right side.}$$

$$y = -\frac{2}{3}x + 1 \qquad \text{Add } -1 \text{ to each side.}$$

Figure 5 shows the graph of the line that passes through the points $(-3, 3)$ and $(3, -1)$. As you can see, the slope and y-intercept are $-\frac{2}{3}$ and 1, respectively.

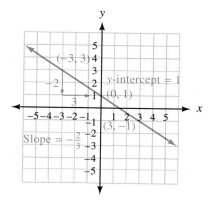

FIGURE 5

Note: We could have used the point $(-3, 3)$ instead of $(3, -1)$ and obtained the same equation. That is, using $(x_1, y_1) = (-3, 3)$ and $m = -\frac{2}{3}$ in $y - y_1 = b(x - x_1)$ gives us

$$y - 3 = -\frac{2}{3}(x + 3)$$

$$y - 3 = -\frac{2}{3}x - 2$$

$$y = -\frac{2}{3}x + 1$$

which is the same result we obtained using $(3, -1)$.

The last form of the equation of a line that we will consider in this section is called the standard form. It is used mainly to write equations in a form that is free of fractions and is easy to compare with other equations.

STANDARD FORM FOR THE EQUATION OF A LINE

If a, b, and c are integers, then the equation of a line is in standard form when it has the form

$$ax + by = c$$

If we were to write the equation

$$y = -\frac{2}{3}x + 1$$

in standard form, we would first multiply both sides by 3 to obtain

$$3y = -2x + 3$$

Then we would add $2x$ to each side, yielding

$$2x + 3y = 3$$

which is a linear equation in standard form.

EXAMPLE 6   Give the equation of the line through $(-1, 4)$ whose graph is perpendicular to the graph of $2x - y = -3$. Write the answer in standard form.

Solution   To find the slope of $2x - y = -3$, we solve for y:

$$2x - y = -3$$
$$y = 2x + 3$$

The slope of this line is 2. The line we are interested in is perpendicular to the line with slope 2 and must, therefore, have a slope of $-\frac{1}{2}$.

Using $(x_1, y_1) = (-1, 4)$ and $m = -\frac{1}{2}$, we have

$$y - y_1 = m(x - x_1)$$

$$y - 4 = -\frac{1}{2}(x + 1)$$

Since we want our answer in standard form, we multiply each side by 2.

$$2y - 8 = -1(x + 1)$$
$$2y - 8 = -x - 1$$
$$x + 2y - 8 = -1$$
$$x + 2y = 7$$

The last equation is in standard form.

As a final note, we should mention again that all horizontal lines have equations of the form $y = b$ and slopes of 0. Vertical lines have no slope and have equations of the form $x = a$. These two special cases do not lend themselves well to either the slope-intercept form or the point-slope form of the equation of a line.

Using TECHNOLOGY

GRAPHING CALCULATORS

One advantage of using a graphing calculator to graph lines is that a calculator does not care whether the equation has been simplified or not. To illustrate, in Example 5 we found that the equation of the line with slope $-\frac{2}{3}$ that passes through the point $(3, -1)$ is

$$y + 1 = -\frac{2}{3}(x - 3)$$

Normally, to graph this equation we would simplify it first. With a graphing calculator we add -1 to each side and enter the equation this way:

$$Y_1 = -(2/3)(X - 3) - 1$$

(continued)

No simplification is necessary. We can graph the equation in this form, and the graph will be the same as the simplified form of the equation, which is $y = -\frac{2}{3}x + 1$. To convince yourself that this is true, graph both the simplified form for the equation and the unsimplified form in the same window. As you will see, the two graphs coincide.

Getting Ready for Class

After reading through the preceding section, respond in your own words and in complete sentences.

A. How would you graph the line $y = \frac{1}{2}x + 3$?

B. What is the slope-intercept form of the equation of a line?

C. Describe how you would find the equation of a line if you knew the slope and the y-intercept of the line.

D. If you had the graph of a line, how would you use it to find the equation of the line?

PROBLEM SET 8.2

Give the equation of the line with the following slope and y-intercept.

1. $m = 2, b = 3$

2. $m = -4, b = 2$

3. $m = 1, b = -5$

4. $m = -5, b = -3$

5. $m = \dfrac{1}{2}, b = \dfrac{3}{2}$

6. $m = \dfrac{2}{3}, b = \dfrac{5}{6}$

7. $m = 0, b = 4$

8. $m = 0, b = -2$

Give the slope and y-intercept for each of the following equations. Sketch the graph using the slope and y-intercept. Give the slope of any line perpendicular to the given line.

9. $y = 3x - 2$

10. $y = 2x + 3$

11. $2x - 3y = 12$

12. $3x - 2y = 12$

13. $4x + 5y = 20$

14. $5x - 4y = 20$

For each of the following lines, name the slope and y-intercept. Then write the equation of the line in slope-intercept form.

15.

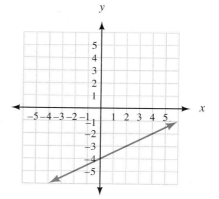

16.

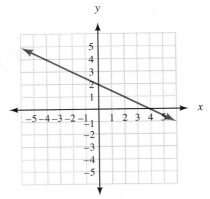

For each of the following problems, the slope and one point on a line are given. In each case, find the equation of that line. (Write the equation for each line in slope-intercept form.)

19. $(-2, -5); m = 2$ **20.** $(-1, -5); m = 2$

21. $(-4, 1); m = -\frac{1}{2}$ **22.** $(-2, 1); m = -\frac{1}{2}$

23. $(-\frac{1}{3}, 2); m = -3$ **24.** $(-\frac{2}{3}, 5); m = -3$

Find the equation of the line that passes through each pair of points. Write your answers in standard form.

25. $(-2, -4), (1, -1)$ **26.** $(2, 4), (-3, -1)$

27. $(-1, -5), (2, 1)$ **28.** $(-1, 6), (1, 2)$

29. $(\frac{1}{3}, -\frac{1}{5}), (-\frac{1}{3}, -1)$ **30.** $(-\frac{1}{2}, -\frac{1}{2}), (\frac{1}{2}, \frac{1}{10})$

17.

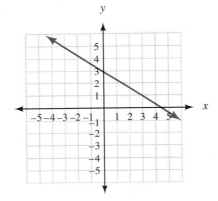

For each of the following lines, name the coordinates of any two points on the line. Then use those two points to find the equation of the line.

31.

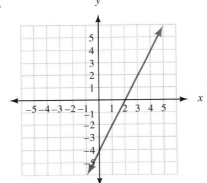

18.

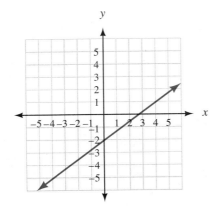

32.

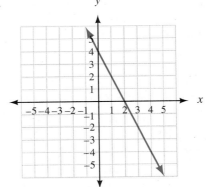

33.

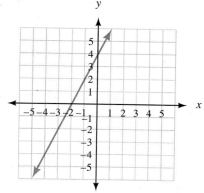

34.

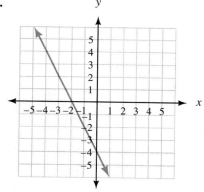

35. Give the slope and y-intercept of $y = -2$. Sketch the graph.

36. For the line $x = -3$, sketch the graph, give the slope, and name any intercepts.

37. Find the equation of the line parallel to the graph of $3x - y = 5$ that contains the point $(-1, 4)$.

38. Find the equation of the line parallel to the graph of $2x - 4y = 5$ that contains the point $(0, 3)$.

39. Line l is perpendicular to the graph of the equation $2x - 5y = 10$ and contains the point $(-4, -3)$. Find the equation for l.

40. Line l is perpendicular to the graph of the equation $-3x - 5y = 2$ and contains the point $(2, -6)$. Find the equation for l.

41. Give the equation of the line perpendicular to the graph of $y = -4x + 2$ that has an x-intercept of -1.

42. Write the equation of the line parallel to the graph of $7x - 2y = 14$ that has an x-intercept of 5.

43. Find the equation of the line with x-intercept 3 and y-intercept 2.

44. Find the equation of the line with x-intercept 2 and y-intercept 3.

Applying the Concepts

45. Deriving the Temperature Equation   The table below resembles the table from the introduction to this section. The rows of the table give us ordered pairs (C, F).

Degrees Celsius	Degrees Fahrenheit
C	F
0	32
25	77
50	122
75	167
100	212

(a) Use any two of the ordered pairs from the table to derive the equation $F = \frac{9}{5}C + 32$.

(b) Use the equation from part (a) to find the Fahrenheit temperature that corresponds to a Celsius temperature of $30°$.

46. Maximum Heart Rate   The table gives the maximum heart rate for adults 30, 40, 50, and 60 years old. Each row of the table below gives us an ordered pair (A, M).

Age (years)	Maximum Heart Rate (beats per minute)
A	M
30	190
40	180
50	170
60	160

(a) Use any two of the ordered pairs from the table to derive the equation $M = 220 - A$, which gives the maximum heart rate M for an adult whose age is A.

(b) Use the equation from part (a) to find the maximum heart rate for a 25-year-old adult.

47. **Oxygen Consumption and Exercise** A person's oxygen consumption (in liters per minute) is linearly related to that person's heart rate (in beats per minute). Suppose that at 98 beats per minute a person consumes 1 liter of oxygen per minute, and after exercising with a heart rate of 155 beats per minute consumes 1.5 liters of oxygen per minute.
   (a) Find an equation that relates oxygen consumption to heart rate.
   (b) What restrictions seem reasonable on values of the independent variable?

48. **Exercise Heart Rate** In an aerobics class, the instructor indicates that her students' exercise heart rate is 60% of their maximum heart rate, where maximum heart rate is 220 minus their age.
   (a) Determine the equation that gives exercise heart rate E in terms of age A.
   (b) Use the equation to find the exercise heart rate of a 22-year-old student.
   (c) Sketch the graph of the equation for students from 18 to 80 years of age.

49. **AIDS Cases** The number of AIDS cases in the United States from 1984 through 1988 increased in a linear relationship. In 1984 there were 3,000 known cases, and in 1988 the number of cases had risen to 20,000.
   (a) Write the equation of the line that relates the number of AIDS cases to the year for the years 1984 through 1988.
   (b) Use the equation found in part (a) to determine how many AIDS cases were reported in 1986.

50. **Cell Phone Rates** Suppose you have a cellular phone with a rate plan that costs $20 per month, with no additional charge for the first 30 minutes of use, and then $0.40 for each minute after the first 30 minutes.
   (a) Write an equation that gives the monthly cost C to talk for x minutes, if x is less than 30 minutes.
   (b) Write an equation that gives the monthly cost C to talk for x minutes, if x is greater than 30 minutes.

(c) On the same coordinate system, graph the first equation for $x < 30$ and the second equation for $x > 30$.

Review Problems

The following problems review material from Section 7.7.

Solve each system.

51. $\begin{aligned} x + y + z &= 6 \\ 2x - y + z &= 3 \\ x + 2y - 3z &= -4 \end{aligned}$

52. $\begin{aligned} x + y + z &= 6 \\ x - y + 2z &= 7 \\ 2x - y - z &= 0 \end{aligned}$

53. $\begin{aligned} 3x + 4y &= 15 \\ 2x - 5z &= -3 \\ 4y - 3z &= 9 \end{aligned}$

54. $\begin{aligned} x + 3y &= 5 \\ 6y + z &= 12 \\ x - 2z &= -10 \end{aligned}$

Extending the Concepts

55. Label the units on a sheet of graph paper in multiples of 10, and graph the line $2x + 5y = 100$.

56. Label the units on a sheet of graph paper in multiples of 20, and graph the line $-4x + 10y = 100$.

57. Label the y-axis in multiples of 10 and the x-axis in multiples of 1, and graph the equation $y = 20x - 50$.

58. Label the y-axis in multiples of 10 and the x-axis in multiples of 1, and graph the equation $y = -20x + 30$.

Write each equation in slope-intercept form. Then name the slope, the y-intercept, and the x-intercept.

59. $\dfrac{x}{2} + \dfrac{y}{3} = 1$

60. $\dfrac{x}{5} + \dfrac{y}{4} = 1$

61. $\dfrac{x}{-2} + \dfrac{y}{3} = 1$

62. $\dfrac{x}{2} + \dfrac{y}{-3} = 1$

63. When a linear equation is written in the form

$$\frac{x}{a} + \frac{y}{b} = 1$$

it is said to be in *two-intercept form*. Find the x-intercept, the y-intercept, and the slope of this line.

A small movie theater holds 100 people. The owner charges more for adults than for children, so it is important to know the different combinations of adults and children that can be seated at one time. The shaded region in Figure 1 contains all the seating combinations. The line $x + y = 100$ shows the combinations for a full theater: The y-intercept corresponds to a theater full of adults, and the x-intercept corresponds to a theater full of children. In the shaded region below the line $x + y = 100$ are the combinations that occur if the theater is not full.

Shaded regions like the one shown in Figure 1 are produced by linear inequalities in two variables, which is the topic of this section.

 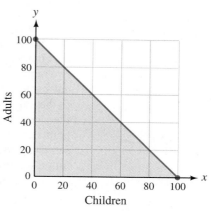

FIGURE I

A *linear inequality in two variables* is any expression that can be put in the form

$$ax + by < c$$

where a, b, and c are real numbers (a and b not both 0). The inequality symbol can be any one of the following four: $<, \leq, >, \geq$.

Some examples of linear inequalities are

$$2x + 3y < 6 \qquad y \geq 2x + 1 \qquad x - y \leq 0$$

Although not all of these examples have the form $ax + by < c$, each one can be put in that form.

The solution set for a linear inequality is a *section of the coordinate plane*. The *boundary* for the section is found by replacing the inequality symbol with an equal sign and graphing the resulting equation. The boundary is included in the solution set (and is represented with a *solid line*) if the inequality symbol used originally is $\leq$ or $\geq$. The boundary is not included (and is represented with a *broken line*) if the original symbol is $<$ or $>$.

EXAMPLE I Graph the solution set for $x + y \leq 4$.

Solution The boundary for the graph is the graph of $x + y = 4$. The boundary is included in the solution set because the inequality symbol is $\leq$.

Figure 2 is the graph of the boundary:

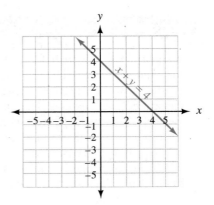

FIGURE 2

The boundary separates the coordinate plane into two regions: the region above the boundary and the region below it. The solution set for $x + y \leq 4$ is one of these two regions along with the boundary. To find the correct region, we simply choose any convenient point that is *not* on the boundary. We then substitute the coordinates of the point into the original inequality $x + y \leq 4$. If the point we choose satisfies the inequality, then it is a member of the solution set, and we can assume that all points on the same side of the boundary as the chosen point are also in the solution set. If the coordinates of our point do not satisfy the original inequality, then the solution set lies on the other side of the boundary.

In this example, a convenient point that is not on the boundary is the origin.

$$\text{Substituting} \quad (0, 0)$$
$$\text{into} \quad x + y \leq 4$$
$$\text{gives us} \quad 0 + 0 \leq 4$$
$$0 \leq 4 \quad \text{A true statement}$$

Since the origin is a solution to the inequality $x + y \leq 4$, and the origin is below the boundary, all other points below the boundary are also solutions.

Figure 3 is the graph of $x + y \leq 4$.

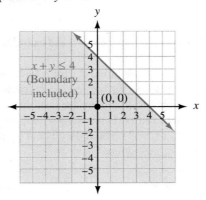

FIGURE 3

The region above the boundary is described by the inequality $x + y > 4$.

Here is a list of steps to follow when graphing the solution set for linear inequalities in two variables.

TO GRAPH A LINEAR INEQUALITY IN TWO VARIABLES

Step 1: Replace the inequality symbol with an equal sign. The resulting equation represents the boundary for the solution set.

Step 2: Graph the boundary found in step 1 using a *solid line* if the boundary is included in the solution set (i.e., if the original inequality symbol was either ≤ or ≥). Use a *broken line* to graph the boundary if it is *not* included in the solution set. (It is not included if the original inequality was either < or >.)

Step 3: Choose any convenient point not on the boundary and substitute the coordinates into the *original* inequality. If the resulting statement is *true*, the graph lies on the *same* side of the boundary as the chosen point. If the resulting statement is *false*, the solution set lies on the *opposite* side of the boundary.

EXAMPLE 2   Graph the solution set for $y < 2x - 3$.

Solution   The boundary is the graph of $y = 2x - 3$, a line with slope 2 and y-intercept -3. The boundary is not included since the original inequality symbol is <. We therefore use a broken line to represent the boundary in Figure 4.

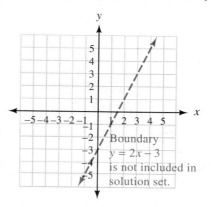

FIGURE 4

A convenient test point is again the origin:

$$\text{Using} \qquad (0, 0)$$

$$\text{in} \qquad y < 2x - 3$$

$$\text{we have} \qquad 0 < 2(0) - 3$$

$$0 < -3 \qquad \text{A false statement}$$

Since our test point gives us a false statement and it lies above the boundary, the solution set must lie on the other side of the boundary (Figure 5).

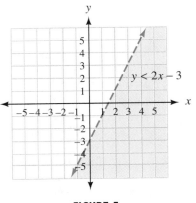

FIGURE 5

Using TECHNOLOGY

GRAPHING CALCULATORS

Most graphing calculators have a Shade command that allows a portion of a graphing screen to be shaded. With this command we can visualize the solution sets to linear inequalities in two variables. Since most graphing calculators cannot draw a dotted line, however, we are not actually "graphing" the solution set, only visualizing it.

STRATEGY FOR VISUALIZING A LINEAR INEQUALITY IN TWO VARIABLES ON A GRAPHING CALCULATOR

Step 1: Solve the inequality for y.
Step 2: Replace the inequality symbol with an equal sign. The resulting equation represents the boundary for the solution set.
Step 3: Graph the equation in an appropriate viewing window.
Step 4: Use the Shade command to indicate the solution set:
For inequalities having the $<$ or $\leq$ sign, use Shade(Xmin, Y_1).
For inequalities having the $>$ or $\geq$ sign, use Shade(Y_1, Xmax).

Note: On the TI-83, step 4 can be done by manipulating the icons in the left column in the list of Y variables.

Figures 6 and 7 show the graphing calculator screens that help us visualize the solution set to the inequality $y < 2x - 3$ that we graphed in Example 2.

Windows: X from -5 to 5, Y from -5 to 5

(continued)

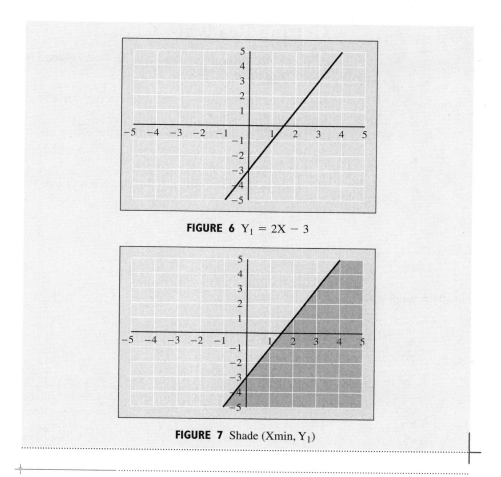

FIGURE 6 $Y_1 = 2X - 3$

FIGURE 7 Shade (Xmin, Y_1)

EXAMPLE 3   Graph the solution set for $x \leq 5$.

Solution   The boundary is $x = 5$, which is a vertical line. All points in Figure 8 to the left have x-coordinates less than 5 and all points to the right have x-coordinates greater than 5.

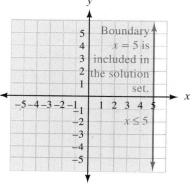

FIGURE 8

Getting Ready for Class

After reading through the preceding section, respond in your own words and in complete sentences.

A. When graphing a linear inequality in two variables, how do you find the equation of the boundary line?

B. What is the significance of a broken line in the graph of an inequality?

C. When graphing a linear inequality in two variables, how do you know which side of the boundary line to shade?

D. Describe the set of ordered pairs that are solutions to $x + y < 6$.

PROBLEM SET 8.3

Graph the solution set for each of the following.

1. $x + y < 5$ **2.** $x + y \leq 5$

3. $x - y \geq -3$ **4.** $x - y > -3$

5. $2x + 3y < 6$ **6.** $2x - 3y > -6$

7. $-x + 2y > -4$ **8.** $-x - 2y < 4$

9. $2x + y < 5$ **10.** $2x + y < -5$

11. $y < 2x - 1$ **12.** $y \leq 2x - 1$

14.

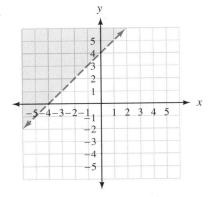

For each graph shown here, name the linear inequality in two variables that is represented by the shaded region.

13.

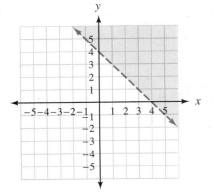

15.

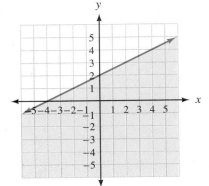

16.

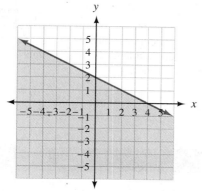

Graph each inequality.

17. $x \geq 3$ **18.** $x > -2$

19. $y \leq 4$ **20.** $y > -5$

21. $y < 2x$ **22.** $y > -3x$

23. $y \geq \frac{1}{2}x$ **24.** $y \leq \frac{1}{3}x$

25. $y \geq \frac{3}{4}x - 2$ **26.** $y > -\frac{2}{3}x + 3$

27. $\frac{x}{3} + \frac{y}{2} > 1$ **28.** $\frac{x}{5} + \frac{y}{4} < 1$

Applying the Concepts

29. Number of People in a Dance Club A dance club holds a maximum of 200 people. The club charges one price for students and a higher price for nonstudents. If the number of students in the club at any time is x and the number of nonstudents is y, shade the region in the first quadrant that contains all combinations of students and nonstudents that are in the club at any time.

30. Many Perimeters Suppose you have 500 feet of fencing that you will use to build a rectangular livestock pen. Let x represent the length of the pen and y represent the width. Shade the region in the first quadrant that contains all possible values of x and y that will give you a rectangle from 500 feet of fencing. (You don't have to use all of the fencing, so the perimeter of the pen could be less than 500 feet.)

31. Gas Mileage You have two cars. The first car travels an average of 12 miles on a gallon of gasoline, and the second averages 22 miles per gallon. Suppose you can afford to buy up to 30 gallons of gasoline this month. If the first car is driven x miles this month, and the second car is driven y miles this

month, shade the region in the first quadrant that gives all the possible values of x and y that will keep you from buying more than 30 gallons of gasoline this month.

32. Number Problem The sum of two positive numbers is at most 20. If the two numbers are represented by x and y, shade the region in the first quadrant that shows all the possibilities for the two numbers.

33. Student Loan Payments When considering how much debt to incur in student loans, it is advisable to keep your student loan payment after graduation to 8% or less of your starting monthly income. Let x represent your starting monthly salary and let y represent your monthly student loan payment, and write an inequality that describes this situation. Shade the region in the first quadrant that is a solution to your inequality.

34. Student Loan Payments During your college career you anticipate borrowing money two times: once as an undergraduate student and once as a graduate student. You anticipate a starting monthly salary after graduate school of approximately $3,500. You are advised to keep your total in student loan payments to 8% or less of your starting monthly salary. If x represents the monthly payment on the first loan and y represents the monthly payment on the second loan, shade the region in the first quadrant representing all combinations of possible monthly payments on the two loans that will total 8% or less of your starting monthly salary.

Review Problems

The problems that follow review material we covered in Sections 2.8 and 7.3.

Solve each of the following inequalities.

35. $5t - 4 > 3t - 8$

36. $-3(t - 2) < 6 - 5(t + 1)$

37. $2x < -8$ or $x + 2 > 6$

38. $3x - 1 \leq 2$ or $2x + 3 \geq 11$

39. $-9 < -4 + 5t < 6$

40. $-3 < 2t + 1 < 3$

Extending the Concepts

Graph each inequality.

41. $y < |x + 2|$ **42.** $y > |x - 2|$

43. $y > |x - 3|$ **44.** $y < |x + 3|$

45. The Associated Students organization holds a *Night at the Movies* fund-raiser. Students tickets are $1.00 each and nonstudent tickets are $2.00 each. The theater holds a maximum of 200 people. The club needs to collect at least $100 to make money. Shade the region in the first quadrant that contains all combinations of students and non-students that could attend the movie night so that the club makes money.

8.4   Introduction to Functions

The ad shown in the margin appeared in the help wanted section of the local newspaper the day I was writing this section of the book. We can use the information in the ad to start an informal discussion of our next topic: functions.

312 Help Wanted

YARD PERSON
Full-time 40 hrs. with weekend work required. Cleaning & loading trucks. $7.50/hr. Valid CDL with clean record & drug screen required. Submit current MVR to KCI, 225 Suburban Rd., SLO. 805-555-3304.

AN INFORMAL LOOK AT FUNCTIONS

To begin with, suppose you have a job that pays $7.50 per hour and that you work anywhere from 0 to 40 hours per week. The amount of money you make in one week depends on the number of hours you work that week. In mathematics we say that your weekly earnings are a *function* of the number of hours you work. If we let the variable x represent hours and the variable y represent the money you make, then the relationship between x and y can be written as

$$y = 7.5x \quad \text{for} \quad 0 \le x \le 40$$

EXAMPLE I   Construct a table and graph for the function

$$y = 7.5x \quad \text{for} \quad 0 \le x \le 40$$

Solution   Table 1 gives some of the paired data that satisfy the equation $y = 7.5x$. Figure 1 is the graph of the equation with the restriction $0 \le x \le 40$.

TABLE I   Weekly Wages

Hours Worked	Rule	Pay
x	$y = 7.5x$	y
0	$y = 7.5(0)$	0
10	$y = 7.5(10)$	75
20	$y = 7.5(20)$	150
30	$y = 7.5(30)$	225
40	$y = 7.5(40)$	300

Ordered Pairs
(0, 0)
(10, 75)
(20, 150)
(30, 225)
(40, 300)

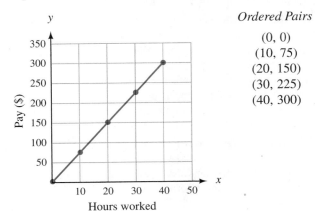

FIGURE I Weekly wages at $7.50 per hour

The equation $y = 7.5x$ with the restriction $0 \leq x \leq 40$, Table 1, and Figure 1 are three ways to describe the same relationship between the number of hours you work in one week and your gross pay for that week. In all three, we *input* values of x, and then use the function rule to *output* values of y.

DOMAIN AND RANGE OF A FUNCTION

We began this discussion by saying that the number of hours worked during the week was from 0 to 40, so these are the values that x can assume. From the line graph in Figure 1, we see that the values of y range from 0 to 300. We call the complete set of values that x can assume the *domain* of the function. The values that are assigned to y are called the *range* of the function.

EXAMPLE 2   State the domain and range for the function

$$y = 7.5x, \quad 0 \leq x \leq 40$$

Solution   From the previous discussion we have

$$\text{Domain} = \{x \mid 0 \leq x \leq 40\}$$

$$\text{Range} = \{y \mid 0 \leq y \leq 300\}$$

FUNCTION MAPS

Another way to visualize the relationship between x and y is with the diagram in Figure 2, which we call a *function map*:

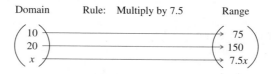

FIGURE 2 A function map

Although the diagram in Figure 2 does not show all the values that x and y can assume, it does give us a visual description of how x and y are related. It shows that values of y in the range come from values of x in the domain according to a specific rule (multiply by 7.5 each time).

A FORMAL LOOK AT FUNCTIONS

What is apparent from the preceding discussion is that we are working with paired data. The solutions to the equation $y = 7.5x$ are pairs of numbers; the points on the line graph in Figure 1 come from paired data; and the diagram in Figure 2 pairs numbers in the domain with numbers in the range. We are now ready for the formal definition of a function.

> **DEFINITION**   A **function** is a rule that pairs each element in one set, called the **domain,** with exactly one element from a second set, called the **range.**

In other words, a function is a rule for which each input is paired with exactly one output.

FUNCTIONS AS ORDERED PAIRS

The function rule $y = 7.5x$ from Example 1 produces ordered pairs of numbers (x, y). The same thing happens with all functions: The function rule produces ordered pairs of numbers. We use this result to write an alternative definition for a function.

> **ALTERNATIVE DEFINITION**   A **function** is a set of ordered pairs in which no two different ordered pairs have the same first coordinate. The set of all first coordinates is called the **domain** of the function. The set of all second co-ordinates is called the **range** of the function.

The restriction on first coordinates in the alternative definition keeps us from assigning a number in the domain to more than one number in the range.

A RELATIONSHIP THAT IS NOT A FUNCTION

You may be wondering if any sets of paired data fail to qualify as functions. The answer is yes, as the next example reveals.

EXAMPLE 3   Table 2 shows the prices of used Ford Mustangs that were listed in the local newspaper. The diagram in Figure 3 is called a *scatter diagram.* It gives a visual representation of the data in Table 2. Why is this data not a function?

TABLE 2 Used Mustang Prices	
Year	**Price ($)**
x	*y*
1997	13,925
1997	11,850
1997	9,995
1996	10,200
1996	9,600
1995	9,525
1994	8,675
1994	7,900
1993	6,975

FIGURE 3 Scatter diagram of data in Table 2

Ordered Pairs

(1997, 13,925)
(1997, 11,850)
(1997, 9,995)
(1996, 10,200)
(1996, 9,600)
(1995, 9,525)
(1994, 8,675)
(1994, 7,900)
(1993, 6,975)

Solution   In Table 2, the year 1997 is paired with three different prices: $13,925, $11,850, and $9,995. That is enough to disqualify the data from belonging to a function. For a set of paired data to be considered a function, each number in the domain must be paired with exactly one number in the range.

Still, there is a relationship between the first coordinates and second coordinates in the used-car data. It is not a function relationship, but it is a relationship. In order to classify all relationships specified by ordered pairs, whether they are functions or not, we include the following two definitions.

> **DEFINITION**   A **relation** is a rule that pairs each element in one set, called the **domain**, with **one or more elements** from a second set, called the **range**.

> **ALTERNATIVE DEFINITION**   A **relation** is a set of ordered pairs. The set of all first coordinates is the **domain** of the relation. The set of all second coordinates is the **range** of the relation.

Here are some facts that will help clarify the distinction between relations and functions.

1. Any rule that assigns numbers from one set to numbers in another set is a relation. If that rule makes the assignment so that no input has more than one output, then it is also a function.
2. Any set of ordered pairs is a relation. If none of the first coordinates of those ordered pairs is repeated, the set of ordered pairs is also a function.
3. Every function is a relation.
4. Not every relation is a function.

GRAPHING RELATIONS AND FUNCTIONS

To give ourselves a wider perspective on functions and relations, we consider some equations whose graphs are not straight lines.

EXAMPLE   4   Kendra is tossing a softball into the air with an underhand motion. The distance of the ball above her hand at any time is given by the function

$$h = 32t - 16t^2 \quad \text{for} \quad 0 \le t \le 2$$

where h is the height of the ball in feet, and t is the time in seconds. Construct a table that gives the height of the ball at quarter-second intervals, starting with $t = 0$ and ending with $t = 2$. Construct a line graph from the table.

Solution   We construct Table 3 using the following values of t: $0, \frac{1}{4}, \frac{1}{2}, \frac{3}{4}, 1, \frac{5}{4}, \frac{3}{2}, \frac{7}{4}$, 2. The values of h come from substituting these values of t into the equation

$h = 32t - 16t^2$. (This equation comes from physics. If you take a physics class, you will learn how to derive this equation.) Then we construct the graph in Figure 4 from the table. The graph appears only in the first quadrant because neither t nor h can be negative.

TABLE 3 Tossing a Softball into the Air		
Time (sec)	**Function Rule**	**Distance (ft)**
t	$h = 32t - 16t^2$	h
0	$h = 32(0) - 16(0)^2 = 0 - 0 = 0$	0
$\frac{1}{4}$	$h = 32\left(\frac{1}{4}\right) - 16\left(\frac{1}{4}\right)^2 = 8 - 1 = 7$	7
$\frac{1}{2}$	$h = 32\left(\frac{1}{2}\right) - 16\left(\frac{1}{2}\right)^2 = 16 - 4 = 12$	12
$\frac{3}{4}$	$h = 32\left(\frac{3}{4}\right) - 16\left(\frac{3}{4}\right)^2 = 24 - 9 = 15$	15
1	$h = 32(1) - 16(1)^2 = 32 - 16 = 16$	16
$\frac{5}{4}$	$h = 32\left(\frac{5}{4}\right) - 16\left(\frac{5}{4}\right)^2 = 40 - 25 = 15$	15
$\frac{3}{2}$	$h = 32\left(\frac{3}{2}\right) - 16\left(\frac{3}{2}\right)^2 = 48 - 36 = 12$	12
$\frac{7}{4}$	$h = 32\left(\frac{7}{4}\right) - 16\left(\frac{7}{4}\right)^2 = 56 - 49 = 7$	7
2	$h = 32(2) - 16(2)^2 = 64 - 64 = 0$	0

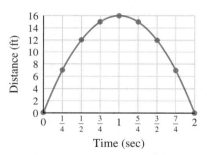

FIGURE 4

Here is a summary of what we know about functions as it applies to this example: We input values of t and output values of h according to the function rule

$$h = 32t - 16t^2 \quad \text{for} \quad 0 \le t \le 2$$

The domain is given by the inequality that follows the equation; it is

$$\text{Domain} = \{t \mid 0 \le t \le 2\}$$

The range is the set of all outputs that are possible by substituting the values of t from the domain into the equation. From our table and graph, it seems that the range is

$$\text{Range} = \{h \mid 0 \le h \le 16\}$$

Using TECHNOLOGY

MORE ABOUT EXAMPLE 4

Most graphing calculators can easily produce the information in Table 3. Simply set Y_1 equal to $32X - 16X^2$. Then set up the table so it starts at 0 and increases by an increment of 0.25 each time. (On a TI-82/83, use the TBLSET key to set up the table.) .

Table Setup

Table minimum = 0
Table increment = .25
Dependent variable: Auto
Independent variable: Auto

Y Variables Setup

$Y_1 = 32X - 16X^2$

The table will look like this:

X	Y_1
0	0
.25	7
.5	12
.75	15
1	16
1.25	15
1.5	12

EXAMPLE 5   Sketch the graph of $x = y^2$.

Solution   Without going into much detail, we graph the equation $x = y^2$ by finding a number of ordered pairs that satisfy the equation, plotting these points, then drawing a smooth curve that connects them. A table of values for x and y that satisfy the equation follows, along with the graph of $x = y^2$ shown in Figure 5.

x	y
0	0
1	1
1	−1
4	2
4	−2
9	3
9	−3

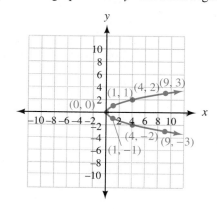

FIGURE 5

As you can see from looking at the table and the graph in Figure 5, several ordered pairs whose graphs lie on the curve have repeated first coordinates, for instance (1, 1) and (1, −1), (4, 2) and (4, −2), as well as (9, 3) and (9, −3). The graph is therefore not the graph of a function.

VERTICAL LINE TEST

Look back at the scatter diagram for used Mustang prices shown in Figure 3. Notice that some of the points on the diagram lie above and below each other along vertical lines. This is an indication that the data do not constitute a function. Two data points that lie on the same vertical line must have come from two ordered pairs with the same first coordinates.

Now, look at the graph shown in Figure 5. The reason this graph is the graph of a relation, but not of a function, is that some points on the graph have the same first coordinates, for example, the points (4, 2) and (4, −2). Furthermore, any time two points on a graph have the same first coordinates, those points must lie on a vertical line. [To convince yourself, connect the points (4, 2) and (4, −2) with a straight line. You will see that it must be a vertical line.] This allows us to write the following test that uses the graph to determine whether a relation is also a function.

VERTICAL LINE TEST

If a vertical line crosses the graph of a relation in more than one place, the relation cannot be a function. If no vertical line can be found that crosses a graph in more than one place, then the graph is the graph of a function.

If we look back to the graph of $h = 32t - 16t^2$ as shown in Figure 4, we see that no vertical line can be found that crosses this graph in more than one place. The graph shown in Figure 4 is therefore the graph of a function.

EXAMPLE 6   Graph $y = |x|$. Use the graph to determine whether we have the graph of a function. State the domain and range.

Solution   We let x take on values of −4, −3, −2, −1, 0, 1, 2, 3, and 4. The corresponding values of y are shown in the table. The graph is shown in Figure 6.

x	y
-4	4
-3	3
-2	2
-1	1
0	0
1	1
2	2
3	3
4	4

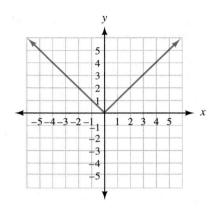

FIGURE 6

Since no vertical line can be found that crosses the graph in more than one place, $y = |x|$ is a function. The domain is all real numbers. The range is $\{y \mid y \geq 0\}$.

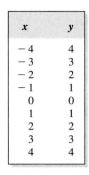 **Getting Ready for Class**

After reading through the preceding section, respond in your own words and in complete sentences.

A. What is a function?

B. What is the vertical line test?

C. Is every line the graph of a function? Explain.

D. Which variable is usually associated with the domain of a function?

PROBLEM SET 8.4

1. Suppose you have a job that pays $8.50 per hour and you work anywhere from 10 to 40 hours per week.

(a) Write an equation, with a restriction on the variable x, that gives the amount of money, y, you will earn for working x hours in one week.

(b) Use the function rule you have written in part (a) to complete Table 4.

TABLE 4   Weekly Wages

Hours Worked	Function Rule	Gross Pay ($)
x		y
10		
20		
30		
40		

(c) Use the template in Figure 7 to construct a line graph from the information in Table 4.

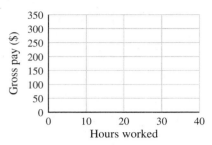

FIGURE 7 Template for line graph

(d) State the domain and range of this function.
(e) What is the minimum amount you can earn in a week with this job? What is the maximum amount?

2. The ad shown here was in the local newspaper. Suppose you are hired for the job described in the ad.

312 Help Wanted
ESPRESSO BAR OPERATOR
Must be dependable, honest, service-oriented. Coffee exp desired. 15–30 hrs per wk. $5.25/hr. Start 5/31. Apply in person: Espresso Yourself, Central Coast Mall. Deadline 5/23.

(a) If *x* is the number of hours you work per week and *y* is your weekly gross pay, write the equation for *y*. (Be sure to include any restrictions on the variable *x* that are given in the ad.)
(b) Use the function rule you have written in part (a) to complete Table 5.

TABLE 5 Weekly Wages

Hours Worked	Function Rule	Gross Pay ($)
x		*y*
15		
20		
25		
30		

(c) Use the template in Figure 8 to construct a line graph from the information in Table 5.

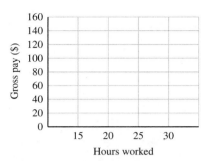

FIGURE 8 Template for line graph

(d) State the domain and range of this function.
(e) What is the minimum amount you can earn in a week with this job? What is the maximum amount?

For each of the following relations, give the domain and range, and indicate which are also functions.

3. $\{(1, 3), (2, 5), (4, 1)\}$ **4.** $\{(3, 1), (5, 7), (2, 3)\}$
5. $\{(-1, 3), (1, 3), (2, -5)\}$
6. $\{(3, -4), (-1, 5), (3, 2)\}$
7. $\{(7, -1), (3, -1), (7, 4)\}$
8. $\{(5, -2), (3, -2), (5, -1)\}$

State whether each of the following graphs represents a function.

9.

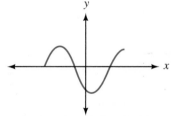

10.

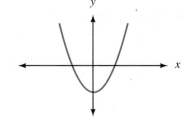

11.

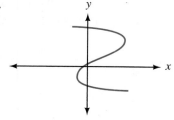

12.

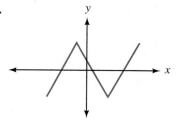

13.

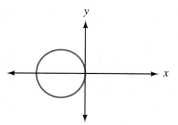

14.

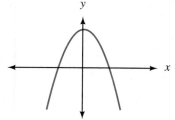

15.

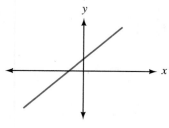

16.

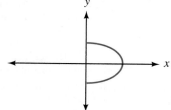

17.

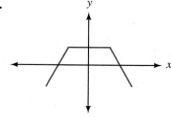

18.

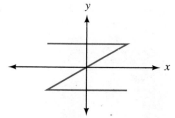

19. Tossing a Coin   Hali is tossing a quarter into the air with an underhand motion. The distance the quarter is above her hand at any time is given by the function

$$h = 16t - 16t^2 \qquad \text{for} \quad 0 \le t \le 1$$

where h is the height of the quarter in feet, and t is the time in seconds.

(a) Fill in the table.

Time (sec)	Function Rule	Distance (ft)
t	$h = 16t - 16t^2$	h
0		
0.1		
0.2		
0.3		
0.4		
0.5		
0.6		
0.7		
0.8		
0.9		
1		

(b) State the domain and range of this function.

(c) Use the data from the table to graph the function.

20. Intensity of Light   The following formula gives the intensity of light that falls on a surface at various distances from a 100-watt light bulb:

$$I = \frac{120}{d^2} \qquad \text{for} \quad d > 0$$

where I is the intensity of light (in lumens per square foot), and d is the distance (in feet) from the light bulb to the surface.

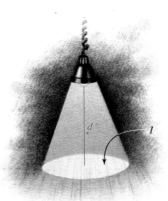

(a) Fill in the table.

Distance (ft)	Function Rule	Intensity
d		I
1		
2		
3		
4		
5		
6		

(b) Use the data from the table in part (a) to construct a line graph of the function.

Determine the domain and range of the following functions. Assume the *entire* function is shown.

21.

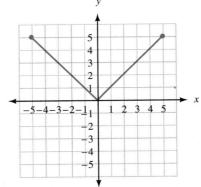

22.

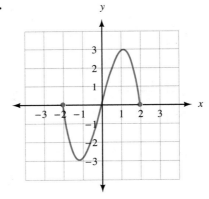

23.

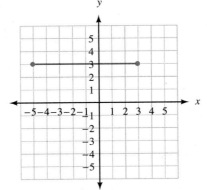

24.

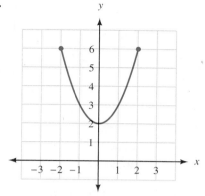

Graph each of the following relations. In each case, use the graph to find the domain and range, and indicate whether the graph is the graph of a function.

25. $y = x^2 - 1$ **26.** $y = x^2 + 1$

27. $y = x^2 + 4$ **28.** $y = x^2 - 9$

29. $x = y^2 - 1$ **30.** $x = y^2 + 1$

31. $x = y^2 + 4$ **32.** $x = y^2 - 9$

33. $y = |x - 2|$ **34.** $y = |x| - 2$

Area of a Circle   The formula for the area A of a circle with radius r is given by $A = \pi r^2$. The formula shows that A is a function of r.

35. Graph the function $A = \pi r^2$ for $0 \le r \le 3$. (On the graph, let the horizontal axis be the r-axis, and let the vertical axis be the A-axis.)

36. State the domain and range of the function $A = \pi r^2, 0 \le r \le 3$. (Use $\pi \approx 3.14$.)

Area and Perimeter of a Rectangle   A rectangle is 2 inches longer than it is wide. Let $x =$ the width, $P =$ the perimeter, and $A =$ the area of the rectangle (Problems 37–40).

37. Write an equation that will give the perimeter P in terms of the width x of the rectangle. Are there any restrictions on the values that x can assume?

38. Graph the relationship between P and x.

39. Write an equation that will give the area A in terms of the width x of the rectangle. Are there any restrictions on the values that x can assume?

40. Graph the relationship between A and x.

41. Tossing a Ball   A ball is thrown straight up into the air from ground level. The relationship between the height h of the ball at any time t is illustrated by the following graph:

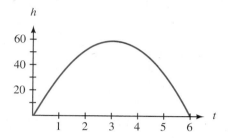

The horizontal axis represents time t, and the vertical axis represents height h.

(a) Is this graph the graph of a function?

(b) State the domain and range.

(c) At what time does the ball reach its maximum height?

(d) What is the maximum height of the ball?

(e) At what time does the ball hit the ground?

42. The following graph shows the relationship between a company's profits P and the number of items it sells x. (P is in dollars.)

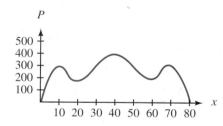

(a) Is this graph the graph of a function?

(b) State the domain and range.

(c) How many items must the company sell to make their maximum profit?

(d) What is their maximum profit?

Reading Graphs

43. Match each of the following statements to the appropriate graph indicated by Figures 9–12.

(a) Sarah works 25 hours in a week to earn $250.

(b) Justin works 35 hours in a week to earn $560.

(c) Rosemary works 30 hours in a week to earn $360.

(d) Marcus works 40 hours in a week to earn $320.

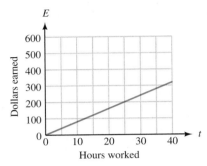

FIGURE 9

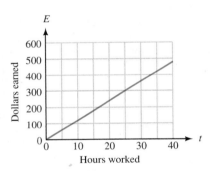

FIGURE 10

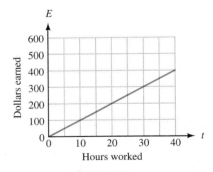

FIGURE 11

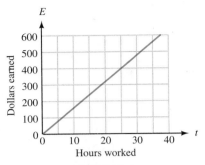

FIGURE 12

44. Find an equation for each of the functions shown in Figures 9–12. Show dollars earned, E, as a function of hours worked, t. Then, indicate the domain and range of each function.

(a) Figure 9: $E =$ _____   Domain $= \{t|$ $\}$;
Range $= \{E|$ $\}$

(b) Figure 10: $E =$ _____   Domain $= \{t|$ $\}$;
Range $= \{E|$ $\}$

(c) Figure 11: $E =$ _____   Domain $= \{t|$ $\}$;
Range $= \{E|$ $\}$

(d) Figure 12: $E =$ _____   Domain $= \{t|$ $\}$;
Range $= \{E|$ $\}$

Review Problems

The problems that follow review material we covered in Section 2.5. Reviewing these problems will help you with the next section.

For the equation $y = 3x - 2$:

45. Find y if x is 4.

46. Find y if x is 0.

47. Find y if x is -4.

48. Find y if x is -2.

For the equation $y = x^2 - 3$:

49. Find y if x is 2.

50. Find y if x is -2.

51. Find y if x is 0.

52. Find y if x is -4.

Let's return to the discussion that introduced us to functions. If a job pays $7.50 per hour for working from 0 to 40 hours a week, then the amount of money y earned in one week is a function of the number of hours worked x. The exact relationship between x and y is written

$$y = 7.5x \quad \text{for} \quad 0 \le x \le 40$$

Since the amount of money earned y depends on the number of hours worked x, we call y the *dependent variable* and x the *independent variable*. Furthermore, if we let f represent all the ordered pairs produced by the equation, then we can write

$$f = \{(x, y) \,|\, y = 7.5x \text{ and } 0 \le x \le 40\}$$

Once we have named a function with a letter, we can use an alternative notation to represent the dependent variable y. The alternative notation for y is $f(x)$. It is read "f of x" and can be used instead of the variable y when working with functions. The notation y and the notation $f(x)$ are equivalent. That is,

$$y = 7.5x \Leftrightarrow f(x) = 7.5x$$

When we use the notation $f(x)$ we are using *function notation*. The benefit of using function notation is that we can write more information with fewer symbols than we can by using just the variable y. For example, asking how much money a person will make for working 20 hours is simply a matter of asking for $f(20)$. Without function notation, we would have to say "find the value of y that corresponds to a value of $x = 20$." To illustrate further, using the variable y, we can say "y is 150 when x is 20." Using the notation $f(x)$, we simply say "$f(20) = 150$." Each expression indicates that you will earn $150 for working 20 hours.

EXAMPLE 1 If $f(x) = 7.5x$, find $f(0)$, $f(10)$, and $f(20)$.

Solution To find $f(0)$ we substitute 0 for x in the expression $7.5x$ and simplify. We find $f(10)$ and $f(20)$ in a similar manner—by substitution.

$$\text{If} \qquad f(x) = 7.5x$$
$$\text{then} \qquad f(\mathbf{0}) = 7.5(\mathbf{0}) = 0$$
$$f(\mathbf{10}) = 7.5(\mathbf{10}) = 75$$
$$f(\mathbf{20}) = 7.5(\mathbf{20}) = 150$$

If we changed the example in the discussion that opened this section so that the hourly wage was $6.50 per hour, we would have a new equation to work with, namely,

$$y = 6.5x \quad \text{for} \quad 0 \le x \le 40$$

Suppose we name this new function with the letter g. Then

$$g = \{(x, y) \,|\, y = 6.5x \text{ and } 0 \le x \le 40\}$$

312 Help Wanted

YARD PERSON
Full-time 40 hrs. with weekend work required. Cleaning & loading trucks. $7.50/hr. Valid CDL with clean record & drug screen required. Submit current MVR to KCI, 225 Suburban Rd., SLO. 805-555-3304.

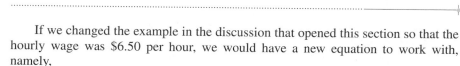

Input x

Function machine

Output $f(x)$

Some students like to think of functions as machines. Values of x are put into the machine, which transforms them into values of $f(x)$, which are then output by the machine.

and

$$g(x) = 6.5x$$

If we want to talk about both functions in the same discussion, having two different letters, f and g, makes it easy to distinguish between them. For example, since $f(x) = 7.5x$ and $g(x) = 6.5x$, asking how much money a person makes for working 20 hours is simply a matter of asking for $f(20)$ or $g(20)$, avoiding any confusion over which hourly wage we are talking about.

The diagrams shown in Figure 1 further illustrate the similarities and differences between the two functions we have been discussing.

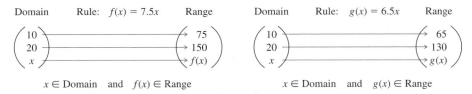

$x \in$ Domain   and $f(x) \in$ Range $x \in$ Domain   and $g(x) \in$ Range

FIGURE 1   Function maps

FUNCTION NOTATION AND GRAPHS

We can visualize the relationship between x and $f(x)$ or $g(x)$ on the graphs of the two functions. Figure 2 shows the graph of $f(x) = 7.5x$ along with two additional line segments. The horizontal line segment corresponds to $x = 20$, and the vertical line segment corresponds to $f(20)$. Figure 3 shows the graph of $g(x) = 6.5x$ along with the horizontal line segment that corresponds to $x = 20$, and the vertical line segment that corresponds to $g(20)$. (Note that the domain in each case is restricted to $0 \leq x \leq 40$.)

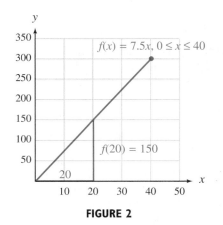

FIGURE 2

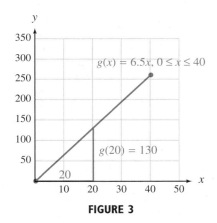

FIGURE 3

USING FUNCTION NOTATION

The remaining examples in this section show a variety of ways to use and interpret function notation.

EXAMPLE 2 If it takes Lorena t minutes to run a mile, then her average speed $s(t)$ in miles per hour, is given by the formula

$$s(t) = \frac{60}{t} \quad \text{for} \quad t > 0$$

Find $s(10)$ and $s(8)$, and then explain what they mean.

Solution To find $s(10)$, we substitute 10 for t in the equation and simplify:

$$s(\mathbf{10}) = \frac{60}{\mathbf{10}} = 6$$

In words: When Lorena runs a mile in 10 minutes, her average speed is 6 miles per hour.

We calculate $s(8)$ by substituting 8 for t in the equation. Doing so gives us

$$s(\mathbf{8}) = \frac{60}{\mathbf{8}} = 7.5$$

In words: Running a mile in 8 minutes is running at a rate of 7.5 miles per hour.

EXAMPLE 3 A painting is purchased as an investment for $125. If its value increases continuously so that it doubles every 5 years, then its value is given by the function

$$V(t) = 125 \cdot 2^{t/5} \quad \text{for} \quad t \geq 0$$

where t is the number of years since the painting was purchased, and $V(t)$ is its value (in dollars) at time t. Find $V(5)$ and $V(10)$, and explain what they mean.

Solution The expression $V(5)$ is the value of the painting when $t = 5$ (5 years after it is purchased). We calculate $V(5)$ by substituting 5 for t in the equation $V(t) = 125 \cdot 2^{t/5}$. Here is our work:

$$V(\mathbf{5}) = 125 \cdot 2^{\mathbf{5}/5} = 125 \cdot 2^1 = 125 \cdot 2 = 250$$

In words: After 5 years, the painting is worth $250.

The expression $V(10)$ is the value of the painting after 10 years. To find this number, we substitute 10 for t in the equation:

$$V(\mathbf{10}) = 125 \cdot 2^{\mathbf{10}/5} = 125 \cdot 2^2 = 125 \cdot 4 = 500$$

In words: The value of the painting 10 years after it is purchased is $500.

EXAMPLE 4 A balloon has the shape of a sphere with a radius of 3 inches. Use the following formulas to find the volume and surface area of the balloon.

$$V(r) = \frac{4}{3} \pi r^3 \qquad S(r) = 4\pi r^2$$

Solution As you can see, we have used function notation to write the two formulas for volume and surface area, because each quantity is a function of the radius.

To find these quantities when the radius is 3 inches, we evaluate $V(3)$ and $S(3)$:

$$V(3) = \frac{4}{3}\pi 3^3 = \frac{4}{3}\pi 27 = 36\pi \text{ cubic inches, or } 113 \text{ cubic inches} \qquad \substack{\text{To the nearest} \\ \text{whole number}}$$

$$S(3) = 4\pi 3^2 = 36\pi \text{ square inches, or } 113 \text{ square inches} \qquad \substack{\text{To the nearest} \\ \text{whole number}}$$

The fact that $V(3) = 36\pi$ means that the ordered pair $(3, 36\pi)$ belongs to the function V. Likewise, the fact that $S(3) = 36\pi$ tells us that the ordered pair $(3, 36\pi)$ is a member of function S.

We can generalize the discussion at the end of Example 4 this way:

$$(a, b) \in f \qquad \text{if and only if} \qquad f(a) = b$$

Using TECHNOLOGY

MORE ABOUT EXAMPLE 4

If we look back at Example 4, we see that when the radius of a sphere is 3, the numerical values of the volume and surface area are equal. How unusual is this? Are there other values of r for which $V(r)$ and $S(r)$ are equal? We can answer this question by looking at the graphs of both V and S.

To graph the function $V(r) = \frac{4}{3}\pi r^3$, set $Y_1 = 4\pi X^3/3$. To graph $S(r) = 4\pi r^2$, set $Y_2 = 4\pi X^2$. Graph the two functions in each of the following windows:

> Window 1: X from -4 to 4, Y from -2 to 10
>
> Window 2: X from 0 to 4, Y from 0 to 50
>
> Window 3: X from 0 to 4, Y from 0 to 150

Then use the Trace and Zoom features of your calculator to locate the point in the first quadrant where the two graphs intersect. How do the coordinates of this point compare with the results in Example 4?

EXAMPLE 5 If $f(x) = 3x^2 + 2x - 1$, find $f(0)$, $f(3)$, and $f(-2)$.

Solution Since $f(x) = 3x^2 + 2x - 1$, we have

$$f(0) = 3(0)^2 + 2(0) - 1 \qquad = 0 + 0 - 1 = -1$$

$$f(3) = 3(3)^2 + 2(3) - 1 \qquad = 27 + 6 - 1 = 32$$

$$f(-2) = 3(-2)^2 + 2(-2) - 1 = 12 - 4 - 1 = 7$$

In Example 5, the function f is defined by the equation $f(x) = 3x^2 + 2x - 1$. We could just as easily have said $y = 3x^2 + 2x - 1$. That is, $y = f(x)$. Saying $f(-2) = 7$ is exactly the same as saying y is 7 when x is -2.

EXAMPLE 6   If $f(x) = 4x - 1$ and $g(x) = x^2 + 2$, then

$$f(5) = 4(5) - 1 = 19 \qquad \text{and} \qquad g(5) = 5^2 + 2 = 27$$

$$f(-2) = 4(-2) - 1 = -9 \qquad \text{and} \qquad g(-2) = (-2)^2 + 2 = 6$$

$$f(0) = 4(0) - 1 = -1 \qquad \text{and} \qquad g(0) = 0^2 + 2 = 2$$

$$f(z) = 4z - 1 \qquad \text{and} \qquad g(z) = z^2 + 2$$

$$f(a) = 4a - 1 \qquad \text{and} \qquad g(a) = a^2 + 2$$

Using TECHNOLOGY

MORE ABOUT EXAMPLE 6

Most graphing calculators can use tables to evaluate functions. To work Example 6 using a graphing calculator table, set Y_1 equal to $4X - 1$ and Y_2 equal to $X^2 + 2$. Then set the independent variable in the table to Ask instead of Auto. Go to your table and input 5, -2, and 0. Under Y_1 in the table, you will find $f(5), f(-2)$, and $f(0)$. Under Y_2, you will find $g(5)$, $g(-2)$, and $g(0)$.

Table Setup   *Y Variables Setup*

Table minimum = 0 $Y_1 = 4X - 1$
Table increment = 1 $Y_2 = X^2 + 2$
Independent variable: Ask
Dependent variable: Ask

The table will look like this:

X	Y_1	Y_2
5	19	27
-2	-9	6
0	-1	2

Although the calculator asks us for a table increment, the increment doesn't matter since we are inputting the X values ourselves.

EXAMPLE 7   If the function f is given by

$$f = \{(-2, 0), (3, -1), (2, 4), (7, 5)\}$$

then $f(-2) = 0$, $f(3) = -1$, $f(2) = 4$, and $f(7) = 5$.

EXAMPLE 8   If $f(x) = 2x^2$ and $g(x) = 3x - 1$, find
(a) $f[g(2)]$ (b) $g[f(2)]$

Solution   The expression $f[g(2)]$ is read "f of g of 2."
(a) Since $g(2) = 3(2) - 1 = 5$,

$$f[g(2)] = f(5) = 2(5)^2 = 50$$

(b) Since $f(2) = 2(2)^2 = 8$,

$$g[f(2)] = g(8) = 3(8) - 1 = 23$$

 Getting Ready for Class

After reading through the preceding section, respond in your own words and in complete sentences.

A. Explain what you are calculating when you find $f(2)$ for a given function f.

B. If $s(t) = \dfrac{60}{t}$, how do you find $s(10)$?

C. If $f(2) = 3$ for a function f, what is the relationship between the numbers 2 and 3 and the graph of f?

D. If $f(6) = 0$ for a particular function f, then you can immediately graph one of the intercepts. Explain.

PROBLEM SET 8.5

Let $f(x) = 2x - 5$ and $g(x) = x^2 + 3x + 4$. Evaluate the following.

1. $f(2)$ **2.** $f(3)$
3. $f(-3)$ **4.** $g(-2)$
5. $g(-1)$ **6.** $f(-4)$
7. $g(-3)$ **8.** $g(2)$
9. $g(4) + f(4)$ **10.** $f(2) - g(3)$
11. $f(3) - g(2)$ **12.** $g(-1) + f(-1)$

Let $f(x) = 3x^2 - 4x + 1$ and $g(x) = 2x - 1$. Evaluate the following.

13. $f(0)$ **14.** $g(0)$
15. $g(-4)$ **16.** $f(1)$
17. $f(-1)$ **18.** $g(-1)$
19. $g(10)$ **20.** $f(10)$
21. $f(3)$ **22.** $g(3)$
23. $g(\frac{1}{2})$ **24.** $g(\frac{1}{4})$

25. $f(a)$ **26.** $g(b)$

If $f = \{(1, 4), (-2, 0), (3, \frac{1}{2}), (\pi, 0)\}$ and $g = \{(1, 1), (-2, 2), (\frac{1}{2}, 0)\}$, find each of the following values of f and g.

27. $f(1)$ **28.** $g(1)$ **29.** $g(\frac{1}{2})$

30. $f(3)$ **31.** $g(-2)$ **32.** $f(\pi)$

Let $f(x) = 2x^2 - 8$ and $g(x) = \frac{1}{2}x + 1$. Evaluate each of the following.

33. $f(0)$ **34.** $g(0)$ **35.** $g(-4)$

36. $f(1)$ **37.** $f(a)$ **38.** $g(z)$

39. $f(b)$ **40.** $g(t)$ **41.** $f[g(2)]$

42. $g[f(2)]$ **43.** $g[f(-1)]$ **44.** $f[g(-2)]$

45. $g[f(0)]$ **46.** $f[g(0)]$

47. Graph the function $f(x) = \frac{1}{2}x + 2$. Then draw and label the line segments that represent $x = 4$ and $f(4)$.

48. Graph the function $f(x) = -\frac{1}{2}x + 6$. Then draw and label the line segments that represent $x = 4$ and $f(4)$.

49. For the function $f(x) = \frac{1}{2}x + 2$, find the value of x for which $f(x) = x$.

50. For the function $f(x) = -\frac{1}{2}x + 6$, find the value of x for which $f(x) = x$.

51. Graph the function $f(x) = x^2$. Then draw and label the line segments that represent $x = 1$ and $f(1)$, $x = 2$ and $f(2)$, and, finally, $x = 3$ and $f(3)$.

52. Graph the function $f(x) = x^2 - 2$. Then draw and label the line segments that represent $x = 2$ and $f(2)$ and the line segments corresponding to $x = 3$ and $f(3)$.

Applying the Concepts

53. Investing in Art A painting is purchased as an investment for $150. If its value increases continuously so that it doubles every 3 years, then its value is given by the function

$$V(t) = 150 \cdot 2^{t/3} \qquad \text{for} \quad t \geq 0$$

where t is the number of years since the painting was purchased, and $V(t)$ is its value (in dollars) at time t. Find $V(3)$ and $V(6)$, and then explain what they mean.

54. Average Speed If it takes Minke t minutes to run a mile, then her average speed $s(t)$, in miles per hour, is given by the formula

$$s(t) = \frac{60}{t} \qquad \text{for} \quad t > 0$$

Find $s(4)$ and $s(5)$, and then explain what they mean.

55. Dimensions of a Rectangle The length of a rectangle is 3 inches more than twice the width. Let x represent the width of the rectangle and $P(x)$ represent the perimeter of the rectangle. Use function notation to write the relationship between x and $P(x)$, noting any restrictions on the variable x.

56. Dimensions of a Rectangle The length of a rectangle is 3 inches more than twice the width. Let x represent the width of the rectangle and $A(x)$ represent the area of the rectangle. Use function notation to write the relationship between x and $A(x)$, noting any restrictions on the variable x.

Area of a Circle The formula for the area A of a circle with radius r can be written with function notation as $A(r) = \pi r^2$.

57. Find $A(2)$, $A(5)$, and $A(10)$. (Use $\pi \approx 3.14$.)

58. Why doesn't it make sense to ask for $A(-10)$?

59. Cost of a Phone Call Suppose a phone company charges 33¢ for the first minute and 24¢ for each additional minute to place a long-distance call between 5 P.M. and 11 P.M. If x is the number of additional minutes and $f(x)$ is the cost of the call, then $f(x) = 24x + 33$.

(a) How much does it cost to talk for 10 minutes?

(b) What does $f(5)$ represent in this problem?

(c) If a call costs $1.29, how long was it?

60. Cost of a Phone Call The same phone company mentioned in Problem 59 charges 52¢ for the first minute and 36¢ for each additional minute to place a long-distance call between 8 A.M. and 5 P.M.

(a) Let $g(x)$ be the total cost of a long-distance call between 8 A.M. and 5 P.M., and write an equation for $g(x)$.

(b) Find $g(5)$.

(c) Find the difference in price between a 10-minute call made between 8 A.M. and 5 P.M. and the same call made between 5 P.M. and 11 P.M.

Straight-Line Depreciation Straight-line depreciation is an accounting method used to help spread the cost of new equipment over a number of years. It takes into account both the cost when new and the salvage value, which is the value of the equipment at the time it gets replaced.

61. Value of a Copy Machine The function $V(t) = -3,300t + 18,000$, where V is value and t is time in years, can be used to find the value of a large copy machine during the first 5 years of use.
(a) What is the value of the copier after 3 years and 9 months?
(b) What is the salvage value of this copier if it is replaced after 5 years of use?
(c) State the domain of this function.
(d) Sketch the graph of this function.

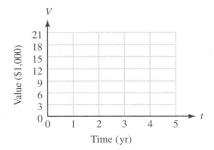

(FPG International/Telegraph Colour Library)

(e) What is the range of this function?
(f) After how many years will the copier be worth only $10,000?

62. Value of a Forklift The function $V(t) = -16,500t + 125,000$, where V is value and t is time in years, can be used to find the value of an electric forklift during the first 6 years of use.
(a) What is the value of the forklift after 2 years and 3 months?

(b) What is the salvage value of this forklift if it is replaced after 6 years of use?
(c) State the domain of this function.
(d) Sketch the graph of this function.

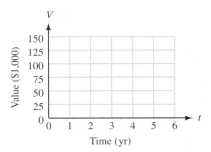

(FPG International/Telegraph Colour Library)

(e) What is the range of this function?
(f) After how many years will the forklift be worth only $45,000?

63. Step Function Figure 4 shows the graph of the step function C that was used to calculate the first-class postage on a letter weighing x ounces in 1997. Use this graph to answer questions (a) through (d).

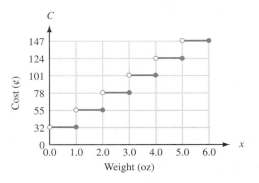

FIGURE 4 The graph of $C(x)$

(a) Fill in the following table:

Weight (ounces)	0.6	1.0	1.1	2.5	3.0	4.8	5.0	5.3
Cost (cents)								

(b) If a letter cost 78 cents to mail, how much does it weigh? State your answer in words. State your answer as an inequality.

(c) If the entire function is shown in Figure 4, state the domain.

(d) State the range of the function shown in Figure 4.

64. Step Function A taxi ride in Boston at the time I am writing this problem is $1.50 for the first $\frac{1}{4}$ mile, and then $0.25 for each additional $\frac{1}{8}$ of a mile. The following graph shows how much you will pay for a taxi ride of 1 mile or less.

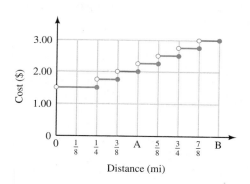

Distance (mi)

FIGURE 5

(a) What is the most you will pay for this taxi ride?

(b) How much does it cost to ride the taxi for $\frac{8}{10}$ of a mile?

(c) Find points A and B on the horizontal axis.

(d) If a taxi ride costs $2.50, what distance was the ride?

(e) If the complete function is shown in Figure 5, find the domain and range of the function.

Review Problems

The problems that follow review material we covered in Section 7.2.

Solve each equation.

65. $|3x - 5| = 7$

66. $|0.04 - 0.03x| = 0.02$

67. $|4y + 2| - 8 = -2$

68. $4 = |3 - 2y| - 5$

69. $5 + |6t + 2| = 3$

70. $7 + |3 - \frac{3}{4}t| = 10$

Extending the Concepts

71. The graphs of two functions are shown in Figures 6 and 7. Use the graphs to find the following.

(a) $f(2)$

(b) $f(-4)$

(c) $g(0)$

(d) $g(3)$

(e) $g(2) - f(2)$

(f) $f(1) + g(1)$

(g) $f[g(3)]$

(h) $g[f(3)]$

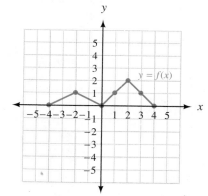

FIGURE 6

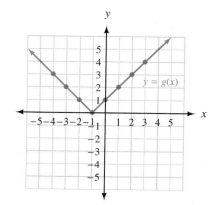

FIGURE 7

A company produces and sells copies of an accounting program for home computers. The price they charge for the program is related to the number of copies sold by the demand function

$$p(x) = 35 - 0.1x$$

We find the revenue for this business by multiplying the number of items sold by the price per item. When we do so, we are forming a new function by combining two existing functions. That is, if $n(x) = x$ is the number of items sold and $p(x) = 35 - 0.1x$ is the price per item, then revenue is

$$R(x) = n(x) \cdot p(x) = x(35 - 0.1x) = 35x - 0.1x^2$$

In this case, the revenue function is the product of two functions. When we combine functions in this manner, we are applying our rules for algebra to functions.

    To carry this situation further, we know the profit function is the difference between two functions. If the cost function for producing x copies of the accounting program is $C(x) = 8x + 500$, then the profit function is

$$P(x) = R(x) - C(x) = (35x - 0.1x^2) - (8x + 500) = -500 + 27x - 0.1x^2$$

The relationship between these last three functions is shown visually in Figure 1.

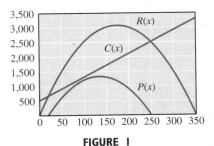

FIGURE 1

    Again, when we combine functions in the manner shown we are applying our rules for algebra to functions. To begin this section we take a formal look at addition, subtraction, multiplication, and division with functions.

    If we are given two functions f and g with a common domain, we can define four other functions as follows.

DEFINITION

$(f + g)(x) = f(x) + g(x)$ The function $f + g$ is the sum of the functions f and g.

$(f - g)(x) = f(x) - g(x)$ The function $f - g$ is the difference of the functions f and g.

$(fg)(x) = f(x)g(x)$ The function fg is the product of the functions f and g.

$\dfrac{f}{g}(x) = \dfrac{f(x)}{g(x)}$ The function f/g is the quotient of the functions f and g, where $g(x) \neq 0$.

EXAMPLE 1   If $f(x) = 4x^2 + 3x + 2$ and $g(x) = 2x^2 - 5x - 6$, write the formula for the functions $f + g$, $f - g$, fg, and f/g.

Solution   The function $f + g$ is defined by

$$(f + g)(x) = f(x) + g(x)$$
$$= (4x^2 + 3x + 2) + (2x^2 - 5x - 6)$$
$$= 6x^2 - 2x - 4$$

The function $f - g$ is defined by

$$(f - g)(x) = f(x) - g(x)$$
$$= (4x^2 + 3x + 2) - (2x^2 - 5x - 6)$$
$$= 4x^2 + 3x + 2 - 2x^2 + 5x + 6$$
$$= 2x^2 + 8x + 8$$

The function fg is defined by

$$(fg)(x) = f(x)g(x)$$
$$= (4x^2 + 3x + 2)(2x^2 - 5x - 6)$$
$$= 8x^4 - 20x^3 - 24x^2 + 6x^3 - 15x^2 - 18x + 4x^2 - 10x - 12$$
$$= 8x^4 - 14x^3 - 35x^2 - 28x - 12$$

The function f/g is defined by

$$\left(\frac{f}{g}\right)(x) = \frac{f(x)}{g(x)}$$

$$= \frac{4x^2 + 3x + 2}{2x^2 - 5x - 6}$$

EXAMPLE 2   Let $f(x) = 4x - 3$, $g(x) = 4x^2 - 7x + 3$, and $h(x) = x - 1$. Find $f + g, fh, fg$, and g/f.

Solution   The function $f + g$, the sum of functions f and g, is defined by

$$(f + g)(x) = f(x) + g(x)$$
$$= (4x - 3) + (4x^2 - 7x + 3)$$
$$= 4x^2 - 3x$$

The function fh, the product of functions f and h, is defined by

$$(fh)(x) = f(x)h(x)$$
$$= (4x - 3)(x - 1)$$
$$= 4x^2 - 7x + 3$$
$$= g(x)$$

The product of the functions f and g, fg, is given by

$$(fg)(x) = f(x)g(x)$$
$$= (4x - 3)(4x^2 - 7x + 3)$$
$$= 16x^3 - 28x^2 + 12x - 12x^2 + 21x - 9$$
$$= 16x^3 - 40x^2 + 33x - 9$$

The quotient of the functions g and f, g/f, is defined as

$$\frac{g}{f}(x) = \frac{g(x)}{f(x)}$$
$$= \frac{4x^2 - 7x + 3}{4x - 3}$$

Factoring the numerator, we can reduce to lowest terms:

$$\frac{g}{f}(x) = \frac{(4x - 3)(x - 1)}{4x - 3}$$
$$= x - 1$$
$$= h(x)$$

EXAMPLE 3   If f, g, and h are the same functions defined in Example 2, evaluate $(f + g)(2)$, $(fh)(-1)$, $(fg)(0)$, and $(g/f)(5)$.

Solution   We use the formulas for $f + g, fh, fg$, and g/f found in Example 2:

$$(f + g)(2) = 4(2)^2 - 3(2)$$
$$= 16 - 6$$
$$= 10$$

$$(fh)(-1) = 4(-1)^2 - 7(-1) + 3$$
$$= 4 + 7 + 3$$
$$= 14$$

$$(fg)(0) = 16(0)^3 - 40(0)^2 + 33(0) - 9$$
$$= 0 - 0 + 0 - 9$$
$$= -9$$

$$\frac{g}{f}(5) = 5 - 1$$
$$= 4$$

COMPOSITION OF FUNCTIONS

In addition to the four operations used to combine functions shown so far in this section, there is a fifth way to combine two functions to obtain a new function. It is called **composition of functions.** To illustrate the concept, we can use the definition of training heart rate: training heart rate, in beats per minute, is resting heart rate plus 60% of the difference between maximum heart rate and resting heart rate. If your resting heart rate is 70 beats per minute, then your training heart rate is a function of your maximum heart rate M

$$T(M) = 70 + 0.6(M - 70) = 70 + 0.6M - 42 = 28 + 0.6M$$

But your maximum heart rate is found by subtracting your age in years from 220. So, if x represents your age in years, then your maximum heart rate is

$$M(x) = 220 - x$$

Therefore, if your resting heart rate is 70 beats per minute and your age in years is x, then your training heart rate can be written as a function of x.

$$T(x) = 28 + 0.6(220 - x)$$

This last line is the composition of functions T and M. We input x into function M, which outputs $M(x)$. Then, we input $M(x)$ into function T, which outputs $T(M(x))$, which is the training heart rate as a function of age x. Here is a diagram of the situation, which is called a function map:

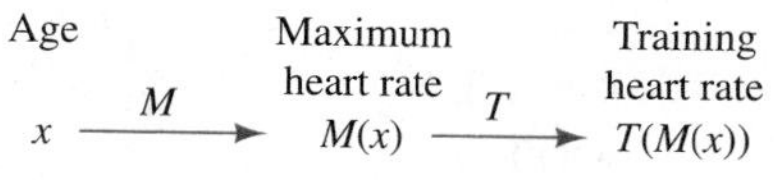

Age		Maximum heart rate		Training heart rate
x	$\xrightarrow{\ M\ }$	$M(x)$	$\xrightarrow{\ T\ }$	$T(M(x))$

FIGURE 2

Now let's generalize the preceding ideas into a formal development of composition of functions. To find the composition of two functions f and g, we first require that the range of g have numbers in common with the domain of f. Then the composition of f with g, is defined this way:

$$(f \circ g)(x) = f(g(x))$$

To understand this new function, we begin with a number x, and we operate on it with g, giving us $g(x)$. Then we take $g(x)$ and operate on it with f, giving us $f(g(x))$. The only numbers we can use for the domain of the composition of f with g are numbers x in the domain of g, for which $g(x)$ is in the domain of f. The diagrams in Figure 3 illustrate the composition of f with g.

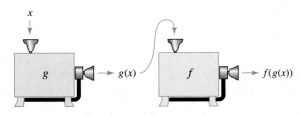

Function machines

FIGURE 3

Composition of functions is not commutative. The composition of f with g, $f \circ g$, may therefore be different from the composition of g with f, $g \circ f$.

$$(g \circ f)(x) = g(f(x))$$

Again, the only numbers we can use for the domain of the composition of g with f are numbers in the domain of f, for which $f(x)$ is in the domain of g. The diagrams in Figure 4 illustrate the composition of g with f.

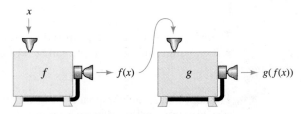

Function machines

FIGURE 4

EXAMPLE 4   If $f(x) = x + 5$ and $g(x) = x^2 - 2x$, find $(f \circ g)(x)$ and $(g \circ f)(x)$.

Solution   The composition of f with g is

$$(f \circ g)(x) = f(g(x))$$
$$= f(x^2 - 2x)$$
$$= (x^2 - 2x) + 5$$
$$= x^2 - 2x + 5$$

The composition of g with f is

$$
\begin{aligned}
(g \circ f)(x) &= g(f(x)) \\
&= g(x + 5) \\
&= (x + 5)^2 - 2(x + 5) \\
&= (x^2 + 10x + 25) - 2(x + 5) \\
&= x^2 + 8x + 15
\end{aligned}
$$

 # Getting Ready for Class

Respond in your own words and in complete sentences.

A. How are profit, revenue, and cost related.

B. How do you find maximum heart rate?

C. For functions f and g, how do you find the composition of f with g?

D. For functions f and g, how do you find the composition of g with f?

PROBLEM SET 8.6

Let $f(x) = 4x - 3$ and $g(x) = 2x + 5$. Write a formula for each of the following functions.

1. $f + g$ **2.** $f - g$

3. $g - f$ **4.** $g + f$

5. fg **6.** f/g

7. g/f **8.** ff

If the functions f, g, and h are defined by $f(x) = 3x - 5$, $g(x) = x - 2$, and $h(x) = 3x^2 - 11x + 10$, write a formula for each of the following functions.

9. $g + f$ **10.** $f + h$

11. $g + h$ **12.** $f - g$

13. $g - f$ **14.** $h - g$

15. fg **16.** gf

17. fh **18.** gh

19. h/f **20.** h/g

21. f/h **22.** g/h

23. $f + g + h$ **24.** $h - g + f$

25. $h + fg$ **26.** $h - fg$

Let $f(x) = 2x + 1$, $g(x) = 4x + 2$, and $h(x) = 4x^2 + 4x + 1$, and find the following.

27. $(f + g)(2)$ **28.** $(f - g)(-1)$

29. $(fg)(3)$ **30.** $(f/g)(-3)$

31. $(h/g)(1)$ **32.** $(hg)(1)$

33. $(fh)(0)$ **34.** $(h - g)(-4)$

35. $(f + g + h)(2)$ **36.** $(h - f + g)(0)$

37. $(h + fg)(3)$ **38.** $(h - fg)(5)$

39. Let $f(x) = x^2$ and $g(x) = x + 4$, and find
  (a) $(f \circ g)(5)$ (c) $(f \circ g)(x)$
  (b) $(g \circ f)(5)$ (d) $(g \circ f)(x)$

40. Let $f(x) = 3 - x$ and $g(x) = x^3 - 1$, and find
  (a) $(f \circ g)(0)$ (c) $(f \circ g)(x)$
  (b) $(g \circ f)(0)$ (d) $(g \circ f)(x)$

41. Let $f(x) = x^2 + 3x$ and $g(x) = 4x - 1$, and find
  (a) $(f \circ g)(0)$ (c) $(f \circ g)(x)$
  (b) $(g \circ f)(0)$ (d) $(g \circ f)(x)$

42. Let $f(x) = (x - 2)^2$ and $g(x) = x + 1$, and find the following
  (a) $(f \circ g)(-1)$ (c) $(f \circ g)(x)$
  (b) $(g \circ f)(-1)$ (d) $(g \circ f)(x)$

For each of the following pairs of functions f and g, show that $(f \circ g)(x) = (g \circ f)(x) = x$.

43. $f(x) = 5x - 4$ and $g(x) = \dfrac{x + 4}{5}$

44. $f(x) = \dfrac{x}{6} - 2$ and $g(x) = 6x + 12$

Applying the Concepts

45. Profit, Revenue, and Cost A company manufactures and sells prerecorded videotapes. Here are the equations they use in connection with their business.

Number of tapes sold each day: $n(x) = x$

Selling price for each tape: $p(x) = 11.5 - 0.05x$

Daily fixed costs: $f(x) = 200$

Daily variable costs: $v(x) = 2x$

Find the following functions.
(a) Revenue $= R(x) =$ the product of the number of tapes sold each day and the selling price of each tape.
(b) Cost $= C(x) =$ the sum of the fixed costs and the variable costs.
(c) Profit $= P(x) =$ the difference between revenue and cost.
(d) Average cost $= \overline{C}(x) =$ the quotient of cost and the number of tapes sold each day.

46. Profit, Revenue, and Cost A company manufactures and sells diskettes for home computers. Here are the equations they use in connection with their business.

Number of diskettes sold each day: $n(x) = x$

Selling price for each diskette: $p(x) = 3 - \dfrac{1}{300}x$

Daily fixed costs: $f(x) = 200$

Daily variable costs: $v(x) = 2x$

Find the following functions.
(a) Revenue $= R(x) =$ the product of the number of diskettes sold each day and the selling price of each diskette.
(b) Cost $= C(x) =$ the sum of the fixed costs and the variable costs.
(c) Profit $= P(x) =$ the difference between revenue and cost.
(d) Average cost $= \overline{C}(x) =$ the quotient of cost and the number of diskettes sold each day.

47. Training Heart Rate Find the training heart rate function, $T(M)$ for a person with a resting heart rate of 62 beats per minute, then find the following.
(a) Find the maximum heart rate function, $M(x)$, for a person x years of age.
(b) What is the maximum heart rate for a 24-year-old person?
(c) What is the training heart rate for a 24-year-old person with a resting heart rate of 62 beats per minute?
(d) What is the training heart rate for a 36-year-old person with a resting heart rate of 62 beats per minute?
(e) What is the training heart rate for a 48-year-old person with a resting heart rate of 62 beats per minute?

48. Training Heart Rate Find the training heart rate function, $T(M)$ for a person with a resting heart rate of 72 beats per minute, then find the following.
(a) Find the maximum heart rate function, $M(x)$, for a person x years of age.
(b) What is the maximum heart rate for a 20-year-old person?
(c) What is the training heart rate for a 20-year-old person with a resting heart rate of 72 beats per minute?
(d) What is the training heart rate for a 30-year-old person with a resting heart rate of 72 beats per minute?
(e) What is the training heart rate for a 40-year-old person with a resting heart rate of 72 beats per minute?

Review Problems

The problems that follow review material we covered in Section 7.6.

Solve each system by the addition method.

49. $4x + 3y = 10$
$2x + y = 4$

50. $3x - 5y = -2$
$2x - 3y = 1$

51. $4x + 5y = 5$
$\dfrac{6}{5}x + y = 2$

52. $4x + 2y = -2$
$\dfrac{1}{2}x + y = 0$

Solve each system by the substitution method.

53. $x + y = 3$
$y = x + 3$

54. $x + y = 6$
$y = x - 4$

55. $2x - 3y = -6$
$y = 3x - 5$

56. $7x - y = 24$
$x = 2y + 9$

Variation

If you are a runner and you average t minutes for every mile you run during one of your workouts, then your speed s in miles per hour is given by the equation and graph shown here [the graph (Figure 1) is shown in the first quadrant only because both t and s are positive]:

$$s = \frac{60}{t}$$

Input t	Output s
4	15
6	10
8	7.5
10	6
12	5

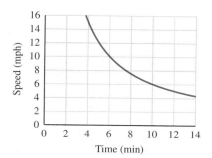

FIGURE 1

You know intuitively that as your average time per mile (t) increases, your speed (s) decreases. Likewise, lowering your time per mile will increase your speed. The equation and Figure 1 also show this to be true: Increasing t decreases s, and decreasing t increases s. Quantities that are connected in this way are said to *vary inversely* with each other. Inverse variation is one of the topics we will study in this section.

There are two main types of variation: *direct variation* and *inverse variation*. Variation problems are most common in the sciences, particularly in chemistry and physics.

DIRECT VARIATION

When we say the variable y *varies directly* with the variable x, we mean that the relationship can be written in symbols as $y = Kx$, where K is a nonzero constant called the *constant of variation* (or *proportionality constant*).

Another way of saying y varies directly with x is to say y is *directly proportional* to x.

Study the following list. It gives the mathematical equivalent of some direct variation statements.

Verbal Phrase	Algebraic Equation
y varies directly with x.	$y = Kx$
s varies directly with the square of t.	$s = Kt^2$
y is directly proportional to the cube of z.	$y = Kz^3$
u is directly proportional to the square root of v.	$u = K\sqrt{v}$

EXAMPLE I y varies directly with x. If y is 15 when x is 5, find y when x is 7.

Solution   The first sentence gives us the general relationship between x and y. The equation equivalent to the statement "y varies directly with x" is

$$y = Kx$$

The first part of the second sentence in our example gives us the information necessary to evaluate the constant K:

$$
\begin{aligned}
\text{When} \qquad & y = 15 \\
\text{and} \qquad & x = 5 \\
\text{the equation} \qquad & y = Kx \\
\text{becomes} \qquad & 15 = K \cdot 5 \\
\text{or} \qquad & K = 3
\end{aligned}
$$

The equation can now be written specifically as

$$y = 3x$$

Letting $x = 7$, we have

$$y = 3 \cdot 7$$
$$y = 21$$

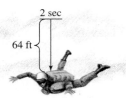

EXAMPLE 2   A skydiver jumps from a plane. Like any object that falls toward Earth, the distance the skydiver falls is directly proportional to the square of the time he has been falling, until he reaches his terminal velocity. If the skydiver falls 64 feet in the first 2 seconds of the jump, then

(a) How far will he have fallen after 3.5 seconds?
(b) Graph the relationship between distance and time.
(c) How long will it take him to fall 256 feet?

Solution   We let t represent the time the skydiver has been falling, then we can let $d(t)$ represent the distance he has fallen.

(a) Since $d(t)$ is directly proportional to the square of t, we have the general function that describes this situation:

$$d(t) = Kt^2$$

Next, we use the fact that $d(2) = 64$ to find K.

$$64 = K(2^2)$$
$$K = 16$$

The specific equation that describes this situation is

$$d(t) = 16t^2$$

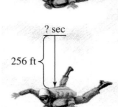

To find how far a skydiver will fall after 3.5 seconds, we find $d(3.5)$,

$$d(3.5) = 16(3.5^2)$$

$$d(3.5) = 196$$

A skydiver will fall 196 feet after 3.5 seconds.

(b) To graph this equation, we use a table:

Input t	Output $d(t)$
0	0
1	16
2	64
3	144
4	256
5	400

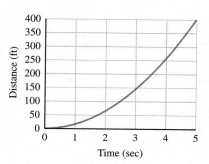

FIGURE 2

(c) From the table or the graph (Figure 2), we see that it will take 4 seconds for the skydiver to fall 256 feet.

INVERSE VARIATION

Running   From the introduction to this section, we know that the relationship between the number of minutes t it takes a person to run a mile and his or her average speed in miles per hour s can be described with the following equation and table, and with Figure 3.

$$s = \frac{60}{t}$$

Input t	Output s
4	15
6	10
8	7.5
10	6
12	5

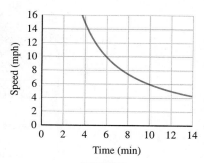

FIGURE 3

If t decreases, then s will increase, and if t increases, then s will decrease. The variable s is *inversely proportional* to the variable t. In this case, the *constant of proportionality* is 60.

Photography If you are familiar with the terminology and mechanics associated with photography, you know that the *f*-stop for a particular lens will increase as the aperture (the maximum diameter of the opening of the lens) decreases. In mathematics we say that *f*-stop and aperture vary inversely with each other. The diagram illustrates this relationship.

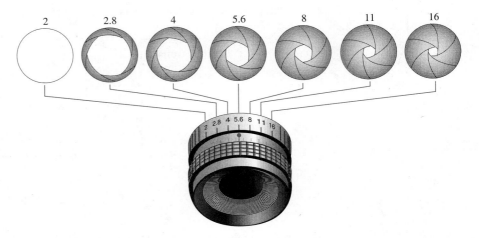

If *f* is the *f*-stop and *d* is the aperture, then their relationship can be written

$$f = \frac{K}{d}$$

In this case, *K* is the constant of proportionality. (Those of you familiar with photography know that *K* is also the focal length of the camera lens.)

In General We generalize this discussion of inverse variation as follows: If *y* varies inversely with *x*, then

$$y = K\frac{1}{x} \qquad \text{or} \qquad y = \frac{K}{x}$$

We can also say *y* is inversely proportional to *x*. The constant *K* is again called the constant of variation or proportionality constant.

Verbal Phrase	Algebraic Equation
y is inversely proportional to *x*.	$y = \dfrac{K}{x}$
s varies inversely with the square of *t*.	$s = \dfrac{K}{t^2}$
y is inversely proportional to x^4.	$y = \dfrac{K}{x^4}$
z varies inversely with the cube root of *t*.	$z = \dfrac{K}{\sqrt[3]{t}}$

EXAMPLE   3   The volume of a gas is inversely proportional to the pressure of the gas on its container. If a pressure of 48 pounds per square inch corresponds to a volume of 50 cubic feet, what pressure is needed to produce a volume of 100 cubic feet?

Solution   We can represent volume with V and pressure with P:

$$V = \frac{K}{P}$$

Using $P = 48$ and $V = 50$, we have

$$50 = \frac{K}{48}$$

$$K = 50(48)$$

$$K = 2{,}400$$

The equation that describes the relationship between P and V is

$$V = \frac{2{,}400}{P}$$

Here is a graph of this relationship.

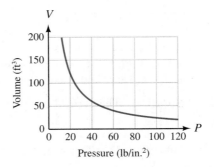

Substituting $V = 100$ into this last equation, we get

$$100 = \frac{2{,}400}{P}$$

$$100P = 2{,}400$$

$$P = \frac{2{,}400}{100}$$

$$P = 24$$

A volume of 100 cubic feet is produced by a pressure of 24 pounds per square inch.

Note: The relationship between pressure and volume as given in this example is known as Boyle's law and applies to situations such as those encountered in a piston-cylinder arrangement. It was Robert Boyle (1627–1691) who, in 1662, published the results of some of his experiments that showed, among other things, that the volume of a gas decreases as the pressure increases. This is an example of inverse variation.

JOINT VARIATION AND OTHER VARIATION COMBINATIONS

Many times relationships among different quantities are described in terms of more than two variables. If the variable y varies directly with *two* other variables, say x and z, then we say y varies *jointly* with x and z. In addition to joint variation, there are many other combinations of direct and inverse variation involving more than two variables. The following table is a list of some variation statements and their equivalent mathematical forms:

Verbal Phrase	Algebraic Equation
y varies jointly with x and z.	$y = Kxz$
z varies jointly with r and the square of s.	$z = Krs^2$
V is directly proportional to T and inversely proportional to P.	$V = \dfrac{KT}{P}$
F varies jointly with m_1 and m_2 and inversely with the square of r.	$F = \dfrac{Km_1m_2}{r^2}$

E X A M P L E   4 y varies jointly with x and the square of z. When x is 5 and z is 3, y is 180. Find y when x is 2 and z is 4.

Solution   The general equation is given by

$$y = Kxz^2$$

Substituting $x = 5$, $z = 3$, and $y = 180$, we have

$$180 = K(5)(3)^2$$

$$180 = 45K$$

$$K = 4$$

The specific equation is

$$y = 4xz^2$$

When $x = 2$ and $z = 4$, the last equation becomes

$$y = 4(2)(4)^2$$

$$y = 128$$

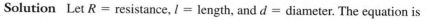

EXAMPLE 5   In electricity, the resistance of a cable is directly proportional to its length and inversely proportional to the square of the diameter. If a 100-foot cable 0.5 inch in diameter has a resistance of 0.2 ohm, what will be the resistance of a cable made from the same material if it is 200 feet long with a diameter of 0.25 inch?

Solution   Let R = resistance, l = length, and d = diameter. The equation is

$$R = \frac{Kl}{d^2}$$

When $R = 0.2$, $l = 100$, and $d = 0.5$, the equation becomes

$$0.2 = \frac{K(100)}{(0.5)^2}$$

or

$$K = 0.0005$$

Using this value of K in our original equation, the result is

$$R = \frac{0.0005l}{d^2}$$

When $l = 200$ and $d = 0.25$, the equation becomes

$$R = \frac{0.0005(200)}{(0.25)^2}$$

$$R = 1.6 \text{ ohms}$$

Getting Ready for Class

After reading through the preceding section, respond in your own words and in complete sentences.

A. Give an example of a direct variation statement, and then translate it into symbols.

B. Translate the equation $y = \dfrac{k}{x}$ into words.

C. For the inverse variation equation $y = \dfrac{3}{x}$, what happens to the values of y as x gets larger?

D. How are direct variation statements and linear equations in two variables related?

PROBLEM SET 8.7

For the following problems, y varies directly with x.

1. If y is 10 when x is 2, find y when x is 6.

2. If y is 20 when x is 5, find y when x is 3.

3. If y is -32 when x is 4, find x when y is -40.

4. If y is -50 when x is 5, find x when y is -70.

For the following problems, r is inversely proportional to s.

5. If r is -3 when s is 4, find r when s is 2.

6. If r is -10 when s is 6, find r when s is -5.

7. If r is 8 when s is 3, find s when r is 48.

8. If r is 12 when s is 5, find s when r is 30.

For the following problems, d varies directly with the square of r.

9. If $d = 10$ when $r = 5$, find d when $r = 10$.

10. If $d = 12$ when $r = 6$, find d when $r = 9$.

11. If $d = 100$ when $r = 2$, find d when $r = 3$.

12. If $d = 50$ when $r = 5$, find d when $r = 7$.

For the following problems, y varies inversely with the square of x.

13. If $y = 45$ when $x = 3$, find y when x is 5.

14. If $y = 12$ when $x = 2$, find y when x is 6.

15. If $y = 18$ when $x = 3$, find y when x is 2.

16. If $y = 45$ when $x = 4$, find y when x is 5.

For the following problems, z varies jointly with x and the square of y.

17. If z is 54 when x and y are 3, find z when $x = 2$ and $y = 4$.

18. If z is 80 when x is 5 and y is 2, find z when $x = 2$ and $y = 5$.

19. If z is 64 when $x = 1$ and $y = 4$, find x when $z = 32$ and $y = 1$.

20. If z is 27 when $x = 6$ and $y = 3$, find x when $z = 50$ and $y = 4$.

Applying the Concepts

21. Length of a Spring The length a spring stretches is directly proportional to the force applied. If a force of 5 pounds stretches a spring 3 inches, how much force is necessary to stretch the same spring 10 inches?

22. Weight and Surface Area The weight of a certain material varies directly with the surface area of that material. If 8 square feet weighs half a pound, how much will 10 square feet weigh?

23. Pressure and Temperature The temperature of a gas varies directly with its pressure. A temperature of 200 K produces a pressure of 50 pounds per square inch.

(a) Find the equation that relates pressure and temperature.

(b) Graph the equation from part (a) in the first quadrant only.

(c) What pressure will the gas have at 280 K?

24. Circumference and Diameter The circumference of a wheel is directly proportional to its diameter. A wheel has a circumference of 8.5 feet and a diameter of 2.7 feet.

(a) Find the equation that relates circumference and diameter.

(b) Graph the equation from part (a) in the first quadrant only.

(c) What is the circumference of a wheel that has a diameter of 11.3 feet?

25. Volume and Pressure The volume of a gas is inversely proportional to the pressure. If a pressure of 36 pounds per square inch corresponds to a volume of 25 cubic feet, what pressure is needed to produce a volume of 75 cubic feet?

26. Wave Frequency The frequency of an electromagnetic wave varies inversely with the wavelength. If a wavelength of 200 meters has a frequency of 800 kilocycles per second, what frequency will be associated with a wavelength of 500 meters?

27. *f*-Stop and Aperture Diameter The relative aperture, or *f*-stop, for a camera lens is inversely proportional to the diameter of the aperture. An *f*-stop of 2 corresponds to an aperture diameter of 40 millimeters for the lens on an automatic camera.

(a) Find the equation that relates *f*-stop and diameter.

(b) Graph the equation from part (a) in the first quadrant only.

(c) What is the *f*-stop of this camera when the aperture diameter is 10 millimeters?

28. ***f*-Stop and Aperture Diameter**   The relative aperture, or *f*-stop, for a camera lens is inversely proportional to the diameter of the aperture. An *f*-stop of 2.8 corresponds to an aperture diameter of 75 millimeters for a certain telephoto lens.
    (a) Find the equation that relates *f*-stop and diameter.
    (b) Graph the equation from part (a) in the first quadrant only.
    (c) What aperture diameter corresponds to an *f*-stop of 5.6?

29. **Surface Area of a Cylinder**   The surface area of a hollow cylinder varies jointly with the height and radius of the cylinder. If a cylinder with radius 3 inches and height 5 inches has a surface area of 94 square inches, what is the surface area of a cylinder with radius 2 inches and height 8 inches?

30. **Capacity of a Cylinder**   The capacity of a cylinder varies jointly with its height and the square of its radius. If a cylinder with a radius of 3 centimeters and a height of 6 centimeters has a capacity of 169.56 cubic centimeters, what will be the capacity of a cylinder with radius 4 centimeters and height 9 centimeters?

31. **Electrical Resistance**   The resistance of a wire varies directly with its length and inversely with the square of its diameter. If 100 feet of wire with diameter 0.01 inch has a resistance of 10 ohms, what is the resistance of 60 feet of the same type of wire if its diameter is 0.02 inch?

32. **Volume and Temperature**   The volume of a gas varies directly with its temperature and inversely with the pressure. If the volume of a certain gas is 30 cubic feet at a temperature of 300 K and a pressure of 20 pounds per square inch, what is the volume of the same gas at 340 K when the pressure is 30 pounds per square inch?

33. **Period of a Pendulum**   The time it takes for a pendulum to complete one period varies directly with the square root of the length of the pendulum. A 100-centimeter pendulum takes 2.1 seconds to complete one period.
    (a) Find the equation that relates period and pendulum length.

(b) Graph the equation from part (a) in quadrant I only.
(c) How long does it take to complete one period if the pendulum hangs 225 centimeters?

34. **Area of a Circle**   The area of a circle varies directly with the square of the radius. A circle with a 5-centimeter radius has an area of 78.5 square centimeters.
    (a) Find the equation that relates area and radius.
    (b) Graph the equation from part (a) in quadrant I only.
    (c) What is the area of a circle that has an 8-centimeter radius?

35. **Music**   A musical tone's pitch varies inversely with its wavelength. If one tone has a pitch of 420 vibrations each second and a wavelength of 2.2 meters, find the wavelength of a tone that has a pitch of 720 vibrations each second.

36. **Hooke's Law**   Hooke's law states that the stress (force per unit area) placed on a solid object varies directly with the strain (deformation) produced.
    (a) Using the variables S_1 for stress and S_2 for strain, state this law in algebraic form.
    (b) Find the constant, K, if for one type of material $S_1 = 24$ and $S_2 = 72$.

37. **Gravity**   In Book Three of his *Principia,* Isaac Newton states that there is a single force in the universe that holds everything together, called the force of universal gravity. Newton stated that the force of universal gravity, F, is directly proportional with the product of two masses, m_1 and m_2, and inversely proportional with the square of the distance d between them. Write the equation for Newton's force of universal gravity, using the symbol G as the constant of proportionality.

38. **Boyle's Law and Charles's Law**   Boyle's law states that for low pressures, the pressure of an ideal gas kept at a constant temperature varies inversely with the volume of the gas. Charles's law states that for low pressures, the density of an ideal gas kept at a constant pressure varies inversely with the absolute temperature of the gas.
    (a) State Boyle's law as an equation using the symbols P, K, and V.
    (b) State Charles's law as an equation using the symbols D, K, and T.

Review Problems

The following problems review material we covered in Section 7.4.

Solve each inequality, and graph the solution set.

39. $\left| \dfrac{x}{5} + 1 \right| \geq \dfrac{4}{5}$ **40.** $|x - 6| \geq 0.01$

41. $|3 - 4t| > -5$ **42.** $|2 - 6t| < -5$

43. $-8 + |3y + 5| < 5$

44. $|6y - 1| - 4 \leq 2$

Extending the Concepts

45. Light Intensity   I found the following diagram while shopping for some track lighting for my home.

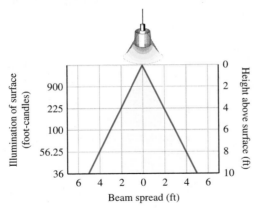

Beam spread (ft)

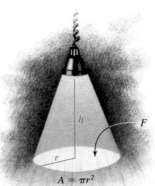

$A = \pi r^2$

I was impressed by the diagram because it displays a lot of useful information in a very efficient manner. As the diagram indicates, the amount of light that falls on a surface depends on how far above the surface the

light is placed and how much the light spreads out on the surface. Assume that this light illuminates a circle on a flat surface, and work the following problems.

(a) Fill in each table.

Height Above Surface (ft)	Illumination (foot-candles)
2	
4	
6	
8	
10	

Height Above Surface (ft)	Area of Illuminated Region (ft²)
2	
4	
6	
8	
10	

(b) Use the templates in Figures 4 and 5 to construct line graphs from the data in the tables.

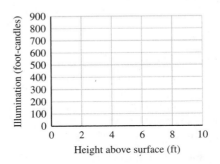

FIGURE 4

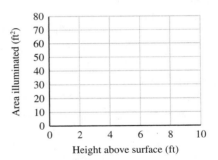

FIGURE 5

(c) Which of the relationships is direct variation, and which is inverse variation?

(d) Let F represent the number of foot-candles that fall on the surface, h the distance the light source is above the surface, and A the area of the illuminated region. Write an equation that shows the relationship between A and h, then write another equation that gives the relationship between F and h.

46. Law of Levers   Inverse variation may also be defined as occurring when the product of two variables is a constant; that is, when $x \cdot y = K$. This definition gives rise to the **Law of Levers,** or "seesaw principle." A seesaw is balanced when the variables on each side of the center, weight and distance from center, have the same inverse variation.

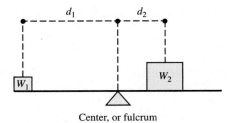

Center, or fulcrum

In the preceding figure, the weight on the left side must vary inversely with its distance from the center in the same way as the weight on the right side with its corresponding distance from the center in order for the seesaw to be balanced. This implies that $W_1 \cdot d_1 = W_2 \cdot d_2$ in order to balance. In a seesaw, an 85-pound girl is 4 feet from the center. How far from the center should her 120-pound brother place himself in order to balance the device?

CHAPTER 8 SUMMARY

1. The slope of the line through $(6, 9)$ and $(1, -1)$ is

$$m = \frac{9 - (-1)}{6 - 1} = \frac{10}{5} = 2$$

The Slope of a Line [8.1]

The *slope* of the line containing points (x_1, y_1) and (x_2, y_2) is given by

$$\text{Slope} = m = \frac{\text{Rise}}{\text{Run}} = \frac{y_2 - y_1}{x_2 - x_1}$$

Horizontal lines have 0 slope, and vertical lines have no slope.

Parallel lines have equal slopes, and perpendicular lines have slopes that are negative reciprocals.

2. The equation of the line with slope 5 and y-intercept 3 is

$$y = 5x + 3$$

The Slope-Intercept Form of a Line [8.2]

The equation of a line with slope m and y-intercept b is given by

$$y = mx + b$$

3. The equation of the line through $(3, 2)$ with slope -4 is

$$y - 2 = -4(x - 3)$$

which can be simplified to

$$y = -4x + 14$$

The Point-Slope Form of a Line [8.2]

The equation of the line through (x_1, y_1) that has slope m can be written as

$$y - y_1 = m(x - x_1)$$

4. The graph of $x - y \leq 3$:

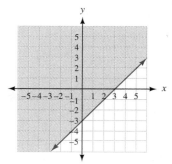

Linear Inequalities in Two Variables [8.3]

An inequality of the form $ax + by < c$ is a *linear inequality in two variables*. The equation for the boundary of the solution set is given by $ax + by = c$. (This equation is found by simply replacing the inequality symbol with an equal sign.)

To graph a linear inequality, first graph the boundary. Next, choose any point not on the boundary and substitute its coordinates into the original inequality. If the resulting statement is true, the graph lies on the same side of the boundary as the test point. A false statement indicates that the solution set lies on the other side of the boundary.

5. The relation

$$\{(8, 1), (6, 1), (-3, 0)\}$$

is also a function since no ordered pairs have the same first coordinates. The domain is $\{8, 6, -3\}$ and the range is $\{1, 0\}$.

Relations and Functions [8.4]

A *function* is a rule that pairs each element in one set, called the *domain*, with exactly one element from a second set, called the *range*.

A *relation* is any set of ordered pairs. The set of all first coordinates is called the *domain* of the relation, and the set of all second coordinates is the *range* of the relation. A function is a relation in which no two different ordered pairs have the same first coordinates.

If the domain for a relation or a function is not specified, it is assumed to be all real numbers for which the relation (or function) is defined. Since we are concerned only with real number functions, a function is not defined for those values of x that give 0 in the denominator or the square root of a negative number.

6. The graph of $x = y^2$ shown in Figure 5 on page 505 fails the vertical line test. It is not the graph of a function.

Vertical Line Test [8.4]

If a vertical line crosses the graph of a relation in more than one place, the relation cannot be a function. If no vertical line can be found that crosses the graph in more than one place, the relation must be a function.

7. If $f(x) = 5x - 3$ then

$$f(0) = 5(0) - 3 = -3$$

$$f(1) = 5(1) - 3 = 2$$

$$f(-2) = 5(-2) - 3 = -13$$

$$f(a) = 5a - 3$$

Function Notation [8.5]

The alternative notation for y is $f(x)$. It is read "f of x," and can be used instead of the variable y when working with functions. The notation y and the notation $f(x)$ are equivalent. That is, $y = f(x)$.

8. If $f(x) = 4x$ and $g(x) = x^2 - 3$, then

$$(f + g)(x) = x^2 + 4x - 3$$

$$(f - g)(x) = -x^2 + 4x + 3$$

$$(fg)(x) = 4x^3 - 12x$$

$$\frac{f}{g}(x) = \frac{4x}{x^2 - 3}$$

Algebra with Functions [8.6]

If f and g are any two functions with a common domain, then:

$(f + g)(x) = f(x) + g(x)$ The function $f + g$ is the sum of the functions f and g.

$(f - g)(x) = f(x) - g(x)$ The function $f - g$ is the difference of the functions f and g.

$(fg)(x) = f(x)g(x)$ The function fg is the product of the functions f and g.

$\dfrac{f}{g}(x) = \dfrac{f(x)}{g(x)}$ The function f/g is the quotient of the functions f and g, where $g(x) \neq 0$.

9. If $f(x) = 4x$ and $g(x) = x^2 - 3$, then

Composition of Functions [8.6]

If f and g are two functions for which the range of each has numbers in common with the domain of the other, then we have the following definitions:

$$(f \circ g)(x) = f(g(x)) = f(x^2 - 3)$$
$$= 4x^2 - 12$$
$$(g \circ f)(x) = g(f(x)) = g(4x)$$
$$= 16x^2 - 3$$

The composition of f with g: $(f \circ g)(x) = f(g(x))$

The composition of g with f: $(g \circ f)(x) = g(f(x))$

10. If y varies directly with x, then

$$y = Kx$$

Then if y is 18 when x is 6,

$$18 = K \cdot 6$$

or

$$K = 3$$

So the equation can be written more specifically as

$$y = 3x$$

If we want to know what y is when x is 4, we simply substitute:

$$y = 3 \cdot 4$$
$$y = 12$$

Variation [8.7]

If y *varies directly* with x (y is directly proportional to x), then we say

$$y = Kx$$

If y *varies inversely* with x (y is inversely proportional to x), then we say

$$y = \frac{K}{x}$$

If z *varies jointly* with x and y (z is directly proportional to both x and y), then we say

$$z = Kxy$$

In each case, K is called the *constant of variation*.

COMMON MISTAKES

1. The two most common mistakes students make when first working with the formula for the slope of a line are
(a) Putting the difference of the x-coordinates over the difference of the y-coordinates.
(b) Subtracting in one order in the numerator and then subtracting in the opposite order in the denominator.

2. When graphing linear inequalities in two variables, remember to graph the boundary with a broken line when the inequality symbol is $<$ or $>$. The only time you use a solid line for the boundary is when the inequality symbol is $\leq$ or $\geq$.

CHAPTER 8 REVIEW

Find the slope of the line through the following pairs of points. [8.1]

1. $(5, 2)$, $(3, 6)$

2. $(-4, 2)$, $(3, 2)$

Graph each line. Then use the graph to find the slope of the line. [8.1]

3. $3x + 2y = 6$

4. $y = -\frac{3}{2}x + 1$

5. $x = 3$

Find x if the line through the two given points has the given slope. [8.1]

6. $(4, x)$, $(1, -3)$; $m = 2$

7. $(-4, 7)$, $(2, x)$; $m = -\frac{1}{3}$

8. Find the slope of any line parallel to the line through $(3, 8)$ and $(5, -2)$. [8.1]

9. The line through $(5, 3y)$ and $(2, y)$ is parallel to a line with slope 4. What is the value of y? [8.1]

Give the equation of the line with the following slope and y-intercept. [8.2]

10. $m = 3, b = 5$ **11.** $m = -2, b = 0$

Give the slope and y-intercept of each equation. [8.2]

12. $3x - y = 6$ **13.** $2x - 3y = 9$

Find the equation of the line that contains the given point and has the given slope. [8.2]

14. $(2, 4)$, $m = 2$

15. $(-3, 1)$, $m = -\frac{1}{3}$

Find the equation of the line that contains the given pair of points. [8.2]

16. $(2, 5)$, $(-3, -5)$

17. $(-3, 7)$, $(4, 7)$

18. $(-5, -1)$, $(-3, -4)$

19. Find the equation of the line that is parallel to $2x - y = 4$ and contains the point $(2, -3)$. [8.2]

20. Find the equation of the line perpendicular to $y = -3x + 1$ that has an x-intercept of 2. [8.2]

Graph each linear inequality. [8.3]

21. $y \le 2x - 3$ **22.** $x \ge -1$

State the domain and range of each relation, and then indicate which relations are also functions. [8.4]

23. $\{(2, 4), (3, 3), (4, 2)\}$

24. $\{(6, 3), (-4, 3), (-2, 0)\}$

If $f = \{(2, -1), (-3, 0), (4, \frac{1}{2}), (\pi, 2)\}$ and $g = \{(2, 2), (-1, 4), (0, 0)\}$, find the following. [8.5]

25. $f(-3)$ **26.** $f(2) + g(2)$

Let $f(x) = 2x^2 - 4x + 1$ and $g(x) = 3x + 2$, and evaluate each of the following. [8.5]

27. $f(0)$ **28.** $g(a)$

29. $f[g(0)]$ **30.** $f[g(1)]$

Let $f(x) = 2x + 1$ and $g(x) = x^2 - 4$, and find. [8.6]

31. $(f + g)(x)$ **32.** $(f - g)(0)$

33. $(fg)(1)$ **34.** $(fg)(x)$

35. $(f \circ g)(-1)$ **36.** $(g \circ f)(x)$

For the following problems, y varies directly with x. [8.7]

37. If y is 6 when x is 2, find y when x is 8.

38. If y is -3 when x is 5, find y when x is -10.

For the following problems, y varies inversely with the square of x. [8.7]

39. If y is 9 when x is 2, find y when x is 3.

40. If y is 4 when x is 5, find y when x is 2.

Solve each application problem. [8.7]

41. Tension in a Spring   The tension t in a spring varies directly with the distance d the spring is stretched. If the tension is 42 pounds when the spring is stretched 2 inches, find the tension when the spring is stretched twice as far.

42. Light Intensity   The intensity of a light source varies inversely with the square of the distance from the source. Four feet from the source the intensity is 9 foot-candles. What is the intensity 3 feet from the source?

C H A P T E R   8   PROJECTS

EQUATIONS AND INEQUALITIES IN TWO VARIABLES

G R O U P   P R O J E C T

PHONE BILL

Number of People: 2–3

Time Needed: 20–30 minutes

Equipment: Paper, pencil, and graphing calculator

Background: Step functions, or greatest integer functions, are used in situations such as phone bills and postage rates, where the function outputs discrete numbers that occur in steps or jumps. On a graphing calculator, the notation int(X), found under the MATH menu on a TI-83, gives the greatest integer less than or equal to x. In other words, it "rounds down." For the calculator to "round up," we use the notation $-\text{int}(-X)$. Figures 1 and 2 show the graphs of $y = \text{int}(x)$ and $y = -\text{int}(-x)$. Their differences are subtle, but important. (To obtain these graphs on a TI-83, press the MODE key and then select DOT.)

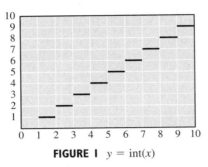

FIGURE 1 $y = \text{int}(x)$

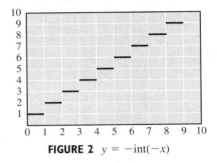

FIGURE 2 $y = -\text{int}(-x)$

Procedure: Suppose a wireless phone company charges $.50 per phone call plus $.35 per minute. Reading the fine print of your contract, you discover that you are actually paying "$.35 per minute with fractions of a minute rounded up to the next full minute."

1. Find a step function that gives the cost of a call as a function of time.

2. Sketch the graph of this function on the given grid.

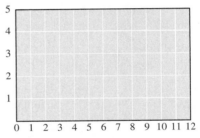

FIGURE 3

3. Fill in the following table using this step function.

Time (minutes)	0.6	1.0	1.1	2.5	6.0	6.8	7.0	8.3
Cost (dollars)								

4. If one phone call costs $5.05, how long was the call?

RESEARCH PROJECT

DESCARTES AND PASCAL

In this chapter we mentioned that René Descartes, the inventor of the rectangular coordinate system, is the person who made the statement "I think, therefore, I am." Blaise Pascal, is responsible for the statement "The heart has its reasons which reason does not know." Although Pascal and Descartes were contemporaries, the philosophies of the two men differed greatly. Research the philosophy of both Descartes and Pascal, and then write an essay that gives the main points of each man's philosophy. In the essay, show how the quotations given here fit in with the philosophy of the man responsible for the quotation.

René Descartes, 1596–1650
(David Eugene Smith
Collection/Columbia
University)

Blaise Pascal, 1623–1662
(David Eugene Smith
Collection/Columbia
University)

C H A P T E R   8   **TEST**

For each of the following straight lines, identify the x-intercept, y-intercept, and slope, and sketch the graph. [8.1]

1. $2x + y = 6$

2. $y = -2x - 3$

3. $y = \dfrac{3}{2}x + 4$

4. $x = -2$

Find the equation for each line. [8.2]

5. Give the equation of the line through $(-1, 3)$ that has slope $m = 2$.

6. Give the equation of the line through $(-3, 2)$ and $(4, -1)$.

7. Line l contains the point $(5, -3)$ and has a graph parallel to the graph of $2x - 5y = 10$. Find the equation for l.

8. Line l contains the point $(-1, -2)$ and has a graph perpendicular to the graph of $y = 3x - 1$. Find the equation for l.

9. Give the equation of the vertical line through $(4, -7)$.

Graph the following linear inequalities. [8.3]

10. $3x - 4y < 12$

11. $y \le -x + 2$

State the domain and range for the following relations, and indicate which relations are also functions. [8.4]

12. $\{(-2, 0), (-3, 0), (-2, 1)\}$

13. $y = x^2 - 9$

Let $f(x) = x - 2$, $g(x) = 3x + 4$, and $h(x) = 3x^2 - 2x - 8$, and find the following. [8.5]

14. $f(3) + g(2)$

15. $h(0) + g(0)$

16. $f[g(2)]$

17. $g[f(2)]$

Constructing a Box A piece of cardboard is in the shape of a square with each side 8 inches long. You are going to make a box out of the piece of cardboard by cutting four equal squares from each corner and then folding up the sides, as shown in Figure 1. Use this information in Problems 18–20. [8.4, 8.5]

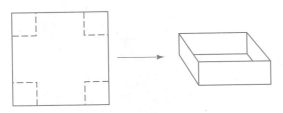

FIGURE I

18. If the length of each side of the squares you are cutting from the corners is represented by the variable x, state the restrictions on x.

19. Use function notation to write the volume of the resulting box $V(x)$ as a function of x.

20. Find $V(2)$, and explain what it represents.

Solve the following variation problems. [8.7]

21. **Direct Variation** Quantity y varies directly with the square of x. If y is 50 when x is 5, find y when x is 3.

22. **Joint Variation** Quantity z varies jointly with x and the cube of y. If z is 15 when x is 5 and y is 2, find z when x is 2 and y is 3.

23. **Maximum Load** The maximum load (L) a horizontal beam can safely hold varies jointly with the width (w) and the square of the depth (d) and inversely with the length (l). If a 10-foot beam with width 3 feet and depth 4 feet will safely hold up to 800 pounds, how many pounds will a 12-foot beam with width 3 feet and depth 4 feet hold?

CHAPTERS 1–8 **CUMULATIVE REVIEW**

Simplify.

1. $11 - (-9) - 7 - (-5)$

2. $\left(\dfrac{5}{6}\right)^{-2}$

3. $\dfrac{x^{-5}}{x^{-8}}$

4. $\left(\dfrac{x^{-6}y^3}{x^{-3}y^{-4}}\right)^{-1}$

5. $-3(5x + 4) + 12x$

6. $\dfrac{\dfrac{2a}{3a^3 - 3}}{\dfrac{4a}{6a - 6}}$

7. $(x + 3)^2 - (x - 3)^2$

8. If $f(x) = 3x - 2$ and $g(x) = x^2 - 1$, find $(f \circ g)(x)$.

9. Subtract $\frac{3}{4}$ from the product of -3 and $\frac{5}{12}$.

10. Write in symbols: The difference of $5a$ and $7b$ is greater than their sum.

11. Factor $8x^3 - 27$.

12. Give the next number of the sequence, and state whether arithmetic or geometric: $-3, 2, 7, \ldots$.

13. State the properties that justify: $(3x - 4) + 2y = (3x + 2y) - 4$

Subtract.

14. $\dfrac{6}{y^2 - 9} - \dfrac{5}{y^2 - y - 6}$

15. $\dfrac{y}{x^2 - y^2} - \dfrac{x}{x^2 - y^2}$

Multiply.

16. $\left(4t^2 + \dfrac{1}{3}\right)\left(3t^2 - \dfrac{1}{4}\right)$

17. $\dfrac{x^4 - 16}{x^3 - 8} \cdot \dfrac{x^2 + 2x + 4}{x^2 + 4}$

Divide.

18. $\dfrac{10x^3 - 15x^4}{5x}$

19. $\dfrac{a^4 + a^3 - 1}{a + 2}$

Solve.

20. $-\dfrac{3}{5}a + 3 = 15$

21. $7y - 6 = 2y + 9$

22. $\dfrac{2}{5}(15x - 2) - \dfrac{1}{5} = 5$

23. $|a| - 5 = 7$

24. $x^3 - 3x^2 - 25x + 75 = 0$

25. $\dfrac{3}{y - 2} = \dfrac{2}{y - 3}$

26. $2 - \dfrac{11}{x} = -\dfrac{12}{x^2}$

Solve each system.

27. $5x - 2y = -1$
$\quad\ y = 3x + 2$

28. $5x - 8y = 4$
$\quad\ 3x + 2y = -1$

29. $-5x + 3y = 1$
$\quad\ \dfrac{5}{3}x - y = 2$

30. $\quad x - y + z = -4$
$-4x - 3y - 2z = 2$
$-5x + 4y + z = 2$

31. $\quad 7x - 9y = 2$
$-3x + 11y = 1$

32. $\quad x - 2y = 1$
$3x + 5z = 8$
$4y - 7z = -3$

Solve each inequality, and graph the solution.

33. $-3(3x - 1) \le -2(3x - 3)$

34. $|2x + 3| - 4 < 1$

35. Find the slope of the line $x + y = -3$.

Graph on a rectangular coordinate system.

36. $6x - 5y = 30$

37. $f(x) = \dfrac{1}{2}x - 1$

38. $3x - y < 4$

Factor completely.

39. 168

40. $x^2 - 3x - 70$

41. $x^2 + 10x + 25 - y^2$

42. Reduce to lowest terms: $\dfrac{x^3 + 2x^2 - 9x - 18}{x^2 - x - 6}$

43. Write 9,270,000 in scientific notation.

44. 18 is 8% of what number?

45. Let $f(x) = x^2 - 2x$ and $g(x) = x + 5$. Find: $f(-2) - g(3)$.

46. Find the slope of any line perpendicular to the line through $(-6, -1)$ and $(-3, -5)$.

47. Find the equation of the line through $(-2, 6)$ and $(-2, 3)$.

48. Joint Variation   Suppose z varies jointly with x and the cube of y. If z is -48 when x is 3 and y is 2, find z when x is 2 and y is 3.

49. Geometry   The height of a triangle is 5 feet less than 2 times the base. If the area is 75 square feet, find the base and height.

50. Geometry   Find all three angles in a triangle if the smallest angle is one-sixth the largest angle and the remaining angle is 20 degrees more than the smallest angle.

Rational Exponents and Roots

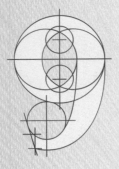

(Chip Porter/Tony Stone Images)

INTRODUCTION

Ecology and conservation are topics that interest most college students. If our rivers and oceans are to be preserved for future generations, we need to work to eliminate pollution from our waters. If a river is flowing at 1 meter per second and a pollutant is entering the river at a constant rate, the shape of the pollution plume can often be modeled by the simple equation

$$y = \sqrt{x}$$

The following table and graph were produced from the equation.

TABLE 1	Width of a Pollutant Plume
Distance From Source (meters) x	**Width of Plume (meters)** y
0	0
1	1
4	2
9	3
16	4

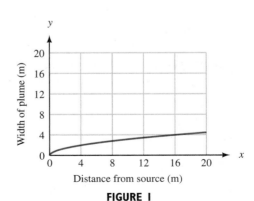

FIGURE 1

To visualize how Figure 1 models the pollutant plume, imagine that the river is flowing from left to right, parallel to the x-axis, with the x-axis as one of its banks. The pollutant is entering the river from the bank at $(0, 0)$.

By modeling pollution with mathematics, we can use our knowledge of mathematics to help control and eliminate pollution.

Figure 1 shows a square in which each of the four sides is 1 inch long. To find the square of the length of the diagonal c, we apply the Pythagorean theorem:

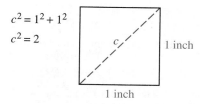

$$c^2 = 1^2 + 1^2$$
$$c^2 = 2$$

FIGURE 1

Because we know that c is positive and that its square is 2, we call c the *positive square root* of 2, and we write $c = \sqrt{2}$. Associating numbers, such as $\sqrt{2}$, with the diagonal of a square or rectangle allows us to analyze some interesting items from geometry. One particularly interesting geometric object that we will study in this section is shown in Figure 2. It is constructed from a right triangle, and the length of the diagonal is found from the Pythagorean theorem. We will come back to this figure at the end of this section.

The Golden Rectangle

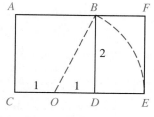

FIGURE 2

In Chapter 4 we developed notation (exponents) to give us the square, cube, or any other power of a number. For instance, if we wanted the square of 3, we wrote $3^2 = 9$. If we wanted the cube of 3, we wrote $3^3 = 27$. In this section we will develop notation that will take us in the reverse direction, that is, from the square of a number, say 25, back to the original number, 5.

DEFINITION If x is a nonnegative real number, then the expression $\sqrt{x}$ is called the **positive square root** of x and is such that

$$(\sqrt{x})^2 = x$$

In words: $\sqrt{x}$ is the positive number we square to get x.

The negative square root of x, $-\sqrt{x}$, is defined in a similar manner.

EXAMPLE 1 The positive square root of 64 is 8 because 8 is the positive number with the property $8^2 = 64$. The negative square root of 64 is -8 since -8 is the negative number whose square is 64. We can summarize both of these facts by saying

$$\sqrt{64} = 8 \quad \text{and} \quad -\sqrt{64} = -8$$

Note: It is a common mistake to assume that an expression like $\sqrt{25}$ indicates both square roots, 5 and -5. The expression $\sqrt{25}$ indicates only the positive square root of 25, which is 5. If we want the negative square root, we must use a negative sign: $-\sqrt{25} = -5$.

The higher roots, cube roots, fourth roots, and so on, are defined by definitions similar to that of square roots.

DEFINITION If x is a real number and n is a positive integer, then

Positive square root of x, $\sqrt{x}$, is such that $(\sqrt{x})^2 = x$ $x \geq 0$

Cube root of x, $\sqrt[3]{x}$ is such that $(\sqrt[3]{x})^3 = x$

Positive fourth root of x, $\sqrt[4]{x}$, is such that $(\sqrt[4]{x})^4 = x$ $x \geq 0$

Fifth root of x, $\sqrt[5]{x}$, is such that $(\sqrt[5]{x})^5 = x$

$\vdots$ $\vdots$

The **nth root of x**, $\sqrt[n]{x}$, is such that $(\sqrt[n]{x})^n = x$ $x \geq 0$ if n is even

---

Note: We have restricted the even roots in this definition to nonnegative numbers. Even roots of negative numbers exist, but are not represented by real numbers. That is, $\sqrt{-4}$ is not a real number since there is no real number whose square is -4.

The following is a table of the most common roots used in this book. Any of the roots that are unfamiliar should be memorized.

Square Roots		Cube Roots	Fourth Roots
$\sqrt{0} = 0$	$\sqrt{49} = 7$	$\sqrt[3]{0} = 0$	$\sqrt[4]{0} = 0$
$\sqrt{1} = 1$	$\sqrt{64} = 8$	$\sqrt[3]{1} = 1$	$\sqrt[4]{1} = 1$
$\sqrt{4} = 2$	$\sqrt{81} = 9$	$\sqrt[3]{8} = 2$	$\sqrt[4]{16} = 2$
$\sqrt{9} = 3$	$\sqrt{100} = 10$	$\sqrt[3]{27} = 3$	$\sqrt[4]{81} = 3$
$\sqrt{16} = 4$	$\sqrt{121} = 11$	$\sqrt[3]{64} = 4$	
$\sqrt{25} = 5$	$\sqrt{144} = 12$	$\sqrt[3]{125} = 5$	
$\sqrt{36} = 6$	$\sqrt{169} = 13$		

Notation An expression like $\sqrt[3]{8}$ that involves a root is called a *radical expression*. In the expression $\sqrt[3]{8}$, the ^{3} is called the *index*, the $\sqrt{}$ is the *radical sign*, and 8 is called the *radicand*. The index of a radical must be a positive integer greater than 1. If no index is written, it is assumed to be 2.

ROOTS AND NEGATIVE NUMBERS

When dealing with negative numbers and radicals, the only restriction concerns negative numbers under even roots. We can have negative signs in front of radicals and negative numbers under odd roots and still obtain real numbers. Here are some examples to help clarify this. In the last section of this chapter we will see how to deal with even roots of negative numbers.

EXAMPLES   Simplify each expression, if possible.

2. $\sqrt[3]{-8} = -2$ because $(-2)^3 = -8$.

3. $\sqrt{-4}$ is not a real number since there is no real number whose square is -4.

4. $-\sqrt{25} = -5$ because -5 is the negative square root of 25.

5. $\sqrt[5]{-32} = -2$ because $(-2)^5 = -32$.

6. $\sqrt[4]{-81}$ is not a real number since there is no real number we can raise to the fourth power and obtain -81.

VARIABLES UNDER A RADICAL

From the preceding examples it is clear that we must be careful that we do not try to take an even root of a negative number. For this reason, we will assume that all variables appearing under a radical sign represent nonnegative numbers.

EXAMPLES   Assume all variables represent nonnegative numbers, and simplify each expression as much as possible.

7. $\sqrt{25a^4b^6} = 5a^2b^3$ because $(5a^2b^3)^2 = 25a^4b^6$.

8. $\sqrt[3]{x^6y^{12}} = x^2y^4$ because $(x^2y^4)^3 = x^6y^{12}$.

9. $\sqrt[4]{81r^8s^{20}} = 3r^2s^5$ because $(3r^2s^5)^4 = 81r^8s^{20}$.

RATIONAL NUMBERS AS EXPONENTS

We will now develop a second kind of notation involving exponents that will allow us to designate square roots, cube roots, and so on in another way.

Consider the equation $x = 8^{1/3}$. Although we have not encountered fractional exponents before, let's assume that all the properties of exponents hold in this case. Cubing both sides of the equation, we have

$$x^3 = (8^{1/3})^3$$

$$x^3 = 8^{(1/3)(3)}$$

$$x^3 = 8^1$$

$$x^3 = 8$$

The last line tells us that x is the number whose cube is 8. It must be true, then, that x is the cube root of 8, $x = \sqrt[3]{8}$. Since we started with $x = 8^{1/3}$, it follows that

$$8^{1/3} = \sqrt[3]{8}$$

It seems reasonable, then, to define fractional exponents as indicating roots. Here is the formal definition.

DEFINITION   If x is a real number and n is a positive integer greater than 1, then

$$x^{1/n} = \sqrt[n]{x} \qquad (x \geq 0 \text{ when } n \text{ is even})$$

In words: The quantity $x^{1/n}$ is the nth root of x.

With this definition we have a way of representing roots with exponents. Here are some examples.

EXAMPLES   Write each expression as a root and then simplify, if possible.

10. $8^{1/3} = \sqrt[3]{8} = 2$

11. $36^{1/2} = \sqrt{36} = 6$

12. $-25^{1/2} = -\sqrt{25} = -5$

13. $(-25)^{1/2} = \sqrt{-25}$, which is not a real number

14. $\left(\dfrac{4}{9}\right)^{1/2} = \sqrt{\dfrac{4}{9}} = \dfrac{2}{3}$

The properties of exponents developed in Chapter 4 were applied to integer exponents only. We will now extend these properties to include rational exponents also. We do so without proof.

PROPERTIES OF EXPONENTS

If a and b are real numbers and r and s are rational numbers, and a and b are nonnegative whenever r and s indicate even roots, then

1. $a^r \cdot a^s = a^{r+s}$ **4.** $a^{-r} = \dfrac{1}{a^r}$ $(a \neq 0)$

2. $(a^r)^s = a^{rs}$ **5.** $\left(\dfrac{a}{b}\right)^r = \dfrac{a^r}{b^r}$ $(b \neq 0)$

3. $(ab)^r = a^r b^r$ **6.** $\dfrac{a^r}{a^s} = a^{r-s}$ $(a \neq 0)$

Sometimes rational exponents can simplify our work with radicals. Here are Examples 8 and 9 again, but this time we will work them using rational exponents.

EXAMPLES   Write each radical with a rational exponent and then simplify.

15. $\sqrt[3]{x^6 y^{12}} = (x^6 y^{12})^{1/3}$

$\qquad\qquad = (x^6)^{1/3}(y^{12})^{1/3}$

$\qquad\qquad = x^2 y^4$

16. $\sqrt[4]{81 r^8 s^{20}} = (81 r^8 s^{20})^{1/4}$

$\qquad\qquad = 81^{1/4}(r^8)^{1/4}(s^{20})^{1/4}$

$\qquad\qquad = 3 r^2 s^5$

So far, the numerators of all the rational exponents we have encountered have been 1. The next theorem extends the work we can do with rational exponents to rational exponents with numerators other than 1.

THEOREM 9.1

If a is a nonnegative real number, m is an integer, and n is a positive integer, then

$$a^{m/n} = (a^{1/n})^m = (a^m)^{1/n}$$

Proof   We can prove Theorem 9.1 using the properties of exponents. Since $\frac{m}{n} = m(\frac{1}{n})$ we have

$$a^{m/n} = a^{m(1/n)} \qquad\qquad a^{m/n} = a^{(1/n)(m)}$$

$$= (a^m)^{1/n} \qquad\qquad\qquad = (a^{1/n})^m$$

Here are some examples that illustrate how we use this theorem.

EXAMPLES   Simplify as much as possible.

17. $8^{2/3} = (8^{1/3})^2$ $\qquad$ Theorem 9.1

$\qquad\; = 2^2$ $\qquad\qquad$ Definition of fractional exponents

$\qquad\; = 4$ $\qquad\qquad$ The square of 2 is 4.

Note: On a scientific calculator, Example 17 would look like this:

$$8 \;\boxed{y^x}\; \boxed{(}\; 2 \;\boxed{\div}\; 3 \;\boxed{)}\; \boxed{=}$$

18. $25^{3/2} = (25^{1/2})^3$ Theorem 9.1

           $= 5^3$ Definition of fractional exponents

           $= 125$ The cube of 5 is 125.

19. $9^{-3/2} = (9^{1/2})^{-3}$ Theorem 9.1

           $= 3^{-3}$ Definition of fractional exponents

           $= \dfrac{1}{3^3}$ Property 4 for exponents

           $= \dfrac{1}{27}$ The cube of 3 is 27.

20. $\left(\dfrac{27}{8}\right)^{-4/3} = \left[\left(\dfrac{27}{8}\right)^{1/3}\right]^{-4}$ Theorem 9.1

           $= \left(\dfrac{3}{2}\right)^{-4}$ Definition of fractional exponents

           $= \left(\dfrac{2}{3}\right)^{4}$ Property 4 for exponents

           $= \dfrac{16}{81}$ The fourth power of $\frac{2}{3}$ is $\frac{16}{81}$.

The following examples show the application of the properties of exponents to rational exponents.

EXAMPLES Assume all variables represent positive quantities, and simplify as much as possible.

21. $x^{1/3} \cdot x^{5/6} = x^{1/3+5/6}$ Property 1

           $= x^{2/6+5/6}$ LCD is 6.

           $= x^{7/6}$ Add fractions.

22. $(y^{2/3})^{3/4} = y^{(2/3)(3/4)}$ Property 2

           $= y^{1/2}$ Multiply fractions: $\frac{2}{3} \cdot \frac{3}{4} = \frac{6}{12} = \frac{1}{2}$

23. $\dfrac{z^{1/3}}{z^{1/4}} = z^{1/3-1/4}$ Property 6

           $= z^{4/12-3/12}$ LCD is 12.

           $= z^{1/12}$ Subtract fractions.

24. $\left(\dfrac{a^{-1/3}}{b^{1/2}}\right)^{6} = \dfrac{(a^{-1/3})^6}{(b^{1/2})^6}$ Property 5

           $= \dfrac{a^{-2}}{b^3}$ Property 2

           $= \dfrac{1}{a^2 b^3}$ Property 4

25. $\dfrac{(x^{-3}y^{1/2})^4}{x^{10}y^{3/2}} = \dfrac{(x^{-3})^4(y^{1/2})^4}{x^{10}y^{3/2}}$ Property 3

$\qquad\qquad = \dfrac{x^{-12}y^2}{x^{10}y^{3/2}}$ Property 2

$\qquad\qquad = x^{-22}y^{1/2}$ Property 6

$\qquad\qquad = \dfrac{y^{1/2}}{x^{22}}$ Property 4

FACTS FROM
Geometry

The Pythagorean Theorem (Again) and the Golden Rectangle

Now that we have had some experience working with square roots, we can rewrite the Pythagorean theorem using a square root. If triangle *ABC* is a right triangle with $C = 90°$, then the length of the longest side is the *positive square root* of the sum of the squares of the other two sides (see Figure 3).

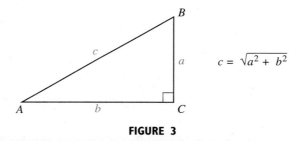

FIGURE 3

In the introduction to this section we mentioned the golden rectangle. Its origins can be traced back over 2,000 years to the Greek civilization that produced Pythagoras, Socrates, Plato, Aristotle, and Euclid. The most important mathematical work to come from that Greek civilization was Euclid's *Elements,* an elegantly written summary of all that was known about geometry at that time in history. Euclid's *Elements,* according to Howard Eves, an authority on the history of mathematics, exercised a greater influence on scientific thinking than any other work. Here is how we construct a golden rectangle from a square of side 2, using the same method that Euclid used in his *Elements.*

CONSTRUCTING A GOLDEN RECTANGLE
FROM A SQUARE OF SIDE 2

Step 1: Draw a square with a side of length 2. Connect the midpoint of side *CD* to corner *B* as shown on the next page. (Note that we have labeled the midpoint of segment *CD* with the letter *O.*)

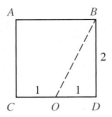

Step 2: Drop the diagonal from step 1 down so it aligns with side *CD*.

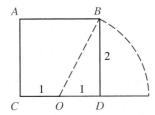

Step 3: Form rectangle *ACEF.* This is a golden rectangle.

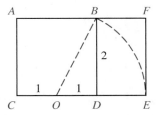

All golden rectangles are constructed from squares. Every golden rectangle, no matter how large or small it is, will have the same shape. To associate a number with the shape of the golden rectangle, we use the ratio of its length to its width. This ratio is called the *golden ratio.* To calculate the golden ratio, we must first find the length of the diagonal we used to construct the golden rectangle. Figure 4 shows the golden rectangle that we constructed from a square of side 2. The length of the diagonal *OB* is found by applying the Pythagorean theorem to triangle *OBD.*

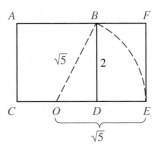

FIGURE 4

The length of segment OE is equal to the length of diagonal OB; both are $\sqrt{5}$. Since the distance from C to O is 1, the length CE of the golden rectangle is $1 + \sqrt{5}$. Now we can find the golden ratio:

$$\text{Golden ratio} = \frac{\text{length}}{\text{width}} = \frac{CE}{EF} = \frac{1 + \sqrt{5}}{2}$$

Using TECHNOLOGY

GRAPHING CALCULATORS—A WORD OF CAUTION

Some graphing calculators give surprising results when evaluating expressions such as $(-8)^{2/3}$. As you know from reading this section, the expression $(-8)^{2/3}$ simplifies to 4, either by taking the cube root first and then squaring the result, or by squaring the base first and then taking the cube root of the result. Here are three different ways to evaluate this expression on your calculator:

1. $(-8)^{\wedge}(2/3)$ To evaluate $(-8)^{2/3}$
2. $((-8)^{\wedge}2)^{\wedge}(1/3)$ To evaluate $((-8)^2)^{1/3}$
3. $((-8)^{\wedge}(1/3))^{\wedge}2$ To evaluate $((-8)^{1/3})^2$

Note any differences in the results.

    Next, graph each of the following functions, one at a time.

1. $Y_1 = X^{2/3}$ **2.** $Y_2 = (X^2)^{1/3}$ **3.** $Y_3 = (X^{1/3})^2$

The correct graph is shown in Figure 5. Note which of your graphs match the correct graph.

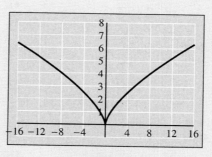

FIGURE 5

    Different calculators evaluate exponential expressions in different ways. You should use the method (or methods) that gave you the correct graph.

Getting Ready for Class

After reading through the preceding section, respond in your own words and in complete sentences.

A. Every real number has two square roots. Explain the notation we use to tell them apart. Use the square roots of 3 for examples.

B. Explain why a square root of -4 is not a real number.

C. We use the notation $\sqrt{2}$ to represent the positive square root of 2. Explain why there isn't a simpler way to express the positive square root of 2.

D. For the expression $a^{m/n}$, explain the significance of the numerator m and the significance of the denominator n in the exponent.

PROBLEM SET 9.1

Find each of the following roots, if possible.

1. $\sqrt{144}$　　　　　　**2.** $-\sqrt{144}$

3. $\sqrt{-144}$　　　　　**4.** $\sqrt{-49}$

5. $-\sqrt{49}$　　　　　　**6.** $\sqrt{49}$

7. $\sqrt[3]{-27}$　　　　　**8.** $-\sqrt[3]{27}$

9. $\sqrt[4]{16}$　　　　　　**10.** $-\sqrt[4]{16}$

11. $\sqrt[4]{-16}$　　　　　**12.** $-\sqrt[4]{-16}$

13. $\sqrt{0.04}$　　　　　　**14.** $\sqrt{0.81}$

15. $\sqrt[3]{0.008}$　　　　　**16.** $\sqrt[3]{0.125}$

Simplify each expression. Assume all variables represent nonnegative numbers.

17. $\sqrt{36a^8}$　　　　　　**18.** $\sqrt{49a^{10}}$

19. $\sqrt[3]{27a^{12}}$　　　　　**20.** $\sqrt[3]{8a^{15}}$

21. $\sqrt[3]{x^3y^6}$　　　　　**22.** $\sqrt[3]{x^6y^3}$

23. $\sqrt[5]{32x^{10}y^5}$　　　　**24.** $\sqrt[5]{32x^5y^{10}}$

25. $\sqrt[4]{16a^{12}b^{20}}$　　　　**26.** $\sqrt[4]{81a^{24}b^8}$

Use the definition of rational exponents to write each of the following with the appropriate root. Then simplify.

27. $36^{1/2}$　　　　　　**28.** $49^{1/2}$

29. $-9^{1/2}$　　　　　　**30.** $-16^{1/2}$

31. $8^{1/3}$　　　　　　**32.** $-8^{1/3}$

33. $(-8)^{1/3}$　　　　　**34.** $-27^{1/3}$

35. $32^{1/5}$　　　　　　**36.** $81^{1/4}$

37. $\left(\dfrac{81}{25}\right)^{1/2}$　　　　**38.** $\left(\dfrac{9}{16}\right)^{1/2}$

39. $\left(\dfrac{64}{125}\right)^{1/3}$　　　　**40.** $\left(\dfrac{8}{27}\right)^{1/3}$

Use Theorem 9.1 to simplify each of the following as much as possible.

41. $27^{2/3}$　　　　　　**42.** $8^{4/3}$

43. $25^{3/2}$　　　　　　**44.** $9^{3/2}$

45. $16^{3/4}$　　　　　　**46.** $81^{3/4}$

Simplify each expression. Remember, negative exponents give reciprocals.

47. $27^{-1/3}$　　　　　　**48.** $9^{-1/2}$

49. $81^{-3/4}$　　　　　　**50.** $4^{-3/2}$

51. $\left(\dfrac{25}{36}\right)^{-1/2}$　　　　**52.** $\left(\dfrac{16}{49}\right)^{-1/2}$

53. $\left(\dfrac{81}{16}\right)^{-3/4}$　　　　**54.** $\left(\dfrac{27}{8}\right)^{-2/3}$

55. $16^{1/2} + 27^{1/3}$　　　　**56.** $25^{1/2} + 100^{1/2}$

57. $8^{-2/3} + 4^{-1/2}$　　　　**58.** $49^{-1/2} + 25^{-1/2}$

Use the properties of exponents to simplify each of the following as much as possible. Assume all bases are positive.

59. $x^{3/5} \cdot x^{1/5}$　　　　**60.** $x^{3/4} \cdot x^{5/4}$

61. $(a^{3/4})^{4/3}$　　　　**62.** $(a^{2/3})^{3/4}$

63. $\dfrac{x^{1/5}}{x^{3/5}}$

64. $\dfrac{x^{2/7}}{x^{5/7}}$

65. $\dfrac{x^{5/6}}{x^{2/3}}$

66. $\dfrac{x^{7/8}}{x^{8/7}}$

67. $(x^{3/5}y^{5/6}z^{1/3})^{3/5}$

68. $(x^{3/4}y^{1/8}z^{5/6})^{4/5}$

69. $\dfrac{a^{3/4}b^2}{a^{7/8}b^{1/4}}$

70. $\dfrac{a^{1/3}b^4}{a^{3/5}b^{1/3}}$

71. $\dfrac{(y^{2/3})^{3/4}}{(y^{1/3})^{3/5}}$

72. $\dfrac{(y^{5/4})^{2/5}}{(y^{1/4})^{4/3}}$

73. $\left(\dfrac{a^{-1/4}}{b^{1/2}}\right)^8$

74. $\left(\dfrac{a^{-1/5}}{b^{1/3}}\right)^{15}$

75. $\dfrac{(r^{-2}s^{1/3})^6}{r^8 s^{3/2}}$

76. $\dfrac{(r^{-5}s^{1/2})^4}{r^{12}s^{5/2}}$

77. $\dfrac{(25a^6b^4)^{1/2}}{(8a^{-9}b^3)^{-1/3}}$

78. $\dfrac{(27a^3b^6)^{1/3}}{(81a^8b^{-4})^{1/4}}$

79. Show that the expression $(a^{1/2} + b^{1/2})^2$ is not equal to $a + b$ by replacing a with 9 and b with 4 in both expressions and then simplifying each.

80. Show that the statement $(a^2 + b^2)^{1/2} = a + b$ is not, in general, true by replacing a with 3 and b with 4 and then simplifying both sides.

81. You may have noticed, if you have been using a calculator to find roots, that you can find the fourth root of a number by pressing the square root button twice. Written in symbols, this fact looks like this:

$$\sqrt{\sqrt{a}} = \sqrt[4]{a} \qquad (a \geq 0)$$

Show that this statement is true by rewriting each side with exponents instead of radical notation and then simplifying the left side.

82. Show that the following statement is true by rewriting each side with exponents instead of radical notation and then simplifying the left side.

$$\sqrt[3]{\sqrt{a}} = \sqrt[6]{a} \qquad (a \geq 0)$$

Applying the Concepts

83. Maximum Speed   The maximum speed (v) that an automobile can travel around a curve of radius r without skidding is given by the equation

$$v = \left(\dfrac{5r}{2}\right)^{1/2}$$

where v is in miles per hour and r is measured in feet. What is the maximum speed a car can travel around a curve with a radius of 250 feet without skidding?

84. Relativity   The equation

$$L = \left(1 - \dfrac{v^2}{c^2}\right)^{1/2}$$

gives the relativistic length of a 1-foot ruler traveling with velocity v. Find L if

$$\dfrac{v}{c} = \dfrac{3}{5}$$

85. Golden Ratio   The golden ratio is the ratio of the length to the width in any golden rectangle. The exact value of this number is $\dfrac{1 + \sqrt{5}}{2}$. Use a calculator to find a decimal approximation to this number and round it to the nearest thousandth.

86. Golden Ratio   The reciprocal of the golden ratio is $\dfrac{2}{1 + \sqrt{5}}$. Find a decimal approximation to this number that is accurate to the nearest thousandth.

87. Sequences   Find the next term in the following sequence. Then explain how this sequence is related to the Fibonacci sequence.

$$\dfrac{3}{2}, \dfrac{5}{3}, \dfrac{8}{5}, \ldots$$

88. Sequences   Write the first ten terms in the sequence shown in Problem 87. Then find a decimal approximation to each of the ten terms, rounding each to the nearest thousandth.

89. Chemistry   Figure 6 shows part of a model of a magnesium oxide (MgO) crystal. Each corner of the square is at the center of one oxygen ion (O^{2-}), and the center of the middle ion is at the center of the square. The radius for each oxygen ion is 60 picometers (pm), and the radius for each magnesium ion (Mg^{2+}) is 150 picometers.

  (a) Find the length of the side of the square. Write your answer in picometers.

  (b) Find the length of the diagonal of the square. Write your answer in picometers.

  (c) If 1 meter is 10^{12} picometers, give the length of the diagonal of the square in meters.

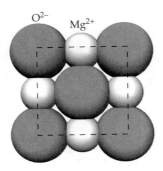

FIGURE 6 *(Susan M. Young)*

90. Chemistry Figure 7 shows part of a model of crystallized aluminum. Each corner of the square is at the center of one aluminum atom, and the center of the middle atom is at the center of the square. The radius for each aluminum atom is 143 picometers (pm).
  (a) Find the length of the side of the square. Write your answer in picometers.
  (b) If 1 meter is 10^{12} picometers, give the length of the side of the square in meters.

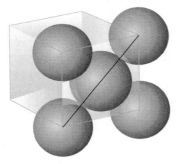

FIGURE 7 *(Susan M. Young)*

91. Geometry The length of each side of the cube shown in Figure 8 is 1 inch.
  (a) Find the length of the diagonal *CH*.
  (b) Find the length of the diagonal *CF*.

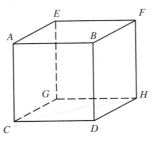

FIGURE 8

92. Chemistry Figure 9 shows part of a model of a crystal containing three atoms. The endpoints of the diagonal of the cube are at the centers of the two outside atoms, and the center of the cube is at the center of the middle atom. The radius of each atom is 100 picometers.
  (a) Find the length of the diagonal of the cube.
  (b) Find the length of the side of the cube.

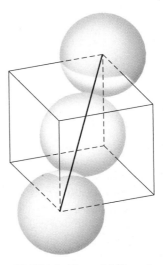

FIGURE 9 *(Susan M. Young)*

93. Comparing Graphs Identify the graph with the correct equation.
  (a) $y = x$
  (b) $y = x^2$
  (c) $y = x^{2/3}$
  (d) What are the two points of intersection of all three graphs?

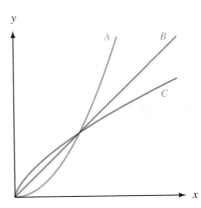

94. Falling Objects The time t it takes an object to fall d feet is given by the equation

$$t = \frac{1}{4}\sqrt{d}$$

(a) The Sears Tower in Chicago is 1,450 feet tall. How long would it take a penny to fall to the ground from the top of the Sears Tower?

(b) An object took 30 seconds to fall to the ground. From what distance must it have been dropped?

Review Problems

The problems that follow review material we covered in Sections 4.5 and 4.6. Reviewing these problems will help you understand the next section.

Multiply.

95. $x^2(x^4 - x)$

96. $5x^2(2x^3 - x)$

97. $(x - 3)(x + 5)$

98. $(x - 2)(x + 2)$

99. $(x^2 - 5)^2$

100. $(x^2 + 5)^2$

101. $(x - 3)(x^2 + 3x + 9)$

102. $(x + 3)(x^2 - 3x + 9)$

Extending the Concepts

Graph the equation $y = x^{3/4}$, and then use the Trace and Zoom features to approximate each of the following to the nearest tenth.

103. $2^{3/4}$

104. $3^{3/4}$

105. $10^{3/4}$

106. $16^{3/4}$

Graph $y = x^{3/4}$ and $y = x^{4/3}$ on the same coordinate system, using a window with X from -4 to 4 and Y from -1 to 4. Use the graphs to answer the following questions.

107. Where do the two graphs intersect?

108. For what values of x is $x^{3/4} \geq x^{4/3}$?

Carbon-14 dating is used extensively in science to find the age of fossils. If at one time a fossilized substance contains an amount A of carbon-14, then t years later the amount of carbon-14 it contains is given by the formula

$$A \cdot 2^{-t/5,600}$$

Use this formula and a calculator to solve the following problems.

109. A fossilized substance contains 3 micrograms of carbon-14. How much carbon-14 will be left at the end of

(a) 5,000 years?   (b) 10,000 years?

(c) 56,000 years?   (d) 112,000 years?

110. A fossilized substance contains 5 micrograms of carbon-14. How much carbon-14 will be left at the end of

(a) 500 years?   (b) 5,000 years?

(c) 56,000 years?   (d) 112,000 years?

Suppose you purchased ten silver proof coin sets in 1997 for $21 each, for a total investment of $210. Three years later, in 2000, you find that each set is worth $30, which means that your ten sets have a total value of $300.

United States Mint Proof Set

You can calculate the annual rate of return on this investment using a formula that involves rational exponents. The annual rate of return will tell you at what interest rate you would have to invest your original $210 in order for it to be worth $300, 3 years later. As you will see at the end of this section, the annual rate of return on this investment is 12.6%, which is a good return on your money.

In this section we will look at multiplication, division, factoring, and simplification of some expressions that resemble polynomials but contain rational exponents. The problems in this section will be of particular interest to you if you are planning to take either an engineering calculus class or a business calculus class. As was the case in the previous section, we will assume all variables represent nonnegative real numbers. That way, we will not have to worry about the possibility of introducing undefined terms—even roots of negative numbers—into any of our examples. Let's begin this section with a look at multiplication of expressions containing rational exponents.

EXAMPLE 1 Multiply $x^{2/3}(x^{4/3} - x^{1/3})$.

Solution Applying the distributive property and then simplifying the resulting terms, we have:

$$x^{2/3}(x^{4/3} - x^{1/3}) = x^{2/3}x^{4/3} - x^{2/3}x^{1/3} \qquad \text{Distributive property}$$
$$= x^{6/3} - x^{3/3} \qquad \text{Add exponents.}$$
$$= x^2 - x \qquad \text{Simplify.}$$

EXAMPLE 2 Multiply $(x^{2/3} - 3)(x^{2/3} + 5)$.

Solution Applying the FOIL method, we multiply as if we were multiplying two binomials:

$$(x^{2/3} - 3)(x^{2/3} + 5) = x^{2/3}x^{2/3} + 5x^{2/3} - 3x^{2/3} - 15$$
$$= x^{4/3} + 2x^{2/3} - 15$$

EXAMPLE 3   Multiply $(3a^{1/3} - 2b^{1/3})(4a^{1/3} - b^{1/3})$.

Solution   Again, we use the FOIL method to multiply:

$$(3a^{1/3} - 2b^{1/3})(4a^{1/3} - b^{1/3}) = 3a^{1/3}4a^{1/3} - 3a^{1/3}b^{1/3} - 2b^{1/3}4a^{1/3} + 2b^{1/3}b^{1/3}$$
$$= 12a^{2/3} - 11a^{1/3}b^{1/3} + 2b^{2/3}$$

EXAMPLE 4   Expand $(t^{1/2} - 5)^2$.

Solution   We can use the definition of exponents and the FOIL method:

$$(t^{1/2} - 5)^2 = (t^{1/2} - 5)(t^{1/2} - 5)$$
$$= t^{1/2}t^{1/2} - 5t^{1/2} - 5t^{1/2} + 25$$
$$= t - 10t^{1/2} + 25$$

We can obtain the same result by using the formula for the square of a binomial, $(a - b)^2 = a^2 - 2ab + b^2$.

$$(t^{1/2} - 5)^2 = (t^{1/2})^2 - 2t^{1/2} \cdot 5 + 5^2$$
$$= t - 10t^{1/2} + 25$$

EXAMPLE 5   Multiply $(x^{3/2} - 2^{3/2})(x^{3/2} + 2^{3/2})$.

Solution   This product has the form $(a - b)(a + b)$, which will result in the difference of two squares, $a^2 - b^2$:

$$(x^{3/2} - 2^{3/2})(x^{3/2} + 2^{3/2}) = (x^{3/2})^2 - (2^{3/2})^2$$
$$= x^3 - 2^3$$
$$= x^3 - 8$$

EXAMPLE 6   Multiply $(a^{1/3} - b^{1/3})(a^{2/3} + a^{1/3}b^{1/3} + b^{2/3})$.

Solution   We can find this product by multiplying in columns:

$$
\begin{array}{l}
a^{2/3} + a^{1/3}b^{1/3} + b^{2/3} \\
\underline{\qquad\qquad a^{1/3} - b^{1/3}} \\
a \quad + a^{2/3}b^{1/3} + a^{1/3}b^{2/3} \\
\underline{\qquad - a^{2/3}b^{1/3} - a^{1/3}b^{2/3} - b} \\
a \qquad\qquad\qquad\qquad\qquad - b
\end{array}
$$

The product is $a - b$.

Our next example involves division with expressions that contain rational exponents. As you will see, this kind of division is very similar to division of a polynomial by a monomial.

EXAMPLE 7 Divide $\dfrac{15x^{2/3}y^{1/3} - 20x^{4/3}y^{2/3}}{5x^{1/3}y^{1/3}}$.

Solution We can approach this problem in the same way we approached division by a monomial. We simply divide each term in the numerator by the term in the denominator:

$$\frac{15x^{2/3}y^{1/3} - 20x^{4/3}y^{2/3}}{5x^{1/3}y^{1/3}} = \frac{15x^{2/3}y^{1/3}}{5x^{1/3}y^{1/3}} - \frac{20x^{4/3}y^{2/3}}{5x^{1/3}y^{1/3}}$$

$$= 3x^{1/3} - 4xy^{1/3}$$

The next three examples involve factoring. In the first example, we are told what to factor from each term of an expression.

EXAMPLE 8 Factor $3(x - 2)^{1/3}$ from $12(x - 2)^{4/3} - 9(x - 2)^{1/3}$, and then simplify, if possible.

Solution This solution is similar to factoring out the greatest common factor:

$$12(x - 2)^{4/3} - 9(x - 2)^{1/3} = 3(x - 2)^{1/3}[4(x - 2) - 3]$$

$$= 3(x - 2)^{1/3}(4x - 11)$$

Although an expression containing rational exponents is not a polynomial—remember, a polynomial must have exponents that are whole numbers—we are going to treat the expressions that follow as if they were polynomials.

EXAMPLE 9 Factor $x^{2/3} - 3x^{1/3} - 10$ as if it were a trinomial.

Solution We can think of $x^{2/3} - 3x^{1/3} - 10$ as if it is a trinomial in which the variable is $x^{1/3}$. To see this, replace $x^{1/3}$ with y to get

$$y^2 - 3y - 10$$

Since this trinomial in y factors as $(y - 5)(y + 2)$, we can factor our original expression similarly:

$$x^{2/3} - 3x^{1/3} - 10 = (x^{1/3} - 5)(x^{1/3} + 2)$$

Remember, with factoring, we can always multiply our factors to check that we have factored correctly.

EXAMPLE 10   Factor $6x^{2/5} + 11x^{1/5} - 10$ as if it were a trinomial.

Solution   We can think of the expression in question as a trinomial in $x^{1/5}$.

$$6x^{2/5} + 11x^{1/5} - 10 = (3x^{1/5} - 2)(2x^{1/5} + 5)$$

In our next example, we combine two expressions by applying the methods we used to add and subtract fractions or rational expressions in Chapter 6.

EXAMPLE 11   Subtract $(x^2 + 4)^{1/2} - \dfrac{x^2}{(x^2 + 4)^{1/2}}$.

Solution   To combine these two expressions, we need to find a least common denominator, change to equivalent fractions, and subtract numerators. The least common denominator is $(x^2 + 4)^{1/2}$.

$$(x^2 + 4)^{1/2} - \frac{x^2}{(x^2 + 4)^{1/2}} = \frac{(x^2 + 4)^{1/2}}{1} \cdot \frac{(x^2 + 4)^{1/2}}{(x^2 + 4)^{1/2}} - \frac{x^2}{(x^2 + 4)^{1/2}}$$

$$= \frac{x^2 + 4 - x^2}{(x^2 + 4)^{1/2}}$$

$$= \frac{4}{(x^2 + 4)^{1/2}}$$

EXAMPLE 12   If you purchase an investment for P dollars and t years later it is worth A dollars, then the annual rate of return r on that investment is given by the formula

$$r = \left(\frac{A}{P} \right)^{1/t} - 1$$

Find the annual rate of return on a coin collection that was purchased for $210 and sold 3 years later for $300.

Solution   Using $A = 300$, $P = 210$, and $t = 3$ in the formula, we have

$$r = \left(\frac{300}{210} \right)^{1/3} - 1$$

The easiest way to simplify this expression is with a calculator.

$$\boxed{(}\ 300\ \boxed{\div}\ 210\ \boxed{)}\ \boxed{\wedge}\ \boxed{(}\ 1\ \boxed{\div}\ 3\ \boxed{)}\ \boxed{-}\ 1\ \boxed{=}$$

Allowing three decimal places, the result is 0.126. The annual return on the coin collection is approximately 12.6%. To do as well with a savings account, we would have to invest the original $210 in an account that paid 12.6%, compounded annually.

Getting Ready for Class

After reading through the preceding section, respond in your own words and in complete sentences.

A. When multiplying expressions with fractional exponents, when do we add the fractional exponents?

B. Is it possible to multiply two expressions with fractional exponents and end up with an expression containing only integer exponents? Support your answer with examples.

C. Write an application modeled by the equation $r = (\frac{1,000}{600})^{1/8} - 1$.

D. When can you use the FOIL method with expressions that contain rational exponents?

PROBLEM SET 9.2

Multiply. (Assume all variables in this problem set represent nonnegative real numbers.)

1. $x^{2/3}(x^{1/3} + x^{4/3})$ **2.** $x^{2/5}(x^{3/5} - x^{8/5})$

3. $a^{1/2}(a^{3/2} - a^{1/2})$ **4.** $a^{1/4}(a^{3/4} + a^{7/4})$

5. $2x^{1/3}(3x^{8/3} - 4x^{5/3} + 5x^{2/3})$

6. $5x^{1/2}(4x^{5/2} + 3x^{3/2} + 2x^{1/2})$

7. $4x^{1/2}y^{3/5}(3x^{3/2}y^{-3/5} - 9x^{-1/2}y^{7/5})$

8. $3x^{4/5}y^{1/3}(4x^{6/5}y^{-1/3} - 12x^{-4/5}y^{5/3})$

9. $(x^{2/3} - 4)(x^{2/3} + 2)$ **10.** $(x^{2/3} - 5)(x^{2/3} + 2)$

11. $(a^{1/2} - 3)(a^{1/2} - 7)$ **12.** $(a^{1/2} - 6)(a^{1/2} - 2)$

13. $(4y^{1/3} - 3)(5y^{1/3} + 2)$

14. $(5y^{1/3} - 2)(4y^{1/3} + 3)$

15. $(5x^{2/3} + 3y^{1/2})(2x^{2/3} + 3y^{1/2})$

16. $(4x^{2/3} - 2y^{1/2})(5x^{2/3} - 3y^{1/2})$

17. $(t^{1/2} + 5)^2$ **18.** $(t^{1/2} - 3)^2$

19. $(x^{3/2} + 4)^2$ **20.** $(x^{3/2} - 6)^2$

21. $(a^{1/2} - b^{1/2})^2$ **22.** $(a^{1/2} + b^{1/2})^2$

23. $(2x^{1/2} - 3y^{1/2})^2$ **24.** $(5x^{1/2} + 4y^{1/2})^2$

25. $(a^{1/2} - 3^{1/2})(a^{1/2} + 3^{1/2})$

26. $(a^{1/2} - 5^{1/2})(a^{1/2} + 5^{1/2})$

27. $(x^{3/2} + y^{3/2})(x^{3/2} - y^{3/2})$

28. $(x^{5/2} + y^{5/2})(x^{5/2} - y^{5/2})$

29. $(t^{1/2} - 2^{3/2})(t^{1/2} + 2^{3/2})$

30. $(t^{1/2} - 5^{3/2})(t^{1/2} + 5^{3/2})$

31. $(2x^{3/2} + 3^{1/2})(2x^{3/2} - 3^{1/2})$

32. $(3x^{1/2} + 2^{3/2})(3x^{1/2} - 2^{3/2})$

33. $(x^{1/3} + y^{1/3})(x^{2/3} - x^{1/3}y^{1/3} + y^{2/3})$

34. $(x^{1/3} - y^{1/3})(x^{2/3} + x^{1/3}y^{1/3} + y^{2/3})$

35. $(a^{1/3} - 2)(a^{2/3} + 2a^{1/3} + 4)$

36. $(a^{1/3} + 3)(a^{2/3} - 3a^{1/3} + 9)$

37. $(2x^{1/3} + 1)(4x^{2/3} - 2x^{1/3} + 1)$

38. $(3x^{1/3} - 1)(9x^{2/3} + 3x^{1/3} + 1)$

39. $(t^{1/4} - 1)(t^{1/4} + 1)(t^{1/2} + 1)$

40. $(t^{1/4} - 2)(t^{1/4} + 2)(t^{1/2} + 4)$

Divide.

41. $\dfrac{18x^{3/4} + 27x^{1/4}}{9x^{1/4}}$ **42.** $\dfrac{25x^{1/4} + 30x^{3/4}}{5x^{1/4}}$

43. $\dfrac{12x^{2/3}y^{1/3} - 16x^{1/3}y^{2/3}}{4x^{1/3}y^{1/3}}$

44. $\dfrac{12x^{4/3}y^{1/3} - 18x^{1/3}y^{4/3}}{6x^{1/3}y^{1/3}}$

45. $\dfrac{21a^{7/5}b^{3/5} - 14a^{2/5}b^{8/5}}{7a^{2/5}b^{3/5}}$

46. $\dfrac{24a^{9/5}b^{3/5} - 16a^{4/5}b^{8/5}}{8a^{4/5}b^{3/5}}$

47. Factor $3(x - 2)^{1/2}$ from
$12(x - 2)^{3/2} - 9(x - 2)^{1/2}$.

48. Factor $4(x + 1)^{1/3}$ from
$4(x + 1)^{4/3} + 8(x + 1)^{1/3}$.

49. Factor $5(x - 3)^{7/5}$ from
$5(x - 3)^{12/5} - 15(x - 3)^{7/5}$.

50. Factor $6(x + 3)^{8/7}$ from
$6(x + 3)^{15/7} - 12(x + 3)^{8/7}$.

51. Factor $3(x + 1)^{1/2}$ from
$9x(x + 1)^{3/2} + 6(x + 1)^{1/2}$.

52. Factor $4x(x + 1)^{1/2}$ from
$4x^2(x + 1)^{1/2} + 8x(x + 1)^{3/2}$.

Factor each of the following as if it were a trinomial.

53. $x^{2/3} - 5x^{1/3} + 6$ **54.** $x^{2/3} - x^{1/3} - 6$

55. $a^{2/5} - 2a^{1/5} - 8$ **56.** $a^{2/5} + 2a^{1/5} - 8$

57. $2y^{2/3} - 5y^{1/3} - 3$ **58.** $3y^{2/3} + 5y^{1/3} - 2$

59. $9t^{2/5} - 25$ **60.** $16t^{2/5} - 49$

61. $4x^{2/7} + 20x^{1/7} + 25$ **62.** $25x^{2/7} - 20x^{1/7} + 4$

Simplify each of the following to a single fraction.

63. $\dfrac{3}{x^{1/2}} + x^{1/2}$ **64.** $\dfrac{2}{x^{1/2}} - x^{1/2}$

65. $x^{2/3} + \dfrac{5}{x^{1/3}}$ **66.** $x^{3/4} - \dfrac{7}{x^{1/4}}$

67. $\dfrac{3x^2}{(x^3 + 1)^{1/2}} + (x^3 + 1)^{1/2}$

68. $\dfrac{x^3}{(x^2 - 1)^{1/2}} + 2x(x^2 - 1)^{1/2}$

69. $\dfrac{x^2}{(x^2 + 4)^{1/2}} - (x^2 + 4)^{1/2}$

70. $\dfrac{x^5}{(x^2 - 2)^{1/2}} + 4x^3(x^2 - 2)^{1/2}$

Use a calculator to find approximations to each of the following. Round your answers for Problems 75 and 76 to three places past the decimal point.

71. $16^{0.25}$ **72.** $81^{0.25}$

73. $9^{1.5}$ **74.** $32^{0.4}$

75. $\left(\dfrac{1}{2}\right)^{1/5}$ **76.** $\left(\dfrac{1}{2}\right)^{1/10}$

Applying the Concepts

77. Investing   A coin collection is purchased as an investment for $500 and sold 4 years later for $900. Find the annual rate of return on the investment.

78. Investing   An investor buys stock in a company for $800. Five years later, the same stock is worth $1,600. Find the annual rate of return on the stocks.

79. Investing   Find the annual rate of return on a home that is purchased for $60,000 and is sold 5 years later for $80,000.

80. Investing   Find the annual rate of return on a home that is purchased for $75,000 and is sold 10 years later for $150,000.

Review Problems

The problems that follow review material we covered in Sections 8.1 and 8.2.

81. Find the slope of the line that contains $(-4, -1)$ and $(-2, 5)$.

82. A line has a slope of $\frac{2}{3}$. Find the slope of any line
(a) Parallel to it.
(b) Perpendicular to it.

83. Give the slope and y-intercept of the line $2x - 3y = 6$.

84. Give the equation of the line with slope -3 and y-intercept 5.

85. Find the equation of the line with slope $\frac{2}{3}$ that contains the point $(-6, 2)$.

86. Find the equation of the line through $(1, 3)$ and $(-1, -5)$.

Earlier in this chapter we showed how the Pythagorean theorem can be used to construct a golden rectangle. In a similar manner, the Pythagorean theorem can be used to construct the attractive spiral shown here.

The Spiral of Roots

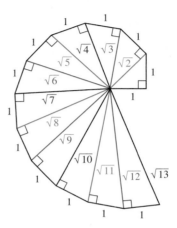

This spiral is called the Spiral of Roots because each of the diagonals is the positive square root of one of the positive integers. At the end of this section, we will use the Pythagorean theorem and some of the material in this section to construct this spiral.

In this section we will use radical notation instead of rational exponents. We will begin by stating two properties of radicals. Following this, we will give a definition for simplified form for radical expressions. The examples in this section show how we use the properties of radicals to write radical expressions in simplified form.

Here are our first two properties of radicals. For these two properties, we will assume a and b are nonnegative real numbers whenever n is an even number.

PROPERTY 1 FOR RADICALS

$$\sqrt[n]{ab} = \sqrt[n]{a}\,\sqrt[n]{b}$$

In words: The nth root of a product is the product of the nth roots.

Proof of Property 1

$\sqrt[n]{ab} = (ab)^{1/n}$	Definition of fractional exponents
$= a^{1/n}b^{1/n}$	Exponents distribute over products.
$= \sqrt[n]{a}\,\sqrt[n]{b}$	Definition of fractional exponents

Note: There is no property for radicals that says the nth root of a sum is the sum of the nth roots. That is,

$$\sqrt[n]{a + b} \neq \sqrt[n]{a} + \sqrt[n]{b}$$

PROPERTY 2 FOR RADICALS

$$\sqrt[n]{\frac{a}{b}} = \frac{\sqrt[n]{a}}{\sqrt[n]{b}} \qquad (b \neq 0)$$

In words: The nth root of a quotient is the quotient of the nth roots.

The proof of Property 2 is similar to the proof of Property 1.

These two properties of radicals allow us to change the form of and simplify radical expressions without changing their value.

SIMPLIFIED FORM FOR RADICAL EXPRESSIONS

A radical expression is in *simplified form* if

1. None of the factors of the radicand (the quantity under the radical sign) can be written as powers greater than or equal to the index—that is, no perfect squares can be factors of the quantity under a square root sign, no perfect cubes can be factors of what is under a cube root sign, and so forth;
2. There are no fractions under the radical sign; and
3. There are no radicals in the denominator.

Satisfying the first condition for simplified form actually amounts to taking as much out from under the radical sign as possible. The following examples illustrate the first condition for simplified form.

EXAMPLE 1   Write $\sqrt{50}$ in simplified form.

Solution   The largest perfect square that divides 50 is 25. We write 50 as $25 \cdot 2$ and apply Property 1 for radicals:

$$\begin{aligned}
\sqrt{50} &= \sqrt{25 \cdot 2} & 50 &= 25 \cdot 2 \\
&= \sqrt{25}\,\sqrt{2} & &\text{Property 1} \\
&= 5\sqrt{2} & \sqrt{25} &= 5
\end{aligned}$$

We have taken as much as possible out from under the radical sign—in this case, factoring 25 from 50 and then writing $\sqrt{25}$ as 5.

EXAMPLE 2   Write in simplified form: $\sqrt{48x^4y^3}$, where $x, y \geq 0$

Solution   The largest perfect square that is a factor of the radicand is $16x^4y^2$. Applying Property 1 again, we have

$$\sqrt{48x^4y^3} = \sqrt{16x^4y^2 \cdot 3y}$$
$$= \sqrt{16x^4y^2}\,\sqrt{3y}$$
$$= 4x^2y\sqrt{3y}$$

EXAMPLE 3   Write $\sqrt[3]{40a^5b^4}$ in simplified form.

Solution   We now want to factor the largest perfect cube from the radicand. We write $40a^5b^4$ as $8a^3b^3 \cdot 5a^2b$ and proceed as we did in Examples 1 and 2.

$$\sqrt[3]{40a^5b^4} = \sqrt[3]{8a^3b^3 \cdot 5a^2b}$$
$$= \sqrt[3]{8a^3b^3}\,\sqrt[3]{5a^2b}$$
$$= 2ab\sqrt[3]{5a^2b}$$

Here are some further examples concerning the first condition for simplified form.

EXAMPLES   Write each expression in simplified form.

4. $\sqrt{12x^7y^6} = \sqrt{4x^6y^6 \cdot 3x}$
$$\phantom{\sqrt{12x^7y^6}} = \sqrt{4x^6y^6}\,\sqrt{3x}$$
$$\phantom{\sqrt{12x^7y^6}} = 2x^3y^3\sqrt{3x}$$

5. $\sqrt[3]{54a^6b^2c^4} = \sqrt[3]{27a^6c^3 \cdot 2b^2c}$
$$\phantom{\sqrt[3]{54a^6b^2c^4}} = \sqrt[3]{27a^6c^3}\,\sqrt[3]{2b^2c}$$
$$\phantom{\sqrt[3]{54a^6b^2c^4}} = 3a^2c\sqrt[3]{2b^2c}$$

The second property of radicals is used to simplify a radical that contains a fraction.

EXAMPLE 6   Simplify $\sqrt{\frac{3}{4}}$.

Solution   Applying Property 2 for radicals, we have

$$\sqrt{\frac{3}{4}} = \frac{\sqrt{3}}{\sqrt{4}} \qquad \text{Property 2}$$

$$= \frac{\sqrt{3}}{2} \qquad \sqrt{4} = 2$$

The last expression is in simplified form because it satisfies all three conditions for simplified form.

E X A M P L E   7   Write $\sqrt{\frac{5}{6}}$ in simplified form.

Solution   Proceeding as in Example 6, we have

$$\sqrt{\frac{5}{6}} = \frac{\sqrt{5}}{\sqrt{6}}$$

The resulting expression satisfies the second condition for simplified form since neither radical contains a fraction. It does, however, violate Condition 3 since it has a radical in the denominator. Getting rid of the radical in the denominator is called *rationalizing the denominator* and is accomplished, in this case, by multiplying the numerator and denominator by $\sqrt{6}$:

$$\frac{\sqrt{5}}{\sqrt{6}} = \frac{\sqrt{5}}{\sqrt{6}} \cdot \frac{\sqrt{6}}{\sqrt{6}}$$

$$= \frac{\sqrt{30}}{\sqrt{6^2}}$$

$$= \frac{\sqrt{30}}{6}$$

E X A M P L E S   Rationalize the denominator.

8. $\dfrac{4}{\sqrt{3}} = \dfrac{4}{\sqrt{3}} \cdot \dfrac{\sqrt{3}}{\sqrt{3}}$

$\qquad = \dfrac{4\sqrt{3}}{\sqrt{3^2}}$

$\qquad = \dfrac{4\sqrt{3}}{3}$

9. $\dfrac{2\sqrt{3x}}{\sqrt{5y}} = \dfrac{2\sqrt{3x}}{\sqrt{5y}} \cdot \dfrac{\sqrt{5y}}{\sqrt{5y}}$

$\qquad = \dfrac{2\sqrt{15xy}}{\sqrt{(5y)^2}}$

$\qquad = \dfrac{2\sqrt{15xy}}{5y}$

When the denominator involves a cube root, we must multiply by a radical that will produce a perfect cube under the cube root sign in the denominator, as our next example illustrates.

E X A M P L E   1 0   Rationalize the denominator in $\dfrac{7}{\sqrt[3]{4}}$.

Solution   Since $4 = 2^2$, we can multiply both numerator and denominator by $\sqrt[3]{2}$ and obtain $\sqrt[3]{2^3}$ in the denominator.

$$\frac{7}{\sqrt[3]{4}} = \frac{7}{\sqrt[3]{2^2}}$$

$$= \frac{7}{\sqrt[3]{2^2}} \cdot \frac{\sqrt[3]{2}}{\sqrt[3]{2}}$$

$$= \frac{7\sqrt[3]{2}}{\sqrt[3]{2^3}}$$

$$= \frac{7\sqrt[3]{2}}{2}$$

EXAMPLE 11   Simplify $\sqrt{\dfrac{12x^5y^3}{5z}}$.

Solution   We use Property 2 to write the numerator and denominator as two separate radicals:

$$\sqrt{\frac{12x^5y^3}{5z}} = \frac{\sqrt{12x^5y^3}}{\sqrt{5z}}$$

Simplifying the numerator, we have

$$\frac{\sqrt{12x^5y^3}}{\sqrt{5z}} = \frac{\sqrt{4x^4y^2}\,\sqrt{3xy}}{\sqrt{5z}}$$

$$= \frac{2x^2y\sqrt{3xy}}{\sqrt{5z}}$$

To rationalize the denominator, we multiply the numerator and denominator by $\sqrt{5z}$:

$$\frac{2x^2y\sqrt{3xy}}{\sqrt{5z}} \cdot \frac{\sqrt{5z}}{\sqrt{5z}} = \frac{2x^2y\sqrt{15xyz}}{\sqrt{(5z)^2}}$$

$$= \frac{2x^2y\sqrt{15xyz}}{5z}$$

THE SQUARE ROOT OF A PERFECT SQUARE

So far in this chapter, we have assumed that all our variables are nonnegative when they appear under a square root symbol. There are times, however, when this is not the case.

Consider the following two statements:

$$\sqrt{3^2} = \sqrt{9} = 3 \qquad \text{and} \qquad \sqrt{(-3)^2} = \sqrt{9} = 3$$

Whether we operate on 3 or -3, the result is the same: Both expressions simplify to 3. The other operation we have worked with in the past that produces the same result is absolute value. That is,

$$|3| = 3 \quad \text{and} \quad |-3| = 3$$

This leads us to the next property of radicals.

PROPERTY 3 FOR RADICALS

If a is a real number, then $\sqrt{a^2} = |a|$.

The result of this discussion and Property 3 is simply this:

If we know a is positive, then $\sqrt{a^2} = a$.

If we know a is negative, then $\sqrt{a^2} = |a|$.

If we don't know if a is positive or negative, then $\sqrt{a^2} = |a|$.

EXAMPLES   Simplify each expression. Do *not* assume the variables represent positive numbers.

12. $\sqrt{9x^2} = 3|x|$

13. $\sqrt{x^3} = |x|\sqrt{x}$

14. $\sqrt{x^2 - 6x + 9} = \sqrt{(x - 3)^2} = |x - 3|$

15. $\sqrt{x^3 - 5x^2} = \sqrt{x^2(x - 5)} = |x|\sqrt{x - 5}$

As you can see, we must use absolute value symbols when we take a square root of a perfect square, unless we know the base of the perfect square is a positive number. The same idea holds for higher even roots, but not for odd roots. With odd roots, no absolute value symbols are necessary.

EXAMPLES   Simplify each expression.

16. $\sqrt[3]{(-2)^3} = \sqrt[3]{-8} = -2$

17. $\sqrt[3]{(-5)^3} = \sqrt[3]{-125} = -5$

We can extend this discussion to all roots as follows:

EXTENDING PROPERTY 3 FOR RADICALS

If a is a real number, then

$$\sqrt[n]{a^n} = |a| \quad \text{if} \quad n \text{ is even}$$

$$\sqrt[n]{a^n} = a \quad \text{if} \quad n \text{ is odd}$$

THE SPIRAL OF ROOTS

In order to visualize the square roots of the positive integers, we can construct the spiral of roots that we mentioned in the introduction to this section. To begin, we draw two line segments, each of length 1, at right angles to each other. Then we use the Pythagorean theorem to find the length of the diagonal. Figure 1 illustrates this procedure.

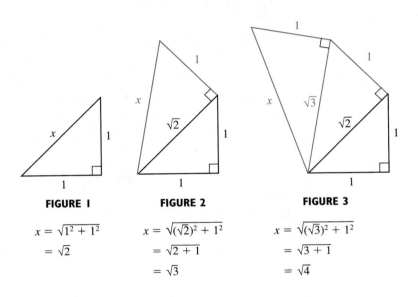

FIGURE 1

$x = \sqrt{1^2 + 1^2}$

$\quad = \sqrt{2}$

FIGURE 2

$x = \sqrt{(\sqrt{2})^2 + 1^2}$

$\quad = \sqrt{2 + 1}$

$\quad = \sqrt{3}$

FIGURE 3

$x = \sqrt{(\sqrt{3})^2 + 1^2}$

$\quad = \sqrt{3 + 1}$

$\quad = \sqrt{4}$

Next, we construct a second triangle by connecting a line segment of length 1 to the end of the first diagonal so that the angle formed is a right angle. We find the length of the second diagonal using the Pythagorean theorem. Figure 2 illustrates this procedure. Continuing to draw new triangles by connecting line segments of length 1 to the end of each new diagonal, so that the angle formed is a right angle, the spiral of roots begins to appear (Figure 3).

THE SPIRAL OF ROOTS AND FUNCTION NOTATION

Looking over the diagrams and calculations in the preceding discussion, we see that each diagonal in the spiral of roots is found by using the length of the previous diagonal.

First diagonal: $\quad \sqrt{1^2 + 1^2} = \sqrt{2}$

Second diagonal: $\quad \sqrt{(\sqrt{2})^2 + 1^2} = \sqrt{3}$

Third diagonal: $\quad \sqrt{(\sqrt{3})^2 + 1^2} = \sqrt{4}$

Fourth diagonal: $\quad \sqrt{(\sqrt{4})^2 + 1^2} = \sqrt{5}$

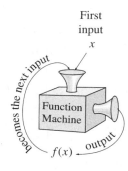

First
input
x

A process like this one, in which the answer to one calculation is used to find the answer to the next calculation, is called a *recursive* process. In this particular case, we can use function notation to model the process. If we let x represent the length of any diagonal, then the length of the next diagonal is given by

$$f(x) = \sqrt{x^2 + 1}$$

To begin the process of finding the diagonals, we let $x = 1$:

$$f(1) = \sqrt{1^2 + 1} = \sqrt{2}$$

To find the next diagonal, we substitute $\sqrt{2}$ for x to obtain

$$f[f(1)] = f(\sqrt{2}) = \sqrt{(\sqrt{2})^2 + 1} = \sqrt{3}$$
$$f(f[f(1)]) = f(\sqrt{3}) = \sqrt{(\sqrt{3})^2 + 1} = \sqrt{4}$$

We can describe this process of finding the diagonals of the spiral of roots very concisely this way:

$$f(1), f[f(1)], f(f[f(1)]), \ldots \qquad \text{where } f(x) = \sqrt{x^2 + 1}$$

This sequence of function values is a special case of a general category of similar sequences that are closely connected to *fractals* and *chaos*, two topics in mathematics that are currently receiving a good deal of attention.

Using TECHNOLOGY

As our preceding discussion indicates, the length of each diagonal in the spiral of roots is used to calculate the length of the next diagonal. The $\boxed{\text{ANS}}$ key on a graphing calculator can be used very effectively in a situation like this. To begin, we store the number 1 in the variable ANS. Next, we key in the formula used to produce each diagonal using ANS for the variable. After that, it is simply a matter of pressing $\boxed{\text{ENTER}}$, as many times as we like, to produce the lengths of as many diagonals as we like. Here is a summary of what we do:

Enter This	Display Shows
1 $\boxed{\text{ENTER}}$	1.000
$\sqrt{\ }$ (ANS^2 + 1) $\boxed{\text{ENTER}}$	1.414
$\boxed{\text{ENTER}}$	1.732
$\boxed{\text{ENTER}}$	2.000
$\boxed{\text{ENTER}}$	2.236

If you continue to press the $\boxed{\text{ENTER}}$ key, you will produce decimal approximations for as many of the diagonals in the spiral of roots as you like.

Getting Ready for Class

After reading through the preceding section, respond in your own words and in complete sentences.

A. Explain why this statement is false: "The square root of a sum is the sum of the square roots."

B. What is simplified form for an expression that contains a square root?

C. Why is it not necessarily true that $\sqrt{a^2} = a$?

D. What does it mean to rationalize the denominator in an expression?

PROBLEM SET 9.3

Use Property 1 for radicals to write each of the following expressions in simplified form. (Assume all variables are nonnegative through Problem 70.)

1. $\sqrt{8}$ **2.** $\sqrt{32}$ **3.** $\sqrt{98}$

4. $\sqrt{75}$ **5.** $\sqrt{288}$ **6.** $\sqrt{128}$

7. $\sqrt{80}$ **8.** $\sqrt{200}$ **9.** $\sqrt{48}$

10. $\sqrt{27}$ **11.** $\sqrt{675}$ **12.** $\sqrt{972}$

13. $\sqrt[3]{54}$ **14.** $\sqrt[3]{24}$ **15.** $\sqrt[3]{128}$

16. $\sqrt[3]{162}$ **17.** $\sqrt[3]{432}$ **18.** $\sqrt[3]{1{,}536}$

19. $\sqrt[5]{64}$ **20.** $\sqrt[4]{48}$ **21.** $\sqrt{18x^3}$

22. $\sqrt{27x^5}$ **23.** $\sqrt[4]{32y^7}$ **24.** $\sqrt[5]{32y^7}$

25. $\sqrt[3]{40x^4y^7}$ **26.** $\sqrt[3]{128x^6y^2}$

27. $\sqrt{48a^2b^3c^4}$ **28.** $\sqrt{72a^4b^3c^2}$

29. $\sqrt[3]{48a^2b^3c^4}$ **30.** $\sqrt[3]{72a^4b^3c^2}$

31. $\sqrt[5]{64x^8y^{12}}$ **32.** $\sqrt[4]{32x^9y^{10}}$

33. $\sqrt[5]{243x^7y^{10}z^5}$ **34.** $\sqrt[4]{64x^8y^4z^{11}}$

Substitute the given numbers into the expression $\sqrt{b^2 - 4ac}$, and then simplify.

35. $a = 2, b = -6, c = 3$

36. $a = 6, b = 7, c = -5$

37. $a = 1, b = 2, c = 6$

38. $a = 2, b = 5, c = 3$

39. $a = \dfrac{1}{2}, b = -\dfrac{1}{2}, c = -\dfrac{5}{4}$

40. $a = \dfrac{7}{4}, b = -\dfrac{3}{4}, c = -2$

Rationalize the denominator in each of the following expressions.

41. $\dfrac{2}{\sqrt{3}}$ **42.** $\dfrac{3}{\sqrt{2}}$ **43.** $\dfrac{5}{\sqrt{6}}$

44. $\dfrac{7}{\sqrt{5}}$ **45.** $\sqrt{\dfrac{1}{2}}$ **46.** $\sqrt{\dfrac{1}{3}}$

47. $\sqrt{\dfrac{1}{5}}$ **48.** $\sqrt{\dfrac{1}{6}}$ **49.** $\dfrac{4}{\sqrt[3]{2}}$

50. $\dfrac{5}{\sqrt[3]{3}}$ **51.** $\dfrac{2}{\sqrt[3]{9}}$ **52.** $\dfrac{3}{\sqrt[3]{4}}$

53. $\sqrt[4]{\dfrac{3}{2x^2}}$ **54.** $\sqrt[4]{\dfrac{5}{3x^2}}$ **55.** $\sqrt[4]{\dfrac{8}{y}}$

56. $\sqrt[4]{\dfrac{27}{y}}$ **57.** $\sqrt[3]{\dfrac{4x}{3y}}$ **58.** $\sqrt[3]{\dfrac{7x}{6y}}$

59. $\sqrt[3]{\dfrac{2x}{9y}}$ **60.** $\sqrt[3]{\dfrac{5x}{4y}}$ **61.** $\sqrt[4]{\dfrac{1}{8x^3}}$

62. $\sqrt[4]{\dfrac{8}{9x^3}}$

Write each of the following in simplified form.

63. $\sqrt{\dfrac{27x^3}{5y}}$ **64.** $\sqrt{\dfrac{12x^5}{7y}}$ **65.** $\sqrt{\dfrac{75x^3y^2}{2z}}$

66. $\sqrt{\dfrac{50x^2y^3}{3z}}$ **67.** $\sqrt[3]{\dfrac{16a^4b^3}{9c}}$ **68.** $\sqrt[3]{\dfrac{54a^5b^4}{25c^2}}$

69. $\sqrt[3]{\dfrac{8x^3y^6}{9z}}$ **70.** $\sqrt[3]{\dfrac{27x^6y^3}{2z^2}}$

Simplify each expression. Do *not* assume the variables represent positive numbers.

71. $\sqrt{25x^2}$

72. $\sqrt{49x^2}$

73. $\sqrt{27x^3y^2}$

74. $\sqrt{40x^3y^2}$

75. $\sqrt{x^2 - 10x + 25}$

76. $\sqrt{x^2 - 16x + 64}$

77. $\sqrt{4x^2 + 12x + 9}$

78. $\sqrt{16x^2 + 40x + 25}$

79. $\sqrt{4a^4 + 16a^3 + 16a^2}$

80. $\sqrt{9a^4 + 18a^3 + 9a^2}$

81. $\sqrt{4x^3 - 8x^2}$

82. $\sqrt{18x^3 - 9x^2}$

83. Show that the statement $\sqrt{a + b} = \sqrt{a} + \sqrt{b}$ is not true by replacing a with 9 and b with 16 and simplifying both sides.

84. Find a pair of values for a and b that will make the statement $\sqrt{a + b} = \sqrt{a} + \sqrt{b}$ true.

Applying the Concepts

85. Diagonal Distance   The distance d between opposite corners of a rectangular room with length l and width w is given by

$$d = \sqrt{l^2 + w^2}$$

How far is it between opposite corners of a living room that measures 10 by 15 feet?

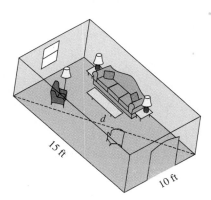

86. Radius of a Sphere   The radius r of a sphere with volume V can be found by using the formula

$$r = \sqrt[3]{\frac{3V}{4\pi}}$$

Find the radius of a sphere with volume 9 cubic feet. Write your answer in simplified form. (Use $\frac{22}{7}$ for π.)

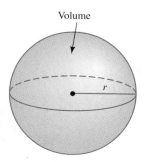

Volume

87. Spiral of Roots   Construct your own spiral of roots by using a ruler. Draw the first triangle by using two 1-inch lines. The first diagonal will have a length of $\sqrt{2}$ inches. Each new triangle will be formed by drawing a 1-inch line segment at the end of the previous diagonal so that the angle formed is 90°.

88. Spiral of Roots   Construct a spiral of roots by using line segments of length 2 inches. The length of the first diagonal will be $2\sqrt{2}$ inches. The length of the second diagonal will be $2\sqrt{3}$ inches.

89. Spiral of Roots   If $f(x) = \sqrt{x^2 + 1}$, find the first six terms in the following sequence. Use your results to predict the value of the 10th term and the 100th term.

$$f(1), f[f(1)], f(f[f(1)]), \ldots$$

90. Spiral of Roots   If $f(x) = \sqrt{x^2 + 4}$, find the first six terms in the following sequence. Use your results to predict the value of the 10th term and the 100th term. (The numbers in this sequence are the lengths of the diagonals of the spiral you drew in Problem 88.)

$$f(2), f[f(2)], f(f[f(2)]), \ldots$$

Review Problems

The following problems review material we covered in Sections 8.5 and 8.6.

Let $f(x) = \frac{1}{2}x + 3$ and $g(x) = x^2 - 4$, and find

91. $f(0)$

92. $g(0)$

93. $g(2)$

94. $f(2)$

95. $f(-4)$

96. $g(-6)$

97. $f[g(2)]$ **98.** $g[f(2)]$
99. $(f + g)(4)$ **100.** $(fg)(x)$

Extending the Concepts

Assume a is a positive number, and rationalize each denominator.

101. $\dfrac{1}{\sqrt[10]{a^3}}$ **102.** $\dfrac{1}{\sqrt[12]{a^7}}$

103. $\dfrac{1}{\sqrt[20]{a^{11}}}$ **104.** $\dfrac{1}{\sqrt[15]{a^{13}}}$

105. Show that the two expressions $\sqrt{x^2 + 1}$ and $x + 1$ are not, in general, equal to each other by graphing $y = \sqrt{x^2 + 1}$ and $y = x + 1$ in the same viewing window.

106. Show that the two expressions $\sqrt{x^2 + 9}$ and $x + 3$ are not, in general, equal to each other by graphing $y = \sqrt{x^2 + 9}$ and $y = x + 3$ in the same viewing window.

107. Approximately how far apart are the graphs in Problem 105 when $x = 2$?

108. Approximately how far apart are the graphs in Problem 106 when $x = 2$?

109. For what value of x are the expressions $\sqrt{x^2 + 1}$ and $x + 1$ equal?

110. For what value of x are the expressions $\sqrt{x^2 + 9}$ and $x + 3$ equal?

111. Hero's Formula Hero's formula for the area of a triangle is

$$A = \sqrt{s(s - a)(s - b)(s - c)}$$

in which a, b, and c are the lengths of the sides of the triangle and s is one-half the perimeter of the triangle.

(a) Write a formula to find s in terms of a, b, and c.

(b) Use the result from part (a) along with Hero's formula to find the area of a triangle that has sides of 5, 6, and 7.

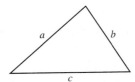

112. Brahmagupta's Formula A cyclic quadrilateral is a quadrilateral that has each of its vertices on a circle. The area of a cyclic quadrilateral can be found with Brahmagupta's formula

$$A = \sqrt{(s - a)(s - b)(s - c)(s - d)}$$

where a, b, c, and d are the lengths of the sides of the cyclic quadrilateral and s is one-half the perimeter of the quadrilateral.

(a) Write a formula for s in terms of a, b, c, and d.

(b) Find the area of a cyclic quadrilateral with sides of 4, 6, 9, and 3.

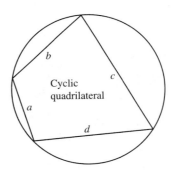

In Chapter 4 we found that we could add similar terms when combining polynomials. The same idea applies to addition and subtraction of radical expressions.

> **DEFINITION**   Two radicals are said to be **similar radicals** if they have the same index and the same radicand.

The expressions $5\sqrt[3]{7}$ and $-8\sqrt[3]{7}$ are similar since the index is 3 in both cases and the radicands are 7. The expressions $3\sqrt[4]{5}$ and $7\sqrt[3]{5}$ are not similar because they have different indices, and the expressions $2\sqrt[5]{8}$ and $3\sqrt[5]{9}$ are not similar because the radicands are not the same.

We add and subtract radical expressions in the same way we add and subtract polynomials—by combining similar terms under the distributive property.

EXAMPLE 1   Combine $5\sqrt{3} - 4\sqrt{3} + 6\sqrt{3}$.

Solution   All three radicals are similar. We apply the distributive property to get

$$5\sqrt{3} - 4\sqrt{3} + 6\sqrt{3} = (5 - 4 + 6)\sqrt{3}$$
$$= 7\sqrt{3}$$

EXAMPLE 2   Combine $3\sqrt{8} + 5\sqrt{18}$.

Solution   The two radicals do not seem to be similar. We must write each in simplified form before applying the distributive property.

$$3\sqrt{8} + 5\sqrt{18} = 3\sqrt{4 \cdot 2} + 5\sqrt{9 \cdot 2}$$
$$= 3\sqrt{4}\,\sqrt{2} + 5\sqrt{9}\,\sqrt{2}$$
$$= 3 \cdot 2\sqrt{2} + 5 \cdot 3\sqrt{2}$$
$$= 6\sqrt{2} + 15\sqrt{2}$$
$$= (6 + 15)\sqrt{2}$$
$$= 21\sqrt{2}$$

The result of Example 2 can be generalized to the following rule for sums and differences of radical expressions.

RULE

To add or subtract radical expressions, put each in simplified form, and apply the distributive property if possible. We can add only similar radicals. We must write each expression in simplified form for radicals before we can tell if the radicals are similar.

EXAMPLE 3   Combine $7\sqrt{75xy^3} - 4y\sqrt{12xy}$, where $x, y \geq 0$.

Solution   We write each expression in simplified form and combine similar radicals:

$$7\sqrt{75xy^3} - 4y\sqrt{12xy} = 7\sqrt{25y^2}\,\sqrt{3xy} - 4y\sqrt{4}\,\sqrt{3xy}$$
$$= 35y\sqrt{3xy} - 8y\sqrt{3xy}$$
$$= (35y - 8y)\sqrt{3xy}$$
$$= 27y\sqrt{3xy}$$

EXAMPLE 4   Combine $10\sqrt[3]{8a^4b^2} + 11a\sqrt[3]{27ab^2}$.

Solution   Writing each radical in simplified form and combining similar terms, we have

$$10\sqrt[3]{8a^4b^2} + 11a\sqrt[3]{27ab^2} = 10\sqrt[3]{8a^3}\,\sqrt[3]{ab^2} + 11a\sqrt[3]{27}\,\sqrt[3]{ab^2}$$
$$= 20a\sqrt[3]{ab^2} + 33a\sqrt[3]{ab^2}$$
$$= 53a\sqrt[3]{ab^2}$$

EXAMPLE 5   Combine $\dfrac{\sqrt{3}}{2} + \dfrac{1}{\sqrt{3}}$.

Solution   We begin by writing the second term in simplified form.

$$\frac{\sqrt{3}}{2} + \frac{1}{\sqrt{3}} = \frac{\sqrt{3}}{2} + \frac{1}{\sqrt{3}} \cdot \frac{\sqrt{3}}{\sqrt{3}}$$
$$= \frac{\sqrt{3}}{2} + \frac{\sqrt{3}}{3}$$
$$= \frac{1}{2}\sqrt{3} + \frac{1}{3}\sqrt{3}$$
$$= \left(\frac{1}{2} + \frac{1}{3}\right)\sqrt{3}$$
$$= \frac{5}{6}\sqrt{3} = \frac{5\sqrt{3}}{6}$$

E X A M P L E   6   Construct a golden rectangle from a square of side 4. Then show that the ratio of the length to the width is the golden ratio $\dfrac{1 + \sqrt{5}}{2}$.

Solution   Figure 1 shows the golden rectangle constructed from a square of side 4.

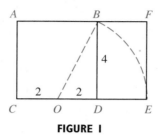

FIGURE 1

The length of the diagonal OB is found from the Pythagorean theorem.

$$OB = \sqrt{2^2 + 4^2} = \sqrt{4 + 16} = \sqrt{20} = 2\sqrt{5}$$

The ratio of the length to the width for the rectangle is the golden ratio.

$$\text{Golden ratio} = \frac{CE}{EF} = \frac{2 + 2\sqrt{5}}{4} = \frac{2(1 + \sqrt{5})}{2 \cdot 2} = \frac{1 + \sqrt{5}}{2}$$

As you can see, showing that the ratio of length to width in this rectangle is the golden ratio depends on our ability to write $\sqrt{20}$ as $2\sqrt{5}$ and our ability to reduce to lowest terms by factoring and then dividing out the common factor 2 from the numerator and denominator.

Getting Ready for Class

After reading through the preceding section, respond in your own words and in complete sentences.

A. What are similar radicals?

B. When can we add two radical expressions?

C. What is the first step when adding or subtracting expressions containing radicals?

D. What is the golden ratio, and where does it come from?

PROBLEM SET 9.4

Combine the following expressions. (Assume any variables under an even root are nonnegative.)

1. $3\sqrt{5} + 4\sqrt{5}$

2. $6\sqrt{3} - 5\sqrt{3}$

3. $3x\sqrt{7} - 4x\sqrt{7}$

4. $6y\sqrt{a} + 7y\sqrt{a}$

5. $5\sqrt[3]{10} - 4\sqrt[3]{10}$

6. $6\sqrt[4]{2} + 9\sqrt[4]{2}$

7. $8\sqrt[5]{6} - 2\sqrt[5]{6} + 3\sqrt[5]{6}$

8. $7\sqrt[6]{7} - \sqrt[6]{7} + 4\sqrt[6]{7}$

9. $3x\sqrt{2} - 4x\sqrt{2} + x\sqrt{2}$

10. $5x\sqrt{6} - 3x\sqrt{6} - 2x\sqrt{6}$

11. $\sqrt{20} - \sqrt{80} + \sqrt{45}$

12. $\sqrt{8} - \sqrt{32} - \sqrt{18}$

13. $4\sqrt{8} - 2\sqrt{50} - 5\sqrt{72}$

14. $\sqrt{48} - 3\sqrt{27} + 2\sqrt{75}$

15. $5x\sqrt{8} + 3\sqrt{32x^2} - 5\sqrt{50x^2}$

16. $2\sqrt{50x^2} - 8x\sqrt{18} - 3\sqrt{72x^2}$

17. $5\sqrt[3]{16} - 4\sqrt[3]{54}$

18. $\sqrt[3]{81} + 3\sqrt[3]{24}$

19. $\sqrt[3]{x^4y^2} + 7x\sqrt[3]{xy^2}$

20. $2\sqrt[3]{x^8y^6} - 3y^2\sqrt[3]{8x^8}$

21. $5a^2\sqrt{27ab^3} - 6b\sqrt{12a^5b}$

22. $9a\sqrt{20a^3b^2} + 7b\sqrt{45a^5}$

23. $b\sqrt[3]{24a^5b} + 3a\sqrt[3]{81a^2b^4}$

24. $7\sqrt[3]{a^4b^3c^2} - 6ab\sqrt[3]{ac^2}$

25. $5x\sqrt[4]{3y^5} + y\sqrt[4]{243x^4y} + \sqrt[4]{48x^4y^5}$

26. $x\sqrt[4]{5xy^8} + y\sqrt[4]{405x^5y^4} + y^2\sqrt[4]{80x^5}$

27. $\dfrac{\sqrt{2}}{2} + \dfrac{1}{\sqrt{2}}$

28. $\dfrac{\sqrt{3}}{3} + \dfrac{1}{\sqrt{3}}$

29. $\dfrac{\sqrt{5}}{3} + \dfrac{1}{\sqrt{5}}$

30. $\dfrac{\sqrt{6}}{2} + \dfrac{1}{\sqrt{6}}$

31. $\sqrt{x} - \dfrac{1}{\sqrt{x}}$

32. $\sqrt{x} + \dfrac{1}{\sqrt{x}}$

33. $\dfrac{\sqrt{18}}{6} + \sqrt{\dfrac{1}{2}} + \dfrac{\sqrt{2}}{2}$

34. $\dfrac{\sqrt{12}}{6} + \sqrt{\dfrac{1}{3}} + \dfrac{\sqrt{3}}{3}$

35. $\sqrt{6} - \sqrt{\dfrac{2}{3}} + \sqrt{\dfrac{1}{6}}$

36. $\sqrt{15} - \sqrt{\dfrac{3}{5}} + \sqrt{\dfrac{5}{3}}$

37. $\sqrt[3]{25} + \dfrac{3}{\sqrt[3]{5}}$

38. $\sqrt[4]{8} + \dfrac{1}{\sqrt[4]{2}}$

39. Use a calculator to find a decimal approximation for $\sqrt{12}$ and for $2\sqrt{3}$.

40. Use a calculator to find decimal approximations for $\sqrt{50}$ and $5\sqrt{2}$.

41. Use a calculator to find a decimal approximation for $\sqrt{8} + \sqrt{18}$. Is it equal to the decimal approximation for $\sqrt{26}$ or $\sqrt{50}$?

42. Use a calculator to find a decimal approximation for $\sqrt{3} + \sqrt{12}$. Is it equal to the decimal approximation for $\sqrt{15}$ or $\sqrt{27}$?

Each of the following statements is false. Correct the right side of each one to make the statement true.

43. $3\sqrt{2x} + 5\sqrt{2x} = 8\sqrt{4x}$

44. $5\sqrt{3} - 7\sqrt{3} = -2\sqrt{9}$

45. $\sqrt{9 + 16} = 3 + 4$

46. $\sqrt{36 + 64} = 6 + 8$

Applying the Concepts

47. Golden Rectangle Construct a golden rectangle from a square of side 8. Then show that the ratio of the length to the width is the golden ratio $\dfrac{1 + \sqrt{5}}{2}$.

48. Golden Rectangle Construct a golden rectangle from a square of side 10. Then show that the ratio of the length to the width is the golden ratio $\dfrac{1 + \sqrt{5}}{2}$.

49. Golden Rectangle Use a ruler to construct a golden rectangle from a square of side 1 inch. Then show that the ratio of the length to the width is the golden ratio.

50. Golden Rectangle Use a ruler to construct a golden rectangle from a square of side $\frac{2}{3}$ inch. Then show that the ratio of the length to the width is the golden ratio.

51. Golden Rectangle To show that all golden rectangles have the same ratio of length to width, construct a golden rectangle from a square of side $2x$. Then show that the ratio of the length to the width is the golden ratio.

52. Golden Rectangle To show that all golden rectangles have the same ratio of length to width, construct a golden rectangle from a square of side x. Then show that the ratio of the length to the width is the golden ratio.

53. Isosceles Right Triangles A triangle is isosceles if it has two equal sides, and a triangle is a right triangle if it has a right angle in it. Sketch an isosceles right triangle, and find the ratio of the hypotenuse to a leg.

54. Equilateral Triangles A triangle is equilateral if it has three equal sides. The triangle in the figure is equilateral with each side of length $2x$. Find the ratio of the height to a side.

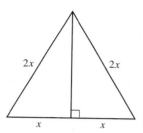

55. Pyramids The following solid is called a regular square pyramid because its base is a square and all eight edges are the same length, 5. It is also true that the vertex, V, is directly above the center of the base.

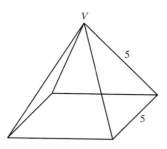

(a) Find the ratio of a diagonal of the base to the length of a side.

(b) Find the ratio of the area of the base to the diagonal of the base.

(c) Find the ratio of the area of the base to the perimeter of the base.

56. Pyramids Refer to the diagram of a square pyramid below. Find the ratio of the height h of the pyramid to the altitude a.

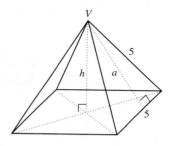

Review Problems

The problems that follow review material we covered in Section 8.3.

Graph each inequality.

57. $2x + 3y < 6$

58. $2x + y < -5$

59. $y \geq -3x - 4$

60. $y \geq 2x - 1$

61. $x \geq 3$

62. $y > -5$

We have worked with the golden rectangle more than once in this chapter. The following is one such golden rectangle.

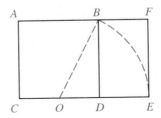

By now you know that in any golden rectangle constructed from a square (of any size), the ratio of the length to the width will be

$$\frac{1 + \sqrt{5}}{2}$$

which we call the golden ratio. What is interesting is that the smaller rectangle on the right, *BFED*, is also a golden rectangle. We will use the mathematics developed in this section to confirm this fact.

In this section we will look at multiplication and division of expressions that contain radicals. As you will see, multiplication of expressions that contain radicals is very similar to multiplication of polynomials. The division problems in this section are just an extension of the work we did previously when we rationalized denominators.

EXAMPLE 1 Multiply $(3\sqrt{5})(2\sqrt{7})$.

Solution We can rearrange the order and grouping of the numbers in this product by applying the commutative and associative properties. Following this, we apply Property 1 for radicals and multiply:

$$(3\sqrt{5})(2\sqrt{7}) = (3 \cdot 2)(\sqrt{5}\,\sqrt{7}) \qquad \text{Commutative and associative properties}$$
$$= (3 \cdot 2)(\sqrt{5 \cdot 7}) \qquad \text{Property 1 for radicals}$$
$$= 6\sqrt{35} \qquad \text{Multiplication}$$

In practice, it is not necessary to show the first two steps.

EXAMPLE 2 Multiply $\sqrt{3}(2\sqrt{6} - 5\sqrt{12})$.

Solution Applying the distributive property, we have

$$\sqrt{3}(2\sqrt{6} - 5\sqrt{12}) = \sqrt{3} \cdot 2\sqrt{6} - \sqrt{3} \cdot 5\sqrt{12}$$
$$= 2\sqrt{18} - 5\sqrt{36}$$

Writing each radical in simplified form gives

$$2\sqrt{18} - 5\sqrt{36} = 2\sqrt{9}\sqrt{2} - 5\sqrt{36}$$
$$= 6\sqrt{2} - 30$$

EXAMPLE 3   Multiply $(\sqrt{3} + \sqrt{5})(4\sqrt{3} - \sqrt{5})$.

Solution   The same principle that applies when multiplying two binomials applies to this product. We must multiply each term in the first expression by each term in the second one. Any convenient method can be used. Let's use the FOIL method.

$$(\sqrt{3} + \sqrt{5})(4\sqrt{3} - \sqrt{5}) = \overset{F}{\sqrt{3}\cdot 4\sqrt{3}} - \overset{O}{\sqrt{3}\sqrt{5}} + \overset{I}{\sqrt{5}\cdot 4\sqrt{3}} - \overset{L}{\sqrt{5}\sqrt{5}}$$
$$= 4\cdot 3 - \sqrt{15} + 4\sqrt{15} - 5$$
$$= 12 + 3\sqrt{15} - 5$$
$$= 7 + 3\sqrt{15}$$

EXAMPLE 4   Expand and simplify $(\sqrt{x} + 3)^2$.

Solution 1   We can write this problem as a multiplication problem and proceed as we did in Example 3:

$$(\sqrt{x} + 3)^2 = (\sqrt{x} + 3)(\sqrt{x} + 3)$$
$$= \overset{F}{\sqrt{x}\cdot\sqrt{x}} + \overset{O}{3\sqrt{x}} + \overset{I}{3\sqrt{x}} + \overset{L}{3\cdot 3}$$
$$= x + 3\sqrt{x} + 3\sqrt{x} + 9$$
$$= x + 6\sqrt{x} + 9$$

Solution 2   We can obtain the same result by applying the formula for the square of a sum: $(a + b)^2 = a^2 + 2ab + b^2$.

$$(\sqrt{x} + 3)^2 = (\sqrt{x})^2 + 2(\sqrt{x})(3) + 3^2$$
$$= x + 6\sqrt{x} + 9$$

EXAMPLE 5   Expand $(3\sqrt{x} - 2\sqrt{y})^2$ and simplify the result.

Solution   Let's apply the formula for the square of a difference, $(a - b)^2 = a^2 - 2ab + b^2$.

$$(3\sqrt{x} - 2\sqrt{y})^2 = (3\sqrt{x})^2 - 2(3\sqrt{x})(2\sqrt{y}) + (2\sqrt{y})^2$$
$$= 9x - 12\sqrt{xy} + 4y$$

EXAMPLE 6 Expand and simplify $(\sqrt{x+2}-1)^2$.

Solution Applying the formula $(a-b)^2 = a^2 - 2ab + b^2$, we have

$$(\sqrt{x+2}-1)^2 = (\sqrt{x+2})^2 - 2\sqrt{x+2}(1) + 1^2$$
$$= x + 2 - 2\sqrt{x+2} + 1$$
$$= x + 3 - 2\sqrt{x+2}$$

EXAMPLE 7 Multiply $(\sqrt{6}+\sqrt{2})(\sqrt{6}-\sqrt{2})$.

Solution We notice the product is of the form $(a+b)(a-b)$, which always gives the difference of two squares, $a^2 - b^2$:

$$(\sqrt{6}+\sqrt{2})(\sqrt{6}-\sqrt{2}) = (\sqrt{6})^2 - (\sqrt{2})^2$$
$$= 6 - 2$$
$$= 4$$

In Example 7 the two expressions $(\sqrt{6}+\sqrt{2})$ and $(\sqrt{6}-\sqrt{2})$ are called *conjugates*. In general, the conjugate of $\sqrt{a}+\sqrt{b}$ is $\sqrt{a}-\sqrt{b}$. If a and b are integers, multiplying conjugates of this form always produces a rational number. That is, if a and b are positive integers, then

$$(\sqrt{a}+\sqrt{b})(\sqrt{a}-\sqrt{b}) = \sqrt{a}\sqrt{a} - \sqrt{a}\sqrt{b} + \sqrt{a}\sqrt{b} - \sqrt{b}\sqrt{b}$$
$$= a - \sqrt{ab} + \sqrt{ab} - b$$
$$= a - b$$

which is rational if a and b are rational.

Division with radical expressions is the same as rationalizing the denominator. In Section 9.3 we were able to divide $\sqrt{3}$ by $\sqrt{2}$ by rationalizing the denominator:

$$\frac{\sqrt{3}}{\sqrt{2}} = \frac{\sqrt{3}}{\sqrt{2}} \cdot \frac{\sqrt{2}}{\sqrt{2}} = \frac{\sqrt{6}}{2}$$

We can accomplish the same result with expressions such as

$$\frac{6}{\sqrt{5}-\sqrt{3}}$$

by multiplying the numerator and denominator by the conjugate of the denominator.

EXAMPLE 8 Divide $\dfrac{6}{\sqrt{5}-\sqrt{3}}$. (Rationalize the denominator.)

Solution Since the product of two conjugates is a rational number, we multiply the numerator and denominator by the conjugate of the denominator.

$$\frac{6}{\sqrt{5}-\sqrt{3}} = \frac{6}{\sqrt{5}-\sqrt{3}} \cdot \frac{(\sqrt{5}+\sqrt{3})}{(\sqrt{5}+\sqrt{3})}$$

$$= \frac{6\sqrt{5} + 6\sqrt{3}}{(\sqrt{5})^2 - (\sqrt{3})^2}$$

$$= \frac{6\sqrt{5} + 6\sqrt{3}}{5 - 3}$$

$$= \frac{6\sqrt{5} + 6\sqrt{3}}{2}$$

The numerator and denominator of this last expression have a factor of 2 in common. We can reduce to lowest terms by factoring 2 from the numerator and then dividing both the numerator and denominator by 2:

$$= \frac{\cancel{2}(3\sqrt{5} + 3\sqrt{3})}{\cancel{2}}$$

$$= 3\sqrt{5} + 3\sqrt{3}$$

EXAMPLE 9   Rationalize the denominator $\dfrac{\sqrt{5} - 2}{\sqrt{5} + 2}$.

Solution   To rationalize the denominator, we multiply the numerator and denominator by the conjugate of the denominator:

$$\frac{\sqrt{5} - 2}{\sqrt{5} + 2} = \frac{\sqrt{5} - 2}{\sqrt{5} + 2} \cdot \frac{(\sqrt{5} - 2)}{(\sqrt{5} - 2)}$$

$$= \frac{5 - 2\sqrt{5} - 2\sqrt{5} + 4}{(\sqrt{5})^2 - 2^2}$$

$$= \frac{9 - 4\sqrt{5}}{5 - 4}$$

$$= \frac{9 - 4\sqrt{5}}{1}$$

$$= 9 - 4\sqrt{5}$$

EXAMPLE 10   A golden rectangle constructed from a square of side 2 is shown in Figure 1. Show that the smaller rectangle *BDEF* is also a golden rectangle by finding the ratio of its length to its width.

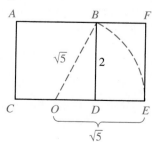

FIGURE I

Solution   First, find expressions for the length and width of the smaller rectangle.

$$\text{Length} = EF = 2$$

$$\text{Width} = DE = \sqrt{5} - 1$$

Next, we find the ratio of length to width.

$$\text{Ratio of length to width} = \frac{EF}{DE} = \frac{2}{\sqrt{5} - 1}$$

To show that the small rectangle is a golden rectangle, we must show that the ratio of length to width is the golden ratio. We do so by rationalizing the denominator.

$$\frac{2}{\sqrt{5} - 1} = \frac{2}{\sqrt{5} - 1} \cdot \frac{\sqrt{5} + 1}{\sqrt{5} + 1}$$

$$= \frac{2(\sqrt{5} + 1)}{5 - 1}$$

$$= \frac{2(\sqrt{5} + 1)}{4}$$

$$= \frac{\sqrt{5} + 1}{2} \qquad\qquad \text{Divide out common factor 2.}$$

Since addition is commutative, this last expression is the golden ratio. Therefore, the small rectangle in Figure 1 is a golden rectangle.

 # Getting Ready for Class

After reading through the preceding section, respond in your own words and in complete sentences.

A. Explain why $(\sqrt{5} + \sqrt{2})^2 \neq 5 + 2$.

B. Explain in words how you would rationalize the denominator in the expression $\dfrac{\sqrt{3}}{\sqrt{5} - \sqrt{2}}$.

C. What are conjugates?

D. What result is guaranteed when multiplying radical expressions that are conjugates?

PROBLEM SET 9.5

Multiply. (Assume all expressions appearing under a square root symbol represent nonnegative numbers throughout this problem set.)

1. $\sqrt{6}\,\sqrt{3}$

2. $\sqrt{6}\,\sqrt{2}$

3. $(2\sqrt{3})(5\sqrt{7})$

4. $(3\sqrt{5})(2\sqrt{7})$

5. $(4\sqrt{6})(2\sqrt{15})(3\sqrt{10})$

6. $(4\sqrt{35})(2\sqrt{21})(5\sqrt{15})$

7. $(3\sqrt[3]{3})(6\sqrt[3]{9})$

8. $(2\sqrt[3]{2})(6\sqrt[3]{4})$

9. $\sqrt{3}(\sqrt{2} - 3\sqrt{3})$

10. $\sqrt{2}(5\sqrt{3} + 4\sqrt{2})$

11. $6\sqrt[3]{4}(2\sqrt[3]{2} + 1)$

12. $7\sqrt[3]{5}(3\sqrt[3]{25} - 2)$

13. $(\sqrt{3} + \sqrt{2})(3\sqrt{3} - \sqrt{2})$

14. $(\sqrt{5} - \sqrt{2})(3\sqrt{5} + 2\sqrt{2})$

15. $(\sqrt{x} + 5)(\sqrt{x} - 3)$

16. $(\sqrt{x} + 4)(\sqrt{x} + 2)$

17. $(3\sqrt{6} + 4\sqrt{2})(\sqrt{6} + 2\sqrt{2})$

18. $(\sqrt{7} - 3\sqrt{3})(2\sqrt{7} - 4\sqrt{3})$

19. $(\sqrt{3} + 4)^2$

20. $(\sqrt{5} - 2)^2$

21. $(\sqrt{x} - 3)^2$

22. $(\sqrt{x} + 4)^2$

23. $(2\sqrt{a} - 3\sqrt{b})^2$

24. $(5\sqrt{a} - 2\sqrt{b})^2$

25. $(\sqrt{x - 4} + 2)^2$

26. $(\sqrt{x - 3} + 2)^2$

27. $(\sqrt{x - 5} - 3)^2$

28. $(\sqrt{x - 3} - 4)^2$

29. $(\sqrt{3} - \sqrt{2})(\sqrt{3} + \sqrt{2})$

30. $(\sqrt{5} - \sqrt{2})(\sqrt{5} + \sqrt{2})$

31. $(\sqrt{a} + 7)(\sqrt{a} - 7)$

32. $(\sqrt{a} + 5)(\sqrt{a} - 5)$

33. $(5 - \sqrt{x})(5 + \sqrt{x})$

34. $(3 - \sqrt{x})(3 + \sqrt{x})$

35. $(\sqrt{x - 4} + 2)(\sqrt{x - 4} - 2)$

36. $(\sqrt{x + 3} + 5)(\sqrt{x + 3} - 5)$

37. $(\sqrt{3} + 1)^3$

38. $(\sqrt{5} - 2)^3$

Rationalize the denominator in each of the following.

39. $\dfrac{\sqrt{2}}{\sqrt{6} - \sqrt{2}}$

40. $\dfrac{\sqrt{5}}{\sqrt{5} + \sqrt{3}}$

41. $\dfrac{\sqrt{5}}{\sqrt{5} + 1}$

42. $\dfrac{\sqrt{7}}{\sqrt{7} - 1}$

43. $\dfrac{\sqrt{x}}{\sqrt{x} - 3}$

44. $\dfrac{\sqrt{x}}{\sqrt{x} + 2}$

45. $\dfrac{\sqrt{5}}{2\sqrt{5} - 3}$

46. $\dfrac{\sqrt{7}}{3\sqrt{7} - 2}$

47. $\dfrac{3}{\sqrt{x} - \sqrt{y}}$

48. $\dfrac{2}{\sqrt{x} + \sqrt{y}}$

49. $\dfrac{\sqrt{6} + \sqrt{2}}{\sqrt{6} - \sqrt{2}}$

50. $\dfrac{\sqrt{5} - \sqrt{3}}{\sqrt{5} + \sqrt{3}}$

51. $\dfrac{\sqrt{7} - 2}{\sqrt{7} + 2}$

52. $\dfrac{\sqrt{11} + 3}{\sqrt{11} - 3}$

53. $\dfrac{\sqrt{a} + \sqrt{b}}{\sqrt{a} - \sqrt{b}}$

54. $\dfrac{\sqrt{a} - \sqrt{b}}{\sqrt{a} + \sqrt{b}}$

55. $\dfrac{\sqrt{x} + 2}{\sqrt{x} - 2}$

56. $\dfrac{\sqrt{x} - 3}{\sqrt{x} + 3}$

57. $\dfrac{2\sqrt{3} - \sqrt{7}}{3\sqrt{3} + \sqrt{7}}$

58. $\dfrac{5\sqrt{6} + 2\sqrt{2}}{\sqrt{6} - \sqrt{2}}$

59. $\dfrac{3\sqrt{x} + 2}{1 + \sqrt{x}}$

60. $\dfrac{5\sqrt{x} - 1}{2 + \sqrt{x}}$

61. Show that the product below is 5.
$(\sqrt[3]{2} + \sqrt[3]{3})(\sqrt[3]{4} - \sqrt[3]{6} + \sqrt[3]{9})$

62. Show that the product below is $x + 8$.
$(\sqrt[3]{x} + 2)(\sqrt[3]{x^2} - 2\sqrt[3]{x} + 4)$

Each of the following statements below is false. Correct the right side of each one to make it true.

63. $5(2\sqrt{3}) = 10\sqrt{15}$

64. $3(2\sqrt{x}) = 6\sqrt{3x}$

65. $(\sqrt{x} + 3)^2 = x + 9$

66. $(\sqrt{x} - 7)^2 = x - 49$

67. $(5\sqrt{3})^2 = 15$

68. $(3\sqrt{5})^2 = 15$

Applying the Concepts

69. Gravity If an object is dropped from the top of a 100-foot building, the amount of time t (in seconds) that it takes for the object to be h feet from the ground is given by the formula

$$t = \frac{\sqrt{100 - h}}{4}$$

How long does it take before the object is 50 feet from the ground? How long does it take to reach the ground? (When it is on the ground, h is 0.)

70. Gravity Use the formula given in Problem 69 to determine h if t is 1.25 seconds.

71. Golden Rectangle   Rectangle *ACEF* in Figure 2 is a golden rectangle. If side *AC* is 6 inches, show that the smaller rectangle *BDEF* is also a golden rectangle.

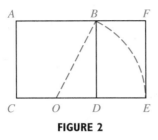

FIGURE 2

72. Golden Rectangle   Rectangle *ACEF* in Figure 2 is a golden rectangle. If side *AC* is 1 inch, show that the smaller rectangle *BDEF* is also a golden rectangle.

73. Golden Rectangle   If side *AC* in Figure 2 is 2*x*, show that rectangle *BDEF* is a golden rectangle.

74. Golden Rectangle   If side *AC* in Figure 2 is *x*, show that rectangle *BDEF* is a golden rectangle.

Review Problems

The following problems review material we covered in Section 8.7.

75. If *y* varies directly with the square of *x*, and *y* is 75 when *x* is 5, find *y* when *x* is 7.

76. Suppose *y* varies directly with the cube of *x*. If *y* is 16 when *x* is 2, find *y* when *x* is 3.

77. Suppose *y* varies inversely with *x*. If *y* is 10 when *x* is 25, find *x* when *y* is 5.

78. If *y* varies inversely with the cube of *x*, and *y* is 2 when *x* is 2, find *y* when *x* is 4.

79. Suppose *z* varies jointly with *x* and the square of *y*. If *z* is 40 when *x* is 5 and *y* is 2, find *z* when *x* is 2 and *y* is 5.

80. Suppose *z* varies jointly with *x* and the cube of *y*. If *z* is 48 when *x* is 3 and *y* is 2, find *z* when *x* is 4 and *y* is $\frac{1}{2}$.

9.6   Equations With Radicals

This section is concerned with solving equations that involve one or more radicals. The first step in solving an equation that contains a radical is to eliminate the radical from the equation. To do so, we need an additional property.

SQUARING PROPERTY OF EQUALITY

If both sides of an equation are squared, the solutions to the original equation are solutions to the resulting equation.

We will never lose solutions to our equations by squaring both sides. We may, however, introduce *extraneous solutions.* Extraneous solutions satisfy the equation obtained by squaring both sides of the original equation, but do not satisfy the original equation.

We know that if two real numbers *a* and *b* are equal, then so are their squares:

$$\text{If} \qquad a = b$$

$$\text{then} \qquad a^2 = b^2$$

On the other hand, extraneous solutions are introduced when we square opposites. That is, even though opposites are not equal, their squares are. For example,

$$5 = -5 \qquad \text{A false statement}$$
$$(5)^2 = (-5)^2 \qquad \text{Square both sides.}$$
$$25 = 25 \qquad \text{A true statement}$$

We are free to square both sides of an equation any time it is convenient. We must be aware, however, that doing so may introduce extraneous solutions. We must, therefore, check all our solutions in the original equation if at any time we square both sides of the original equation.

EXAMPLE   1   Solve for x: $\sqrt{3x + 4} = 5$

Solution   We square both sides and proceed as usual:

$$\sqrt{3x + 4} = 5$$
$$(\sqrt{3x + 4})^2 = 5^2$$
$$3x + 4 = 25$$
$$3x = 21$$
$$x = 7$$

Checking $x = 7$ in the original equation, we have

$$\sqrt{3(7) + 4} \stackrel{?}{=} 5$$
$$\sqrt{21 + 4} = 5$$
$$\sqrt{25} = 5$$
$$5 = 5$$

The solution $x = 7$ satisfies the original equation.

EXAMPLE   2   Solve $\sqrt{4x - 7} = -3$.

Solution   Squaring both sides, we have

$$\sqrt{4x - 7} = -3$$
$$(\sqrt{4x - 7})^2 = (-3)^2$$
$$4x - 7 = 9$$
$$4x = 16$$
$$x = 4$$

Checking $x = 4$ in the original equation gives

$$\sqrt{4(4) - 7} \overset{?}{=} -3$$
$$\sqrt{16 - 7} = -3$$
$$\sqrt{9} = -3$$
$$3 = -3$$

The solution $x = 4$ produces a false statement when checked in the original equation. Since $x = 4$ was the only possible solution, there is no solution to the original equation. The possible solution $x = 4$ is an extraneous solution. It satisfies the equation obtained by squaring both sides of the original equation, but does not satisfy the original equation.

Note: The fact that there is no solution to the equation in Example 2 was obvious to begin with. Notice that the left side of the equation is the *positive* square root of $4x - 7$, which must be a positive number or 0. The right side of the equation is -3. Since we cannot have a number that is either positive or zero equal to a negative number, there is no solution to the equation.

EXAMPLE 3   Solve $\sqrt{5x - 1} + 3 = 7$.

Solution   We must isolate the radical on the left side of the equation. If we attempt to square both sides without doing so, the resulting equation will also contain a radical. Adding -3 to both sides, we have

$$\sqrt{5x - 1} + 3 = 7$$
$$\sqrt{5x - 1} = 4$$

We can now square both sides and proceed as usual:

$$(\sqrt{5x - 1})^2 = 4^2$$
$$5x - 1 = 16$$
$$5x = 17$$
$$x = \frac{17}{5}$$

Checking $x = \frac{17}{5}$, we have

$$\sqrt{5\left(\frac{17}{5}\right) - 1} + 3 \overset{?}{=} 7$$
$$\sqrt{17 - 1} + 3 = 7$$
$$\sqrt{16} + 3 = 7$$
$$4 + 3 = 7$$
$$7 = 7$$

E X A M P L E 4 Solve $t + 5 = \sqrt{t + 7}$.

Solution This time, squaring both sides of the equation results in a quadratic equation:

$$(t + 5)^2 = (\sqrt{t + 7})^2 \qquad \text{Square both sides.}$$
$$t^2 + 10t + 25 = t + 7$$
$$t^2 + 9t + 18 = 0 \qquad \text{Standard form}$$
$$(t + 3)(t + 6) = 0 \qquad \text{Factor the left side.}$$
$$t + 3 = 0 \quad \text{or} \quad t + 6 = 0 \qquad \text{Set factors equal to 0.}$$
$$t = -3 \quad \text{or} \quad t = -6$$

We must check each solution in the original equation:

Check $t = -3$

$$-3 + 5 \overset{?}{=} \sqrt{-3 + 7}$$
$$2 = \sqrt{4}$$
$$2 = 2 \qquad \text{A true statement}$$

Check $t = -6$

$$-6 + 5 \overset{?}{=} \sqrt{-6 + 7}$$
$$-1 = \sqrt{1}$$
$$-1 = 1 \qquad \text{A false statement}$$

Since $t = -6$ does not check, our only solution is $t = -3$.

E X A M P L E 5 Solve $\sqrt{x - 3} = \sqrt{x} - 3$.

Solution We begin by squaring both sides. Note carefully what happens when we square the right side of the equation, and compare the square of the right side with the square of the left side. You must convince yourself that these results are correct. (The note on the next page will help if you are having trouble convincing yourself that what is written below is true.)

$$(\sqrt{x - 3})^2 = (\sqrt{x} - 3)^2$$
$$x - 3 = x - 6\sqrt{x} + 9$$

Now we still have a radical in our equation, so we will have to square both sides again. Before we do, though, let's isolate the remaining radical.

$$x - 3 = x - 6\sqrt{x} + 9$$
$$-3 = -6\sqrt{x} + 9 \qquad \text{Add } -x \text{ to each side.}$$
$$-12 = -6\sqrt{x} \qquad \text{Add } -9 \text{ to each side.}$$
$$2 = \sqrt{x} \qquad \text{Divide each side by } -6.$$
$$4 = x \qquad \text{Square each side.}$$

Our only possible solution is $x = 4$, which we check in our original equation as follows:

$$\sqrt{4 - 3} \overset{?}{=} \sqrt{4} - 3$$
$$\sqrt{1} = 2 - 3$$
$$1 = -1 \qquad \text{A false statement}$$

Substituting 4 for x in the original equation yields a false statement. Since 4 was our only possible solution, there is no solution to our equation.

Note: It is very important that you realize that the square of $(\sqrt{x} - 3)$ is not $x + 9$. Remember, when we square a difference with two terms, we use the formula

$$(a - b)^2 = a^2 - 2ab + b^2$$

Applying this formula to $(\sqrt{x} - 3)^2$, we have

$$(\sqrt{x} - 3)^2 = (\sqrt{x})^2 - 2(\sqrt{x})(3) + 3^2$$
$$= x - 6\sqrt{x} + 9$$

Here is another example of an equation for which we must apply our squaring property twice before all radicals are eliminated.

EXAMPLE 6   Solve $\sqrt{x + 1} = 1 - \sqrt{2x}$.

Solution   This equation has two separate terms involving radical signs.
Squaring both sides gives

$$x + 1 = 1 - 2\sqrt{2x} + 2x$$
$$-x = -2\sqrt{2x} \qquad \text{Add } -2x \text{ and } -1 \text{ to both sides.}$$
$$x^2 = 4(2x) \qquad \text{Square both sides.}$$
$$x^2 - 8x = 0 \qquad \text{Standard form}$$

Our equation is a quadratic equation in standard form. To solve for x, we factor the left side and set each factor equal to 0:

$$x(x - 8) = 0 \qquad \text{Factor left side.}$$
$$x = 0 \quad \text{or} \quad x - 8 = 0 \qquad \text{Set factors equal to 0.}$$
$$x = 8$$

Since we squared both sides of our equation, we have the possibility that one or both of the solutions are extraneous. We must check each one in the original equation:

Check $x = 8$	Check $x = 0$
$\sqrt{8 + 1} \overset{?}{=} 1 - \sqrt{2 \cdot 8}$	$\sqrt{0 + 1} \overset{?}{=} 1 - \sqrt{2 \cdot 0}$
$\sqrt{9} = 1 - \sqrt{16}$	$\sqrt{1} = 1 - \sqrt{0}$
$3 = 1 - 4$	$1 = 1 - 0$
$3 = -3$ A false statement	$1 = 1$ A true statement

Since $x = 8$ does not check, it is an extraneous solution. Our only solution is $x = 0$.

EXAMPLE 7   Solve $\sqrt{x + 1} = \sqrt{x + 2} - 1$.

Solution   Squaring both sides we have

$$(\sqrt{x + 1})^2 = (\sqrt{x + 2} - 1)^2$$
$$x + 1 = x + 2 - 2\sqrt{x + 2} + 1$$

Once again we are left with a radical in our equation. Before we square each side again, we must isolate the radical on the right side of the equation.

$x + 1 = x + 3 - 2\sqrt{x + 2}$	Simplify the right side.
$1 = 3 - 2\sqrt{x + 2}$	Add $-x$ to each side.
$-2 = -2\sqrt{x + 2}$	Add -3 to each side.
$1 = \sqrt{x + 2}$	Divide each side by -2.
$1 = x + 2$	Square both sides.
$-1 = x$	Add -2 to each side.

Checking our only possible solution, $x = -1$, in our original equation, we have

$$\sqrt{-1 + 1} \overset{?}{=} \sqrt{-1 + 2} - 1$$
$$\sqrt{0} = \sqrt{1} - 1$$
$$0 = 1 - 1$$
$$0 = 0 \qquad \text{A true statement}$$

Our solution checks.

It is also possible to raise both sides of an equation to powers greater than 2. We only need to check for extraneous solutions when we raise both sides of an equation to an even power. Raising both sides of an equation to an odd power will not produce extraneous solutions.

EXAMPLE 8   Solve $\sqrt[3]{4x + 5} = 3$.

Solution   Cubing both sides we have

$$(\sqrt[3]{4x + 5})^3 = 3^3$$
$$4x + 5 = 27$$
$$4x = 22$$
$$x = \frac{22}{4}$$
$$x = \frac{11}{2}$$

We do not need to check $x = \frac{11}{2}$ since we raised both sides to an odd power.

We end this section by looking at graphs of some equations that contain radicals.

EXAMPLE 9 Graph $y = \sqrt{x}$ and $y = \sqrt[3]{x}$.

Solution The graphs are shown in Figures 1 and 2. Notice that the graph of $y = \sqrt{x}$ appears in the first quadrant only, because in the equation $y = \sqrt{x}$, x and y cannot be negative.

The graph of $y = \sqrt[3]{x}$ appears in Quadrants 1 and 3 since the cube root of a positive number is also a positive number, and the cube root of a negative number is a negative number. That is, when x is positive, y will be positive, and when x is negative, y will be negative.

The graphs of both equations will contain the origin, since $y = 0$ when $x = 0$ in both equations.

x	y
-4	Undefined
-1	Undefined
0	0
1	1
4	2
9	3
16	4

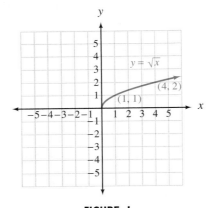

FIGURE 1

x	y
-27	-3
-8	-2
-1	-1
0	0
1	1
8	2
27	3

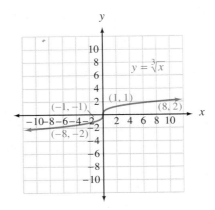

FIGURE 2

Getting Ready for Class

After reading through the preceding section, respond in your own words and in complete sentences.

A. What is the squaring property of equality?

B. Under what conditions do we obtain extraneous solutions to equations that contain radical expressions?

C. If we have raised both sides of an equation to a power, when is it not necessary to check for extraneous solutions?

D. When will you need to apply the squaring property of equality twice in the process of solving an equation containing radicals?

PROBLEM SET 9.6

Solve each of the following equations.

1. $\sqrt{2x + 1} = 3$

2. $\sqrt{3x + 1} = 4$

3. $\sqrt{4x + 1} = -5$

4. $\sqrt{6x + 1} = -5$

5. $\sqrt{2y - 1} = 3$

6. $\sqrt{3y - 1} = 2$

7. $\sqrt{5x - 7} = -1$

8. $\sqrt{8x + 3} = -6$

9. $\sqrt{2x - 3} - 2 = 4$

10. $\sqrt{3x + 1} - 4 = 1$

11. $\sqrt{4a + 1} + 3 = 2$

12. $\sqrt{5a - 3} + 6 = 2$

13. $\sqrt[4]{3x + 1} = 2$

14. $\sqrt[4]{4x + 1} = 3$

15. $\sqrt[3]{2x - 5} = 1$

16. $\sqrt[3]{5x + 7} = 2$

17. $\sqrt[3]{3a + 5} = -3$

18. $\sqrt[3]{2a + 7} = -2$

19. $\sqrt{y - 3} = y - 3$

20. $\sqrt{y + 3} = y - 3$

21. $\sqrt{a + 2} = a + 2$

22. $\sqrt{a + 10} = a - 2$

23. $\sqrt{2x + 4} = \sqrt{1 - x}$

24. $\sqrt{3x + 4} = -\sqrt{2x + 3}$

25. $\sqrt{4a + 7} = -\sqrt{a + 2}$

26. $\sqrt{7a - 1} = \sqrt{2a + 4}$

27. $\sqrt[4]{5x - 8} = \sqrt[4]{4x - 1}$

28. $\sqrt[4]{6x + 7} = \sqrt[4]{x + 2}$

29. $x + 1 = \sqrt{5x + 1}$

30. $x - 1 = \sqrt{6x + 1}$

31. $t + 5 = \sqrt{2t + 9}$

32. $t + 7 = \sqrt{2t + 13}$

33. $\sqrt{y - 8} = \sqrt{8 - y}$

34. $\sqrt{2y + 5} = \sqrt{5y + 2}$

35. $\sqrt[3]{3x + 5} = \sqrt[3]{5 - 2x}$

36. $\sqrt[3]{4x + 9} = \sqrt[3]{3 - 2x}$

The following equations will require that you square both sides twice before all the radicals are eliminated. Solve each equation using the methods shown in Examples 5, 6, and 7.

37. $\sqrt{x - 8} = \sqrt{x} - 2$

38. $\sqrt{x + 3} = \sqrt{x} - 3$

39. $\sqrt{x + 1} = \sqrt{x} + 1$

40. $\sqrt{x - 1} = \sqrt{x} - 1$

41. $\sqrt{x + 8} = \sqrt{x - 4} + 2$

42. $\sqrt{x + 5} = \sqrt{x - 3} + 2$

43. $\sqrt{x - 5} - 3 = \sqrt{x - 8}$

44. $\sqrt{x - 3} - 4 = \sqrt{x - 3}$

45. $\sqrt{x + 4} = 2 - \sqrt{2x}$

46. $\sqrt{5x + 1} = 1 + \sqrt{5x}$

47. $\sqrt{2x + 4} = \sqrt{x + 3} + 1$

48. $\sqrt{2x - 1} = \sqrt{x - 4} + 2$

Applying the Concepts

49. Solving a Formula   Solve the following formula for h:

$$t = \frac{\sqrt{100 - h}}{4}$$

50. Solving a Formula Solve the following formula for h:

$$t = \sqrt{\frac{2h - 40t}{g}}$$

51. Pendulum Clock The length of time (T) in seconds it takes the pendulum of a clock to swing through one complete cycle is given by the formula

$$T = 2\pi\sqrt{\frac{L}{32}}$$

where L is the length, in feet, of the pendulum, and π is approximately $\frac{22}{7}$. How long must the pendulum be if one complete cycle takes 2 seconds?

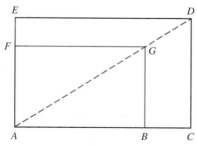

1 sec

52. Pendulum Clock Solve the formula in Problem 51 for L.

53. Similar Rectangles Two rectangles are similar if their vertices lie along the same diagonal, as shown in the following diagram.

Rectangle ABGF is similar to rectangle ACDE.

If two rectangles are similar, then corresponding sides are in proportion, which means that $\frac{ED}{DC} = \frac{FG}{GB}$. Use these facts in the following diagram to express the length of the larger rectangle l in terms of x.

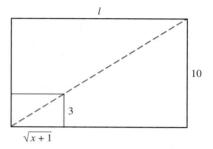

54. Volume Recall that the volume of a box V can be found from the formula $V =$ (length)(width)(height). Find the volume of the following box in terms of x.

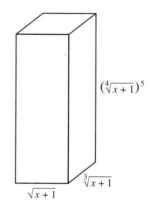

$(\sqrt[4]{x + 1})^5$

$\sqrt[3]{x + 1}$

$\sqrt{x + 1}$

Pollution A long straight river, 100 meters wide, is flowing at 1 meter per second. A pollutant is entering the river at a constant rate, from one of its banks. As the pollutant disperses in the water, it forms a plume that is modeled by the equation $y = \sqrt{x}$. Use this information to answer the following questions.

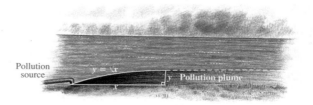

55. How wide is the plume 25 meters downriver from the source of the pollution?

56. How wide is the plume 100 meters downriver from the source of the pollution?

57. How far down river from the source of the pollution does the plume reach halfway across the river?

58. How far down the river from the source of the pollution does the plume reach the other side of the river?

59. For the situation described in the instructions and modeled by the equation $y = \sqrt{x}$, what is the range of values that y can assume?

60. If the river was moving at 2 meters per second, would the plume be larger or smaller 100 meters downstream from the source?

Graph each equation.

61. $y = 2\sqrt{x}$

62. $y = -2\sqrt{x}$

63. $y = \sqrt{x} - 2$

64. $y = \sqrt{x} + 2$

65. $y = \sqrt{x - 2}$

66. $y = \sqrt{x + 2}$

67. $y = 3\sqrt[3]{x}$

68. $y = -3\sqrt[3]{x}$

69. $y = \sqrt[3]{x} + 3$

70. $y = \sqrt[3]{x} - 3$

71. $y = \sqrt[3]{x + 3}$

72. $y = \sqrt[3]{x - 3}$

Review Problems

The problems that follow review material we covered in Section 9.5. Reviewing these problems will help you understand the next section.

Multiply.

73. $\sqrt{2}(\sqrt{3} - \sqrt{2})$

74. $(\sqrt{x} - 4)(\sqrt{x} + 5)$

75. $(\sqrt{x} + 5)^2$

76. $(\sqrt{5} + \sqrt{3})(\sqrt{5} - \sqrt{3})$

Rationalize the denominator.

77. $\dfrac{\sqrt{x}}{\sqrt{x} + 3}$

78. $\dfrac{\sqrt{5} - \sqrt{3}}{\sqrt{5} + \sqrt{3}}$

Extending the Concepts

Solve each equation.

79. $\dfrac{x}{3\sqrt{2x - 3}} - \dfrac{1}{\sqrt{2x - 3}} = \dfrac{1}{3}$

80. $\dfrac{x}{5\sqrt{2x + 10}} + \dfrac{1}{\sqrt{2x + 10}} = \dfrac{1}{5}$

81. $x + 1 = \sqrt[3]{4x + 4}$

82. $x - 1 = \sqrt[3]{4x - 4}$

Solve for y in terms of x.

83. $y + 2 = \sqrt{x^2 + (y - 2)^2}$

84. $y + \dfrac{1}{2} = \sqrt{x^2 + \left(y - \dfrac{1}{2}\right)^2}$

85. Use your Y variables list, or write a program, to graph the family of curves $Y = \sqrt{X} + B$ for $B = -3, -2, -1, 0, 1, 2,$ and 3.

86. Use your Y variables list, or write a program, to graph the family of curves $Y = \sqrt{X + B}$ for $B = -3, -2, -1, 0, 1, 2,$ and 3.

87. Summarize the results of Problem 85 by giving a written description of the effect of b on the graph of $y = \sqrt{x} + b$.

88. Summarize the results of Problem 86 by giving a written description of the effect of b on the graph of $y = \sqrt{x + b}$.

89. Use your Y variables list, or write a program, to graph the family of curves $Y = \sqrt[3]{X} + B$ for $B = -3, -2, -1, 0, 1, 2,$ and 3.

90. Use your Y variables list, or write a program, to graph the family of curves $Y = \sqrt[3]{X + B}$ for $B = -3, -2, -1, 0, 1, 2,$ and 3.

91. Summarize the results of Problem 89 by giving a written description of the effect of b on the graph of $y = \sqrt[3]{x} + b$.

92. Summarize the results of Problem 90 by giving a written description of the effect of b on the graph of $y = \sqrt[3]{x + b}$.

93. Use your Y variables list, or write a program, to graph the family of curves $Y = A\sqrt{X}$ for $A = -3, -2, -1, 0, 1, 2,$ and 3.

94. Use your Y variables list, or write a program, to graph the family of curves $Y = A\sqrt{X}$ for $A = \dfrac{1}{4}, \dfrac{1}{3}, \dfrac{1}{2}, 1, 2,$ and 3.

95. Summarize the results of Problems 93 and 94 by giving a written description of the effect of a on the graph of $y = a\sqrt{x}$.

Complex Numbers

The equation $x^2 = -9$ has no real number solutions since the square of a real number is always positive. We have been unable to work with square roots of negative numbers like $\sqrt{-25}$ and $\sqrt{-16}$ for the same reason. Complex numbers allow us to expand our work with radicals to include square roots of negative numbers and to solve equations like $x^2 = -9$ and $x^2 = -64$. Our work with complex numbers is based on the following definition.

> **DEFINITION**   The **number** i is such that $i = \sqrt{-1}$ (which is the same as saying $i^2 = -1$).

The number i, as we have defined it here, is not a real number. Because of the way we have defined i, we can use it to simplify square roots of negative numbers.

SQUARE ROOTS OF NEGATIVE NUMBERS

If a is a positive number, then $\sqrt{-a}$ can always be written as $i\sqrt{a}$. That is,

$$\sqrt{-a} = i\sqrt{a} \qquad \text{if } a \text{ is a positive number}$$

To justify our rule, we simply square the quantity $i\sqrt{a}$ to obtain $-a$. Here is what it looks like when we do so:

$$(i\sqrt{a})^2 = i^2 \cdot (\sqrt{a})^2$$
$$= -1 \cdot a$$
$$= -a$$

Here are some examples that illustrate the use of our new rule.

EXAMPLES   Write each square root in terms of the number i.

1. $\sqrt{-25} = i\sqrt{25} = i \cdot 5 = 5i$

2. $\sqrt{-49} = i\sqrt{49} = i \cdot 7 = 7i$

3. $\sqrt{-12} = i\sqrt{12} = i \cdot 2\sqrt{3} = 2i\sqrt{3}$

4. $\sqrt{-17} = i\sqrt{17}$

Note: In Examples 3 and 4 we wrote i before the radical simply to avoid confusion. If we were to write the answer to 3 as $2\sqrt{3}i$, some people would think the i was under the radical sign, and it is not.

If we assume all the properties of exponents hold when the base is i, we can write any power of i as i, -1, $-i$, or 1. Using the fact that $i^2 = -1$, we have

$$i^1 = i$$

$$i^2 = -1$$

$$i^3 = i^2 \cdot i = -1(i) = -i$$

$$i^4 = i^2 \cdot i^2 = -1(-1) = 1$$

Since $i^4 = 1$, i^5 will simplify to i, and we will begin repeating the sequence i, -1, $-i$, 1 as we simplify higher powers of i: Any power of i simplifies to i, -1, $-i$, or 1. The easiest way to simplify higher powers of i is to write them in terms of i^2. For instance, to simplify i^{21}, we would write it as

$$(i^2)^{10} \cdot i \qquad \text{because } 2 \cdot 10 + 1 = 21$$

Then, since $i^2 = -1$, we have

$$(-1)^{10} \cdot i = 1 \cdot i = i$$

E X A M P L E S   Simplify as much as possible.

5. $i^{30} = (i^2)^{15} = (-1)^{15} = -1$

6. $i^{11} = (i^2)^5 \cdot i = (-1)^5 \cdot i = (-1)i = -i$

7. $i^{40} = (i^2)^{20} = (-1)^{20} = 1$

DEFINITION   A **complex number** is any number that can be put in the form

$$a + bi$$

where a and b are real numbers and $i = \sqrt{-1}$. The form $a + bi$ is called *standard form* for complex numbers. The number a is called the *real part* of the complex number. The number b is called the *imaginary part* of the complex number.

Every real number is also a complex number. The real number 8, for example, can be written as $8 + 0i$; therefore, 8 is also considered a complex number.

EQUALITY FOR COMPLEX NUMBERS

Two complex numbers are equal if and only if their real parts are equal and their imaginary parts are equal. That is, for real numbers a, b, c, and d,

$$a + bi = c + di \qquad \text{if and only if} \qquad a = c \quad \text{and} \quad b = d$$

EXAMPLE 8 Find x and y if $3x + 4i = 12 - 8yi$.

Solution Since the two complex numbers are equal, their real parts are equal, and their imaginary parts are equal:

$$3x = 12 \quad \text{and} \quad 4 = -8y$$

$$x = 4 \qquad\qquad y = -\frac{1}{2}$$

EXAMPLE 9 Find x and y if $(4x - 3) + 7i = 5 + (2y - 1)i$.

Solution The real parts are $4x - 3$ and 5. The imaginary parts are 7 and $2y - 1$:

$$4x - 3 = 5 \quad \text{and} \quad 7 = 2y - 1$$

$$4x = 8 \qquad\qquad 8 = 2y$$

$$x = 2 \qquad\qquad y = 4$$

ADDITION AND SUBTRACTION OF COMPLEX NUMBERS

To add two complex numbers, add their real parts and add their imaginary parts. That is, if a, b, c, and d are real numbers, then

$$(a + bi) + (c + di) = (a + c) + (b + d)i$$

If we assume that the commutative, associative, and distributive properties hold for the number i, then the definition of addition is simply an extension of these properties.

We define subtraction in a similar manner. If a, b, c, and d are real numbers, then

$$(a + bi) - (c + di) = (a - c) + (b - d)i$$

EXAMPLES Add or subtract as indicated.

10. $(3 + 4i) + (7 - 6i) = (3 + 7) + (4 - 6)i = 10 - 2i$

11. $(7 + 3i) - (5 + 6i) = (7 - 5) + (3 - 6)i = 2 - 3i$

12. $(5 - 2i) - (9 - 4i) = (5 - 9) + (-2 + 4)i = -4 + 2i$

MULTIPLICATION OF COMPLEX NUMBERS

Since complex numbers have the same form as binomials, we find the product of two complex numbers the same way we find the product of two binomials.

EXAMPLE 13   Multiply $(3 - 4i)(2 + 5i)$.

Solution   Multiplying each term in the second complex number by each term in the first, we have

$$\overset{\text{F} \qquad \text{O} \qquad \text{I} \qquad \text{L}}{(3 - 4i)(2 + 5i) = 3 \cdot 2 + 3 \cdot 5i - 2 \cdot 4i - 4i(5i)}$$

$$= 6 + 15i - 8i - 20i^2$$

Combining similar terms and using the fact that $i^2 = -1$, we can simplify as follows:

$$6 + 15i - 8i - 20i^2 = 6 + 7i - 20(-1)$$

$$= 6 + 7i + 20$$

$$= 26 + 7i$$

The product of the complex numbers $3 - 4i$ and $2 + 5i$ is the complex number $26 + 7i$.

EXAMPLE 14   Multiply $2i(4 - 6i)$.

Solution   Applying the distributive property gives us

$$2i(4 - 6i) = 2i \cdot 4 - 2i \cdot 6i$$

$$= 8i - 12i^2$$

$$= 12 + 8i$$

EXAMPLE 15   Expand $(3 + 5i)^2$.

Solution   We treat this like the square of a binomial. Remember, $(a + b)^2 = a^2 + 2ab + b^2$:

$$(3 + 5i)^2 = 3^2 + 2(3)(5i) + (5i)^2$$

$$= 9 + 30i + 25i^2$$

$$= 9 + 30i - 25$$

$$= -16 + 30i$$

EXAMPLE 16   Multiply $(2 - 3i)(2 + 3i)$.

Solution   This product has the form $(a - b)(a + b)$, which we know results in the difference of two squares, $a^2 - b^2$:

$$(2 - 3i)(2 + 3i) = 2^2 - (3i)^2$$
$$= 4 - 9i^2$$
$$= 4 + 9$$
$$= 13$$

The product of the two complex numbers $2 - 3i$ and $2 + 3i$ is the real number 13. The two complex numbers $2 - 3i$ and $2 + 3i$ are called complex conjugates. The fact that their product is a real number is very useful.

DEFINITION   The complex numbers $a + bi$ and $a - bi$ are called **complex conjugates.** One important property they have is that their product is the real number $a^2 + b^2$. Here's why:

$$(a + bi)(a - bi) = a^2 - (bi)^2$$
$$= a^2 - b^2i^2$$
$$= a^2 - b^2(-1)$$
$$= a^2 + b^2$$

DIVISION WITH COMPLEX NUMBERS

The fact that the product of two complex conjugates is a real number is the key to division with complex numbers.

EXAMPLE 17   Divide $\dfrac{2 + i}{3 - 2i}$.

Solution   We want a complex number in standard form that is equivalent to the quotient $\dfrac{2 + i}{3 - 2i}$. We need to eliminate i from the denominator. Multiplying the numerator and denominator by $3 + 2i$ will give us what we want:

$$\frac{2 + i}{3 - 2i} = \frac{2 + i}{3 - 2i} \cdot \frac{(3 + 2i)}{(3 + 2i)}$$
$$= \frac{6 + 4i + 3i + 2i^2}{9 - 4i^2}$$
$$= \frac{6 + 7i - 2}{9 + 4}$$

$$= \frac{4 + 7i}{13}$$

$$= \frac{4}{13} + \frac{7}{13} i$$

Dividing the complex number $2 + i$ by $3 - 2i$ gives the complex number $\frac{4}{13} + \frac{7}{13}i$.

EXAMPLE 18 Divide $\dfrac{7 - 4i}{i}$.

Solution The conjugate of the denominator is $-i$. Multiplying numerator and denominator by this number, we have

$$\frac{7 - 4i}{i} = \frac{7 - 4i}{i} \cdot \frac{-i}{-i}$$

$$= \frac{-7i + 4i^2}{-i^2}$$

$$= \frac{-7i + 4(-1)}{-(-1)}$$

$$= -4 - 7i$$

Getting Ready for Class

After reading through the preceding section, respond in your own words and in complete sentences.

A. What is the number i?

B. What is a complex number?

C. What kind of number will always result when we multiply complex conjugates?

D. Explain how to divide complex numbers.

PROBLEM SET 9.7

Write the following in terms of i, and simplify as much as possible.

1. $\sqrt{-36}$ **2.** $\sqrt{-49}$ **3.** $-\sqrt{-25}$

4. $-\sqrt{-81}$ **5.** $\sqrt{-72}$ **6.** $\sqrt{-48}$

7. $-\sqrt{-12}$ **8.** $-\sqrt{-75}$

Write each of the following as i, -1, $-i$, or 1.

9. i^{28} **10.** i^{31} **11.** i^{26}

12. i^{37} **13.** i^{75} **14.** i^{42}

Find x and y so that each of the following equations is true.

15. $2x + 3yi = 6 - 3i$ **16.** $4x - 2yi = 4 + 8i$

17. $2 - 5i = -x + 10yi$ **18.** $4 + 7i = 6x - 14yi$

19. $2x + 10i = -16 - 2yi$

20. $4x - 5i = -2 + 3yi$

21. $(2x - 4) - 3i = 10 - 6yi$

22. $(4x - 3) - 2i = 8 + yi$

23. $(7x - 1) + 4i = 2 + (5y + 2)i$

24. $(5x + 2) - 7i = 4 + (2y + 1)i$

Combine the following complex numbers.

25. $(2 + 3i) + (3 + 6i)$ **26.** $(4 + i) + (3 + 2i)$

27. $(3 - 5i) + (2 + 4i)$ **28.** $(7 + 2i) + (3 - 4i)$

29. $(5 + 2i) - (3 + 6i)$ **30.** $(6 + 7i) - (4 + i)$

31. $(3 - 5i) - (2 + i)$

32. $(7 - 3i) - (4 + 10i)$

33. $[(3 + 2i) - (6 + i)] + (5 + i)$

34. $[(4 - 5i) - (2 + i)] + (2 + 5i)$

35. $[(7 - i) - (2 + 4i)] - (6 + 2i)$

36. $[(3 - i) - (4 + 7i)] - (3 - 4i)$

37. $(3 + 2i) - [(3 - 4i) - (6 + 2i)]$

38. $(7 - 4i) - [(-2 + i) - (3 + 7i)]$

39. $(4 - 9i) + [(2 - 7i) - (4 + 8i)]$

40. $(10 - 2i) - [(2 + i) - (3 - i)]$

Find the following products.

41. $3i(4 + 5i)$ **42.** $2i(3 + 4i)$

43. $6i(4 - 3i)$ **44.** $11i(2 - i)$

45. $(3 + 2i)(4 + i)$ **46.** $(2 - 4i)(3 + i)$

47. $(4 + 9i)(3 - i)$ **48.** $(5 - 2i)(1 + i)$

49. $(1 + i)^3$ **50.** $(1 - i)^3$

51. $(2 - i)^3$ **52.** $(2 + i)^3$

53. $(2 + 5i)^2$ **54.** $(3 + 2i)^2$

55. $(1 - i)^2$ **56.** $(1 + i)^2$

57. $(3 - 4i)^2$ **58.** $(6 - 5i)^2$

59. $(2 + i)(2 - i)$ **60.** $(3 + i)(3 - i)$

61. $(6 - 2i)(6 + 2i)$ **62.** $(5 + 4i)(5 - 4i)$

63. $(2 + 3i)(2 - 3i)$ **64.** $(2 - 7i)(2 + 7i)$

65. $(10 + 8i)(10 - 8i)$ **66.** $(11 - 7i)(11 + 7i)$

Find the following quotients. Write all answers in standard form for complex numbers.

67. $\dfrac{2 - 3i}{i}$ **68.** $\dfrac{3 + 4i}{i}$

69. $\dfrac{5 + 2i}{-i}$ **70.** $\dfrac{4 - 3i}{-i}$ **71.** $\dfrac{4}{2 - 3i}$

72. $\dfrac{3}{4 - 5i}$ **73.** $\dfrac{6}{-3 + 2i}$ **74.** $\dfrac{-1}{-2 - 5i}$

75. $\dfrac{2 + 3i}{2 - 3i}$ **76.** $\dfrac{4 - 7i}{4 + 7i}$ **77.** $\dfrac{5 + 4i}{3 + 6i}$

78. $\dfrac{2 + i}{5 - 6i}$

Applying the Concepts

79. Electric Circuits Complex numbers may be applied to electrical circuits. Electrical engineers use the fact that resistance R to electrical flow in a circuit is related to the electrical current I and the voltage V by the formula $V = RI$. (Voltage is measured in volts, resistance in ohms, and current in amperes.) Find the resistance to electrical flow in a circuit that has a voltage $V = (80 + 20i)$ volts and current $I = (-6 + 2i)$ amps.

80. Electric Circuits Refer to the information about electrical circuits in Problem 79, and find the current in a circuit that has a resistance of $(4 + 10i)$ ohms and a voltage of $(5 - 7i)$ volts.

Review Problems

The problems below review material we covered in Section 8.4.

For each relation below, state the domain and the range, and indicate which are also functions.

81. $\{(1, 2), (3, 4), (4, 2)\}$

82. $\{(0, 0), (1, 1), (0, 1)\}$

83. $\{(3, 1), (2, 3), (1, 2)\}$

84. $\{(-1, 1), (2, -2), (-3, -3)\}$

State whether each of the following graphs is the graph of a function.

85.

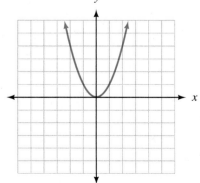

86.

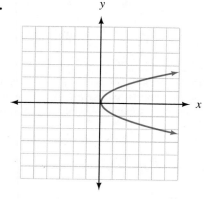

Extending the Concepts

87. Show that $-i$ and $\dfrac{1}{i}$ (the opposite and the reciprocal of i) are the same number.

88. Show that i^{2n+1} is the same as i for all positive even integers n.

89. Show that $x = 1 + i$ is a solution to the equation $x^2 - 2x + 2 = 0$.

90. Show that $x = 1 - i$ is a solution to the equation $x^2 - 2x + 2 = 0$.

91. Show that $x = 2 + i$ is a solution to the equation $x^3 - 11x + 20 = 0$.

92. Show that $x = 2 - i$ is a solution to the equation $x^3 - 11x + 20 = 0$.

Examples

1. The number 49 has two square roots, 7 and -7. They are written like this:

$$\sqrt{49} = 7 \qquad -\sqrt{49} = -7$$

Square Roots [9.1]

Every positive real number x has two square roots. The *positive square root* of x is written $\sqrt{x}$, and the *negative square root* of x is written $-\sqrt{x}$. Both the positive and the negative square roots of x are numbers we square to get x. That is,

$$\left. \begin{array}{c} (\sqrt{x})^2 = x \\ \text{and} \qquad (-\sqrt{x})^2 = x \end{array} \right\} \qquad \text{for } x \geq 0$$

2. $\sqrt[3]{8} = 2$
$\sqrt[3]{-27} = -3$

Higher Roots [9.1]

Consider the expression $\sqrt[n]{a}$: n is the *index,* a is the *radicand,* and $\sqrt{}$ is the *radical sign.* The expression $\sqrt[n]{a}$ is such that

$$(\sqrt[n]{a})^n = a \qquad a \geq 0 \text{ when } n \text{ is even}$$

3. $25^{1/2} = \sqrt{25} = 5$
$8^{2/3} = (\sqrt[3]{8})^2 = 2^2 = 4$
$9^{3/2} = (\sqrt{9})^3 = 3^3 = 27$

Rational Exponents [9.1, 9.2]

Rational exponents are used to indicate roots. The relationship between rational exponents and roots is as follows:

$$a^{1/n} = \sqrt[n]{a} \qquad \text{and} \qquad a^{m/n} = (a^{1/n})^m = (a^m)^{1/n}$$

$$a \geq 0 \text{ when } n \text{ is even}$$

4. $\sqrt{4 \cdot 5} = \sqrt{4}\,\sqrt{5} = 2\sqrt{5}$
$\sqrt{\dfrac{7}{9}} = \dfrac{\sqrt{7}}{\sqrt{9}} = \dfrac{\sqrt{7}}{3}$

Properties of Radicals [9.3]

If a and b are nonnegative real numbers whenever n is even, then

1. $\sqrt[n]{ab} = \sqrt[n]{a}\,\sqrt[n]{b}$

2. $\sqrt[n]{\dfrac{a}{b}} = \dfrac{\sqrt[n]{a}}{\sqrt[n]{b}} \qquad (b \neq 0)$

5. $\sqrt{\dfrac{4}{5}} = \dfrac{\sqrt{4}}{\sqrt{5}}$

$\phantom{\sqrt{\dfrac{4}{5}}} = \dfrac{2}{\sqrt{5}} \cdot \dfrac{\sqrt{5}}{\sqrt{5}}$

$\phantom{\sqrt{\dfrac{4}{5}}} = \dfrac{2\sqrt{5}}{5}$

Simplified Form for Radicals [9.3]

A radical expression is said to be in *simplified form*

1. If there is no factor of the radicand that can be written as a power greater than or equal to the index;

2. If there are no fractions under the radical sign; and

3. If there are no radicals in the denominator.

6. $5\sqrt{3} - 7\sqrt{3} = (5 - 7)\sqrt{3}$
$\phantom{5\sqrt{3} - 7\sqrt{3}} = -2\sqrt{3}$
$\sqrt{20} + \sqrt{45} = 2\sqrt{5} + 3\sqrt{5}$
$\phantom{\sqrt{20} + \sqrt{45}} = (2 + 3)\sqrt{5}$
$\phantom{\sqrt{20} + \sqrt{45}} = 5\sqrt{5}$

Addition and Subtraction of Radical Expressions [9.4]

We add and subtract radical expressions by using the distributive property to combine similar radicals. Similar radicals are radicals with the same index and the same radicand.

7. $(\sqrt{x} + 2)(\sqrt{x} + 3)$
$= \sqrt{x}\,\sqrt{x} + 3\sqrt{x} + 2\sqrt{x} + 2\cdot 3$
$= x + 5\sqrt{x} + 6$

Multiplication of Radical Expressions [9.5]
We multiply radical expressions in the same way that we multiply polynomials. We can use the distributive property and the FOIL method.

8. $\dfrac{3}{\sqrt{2}} = \dfrac{3}{\sqrt{2}} \cdot \dfrac{\sqrt{2}}{\sqrt{2}} = \dfrac{3\sqrt{2}}{2}$

$\dfrac{3}{\sqrt{5} - \sqrt{3}} = \dfrac{3}{\sqrt{5} - \sqrt{3}} \cdot \dfrac{\sqrt{5} + \sqrt{3}}{\sqrt{5} + \sqrt{3}}$

$\phantom{\dfrac{3}{\sqrt{5} - \sqrt{3}}} = \dfrac{3\sqrt{5} + 3\sqrt{3}}{5 - 3}$

$\phantom{\dfrac{3}{\sqrt{5} - \sqrt{3}}} = \dfrac{3\sqrt{5} + 3\sqrt{3}}{2}$

Rationalizing the Denominator [9.3, 9.5]
When a fraction contains a square root in the denominator, we rationalize the denominator by multiplying numerator and denominator by

1. The square root itself if there is only one term in the denominator, or

2. The conjugate of the denominator if there are two terms in the denominator.

Rationalizing the denominator can also be called division of radical expressions.

9. $\sqrt{2x + 1} = 3$
$(\sqrt{2x + 1})^2 = 3^2$
$2x + 1 = 9$
$x = 4$

Squaring Property of Equality [9.6]
We may square both sides of an equation any time it is convenient to do so, as long as we check all resulting solutions in the original equation.

10. $3 + 4i$ is a complex number.

Addition

$(3 + 4i) + (2 - 5i) = 5 - i$

Multiplication

$(3 + 4i)(2 - 5i)$
$ = 6 - 15i + 8i - 20i^2$
$ = 6 - 7i + 20$
$ = 26 - 7i$

Division

$\dfrac{2}{3 + 4i} = \dfrac{2}{3 + 4i} \cdot \dfrac{3 - 4i}{3 - 4i}$

$\phantom{\dfrac{2}{3 + 4i}} = \dfrac{6 - 8i}{9 + 16}$

$\phantom{\dfrac{2}{3 + 4i}} = \dfrac{6}{25} - \dfrac{8}{25}i$

Complex Numbers [9.7]
A *complex number* is any number that can be put in the form

$$a + bi$$

where a and b are real numbers and $i = \sqrt{-1}$. The *real part* of the complex number is a, and b is the *imaginary part*.

If a, b, c, and d are real numbers, then we have the following definitions associated with complex numbers:

1. Equality

$\quad a + bi = c + di \qquad$ if and only if $\qquad a = c$ and $b = d$

2. Addition and subtraction

$$(a + bi) + (c + di) = (a + c) + (b + d)i$$
$$(a + bi) - (c + di) = (a - c) + (b - d)i$$

3. Multiplication

$$(a + bi)(c + di) = (ac - bd) + (ad + bc)i$$

4. Division is similar to rationalizing the denominator.

Simplify each expression as much as possible. [9.1]

1. $\sqrt{49}$ **2.** $(-27)^{1/3}$ **3.** $16^{1/4}$

4. $9^{3/2}$ **5.** $\sqrt[5]{32x^{15}y^{10}}$ **6.** $8^{-4/3}$

Use the properties of exponents to simplify each expression. Assume all bases represent positive numbers. [9.1]

7. $x^{2/3} \cdot x^{4/3}$ **8.** $(a^{2/3}b^{4/3})^3$ **9.** $\dfrac{a^{3/5}}{a^{1/4}}$

10. $\dfrac{a^{2/3}b^3}{a^{1/4}b^{1/3}}$

Multiply. [9.2]

11. $(3x^{1/2} + 5y^{1/2})(4x^{1/2} - 3y^{1/2})$

12. $(a^{1/3} - 5)^2$

13. Divide: $\dfrac{28x^{5/6} + 14x^{7/6}}{7x^{1/3}}$ (Assume $x > 0$.) [9.2]

14. Factor $2(x - 3)^{1/4}$ from $8(x - 3)^{5/4} - 2(x - 3)^{1/4}$. [9.2]

15. Simplify $x^{3/4} + \dfrac{5}{x^{1/4}}$ into a single fraction. (Assume $x > 0$.) [9.2]

Write each expression in simplified form for radicals. (Assume all variables represent nonnegative numbers.) [9.3]

16. $\sqrt{12}$ **17.** $\sqrt{50}$ **18.** $\sqrt[3]{16}$

19. $\sqrt{18x^2}$ **20.** $\sqrt{80a^3b^4c^2}$ **21.** $\sqrt[4]{32a^4b^5c^6}$

Rationalize the denominator in each expression. [9.3]

22. $\dfrac{3}{\sqrt{2}}$ **23.** $\dfrac{6}{\sqrt[3]{2}}$

Write each expression in simplified form. (Assume all variables represent positive numbers.) [9.3]

24. $\sqrt{\dfrac{48x^3}{7y}}$ **25.** $\sqrt[3]{\dfrac{40x^2y^3}{3z}}$

Combine the following expressions. (Assume all variables represent positive numbers.) [9.4]

26. $5x\sqrt{6} + 2x\sqrt{6} - 9x\sqrt{6}$

27. $\sqrt{12} + \sqrt{3}$ **28.** $\dfrac{3}{\sqrt{5}} + \sqrt{5}$

29. $3\sqrt{8} - 4\sqrt{72} + 5\sqrt{50}$

30. $3b\sqrt{27a^5b} + 2a\sqrt{3a^3b^3}$

31. $2x\sqrt[3]{xy^3z^2} - 6y\sqrt[3]{x^4z^2}$

Multiply. [9.5]

32. $\sqrt{2}(\sqrt{3} - 2\sqrt{2})$ **33.** $(\sqrt{x} - 2)(\sqrt{x} - 3)$

Rationalize the denominator. (Assume $x, y > 0$.) [9.5]

34. $\dfrac{3}{\sqrt{5} - 2}$ **35.** $\dfrac{\sqrt{7} + \sqrt{5}}{\sqrt{7} - \sqrt{5}}$ **36.** $\dfrac{3\sqrt{7}}{3\sqrt{7} - 4}$

Solve each equation. [9.6]

37. $\sqrt{4a + 1} = 1$ **38.** $\sqrt[3]{3x - 8} = 1$

39. $\sqrt{3x + 1} - 3 = 1$ **40.** $\sqrt{x + 4} = \sqrt{x} - 2$

Graph each equation. [9.6]

41. $y = 3\sqrt{x}$ **42.** $y = \sqrt[3]{x} + 2$

Write each of the following as $i, -1, -i$, or 1. [9.7]

43. i^{24} **44.** i^{27}

Find x and y so that each of the following equations is true. [9.7]

45. $3 - 4i = -2x + 8yi$

46. $(3x + 2) - 8i = -4 + 2yi$

Combine the following complex numbers. [9.7]

47. $(3 + 5i) + (6 - 2i)$

48. $(2 + 5i) - [(3 + 2i) + (6 - i)]$

Multiply. [9.7]

49. $3i(4 + 2i)$ **50.** $(2 + 3i)(4 + i)$

51. $(4 + 2i)^2$ **52.** $(4 + 3i)(4 - 3i)$

Divide. Write all answers in standard form for complex numbers. [9.7]

53. $\dfrac{3 + i}{i}$ **54.** $\dfrac{-3}{2 + i}$

55. Construction The roof of a house shown in Figure 1 is to extend up 13.5 feet above the ceiling, which is 36 feet across. Find the length of one side of the roof.

FIGURE 1

56. Surveying A surveyor is attempting to find the distance across a pond. From a point on one side of the pond he walks 25 yards to the end of the pond and then makes a 90-degree turn and walks another 60 yards before coming to a point directly across the pond from the point at which he started. What is the distance across the pond? (See Figure 2.)

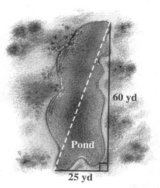

FIGURE 2

CHAPTER 9 PROJECTS

RATIONAL EXPONENTS AND ROOTS

GROUP PROJECT

CONSTRUCTING THE SPIRAL OF ROOTS

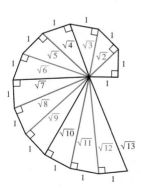

Number of People: 3

Time Needed: 20 minutes

Equipment: Two sheets of graph paper (4 or 5 squares per inch) and pencils

Background: The spiral of roots gives us a way to visualize the positive square roots of the counting numbers, and in so doing, we see many line segments whose lengths are irrational numbers.

Procedure: You are to construct a spiral of roots from a line segment 1 inch long. The graph paper you have contains either 4 or 5 squares per inch, allowing you to accurately draw 1-inch line segments. Since the lines on the graph paper are perpendicular to one another, if you are careful, you can also use the graph paper to connect one line segment to another so that they form a right angle.

1. Fold one of the pieces of graph paper so it can be used as a ruler.

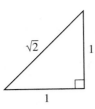

FIGURE 1

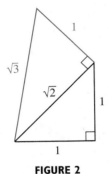

FIGURE 2

2. Use the folded paper to draw a line segment 1 inch long, just to the right of the middle of the unfolded paper. On the end of this segment, attach another segment of 1-inch length at a right angle to the first one. Connect the endpoints of the segments to form a right triangle. Label each of the sides of this triangle. When you are finished, your work should resemble Figure 1.

3. On the end of the hypotenuse of the triangle, attach a 1-inch line segment so that the two segments form a right angle. (Use the folded paper to do this.) Draw the hypotenuse of this triangle. Label all the sides of this second triangle. Your work should resemble Figure 2.

4. Continue to draw a new right triangle by attaching 1-inch line segments at right angles to the previous 1-inch line segment. Label all the sides of each triangle.

5. Stop when you have drawn a hypotenuse $\sqrt{8}$ inches long.

RESEARCH PROJECT

CONNECTIONS

Although it may not look like it, the three items shown here are related very closely to one another. Your job is to find the connection.

A Continued Fraction   *The Fibonacci Sequence*   *The Golden Rectangle*

$$1 + \cfrac{1}{1 + \cfrac{1}{1 + \cfrac{1}{1 + \cdots}}}$$

1, 1, 2, 3, 5, . . .

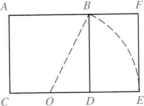

Step 1: The dots in the continued fraction indicate that the pattern shown continues indefinitely. This means that there is no way for us to simplify this expression, as we have simplified the expressions in this chapter. However, we can begin to understand the continued fraction, and what it simplifies to, by working with the following sequence of expressions. Simplify each expression. Write each answer as a fraction, in lowest terms.

$$1 + \cfrac{1}{1+1} \qquad 1 + \cfrac{1}{1+\cfrac{1}{1+1}} \qquad 1 + \cfrac{1}{1+\cfrac{1}{1+\cfrac{1}{1+1}}} \qquad 1 + \cfrac{1}{1+\cfrac{1}{1+\cfrac{1}{1+\cfrac{1}{1+1}}}}$$

Step 2: Compare the fractional answers to step 1 with the numbers in the Fibonacci sequence. Based on your observation, give the answer to the following problem, without actually doing any arithmetic.

$$1 + \cfrac{1}{1+\cfrac{1}{1+\cfrac{1}{1+\cfrac{1}{1+\cfrac{1}{1+1}}}}}$$

Step 3: Continue the sequence of simplified fractions you have written in steps 1 and 2, until you have nine numbers in the sequence. Convert each of these numbers to a decimal, accurate to four places past the decimal point.

Step 4: Find a decimal approximation to the golden ratio $\dfrac{1+\sqrt{5}}{2}$, accurate to four places past the decimal point.

Step 5: Compare the results in steps 3 and 4, and then make a conjecture about what number the continued fraction would simplify to, if it was actually possible to simplify it.

C H A P T E R   9   **TEST**

Simplify each of the following. (Assume all variable bases are positive integers and all variable exponents are positive real numbers throughout this test.) [9.1]

1. $27^{-2/3}$

2. $\left(\dfrac{25}{49}\right)^{-1/2}$

3. $a^{3/4} \cdot a^{-1/3}$

4. $\dfrac{(x^{2/3}y^{-3})^{1/2}}{(x^{3/4}y^{1/2})^{-1}}$

5. $\sqrt{49x^8 y^{10}}$

6. $\sqrt[5]{32x^{10}y^{20}}$

7. $\dfrac{(36a^8 b^4)^{1/2}}{(27a^9 b^6)^{1/3}}$

8. $\dfrac{(x^n y^{1/n})^n}{(x^{1/n}y^n)^{n^2}}$

Multiply. [9.2]

9. $2a^{1/2}(3a^{3/2} - 5a^{1/2})$ **10.** $(4a^{3/2} - 5)^2$

Factor. [9.2]

11. $3x^{2/3} + 5x^{1/3} - 2$ **12.** $9x^{2/3} - 49$

Combine. [9.4]

13. $\dfrac{4}{x^{1/2}} + x^{1/2}$

14. $\dfrac{x^2}{(x^2 - 3)^{1/2}} - (x^2 - 3)^{1/2}$

Write in simplified form. [9.3]

15. $\sqrt{125x^3y^5}$

16. $\sqrt[3]{40x^7y^8}$

17. $\sqrt{\dfrac{2}{3}}$

18. $\sqrt{\dfrac{12a^4b^3}{5c}}$

Combine. [9.4]

19. $3\sqrt{12} - 4\sqrt{27}$

20. $\sqrt[3]{24a^3b^3} - 5a\sqrt[3]{3b^3}$

Multiply. [9.5]

21. $(\sqrt{x} + 7)(\sqrt{x} - 4)$

22. $(3\sqrt{2} - \sqrt{3})^2$

Rationalize the denominator. [9.5]

23. $\dfrac{5}{\sqrt{3} - 1}$

24. $\dfrac{\sqrt{x} - \sqrt{2}}{\sqrt{x} + \sqrt{2}}$

Solve for x. [9.6]

25. $\sqrt{3x + 1} = x - 3$

26. $\sqrt[3]{2x + 7} = -1$

27. $\sqrt{x + 3} = \sqrt{x + 4} - 1$

Graph. [9.6]

28. $y = \sqrt{x} - 2$

29. $y = \sqrt[3]{x} + 3$

30. Solve for x and y so that the following equation is true [9.7]:

$$(2x + 5) - 4i = 6 - (y - 3)i$$

Perform the indicated operations. [9.7]

31. $(3 + 2i) - [(7 - i) - (4 + 3i)]$

32. $(2 - 3i)(4 + 3i)$

33. $(5 - 4i)^2$

34. $\dfrac{2 - 3i}{2 + 3i}$

35. Show that i^{38} can be written as -1. [9.7]

Simplify.

1. $33 - 22 - (-11) + 1$
2. $12 - 20 \div 4 - 3 \cdot 2$
3. $-6 + 5[3 - 2(-4 - 1)]$
4. $3(2x + 5) + 4(4x - 1)$
5. $(2y^{-3})^{-1}(4y^{-3})^2$
6. $8^{2/3}$
7. $\sqrt{72y^5}$
8. $\sqrt{15} - \sqrt{\dfrac{3}{5}}$
9. Give the opposite and reciprocal of 12.
10. Specify the domain and range for the relation $\{(-1, 2), (3, -1), (-1, 0)\}$. Is this relation also a function?
11. Write 41,500 in scientific notation.
12. Let $f(x) = x^2 - 3x$ and $g(x) = x - 1$. Find: $f(3) - g(0)$

Factor completely.

13. $625a^4 - 16b^4$
14. $24a^4 + 10a^2 - 6$

Reduce to lowest terms.

15. $\dfrac{246}{861}$
16. $\dfrac{28xy^3z^2}{14x^2y^3z}$
17. $\dfrac{x^2 - 9x + 20}{x^2 - 7x + 12}$

Multiply.

18. $\dfrac{6}{7} \cdot \dfrac{21}{35} \cdot 5$
19. $(2x - 5y)(3x - 2y)$
20. $(9x^2 - 25) \cdot \dfrac{x + 5}{3x - 5}$
21. $(\sqrt{x} - 2)^2$

Divide.

22. $\dfrac{18a^4b^2 - 9a^2b^2 + 27a^2b^4}{-9a^2b^2}$
23. $\dfrac{27x^3y^2}{13x^2y^4} \div \dfrac{9xy}{26y}$
24. $\dfrac{\sqrt{x} + \sqrt{y}}{\sqrt{x} - \sqrt{y}}$

Subtract.

25. $\left(\dfrac{2}{3}x^3 + \dfrac{1}{6}x^2 + \dfrac{1}{2}\right) - \left(\dfrac{1}{4}x^2 - \dfrac{1}{3}x + \dfrac{1}{12}\right)$
26. $\dfrac{3}{x^2 + 8x + 15} - \dfrac{1}{x^2 + 7x + 12} - \dfrac{1}{x^2 + 9x + 20}$

Solve.

27. $3y - 8 = -4y + 6$
28. $\dfrac{2}{3}(9x - 2) + \dfrac{1}{3} = 4$
29. $|3x - 1| - 2 = 6$
30. $x^3 + 2x^2 - 16x - 32 = 0$
31. $\dfrac{x + 2}{x + 1} - 2 = \dfrac{1}{x + 1}$
32. $\sqrt[3]{8 - 3x} = -1$
33. $(x + 3)^2 - 3(x + 3) - 70 = 0$
34. $\sqrt{y + 7} - \sqrt{y + 2} = 1$
35. Solve for L if $P = 2L + 2W$, $P = 18$ and $W = 3$.

Solve each inequality, and graph the solution.

36. $4y - 2 \geq 10$ or $4y - 2 \leq -10$
37. $|5x - 4| \geq 6$

Solve each system.

38. $\begin{aligned} -7x + 8y &= 7 \\ 6x - 5y &= -19 \end{aligned}$
39. $\begin{aligned} 4x + 9y &= 2 \\ y &= 2x - 12 \end{aligned}$
40. $\begin{aligned} x + y + z &= -2 \\ 2x + y - 3z &= 1 \\ -2x - 3y + 4z &= 9 \end{aligned}$
41. $\begin{aligned} x + y &= -2 \\ y + 10z &= -1 \\ 2x - 13z &= 5 \end{aligned}$
42. Evaluate $x^2 - 10x + 25$ and $(x - 5)^2$ when $x = 10$.

Graph on a rectangular coordinate system.

43. $y = \dfrac{1}{3}x - 3$
44. $y = \sqrt{x} + 3$
45. $y = \dfrac{4}{x}$

46. Find y if the line through $(-1, 2)$ and $(-3, y)$ has a slope of -2.
47. Find the slope and y-intercept of $4x - 5y = 15$.
48. Give the equation of a line with slope $-\dfrac{2}{5}$ and y-intercept $= 5$.
49. Write the equation of the line in slope-intercept form if the slope is $\dfrac{2}{3}$ and $(9, 3)$ is a point on the line.
50. **Geometry** The length of a rectangle is 4 feet more than twice the width. The perimeter is 68 feet. Find the dimensions.

Quadratic Functions

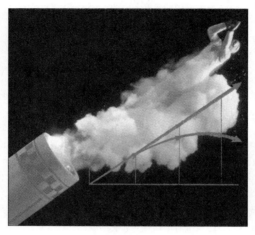

(Jerry Yulsman/The Image Bank)

INTRODUCTION

If you have been to the circus or the county fair recently, you may have witnessed one of the more spectacular acts, the human cannonball. The human cannonball shown in the photograph will reach a height of 70 feet, and travel a distance of 160 feet, before landing in a safety net. In this chapter we use this information to derive the equation

$$y = -\frac{7}{640}(x - 80)^2 + 70 \quad \text{for } 0 \le x \le 160$$

which describes the path flown by this particular cannonball. The table and graph below were constructed from this equation.

TABLE I	Path of a Human Cannonball
x (feet)	**y** (nearest foot)
0	0
40	53
80	70
120	53
160	0

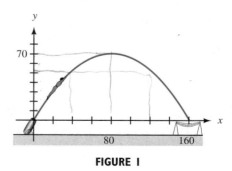

FIGURE I

All objects that are projected into the air, whether they are basketballs, bullets, arrows, or coins, follow parabolic paths like the one shown in Figure 1. Studying the material in this chapter will give you a more mathematical hold on the world around you.

Table 2 is taken from the trail map given to skiers at the Northstar at Tahoe Ski Resort in Lake Tahoe, California. The table gives the length of each chair lift at Northstar, along with the change in elevation from the beginning of the lift to the end of the lift.

Right triangles are good mathematical models for chair lifts. In this section we will use our knowledge of right triangles, along with the new material developed in the section, to solve problems involving chair lifts and a variety of other examples.

TABLE 2 From the Trail Map for the Northstar at Tahoe Ski Resort

Lift Information

Lift	Vertical Rise (feet)	Length (feet)
Big Springs Gondola	480	4,100
Bear Paw Double	120	790
Echo Triple	710	4,890
Aspen Express Quad	900	5,100
Forest Double	1,170	5,750
Lookout Double	960	4,330
Comstock Express Quad	1,250	5,900
Rendezvous Triple	650	2,900
Schaffer Camp Triple	1,860	6,150
Chipmunk Tow Lift	28	280
Bear Cub Tow Lift	120	750

In this section we will develop the first of our new methods of solving quadratic equations. The new method is called *completing the square*. Completing the square on a quadratic equation allows us to obtain solutions, regardless of whether the equation can be factored. Before we solve equations by completing the square, we need to learn how to solve equations by taking square roots of both sides.

Consider the equation

$$x^2 = 16$$

We could solve it by writing it in standard form, factoring the left side, and proceeding as we did in Chapter 5. We can shorten our work considerably, however, if

617

we simply notice that x must be either the positive square root of 16 or the negative square root of 16. That is,

$$\text{If} \quad x^2 = 16$$

$$\text{then} \quad x = \sqrt{16} \quad \text{or} \quad x = -\sqrt{16}$$

$$x = 4 \quad \text{or} \quad x = -4$$

We can generalize this result into a theorem as follows.

THEOREM 10.1

If $a^2 = b$, where b is a real number, then $a = \sqrt{b}$ or $a = -\sqrt{b}$.

Notation   The expression $a = \sqrt{b}$ or $a = -\sqrt{b}$ can be written in shorthand form as $a = \pm\sqrt{b}$. The symbol $\pm$ is read "plus or minus."

We can apply Theorem 10.1 to some fairly complicated quadratic equations.

EXAMPLE   1   Solve $(2x - 3)^2 = 25$.

Solution

$$(2x - 3)^2 = 25$$

$$2x - 3 = \pm\sqrt{25} \qquad \text{Theorem 10.1}$$

$$2x - 3 = \pm 5 \qquad \sqrt{25} = 5$$

$$2x = 3 \pm 5 \qquad \text{Add 3 to both sides.}$$

$$x = \frac{3 \pm 5}{2} \qquad \text{Divide both sides by 2.}$$

The last equation can be written as two separate statements:

$$x = \frac{3 + 5}{2} \quad \text{or} \quad x = \frac{3 - 5}{2}$$

$$= \frac{8}{2} \qquad\qquad = \frac{-2}{2}$$

$$= 4 \quad \text{or} \quad = -1$$

The solution set is $\{4, -1\}$.

Notice that we could have solved the equation in Example 1 by expanding the left side, writing the resulting equation in standard form, and then factoring. The problem would look like this:

$$(2x - 3)^2 = 25 \qquad \text{Original equation}$$

$$4x^2 - 12x + 9 = 25 \qquad \text{Expand the left side.}$$

$$4x^2 - 12x - 16 = 0 \qquad \text{Add } -25 \text{ to each side.}$$

$$4(x^2 - 3x - 4) = 0 \qquad \text{Begin factoring.}$$

$$4(x - 4)(x + 1) = 0 \qquad \text{Factor completely.}$$

$$x - 4 = 0 \quad \text{or} \quad x + 1 = 0 \qquad \text{Set variable factors equal to 0.}$$

$$x = 4 \quad \text{or} \quad x = -1$$

As you can see, solving the equation by factoring leads to the same two solutions.

E X A M P L E   2   Solve for x: $(3x - 1)^2 = -12$

Solution

$$(3x - 1)^2 = -12$$

$$3x - 1 = \pm\sqrt{-12} \qquad \text{Theorem 10.1}$$

$$3x - 1 = \pm 2i\sqrt{3} \qquad \sqrt{-12} = 2i\sqrt{3}$$

$$3x = 1 \pm 2i\sqrt{3} \qquad \text{Add 1 to both sides.}$$

$$x = \frac{1 \pm 2i\sqrt{3}}{3} \qquad \text{Divide both sides by 3.}$$

The solution set is $\left\{ \dfrac{1 + 2i\sqrt{3}}{3}, \dfrac{1 - 2i\sqrt{3}}{3} \right\}$.

Both solutions are complex. Here is a check of the first solution:

When $\qquad\qquad\qquad\qquad\qquad\qquad\qquad x = \dfrac{1 + 2i\sqrt{3}}{3}$

the equation $\qquad\qquad\qquad\qquad\qquad (3x - 1)^2 = -12$

becomes $\qquad\qquad\qquad\qquad \left(3 \cdot \dfrac{1 + 2i\sqrt{3}}{3} - 1 \right)^2 \overset{?}{=} -12$

or $\qquad\qquad\qquad\qquad\qquad (1 + 2i\sqrt{3} - 1)^2 \overset{?}{=} -12$

$$(2i\sqrt{3})^2 \overset{?}{=} -12$$

$$4 \cdot i^2 \cdot 3 \overset{?}{=} -12$$

$$12(-1) \overset{?}{=} -12$$

$$-12 = -12$$

Note: We cannot solve the equation in Example 2 by factoring. If we expand the left side and write the resulting equation in standard form, we are left with a quadratic equation that does not factor:

$$(3x - 1)^2 = -12 \qquad \text{Equation from Example 2}$$

$$9x^2 - 6x + 1 = -12 \qquad \text{Expand the left side.}$$

$$9x^2 - 6x + 13 = 0 \qquad \text{Standard form, but not factorable}$$

EXAMPLE 3 Solve $x^2 + 6x + 9 = 12$.

Solution We can solve this equation as we have the equations in Examples 1 and 2 if we first write the left side as $(x + 3)^2$.

$$x^2 + 6x + 9 = 12 \qquad \text{Original equation}$$

$$(x + 3)^2 = 12 \qquad \text{Write } x^2 + 6x + 9 \text{ as } (x + 3)^2.$$

$$x + 3 = \pm 2\sqrt{3} \qquad \text{Theorem 10.1}$$

$$x = -3 \pm 2\sqrt{3} \qquad \text{Add } -3 \text{ to each side.}$$

We have two irrational solutions: $-3 + 2\sqrt{3}$ and $-3 - 2\sqrt{3}$. What is important about this problem, however, is the fact that the equation was easy to solve because the left side was a perfect square trinomial.

METHOD OF COMPLETING THE SQUARE

The method of completing the square is simply a way of transforming any quadratic equation into an equation of the form found in the preceding three examples.

The key to understanding the method of completing the square lies in recognizing the relationship between the last two terms of any perfect square trinomial whose leading coefficient is 1.

Consider the following list of perfect square trinomials and their corresponding binomial squares:

$$x^2 - 6x + 9 = (x - 3)^2$$

$$x^2 + 8x + 16 = (x + 4)^2$$

$$x^2 - 10x + 25 = (x - 5)^2$$

$$x^2 + 12x + 36 = (x + 6)^2$$

In each case the leading coefficient is 1. A more important observation comes from noticing the relationship between the linear and constant terms (middle and last terms) in each trinomial. Observe that the constant term in each case is the square of half the coefficient of x in the middle term. For example, in the last expression, the constant term 36 is the square of half of 12, where 12 is the coefficient of x in the middle term. (Notice also that the second terms in all the binomials on the right side are half the coefficients of the middle terms of the trinomials on the left side.) We can use these observations to build our own perfect square trinomials and, in doing so, solve some quadratic equations.

Consider the following equation:

$$x^2 + 6x = 3$$

We can think of the left side as having the first two terms of a perfect square trinomial. We need only add the correct constant term. If we take half the coefficient of x, we get 3. If we then square this quantity, we have 9. Adding the 9 to both sides, the equation becomes

$$x^2 + 6x + \mathbf{9} = 3 + \mathbf{9}$$

The left side is the perfect square $(x + 3)^2$; the right side is 12:

$$(x + 3)^2 = 12$$

The equation is now in the correct form. We can apply Theorem 10.1 and finish the solution:

$$(x + 3)^2 = 12$$
$$x + 3 = \pm\sqrt{12} \qquad \text{Theorem 10.1}$$
$$x + 3 = \pm 2\sqrt{3}$$
$$x = -3 \pm 2\sqrt{3}$$

The solution set is $\{-3 + 2\sqrt{3}, -3 - 2\sqrt{3}\}$. The method just used is called *completing the square,* since we complete the square on the left side of the original equation by adding the appropriate constant term.

EXAMPLE 4   Solve by completing the square: $x^2 + 5x - 2 = 0$

Solution   We must begin by adding 2 to both sides. (The left side of the equation, as it is, is not a perfect square, because it does not have the correct constant term. We will simply "move" that term to the other side and use our own constant term.)

$$x^2 + 5x = 2 \qquad \text{Add 2 to each side.}$$

We complete the square by adding the square of half the coefficient of the linear term to both sides:

$$x^2 + 5x + \frac{25}{4} = 2 + \frac{25}{4} \qquad \text{Half of 5 is } \tfrac{5}{2}, \text{ the square of which is } \tfrac{25}{4}.$$

$$\left(x + \frac{5}{2}\right)^2 = \frac{33}{4} \qquad\qquad 2 + \frac{25}{4} = \frac{8}{4} + \frac{25}{4} = \frac{33}{4}$$

$$x + \frac{5}{2} = \pm\sqrt{\frac{33}{4}} \qquad\qquad \text{Theorem 10.1}$$

$$x + \frac{5}{2} = \pm\frac{\sqrt{33}}{2} \qquad\qquad \text{Simplify the radical.}$$

$$x = -\frac{5}{2} \pm \frac{\sqrt{33}}{2} \qquad\qquad \text{Add } -\frac{5}{2} \text{ to both sides.}$$

$$x = \frac{-5 \pm \sqrt{33}}{2}$$

The solution set is $\left\{ \dfrac{-5 + \sqrt{33}}{2}, \dfrac{-5 - \sqrt{33}}{2} \right\}$.

We can use a calculator to get decimal approximations to these solutions. If $\sqrt{33} \approx 5.74$, then

$$\frac{-5 + 5.74}{2} = 0.37$$

$$\frac{-5 - 5.74}{2} = -5.37$$

E X A M P L E   5   Solve for x: $3x^2 - 8x + 7 = 0$

Solution

$$3x^2 - 8x + 7 = 0$$

$$3x^2 - 8x = -7 \qquad \text{Add } -7 \text{ to both sides.}$$

We cannot complete the square on the left side because the leading coefficient is not 1. We take an extra step and divide both sides by 3:

$$\frac{3x^2}{3} - \frac{8x}{3} = -\frac{7}{3}$$

$$x^2 - \frac{8}{3}x = -\frac{7}{3}$$

Half of $\frac{8}{3}$ is $\frac{4}{3}$, the square of which is $\frac{16}{9}$:

$$x^2 - \frac{8}{3}x + \mathbf{\frac{16}{9}} = -\frac{7}{3} + \mathbf{\frac{16}{9}} \qquad \text{Add } \mathbf{\frac{16}{9}} \text{ to both sides.}$$

$$\left(x - \frac{4}{3}\right)^2 = -\frac{5}{9} \qquad \text{Simplify right side.}$$

$$x - \frac{4}{3} = \pm\sqrt{-\frac{5}{9}} \qquad \text{Theorem 10.1}$$

$$x - \frac{4}{3} = \pm\frac{i\sqrt{5}}{3} \qquad \sqrt{-\frac{5}{9}} = \frac{\sqrt{-5}}{3} = \frac{i\sqrt{5}}{3}$$

$$x = \frac{4}{3} \pm \frac{i\sqrt{5}}{3} \qquad \text{Add } \frac{4}{3} \text{ to both sides.}$$

$$x = \frac{4 \pm i\sqrt{5}}{3}$$

The solution set is $\left\{ \dfrac{4 + i\sqrt{5}}{3}, \dfrac{4 - i\sqrt{5}}{3} \right\}$.

TO SOLVE A QUADRATIC EQUATION BY COMPLETING THE SQUARE

To summarize the method used in the preceding two examples, we list the following steps:

Step 1: Write the equation in the form $ax^2 + bx = c$.

Step 2: If the leading coefficient is not 1, divide both sides by the coefficient so that the resulting equation has a leading coefficient of 1. That is, if $a \neq 1$, then divide both sides by a.

Step 3: Add the square of half the coefficient of the linear term to both sides of the equation.

Step 4: Write the left side of the equation as the square of a binomial, and simplify the right side if possible.

Step 5: Apply Theorem 10.1, and solve as usual.

FACTS FROM
Geometry

More Special Triangles

The triangles shown in Figures 1 and 2 occur frequently in mathematics.

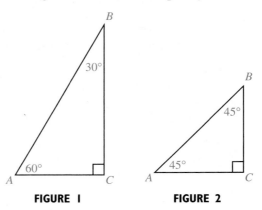

FIGURE 1   **FIGURE 2**

Note that both of the triangles are right triangles. We refer to the triangle in Figure 1 as a $30°–60°–90°$ triangle, and the triangle in Figure 2 as a $45°–45°–90°$ triangle.

EXAMPLE 6   If the shortest side in a $30°-60°-90°$ triangle is 1 inch, find the lengths of the other two sides.

Solution   In Figure 3 triangle ABC is a $30°-60°-90°$ triangle in which the shortest side AC is 1 inch long. Triangle DBC is also a $30°-60°-90°$ triangle in which the shortest side DC is 1 inch long.

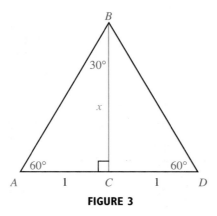

FIGURE 3

Notice that the large triangle ABD is an equilateral triangle because each of its interior angles is $60°$. Each side of triangle ABD is 2 inches long. Side AB in triangle ABC is therefore 2 inches. To find the length of side BC, we use the Pythagorean theorem.

$$BC^2 + AC^2 = AB^2$$
$$x^2 + 1^2 = 2^2$$
$$x^2 + 1 = 4$$
$$x^2 = 3$$
$$x = \sqrt{3} \text{ inches}$$

Note that we write only the positive square root because x is the length of a side in a triangle and is therefore a positive number.

EXAMPLE 7   Table 2 in the introduction to this section gives the vertical rise of the Forest Double chair lift as 1,170 feet and the length of the chair lift as 5,750 feet. To the nearest foot, find the horizontal distance covered by a person riding this lift.

Solution   Figure 4 is a model of the Forest Double chair lift. A rider gets on the lift at point A and exits at point B. The length of the lift is AB.

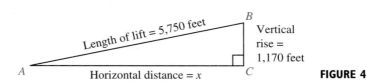

FIGURE 4

To find the horizontal distance covered by a person riding the chair lift, we use the Pythagorean theorem.

$$5{,}750^2 = x^2 + 1{,}170^2$$ Pythagorean theorem

$$33{,}062{,}500 = x^2 + 1{,}368{,}900$$ Simplify squares.

$$x^2 = 33{,}062{,}500 - 1{,}368{,}900$$ Solve for x^2.

$$x^2 = 31{,}693{,}600$$ Simplify the right side.

$$x = \sqrt{31{,}693{,}600}$$ Theorem 10.1

$$= 5{,}630 \text{ feet}$$ to the nearest foot

A rider getting on the lift at point A and riding to point B will cover a horizontal distance of approximately 5,630 feet.

Getting Ready for Class

After reading through the preceding section, respond in your own words and in complete sentences.

A. What kind of equation do we solve using the method of completing the square?

B. Explain in words how you would complete the square on $x^2 - 16x = 4$.

C. What is the relationship between the shortest side and the longest side in a $30°-60°-90°$ triangle?

D. What two expressions together are equivalent to $x = \pm 4$?

PROBLEM SET 10.1

Solve the following equations.

1. $x^2 = 25$

2. $x^2 = 16$

3. $a^2 = -9$

4. $a^2 = -49$

5. $y^2 = \dfrac{3}{4}$

6. $y^2 = \dfrac{5}{9}$

7. $x^2 + 12 = 0$

8. $x^2 + 8 = 0$

9. $4a^2 - 45 = 0$

10. $9a^2 - 20 = 0$

11. $(2y - 1)^2 = 25$

12. $(3y + 7)^2 = 1$

13. $(2a + 3)^2 = -9$

14. $(3a - 5)^2 = -49$

15. $(5x + 2)^2 = -8$

16. $(6x - 7)^2 = -75$

17. $x^2 + 8x + 16 = -27$

18. $x^2 - 12x + 36 = -8$

19. $4a^2 - 12a + 9 = -4$

20. $9a^2 - 12a + 4 = -9$

Simplify the left side of each equation, and then solve for x.

21. $(x + 5)^2 + (x - 5)^2 = 52$

22. $(2x + 1)^2 + (2x - 1)^2 = 10$

23. $(2x + 3)^2 + (2x - 3)^2 = 26$

24. $(3x + 2)^2 + (3x - 2)^2 = 26$

25. $(3x + 4)(3x - 4) - (x + 2)(x - 2) = -4$

26. $(5x + 2)(5x - 2) - (x + 3)(x - 3) = 29$

Copy each of the following, and fill in the blanks so that the left side of each is a perfect square trinomial. That is, complete the square.

27. $x^2 + 12x + \underline{} = (x + \underline{})^2$

28. $x^2 + 6x + \underline{} = (x + \underline{})^2$

29. $x^2 - 4x + \underline{\hspace{0.4cm}} = (x - \underline{\hspace{0.4cm}})^2$

30. $x^2 - 2x + \underline{\hspace{0.4cm}} = (x - \underline{\hspace{0.4cm}})^2$

31. $a^2 - 10a + \underline{\hspace{0.4cm}} = (a - \underline{\hspace{0.4cm}})^2$

32. $a^2 - 8a + \underline{\hspace{0.4cm}} = (a - \underline{\hspace{0.4cm}})^2$

33. $x^2 + 5x + \underline{\hspace{0.4cm}} = (x + \underline{\hspace{0.4cm}})^2$

34. $x^2 + 3x + \underline{\hspace{0.4cm}} = (x + \underline{\hspace{0.4cm}})^2$

35. $y^2 - 7y + \underline{\hspace{0.4cm}} = (y - \underline{\hspace{0.4cm}})^2$

36. $y^2 - y + \underline{\hspace{0.4cm}} = (y - \underline{\hspace{0.4cm}})^2$

Solve each of the following quadratic equations by completing the square.

37. $x^2 + 4x = 12$ **38.** $x^2 - 2x = 8$

39. $x^2 + 12x = -27$ **40.** $x^2 - 6x = 16$

41. $a^2 - 2a + 5 = 0$ **42.** $a^2 + 10a + 22 = 0$

43. $y^2 - 8y + 1 = 0$ **44.** $y^2 + 6y - 1 = 0$

45. $x^2 - 5x - 3 = 0$ **46.** $x^2 - 5x - 2 = 0$

47. $2x^2 - 4x - 8 = 0$ **48.** $3x^2 - 9x - 12 = 0$

49. $3t^2 - 8t + 1 = 0$ **50.** $5t^2 + 12t - 1 = 0$

51. $4x^2 - 3x + 5 = 0$ **52.** $7x^2 - 5x + 2 = 0$

Applying the Concepts

53. Geometry If the shortest side in a $30°-60°-90°$ triangle is $\frac{1}{2}$ inch long, find the lengths of the other two sides.

54. Geometry If the shortest side in a $30°-60°-90°$ triangle is 3 feet long, find the lengths of the other two sides.

55. Geometry If the length of the shortest side of a $30°-60°-90°$ triangle is x, find the lengths of the other two sides in terms of x.

56. Geometry If the length of the longest side of a $30°-60°-90°$ triangle is x, find the lengths of the other two sides in terms of x.

57. Geometry If the length of the shorter sides of a $45°-45°-90°$ triangle is 1 inch, find the length of the hypotenuse.

58. Geometry If the length of the shorter sides of a $45°-45°-90°$ triangle is 3 feet, find the length of the hypotenuse.

59. Geometry If the length of the hypotenuse of a $45°-45°-90°$ triangle is 1 inch, find the length of the shorter sides.

60. Geometry If the length of the hypotenuse of a $45°-45°-90°$ triangle is 2 feet, find the length of the shorter sides.

61. Geometry If the length of the shorter sides of a $45°-45°-90°$ triangle is x, find the length of the hypotenuse, in terms of x.

62. Geometry If the length of the hypotenuse of a $45°-45°-90°$ triangle is x, find the length of the shorter sides, in terms of x.

63. Chair Lift Use Table 2 from the introduction to this section to find the horizontal distance covered by a person riding the Bear Paw Double chair lift. Round your answer to the nearest foot.

64. Chair Lift Use Table 2 from the introduction to this section to find the horizontal distance covered by a person riding the Big Springs Gondola lift. Round your answer to the nearest foot.

65. Chair Lift Using a right triangle to model the Forest Double chair lift, find the slope of the lift to the nearest hundredth.

66. Chair Lift Using a right triangle to model the Echo Triple chair lift, find the slope of the lift to the nearest hundredth.

67. Length of an Escalator An escalator in a department store is to carry people a vertical distance of 20 feet between floors. How long is the escalator if it makes an angle of $45°$ with the ground? (See Figure 5.)

FIGURE 5

68. Dimensions of a Tent A two-person tent is to be made so that the height at the center is 4 feet. If the sides of the tent are to meet the ground at an angle of $60°$, and the tent is to be 6 feet in length, how many square feet of material will be needed to make the tent? (Figure 6; assume that the tent has a floor and is closed at both ends.) Give your answer to the nearest tenth of a square foot.

FIGURE 6

69. Interest Rate Suppose a deposit of $3,000 in a savings account that paid an annual interest rate r (compounded yearly) is worth $3,456 after 2 years. Using the formula $A = P(1 + r)^t$, we have

$$3,456 = 3,000(1 + r)^2$$

Solve for r to find the annual interest rate.

70. Special Triangles In Figure 7, triangle ABC has angles 45° and 30°, and height x. Find the lengths of sides AB, BC, and AC, in terms of x.

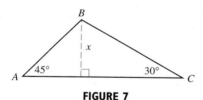

FIGURE 7

71. Special Triangles In Figure 8, AB has a length of 12 inches. Find the length of AD.

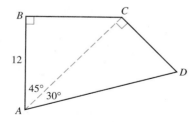

FIGURE 8

72. Wheelchair Access Most local laws state that ramps for wheelchairs may not have more than a 1/10 slope; that is, for every foot in the vertical direction the ramp must travel at least 10 feet in the horizontal

direction. A wheelchair ramp will take the place of four steps that are 8 inches deep and 8 inches high (Figure 9). What is the minimum length of such a ramp?

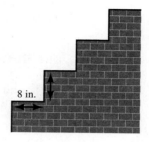

FIGURE 9

Review Problems

The problems that follow review material we covered in Section 9.3. Reviewing these problems will help you with the next section.

Write each of the following in simplified form for radicals.

73. $\sqrt{45}$ **74.** $\sqrt{24}$ **75.** $\sqrt{27y^5}$

76. $\sqrt{8y^3}$ **77.** $\sqrt[3]{54x^6y^5}$ **78.** $\sqrt[3]{16x^9y^7}$

79. Simplify $\sqrt{b^2 - 4ac}$ when $a = 6$, $b = 7$, and $c = -5$.

80. Simplify $\sqrt{b^2 - 4ac}$ when $a = 2$, $b = -6$, and $c = 3$.

Rationalize the denominator.

81. $\dfrac{3}{\sqrt{2}}$ **82.** $\dfrac{5}{\sqrt{3}}$ **83.** $\dfrac{2}{\sqrt[3]{4}}$ **84.** $\dfrac{3}{\sqrt[3]{2}}$

Extending the Concepts

Solve for x.

85. $(x + a)^2 + (x - a)^2 = 10a^2$

86. $(ax + 1)^2 + (ax - 1)^2 = 10$

Assume a is a positive number and solve for x by completing the square on x.

87. $x^2 + 2ax = -a^2$ **88.** $x^2 + 2ax = -4a^2$

89. $x^2 + 2ax = 0$ **90.** $x^2 + ax = 0$

Assume p and q are positive numbers and solve for x by completing the square on x.

91. $x^2 + px + q = 0$ **92.** $x^2 - px + q = 0$

The Quadratic Formula

In this section we will use the method of completing the square from the preceding section to derive the quadratic formula. The *quadratic formula* is a very useful tool in mathematics. It allows us to solve all types of quadratic equations.

THE QUADRATIC THEOREM

For any quadratic equation in the form $ax^2 + bx + c = 0$, where $a \neq 0$, the two solutions are

$$x = \frac{-b + \sqrt{b^2 - 4ac}}{2a} \qquad \text{and} \qquad x = \frac{-b - \sqrt{b^2 - 4ac}}{2a}$$

Proof We will prove the quadratic theorem by completing the square on $ax^2 + bx + c = 0$:

$$ax^2 + bx + c = 0$$

$$ax^2 + bx = -c \qquad \text{Add } -c \text{ to both sides.}$$

$$x^2 + \frac{b}{a}x = -\frac{c}{a} \qquad \text{Divide both sides by } a.$$

To complete the square on the left side, we add the square of $\frac{1}{2}$ of $\frac{b}{a}$ to both sides. $\left(\frac{1}{2} \text{ of } \frac{b}{a} \text{ is } \frac{b}{2a}.\right)$

$$x^2 + \frac{b}{a}x + \left(\frac{b}{2a}\right)^2 = -\frac{c}{a} + \left(\frac{b}{2a}\right)^2$$

We now simplify the right side as a separate step. We square the second term and combine the two terms by writing each with the least common denominator $4a^2$:

$$-\frac{c}{a} + \left(\frac{b}{2a}\right)^2 = -\frac{c}{a} + \frac{b^2}{4a^2} = \frac{4a}{4a}\left(\frac{-c}{a}\right) + \frac{b^2}{4a^2} = \frac{-4ac + b^2}{4a^2}$$

It is convenient to write this last expression as

$$\frac{b^2 - 4ac}{4a^2}$$

Continuing with the proof, we have

$$x^2 + \frac{b}{a}x + \left(\frac{b}{2a}\right)^2 = \frac{b^2 - 4ac}{4a^2}$$

$$\left(x + \frac{b}{2a}\right)^2 = \frac{b^2 - 4ac}{4a^2} \qquad \text{Write left side as a binomial square.}$$

$$x + \frac{b}{2a} = \pm \frac{\sqrt{b^2 - 4ac}}{2a} \qquad \text{Theorem 10.1}$$

$$x = -\frac{b}{2a} \pm \frac{\sqrt{b^2 - 4ac}}{2a} \qquad \text{Add } -\frac{b}{2a} \text{ to both sides.}$$

$$= \frac{-b \pm \sqrt{b^2 - 4ac}}{2a}$$

Our proof is now complete. What we have is this: If our equation is in the form $ax^2 + bx + c = 0$ (standard form), where $a \neq 0$, the two solutions are always given by the formula

$$x = \frac{-b \pm \sqrt{b^2 - 4ac}}{2a}$$

This formula is known as the *quadratic formula*. If we substitute the coefficients a, b, and c of any quadratic equation in standard form into the formula, we need only perform some basic arithmetic to arrive at the solution set.

EXAMPLE I   Use the quadratic formula to solve $6x^2 + 7x - 5 = 0$.

Solution   Using $a = 6$, $b = 7$, and $c = -5$ in the formula

$$x = \frac{-b \pm \sqrt{b^2 - 4ac}}{2a}$$

we have

$$x = \frac{-7 \pm \sqrt{49 - 4(6)(-5)}}{2(6)}$$

or

$$x = \frac{-7 \pm \sqrt{49 + 120}}{12}$$

$$= \frac{-7 \pm \sqrt{169}}{12}$$

$$= \frac{-7 \pm 13}{12}$$

We separate the last equation into the two statements

$$x = \frac{-7 + 13}{12} \qquad \text{or} \qquad x = \frac{-7 - 13}{12}$$

$$x = \frac{1}{2} \qquad \text{or} \qquad x = -\frac{5}{3}$$

The solution set is $\{\frac{1}{2}, -\frac{5}{3}\}$.

Whenever the solutions to a quadratic equation are rational numbers, as they are in Example 1, it means that the original equation was solvable by factoring. To illustrate, let's solve the equation from Example 1 again, but this time by factoring:

$$6x^2 + 7x - 5 = 0 \qquad \text{Equation in standard form}$$

$$(3x + 5)(2x - 1) = 0 \qquad \text{Factor the left side.}$$

$$3x + 5 = 0 \quad \text{or} \quad 2x - 1 = 0 \qquad \text{Set factors equal to 0.}$$

$$x = -\frac{5}{3} \quad \text{or} \quad x = \frac{1}{2}$$

When an equation can be solved by factoring, then factoring is usually the faster method of solution. It is best to try to factor first, and then if you have trouble factoring, go to the quadratic formula. It always works.

EXAMPLE 2   Solve $\dfrac{x^2}{3} - x = -\dfrac{1}{2}$.

Solution   Multiplying through by 6 and writing the result in standard form, we have

$$2x^2 - 6x + 3 = 0$$

the left side of which is not factorable. Therefore, we use the quadratic formula with $a = 2$, $b = -6$, and $c = 3$. The two solutions are given by

$$x = \frac{-(-6) \pm \sqrt{36 - 4(2)(3)}}{2(2)}$$

$$= \frac{6 \pm \sqrt{12}}{4}$$

$$= \frac{6 \pm 2\sqrt{3}}{4} \qquad \sqrt{12} = \sqrt{4 \cdot 3} = \sqrt{4}\,\sqrt{3} = 2\sqrt{3}$$

We can reduce this last expression to lowest terms by factoring 2 from the numerator and denominator and then dividing the numerator and denominator by 2:

$$x = \frac{2(3 \pm \sqrt{3})}{2 \cdot 2} = \frac{3 \pm \sqrt{3}}{2}$$

EXAMPLE 3   Solve $\dfrac{1}{x + 2} - \dfrac{1}{x} = \dfrac{1}{3}$.

Solution   To solve this equation, we must first put it in standard form. To do so, we must clear the equation of fractions by multiplying each side by the LCD for all the denominators, which is $3x(x + 2)$. Multiplying both sides by the LCD, we have

$$3x(x + 2)\left(\frac{1}{x + 2} - \frac{1}{x}\right) = \frac{1}{3} \cdot 3x(x + 2) \qquad \text{Multiply each} \\ \text{by the LCD.}$$

$$3x(x + 2) \cdot \frac{1}{x + 2} - 3x(x + 2) \cdot \frac{1}{x} = \frac{1}{3} \cdot 3x(x + 2)$$

$$3x - 3(x + 2) = x(x + 2)$$

$$3x - 3x - 6 = x^2 + 2x \qquad \text{Multiplication}$$

$$-6 = x^2 + 2x \qquad \text{Simplify left side.}$$

$$0 = x^2 + 2x + 6 \qquad \text{Add 6 to each side.}$$

Since the right side of our last equation is not factorable, we use the quadratic formula. From our last equation, we have $a = 1$, $b = 2$, and $c = 6$. Using these numbers for a, b, and c in the quadratic formula gives us

$$x = \frac{-2 \pm \sqrt{4 - 4(1)(6)}}{2(1)}$$

$$= \frac{-2 \pm \sqrt{4 - 24}}{2} \qquad \text{Simplify inside the radical.}$$

$$= \frac{-2 \pm \sqrt{-20}}{2} \qquad 4 - 24 = -20$$

$$= \frac{-2 \pm 2i\sqrt{5}}{2} \qquad \sqrt{-20} = i\sqrt{20} = i\sqrt{4}\,\sqrt{5} = 2i\sqrt{5}$$

$$= \frac{2(-1 \pm i\sqrt{5})}{2} \qquad \text{Factor 2 from the numerator.}$$

$$= -1 \pm i\sqrt{5} \qquad \text{Divide numerator and denominator by 2.}$$

Since neither of the two solutions, $-1 + i\sqrt{5}$ nor $-1 - i\sqrt{5}$, will make any of the denominators in our original equation 0, they are both solutions.

Although the equation in our next example is not a quadratic equation, we solve it by using both factoring and the quadratic formula.

EXAMPLE 4   Solve $27t^3 - 8 = 0$.

Solution   It would be a mistake to add 8 to each side of this equation and then take the cube root of each side because we would lose two of our solutions. Instead, we factor the left side, and then set the factors equal to 0:

$$27t^3 - 8 = 0 \qquad \text{Equation in standard form}$$

$$(3t - 2)(9t^2 + 6t + 4) = 0 \qquad \text{Factor as the difference} \\ \text{of two cubes.}$$

$$3t - 2 = 0 \quad \text{or} \quad 9t^2 + 6t + 4 = 0 \qquad \text{Set each factor equal to 0.}$$

The first equation leads to a solution of $t = \frac{2}{3}$. The second equation does not factor, so we use the quadratic formula with $a = 9$, $b = 6$, and $c = 4$:

$$t = \frac{-6 \pm \sqrt{36 - 4(9)(4)}}{2(9)}$$

$$= \frac{-6 \pm \sqrt{36 - 144}}{18}$$

$$= \frac{-6 \pm \sqrt{-108}}{18}$$

$$= \frac{-6 \pm 6i\sqrt{3}}{18} \qquad \sqrt{-108} = i\sqrt{36 \cdot 3} = 6i\sqrt{3}$$

$$= \frac{\cancel{6}(-1 \pm i\sqrt{3})}{\cancel{6} \cdot 3} \qquad \text{Factor 6 from the numerator and denominator.}$$

$$= \frac{-1 \pm i\sqrt{3}}{3} \qquad \text{Divide out common factor 6.}$$

The three solutions to our original equation are

$$\frac{2}{3}, \qquad \frac{-1 + i\sqrt{3}}{3}, \qquad \text{and} \qquad \frac{-1 - i\sqrt{3}}{3}$$

EXAMPLE 5   If an object is thrown downward with an initial velocity of 20 feet per second, the distance $s(t)$, in feet, it travels in t seconds is given by the function $s(t) = 20t + 16t^2$. How long does it take the object to fall 40 feet?

Solution   We let $s(t) = 40$, and solve for t:

When $s(t) = 40$

the function $s(t) = 20t + 16t^2$

becomes $40 = 20t + 16t^2$

or $16t^2 + 20t - 40 = 0$

$4t^2 + 5t - 10 = 0$ Divide by 4.

Using the quadratic formula, we have

$$t = \frac{-5 \pm \sqrt{25 - 4(4)(-10)}}{2(4)}$$

$$t = \frac{-5 \pm \sqrt{185}}{8}$$

$$t = \frac{-5 + \sqrt{185}}{8} \qquad \text{or} \qquad t = \frac{-5 - \sqrt{185}}{8}$$

The second solution is impossible since it is a negative number and time t must be positive. It takes

$$t = \frac{-5 + \sqrt{185}}{8} \qquad \text{or approximately} \qquad \frac{-5 + 13.60}{8} \approx 1.08 \text{ seconds}$$

for the object to fall 40 feet.

The relationship between profit, revenue, and cost is given by the formula

$$P(x) = R(x) - C(x)$$

where $P(x)$ is the profit, $R(x)$ is the total revenue, and $C(x)$ is the total cost of producing and selling x items.

EXAMPLE 6   A company produces and sells copies of an accounting program for home computers. The total weekly cost (in dollars) to produce x copies of the program is $C(x) = 8x + 500$, and the weekly revenue for selling all x copies of the program is $R(x) = 35x - 0.1x^2$. How many programs must be sold each week for the weekly profit to be $1,200?

Solution   Substituting the given expressions for $R(x)$ and $C(x)$ in the equation $P(x) = R(x) - C(x)$, we have a polynomial in x that represents the weekly profit $P(x)$:

$$\begin{aligned}
P(x) &= R(x) - C(x) \\
&= 35x - 0.1x^2 - (8x + 500) \\
&= 35x - 0.1x^2 - 8x - 500 \\
&= -500 + 27x - 0.1x^2
\end{aligned}$$

Setting this expression equal to 1,200, we have a quadratic equation to solve that gives us the number of programs x that need to be sold each week to bring in a profit of $1,200:

$$1,200 = -500 + 27x - 0.1x^2$$

We can write this equation in standard form by adding the opposite of each term on the right side of the equation to both sides of the equation. Doing so produces the following equation:

$$0.1x^2 - 27x + 1,700 = 0$$

Applying the quadratic formula to this equation with $a = 0.1$, $b = -27$, and $c = 1,700$, we have

$$\begin{aligned}
x &= \frac{27 \pm \sqrt{(-27)^2 - 4(0.1)(1,700)}}{2(0.1)} \\
&= \frac{27 \pm \sqrt{729 - 680}}{0.2} \\
&= \frac{27 \pm \sqrt{49}}{0.2} \\
&= \frac{27 \pm 7}{0.2}
\end{aligned}$$

Writing this last expression as two separate expressions, we have our two solutions:

$$x = \frac{27 + 7}{0.2} \quad \text{or} \quad x = \frac{27 - 7}{0.2}$$

$$= \frac{34}{0.2} \qquad\qquad\qquad = \frac{20}{0.2}$$

$$= 170 \qquad\qquad\qquad = 100$$

The weekly profit will be $1,200 if the company produces and sells 100 programs or 170 programs.

What is interesting about the equation we solved in Example 6 is that it has rational solutions, meaning it could have been solved by factoring. But looking back at the equation, factoring does not seem like a reasonable method of solution because the coefficients are either very large or very small. So, there are times when using the quadratic formula is a faster method of solution, even though the equation you are solving is factorable.

Using TECHNOLOGY

GRAPHING CALCULATORS

More About Example 5

We can solve the problem discussed in Example 5 by graphing the function $Y_1 = 20X + 16X^2$ in a window with X from 0 to 2 (because X is taking the place of t and we know t is a positive quantity) and Y from 0 to 50 (because we are looking for X when Y_1 is 40). Graphing Y_1 gives a graph similar to the graph in Figure 1. Using the Zoom and Trace features at $Y_1 = 40$ gives us X = 1.08 to the nearest hundredth, matching the results we obtained by solving the original equation algebraically.

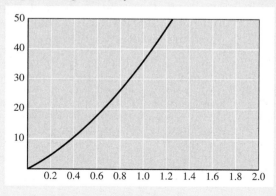

FIGURE 1

(continued)

More About Example 6

To visualize the functions in Example 6, we set up our calculator this way:

$$Y_1 = 35X - .1X^2 \qquad \text{Revenue function}$$
$$Y_2 = 8X + 500 \qquad \text{Cost function}$$
$$Y_3 = Y_1 - Y_2 \qquad \text{Profit function}$$

Window: X from 0 to 350, Y from 0 to 3,500

Graphing these functions produces graphs similar to the ones shown in Figure 2. The lowest graph is the graph of the profit function. Using the Zoom and Trace features on the lowest graph at $Y_3 = 1,200$ produces two corresponding values of X, 170 and 100, which match the results in Example 6.

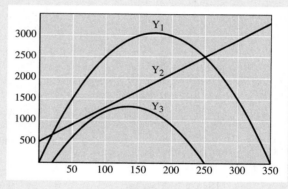

FIGURE 2

We will continue this discussion of the relationship between graphs of functions and solutions to equations in the Using Technology material in the next section.

 # Getting Ready for Class

After reading through the preceding section, respond in your own words and in complete sentences.

A. What is the quadratic formula?

B. Under what circumstances should the quadratic formula be applied?

C. When would the quadratic formula result in complex solutions?

D. When will the quadratic formula result in only one solution?

PROBLEM SET 10.2

Solve each equation. Use factoring or the quadratic formula, whichever is appropriate. (Try factoring first. If you have any difficulty factoring, then go right to the quadratic formula.)

1. $x^2 + 5x + 6 = 0$ **2.** $x^2 + 5x - 6 = 0$

3. $a^2 - 4a + 1 = 0$ **4.** $a^2 + 4a + 1 = 0$

5. $\frac{1}{6}x^2 - \frac{1}{2}x + \frac{1}{3} = 0$

6. $\frac{1}{4}x^2 + \frac{1}{4}x - \frac{1}{2} = 0$

7. $\frac{x^2}{2} + 1 = \frac{2x}{3}$ **8.** $\frac{x^2}{2} + \frac{2}{3} = -\frac{2x}{3}$

9. $y^2 - 5y = 0$ **10.** $2y^2 + 10y = 0$

11. $30x^2 + 40x = 0$ **12.** $50x^2 - 20x = 0$

13. $\frac{2t^2}{3} - t = -\frac{1}{6}$ **14.** $\frac{t^2}{3} - \frac{t}{2} = -\frac{3}{2}$

15. $0.01x^2 + 0.06x - 0.08 = 0$

16. $0.02x^2 - 0.03x + 0.05 = 0$

17. $2x + 3 = -2x^2$ **18.** $2x - 3 = 3x^2$

19. $100x^2 - 200x + 100 = 0$

20. $100x^2 - 600x + 900 = 0$

21. $\frac{1}{2}r^2 = \frac{1}{6}r - \frac{2}{3}$ **22.** $\frac{1}{4}r^2 = \frac{2}{5}r + \frac{1}{10}$

23. $(x - 3)(x - 5) = 1$

24. $(x - 3)(x + 1) = -6$

25. $(x + 3)^2 + (x - 8)(x - 1) = 16$

26. $(x - 4)^2 + (x + 2)(x + 1) = 9$

27. $\frac{x^2}{3} - \frac{5x}{6} = \frac{1}{2}$ **28.** $\frac{x^2}{6} + \frac{5}{6} = -\frac{x}{3}$

Multiply both sides of each equation by its LCD. Then solve the resulting equation.

29. $\frac{1}{x + 1} - \frac{1}{x} = \frac{1}{2}$ **30.** $\frac{1}{x + 1} + \frac{1}{x} = \frac{1}{3}$

31. $\frac{1}{y - 1} + \frac{1}{y + 1} = 1$ **32.** $\frac{2}{y + 2} + \frac{3}{y - 2} = 1$

33. $\frac{1}{x + 2} + \frac{1}{x + 3} = 1$

34. $\frac{1}{x + 3} + \frac{1}{x + 4} = 1$

35. $\frac{6}{r^2 - 1} - \frac{1}{2} = \frac{1}{r + 1}$

36. $2 + \frac{5}{r - 1} = \frac{12}{(r - 1)^2}$

Solve each equation. In each case you will have three solutions.

37. $x^3 - 8 = 0$ **38.** $x^3 - 27 = 0$

39. $8a^3 + 27 = 0$ **40.** $27a^3 + 8 = 0$

41. $125t^3 - 1 = 0$ **42.** $64t^3 + 1 = 0$

Each of the following equations has three solutions. Look for the greatest common factor; then use the quadratic formula to find all solutions.

43. $2x^3 + 2x^2 + 3x = 0$ **44.** $6x^3 - 4x^2 + 6x = 0$

45. $3y^4 = 6y^3 - 6y^2$ **46.** $4y^4 = 16y^3 - 20y^2$

47. $6t^5 + 4t^4 = -2t^3$ **48.** $8t^5 + 2t^4 = -10t^3$

49. One solution to a quadratic equation is $\frac{-3 + 2i}{5}$. What is the other solution?

50. One solution to a quadratic equation is $\frac{-2 + 3i\sqrt{2}}{5}$. What is the other solution?

Applying the Concepts

51. Falling Object An object is thrown downward with an initial velocity of 5 feet per second. The relationship between the distance s it travels and time t is given by $s = 5t + 16t^2$. How long does it take the object to fall 74 feet?

52. Falling Object The distance an object falls from rest is given by the equation $s = 16t^2$, where $s =$

distance and t = time. How long does it take an object dropped from a 100-foot cliff to hit the ground?

53. **Ball Toss**   A ball is thrown upward with an initial velocity of 20 feet per second. The equation that gives the height h of the object at any time t is $h = 20t - 16t^2$. At what times will the object be 4 feet off the ground?

54. **Coin Toss**   A coin is tossed upward with an initial velocity of 32 feet per second from a height of 16 feet above the ground. The equation giving the object's height h at any time t is $h = 16 + 32t - 16t^2$. Does the object ever reach a height of 32 feet?

55. **Profit**   The total cost (in dollars) for a company to manufacture and sell x items per week is $C = 60x + 300$, whereas the revenue brought in by selling all x items is $R = 100x - 0.5x^2$. How many items must be sold to obtain a weekly profit of $300?

56. **Profit**   The total cost (in dollars) for a company to produce and sell x items per week is $C = 200x + 1,600$, whereas the revenue brought in by selling all x items is $R = 300x - 0.5x^2$. How many items must be sold in order for the weekly profit to be $2,150?

57. **Profit**   Suppose it costs a company selling patterns $C = 800 + 6.5x$ dollars to produce and sell x patterns a month. If the revenue obtained by selling x patterns is $R = 10x - 0.002x^2$, how many patterns must it sell each month if it wants a monthly profit of $700?

58. **Profit**   Suppose a company manufactures and sells x picture frames each month with a total cost of $C = 1,200 + 3.5x$ dollars. If the revenue obtained by selling x frames is $R = 9x - 0.002x^2$, find the number of frames it must sell each month if its monthly profit is to be $2,300.

59. **Photograph Cropping**   The following figure shows a photographic image on a 10.5-centimeter by 8.2-centimeter background. The overall area of the background is to be reduced to 80% of its original area by cutting off (cropping) equal strips on all four sides. What is the width of the strip that is cut from each side?

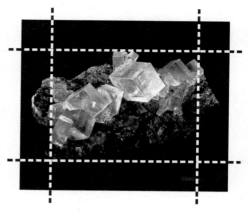

(*American Museum of Natural History*)

60. **Area of a Garden**   A garden measures 20.3 meters by 16.4 meters. In order to double the area of the garden, strips of equal width are added to all four sides.
    (a) Draw a diagram that illustrates these conditions.
    (b) What are the new overall dimensions of the garden?

61. **Area and Perimeter**   A rectangle has a perimeter of 20 yards and an area of 15 square yards.
    (a) Write two equations that state these facts in terms of the rectangle's length, l, and its width, w.
    (b) Solve the two equations from part (a) to determine the actual length and width of the rectangle.
    (c) Explain why two answers are possible to part (b).

62. **Population Size**   Writing in 1829, former President James Madison made some predictions about the growth of the population of the United States. The populations he predicted fit the equation

$$y = 0.029x^2 - 1.39x + 42$$

where y is the population in millions of people x years from 1829.
    (a) Use the equation to determine the approximate year President Madison would have predicted that the U.S. population would reach 100,000,000.
    (b) If the U.S. population in 1990 was approximately 200,000,000, were President Madison's predictions accurate in the long term? Explain why or why not.

Review Problems

The problems that follow review material we covered in Sections 4.8 and 9.1. Reviewing these problems will help you with the next section.

Divide, using long division. [4.8]

63. $\dfrac{8y^2 - 26y - 9}{2y - 7}$

64. $\dfrac{6y^2 + 7y - 18}{3y - 4}$

65. $\dfrac{x^3 + 9x^2 + 26x + 24}{x + 2}$

66. $\dfrac{x^3 + 6x^2 + 11x + 6}{x + 3}$

Simplify each expression. (Assume $x, y > 0$.) [9.1]

67. $25^{1/2}$

68. $8^{1/3}$

69. $\left(\dfrac{9}{25}\right)^{3/2}$

70. $\left(\dfrac{16}{81}\right)^{3/4}$

71. $8^{-2/3}$

72. $4^{-3/2}$

73. $\dfrac{(49x^8y^{-4})^{1/2}}{(27x^{-3}y^9)^{-1/3}}$

74. $\dfrac{(x^{-2}y^{1/3})^6}{x^{-10}y^{3/2}}$

Extending the Concepts

So far, all the equations we have solved have had coefficients that were rational numbers. Here are some equations that have irrational coefficients and some that have complex coefficients. Solve each equation. (Remember, $i^2 = -1$.)

75. $x^2 + \sqrt{3}x - 6 = 0$

76. $x^2 - \sqrt{5}x - 5 = 0$

77. $\sqrt{2}x^2 + 2x - \sqrt{2} = 0$

78. $\sqrt{7}x^2 + 2\sqrt{2}x - \sqrt{7} = 0$

79. $x^2 + ix + 2 = 0$

80. $x^2 + 3ix - 2 = 0$

81. $ix^2 + 3x + 4i = 0$

82. $4ix^2 + 5x + 9i = 0$

10.3   Additional Items Involving Solutions to Equations

In this section we will do two things. First, we will define the discriminant and use it to find the kind of solutions a quadratic equation has without solving the equation. Second, we will use the zero-factor property to build equations from their solutions.

THE DISCRIMINANT

The quadratic formula

$$x = \frac{-b \pm \sqrt{b^2 - 4ac}}{2a}$$

gives the solutions to any quadratic equation in standard form. There are times, when working with quadratic equations, when it is important only to know what kind of solutions the equation has.

> **DEFINITION** The expression under the radical in the quadratic formula is called the **discriminant**:
>
> $$\text{Discriminant} = D = b^2 - 4ac$$

The discriminant indicates the number and type of solutions to a quadratic equation, when the original equation has integer coefficients. For example, if we were to use the quadratic formula to solve the equation $2x^2 + 2x + 3 = 0$, we would find the discriminant to be

$$b^2 - 4ac = 2^2 - 4(2)(3) = -20$$

Since the discriminant appears under a square root symbol, we have the square root of a negative number in the quadratic formula. Our solutions would therefore be complex numbers. Similarly, if the discriminant were 0, the quadratic formula would yield

$$x = \frac{-b \pm \sqrt{0}}{2a} = \frac{-b \pm 0}{2a} = \frac{-b}{2a}$$

and the equation would have one rational solution, the number $\dfrac{-b}{2a}$.

The following table gives the relationship between the discriminant and the type of solutions to the equation.

For the equation $ax^2 + bx + c = 0$ where a, b, and c are integers and $a \neq 0$:

If the Discriminant $b^2 - 4ac$ Is	Then the Equation Will Have
Negative	Two complex solutions containing i
Zero	One rational solution
A positive number that is also a perfect square	Two rational solutions
A positive number that is not a perfect square	Two irrational solutions

In the second and third cases, when the discriminant is 0 or a positive perfect square, the solutions are rational numbers. The quadratic equations in these two cases are the ones that can be factored.

E X A M P L E S   For each equation, give the number and kind of solutions.

1. $x^2 - 3x - 40 = 0$

Solution   Using $a = 1$, $b = -3$, and $c = -40$ in $b^2 - 4ac$, we have $(-3)^2 - 4(1)(-40) = 9 + 160 = 169$.

The discriminant is a perfect square. The equation therefore has two rational solutions.

2. $2x^2 - 3x + 4 = 0$

Solution   Using $a = 2$, $b = -3$, and $c = 4$, we have

$$b^2 - 4ac = (-3)^2 - 4(2)(4) = 9 - 32 = -23$$

The discriminant is negative, implying the equation has two complex solutions that contain i.

3. $4x^2 - 12x + 9 = 0$

Solution Using $a = 4$, $b = -12$, and $c = 9$, the discriminant is

$$b^2 - 4ac = (-12)^2 - 4(4)(9) = 144 - 144 = 0$$

Since the discriminant is 0, the equation will have one rational solution.

4. $x^2 + 6x = 8$

Solution We must first put the equation in standard form by adding -8 to each side. If we do so, the resulting equation is

$$x^2 + 6x - 8 = 0$$

Now we identify *a, b,* and *c* as 1, 6, and -8, respectively:

$$b^2 - 4ac = 6^2 - 4(1)(-8) = 36 + 32 = 68$$

The discriminant is a positive number, but not a perfect square. The equation will therefore have two irrational solutions.

E X A M P L E 5 Find an appropriate k so that the equation $4x^2 - kx = -9$ has exactly one rational solution.

Solution We begin by writing the equation in standard form:

$$4x^2 - kx + 9 = 0$$

Using $a = 4$, $b = -k$, and $c = 9$, we have

$$b^2 - 4ac = (-k)^2 - 4(4)(9)$$
$$= k^2 - 144$$

An equation has exactly one rational solution when the discriminant is 0. We set the discriminant equal to 0 and solve:

$$k^2 - 144 = 0$$
$$k^2 = 144$$
$$k = \pm 12$$

Choosing k to be 12 or -12 will result in an equation with one rational solution.

BUILDING EQUATIONS FROM THEIR SOLUTIONS

Suppose we know that the solutions to an equation are $x = 3$ and $x = -2$. We can find equations with these solutions by using the zero-factor property. First, let's write our solutions as equations with 0 on the right side:

If	$x = 3$	First solution
then	$x - 3 = 0$	Add -3 to each side.

and if $\qquad\qquad x = -2 \qquad$ Second solution

then $\qquad x + 2 = 0 \qquad$ Add 2 to each side.

Now, since both $x - 3$ and $x + 2$ are 0, their product must be 0 also. We can therefore write

$$(x - 3)(x + 2) = 0 \qquad \text{Zero-factor property}$$

$$x^2 - x - 6 = 0 \qquad \text{Multiply out the left side.}$$

Many other equations have 3 and -2 as solutions. For example, any constant multiple of $x^2 - x - 6 = 0$, such as $5x^2 - 5x - 30 = 0$, also has 3 and -2 as solutions. Similarly, any equation built from positive integer powers of the factors $x - 3$ and $x + 2$ will also have 3 and -2 as solutions. One such equation is

$$(x - 3)^2(x + 2) = 0$$

$$(x^2 - 6x + 9)(x + 2) = 0$$

$$x^3 - 4x^2 - 3x + 18 = 0$$

In mathematics we distinguish between the solutions to this last equation and those to the equation $x^2 - x - 6 = 0$ by saying $x = 3$ is a solution of *multiplicity 2* in the equation $x^3 - 4x^2 - 3x + 18 = 0$, and a solution of *multiplicity 1* in the equation $x^2 - x - 6 = 0$.

- - - - - - - - - - - -

E X A M P L E 6 Find an equation that has solutions $t = 5$, $t = -5$, and $t = 3$.

Solution First, we use the given solutions to write equations that have 0 on their right sides:

If $\qquad\qquad t = 5 \qquad\quad t = -5 \qquad\quad t = 3$

then $\qquad t - 5 = 0 \qquad t + 5 = 0 \qquad t - 3 = 0$

Since $t - 5$, $t + 5$, and $t - 3$ are all 0, their product is also 0 by the zero-factor property. An equation with solutions of 5, -5, and 3 is

$$(t - 5)(t + 5)(t - 3) = 0 \qquad \text{Zero-factor property}$$

$$(t^2 - 25)(t - 3) = 0 \qquad \text{Multiply first two binomials.}$$

$$t^3 - 3t^2 - 25t + 75 = 0 \qquad \text{Complete the multiplication.}$$

The last line gives us an equation with solutions of 5, -5, and 3. Remember, many other equations have these same solutions.

- - - - - - - - - - - -

E X A M P L E 7 Find an equation with solutions $x = -\frac{2}{3}$ and $x = \frac{4}{5}$.

Solution The solution $x = -\frac{2}{3}$ can be rewritten as $3x + 2 = 0$ as follows:

$$x = -\frac{2}{3} \qquad \text{The first solution}$$

$$3x = -2 \qquad \text{Multiply each side by 3.}$$

$$3x + 2 = 0 \qquad \text{Add 2 to each side.}$$

Similarly, the solution $x = \frac{4}{5}$ can be rewritten as $5x - 4 = 0$:

$$x = \frac{4}{5} \qquad \text{The second solution}$$

$$5x = 4 \qquad \text{Multiply each side by 5.}$$

$$5x - 4 = 0 \qquad \text{Add } -4 \text{ to each side.}$$

Since both $3x + 2$ and $5x - 4$ are 0, their product is 0 also, giving us the equation we are looking for:

$$(3x + 2)(5x - 4) = 0 \qquad \text{Zero-factor property}$$

$$15x^2 - 2x - 8 = 0 \qquad \text{Multiplication}$$

Using TECHNOLOGY

GRAPHING CALCULATORS

Solving Equations

Now that we have explored the relationship between equations and their solutions, we can look at how a graphing calculator can be used in the solution process. To begin, let's solve the equation $x^2 = x + 2$ using techniques from algebra: writing it in standard form, factoring, and then setting each factor equal to 0.

$$x^2 - x - 2 = 0 \qquad \text{Standard form}$$

$$(x - 2)(x + 1) = 0 \qquad \text{Factor.}$$

$$x - 2 = 0 \quad \text{or} \quad x + 1 = 0 \qquad \text{Set each factor equal to 0.}$$

$$x = 2 \quad \text{or} \quad x = -1 \qquad \text{Solve.}$$

Our original equation, $x^2 = x + 2$, has two solutions: $x = 2$ and $x = -1$. To solve the equation using a graphing calculator, we need to associate it with an equation (or equations) in two variables. One way to do this is to associate the left side with the equation $y = x^2$ and the right side of the equation with $y = x + 2$. To do so, we set up the functions list in our calculator this way:

$$Y_1 = X^2$$

$$Y_2 = X + 2$$

Window: X from -5 to 5, Y from -5 to 5

Graphing these functions in this window will produce a graph similar to the one shown in Figure 1.

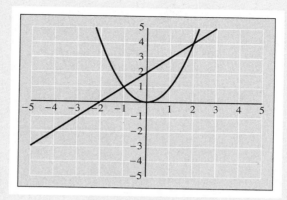

FIGURE 1

If we use the Trace feature to find the coordinates of the points of inter-section, we find that the two curves intersect at $(-1, 1)$ and $(2, 4)$. We note that the x-coordinates of these two points match the solutions to the equation $x^2 = x + 2$, which we found using algebraic techniques. This makes sense because if two graphs intersect at a point (x, y), then the coordinates of that point satisfy both equations. If a point (x, y) satisfies both $y = x^2$ and $y = x + 2$, then, for that particular point, $x^2 = x + 2$. From this we conclude that the x-coordinates of the points of intersection are solutions to our origi-nal equation. Here is a summary of what we have discovered:

Conclusion 1   If the graphs of two functions $y = f(x)$ and $y = g(x)$ intersect in the coordinate plane, then the x-coordinates of the points of intersection are solutions to the equation $f(x) = g(x)$.

A second method of solving our original equation $x^2 = x + 2$ graphi-cally requires the use of one function instead of two. To begin, we write the equation in standard form as $x^2 - x - 2 = 0$. Next, we graph the function $y = x^2 - x - 2$. The x-intercepts of the graph are the points with y-coordi-nates of 0. They therefore satisfy the equation $0 = x^2 - x - 2$, which is equivalent to our original equation. The graph in Figure 2 shows $Y_1 = X^2 - X - 2$ in a window with X from -5 to 5 and Y from -5 to 5.

Using the Trace feature, we find that the x-intercepts of the graph are $x = -1$ and $x = 2$, which match the solutions to our original equation $x^2 = x + 2$. We can summarize the relationship between solutions to an equa-tion and the intercepts of its associated graph this way:

Conclusion 2   If $y = f(x)$ is a function, then any x-intercept on the graph of $y = f(x)$ is a solution to the equation $f(x) = 0$.

(continued)

(continued)

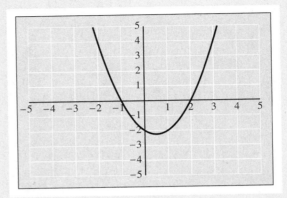

FIGURE 2

Solving Equations With Complex Solutions

There are limitations to using a graph to solve an equation. To illustrate, suppose we solve the equation $4x^3 - 6x^2 + 2x - 3 = 0$ by factoring:

$$4x^3 - 6x^2 + 2x - 3 = 0 \qquad \text{Original equation}$$

$$2x^2(2x - 3) + (2x - 3) = 0$$

$$(2x^2 + 1)(2x - 3) = 0 \qquad \text{Factor by grouping.}$$

$$2x^2 + 1 = 0 \qquad \text{or} \qquad 2x - 3 = 0 \qquad \text{Set factors equal to 0.}$$

$$2x^2 = -1 \qquad \text{or} \qquad 2x = 3$$

$$x^2 = -\frac{1}{2} \qquad \text{or} \qquad x = \frac{3}{2} \qquad \text{Solve the resulting equations.}$$

$$x = \pm\frac{1}{\sqrt{2}}\,i$$

$$= \pm\frac{\sqrt{2}}{2}\,i$$

We have three solutions:

$$-\frac{\sqrt{2}}{2}\,i, \qquad \frac{\sqrt{2}}{2}\,i, \qquad \frac{3}{2}$$

Figure 3 shows the graph of $y = 4x^3 - 6x^2 + 2x - 3$. As you can see, the graph crosses the x-axis exactly once at $x = \frac{3}{2}$, which we expect. The rectangular coordinate system consists of ordered pairs of *real numbers,* so our complex solutions cannot appear on the graph.

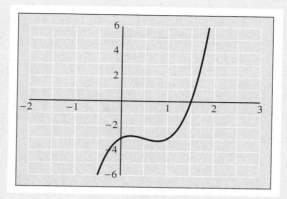

FIGURE 3

Every cubic equation will have at least one real number solution. If we are interested in exact values for all solutions to our equation, including complex solutions as well as irrational solutions, we can take our one real solution and, with the aid of long division and factoring or the quadratic formula, find the other solutions. But how do we find the one real solution in the first place? We use a graphing calculator.

Suppose we want to solve the equation $x^3 = 15x + 4$ completely. We can use a graphing calculator to graph the function $y = x^3 - 15x - 4$ and note that it crosses the x-axis at $x = 4$. The graph is shown in Figure 4.

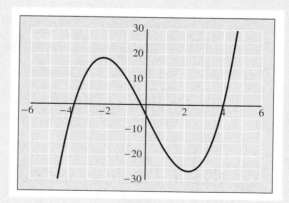

FIGURE 4

If we zoom and trace on the other two solutions, we find they are approximated by -3.732 and -0.268 to the nearest thousandth. To find exact values for these last two solutions, we reason that if $x = 4$ is a solution to our equation, then $x - 4$ must divide $x^3 - 15x - 4$ without a remainder. Here is the

(continued)

(continued)

division problem:

$$x - 4 \overline{)x^3 + 0x^2 - 15x - 4}$$
$$\begin{array}{r} x^2 + 4x + 1 \\ x - 4 \overline{)x^3 + 0x^2 - 15x - 4} \\ \underline{x^3 - 4x^2} \\ 4x^2 - 15x \\ \underline{4x^2 - 16x} \\ x - 4 \\ \underline{x - 4} \\ 0 \end{array}$$

We find that $x^3 - 15x - 4$ factors into $(x - 4)(x^2 + 4x + 1)$. This allows us to find all solutions to our original equation:

$$x^3 - 15x - 4 = 0$$

$$(x - 4)(x^2 + 4x + 1) = 0$$

$$x - 4 = 0 \quad \text{or} \quad x^2 + 4x + 1 = 0$$

$$x = 4 \quad \text{or} \quad x = \frac{-4 \pm \sqrt{16 - 4(1)(1)}}{2(1)}$$

$$= \frac{-4 \pm \sqrt{12}}{2}$$

$$= \frac{-4 \pm 2\sqrt{3}}{2}$$

$$= -2 \pm \sqrt{3}$$

The three exact solutions to our equation are 4, $-2 + \sqrt{3}$, and $-2 - \sqrt{3}$.

 ## Getting Ready for Class

After reading through the preceding section, respond in your own words and in complete sentences.

A. What is the discriminant?

B. What kind of solutions do we get to a quadratic equation when the discriminant is negative?

C. What does it mean for a solution to have multiplicity 3?

D. When will a quadratic equation have two rational solutions?

PROBLEM SET 10.3

Use the discriminant to find the number and kind of solutions for each of the following equations.

1. $x^2 - 6x + 5 = 0$ **2.** $x^2 - x - 12 = 0$

3. $4x^2 - 4x = -1$ **4.** $9x^2 + 12x = -4$

5. $x^2 + x - 1 = 0$ **6.** $x^2 - 2x + 3 = 0$

7. $2y^2 = 3y + 1$ **8.** $3y^2 = 4y - 2$

9. $x^2 - 9 = 0$ **10.** $4x^2 - 81 = 0$

11. $5a^2 - 4a = 5$ **12.** $3a = 4a^2 - 5$

Determine k so that each of the following has exactly one real solution.

13. $x^2 - kx + 25 = 0$ **14.** $x^2 + kx + 25 = 0$

15. $x^2 = kx - 36$ **16.** $x^2 = kx - 49$

17. $4x^2 - 12x + k = 0$

18. $9x^2 + 30x + k = 0$

19. $kx^2 - 40x = 25$ **20.** $kx^2 - 2x = -1$

21. $3x^2 - kx + 2 = 0$ **22.** $5x^2 + kx + 1 = 0$

For each of the following problems, find an equation that has the given solutions.

23. $x = 5, x = 2$ **24.** $x = -5, x = -2$

25. $t = -3, t = 6$ **26.** $t = -4, t = 2$

27. $y = 2, y = -2, y = 4$

28. $y = 1, y = -1, y = 3$

29. $x = \frac{1}{2}, x = 3$ **30.** $x = \frac{1}{3}, x = 5$

31. $t = -\frac{3}{4}, t = 3$ **32.** $t = -\frac{4}{5}, t = 2$

33. $x = 3, x = -3, x = \frac{5}{6}$

34. $x = 5, x = -5, x = \frac{2}{3}$

35. $a = -\frac{1}{2}, a = \frac{3}{5}$ **36.** $a = -\frac{1}{3}, a = \frac{4}{7}$

37. $x = -\frac{2}{3}, x = \frac{2}{3}, x = 1$

38. $x = -\frac{4}{5}, x = \frac{4}{5}, x = -1$

39. $x = 2, x = -2, x = 3, x = -3$

40. $x = 1, x = -1, x = 5, x = -5$

41. Find an equation that has a solution of $x = 3$ of multiplicity 1 and a solution $x = -5$ of multiplicity 2.

42. Find an equation that has a solution of $x = 5$ of multiplicity 1 and a solution $x = -3$ of multiplicity 2.

43. Find an equation that has solutions $x = 3$ and $x = -3$, both of multiplicity 2.

44. Find an equation that has solutions $x = 4$ and $x = -4$, both of multiplicity 2.

45. Find all solutions to $x^3 + 6x^2 + 11x + 6 = 0$ if $x = -3$ is one of its solutions.

46. Find all solutions to $x^3 + 10x^2 + 29x + 20 = 0$ if $x = -4$ is one of its solutions.

47. One solution to $y^3 + 5y^2 - 2y - 24 = 0$ is $y = -3$. Find all solutions.

48. One solution to $y^3 + 3y^2 - 10y - 24 = 0$ is $y = -2$. Find all solutions.

49. If $x = 3$ is one solution to $x^3 - 5x^2 + 8x = 6$, find the other solutions.

50. If $x = 2$ is one solution to $x^3 - 6x^2 + 13x = 10$, find the other solutions.

51. Find all solutions to $t^3 = 13t^2 - 65t + 125$ if $t = 5$ is one of the solutions.

52. Find all solutions to $t^3 = 8t^2 - 25t + 26$ if $t = 2$ is one of the solutions.

Review Problems

Problems 53–62 review material we covered in Section 9.2. Reviewing these problems will help you with the next section.

Multiply.

53. $a^4(a^{3/2} - a^{1/2})$

54. $(a^{1/2} - 5)(a^{1/2} + 3)$

55. $(x^{3/2} - 3)^2$

56. $(x^{1/2} - 8)(x^{1/2} + 8)$

Divide.

57. $\dfrac{30x^{3/4} - 25x^{5/4}}{5x^{1/4}}$

58. $\dfrac{45x^{5/3}y^{7/3} - 36x^{8/3}y^{4/3}}{9x^{2/3}y^{1/3}}$

59. Factor $5(x - 3)^{1/2}$ from

$$10(x - 3)^{3/2} - 15(x - 3)^{1/2}$$

60. Factor $2(x + 1)^{1/3}$ from

$$8(x + 1)^{4/3} - 2(x + 1)^{1/3}$$

Factor each of the following as if they were trinomials.

61. $2x^{2/3} - 11x^{1/3} + 12$

62. $9x^{2/3} + 12x^{1/3} + 4$

Extending the Concepts

63. Find all solutions to $x^4 + x^3 - x^2 + x - 2 = 0$ if $x = -2$ is one solution.

64. Find all solutions to $x^4 - x^3 + 2x^2 - 4x - 8 = 0$ if $x = 2$ is one solution.

65. Find all solutions to $x^3 + 3ax^2 + 3a^2x + a^3 = 0$ if $x = -a$ is one solution.

66. If $x = -2a$ is one solution to $x^3 + 6ax^2 + 12a^2x + 8a^3 = 0$, find the other solutions.

Find all solutions to the following equations. Solve using algebra and by graphing. If rounding is necessary, round to the nearest hundredth.

67. $x^2 = 4x + 5$ **68.** $4x^2 = 8x + 5$

69. $x^2 - 1 = 2x$ **70.** $4x^2 - 1 = 4x$

Find all solutions to each equation. If rounding is necessary, round to the nearest hundredth.

71. $2x^3 - x^2 - 2x + 1 = 0$

72. $3x^3 - 2x^2 - 3x + 2 = 0$

73. $2x^3 + 2 = x^2 + 4x$

74. $3x^3 - 9x = 2x^2 - 6$

Each of the following equations has only one real solution. Find it by graphing. Then use long division and the quadratic formula to find the remaining solutions.

75. $3x^3 - 8x^2 + 10x - 4 = 0$

76. $10x^3 + 6x^2 + x - 2 = 0$

10.4   ## Equations Quadratic in Form

We are now in a position to put our knowledge of quadratic equations to work to solve a variety of equations.

EXAMPLE 1   Solve $(x + 3)^2 - 2(x + 3) - 8 = 0$.

Solution   We can see that this equation is quadratic in form by replacing $x + 3$ with another variable, say, y. Replacing $x + 3$ with y we have

$$y^2 - 2y - 8 = 0$$

We can solve this equation by factoring the left side and then setting each factor equal to 0.

$$y^2 - 2y - 8 = 0$$

$$(y - 4)(y + 2) = 0 \qquad \text{Factor.}$$

$$y - 4 = 0 \quad \text{or} \quad y + 2 = 0 \qquad \text{Set factors to 0.}$$

$$y = 4 \quad \text{or} \qquad y = -2$$

Since our original equation was written in terms of the variable x, we would like our solutions in terms of x also. Replacing y with $x + 3$ and then solving for x we have

$$x + 3 = 4 \qquad \text{or} \qquad x + 3 = -2$$

$$x = 1 \qquad \text{or} \qquad x = -5$$

The solutions to our original equation are 1 and -5.

The method we have just shown lends itself well to other types of equations that are quadratic in form, as we will see. In this example, however, there is another method that works just as well. Let's solve our original equation again, but this time, let's begin by expanding $(x + 3)^2$ and $2(x + 3)$.

$$(x + 3)^2 - 2(x + 3) - 8 = 0$$

$$x^2 + 6x + 9 - 2x - 6 - 8 = 0 \qquad \text{Multiply.}$$

$$x^2 + 4x - 5 = 0 \qquad \text{Combine similar terms.}$$

$$(x - 1)(x + 5) = 0 \qquad \text{Factor.}$$

$$x - 1 = 0 \qquad \text{or} \qquad x + 5 = 0 \qquad \text{Set factors to 0.}$$

$$x = 1 \qquad \text{or} \qquad x = -5$$

As you can see, either method produces the same result.

EXAMPLE 2   Solve $4x^4 + 7x^2 = 2$.

Solution   This equation is quadratic in x^2. We can make it easier to look at by using the substitution $y = x^2$. (The choice of the letter y is arbitrary. We could just as easily use the substitution $m = x^2$.) Making the substitution $y = x^2$ and then solving the resulting equation we have

$$4y^2 + 7y = 2$$

$$4y^2 + 7y - 2 = 0 \qquad \text{Standard form}$$

$$(4y - 1)(y + 2) = 0 \qquad \text{Factor.}$$

$$4y - 1 = 0 \qquad \text{or} \qquad y + 2 = 0 \qquad \text{Set factors to 0.}$$

$$y = \frac{1}{4} \qquad \text{or} \qquad y = -2$$

Now we replace y with x^2 in order to solve for x:

$$x^2 = \frac{1}{4} \qquad \text{or} \qquad x^2 = -2$$

$$x = \pm\sqrt{\frac{1}{4}} \qquad \text{or} \qquad x = \pm\sqrt{-2} \qquad \text{Theorem 10.1}$$

$$= \pm\frac{1}{2} \qquad \text{or} \qquad = \pm i\sqrt{2}$$

The solution set is $\{\frac{1}{2}, -\frac{1}{2}, i\sqrt{2}, -i\sqrt{2}\}$.

EXAMPLE 3   Solve for x: $x + \sqrt{x} - 6 = 0$

Solution   To see that this equation is quadratic in form, we have to notice that $(\sqrt{x})^2 = x$. That is, the equation can be rewritten as

$$(\sqrt{x})^2 + \sqrt{x} - 6 = 0$$

Replacing $\sqrt{x}$ with y and solving as usual, we have

$$y^2 + y - 6 = 0$$

$$(y + 3)(y - 2) = 0$$

$$y + 3 = 0 \qquad \text{or} \qquad y - 2 = 0$$

$$y = -3 \qquad \text{or} \qquad y = 2$$

Again, to find x, we replace y with $\sqrt{x}$ and solve:

$$\sqrt{x} = -3 \qquad \text{or} \qquad \sqrt{x} = 2$$

$$x = 9 \qquad\qquad\qquad x = 4 \qquad \text{Square both sides of each equation.}$$

Since we squared both sides of each equation, we have the possibility of obtaining extraneous solutions. We have to check both solutions in our original equation.

When	$x = 9$	When	$x = 4$
the equation	$x + \sqrt{x} - 6 = 0$	the equation	$x + \sqrt{x} - 6 = 0$
becomes	$9 + \sqrt{9} - 6 \overset{?}{=} 0$	becomes	$4 + \sqrt{4} - 6 \overset{?}{=} 0$
	$9 + 3 - 6 \overset{?}{=} 0$		$4 + 2 - 6 \overset{?}{=} 0$
	$6 \neq 0$		$0 = 0$
	This means 9 is extraneous.		This means 4 is a solution.

The only solution to the equation $x + \sqrt{x} - 6 = 0$ is $x = 4$.

We should note here that the two possible solutions, 9 and 4, to the equation in Example 3 can be obtained by another method. Instead of substituting for $\sqrt{x}$, we can isolate it on one side of the equation and then square both sides to clear the equation of radicals.

$$x + \sqrt{x} - 6 = 0$$

$$\sqrt{x} = -x + 6 \qquad\qquad \text{Isolate } \sqrt{x}.$$

$$x = x^2 - 12x + 36 \qquad\qquad \text{Square both sides.}$$

$$0 = x^2 - 13x + 36 \qquad\qquad \text{Add } -x \text{ to both sides.}$$

$$0 = (x - 4)(x - 9) \qquad\qquad \text{Factor.}$$

$$x - 4 = 0 \qquad \text{or} \qquad x - 9 = 0$$

$$x = 4 \qquad\qquad\qquad x = 9$$

We obtain the same two possible solutions. Since we squared both sides of the equation to find them, we would have to check each one in the original equation. As was the case in Example 3, only $x = 4$ is a solution; $x = 9$ is extraneous.

E X A M P L E   4   If an object is tossed into the air with an upward velocity of 12 feet per second from the top of a building h feet high, the time it takes for the object to hit the ground below is given by the formula

$$16t^2 - 12t - h = 0$$

Solve this formula for t.

Solution   The formula is in standard form and is quadratic in t. The coefficients a, b, and c that we need to apply to the quadratic formula are $a = 16$, $b = -12$, and $c = -h$. Substituting these quantities into the quadratic formula, we have

$$t = \frac{12 \pm \sqrt{144 - 4(16)(-h)}}{2(16)}$$

$$= \frac{12 \pm \sqrt{144 + 64h}}{32}$$

We can factor the perfect square 16 from the two terms under the radical and simplify our radical somewhat:

$$t = \frac{12 \pm \sqrt{16(9 + 4h)}}{32}$$

$$= \frac{12 \pm 4\sqrt{9 + 4h}}{32}$$

Now we can reduce to lowest terms by factoring a 4 from the numerator and denominator:

$$t = \frac{\cancel{4}(3 \pm \sqrt{9 + 4h})}{\cancel{4} \cdot 8}$$

$$= \frac{3 \pm \sqrt{9 + 4h}}{8}$$

If we were given a value of h, we would find that one of the solutions to this last formula would be a negative number. Since time is always measured in positive units, we wouldn't use that solution.

More About the Golden Ratio

In Section 9.1 we derived the golden ratio $\dfrac{1 + \sqrt{5}}{2}$ by finding the ratio of length to width for a golden rectangle. The golden ratio was actually discovered before the golden rectangle by the Greeks who lived before Euclid. The early Greeks found the golden ratio by dividing a line segment into two parts so that the ratio of the shorter part to the longer part was the same as the ratio of the longer part to the whole segment. When they divided a line segment in this manner, they said it was divided in "extreme and mean ratio." Figure 1 illustrates a line segment divided this way.

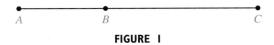

$$A \qquad\qquad B \qquad\qquad\qquad\qquad C$$

FIGURE 1

If point B divides segment AC in "extreme and mean ratio," then

$$\frac{\text{Length of shorter segment}}{\text{Length of longer segment}} = \frac{\text{length of longer segment}}{\text{length of whole segment}}$$

$$\frac{AB}{BC} = \frac{BC}{AC}$$

EXAMPLE 5   If the length of segment AB in Figure 1 is 1 inch, find the length of BC so that the whole segment AC is divided in "extreme and mean ratio."

Solution   Using Figure 1 as a guide, if we let $x =$ the length of segment BC, then the length of AC is $x + 1$. If B divides AC into "extreme and mean ratio," then the ratio of AB to BC must equal the ratio of BC to AC. Writing this relationship using the variable x, we have

$$\frac{1}{x} = \frac{x}{x + 1}$$

If we multiply both sides of this equation by the LCD $x(x + 1)$ we have

$$x + 1 = x^2$$

$$0 = x^2 - x - 1 \qquad \text{Write equation in standard form.}$$

Since this last equation is not factorable, we apply the quadratic formula.

$$x = \frac{1 \pm \sqrt{(-1)^2 - 4(1)(-1)}}{2}$$

$$= \frac{1 \pm \sqrt{5}}{2}$$

Our equation has two solutions, which we approximate using decimals:

$$\frac{1 + \sqrt{5}}{2} \approx 1.618 \qquad \frac{1 - \sqrt{5}}{2} \approx -0.618$$

Since we originally let x equal the length of segment BC, we use only the positive solution to our equation. As you can see, the positive solution is the golden ratio.

GRAPHING CALCULATORS

More About Example 1

As we mentioned before, algebraic expressions entered into a graphing calculator do not have to be simplified in order to be evaluated. This fact applies to equations as well. We can graph the equation $y = (x + 3)^2 - 2(x + 3) - 8$ to assist us in solving the equation in Example 1. The graph is shown in Figure 2. Using the Zoom and Trace features at the x-intercepts gives us $x = 1$ and $x = -5$ as the solutions to the equation $0 = (x + 3)^2 - 2(x + 3) - 8$.

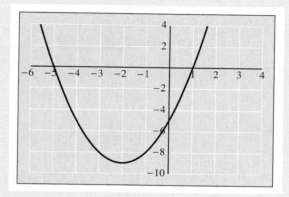

FIGURE 2

More About Example 2

Figure 3 shows the graph of $y = 4x^4 + 7x^2 - 2$. As we expect, the x-intercepts give the real number solutions to the equation $0 = 4x^4 + 7x^2 - 2$. The complex solutions do not appear on the graph.

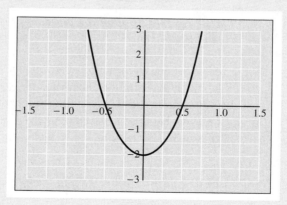

FIGURE 3

(continued)

(continued)

More About Example 3

In solving the equation in Example 3, we found that one of the possible solutions was an extraneous solution. If we solve the equation $x + \sqrt{x} - 6 = 0$ by graphing the function $y = x + \sqrt{x} - 6$, we find that the extraneous solution, 9, is not an x-intercept. Figure 4 shows that the only solution to the equation occurs at the x-intercept 4.

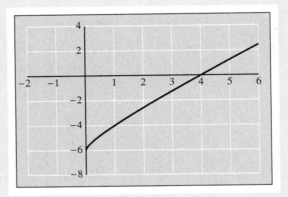

FIGURE 4

 # Getting Ready for Class

After reading through the preceding section, respond in your own words and in complete sentences.

A. What does it mean for an equation to be quadratic in form?

B. What are all of the circumstances in solving equations (that we have studied) in which it is necessary to check for extraneous solutions?

C. How would you start to solve the equation $x + \sqrt{x} - 6 = 0$?

D. What does it mean for a line segment to be divided in "extreme and mean ratio"?

PROBLEM SET 10.4

Solve each equation.

1. $(x - 3)^2 + 3(x - 3) + 2 = 0$
2. $(x + 4)^2 - (x + 4) - 6 = 0$
3. $2(x + 4)^2 + 5(x + 4) - 12 = 0$
4. $3(x - 5)^2 + 14(x - 5) - 5 = 0$
5. $x^4 - 6x^2 - 27 = 0$ 6. $x^4 + 2x^2 - 8 = 0$
7. $x^4 + 9x^2 = -20$ 8. $x^4 - 11x^2 = -30$
9. $(2a - 3)^2 - 9(2a - 3) = -20$
10. $(3a - 2)^2 + 2(3a - 2) = 3$
11. $2(4a + 2)^2 = 3(4a + 2) + 20$
12. $6(2a + 4)^2 = (2a + 4) + 2$
13. $6t^4 = -t^2 + 5$ 14. $3t^4 = -2t^2 + 8$
15. $9x^4 - 49 = 0$ 16. $25x^4 - 9 = 0$

Solve each of the following equations.

Remember, if you square both sides of an equation in the process of solving it, you have to check all solutions in the original equation.

17. $x - 7\sqrt{x} + 10 = 0$ 18. $x - 6\sqrt{x} + 8 = 0$
19. $t - 2\sqrt{t} - 15 = 0$ 20. $t - 3\sqrt{t} - 10 = 0$
21. $6x + 11\sqrt{x} = 35$ 22. $2x + \sqrt{x} = 15$
23. $(a - 2) - 11\sqrt{a - 2} + 30 = 0$
24. $(a - 3) - 9\sqrt{a - 3} + 20 = 0$
25. $(2x + 1) - 8\sqrt{2x + 1} + 15 = 0$
26. $(2x - 3) - 7\sqrt{2x - 3} + 12 = 0$
27. Solve the formula $16t^2 - vt - h = 0$ for t.
28. Solve the formula $16t^2 + vt + h = 0$ for t.
29. Solve the formula $kx^2 + 8x + 4 = 0$ for x.
30. Solve the formula $k^2x^2 + kx + 4 = 0$ for x.
31. Solve $x^2 + 2xy + y^2 = 0$ for x by using the quadratic formula with $a = 1$, $b = 2y$, and $c = y^2$.
32. Solve $x^2 - 2xy + y^2 = 0$ for x by using the quadratic formula, with $a = 1$, $b = -2y$, and $c = y^2$.

Applying the Concepts

For Problems 33–36, t is in seconds.

33. **Falling Object** An object is tossed into the air with an upward velocity of 8 feet per second from the top of a building h feet high. The time it takes for the object to hit the ground below is given by the formula $16t^2 - 8t - h = 0$. Solve this formula for t.

34. **Falling Object** An object is tossed into the air with an upward velocity of 6 feet per second from the top of a building h feet high. The time it takes for the object to hit the ground below is given by the formula $16t^2 - 6t - h = 0$. Solve this formula for t.

35. **Falling Object** An object is tossed into the air with an upward velocity of v feet per second from the top of a building 20 feet high. The time it takes for the object to hit the ground below is given by the formula $16t^2 - vt - 20 = 0$. Solve this formula for t.

36. **Falling Object** An object is tossed into the air with an upward velocity of v feet per second from the top of a building 40 feet high. The time it takes for the object to hit the ground below is given by the formula $16t^2 - vt - 40 = 0$. Solve this formula for t.

Use Figure 1 from this section as a guide to working Problems 37–40.

37. **Golden Ratio** If AB in Figure 1 is 4 inches, and B divides AC in "extreme and mean ratio," find BC, and then show that BC is 4 times the golden ratio.

38. **Golden Ratio** If AB in Figure 1 is $\frac{1}{2}$ inch, and B divides AC in "extreme and mean ratio," find BC, and then show that BC is half the golden ratio.

39. **Golden Ratio** If AB in Figure 1 is 2 inches, and B divides AC in "extreme and mean ratio," find BC, and then show that the ratio of BC to AB is the golden ratio.

40. **Golden Ratio** If AB in Figure 1 is $\frac{1}{2}$ inch, and B divides AC in "extreme and mean ratio," find BC, and then show that the ratio of BC to AB is the golden ratio.

41. **Saint Louis Arch** The shape of the famous "Gateway to the West" arch in Saint Louis can be modeled by a parabola. The equation for one such parabola is:

$$y = -\frac{1}{150}x^2 + \frac{21}{5}x$$

(a) Sketch the graph of the arch's equation on a coordinate axis.

(b) Approximately how far do you have to walk to get from one side of the arch to the other?

(Tom Tracy/Tony Stone Images)

42. **Area** In the following diagram, *ABCD* is a rectangle with diagonal *AC*. Find its area.

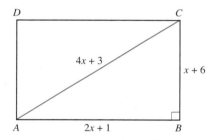

43. **Area and Perimeter** A total of 160 yards of fencing is to be used to enclose part of a lot that borders on a river. This situation is shown in the following diagram.

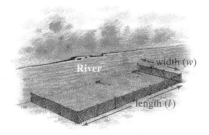

(a) Write an equation that gives the relationship between the length and width and the 160 yards of fencing.

(b) The formula for the area that is enclosed by the fencing and the river is $A = lw$. Solve the equation in part (a) for *l*, and then use the result to write the area in terms of *w* only.

(c) Make a table that gives at least five possible values of *w* and associated area *A*.

(d) From the pattern in your table shown in part (c), what is the largest area that can be enclosed by the 160 yards of fencing? (Try some other table values if necessary.)

44. **Area and Perimeter** Rework all four parts of the preceding problem if it is desired to have an opening 2 yards wide in one of the shorter sides, as shown in the following diagram.

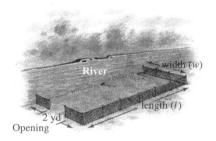

Review Problems

The problems that follow review material we covered in Sections 9.4 and 9.5.

Combine, if possible. [9.4]

45. $5\sqrt{7} - 2\sqrt{7}$ **46.** $6\sqrt{2} - 9\sqrt{2}$

47. $\sqrt{18} - \sqrt{8} + \sqrt{32}$ **48.** $\sqrt{50} + \sqrt{72} - \sqrt{8}$

49. $9x\sqrt{20x^3y^2} + 7y\sqrt{45x^5}$

50. $5x^2\sqrt{27xy^3} - 6y\sqrt{12x^5y}$

Multiply. [9.5]

51. $(\sqrt{5} - 2)(\sqrt{5} + 8)$ **52.** $(2\sqrt{3} - 7)(2\sqrt{3} + 7)$

53. $(\sqrt{x} + 2)^2$ **54.** $(3 - \sqrt{x})(3 + \sqrt{x})$

Rationalize the denominator. [9.5]

55. $\dfrac{\sqrt{7}}{\sqrt{7} - 2}$ **56.** $\dfrac{\sqrt{5} - \sqrt{2}}{\sqrt{5} + \sqrt{2}}$

Extending the Concepts

Find the x- and y-intercepts.

57. $y = x^3 - 4x$

58. $y = x^4 - 10x^2 + 9$

59. $y = 3x^3 + x^2 - 27x - 9$

60. $y = 2x^3 + x^2 - 8x - 4$

61. The graph of $y = 2x^3 - 7x^2 - 5x + 4$ crosses the x-axis at $x = 4$. Where else does it cross the x-axis?

62. The graph of $y = 6x^3 + x^2 - 12x + 5$ crosses the x-axis at $x = 1$. Where else does it cross the x-axis?

10.5   Graphing Parabolas

The solution set to the equation

$$y = x^2 - 3$$

consists of ordered pairs. One method of graphing the solution set is to find a number of ordered pairs that satisfy the equation and to graph them. We can obtain some ordered pairs that are solutions to $y = x^2 - 3$ by use of a table as follows:

x	$y = x^2 - 3$	y	Solutions
-3	$y = (-3)^2 - 3 = 9 - 3 = 6$	6	$(-3, 6)$
-2	$y = (-2)^2 - 3 = 4 - 3 = 1$	1	$(-2, 1)$
-1	$y = (-1)^2 - 3 = 1 - 3 = -2$	-2	$(-1, -2)$
0	$y = 0^2 \quad\; - 3 = 0 - 3 = -3$	-3	$(0, -3)$
1	$y = 1^2 \quad\; - 3 = 1 - 3 = -2$	-2	$(1, -2)$
2	$y = 2^2 \quad\; - 3 = 4 - 3 = 1$	1	$(2, 1)$
3	$y = 3^2 \quad\; - 3 = 9 - 3 = 6$	6	$(3, 6)$

Graphing these solutions and then connecting them with a smooth curve, we have the graph of $y = x^2 - 3$. (See Figure 1.)

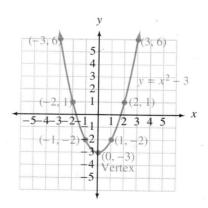

FIGURE 1

This graph is an example of a *parabola*. All equations of the form $y = ax^2 + bx + c$, $a \neq 0$, have parabolas for graphs.

Although it is always possible to graph parabolas by making a table of values of x and y that satisfy the equation, there are other methods that are faster and, in some cases, more accurate.

The important points associated with the graph of a parabola are the highest (or lowest) point on the graph and the x-intercepts. The y-intercepts can also be useful.

INTERCEPTS FOR PARABOLAS

The graph of the equation $y = ax^2 + bx + c$ crosses the y-axis at $y = c$, since substituting $x = 0$ into $y = ax^2 + bx + c$ yields $y = c$.

Since the graph crosses the x-axis when $y = 0$, the x-intercepts are those values of x that are solutions to the quadratic equation $0 = ax^2 + bx + c$.

THE VERTEX OF A PARABOLA

The highest or lowest point on a parabola is called the *vertex*. The vertex for the graph of $y = ax^2 + bx + c$ will always occur when

$$x = \frac{-b}{2a}$$

To see this, we must transform the right side of $y = ax^2 + bx + c$ into an expression that contains x in just one of its terms. This is accomplished by completing the square on the first two terms. Here is what it looks like:

$$y = ax^2 + bx + c$$

$$= a\left(x^2 + \frac{b}{a}x\right) + c$$

$$= a\left[x^2 + \frac{b}{a}x + \left(\frac{b}{2a}\right)^2\right] + c - a\left(\frac{b}{2a}\right)^2$$

$$= a\left(x + \frac{b}{2a}\right)^2 + \frac{4ac - b^2}{4a}$$

It may not look like it, but this last line indicates that the vertex of the graph of $y = ax^2 + bx + c$ has an x-coordinate of $\frac{-b}{2a}$. Since a, b, and c are constants, the only quantity that is varying in the last expression is the x in $\left(x + \frac{b}{2a}\right)^2$.

Since the quantity $\left(x + \frac{b}{2a}\right)^2$ is the square of $x + \frac{b}{2a}$, the smallest it will ever be is 0, and that will happen when $x = \frac{-b}{2a}$.

We can use the vertex point along with the x- and y-intercepts to sketch the graph of any equation of the form $y = ax^2 + bx + c$. Here is a summary of the preceding information.

GRAPHING PARABOLAS

The graph of $y = ax^2 + bx + c$, $a \neq 0$, will have

1. A y-intercept at $y = c$
2. x-intercepts (if they exist) at

$$x = \frac{-b \pm \sqrt{b^2 - 4ac}}{2a}$$

3. A vertex when $x = \dfrac{-b}{2a}$

E X A M P L E   1   Sketch the graph of $y = x^2 - 6x + 5$.

Solution   To find the x-intercepts, we let $y = 0$ and solve for x:

$$0 = x^2 - 6x + 5$$
$$0 = (x - 5)(x - 1)$$
$$x = 5 \quad \text{or} \quad x = 1$$

To find the coordinates of the vertex, we first find

$$x = \frac{-b}{2a} = \frac{-(-6)}{2(1)} = 3$$

The x-coordinate of the vertex is 3. To find the y-coordinate, we substitute 3 for x in our original equation:

$$y = 3^2 - 6(3) + 5 = 9 - 18 + 5 = -4$$

The graph crosses the x-axis at 1 and 5 and has its vertex at $(3, -4)$. Plotting these points and connecting them with a smooth curve, we have the graph shown in Figure 2. The graph is a parabola that opens up, so we say the graph is *concave up*. The

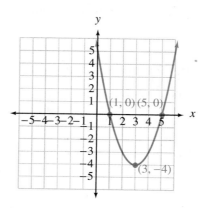

FIGURE 2

vertex is the lowest point on the graph. (Note that the graph crosses the y-axis at 5, which is the value of y we obtain when we let $x = 0$.)

FINDING THE VERTEX BY COMPLETING THE SQUARE

Another way to locate the vertex of the parabola in Example 1 is by completing the square on the first two terms on the right side of the equation $y = x^2 - 6x + 5$. In this case, we would do so by adding 9 to and subtracting 9 from the right side of the equation. This amounts to adding 0 to the equation, so we know we haven't changed its solutions. This is what it looks like:

$$y = (x^2 - 6x \quad) + 5$$
$$= (x^2 - 6x + 9) + 5 - 9$$
$$= (x - 3)^2 - 4$$

You may have to look at this last equation awhile to see this, but when $x = 3$, then $y = (x - 3)^2 - 4 = 0^2 - 4 = -4$ is the smallest y will ever be. And that is why the vertex is at $(3, -4)$. As a matter of fact, this is the same kind of reasoning we used when we derived the formula $x = \dfrac{-b}{2a}$ for the x-coordinate of the vertex.

EXAMPLE 2   Graph $y = -x^2 - 2x + 3$.

Solution   To find the x-intercepts, we let $y = 0$:

$$0 = -x^2 - 2x + 3$$
$$0 = x^2 + 2x - 3 \qquad \text{Multiply each side by } -1.$$
$$0 = (x + 3)(x - 1)$$
$$x = -3 \quad \text{or} \quad x = 1$$

The x-coordinate of the vertex is given by

$$x = \frac{-b}{2a} = \frac{-(-2)}{2(-1)} = \frac{2}{-2} = -1$$

To find the y-coordinate of the vertex, we substitute -1 for x in our original equation to get

$$y = -(-1)^2 - 2(-1) + 3 = -1 + 2 + 3 = 4$$

Our parabola has x-intercepts at -3 and 1, and a vertex at $(-1, 4)$. Figure 3 shows the graph. We say the graph is *concave down* since it opens downward. Again, we could have obtained the coordinates of the vertex by completing the square on the first two terms on the right side of our equation. To do so, we must first factor -1 from the first two terms. (Remember, the leading coefficient must be 1 in order to

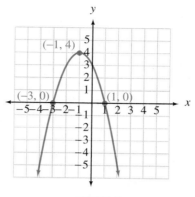

FIGURE 3

complete the square.) When we complete the square, we add 1 inside the parentheses, which actually decreases the right side of the equation by -1 since everything in the parentheses is multiplied by -1. To make up for it, we add 1 outside the parentheses.

$$y = -1(x^2 + 2x \qquad) + 3$$
$$= -1(x^2 + 2x + \mathbf{1}) + 3 + \mathbf{1}$$
$$= -1(x + 1)^2 + 4$$

The last line tells us that the *largest* value of y will be 4, and that will occur when $x = -1$.

E X A M P L E   3   Graph $y = 3x^2 - 6x + 1$.

Solution   To find the x-intercepts, we let $y = 0$ and solve for x:

$$0 = 3x^2 - 6x + 1$$

Since the right side of this equation does not factor, we can look at the discriminant to see what kind of solutions are possible. The discriminant for this equation is

$$b^2 - 4ac = 36 - 4(3)(1) = 24$$

Since the discriminant is a positive number but not a perfect square, the equation will have irrational solutions. This means that the x-intercepts are irrational numbers and will have to be approximated with decimals using the quadratic formula. Rather than use the quadratic formula, we will find some other points on the graph, but first let's find the vertex.

Here are both methods of finding the vertex:

Using the formula that gives us the x-coordinate of the vertex, we have:

$$x = \frac{-b}{2a} = \frac{-(-6)}{2(3)} = 1$$

Substituting 1 for x in the equation gives us the y-coordinate of the vertex:

$$y = 3 \cdot 1^2 - 6 \cdot 1 + 1 = -2$$

To complete the square on the right side of the equation, we factor 3 from the first two terms, add 1 inside the parentheses, and add -3 outside the parentheses (this amounts to adding 0 to the right side):

$$y = 3(x^2 - 2x \quad) + 1$$
$$= 3(x^2 - 2x + \mathbf{1}) + 1 - \mathbf{3}$$
$$= 3(x - 1)^2 - 2$$

In either case, the vertex is $(1, -2)$.

If we can find two points, one on each side of the vertex, we can sketch the graph. Let's let $x = 0$ and $x = 2$, since each of these numbers is the same distance from $x = 1$, and $x = 0$ will give us the y-intercept.

When $x = 0$

$$y = 3(0)^2 - 6(0) + 1$$
$$= 0 - 0 + 1$$
$$= 1$$

When $x = 2$

$$y = 3(2)^2 - 6(2) + 1$$
$$= 12 - 12 + 1$$
$$= 1$$

The two points just found are $(0, 1)$ and $(2, 1)$. Plotting these two points along with the vertex $(1, -2)$, we have the graph shown in Figure 4.

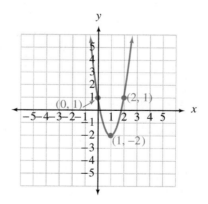

FIGURE 4

EXAMPLE 4   Graph $y = -2x^2 + 6x - 5$.

Solution   Letting $y = 0$, we have

$$0 = -2x^2 + 6x - 5$$

Again, the right side of this equation does not factor. The discriminant is $b^2 - 4ac = 36 - 4(-2)(-5) = -4$, which indicates that the solutions are complex numbers. This means that our original equation does not have x-intercepts. The graph does not cross the x-axis.

Let's find the vertex.

Using our formula for the x-coordinate of the vertex, we have

$$x = \frac{-b}{2a} = \frac{-6}{2(-2)} = \frac{6}{4} = \frac{3}{2}$$

To find the y-coordinate, we let $x = \frac{3}{2}$:

$$y = -2\left(\frac{3}{2}\right)^2 + 6\left(\frac{3}{2}\right) - 5$$

$$= \frac{-18}{4} + \frac{18}{2} - 5$$

$$= \frac{-18 + 36 - 20}{4}$$

$$= -\frac{1}{2}$$

Finding the vertex by completing the square is a more complicated matter. In order to make the coefficient of x^2 a 1, we must factor -2 from the first two terms. To complete the square inside the parentheses, we add $\frac{9}{4}$. Since each term inside the parentheses is multiplied by -2, we add $\frac{9}{2}$ outside the parentheses so that the net result is the same as adding 0 to the right side:

$$y = -2(x^2 - 3x \quad) - 5$$

$$= -2\left(x^2 - 3x + \frac{9}{4}\right) - 5 + \frac{9}{2}$$

$$= -2\left(x - \frac{3}{2}\right)^2 - \frac{1}{2}$$

The vertex is $\left(\frac{3}{2}, -\frac{1}{2}\right)$. Since this is the only point we have so far, we must find two others. Let's let $x = 3$ and $x = 0$, since each point is the same distance from $x = \frac{3}{2}$ and on either side:

When $x = 3$

$$y = -2(3)^2 + 6(3) - 5$$

$$= -18 + 18 - 5$$

$$= -5$$

When $x = 0$

$$y = -2(0)^2 + 6(0) - 5$$

$$= 0 + 0 - 5$$

$$= -5$$

The two additional points on the graph are $(3, -5)$ and $(0, -5)$. Figure 5 shows the graph.

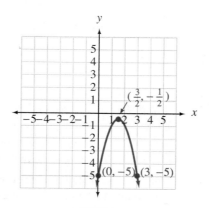

FIGURE 5

The graph is concave down. The vertex is the highest point on the graph.

By looking at the equations and graphs in Examples 1 through 4, we can conclude that the graph of $y = ax^2 + bx + c$ will be concave up when a is positive, and concave down when a is negative. Taking this even further, if $a > 0$, then the vertex is the lowest point on the graph, and if $a < 0$, the vertex is the highest point on the graph. Finally, if we complete the square on x, we can rewrite the equation of our parabola as $y = a(x - h)^2 + k$ where (h, k) is the vertex of our parabola.

EXAMPLE 5   A company selling copies of an accounting program for home computers finds that it will make a weekly profit of P dollars from selling x copies of the program, according to the equation

$$P(x) = -0.1x^2 + 27x - 500$$

How many copies of the program should it sell to make the largest possible profit, and what is the largest possible profit?

Solution   Since the coefficient of x^2 is negative, we know the graph of this parabola will be concave down, meaning that the vertex is the highest point of the curve. We find the vertex by first finding its x-coordinate:

$$x = \frac{-b}{2a} = \frac{-27}{2(-0.1)} = \frac{27}{0.2} = 135$$

This represents the number of programs the company needs to sell each week in order to make a maximum profit. To find the maximum profit, we substitute 135 for x in the original equation. (A calculator is helpful for these kinds of calculations.)

$$P(135) = -0.1(135)^2 + 27(135) - 500$$

$$= -0.1(18,225) + 3,645 - 500$$

$$= -1,822.5 + 3,645 - 500$$

$$= 1,322.5$$

The maximum weekly profit is $1,322.50 and is obtained by selling 135 programs a week.

EXAMPLE 6   An art supply store finds that they can sell x sketch pads each week at p dollars each, according to the equation $x = 900 - 300p$. Graph the revenue equation $R = xp$. Then use the graph to find the price p that will bring in the maximum revenue. Finally, find the maximum revenue.

Solution   As it stands, the revenue equation contains three variables. Since we are asked to find the value of p that gives us the maximum value of R, we rewrite the equation using just the variables R and p. Since $x = 900 - 300p$, we have

$$R = xp = (900 - 300p)p$$

The graph of this equation is shown in Figure 6. The graph appears in the first quadrant only, since R and p are both positive quantities.

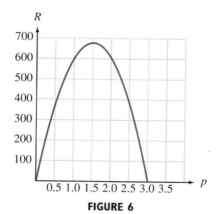

FIGURE 6

From the graph we see that the maximum value of R occurs when $p = \$1.50$. We can calculate the maximum value of R from the equation:

When $\qquad\qquad p = 1.5$

the equation $\qquad R = (900 - 300p)p$

becomes $\qquad R = (900 - 300 \cdot 1.5)1.5$

$$= (900 - 450)1.5$$

$$= 450 \cdot 1.5$$

$$= 675$$

The maximum revenue is \$675. It is obtained by setting the price of each sketch pad at $p = \$1.50$.

 Using TECHNOLOGY

GRAPHING CALCULATORS

If you have been using a graphing calculator for some of the material in this course, you are well aware that your calculator can draw all the graphs in this section very easily. It is important, however, that you be able to recognize and sketch the graph of any parabola by hand. It is a skill that all successful intermediate algebra students should possess, even if they are proficient in the use of a graphing calculator. My suggestion is that you work the problems in this section and problem set without your calculator. Then use your calculator to check your results.

FINDING THE EQUATION FROM THE GRAPH

EXAMPLE 7   At the 1997 Washington County Fair in Oregon, David Smith, Jr., The Bullet, was shot from a cannon. As a human cannonball, he reached a height of 70 feet before landing in a net 160 feet from the cannon. Sketch the graph of his path, and then find the equation of the graph.

Solution   We assume that the path taken by the human cannonball is a parabola. If the origin of the coordinate system is at the opening of the cannon, then the net that catches him will be at 160 on the *x*-axis. Figure 7 shows the graph:

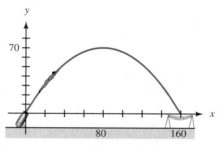

FIGURE 7

Since the curve is a parabola, we know the equation will have the form

$$y = a(x - h)^2 + k$$

Since the vertex of the parabola is at (80, 70), we can fill in two of the three constants in our equation, giving us

$$y = a(x - 80)^2 + 70$$

To find *a*, we note that the landing point will be (160, 0). Substituting the coordinates of this point into the equation, we solve for *a*:

$$0 = a(160 - 80)^2 + 70$$

$$0 = a(80)^2 + 70$$

$$0 = 6{,}400a + 70$$

$$a = -\frac{70}{6{,}400} = -\frac{7}{640}$$

The equation that describes the path of the human cannonball is

$$y = -\frac{7}{640}(x - 80)^2 + 70 \quad \text{for } 0 \le x \le 160$$

Using TECHNOLOGY

GRAPHING CALCULATORS

Graph the equation found in Example 7 by graphing it on a graphing calculator using the window shown here. (We will use this graph later in the book to find the angle between the cannon and the horizontal.)

Window: X from 0 to 180, increment 20

Y from 0 to 80, increment 10

On the TI-83, an increment of 20 for X means Xscl=20.

Getting Ready for Class

After reading through the preceding section, respond in your own words and in complete sentences.

A. What is a parabola?

B. What part of the equation of a parabola determines whether the graph is concave up or concave down?

C. Suppose $f(x) = ax^2 + bx + c$ is the equation of a parabola. Explain how $f(4) = 1$ relates to the graph of the parabola.

D. A line can be graphed with two points. How many points are necessary to get a reasonable sketch of a parabola? Explain.

PROBLEM SET 10.5

For each of the following equations, give the x-intercepts and the coordinates of the vertex, and sketch the graph.

1. $y = x^2 + 2x - 3$ **2.** $y = x^2 - 2x - 3$

3. $y = -x^2 - 4x + 5$ **4.** $y = x^2 + 4x - 5$

5. $y = x^2 - 1$ **6.** $y = x^2 - 4$

7. $y = -x^2 + 9$ **8.** $y = -x^2 + 1$

9. $y = 2x^2 - 4x - 6$ **10.** $y = 2x^2 + 4x - 6$

11. $y = x^2 - 2x - 4$ **12.** $y = x^2 - 2x - 2$

Find the vertex and any two convenient points to sketch the graphs of the following equations.

13. $y = x^2 - 4x - 4$ **14.** $y = x^2 - 2x + 3$

15. $y = -x^2 + 2x - 5$ **16.** $y = -x^2 + 4x - 2$

17. $y = x^2 + 1$ **18.** $y = x^2 + 4$

19. $y = -x^2 - 3$ **20.** $y = -x^2 - 2$

21. $y = 3x^2 + 4x + 1$ **22.** $y = 2x^2 + 4x + 3$

For each of the following equations, find the coordinates of the vertex, and indicate whether the vertex is the highest point on the graph or the lowest point on the graph. (Do not graph.)

23. $y = x^2 - 6x + 5$ **24.** $y = -x^2 + 6x - 5$

25. $y = -x^2 + 2x + 8$ **26.** $y = x^2 - 2x - 8$

27. $y = 12 + 4x - x^2$ **28.** $y = -12 - 4x + x^2$

29. $y = -x^2 - 8x$ **30.** $y = x^2 + 8x$

Applying the Concepts

31. Maximum Profit A company earns a weekly profit of P dollars by selling x items, according to the equation $P(x) = -0.5x^2 + 40x - 300$. Find the number of items the company must sell each week in order to obtain the largest possible profit. Then, find the largest possible profit.

32. Maximum Profit A company earns a weekly profit of P dollars by selling x items, according to the equation $P(x) = -0.5x^2 + 100x - 1,600$. Find the number of items the company must sell each week in order to obtain the largest possible profit. Then, find the largest possible profit.

33. Maximum Profit A company finds that it can make a profit of P dollars each month by selling x patterns, according to the formula $P(x) = -0.002x^2 + 3.5x - 800$. How many patterns must it sell each month in order to have a maximum profit? What is the maximum profit?

34. Maximum Profit A company selling picture frames finds that it can make a profit of P dollars each month by selling x frames, according to the formula $P(x) = -0.002x^2 + 5.5x - 1,200$. How many frames must it sell each month in order to have a maximum profit? What is the maximum profit?

35. Maximum Height Chaudra is tossing a softball into the air with an underhand motion. The distance of the ball above her hand at any time is given by the function

$$h(t) = 32t - 16t^2 \quad \text{for } 0 \le t \le 2$$

where $h(t)$ is the height of the ball (in feet) and t is the time (in seconds). Find the times at which the ball is in her hand, and the maximum height of the ball.

36. Maximum Height Hali is tossing a quarter into the air with an underhand motion. The distance of the quarter above her hand at any time is given by the function

$$h(t) = 16t - 16t^2 \quad \text{for } 0 \le t \le 1$$

where $h(t)$ is the height of the quarter (in feet) and t is the time (in seconds). Find the times at which the quarter is in her hand, and the maximum height of the quarter.

37. Maximum Height An arrow is shot straight up into the air with an initial velocity of 128 feet per second. If h represents the height (in feet) of the arrow at any time t (in seconds), then the equation that gives h in terms of t is $h(t) = 128t - 16t^2$. Find the maximum height attained by the arrow.

38. Maximum Height A ball is projected into the air with an upward velocity of 64 feet per second. The equation that gives the height h (in feet) of the ball at any time t (in seconds) is $h(t) = 64t - 16t^2$. Find the maximum height attained by the ball.

39. Maximum Area Justin wants to fence three sides of a rectangular exercise yard for his dog. The fourth side of the exercise yard will be a side of the house. He has 80 feet of fencing available. Find the dimensions of the exercise yard that will enclose the maximum area.

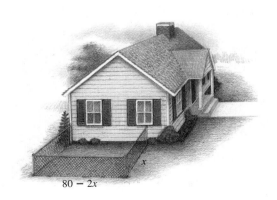

$80 - 2x$

40. Maximum Area Repeat Problem 39, assuming that Justin has 60 feet of fencing available.

41. Maximum Revenue A company that manufactures typewriter ribbons knows that the number of ribbons x it can sell each week is related to the price p of each ribbon by the equation $x = 1,200 - 100p$. Graph the revenue equation $R = xp$. Then use the graph to find the price p that will bring in the maximum revenue. Finally, find the maximum revenue.

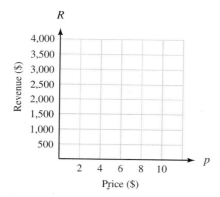

42. Maximum Revenue A company that manufactures diskettes for home computers finds that it can sell x diskettes each day at p dollars per diskette, according to the equation $x = 800 - 100p$. Graph the revenue equation $R = xp$. Then use the graph to find the price p that will bring in the maximum revenue. Finally, find the maximum revenue.

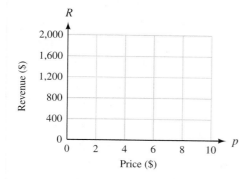

43. Maximum Revenue The relationship between the number of calculators x a company sells each day and the price p of each calculator is given by the equation $x = 1,700 - 100p$. Graph the revenue equation $R = xp$, and use the graph to find the

price p that will bring in the maximum revenue. Then find the maximum revenue.

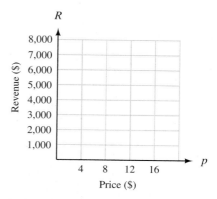

44. Maximum Revenue The relationship between the number x of pencil sharpeners a company sells each week and the price p of each sharpener is given by the equation $x = 1,800 - 100p$. Graph the revenue equation $R = xp$, and use the graph to find the price p that will bring in the maximum revenue. Then find the maximum revenue.

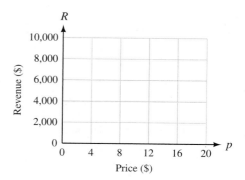

Review Problems

The problems that follow review material we covered in Section 9.7.

Perform the indicated operations.

45. $(3 - 5i) - (2 - 4i)$ **46.** $2i(5 - 6i)$

47. $(3 + 2i)(7 - 3i)$ **48.** $(4 + 5i)^2$

49. $\dfrac{i}{3 + i}$ **50.** $\dfrac{2 + 3i}{2 - 3i}$

Extending the Concepts

Finding the Equation From the Graph For each of the following problems, the graph is a parabola. In each case, find an equation in the form $y = a(x - h)^2 + k$ that describes the graph.

51.

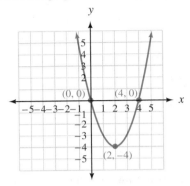

52.

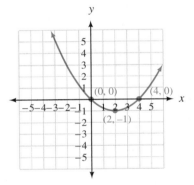

53. Human Cannonball A human cannonball is shot from a cannon at the county fair. He reaches a height of 60 feet before landing in a net 180 feet from the cannon. Sketch the graph of his path, and then find the equation of the graph.

54. Human Cannonball Referring to Problem 53, find the height reached by the human cannonball after he has traveled 30 feet horizontally, and after he has traveled 150 feet horizontally.

55. Comparing Expressions, Equations, and Functions Four problems follow. The solution to Problem 3 is shown in Figure 8. Solve the other three problems, and then explain how the solutions to the four problems are related.

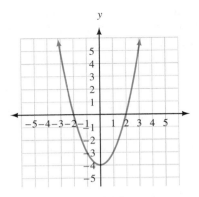

FIGURE 8

Problem 1 Factor the expression $x^2 - 4$.
Problem 2 Solve the equation $x^2 - 4 = 0$.
Problem 3 Graph the function $y = x^2 - 4$.
Problem 4 If $f(x) = x^2 - 4$, find the value of x for which $f(x) = 0$.

56. Comparing Expressions, Equations, and Functions Four problems are shown here. The solution to Problem 3 is shown in Figure 9. Solve the other three problems, and then explain how the solutions to the four problems are related.

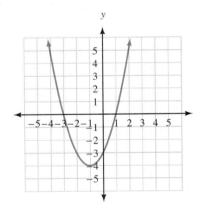

FIGURE 9

Problem 1 Factor the expression $x^2 + 2x - 3$.
Problem 2 Solve the equation $x^2 + 2x - 3 = 0$.
Problem 3 Graph the function $y = x^2 + 2x - 3$.
Problem 4 If $f(x) = x^2 + 2x - 3$, find the value of x for which $f(x) = 0$.

Quadratic inequalities in one variable are inequalities of the form

$$ax^2 + bx + c < 0 \qquad ax^2 + bx + c > 0$$
$$ax^2 + bx + c \le 0 \qquad ax^2 + bx + c \ge 0$$

where a, b, and c are constants, with $a \ne 0$. The technique we will use to solve inequalities of this type involves graphing. Suppose, for example, we wish to find the solution set for the inequality $x^2 - x - 6 > 0$. We begin by factoring the left side to obtain

$$(x - 3)(x + 2) > 0$$

We have two real numbers $x - 3$ and $x + 2$ whose product $(x - 3)(x + 2)$ is greater than zero. That is, their product is positive. The only way the product can be positive is either if both factors, $(x - 3)$ and $(x + 2)$, are positive or if they are both negative. To help visualize where $x - 3$ is positive and where it is negative, we draw a real number line and label it accordingly:

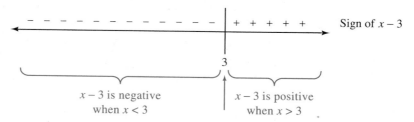

Here is a similar diagram showing where the factor $x + 2$ is positive and where it is negative:

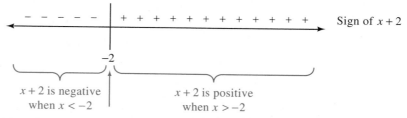

Drawing the two number lines together and eliminating the unnecessary numbers, we have

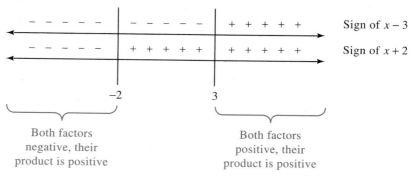

We can see from the preceding diagram that the graph of the solution to $x^2 - x - 6 > 0$ is

$$x < -2 \quad \text{or} \quad x > 3$$

GRAPHICAL SOLUTIONS TO QUADRATIC INEQUALITIES

We can solve the preceding problem by using a graphing calculator to visualize where the product $(x - 3)(x + 2)$ is positive. First, we graph the function $y = (x - 3)(x + 2)$ as shown in Figure 1.

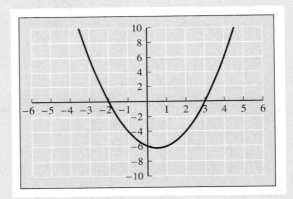

FIGURE 1

Next, we observe where the graph is above the x-axis. As you can see, the graph is above the x-axis to the right of 3 and to the left of -2, as shown in Figure 2.

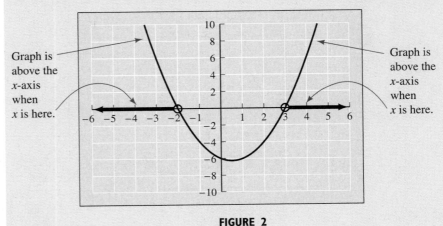

Graph is above the x-axis when x is here.

Graph is above the x-axis when x is here.

FIGURE 2

When the graph is above the x-axis, we have points whose y-coordinates are positive. Since these y-coordinates are the same as the expression $(x - 3)(x + 2)$, the values of x for which the graph of $y = (x - 3)(x + 2)$ is above the x-axis are the values of x for which the inequality $(x - 3)(x + 2) > 0$ is true. Our solution set is therefore

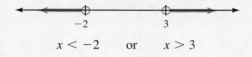

$$x < -2 \quad \text{or} \quad x > 3$$

EXAMPLE   1   Solve for x: $x^2 - 2x - 8 \leq 0$

Algebraic Solution   We begin by factoring:

$$x^2 - 2x - 8 \leq 0$$

$$(x - 4)(x + 2) \leq 0$$

The product $(x - 4)(x + 2)$ is negative or zero. The factors must have opposite signs. We draw a diagram showing where each factor is positive and where each factor is negative:

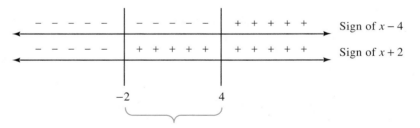

From the diagram we have the graph of the solution set:

$$-2 \leq x \leq 4$$

Graphical Solution

To solve this inequality with a graphing calculator, we graph the function $y = (x - 4)(x + 2)$ and observe where the graph is below the x-axis. These points have negative y-coordinates, which means that the product $(x - 4)(x + 2)$ is negative for these points. Figure 3 shows the graph of $y = (x - 4)(x + 2)$, along with the region on the x-axis where the graph contains points with negative y-coordinates.

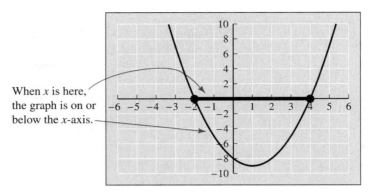

FIGURE 3

When x is here, the graph is on or below the x-axis.

As you can see, the graph is below the x-axis when x is between -2 and 4. Since our original inequality includes the possibility that $(x - 4)(x + 2)$ is 0, we include the endpoints, -2 and 4, with our solution set.

$$-2 \leq x \leq 4$$

E X A M P L E 2   Solve for x: $6x^2 - x \geq 2$

Algebraic Solution

$$6x^2 - x \geq 2$$

$$6x^2 - x - 2 \geq 0 \leftarrow \text{Standard form}$$

$$(3x - 2)(2x + 1) \geq 0$$

The product is positive, so the factors must agree in sign. Here is the diagram showing where that occurs:

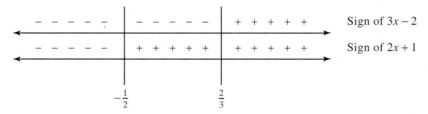

Since the factors agree in sign below $-\frac{1}{2}$ and above $\frac{2}{3}$, the graph of the solution set is

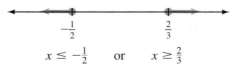

$$x \leq -\frac{1}{2} \quad \text{or} \quad x \geq \frac{2}{3}$$

Graphical Solution   To solve this inequality with a graphing calculator, we graph the function $y = (3x - 2)(2x + 1)$ and observe where the graph is above the x-axis. These are the points that have positive y-coordinates, which means that the product $(3x - 2)(2x + 1)$ is positive for these points. Figure 4 shows the graph of $y = (3x - 2)(2x + 1)$, along with the regions on the x-axis where the graph is on or above the x-axis.

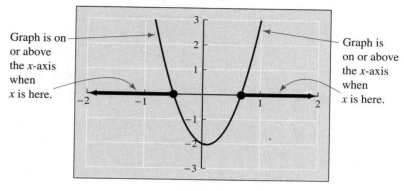

Graph is on or above the x-axis when x is here.

Graph is on or above the x-axis when x is here.

FIGURE 4

To find the points where the graph crosses the x-axis, we need to use either the Trace and Zoom features to zoom in on each point, or the calculator function that finds the intercepts automatically (on the TI-82/83 this is the root/zero function under the CALC key). Whichever method we use, we will obtain the following result:

$$x \le -0.5 \qquad \text{or} \qquad x \ge 0.67$$

−0.5   0.67

EXAMPLE 3   Solve $x^2 - 6x + 9 \ge 0$.

Algebraic Solution

$$x^2 - 6x + 9 \ge 0$$

$$(x - 3)^2 \ge 0$$

This is a special case in which both factors are the same. Since $(x - 3)^2$ is always positive or zero, the solution set is all real numbers. That is, any real number that is used in place of x in the original inequality will produce a true statement.

Graphical Solution   The graph of $y = (x - 3)^2$ is shown in Figure 5. Notice that it touches the x-axis at 3 and is above the x-axis everywhere else. This means that every point on the graph has a y-coordinate greater than or equal to 0, no matter what the value of x. The conclusion that we draw from the graph is that the inequality $(x - 3)^2 \ge 0$ is true for all values of x.

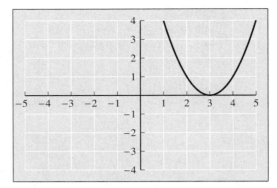

FIGURE 5

Our next two examples involve inequalities that contain rational expressions.

EXAMPLE 4   Solve: $\dfrac{x - 4}{x + 1} \leq 0$

Solution   The inequality indicates that the quotient of $(x - 4)$ and $(x + 1)$ is negative or 0 (less than or equal to 0). We can use the same reasoning we used to solve the first three examples, because quotients are positive or negative under the same conditions that products are positive or negative. Here is the diagram that shows where each factor is positive and where each factor is negative:

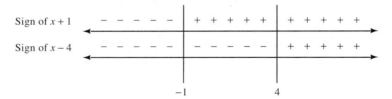

Between -1 and 4 the factors have opposite signs, making the quotient negative. Thus, the region between -1 and 4 is where the solutions lie, since the original inequality indicates the quotient $\dfrac{x - 4}{x + 1}$ is negative. The solution set and its graph are shown here:

$$-1 < x \leq 4$$

Notice that the left endpoint is open—that is, it is not included in the solution set—because $x = -1$ would make the denominator in the original inequality 0. It is important to check all endpoints of solution sets to inequalities that involve rational expressions.

EXAMPLE 5   Solve: $\dfrac{3}{x-2} - \dfrac{2}{x-3} > 0$

Solution   We begin by adding the two rational expressions on the left side. The common denominator is $(x-2)(x-3)$:

$$\frac{3}{x-2}\cdot\frac{(x-3)}{(x-3)} - \frac{2}{x-3}\cdot\frac{(x-2)}{(x-2)} > 0$$

$$\frac{3x-9-2x+4}{(x-2)(x-3)} > 0$$

$$\frac{x-5}{(x-2)(x-3)} > 0$$

This time the quotient involves three factors. Here is the diagram that shows the signs of the three factors:

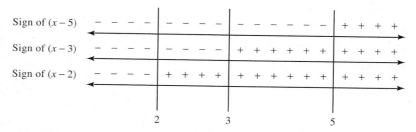

The original inequality indicates that the quotient is positive. In order for this to happen, either all three factors must be positive, or exactly two factors must be negative. Looking back at the diagram, we see the regions that satisfy these conditions are between 2 and 3 or above 5. Here is our solution set:

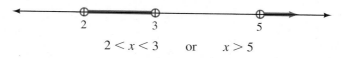

$$2 < x < 3 \quad \text{or} \quad x > 5$$

Getting Ready for Class

After reading through the preceding section, respond in your own words and in complete sentences.

A.   What is the first step in solving a quadratic inequality?

B.   How do you show that the endpoint of a line segment is not part of the graph of a quadratic inequality?

C.   How would you use the graph of $y = ax^2 + bx + c$ to help you find the graph of $ax^2 + bx + c < 0$?

D.   Can a quadratic inequality have exactly one solution? Give an example.

PROBLEM SET 10.6

Solve each of the following inequalities, and graph the solution set.

1. $x^2 + x - 6 > 0$

2. $x^2 + x - 6 < 0$

3. $x^2 - x - 12 \leq 0$

4. $x^2 - x - 12 \geq 0$

5. $x^2 + 5x \geq -6$

6. $x^2 - 5x > 6$

7. $6x^2 < 5x - 1$

8. $4x^2 \geq -5x + 6$

9. $x^2 - 9 < 0$

10. $x^2 - 16 \geq 0$

11. $4x^2 - 9 \geq 0$

12. $9x^2 - 4 < 0$

13. $2x^2 - x - 3 < 0$

14. $3x^2 + x - 10 \geq 0$

15. $x^2 - 4x + 4 \geq 0$

16. $x^2 - 4x + 4 < 0$

17. $x^2 - 10x + 25 < 0$

18. $x^2 - 10x + 25 > 0$

19. $(x - 2)(x - 3)(x - 4) > 0$

20. $(x - 2)(x - 3)(x - 4) < 0$

21. $(x + 1)(x + 2)(x + 3) \leq 0$

22. $(x + 1)(x + 2)(x + 3) \geq 0$

23. $\dfrac{x - 1}{x + 4} \leq 0$

24. $\dfrac{x + 4}{x - 1} \leq 0$

25. $\dfrac{3x}{x + 6} - \dfrac{8}{x + 6} < 0$

26. $\dfrac{5x}{x + 1} - \dfrac{3}{x + 1} < 0$

27. $\dfrac{4}{x - 6} + 1 > 0$

28. $\dfrac{2}{x - 3} + 1 \geq 0$

29. $\dfrac{x - 2}{(x + 3)(x - 4)} < 0$

30. $\dfrac{x - 1}{(x + 2)(x - 5)} < 0$

31. $\dfrac{2}{x - 4} - \dfrac{1}{x - 3} > 0$

32. $\dfrac{4}{x + 3} - \dfrac{3}{x + 2} > 0$

33. $\dfrac{x + 7}{2x + 12} + \dfrac{6}{x^2 - 36} \leq 0$

34. $\dfrac{x + 1}{2x - 2} - \dfrac{2}{x^2 - 1} \leq 0$

35. The graph of $y = x^2 - 4$ is shown in Figure 6. Use the graph to write the solution set for each of the following:

(a) $x^2 - 4 < 0$

(b) $x^2 - 4 > 0$

(c) $x^2 - 4 = 0$

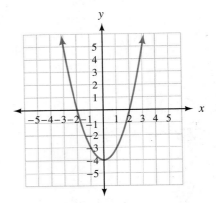

FIGURE 6

36. The graph of $y = 4 - x^2$ is shown in Figure 7. Use the graph to write the solution set for each of the following:

(a) $4 - x^2 < 0$

(b) $4 - x^2 > 0$

(c) $4 - x^2 = 0$

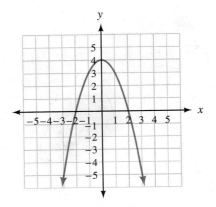

FIGURE 7

37. The graph of $y = x^2 - 3x - 10$ is shown in Figure 8. Use the graph to write the solution set for each of the following:

(a) $x^2 - 3x - 10 < 0$
(b) $x^2 - 3x - 10 > 0$
(c) $x^2 - 3x - 10 = 0$

(a) $x^3 - 3x^2 - x + 3 < 0$
(b) $x^3 - 3x^2 - x + 3 > 0$
(c) $x^3 - 3x^2 - x + 3 = 0$

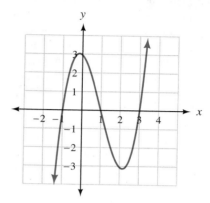

FIGURE 10

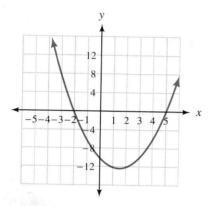

FIGURE 8

38. The graph of $y = x^2 + x - 12$ is shown in Figure 9. Use the graph to write the solution set for each of the following:

(a) $x^2 + x - 12 < 0$
(b) $x^2 + x - 12 > 0$
(c) $x^2 + x - 12 = 0$

40. The graph of $y = x^3 + 4x^2 - 4x - 16$ is shown in Figure 11. Use the graph to write the solution set for each of the following:

(a) $x^3 + 4x^2 - 4x - 16 < 0$
(b) $x^3 + 4x^2 - 4x - 16 > 0$
(c) $x^3 + 4x^2 - 4x - 16 = 0$

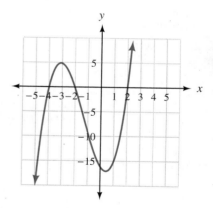

FIGURE 11

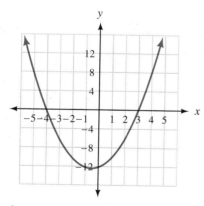

FIGURE 9

39. The graph of $y = x^3 - 3x^2 - x + 3$ is shown in Figure 10. Use the graph to write the solution set for each of the following:

Applying the Concepts

41. Dimensions of a Rectangle The length of a rectangle is 3 inches more than twice the width. If the area is to be at least 44 square inches, what are the possibilities for the width?

42. Dimensions of a Rectangle   The length of a rectangle is 5 inches less than three times the width. If the area is to be less than 12 square inches, what are the possibilities for the width?

43. Revenue   A manufacturer of portable radios knows that the weekly revenue produced by selling x radios is given by the equation $R = 1,300p - 100p^2$, where p is the price of each radio (in dollars). What price should be charged for each radio if the weekly revenue is to be at least $4,000?

44. Revenue   A manufacturer of small calculators knows that the weekly revenue produced by selling x calculators is given by the equation $R = 1,700p - 100p^2$, where p is the price of each calculator (in dollars). What price should be charged for each calculator if the revenue is to be at least $7,000 each week?

45. Union Dues   A labor union has 10,000 members. For every $10 increase in union dues, membership is decreased by 200 people. If the current dues are $100, what should be the new dues (to the nearest multiple of $10) so that income from dues is greatest, and what is that income? *Hint:* Since Income = (membership)(dues), we can let $x =$ the number of $10 increases in dues, and then we can rewrite the income equation as $y = (10,000 - 200x)(100 + 10x)$.

46. Bookstore Receipts   The owner of a used bookstore charges $2 for quality paperbacks and usually sells 40 per day. For every 10-cent increase in the price of these paperbacks, he thinks that he will sell two fewer per day. What is the price he should charge (to the nearest 10 cents) for these books, in order to maximize his income, and what would be that income? *Hint:* let $x =$ the number of 10-cent increases in price.

47. Jiffy-Lube   The owner of a quick oil-change business charges $20 per oil change and has 40 customers per day. If each increase of $2 results in 2 fewer daily customers, what price should the owner charge (to the nearest $2) for an oil change if the income from this business is to be as great as possible?

48. Computer Sales   A computer manufacturer charges $2,200 for its basic model and sells 1,500 computers per month at this price. For every $200 increase in price, it is believed that 75 fewer computers will be sold. What price should the company place on its basic model of computer (to the nearest $100) to have the greatest income?

Review Problems

Problems 49–54 review material we covered in Section 9.6.

Solve each equation.

49. $\sqrt{3t - 1} = 2$

50. $\sqrt{4t + 5} + 7 = 3$

51. $\sqrt{x + 3} = x - 3$

52. $\sqrt{x + 3} = \sqrt{x} - 3$

Graph each equation.

53. $y = \sqrt[3]{x - 1}$

54. $y = \sqrt[3]{x} - 1$

Extending the Concepts

Graph the solution set for each inequality.

55. $x^2 - 2x - 1 < 0$

56. $x^2 - 6x + 7 < 0$

57. $x^2 - 8x + 13 > 0$

58. $x^2 - 10x + 18 > 0$

CHAPTER 10 SUMMARY

Examples

1. If $(x - 3)^2 = 25$

then $x - 3 = \pm 5$

$x = 3 \pm 5$

$x = 8$ or $x = -2$

Theorem 10.1 [10.1]

If $a^2 = b$, where b is a real number, then

$$a = \sqrt{b} \quad \text{or} \quad a = -\sqrt{b} \quad \text{or} \quad a = \pm\sqrt{b}$$

2. Solve $x^2 - 6x - 6 = 0$.

$x^2 - 6x = 6$

$x^2 - 6x + \mathbf{9} = 6 + \mathbf{9}$

$(x - 3)^2 = 15$

$x - 3 = \pm\sqrt{15}$

$x = 3 \pm \sqrt{15}$

To Solve a Quadratic Equation by Completing the Square [10.1]

> **Step 1:** Write the equation in the form $ax^2 + bx = c$.
>
> **Step 2:** If $a \neq 1$, divide through by the constant a so the coefficient of x^2 is 1.
>
> **Step 3:** Complete the square on the left side by adding the square of $\frac{1}{2}$ the coefficient of x to both sides.
>
> **Step 4:** Write the left side of the equation as the square of a binomial. Simplify the right side if possible.
>
> **Step 5:** Apply Theorem 10.1, and solve as usual.

3. If $2x^2 + 3x - 4 = 0$, then

$$x = \frac{-3 \pm \sqrt{9 - 4(2)(-4)}}{2(2)}$$

$$= \frac{-3 \pm \sqrt{41}}{4}$$

The Quadratic Theorem [10.2]

For any quadratic equation in the form $ax^2 + bx + c = 0, a \neq 0$, the two solutions are

$$x = \frac{-b \pm \sqrt{b^2 - 4ac}}{2a}$$

This last expression is known as the *quadratic formula*.

4. The discriminant for

$$x^2 + 6x + 9 = 0$$

is $D = 36 - 4(1)(9) = 0$, which means the equation has one rational solution.

The Discriminant [10.3]

The expression $b^2 - 4ac$ that appears under the radical sign in the quadratic formula is known as the *discriminant*.

We can classify the solutions to $ax^2 + bx + c = 0$:

The Solutions Are	When the Discriminant Is
Two complex numbers containing i	Negative
One rational number	Zero
Two rational numbers	A positive perfect square
Two irrational numbers	A positive number, but not a perfect square

5. The equation $x^4 - x^2 - 12 = 0$ is quadratic in x^2. Letting $y = x^2$ we have

$$y^2 - y - 12 = 0$$

$$(y - 4)(y + 3) = 0$$

$$y = 4 \quad \text{or} \quad y = -3$$

Resubstituting x^2 for y, we have

$$x^2 = 4 \quad \text{or} \quad x^2 = -3$$

$$x = \pm 2 \quad \text{or} \quad x = \pm i\sqrt{3}$$

6. The graph of $y = x^2 - 4$ will be a parabola. It will cross the x-axis at 2 and -2, and the vertex will be $(0, -4)$.

7. Solve $x^2 - 2x - 8 > 0$. We factor and draw the sign diagram:

$$(x - 4)(x + 2) > 0$$

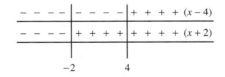

The solution is $x < -2$ or $x > 4$.

Equations Quadratic in Form [10.4]

There are a variety of equations whose form is quadratic. We solve most of them by making a substitution so that the equation becomes quadratic, and then solving that equation by factoring or the quadratic formula. For example,

The Equation	Is Quadratic in
$(2x - 3)^2 + 5(2x - 3) - 6 = 0$	$2x - 3$
$4x^4 - 7x^2 - 2 = 0$	x^2
$2x - 7\sqrt{x} + 3 = 0$	$\sqrt{x}$

Graphing Parabolas [10.5]

The graph of any equation of the form

$$y = ax^2 + bx + c \quad a \neq 0$$

is a *parabola*. The graph is *concave up* if $a > 0$, and *concave down* if $a < 0$. The highest or lowest point on the graph is called the *vertex* and always has an x-coordinate of $x = \dfrac{-b}{2a}$.

Quadratic Inequalities [10.6]

We solve quadratic inequalities by manipulating the inequality to get 0 on the right side and then factoring the left side. We then make a diagram that indicates where the factors are positive and where they are negative. From this sign diagram and the original inequality we graph the appropriate solution set.

CHAPTER 10 REVIEW

Solve each equation. [10.1]

1. $(2t - 5)^2 = 25$ **2.** $(3t - 2)^2 = 4$

3. $(3y - 4)^2 = -49$ **4.** $(2x + 6)^2 = 12$

Solve by completing the square. [10.1]

5. $2x^2 + 6x - 20 = 0$ **6.** $3x^2 + 15x = -18$

7. $a^2 + 9 = 6a$ **8.** $a^2 + 4 = 4a$

9. $2y^2 + 6y = -3$ **10.** $3y^2 + 3 = 9y$

Solve each equation. [10.2]

11. $\frac{1}{6}x^2 + \frac{1}{2}x - \frac{5}{3} = 0$ **12.** $8x^2 - 18x = 0$

13. $4t^2 - 8t + 19 = 0$

14. $100x^2 - 200x = 100$

15. $0.06a^2 + 0.05a = 0.04$

16. $9 - 6x = -x^2$

17. $(2x + 1)(x - 5) - (x + 3)(x - 2) = -17$

18. $2y^3 + 2y = 10y^2$ **19.** $5x^2 = -2x + 3$

20. $x^3 - 27 = 0$ **21.** $3 - \frac{2}{x} + \frac{1}{x^2} = 0$

22. $\dfrac{1}{x - 3} + \dfrac{1}{x + 2} = 1$

23. Profit The total cost (in dollars) for a company to produce x items per week is $C = 7x + 400$. The revenue for selling all x items is $R = 34x - 0.1x^2$. How many items must it produce and sell each week for its weekly profit to be $1,300? [10.2]

24. Profit The total cost (in dollars) for a company to produce x items per week is $C = 70x + 300$. The revenue for selling all x items is $R = 110x - 0.5x^2$. How many items must it produce and sell each week for its weekly profit to be $300? [10.2]

Use the discriminant to find the number and kind of solutions for each equation. [10.3]

25. $2x^2 - 8x = -8$ **26.** $4x^2 - 8x = -4$

27. $2x^2 + x - 3 = 0$ **28.** $5x^2 + 11x = 12$

29. $x^2 - x = 1$ **30.** $x^2 - 5x = -5$

31. $3x^2 + 5x = -4$ **32.** $4x^2 - 3x = -6$

Determine k so that each equation has exactly one real solution. [10.3]

33. $25x^2 - kx + 4 = 0$ **34.** $4x^2 + kx + 25 = 0$

35. $kx^2 + 12x + 9 = 0$ **36.** $kx^2 - 16x + 16 = 0$

37. $9x^2 + 30x + k = 0$ **38.** $4x^2 + 28x + k = 0$

For each of the following problems, find an equation that has the given solutions. [10.3]

39. $x = 3, x = 5$ **40.** $x = -2, x = 4$

41. $y = \frac{1}{2}, y = -4$ **42.** $t = 3, t = -3, t = 5$

Find all solutions. [10.4]

43. $(x - 2)^2 - 4(x - 2) - 60 = 0$

44. $6(2y + 1)^2 - (2y + 1) - 2 = 0$

45. $x^4 - x^2 = 12$ **46.** $x - \sqrt{x} - 2 = 0$

47. $2x - 11\sqrt{x} = -12$ **48.** $\sqrt{x + 5} = \sqrt{x} + 1$

49. $\sqrt{y + 21} + \sqrt{y} = 7$

50. $\sqrt{y + 9} - \sqrt{y - 6} = 3$

51. Projectile Motion An object is tossed into the air with an upward velocity of 10 feet per second from the top of a building h feet high. The time it takes for the object to hit the ground below is given by the formula $16t^2 - 10t - h = 0$. Solve this formula for t. [10.4]

52. Projectile Motion An object is tossed into the air with an upward velocity of v feet per second from the top of a 10-foot wall. The time it takes for the object to hit the ground below is given by the formula $16t^2 - vt - 10 = 0$. Solve this formula for t. [10.4]

Solve each inequality and graph the solution set. [10.6]

53. $x^2 - x - 2 < 0$ **54.** $3x^2 - 14x + 8 \le 0$

55. $2x^2 + 5x - 12 \ge 0$

56. $(x + 2)(x - 3)(x + 4) > 0$

Find the x-intercepts, if they exist, and the vertex for each parabola. Then use them to sketch the graph. [10.5]

57. $y = x^2 - 6x + 8$ **58.** $y = x^2 - 4$

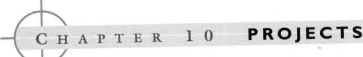

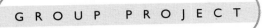

GROUP PROJECT

MAXIMUM VOLUME OF A BOX

Number of People: 5

Time Needed: 30 minutes

Equipment: Graphing calculator and five pieces of graph paper

Background: For many people, having a concrete model to work with allows them to visualize situations that they would have difficulty with if they had only a written description to work with. The purpose of this project is to rework a problem we have worked previously, but this time with a concrete model.

Procedure: You are going to make boxes of varying dimensions from rectangles that are 11 centimeters wide and 17 centimeters long.

1. Cut a rectangle from your graph paper that is 11 squares for the width and 17 squares for the length. Pretend that each small square is 1 centimeter by 1 centimeter. Do this with five pieces of paper.

2. One person tears off one square from each corner of their paper, then folds up the sides to form a box. Write down the length, width, and height of this box. Then calculate its volume.

3. The next person tears a square that is two units on a side from each corner of their paper, then folds up the sides to form a box. Write down the length, width, and height of this box. Then calculate its volume.

4. The next person follows the same procedure, tearing a still larger square from each corner of their paper. This continues until the squares that are to be torn off are larger than the original piece of paper.

5. Enter the data from each box you have created into the table below. Then graph all the points (x, V) from the table.

TABLE I	Volume of a Box			
Side of Square x (cm)	Length of Box L	Width of Box W	Height of Box H	Volume of Box V
1				
2				
3				
4				
5				

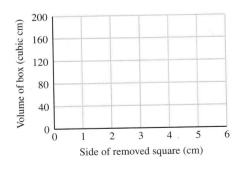

6. Using what you have learned from filling in the table, write a formula for the volume of the box that is created when a square of side x is cut from each corner of the original piece of paper.

7. Graph the equation you found in Problem 6 on a graphing calculator. Use that result to connect the points you plotted in the graph.

8. Use the graphing calculator to find the value of x that will give the maximum volume of the box. Your answer should be accurate to the nearest hundredth.

RESEARCH PROJECT

A CONTINUED FRACTION AND THE GOLDEN RATIO

If you successfully completed the research project in Chapter 9, then you used number sequences and inductive reasoning to conclude that the continued fraction

$$1 + \cfrac{1}{1 + \cfrac{1}{1 + \cfrac{1}{1 + \cdots}}}$$

is equal to the golden ratio.

The same conclusion can be found with the quadratic formula by using the following conditional statement

$$\text{If } x = 1 + \cfrac{1}{1 + \cfrac{1}{1 + \cfrac{1}{1 + \cdots}}}, \text{ then } x = 1 + \frac{1}{x}$$

Work with this conditional statement until you see that it is true. Then solve the equation $x = 1 + \dfrac{1}{x}$. Write an essay in which you explain in your own words why the conditional statement is true. Then show the details of the solution to the equation $x = 1 + \dfrac{1}{x}$. The message that you want to get across in your essay is that the continued fraction shown here is actually the same as the golden ratio.

CHAPTER 10   TEST

Solve each equation. [10.1, 10.2]

1. $(2x + 4)^2 = 25$ **2.** $(2x - 6)^2 = -8$

3. $y^2 - 10y + 25 = -4$

4. $(y + 1)(y - 3) = -6$

5. $8t^3 - 125 = 0$ **6.** $\dfrac{1}{a + 2} - \dfrac{1}{3} = \dfrac{1}{a}$

7. Solve the formula $64(1 + r)^2 = A$ for r. [10.1]

8. Solve $x^2 - 4x = -2$ by completing the square. [10.1]

9. Projectile Motion An object projected upward with an initial velocity of 32 feet per second will rise and fall according to the equation $s(t) = 32t - 16t^2$, where s is its distance above the ground at time t. At what times will the object be 12 feet above the ground? [10.2]

10. Revenue The total weekly cost for a company to make x ceramic coffee cups is given by the formula $C(x) = 2x + 100$. If the weekly revenue from selling all x cups is $R(x) = 25x - 0.2x^2$, how many cups must it sell a week to make a profit of $200 a week? [10.2]

11. Find k so that $kx^2 = 12x - 4$ has one rational solution. [10.3]

12. Use the discriminant to identify the number and kind of solutions to $2x^2 - 5x = 7$. [10.3]

Find equations that have the given solutions. [10.3]

13. $x = 5, x = -\frac{2}{3}$

14. $x = 2, x = -2, x = 7$

Solve each equation. [10.4]

15. $4x^4 - 7x^2 - 2 = 0$

16. $(2t + 1)^2 - 5(2t + 1) + 6 = 0$

17. $2t - 7\sqrt{t} + 3 = 0$

18. Projectile Motion An object is tossed into the air with an upward velocity of 14 feet per second from the top of a building h feet high. The time it takes for the object to hit the ground below is given by the formula $16t^2 - 14t - h = 0$. Solve this formula for t. [10.4]

Sketch the graph of each of the following equations. Give the coordinates of the vertex in each case. [10.5]

19. $y = x^2 - 2x - 3$ **20.** $y = -x^2 + 2x + 8$

Graph each of the following inequalities. [10.6]

21. $x^2 - x - 6 \le 0$ **22.** $2x^2 + 5x > 3$

23. Profit Find the maximum weekly profit for a company with weekly costs of $C = 5x + 100$ and weekly revenue of $R = 25x - 0.1x^2$. [10.5]

Simplify.

1. $11 + 20 \div 5 - 3 \cdot 5$

2. Evaluate: $\left(-\dfrac{2}{3}\right)^3$.

3. $4(15 - 19)^2 - 3(17 - 19)^3$

4. $4 + 8x - 3(5x - 2)$

5. $3 - 5[2x - 4(x - 2)]$

6. $\left(\dfrac{x^{-5}y^4}{x^{-2}y^{-3}}\right)^{-1}$

7. $\sqrt[3]{32}$

8. $8^{-2/3} + 25^{-1/2}$

9. $\dfrac{1 - \dfrac{3}{4}}{1 + \dfrac{3}{4}}$

Reduce.

10. $\dfrac{468}{585}$

11. $\dfrac{5x^2 - 26xy - 24y^2}{5x + 4y}$

12. $\dfrac{x^2 - x - 6}{x + 2}$

Multiply.

13. $(3x - 2)(x^2 - 3x - 2)$

14. $(1 + i)^2$ \quad (Remember, $i^2 = -1$.)

15. Divide $\dfrac{7 - i}{3 - 2i}$ \quad $(i^2 = -1)$

16. Subtract $\dfrac{3}{4}$ from the product of -7 and $\dfrac{5}{28}$.

Solve.

17. $\dfrac{7}{5}a - 6 = 15$

18. $|a| - 6 = 3$

19. $\dfrac{a}{2} + \dfrac{3}{a - 3} = \dfrac{a}{a - 3}$

20. $\sqrt{y + 3} = y + 3$

21. $(3x - 4)^2 = 18$

22. $\dfrac{2}{15}x^2 + \dfrac{1}{3}x + \dfrac{1}{5} = 0$

23. $3y^3 - y = 5y^2$

24. $0.06a^2 - 0.01a = 0.02$

25. $\sqrt{x - 2} = 2 - \sqrt{x}$

26. Solve for x: $ax - 3 = bx + 5$

Solve each inequality, and graph the solution.

27. $5 \le \dfrac{1}{4}x + 3 \le 8$

28. $|4x - 3| \ge 5$

Solve each system.

29. $\begin{aligned} 3x - y &= 2 \\ -6x + 2y &= -4 \end{aligned}$

30. $\begin{aligned} 4x - 8y &= 6 \\ 6x - 12y &= 6 \end{aligned}$

31. $\begin{aligned} 3x - 2y &= 5 \\ y &= 3x - 7 \end{aligned}$

32. $\begin{aligned} 2x + 3y - 8z &= 2 \\ 3x - y + 2z &= 10 \\ 4x + y + 8z &= 16 \end{aligned}$

Graph on a rectangular coordinate system.

33. $2x - 3y = 12$

34. $y = x^2 - x - 2$

35. Find the equation of the line passing through the two points $(\frac{3}{2}, \frac{4}{3})$, $(\frac{1}{4}, -\frac{1}{3})$.

36. Factor completely: $x^2 + 8x + 16 - y^2$.

37. Rationalize the denominator $\dfrac{7}{\sqrt[3]{9}}$.

38. Rationalize the denominator: $\dfrac{2}{\sqrt{5} - 1}$.

39. **Geometry** Find all three angles in a triangle if the smallest angle is one-fourth the largest angle and the remaining angle is 30° more than the smallest angle.

40. **Inverse Variation** y varies inversely with the square of x. If $y = 4$ when $x = \frac{5}{3}$, find y when $x = \frac{8}{3}$.

Exponential and Logarithmic Functions

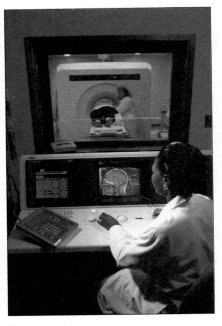

(Index Stock)

INTRODUCTION

If you have had any problems with or had testing done on your thyroid gland, then you may have come in contact with radioactive iodine-131. Like all radioactive elements, iodine-131 decays naturally. The half-life of iodine-131 is 8 days, which means that every 8 days a sample of iodine-131 will decrease to half of its original amount. The following table and graph show what happens to a 1,600-microgram sample of iodine-131 over time.

TABLE 1	Iodine-131 as a Function of Time
t (days)	A (micrograms)
0	1,600
8	800
16	400
24	200
32	100

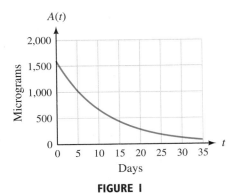

FIGURE 1

The function represented by the information in Table 1 and Figure 1 is

$$A(t) = 1,600 \cdot 2^{-t/8}$$

It is one of the types of functions we will study in this chapter.

To obtain an intuitive idea of how exponential functions behave, we can consider the heights attained by a bouncing ball. When a ball used in the game of racquetball is dropped from any height, the first bounce will reach a height that is $\frac{2}{3}$ of the original height. The second bounce will reach $\frac{2}{3}$ of the height of the first bounce, and so on, as shown in Figure 1.

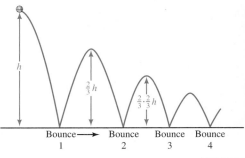
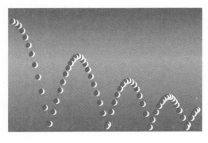

FIGURE 1

If the ball is initially dropped from a height of 1 meter, then during the first bounce it will reach a height of $\frac{2}{3}$ meter. The height of the second bounce will reach $\frac{2}{3}$ of the height reached on the first bounce. The maximum height of any bounce is $\frac{2}{3}$ of the height of the previous bounce.

$$
\begin{aligned}
\text{Initial height:} \quad & h = 1 \\
\text{Bounce 1:} \quad & h = \tfrac{2}{3}(1) = \tfrac{2}{3} \\
\text{Bounce 2:} \quad & h = \tfrac{2}{3}(\tfrac{2}{3}) = (\tfrac{2}{3})^2 \\
\text{Bounce 3:} \quad & h = \tfrac{2}{3}(\tfrac{2}{3})^2 = (\tfrac{2}{3})^3 \\
\text{Bounce 4:} \quad & h = \tfrac{2}{3}(\tfrac{2}{3})^3 = (\tfrac{2}{3})^4 \\
& \quad \vdots \\
\text{Bounce } n: \quad & h = \tfrac{2}{3}(\tfrac{2}{3})^{n-1} = (\tfrac{2}{3})^n
\end{aligned}
$$

This last equation is exponential in form. We classify all exponential functions together with the following definition.

> **DEFINITION**   An **exponential function** is any function that can be written in the form
> $$ f(x) = b^x $$
> where b is a positive real number other than 1.

Each of the following is an exponential function:

$$f(x) = 2^x \qquad y = 3^x \qquad f(x) = \left(\frac{1}{4}\right)^x$$

The first step in becoming familiar with exponential functions is to find some values for specific exponential functions.

EXAMPLE 1   If the exponential functions f and g are defined by

$$f(x) = 2^x \qquad \text{and} \qquad g(x) = 3^x$$

then

$$f(0) = 2^0 = 1 \qquad\qquad g(0) = 3^0 = 1$$

$$f(1) = 2^1 = 2 \qquad\qquad g(1) = 3^1 = 3$$

$$f(2) = 2^2 = 4 \qquad\qquad g(2) = 3^2 = 9$$

$$f(3) = 2^3 = 8 \qquad\qquad g(3) = 3^3 = 27$$

$$f(-2) = 2^{-2} = \frac{1}{2^2} = \frac{1}{4} \qquad g(-2) = 3^{-2} = \frac{1}{3^2} = \frac{1}{9}$$

$$f(-3) = 2^{-3} = \frac{1}{2^3} = \frac{1}{8} \qquad g(-3) = 3^{-3} = \frac{1}{3^3} = \frac{1}{27}$$

In the introduction to this chapter we indicated that the half-life of iodine-131 is 8 days, which means that every 8 days a sample of iodine-131 will decrease to half of its original amount. If we start with A_0 micrograms of iodine-131, then after t days the sample will contain

$$A(t) = A_0 \cdot 2^{-t/8}$$

micrograms of iodine-131.

EXAMPLE 2   A patient is administered a 1,200-microgram dose of iodine-131. How much iodine-131 will be in the patient's system after 10 days, and after 16 days?

Solution   The initial amount of iodine-131 is $A_0 = 1,200$, so the function that gives the amount left in the patient's system after t days is

$$A(t) = 1,200 \cdot 2^{-t/8}$$

After 10 days, the amount left in the patient's system is

$$A(10) = 1,200 \cdot 2^{-10/8} = 1,200 \cdot 2^{-1.25} \approx 504.5 \text{ micrograms}$$

After 16 days, the amount left in the patient's system is

$$A(16) = 1,200 \cdot 2^{-16/8} = 1,200 \cdot 2^{-2} = 300 \text{ micrograms}$$

We will now turn our attention to the graphs of exponential functions. Since the notation y is easier to use when graphing, and $y = f(x)$, for convenience we will write the exponential functions as

$$y = b^x$$

EXAMPLE 3   Sketch the graph of the exponential function $y = 2^x$.

Solution   Using the results of Example 1, we produce the following table. Graphing the ordered pairs given in the table and connecting them with a smooth curve, we have the graph of $y = 2^x$ shown in Figure 2.

x	y
-3	$\frac{1}{8}$
-2	$\frac{1}{4}$
-1	$\frac{1}{2}$
0	1
1	2
2	4
3	8

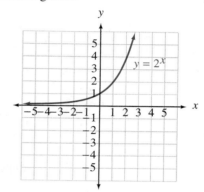

FIGURE 2

Notice that the graph does not cross the x-axis. It *approaches* the x-axis—in fact, we can get it as close to the x-axis as we want without it actually intersecting the x-axis. In order for the graph of $y = 2^x$ to intersect the x-axis, we would have to find a value of x that would make $2^x = 0$. Because no such value of x exists, the graph of $y = 2^x$ cannot intersect the x-axis.

EXAMPLE 4   Sketch the graph of $y = (\frac{1}{3})^x$.

Solution   The table beside Figure 3 gives some ordered pairs that satisfy the equation. Using the ordered pairs from the table, we have the graph shown in Figure 3.

x	y
-3	27
-2	9
-1	3
0	1
1	$\frac{1}{3}$
2	$\frac{1}{9}$
3	$\frac{1}{27}$

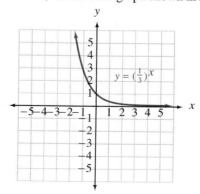

FIGURE 3

The graphs of all exponential functions have two things in common: (1) Each crosses the y-axis at $(0, 1)$, since $b^0 = 1$; and (2) none can cross the x-axis, since $b^x = 0$ is impossible because of the restrictions on b.

Figures 4 and 5 show some families of exponential curves to help you become more familiar with them on an intuitive level.

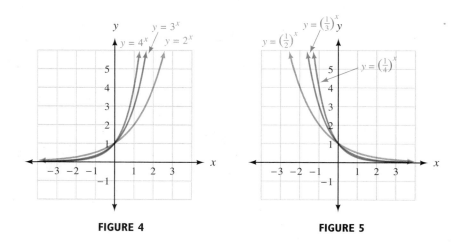

FIGURE 4   **FIGURE 5**

Among the many applications of exponential functions are the applications having to do with interest-bearing accounts. Here are the details.

Compound Interest

If P dollars are deposited in an account with annual interest rate r, compounded n times per year, then the amount of money in the account after t years is given by the formula

$$A(t) = P\left(1 + \frac{r}{n}\right)^{nt}$$

EXAMPLE 5 Suppose you deposit $500 in an account with an annual interest rate of 8% compounded quarterly. Find an equation that gives the amount of money in the account after t years. Then find

(a) The amount of money in the account after 5 years.

(b) The number of years it will take for the account to contain $1,000.

Solution First, we note that $P = 500$ and $r = 0.08$. Interest that is compounded quarterly is compounded four times a year, giving us $n = 4$. Substituting these numbers into the preceding formula, we have our function

$$A(t) = 500\left(1 + \frac{0.08}{4}\right)^{4t} = 500(1.02)^{4t}$$

(a) To find the amount after 5 years, we let $t = 5$:

$$A(5) = 500(1.02)^{4 \cdot 5} = 500(1.02)^{20} \approx \$742.97$$

Our answer is found on a calculator, and then rounded to the nearest cent.

(b) To see how long it will take for this account to total $1,000, we graph the equation $Y_1 = 500(1.02)^{4X}$ on a graphing calculator, and then look to see where it intersects the line $Y_2 = 1,000$. The two graphs are shown in Figure 6.

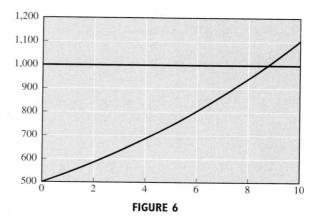

FIGURE 6

Using Zoom and Trace, or the Intersect function on the graphing calculator, we find that the two curves intersect at $X \approx 8.75$ and $Y = 1,000$. This means that our account will contain $1,000 after the money has been on deposit for 8.75 years.

The Natural Exponential Function

A very commonly occurring exponential function is based on a special number we denote with the letter e. The number e is a number like π. It is irrational and occurs in many formulas that describe the world around us. Like π, it can be approximated with a decimal number. Whereas π is approximately 3.1416, e is approximately 2.7183. (If you have a calculator with a key labeled $\boxed{e^x}$, you can use it to find e^1 to find a more accurate approximation to e.) We cannot give a more precise definition of the number e without using some of the topics taught in calculus. For the work we are going to do with the number e, we only need to know that it is an irrational number that is approximately 2.7183.

Here are a table and graph (Figure 7) for the natural exponential function

$$y = f(x) = e^x$$

x	$f(x) = e^x$
-2	$f(-2) = e^{-2} = \dfrac{1}{e^2} \approx 0.135$
-1	$f(-1) = e^{-1} = \dfrac{1}{e} \approx 0.368$
0	$f(0) = e^0 = 1$
1	$f(1) = e^1 = e \approx 2.72$
2	$f(2) = e^2 \approx 7.39$
3	$f(3) = e^3 \approx 20.09$

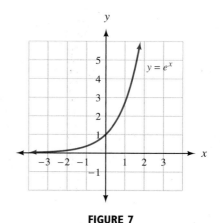

FIGURE 7

One common application of natural exponential functions is with interest-bearing accounts. In Example 5 we worked with the formula

$$A = P\left(1 + \frac{r}{n}\right)^{nt}$$

that gives the amount of money in an account if P dollars are deposited for t years at annual interest rate r, compounded n times per year. In Example 5 the number of compounding periods was four. What would happen if we let the number of compounding periods become larger and larger, so that we compounded the interest every day, then every hour, then every second, and so on? If we take this as far as it can go, we end up compounding the interest every moment. When this happens, we have an account with interest that is compounded continuously, and the amount of money in such an account depends on the number e. Here are the details.

Continuously Compounded Interest

If P dollars are deposited in an account with annual interest rate r, compounded continuously, then the amount of money in the account after t years is given by the formula

$$A(t) = Pe^{rt}$$

EXAMPLE 6 Suppose you deposit \$500 in an account with an annual interest rate of 8% compounded continuously. Find an equation that gives the amount of money in the account after t years. Then find the amount of money in the account after 5 years.

Solution Since the interest is compounded continuously, we use the formula $A(t) = Pe^{rt}$. Substituting $P = 500$ and $r = 0.08$ into this formula we have

$$A(t) = 500e^{0.08t}$$

After 5 years, this account will contain

$$A(5) = 500e^{0.08 \cdot 5} = 500e^{0.4} \approx \$745.91$$

to the nearest cent. Compare this result with the answer to Example 5a.

 # Getting Ready for Class

After reading through the preceding section, respond in your own words and in complete sentences.

A. What is an exponential function?

B. In an exponential function, explain why the base b cannot equal 1. (What kind of function would you get if the base was equal to 1?)

C. Explain continuously compounded interest.

D. What characteristics do the graphs of $y = 2^x$ and $y = (\frac{1}{2})^x$ have in common?

PROBLEM SET 11.1

Let $f(x) = 3^x$ and $g(x) = (\frac{1}{2})^x$, and evaluate each of the following.

1. $g(0)$

2. $f(0)$

3. $g(-1)$

4. $g(-4)$

5. $f(-3)$

6. $f(-1)$

7. $f(2) + g(-2)$

8. $f(2) - g(-2)$

Graph each of the following functions.

9. $y = 4^x$

10. $y = 2^{-x}$

11. $y = 3^{-x}$

12. $y = (\frac{1}{3})^{-x}$

13. $y = 2^{x+1}$

14. $y = 2^{x-3}$

15. $y = e^x$

16. $y = e^{-x}$

Graph each of the following functions on the same coordinate system for positive values of x only.

17. $y = 2x, y = x^2, y = 2^x$

18. $y = 3x, y = x^3, y = 3^x$

19. On a graphing calculator, graph the family of curves $y = b^x$, $b = 2, 4, 6, 8$.

20. On a graphing calculator, graph the family of curves $y = b^x$, $b = \frac{1}{2}, \frac{1}{4}, \frac{1}{6}, \frac{1}{8}$.

Applying the Concepts

21. Bouncing Ball Suppose the ball mentioned in the introduction to this section is dropped from a height of 6 feet above the ground. Find an exponential equation that gives the height h the ball will attain during the nth bounce. How high will it bounce on the fifth bounce?

22. Bouncing Ball A golf ball is manufactured so that if it is dropped from A feet above the ground onto a hard surface, the maximum height of each bounce will be one half of the height of the previous bounce. Find an exponential equation that gives the height h the ball will attain during the nth bounce. If the ball is dropped from 10 feet above the ground onto a hard surface, how high will it bounce on the eighth bounce?

23. Exponential Decay Twinkies on the shelf of a convenience store lose their fresh tastiness over time. We say that the taste quality is 1 when the Twinkies are first put on the shelf at the store, and that the quality of tastiness declines according to the function $Q(t) = 0.85^t$. Graph this function on a graphing calculator, and determine when the taste quality will be one half of its original value.

24. Exponential Growth Automobiles built before 1993 use Freon in their air conditioners. The federal government now prohibits the manufacture of Freon. Because the supply of Freon is decreasing, the price per pound is increasing exponentially. Current estimates put the formula for the price per pound of Freon at $p(t) = 1.89(1.25)^t$, where t is the number of years since 1990. Find the price of Freon in 1995 and 1990. How much will Freon cost in the year 2000?

25. **Compound Interest** Suppose you deposit $1,200 in an account with an annual interest rate of 6% compounded quarterly.
    (a) Find an equation that gives the amount of money in the account after t years.
    (b) Find the amount of money in the account after 8 years.
    (c) How many years will it take for the account to contain $2,400?
    (d) If the interest were compounded continuously, how much money would the account contain after 8 years?

26. **Compound Interest** Suppose you deposit $500 in an account with an annual interest rate of 8% compounded monthly.
    (a) Find an equation that gives the amount of money in the account after t years.
    (b) Find the amount of money in the account after 5 years.
    (c) How many years will it take for the account to contain $1,000?
    (d) If the interest were compounded continuously, how much money would the account contain after 5 years?

Declining-Balance Depreciation The declining-balance method of depreciation is an accounting method businesses use to deduct most of the cost of new equipment during the first few years of purchase. Unlike other methods, the declining-balance formula does not consider salvage value.

27. **Value of a Crane** The function $V(t) = 450,000(1 - 0.30)^t$, where V is value and t is time in years, can be used to find the value of a crane for the first 6 years of use.
    (a) What is the value of the crane after 3 years and 6 months?
    (b) State the domain of this function.
    (c) Sketch the graph of this function using the template below.
    (d) State the range of this function.

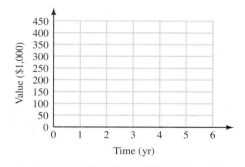

(SuperStock)

(e) After how many years will the crane be worth only $85,000?

28. **Value of a Printing Press** The function $V(t) = 375,000(1 - 0.25)^t$, where V is value and t is time in years, can be used to find the value of a printing press during the first 7 years of use.
    (a) What is the value of the printing press after 4 years and 9 months?
    (b) State the domain of this function.
    (c) Sketch the graph of this function using the template below.
    (d) State the range of this function.

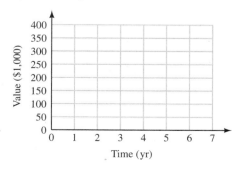

(Jim McCrary/Tony Stone Images)

(Kactus Foto, Santiago, Chile/SuperStock)

(e) After how many years will the printing press be worth only $65,000?

29. Bacteria Growth Suppose it takes 12 hours for a certain strain of bacteria to reproduce by dividing in half. If 50 bacteria are present to begin with, then the total number present after x days will be $f(x) = 50 \cdot 4^x$. Find the total number present after 1 day, 2 days, and 3 days.

30. Bacteria Growth Suppose it takes 1 day for a certain strain of bacteria to reproduce by dividing in half. If 100 bacteria are present to begin with, then the total number present after x days will be $f(x) = 100 \cdot 2^x$. Find the total number present after 1 day, 2 days, 3 days, and 4 days. How many days must elapse before over 100,000 bacteria are present?

31. Value of a Painting A painting is purchased as an investment for $150. If the painting's value doubles every 3 years, then its value is given by the function

$$V(t) = 150 \cdot 2^{t/3} \quad \text{for } t \geq 0$$

where t is the number of years since it was purchased, and $V(t)$ is its value (in dollars) at that time. Graph this function.

32. Value of a Painting A painting is purchased as an investment for $125. If the painting's value doubles every 5 years, then its value is given by the function

$$V(t) = 125 \cdot 2^{t/5} \quad \text{for } t \geq 0$$

where t is the number of years since it was pur-

chased, and $V(t)$ is its value (in dollars) at that time. Graph this function.

Review Problems

The following problems review material from Sections 8.4, 8.5 and 8.6.

For each of the following relations, specify the domain and range; then indicate which are also functions. [8.4]

33. $\{(-2, 6), (-2, 8), (2, 3)\}$
34. $\{(1, 2), (3, 4), (4, 1)\}$

State the domain for each of the following functions. [8.4]

35. $y = \dfrac{-4}{x^2 + 2x - 35}$ **36.** $y = \sqrt{3x + 1}$

If $f(x) = 2x^2 - 18$ and $g(x) = 2x - 6$, find [8.5, 8.6]

37. $f(0)$ **38.** $(g \circ f)(0)$

39. $\dfrac{g(x + h) - g(x)}{h}$ **40.** $\dfrac{g}{f}(x)$

Extending the Concepts

41. Reading Graphs The graphs of two exponential functions are given in Figures 8 and 9. Use the graphs to find the following:

(a) $f(0)$ (b) $f(-1)$
(c) $f(1)$ (d) $g(0)$
(e) $g(1)$ (f) $g(-1)$
(g) $f[g(0)]$ (h) $g[f(0)]$

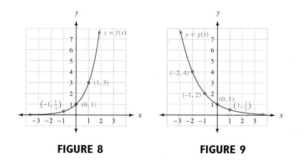

FIGURE 8   **FIGURE 9**

42. **Drag Racing** In Chapter 3 we mentioned the dragster equipped with a computer. Table 1 gives the speed of the dragster every second during one

TABLE 1 Speed of a Dragster	
Elapsed Time (sec)	**Speed (mi/hr)**
0	0.0
1	72.7
2	129.9
3	162.8
4	192.2
5	212.4
6	228.1

FIGURE 10

race at the 1993 Winternationals. Figure 10 is a line graph constructed from the data in Table 1. The graph of the function $s(t) = 250(1 - 1.5^{-t})$ contains the first point and the last point shown in Figure 10. That is, both $(0, 0)$ and $(6, 228.1)$ satisfy the function. Graph the function to see how close it comes to the other points in Figure 10.

43. **Analyzing Graphs** The goal of this problem is to obtain a sense of the growth rate of an exponential function. We will compare the exponential function $y = 2^x$ to the familiar function $y = x^2$.
(a) Graph $y = 2^x$ and $y = x^2$ using the window X: 0 to 3 and Y: 0 to 10. Which function appears to be taking over as x grows larger?
(b) Show algebraically that both graphs contain the points $(2, 4)$ and $(4, 16)$. You may also want to confirm this on your graph by enlarging the window.
(c) Graph $y = 2^x$ and $y = x^2$ with the window X: 0 to 15 and Y: 0 to 2,500. Which function dominates (grows the fastest) as x gets larger in the positive direction? Confirm your answer by evaluating $2^{(100)}$ and $(100)^2$.

11.2 The Inverse of a Function

The following diagram (Figure 1) shows the route Justin takes to school. He leaves his home and drives 3 miles east, and then turns left and drives 2 miles north. When he leaves school to drive home, he drives the same two segments, but in the reverse order and the opposite direction; that is, he drives 2 miles south, turns right, and drives 3 miles west. When he arrives home from school, he is right where he started. His route home "undoes" his route to school, leaving him where he began.

As you will see, the relationship between a function and its inverse function is similar to the relationship between Justin's route from home to school and his route from school to home.

Suppose the function f is given by

$$f = \{(1, 4), (2, 5), (3, 6), (4, 7)\}$$

The inverse of f is obtained by reversing the order of the coordinates in each or-

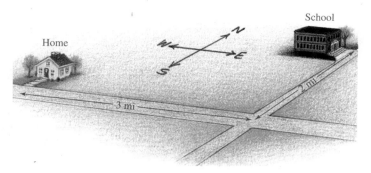

FIGURE 1

dered pair in f. The inverse of f is the relation given by

$$g = \{(4, 1), (5, 2), (6, 3), (7, 4)\}$$

It is obvious that the domain of f is now the range of g, and the range of f is now the domain of g. Every function (or relation) has an inverse that is obtained from the original function by interchanging the components of each ordered pair.

Suppose a function f is defined with an equation instead of a list of ordered pairs. We can obtain the equation of the inverse of f by interchanging the role of x and y in the equation for f.

E X A M P L E   1   If the function f is defined by $f(x) = 2x - 3$, find the equation that represents the inverse of f.

Solution   Since the inverse of f is obtained by interchanging the components of all the ordered pairs belonging to f, and each ordered pair in f satisfies the equation $y = 2x - 3$, we simply exchange x and y in the equation $y = 2x - 3$ to get the formula for the inverse of f:

$$x = 2y - 3$$

We now solve this equation for y in terms of x:

$$x + 3 = 2y$$

$$\frac{x + 3}{2} = y$$

$$y = \frac{x + 3}{2}$$

The last line gives the equation that defines the inverse of f. Let's compare the graphs of f and its inverse as given here. (See Figure 2.)

The graphs of f and its inverse have symmetry about the line $y = x$. This is a reasonable result since the one function was obtained from the other by interchang-

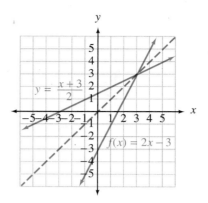

FIGURE 2

ing x and y in the equation. The ordered pairs (a, b) and (b, a) always have symmetry about the line $y = x$.

EXAMPLE 2 Graph the function $y = x^2 - 2$ and its inverse. Give the equation for the inverse.

Solution We can obtain the graph of the inverse of $y = x^2 - 2$ by graphing $y = x^2 - 2$ by the usual methods, and then reflecting the graph about the line $y = x$.

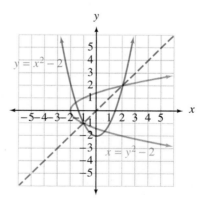

FIGURE 3

The equation that corresponds to the inverse of $y = x^2 - 2$ is obtained by interchanging x and y to get $x = y^2 - 2$.

We can solve the equation $x = y^2 - 2$ for y in terms of x as follows:

$$x = y^2 - 2$$

$$x + 2 = y^2$$

$$y = \pm\sqrt{x + 2}$$

Comparing the graphs from Examples 1 and 2, we observe that the inverse of a function is not always a function. In Example 1, both f and its inverse have graphs that are nonvertical straight lines and therefore both represent functions. In Example 2, the inverse of function f is not a function, since a vertical line crosses it in more than one place.

ONE-TO-ONE FUNCTIONS

We can distinguish between those functions with inverses that are also functions and those functions with inverses that are not functions with the following definition.

> **DEFINITION**   A function is a **one-to-one function** if every element in the range comes from exactly one element in the domain.

This definition indicates that a one-to-one function will yield a set of ordered pairs in which no two different ordered pairs have the same second coordinates. For example, the function

$$f = \{(2, 3), (-1, 3), (5, 8)\}$$

is not one-to-one because the element 3 in the range comes from both 2 and -1 in the domain. On the other hand, the function

$$g = \{(5, 7), (3, -1), (4, 2)\}$$

is a one-to-one function because every element in the range comes from only one element in the domain.

HORIZONTAL LINE TEST

If we have the graph of a function, we can determine if the function is one-to-one with the following test. If a horizontal line crosses the graph of a function in more than one place, then the function is not a one-to-one function because the points at which the horizontal line crosses the graph will be points with the same y-coordinates, but different x-coordinates. Therefore, the function will have an element in the range (the y-coordinate) that comes from more than one element in the domain (the x-coordinates).

Of the functions we have covered previously, all the linear functions and exponential functions are one-to-one functions because no horizontal lines can be found that will cross their graphs in more than one place.

FUNCTIONS WHOSE INVERSES ARE ALSO FUNCTIONS

Because one-to-one functions do not repeat second coordinates, when we reverse the order of the ordered pairs in a one-to-one function, we obtain a relation in which no two ordered pairs have the same first coordinate—by definition, this relation must be a function. In other words, every one-to-one function has an inverse that is itself a function. Because of this, we can use function notation to represent that inverse.

INVERSE FUNCTION NOTATION

If $y = f(x)$ is a one-to-one function, then the inverse of f is also a function and can be denoted by $y = f^{-1}(x)$.

To illustrate, in Example 1 we found that the inverse of $f(x) = 2x - 3$ was the function $y = \dfrac{x + 3}{2}$. We can write this inverse function with inverse function notation as

$$f^{-1}(x) = \frac{x + 3}{2}$$

On the other hand, the inverse of the function in Example 2 is not itself a function, so we do not use the notation $f^{-1}(x)$ to represent it.

Note: The notation f^{-1} does not represent the reciprocal of f. That is, the -1 in this notation is not an exponent. The notation f^{-1} is defined as representing the inverse function for a one-to-one function.

EXAMPLE 3 Find the inverse of $g(x) = \dfrac{x - 4}{x - 2}$.

Solution To find the inverse for g, we begin by replacing $g(x)$ with y to obtain

$$y = \frac{x - 4}{x - 2} \qquad \text{The original function}$$

To find an equation for the inverse, we exchange x and y.

$$x = \frac{y - 4}{y - 2} \qquad \text{The inverse of the original function}$$

To solve for y, we first multiply each side by $y - 2$ to obtain

$$x(y - 2) = y - 4$$

$$xy - 2x = y - 4 \qquad \text{Distributive property}$$

$$xy - y = 2x - 4 \qquad \text{Collect all terms containing } y \text{ on the left side.}$$

$$y(x - 1) = 2x - 4 \qquad \text{Factor } y \text{ from each term on the left side.}$$

$$y = \frac{2x - 4}{x - 1} \qquad \text{Divide each side by } x - 1.$$

Figure 4 shows that the graph of this function passes the horizontal line test. Therefore, it is a one-to-one function.

Since our original function is one-to-one, its inverse is also a function. Therefore, we can use inverse function notation to write

$$g^{-1}(x) = \frac{2x - 4}{x - 1}$$

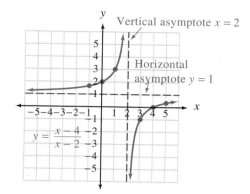

FIGURE 4

EXAMPLE   4   Graph the function $y = 2^x$ and its inverse $x = 2^y$.

Solution   We graphed $y = 2^x$ in the preceding section. We simply reflect its graph about the line $y = x$ to obtain the graph of its inverse $x = 2^y$.

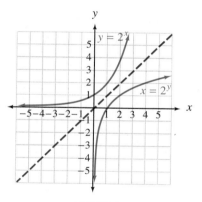

FIGURE 5

As you can see from the graph, $x = 2^y$ is a function. We do not have the mathematical tools to solve this equation for y, however. We are therefore unable to use the inverse function notation to represent this function. In the next section, we will give a definition that solves this problem. For now, we simply leave the equation as $x = 2^y$.

FUNCTIONS, RELATIONS, AND INVERSES—A SUMMARY

Here is a summary of some of the things we know about functions, relations, and their inverses:

1. Every function is a relation, but not every relation is a function.
2. Every function has an inverse, but only one-to-one functions have inverses that are also functions.
3. The domain of a function is the range of its inverse, and the range of a function is the domain of its inverse.
4. If $y = f(x)$ is a one-to-one function, then we can use the notation $y = f^{-1}(x)$ to represent its inverse function.
5. The graph of a function and its inverse have symmetry about the line $y = x$.
6. If (a, b) belongs to the function f, then the point (b, a) belongs to its inverse.

 ## Getting Ready for Class

After reading through the preceding section, respond in your own words and in complete sentences.

A. What is the inverse of a function?
B. What is the relationship between the graph of a function and the graph of its inverse?
C. Explain why only one-to-one functions have inverses that are also functions.
D. Describe the vertical line test, and explain the difference between the vertical line test and the horizontal line test.

PROBLEM SET 11.2

For each of the following one-to-one functions, find the equation of the inverse. Write the inverse using the notation $f^{-1}(x)$.

1. $f(x) = 3x - 1$
2. $f(x) = 2x - 5$
3. $f(x) = x^3$
4. $f(x) = x^3 - 2$
5. $f(x) = \dfrac{x - 3}{x - 1}$
6. $f(x) = \dfrac{x - 2}{x - 3}$
7. $f(x) = \dfrac{x - 3}{4}$
8. $f(x) = \dfrac{x + 7}{2}$
9. $f(x) = \dfrac{1}{2}x - 3$
10. $f(x) = \dfrac{1}{3}x + 1$
11. $f(x) = \dfrac{2x + 1}{3x + 1}$
12. $f(x) = \dfrac{3x + 2}{5x + 1}$

For each of the following relations, sketch the graph of the relation and its inverse, and write an equation for the inverse.

13. $y = 2x - 1$
14. $y = 3x + 1$
15. $y = x^2 - 3$
16. $y = x^2 + 1$
17. $y = x^2 - 2x - 3$
18. $y = x^2 + 2x - 3$
19. $y = 3^x$
20. $y = \left(\dfrac{1}{2}\right)^x$
21. $y = 4$
22. $y = -2$
23. $y = \dfrac{1}{2}x^3$
24. $y = x^3 - 2$
25. $y = \dfrac{1}{2}x + 2$
26. $y = \dfrac{1}{3}x - 1$

27. $y = \sqrt{x} + 2$ **28.** $y = \sqrt{x} + 2$

29. Determine if the following functions are one-to-one.

(a)

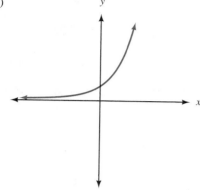

(b)

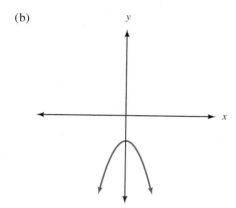

(c)

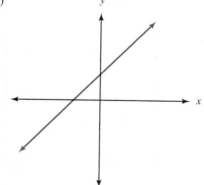

30. Could the following tables of values represent ordered pairs from one-to-one functions? Explain your answer.

(a)

x	y
−2	5
−1	4
0	3
1	4
2	5

(b)

x	y
1.5	0.1
2.0	0.2
2.5	0.3
3.0	0.4
3.5	0.5

31. If $f(x) = 3x - 2$, then $f^{-1}(x) = \dfrac{x + 2}{3}$. Use these two functions to find

(a) $f(2)$ (b) $f^{-1}(2)$

(c) $f[f^{-1}(2)]$ (d) $f^{-1}[f(2)]$

32. If $f(x) = \frac{1}{2}x + 5$, then $f^{-1}(x) = 2x - 10$. Use these two functions to find

(a) $f(-4)$ (b) $f^{-1}(-4)$

(c) $f[f^{-1}(-4)]$ (d) $f^{-1}[f(-4)]$

33. Let $f(x) = \dfrac{1}{x}$, and find $f^{-1}(x)$.

34. Let $f(x) = \dfrac{a}{x}$, and find $f^{-1}(x)$. (a is a real number constant.)

Applying the Concepts

35. Inverse Functions in Words Inverses may also be found by *inverse reasoning*. For example, to find the inverse of $f(x) = 3x + 2$, first list, in order, the operations done to variable x:

(a) Multiply by 3.

(b) Add 2.

Then, to find the inverse, simply apply the inverse operations, in reverse order, to the variable x. That is:

(a) Subtract 2.

(b) Divide by 3.

The inverse function then becomes $f^{-1}(x) = \dfrac{x - 2}{3}$.

Use this method of "inverse reasoning" to find the inverse of the function $f(x) = \dfrac{x}{7} - 2$.

36. Inverse Functions in Words Refer to the method of *inverse reasoning* explained in Problem 35. Use *inverse reasoning* to find the following inverses:

(a) $f(x) = 2x + 7$

(b) $f(x) = \sqrt{x} - 9$

(c) $f(x) = x^3 - 4$
(d) $f(x) = \sqrt{x^3 - 4}$

37. Reading Tables Evaluate each of the following functions using the functions defined by Tables 1 and 2.
(a) $f[g(-3)]$
(b) $g[f(-6)]$
(c) $g[f(2)]$
(d) $f[g(3)]$
(e) $f[g(-2)]$
(f) $g[f(3)]$
(g) What can you conclude about the relationship between functions f and g?

TABLE 1

x	-6	2	3	6
$f(x)$	3	-3	-2	4

TABLE 2

x	-3	-2	3	4
$g(x)$	2	3	-6	6

38. Reading Tables Use the functions defined in Tables 1 and 2 in Problem 37 to answer the following questions.
(a) What are the domain and range of f?
(b) What are the domain and range of g?
(c) How are the domain and range of f related to the domain and range of g?
(d) Is f a one-to-one function?
(e) Is g a one-to-one function?

Review Problems

The problems that follow review material we covered in Section 10.1.

Solve each equation.

39. $(2x - 1)^2 = 25$
40. $(3x + 5)^2 = -12$
41. What number would you add to $x^2 - 10x$ to make it a perfect square trinomial?
42. What number would you add to $x^2 - 5x$ to make it a perfect square trinomial?

Solve by completing the square.

43. $x^2 - 10x + 8 = 0$
44. $x^2 - 5x + 4 = 0$
45. $3x^2 - 6x + 6 = 0$
46. $4x^2 - 16x - 8 = 0$

Extending the Concepts

For each of the following functions, find $f^{-1}(x)$. Then show that $f[f^{-1}(x)] = x$.

47. $f(x) = 3x + 5$
48. $f(x) = 6 - 8x$
49. $f(x) = x^3 + 1$
50. $f(x) = x^3 - 8$
51. $f(x) = \dfrac{x - 4}{x - 2}$
52. $f(x) = \dfrac{x - 3}{x - 1}$

53. Reading Graphs The graphs of a function and its inverse are shown in Figure 6. Use the graphs to find the following:
(a) $f(0)$
(b) $f(1)$
(c) $f(2)$
(d) $f^{-1}(1)$
(e) $f^{-1}(2)$
(f) $f^{-1}(5)$
(g) $f^{-1}[f(2)]$
(h) $f[f^{-1}(5)]$

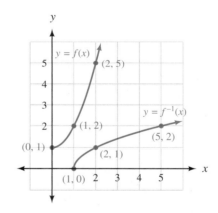

FIGURE 6

54. Domain From a Graph The function f is defined by the equation $f(x) = \sqrt{x + 4}$ for $x \geq -4$, meaning the domain is all x in the interval $[-4, \infty)$.
(a) Graph the function f.
(b) On the same set of axes, graph the line $y = x$ and the inverse function f^{-1}.
(c) State the domain for the inverse function for f^{-1}.
(d) Find the equation for $f^{-1}(x)$.

Logarithms Are Exponents

In January 1999, ABC News reported that an earthquake had occurred in Colombia, causing massive destruction. They reported the strength of the quake by indicating that it measured 6.0 on the Richter scale. For comparison, Table 1 gives the Richter magnitude of a number of other earthquakes.

TABLE 1 Earthquakes		
Year	Earthquake	Richter Magnitude
1971	Los Angeles	6.6
1985	Mexico City	8.1
1989	San Francisco	7.1
1992	Kobe, Japan	7.2
1994	Northridge	6.6
1999	Armenia, Colombia	6.0

(Reuters/José Miguel Gomez/Archives Photos)

Although the size of the numbers in the table do not seem to be very different, the intensity of the earthquakes they measure can be very different. For example, the 1989 San Francisco earthquake was more than 10 times stronger than the 1999 earthquake in Colombia. The reason behind this is that the Richter scale is a **logarithmic scale.** In this section we start our work with logarithms, which will give you an understanding of the Richter scale. Let's begin.

As you know from your work in the previous sections, equations of the form

$$y = b^x \quad b > 0, b \neq 1$$

are called exponential functions. Since the equation of the inverse of a function can be obtained by exchanging x and y in the equation of the original function, the inverse of an exponential function must have the form

$$x = b^y \quad b > 0, b \neq 1$$

Now, this last equation is actually the equation of a logarithmic function, as the following definition indicates:

DEFINITION The expression $y = \log_b x$ is read "y is the logarithm to the base b of x" and is equivalent to the expression

$$x = b^y \quad b > 0, b \neq 1$$

In words, we say "y is the number we raise b to in order to get x."

Notation When an expression is in the form $x = b^y$, it is said to be in exponential form. On the other hand, if an expression is in the form $y = \log_b x$, it is said to be in logarithmic form.

Here are some equivalent statements written in both forms.

Exponential Form		Logarithmic Form
$8 = 2^3$	$\Leftrightarrow$	$\log_2 8 = 3$
$25 = 5^2$	$\Leftrightarrow$	$\log_5 25 = 2$
$0.1 = 10^{-1}$	$\Leftrightarrow$	$\log_{10} 0.1 = -1$
$\dfrac{1}{8} = 2^{-3}$	$\Leftrightarrow$	$\log_2 \dfrac{1}{8} = -3$
$r = z^s$	$\Leftrightarrow$	$\log_z r = s$

EXAMPLE 1   Solve for x: $\log_3 x = -2$

Solution   In exponential form the equation looks like this:

$$x = 3^{-2}$$

or

$$x = \frac{1}{9}$$

The solution is $\frac{1}{9}$.

EXAMPLE 2   Solve $\log_x 4 = 3$.

Solution   Again, we use the definition of logarithms to write the expression in exponential form:

$$4 = x^3$$

Taking the cube root of both sides, we have

$$\sqrt[3]{4} = \sqrt[3]{x^3}$$
$$x = \sqrt[3]{4}$$

The solution set is $\{\sqrt[3]{4}\}$.

EXAMPLE 3   Solve $\log_8 4 = x$.

Solution   We write the expression again in exponential form:

$$4 = 8^x$$

Since both 4 and 8 can be written as powers of 2, we write them in terms of powers of 2:

$$2^2 = (2^3)^x$$
$$2^2 = 2^{3x}$$

The only way the left and right sides of this last line can be equal is if the exponents are equal—that is, if

$$2 = 3x$$

or
$$x = \frac{2}{3}$$

The solution is $\frac{2}{3}$. We check as follows:

$$\log_8 4 = \frac{2}{3} \Leftrightarrow 4 = 8^{2/3}$$
$$4 = (\sqrt[3]{8})^2$$
$$4 = 2^2$$
$$4 = 4$$

The solution checks when used in the original equation.

GRAPHING LOGARITHMIC FUNCTIONS

Graphing logarithmic functions can be done using the graphs of exponential functions and the fact that the graphs of inverse functions have symmetry about the line $y = x$. Here's an example to illustrate.

E X A M P L E   4   Graph the equation $y = \log_2 x$.

Solution   The equation $y = \log_2 x$ is, by definition, equivalent to the exponential equation

$$x = 2^y$$

which is the equation of the inverse of the function

$$y = 2^x$$

The graph of $y = 2^x$ was given in Figure 2 of Section 11.1. We simply reflect the graph of $y = 2^x$ about the line $y = x$ to get the graph of $x = 2^y$, which is also the graph of $y = \log_2 x$. (See Figure 1.)

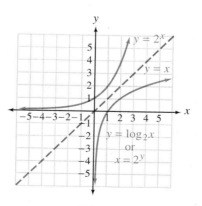

FIGURE 1

It is apparent from the graph that $y = \log_2 x$ is a function, since no vertical line will cross its graph in more than one place. The same is true for all logarithmic equations of the form $y = \log_b x$, where b is a positive number other than 1. Note also that the graph of $y = \log_b x$ will always appear to the right of the y-axis, meaning that x will always be positive in the expression $y = \log_b x$.

TWO SPECIAL IDENTITIES

If b is a positive real number other than 1, then each of the following is a consequence of the definition of a logarithm:

$$(1) \quad b^{\log_b x} = x \qquad \text{and} \qquad (2) \quad \log_b b^x = x$$

The justifications for these identities are similar. Let's consider only the first one. Consider the expression

$$y = \log_b x$$

By definition, it is equivalent to

$$x = b^y$$

Substituting $\log_b x$ for y in the last line gives us

$$x = b^{\log_b x}$$

The next examples in this section show how these two special properties can be used to simplify expressions involving logarithms.

EXAMPLE 5   Simplify $\log_2 8$.

Solution   Substitute 2^3 for 8:

$$\log_2 8 = \log_2 2^3$$
$$= 3$$

EXAMPLE 6   Simplify $\log_{10} 10{,}000$.

Solution $10{,}000$ can be written as 10^4:

$$\log_{10} 10{,}000 = \log_{10} 10^4$$
$$= 4$$

EXAMPLE 7   Simplify $\log_b b \ (b > 0, b \neq 1)$.

Solution   Since $b^1 = b$, we have

$$\log_b b = \log_b b^1$$
$$= 1$$

EXAMPLE   8   Simplify $\log_b 1$ ($b > 0, b \neq 1$).

Solution   Since $1 = b^0$, we have

$$\log_b 1 = \log_b b^0$$
$$= 0$$

EXAMPLE   9   Simplify $\log_4(\log_5 5)$.

Solution   Since $\log_5 5 = 1$,

$$\log_4(\log_5 5) = \log_4 1$$
$$= 0$$

APPLICATION

As we mentioned in the introduction to this section, one application of logarithms is in measuring the magnitude of an earthquake. If an earthquake has a shock wave T times greater than the smallest shock wave that can be measured on a seismograph, then the magnitude M of the earthquake, as measured on the Richter scale, is given by the formula

$$M = \log_{10} T$$

(When we talk about the size of a shock wave, we are talking about its amplitude. The amplitude of a wave is half the difference between its highest point and its lowest point.)

To illustrate the discussion, an earthquake that produces a shock wave that is 10,000 times greater than the smallest shock wave measurable on a seismograph will have a magnitude M on the Richter scale of

$$M = \log_{10} 10,000 = 4$$

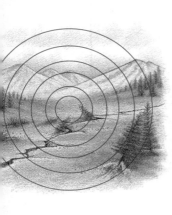

EXAMPLE   10   If an earthquake has a magnitude of $M = 5$ on the Richter scale, what can you say about the size of its shock wave?

Solution   To answer this question, we put $M = 5$ into the formula $M = \log_{10} T$ to obtain

$$5 = \log_{10} T$$

Writing this expression in exponential form, we have

$$T = 10^5 = 100,000$$

We can say that an earthquake that measures 5 on the Richter scale has a shock wave 100,000 times greater than the smallest shock wave measurable on a seismograph.

From Example 10 and the discussion that preceded it, we find that an earthquake of magnitude 5 has a shock wave that is 10 times greater than an earthquake of magnitude 4, because 100,000 is 10 times 10,000.

 Getting Ready for Class

After reading through the preceding section, respond in your own words and in complete sentences.

A. What is a logarithm?

B. What is the relationship between $y = 2^x$ and $y = \log_2 x$? How are their graphs related?

C. Will the graph of $y = \log_b x$ ever appear in the second or third quadrants? Explain why or why not.

D. Explain why $\log_2 0 = x$ has no solution for x.

PROBLEM SET 11.3

Write each of the following expressions in logarithmic form.

1. $2^4 = 16$ **2.** $3^2 = 9$

3. $125 = 5^3$ **4.** $16 = 4^2$

5. $0.01 = 10^{-2}$ **6.** $0.001 = 10^{-3}$

7. $2^{-5} = \dfrac{1}{32}$ **8.** $4^{-2} = \dfrac{1}{16}$

9. $\left(\dfrac{1}{2}\right)^{-3} = 8$ **10.** $\left(\dfrac{1}{3}\right)^{-2} = 9$

11. $27 = 3^3$ **12.** $81 = 3^4$

Write each of the following expressions in exponential form.

13. $\log_{10} 100 = 2$ **14.** $\log_2 8 = 3$

15. $\log_2 64 = 6$ **16.** $\log_2 32 = 5$

17. $\log_8 1 = 0$ **18.** $\log_9 9 = 1$

19. $\log_{10} 0.001 = -3$ **20.** $\log_{10} 0.0001 = -4$

21. $\log_6 36 = 2$ **22.** $\log_7 49 = 2$

23. $\log_5 \dfrac{1}{25} = -2$ **24.** $\log_3 \dfrac{1}{81} = -4$

Solve each of the following equations for x.

25. $\log_3 x = 2$ **26.** $\log_4 x = 3$

27. $\log_5 x = -3$ **28.** $\log_2 x = -4$

29. $\log_2 16 = x$ **30.** $\log_3 27 = x$

31. $\log_8 2 = x$ **32.** $\log_{25} 5 = x$

33. $\log_x 4 = 2$ **34.** $\log_x 16 = 4$

35. $\log_x 5 = 3$ **36.** $\log_x 8 = 2$

Sketch the graph of each of the following logarithmic equations.

37. $y = \log_3 x$ **38.** $y = \log_{1/2} x$

39. $y = \log_{1/3} x$ **40.** $y = \log_4 x$

41. $y = \log_5 x$ **42.** $y = \log_{1/5} x$

43. $y = \log_{10} x$ **44.** $y = \log_{1/4} x$

Each of the following graphs has an equation of the form $y = b^x$ or $y = \log_b x$. Find the equation for each graph.

45.

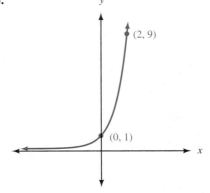

46.

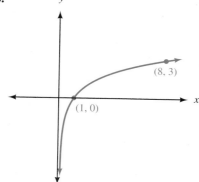

47.

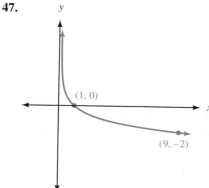

48.

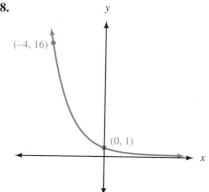

Simplify each of the following.

49. $\log_2 16$

50. $\log_3 9$

51. $\log_{25} 125$

52. $\log_9 27$

53. $\log_{10} 1{,}000$

54. $\log_{10} 10{,}000$

55. $\log_3 3$

56. $\log_4 4$

57. $\log_5 1$

58. $\log_{10} 1$

59. $\log_3(\log_6 6)$

60. $\log_5(\log_3 3)$

61. $\log_4[\log_2(\log_2 16)]$

62. $\log_4[\log_3(\log_2 8)]$

Applying the Concepts

63. Metric System   The metric system uses logical and systematic prefixes for multiplication. For instance, to multiply a unit by 100, the prefix "hecto" is applied, so a hectometer is equal to 100 meters. For each of the prefixes in the following table find the logarithm, base 10, of the multiplying factor.

Prefix	Multiplying Factor	$\log_{10}$ (Multiplying Factor)
Nano	0.000 000 001	
Micro	0.000 001	
Deci	0.1	
Giga	1,000,000,000	
Peta	1,000,000,000,000,000	

64. Domain and Range   Use the graphs of $y = 2^x$ and $y = \log_2 x$ shown in Figure 1 of this section to find the domain and range for each function. Explain how the domain and range found for $y = 2^x$ relate to the domain and range found for $y = \log_2 x$.

65. Magnitude of an Earthquake   Find the magnitude M of an earthquake with a shock wave that measures $T = 100$ on a seismograph.

66. Magnitude of an Earthquake   Find the magnitude M of an earthquake with a shock wave that measures $T = 100{,}000$ on a seismograph.

67. Shock Wave   If an earthquake has a magnitude of 8 on the Richter scale, how many times greater is its shock wave than the smallest shock wave measurable on a seismograph?

68. Shock Wave   If the 1999 Colombia earthquake had a magnitude of 6 on the Richter scale, how many times greater was its shock wave than the smallest shock wave measurable on a seismograph?

Review Problems

The following problems review material we covered in Section 10.2.

Solve.

69. $2x^2 + 4x - 3 = 0$

70. $3x^2 + 4x - 2 = 0$

71. $(2y - 3)(2y - 1) = -4$

72. $(y - 1)(3y - 3) = 10$ **73.** $t^3 - 125 = 0$

74. $8t^3 + 1 = 0$ **75.** $4x^5 - 16x^4 = 20x^3$

76. $3x^4 + 6x^2 = 6x^3$ **77.** $\dfrac{1}{x - 3} + \dfrac{1}{x + 2} = 1$

78. $\dfrac{1}{x + 3} + \dfrac{1}{x - 2} = 1$

Extending the Concepts

79. The graph of the exponential function $y = f(x) = b^x$ is shown below. Use the graph to complete parts a through d.

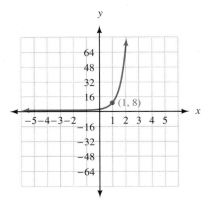

(a) Fill in the table.

x	-1	0	1	2
$f(x)$				

(b) Fill in the table.

x				
$f^{-1}(x)$	-1	0	1	2

(c) Find the equation for $f(x)$.

(d) Find the equation for $f^{-1}(x)$.

80. The graph of the exponential function $y = f(x) = b^x$ is shown below. Use the graph to complete parts a through d.

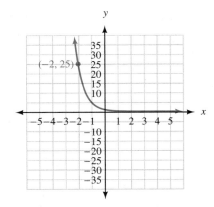

(a) Fill in the table.

x	-1	0	1	2
$f(x)$				

(b) Fill in the table.

x				
$f^{-1}(x)$	-1	0	1	2

(c) Find the equation for $f(x)$.

(d) Find the equation for $f^{-1}(x)$.

If we search for the word *decibel* in *Microsoft Bookshelf 98,* we find the following definition:

A unit used to express relative difference in power or intensity, usually between two acoustic or electric signals, equal to ten times the common logarithm of the ratio of the two levels.

Decibels	Comparable to
10	A light whisper
20	Quiet conversation
30	Normal conversation
40	Light traffic
50	Typewriter, loud conversation
60	Noisy office
70	Normal traffic, quiet train
80	Rock music, subway
90	Heavy traffic, thunder
100	Jet plane at takeoff

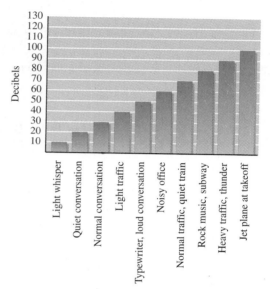

The precise definition for a *decibel* is

$$D = 10 \log_{10}\left(\frac{I}{I_0}\right)$$

where I is the intensity of the sound being measured, and I_0 is the intensity of the least audible sound. (Sound intensity is related to the amplitude of the sound wave that models the sound and is given in units of watts per meter^2.) In this section we will see that the preceding formula can also be written as

$$D = 10(\log_{10} I - \log_{10} I_0)$$

The rules we use to rewrite expressions containing logarithms are called the *properties of logarithms.* There are three of them.

For the following three properties, x, y, and b are all positive real numbers, $b \neq 1$, and r is any real number.

PROPERTY I

$$\log_b(xy) = \log_b x + \log_b y$$

In words: The logarithm of a *product* is the *sum* of the logarithms.

PROPERTY 2

$$\log_b \left(\frac{x}{y} \right) = \log_b x - \log_b y$$

In words: The logarithm of a *quotient* is the *difference* of the logarithms.

PROPERTY 3

$$\log_b x^r = r \log_b x$$

In words: The logarithm of a number raised to a *power* is the *product* of the power and the logarithm of the number.

Proof of Property 1   To prove Property 1, we simply apply the first identity for logarithms given in the preceding section:

$$b^{\log_b xy} = xy = (b^{\log_b x})(b^{\log_b y}) = b^{\log_b x + \log_b y}$$

Since the first and last expressions are equal and the bases are the same, the exponents $\log_b xy$ and $\log_b x + \log_b y$ must be equal. Therefore,

$$\log_b xy = \log_b x + \log_b y$$

The proofs of Properties 2 and 3 proceed in much the same manner, so we will omit them here. The examples that follow show how the three properties can be used.

EXAMPLE 1   Expand, using the properties of logarithms: $\log_5 \dfrac{3xy}{z}$

Solution   Applying Property 2, we can write the quotient of $3xy$ and z in terms of a difference:

$$\log_5 \frac{3xy}{z} = \log_5 3xy - \log_5 z$$

Applying Property 1 to the product $3xy$, we write it in terms of addition:

$$\log_5 \frac{3xy}{z} = \log_5 3 + \log_5 x + \log_5 y - \log_5 z$$

EXAMPLE 2   Expand, using the properties of logarithms:

$$\log_2 \frac{x^4}{\sqrt{y} \cdot z^3}$$

Solution   We write $\sqrt{y}$ as $y^{1/2}$ and apply the properties:

$$\log_2 \frac{x^4}{\sqrt{y} \cdot z^3} = \log_2 \frac{x^4}{y^{1/2} z^3} \qquad\qquad \sqrt{y} = y^{1/2}$$

$$= \log_2 x^4 - \log_2 (y^{1/2} \cdot z^3) \qquad\qquad \text{Property 2}$$

$$= \log_2 x^4 - (\log_2 y^{1/2} + \log_2 z^3) \qquad\qquad \text{Property 1}$$

$$= \log_2 x^4 - \log_2 y^{1/2} - \log_2 z^3 \qquad\qquad \text{Remove parentheses.}$$

$$= 4 \log_2 x - \frac{1}{2} \log_2 y - 3 \log_2 z \qquad\qquad \text{Property 3}$$

We can also use the three properties to write an expression in expanded form as just one logarithm.

E X A M P L E   3   Write as a single logarithm:

$$2 \log_{10} a + 3 \log_{10} b - \frac{1}{3} \log_{10} c$$

Solution   We begin by applying Property 3:

$$2 \log_{10} a + 3 \log_{10} b - \frac{1}{3} \log_{10} c = \log_{10} a^2 + \log_{10} b^3 - \log_{10} c^{1/3} \qquad \text{Property 3}$$

$$= \log_{10} (a^2 \cdot b^3) - \log_{10} c^{1/3} \qquad \text{Property 1}$$

$$= \log_{10} \frac{a^2 b^3}{c^{1/3}} \qquad \text{Property 2}$$

$$= \log_{10} \frac{a^2 b^3}{\sqrt[3]{c}} \qquad c^{1/3} = \sqrt[3]{c}$$

The properties of logarithms along with the definition of logarithms are useful in solving equations that involve logarithms.

E X A M P L E   4   Solve for x: $\log_2(x + 2) + \log_2 x = 3$

Solution   Applying Property 1 to the left side of the equation allows us to write it as a single logarithm:

$$\log_2(x + 2) + \log_2 x = 3$$

$$\log_2[(x + 2)(x)] = 3$$

The last line can be written in exponential form using the definition of logarithms:

$$(x + 2)(x) = 2^3$$

Solve as usual:

$$x^2 + 2x = 8$$

$$x^2 + 2x - 8 = 0$$

$$(x + 4)(x - 2) = 0$$

$$x + 4 = 0 \quad \text{or} \quad x - 2 = 0$$

$$x = -4 \quad \text{or} \quad x = 2$$

In the previous section we noted the fact that x in the expression $y = \log_b x$ cannot be a negative number. Since substitution of $x = -4$ into the original equation gives

$$\log_2(-2) + \log_2(-4) = 3$$

which contains logarithms of negative numbers, we cannot use -4 as a solution. The solution set is $\{2\}$.

Getting Ready for Class

After reading through the preceding section, respond in your own words and in complete sentences.

A. Explain why the following statement is false: "The logarithm of a product is the product of the logarithms."

B. Explain why the following statement is false: "The logarithm of a quotient is the quotient of the logarithms."

C. Explain the difference between $\log_b m + \log_b n$ and $\log_b(m + n)$. Are they equivalent?

D. Explain the difference between $\log_b(mn)$ and $(\log_b m)(\log_b n)$. Are they equivalent?

PROBLEM SET 11.4

Use the three properties of logarithms given in this section to expand each expression as much as possible.

1. $\log_3 4x$

2. $\log_2 5x$

3. $\log_6 \dfrac{5}{x}$

4. $\log_3 \dfrac{x}{5}$

5. $\log_2 y^5$

6. $\log_7 y^3$

7. $\log_9 \sqrt[3]{z}$

8. $\log_8 \sqrt{z}$

9. $\log_6 x^2 y^4$

10. $\log_{10} x^2 y^4$

11. $\log_5 \sqrt{x} \cdot y^4$

12. $\log_8 \sqrt[3]{xy^6}$

13. $\log_b \dfrac{xy}{z}$

14. $\log_b \dfrac{3x}{y}$

15. $\log_{10} \dfrac{4}{xy}$

16. $\log_{10} \dfrac{5}{4y}$

17. $\log_{10} \dfrac{x^2 y}{\sqrt{z}}$

18. $\log_{10} \dfrac{\sqrt{x} \cdot y}{z^3}$

19. $\log_{10} \dfrac{x^3\sqrt{y}}{z^4}$

20. $\log_{10} \dfrac{x^4\sqrt[3]{y}}{\sqrt{z}}$

21. $\log_b \sqrt[3]{\dfrac{x^2 y}{z^4}}$

22. $\log_b \sqrt[4]{\dfrac{x^4 y^3}{z^5}}$

Write each expression as a single logarithm.

23. $\log_b x + \log_b z$

24. $\log_b x - \log_b z$

25. $2 \log_3 x - 3 \log_3 y$

26. $4 \log_2 x + 5 \log_2 y$

27. $\dfrac{1}{2} \log_{10} x + \dfrac{1}{3} \log_{10} y$

28. $\dfrac{1}{3} \log_{10} x - \dfrac{1}{4} \log_{10} y$

29. $3 \log_2 x + \dfrac{1}{2} \log_2 y - \log_2 z$

30. $2 \log_3 x + 3 \log_3 y - \log_3 z$

31. $\dfrac{1}{2} \log_2 x - 3 \log_2 y - 4 \log_2 z$

32. $3 \log_{10} x - \log_{10} y - \log_{10} z$

33. $\dfrac{3}{2} \log_{10} x - \dfrac{3}{4} \log_{10} y - \dfrac{4}{5} \log_{10} z$

34. $3 \log_{10} x - \dfrac{4}{3} \log_{10} y - 5 \log_{10} z$

Solve each of the following equations.

35. $\log_2 x + \log_2 3 = 1$

36. $\log_3 x + \log_3 3 = 1$

37. $\log_3 x - \log_3 2 = 2$

38. $\log_3 x + \log_3 2 = 2$

39. $\log_3 x + \log_3(x - 2) = 1$

40. $\log_6 x + \log_6(x - 1) = 1$

41. $\log_3(x + 3) - \log_3(x - 1) = 1$

42. $\log_4(x - 2) - \log_4(x + 1) = 1$

43. $\log_2 x + \log_2(x - 2) = 3$

44. $\log_4 x + \log_4(x + 6) = 2$

45. $\log_8 x + \log_8(x - 3) = \dfrac{2}{3}$

46. $\log_{27} x + \log_{27}(x + 8) = \dfrac{2}{3}$

✳ **47.** $\log_5 \sqrt{x} + \log_5 \sqrt{6x + 5} = 1$

48. $\log_2 \sqrt{x} + \log_2 \sqrt{6x + 5} = 1$

Applying the Concepts

49. Decibel Formula   Use the properties of logarithms to rewrite the decibel formula $D = 10 \log_{10}(\frac{I}{I_0})$ as $D = 10(\log_{10} I - \log_{10} I_0)$.

50. Decibel Formula   In the decibel formula $D = 10 \log_{10}(\frac{I}{I_0})$, the threshold of hearing, I_0, is

$$I_0 = 10^{-12} \text{ watts/meter}^2$$

Substitute 10^{-12} for I_0 in the decibel formula, and then show that it simplifies to

$$D = 10(\log_{10} I + 12)$$

51. Finding Logarithms   If $\log_{10} 8 = 0.903$ and $\log_{10} 5 = 0.699$, find the following without using a calculator.
(a) $\log_{10} 40$
(b) $\log_{10} 320$
(c) $\log_{10} 1{,}600$

52. Matching   Match each expression in the first column with an equivalent expression in the second column:

(a) $\log_2(ab)$ (i) b
(b) $\log_2(\frac{a}{b})$ (ii) 2
(c) $\log_5 a^b$ (iii) $\log_2 a + \log_2 b$
(d) $\log_a b^a$ (iv) $\log_2 a - \log_2 b$
(e) $\log_a a^b$ (v) $a \log_a b$
(f) $\log_3 9$ (vi) $b \log_5 a$

53. Henderson-Hasselbalch Formula   Doctors use the Henderson-Hasselbalch formula to calculate the pH of a person's blood. pH is a measure of the acidity and/or the alkalinity of a solution. We will say more about this in Section 11.5. This formula is represented as

$$\text{pH} = 6.1 + \log_{10}\left(\frac{x}{y}\right)$$

where x is the base concentration and y is the acidic concentration. Rewrite the Henderson–Hasselbalch formula so that the logarithm of a quotient is not involved.

54. Henderson-Hasselbalch Formula   Refer to the information in the preceding problem about the Henderson-Hasselbalch formula. If most people have a blood pH of 7.4, use the Henderson-Hasselbalch formula to find the ratio of x/y for an average person.

55. Food Processing The formula $M = 0.21(\log_{10} a - \log_{10} b)$ is used in the food processing industry to find the number of minutes M of heat processing a certain food should undergo at 250°F to reduce the probability of survival of *Clostridium botulinum* spores. The letter a represents the number of spores per can before heating, and b represents the number of spores per can after heating. Find M if $a = 1$ and $b = 10^{-12}$. Then find M using the same values for a and b in the formula

$$M = 0.21 \log_{10} \frac{a}{b}.$$

56. Acoustic Powers The formula $N = \log_{10} \dfrac{P_1}{P_2}$ is used in radio electronics to find the ratio of the acoustic powers of two electric circuits in terms of their electric powers. Find N if P_1 is 100 and P_2 is

1. Then use the same two values of P_1 and P_2 to find N in the formula $N = \log_{10} P_1 - \log_{10} P_2$.

Review Problems

The problems that follow review material we covered in Section 10.3.

Use the discriminant to find the number and kind of solutions to the following equations.

57. $2x^2 - 5x + 4 = 0$ **58.** $4x^2 - 12x = -9$

For each of the following problems, find an equation with the given solutions.

59. $x = -3, x = 5$

60. $x = 2, x = -2, x = 1$

61. $y = \dfrac{2}{3}, y = 3$ **62.** $y = -\dfrac{3}{5}, y = 2$

11.5   Common Logarithms and Natural Logarithms

Acid rain was first discovered in the 1960s by Gene Likens and his research team who studied the damage caused by acid rain to Hubbard Brook in New Hampshire. Acid rain is rain with a pH of 5.6 and below. As you will see as you work your way through this section, pH is defined in terms of common logarithms—one of the topics we present in this section. So, when you are finished with this section, you will have a more detailed knowledge of pH and acid rain.

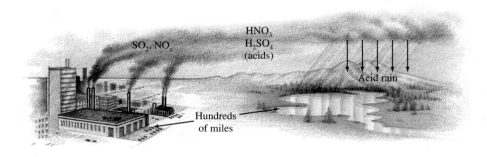

Two kinds of logarithms occur more frequently than other logarithms. Logarithms with a base of 10 are very common because our number system is a base-10 number system. For this reason, we call base-10 logarithms *common logarithms*.

> **DEFINITION**   A **common logarithm** is a logarithm with a base of 10. Since common logarithms are used so frequently, it is customary, in order to save time, to omit notating the base. That is,
>
> $$\log_{10} x = \log x$$
>
> When the base is not shown, it is assumed to be 10.

COMMON LOGARITHMS

Common logarithms of powers of 10 are very simple to evaluate. We need only recognize that $\log 10 = \log_{10} 10 = 1$ and apply the third property of logarithms: $\log_b x^r = r \log_b x$.

$$\log 1{,}000 = \log 10^3 \;\; = 3 \log 10 \;\; = 3(1) \;\; = \;\; 3$$
$$\log 100 \;\; = \log 10^2 \;\; = 2 \log 10 \;\; = 2(1) \;\; = \;\; 2$$
$$\log 10 \;\; = \log 10^1 \;\; = 1 \log 10 \;\; = 1(1) \;\; = \;\; 1$$
$$\log 1 \;\; = \log 10^0 = 0 \log 10 \;\; = 0(1) \;\; = \;\; 0$$
$$\log 0.1 \;\; = \log 10^{-1} = -1 \log 10 = -1(1) = -1$$
$$\log 0.01 \;\; = \log 10^{-2} = -2 \log 10 = -2(1) = -2$$
$$\log 0.001 = \log 10^{-3} = -3 \log 10 = -3(1) = -3$$

To find common logarithms of numbers that are not powers of 10, we use a calculator with a $\boxed{\log}$ key.

Check the following logarithms to be sure you know how to use your calculator. (These answers have been rounded to the nearest ten-thousandth.)

$$\log 7.02 = 0.8463$$
$$\log 1.39 = 0.1430$$
$$\log 6.00 = 0.7782$$
$$\log 9.99 = 0.9996$$

EXAMPLE   1   Use a calculator to find log 2,760.

Solution

$$\log 2{,}760 = 3.4409$$

To work this problem on a scientific calculator, we simply enter the number 2,760 and press the key labeled $\boxed{\log}$. On a graphing calculator we press the $\boxed{\log}$ key first, then the number 2,760.

The 3 in the answer is called the *characteristic,* and the decimal part of the logarithm is called the *mantissa.*

EXAMPLE 2 Find log 0.0391.

Solution $\log 0.0391 = -1.4078$

EXAMPLE 3 Find log 0.00523.

Solution $\log 0.00523 = -2.2815$

EXAMPLE 4 Find x if $\log x = 3.8774$.

Solution We are looking for the number whose logarithm is 3.8774. On a scientific calculator we enter 3.8774 and then press the key labeled $\boxed{10^x}$. On a graphing calculator we press the $\boxed{10^x}$ key first, then the number 3.8774. (Sometimes it is the inverse of the $\boxed{\log}$ key.) The result is 7,540 to four significant digits.

$$\text{If} \qquad \log x = 3.8774$$
$$\text{then} \qquad x = 10^{3.8774}$$
$$= 7,540$$

The number 7,540 is called the *antilogarithm* or just *antilog* of 3.8774. That is, 7,540 is the number whose logarithm is 3.8774.

EXAMPLE 5 Find x if $\log x = -2.4179$.

Solution Using the $\boxed{10^x}$ key, the result is 0.00382. Here is the algebra justifying the use of this calculator key:

$$\text{If} \qquad \log x = -2.4179$$
$$\text{then} \qquad x = 10^{-2.4179}$$
$$= 0.00382$$

The antilog of -2.4179 is 0.00382. That is, the logarithm of 0.00382 is -2.4179.

In Section 11.3 we found that the magnitude M of an earthquake that produces a shock wave T times larger than the smallest shock wave that can be measured on a seismograph is given by the formula

$$M = \log_{10} T$$

We can rewrite this formula using our shorthand notation for common logarithms as

$$M = \log T$$

E X A M P L E   6   The San Francisco earthquake of 1906 is estimated to have measured 8.3 on the Richter scale. The San Fernando earthquake of 1971 measured 6.6 on the Richter scale. Find T for each earthquake, and then give some indication of how much stronger the 1906 earthquake was than the 1971 earthquake.

Solution   For the 1906 earthquake:

$$\text{If } \log T = 8.3, \text{ then } T = 2.00 \times 10^8.$$

For the 1971 earthquake:

$$\text{If } \log T = 6.6, \text{ then } T = 3.98 \times 10^6.$$

Dividing the two values of T and rounding our answer to the nearest whole number, we have

$$\frac{2.00 \times 10^8}{3.98 \times 10^6} = 50$$

The shock wave for the 1906 earthquake was approximately 50 times larger than the shock wave for the 1971 earthquake.

8.3

In chemistry, the pH of a solution is the measure of the acidity of the solution. The definition for pH involves common logarithms. Here it is:

$$\text{pH} = -\log[\text{H}^+]$$

where $[\text{H}^+]$ is the concentration of the hydrogen ion in moles per liter. The range for pH is from 0 to 14. Pure water, a neutral solution, has a pH of 7. An acidic solution, such as vinegar, will have a pH less than 7, and an alkaline solution, such as ammonia, has a pH above 7.

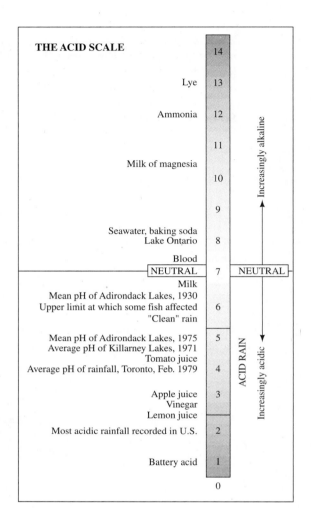

THE ACID SCALE

	14
Lye	13
Ammonia	12
	11
Milk of magnesia	
	10
	9
Seawater, baking soda	
Lake Ontario	8
Blood	
NEUTRAL	7
Milk	
Mean pH of Adirondack Lakes, 1930	
Upper limit at which some fish affected	6
"Clean" rain	
Mean pH of Adirondack Lakes, 1975	5
Average pH of Killarney Lakes, 1971	
Tomato juice	
Average pH of rainfall, Toronto, Feb. 1979	4
Apple juice	3
Vinegar	
Lemon juice	
Most acidic rainfall recorded in U.S.	2
Battery acid	1
	0

Increasingly alkaline

NEUTRAL

ACID RAIN

Increasingly acidic

EXAMPLE 7   Normal rainwater has a pH of 5.6. What is the concentration of the hydrogen ion in normal rainwater?

Solution   Substituting 5.6 for pH in the formula $pH = -\log[H^+]$, we have

$$5.6 = -\log[H^+] \qquad \text{Substitution}$$

$$\log[H^+] = -5.6 \qquad \text{Isolate the logarithm.}$$

$$[H^+] = 10^{-5.6} \qquad \text{Write in exponential form.}$$

$$\approx 2.5 \times 10^{-6} \text{ moles per liter} \qquad \text{Answer in scientific notation.}$$

EXAMPLE 8   The concentration of the hydrogen ion in a sample of acid rain known to kill fish is 3.2×10^{-5} mole per liter. Find the pH of this acid rain to the nearest tenth.

Solution   Substituting 3.2×10^{-5} for $[H^+]$ in the formula $pH = -\log[H^+]$, we have

$$pH = -\log[3.2 \times 10^{-5}] \qquad \text{Substitution}$$
$$\approx -(-4.5) \qquad \text{Evaluate the logarithm.}$$
$$\approx 4.5 \qquad \text{Simplify.}$$

NATURAL LOGARITHMS

> **DEFINITION**   A natural **logarithm** is a logarithm with a base of e. The natural logarithm of x is denoted by $\ln x$. That is,
>
> $$\ln x = \log_e x$$

We can assume that all our properties of exponents and logarithms hold for expressions with a base of e, since e is a real number. Here are some examples intended to make you more familiar with the number e and natural logarithms.

EXAMPLE 9   Simplify each of the following expressions.

(a) $e^0 = 1$
(b) $e^1 = e$
(c) $\ln e = 1$ In exponential form, $e^1 = e$.
(d) $\ln 1 = 0$ In exponential form, $e^0 = 1$.
(e) $\ln e^3 = 3$
(f) $\ln e^{-4} = -4$
(g) $\ln e^t = t$

EXAMPLE 10   Use the properties of logarithms to expand the expression $\ln Ae^{5t}$.

Solution   Since the properties of logarithms hold for natural logarithms, we have

$$\ln Ae^{5t} = \ln A + \ln e^{5t}$$
$$= \ln A + 5t \ln e$$
$$= \ln A + 5t \qquad \text{Because } \ln e = 1$$

EXAMPLE 11   If $\ln 2 = 0.6931$ and $\ln 3 = 1.0986$, find

(a) $\ln 6$ (b) $\ln 0.5$ (c) $\ln 8$

Solution

(a) Since $6 = 2 \cdot 3$, we have

$$\begin{aligned} \ln 6 &= \ln 2 \cdot 3 \\ &= \ln 2 + \ln 3 \\ &= 0.6931 + 1.0986 \\ &= 1.7917 \end{aligned}$$

(b) Writing 0.5 as $\frac{1}{2}$ and applying Property 2 for logarithms gives us

$$\begin{aligned} \ln 0.5 &= \ln \frac{1}{2} \\ &= \ln 1 - \ln 2 \\ &= 0 - 0.6931 \\ &= -0.6931 \end{aligned}$$

(c) Writing 8 as 2^3 and applying Property 3 for logarithms, we have

$$\begin{aligned} \ln 8 &= \ln 2^3 \\ &= 3 \ln 2 \\ &= 3(0.6931) \\ &= 2.0793 \end{aligned}$$

 ## Getting Ready for Class

After reading through the preceding section, respond in your own words and in complete sentences.

A. What is a common logarithm?

B. What is a natural logarithm?

C. Is e a rational number? Explain.

D. Find $\ln e$, and explain how you arrived at your answer.

PROBLEM SET 11.5

Find the following logarithms.

1. $\log 378$

2. $\log 426$

3. $\log 37.8$

4. $\log 42{,}600$

5. $\log 3{,}780$

6. $\log 0.4260$

7. $\log 0.0378$

8. $\log 0.0426$

9. $\log 37{,}800$

10. $\log 4{,}900$

11. $\log 600$

12. $\log 900$

13. $\log 2{,}010$

14. $\log 10{,}200$

15. $\log 0.00971$

16. $\log 0.0312$

17. $\log 0.0314$

18. $\log 0.00052$

19. $\log 0.399$

20. $\log 0.111$

Find x in the following equations.

21. $\log x = 2.8802$

22. $\log x = 4.8802$

23. $\log x = -2.1198$

24. $\log x = -3.1198$

25. $\log x = 3.1553$

26. $\log x = 5.5911$

27. $\log x = -5.3497$

28. $\log x = -1.5670$

29. $\log x = -7.0372$

30. $\log x = -4.2000$

31. $\log x = 10$

32. $\log x = -1$

33. $\log x = -10$

34. $\log x = 1$

35. $\log x = 20$

36. $\log x = -20$

37. $\log x = -2$

38. $\log x = 4$

39. $\log x = \log_2 8$

40. $\log x = \log_3 9$

41. Solve the following for x: $\log x^2 = (\log x)^2$. (*Hint:* Use properties of logarithms and factor.)

42. Solve the following for x: $\ln x^2 = (\ln x)^2$.

Applying the Concepts

43. **Interpreting Calculator Results** Use your calculator to find $\log(-10)$. Explain the result your calculator gives you.

44. **Interpreting Calculator Results** Use your calculator to find $\ln 0$. Explain the result your calculator gives you.

45. **Atomic Bomb Tests** The Bikini Atoll in the Pacific Ocean was used as a location for atomic bomb tests by the United States government in the 1950s. One such test resulted in an earthquake measurement of 5.0 on the Richter scale. Compare the 1906 San Francisco earthquake of estimated magnitude 8.3 on the Richter scale to this atomic bomb test. Use the shock wave T for purposes of comparison.

46. **Atomic Bomb Tests** Today's nuclear weapons are 1,000 times more powerful than the atomic bombs tested in the Bikini Atoll mentioned in Problem 45. Use the shock wave T to determine the Richter scale measurement of a nuclear test today.

47. **Getting Close to *e*** Use a calculator to complete the following table.

x	$(1+x)^{1/x}$
1	
0.5	
0.1	
0.01	
0.001	
0.0001	
0.00001	

What number does the expression $(1+x)^{1/x}$ seem to approach as x gets closer and closer to zero?

48. **Getting Close to *e*** Use a calculator to complete the following table.

x	$\left(1+\frac{1}{x}\right)^x$
1	
10	
50	
100	
500	
1,000	
10,000	
1,000,000	

What number does the expression $\left(1+\frac{1}{x}\right)^x$ seem to approach as x gets larger and larger?

Kepler's Law Johannes Kepler (German astronomer, 1571–1630) found that for every planet in the solar system, the following relationship holds: If R is the radius of the orbit of a planet and T is its period (the time for one complete revolution around the sun), then the quotient R^3/T^2 is a constant.

49. Show that the expression $3 \log R - 2 \log T$ must also be constant if R^3/T^2 is a constant.

50. The same relationship holds for satellites orbiting planets: The quantity R^3/T^2 is a constant. Table 1 gives the orbit radius and the period for five of the satellites of Jupiter. Verify that the expression

3 log R − 2 log T is a constant for the satellites of Jupiter.

(JPL/NASA)

TABLE 1 Satellites of Jupiter		
Name	**Radius of the Orbit (mi)**	**Period (h)**
V	112,000	11.96
I (Io)	262,000	42.46
II (Europa)	417,000	85.22
III (Ganymede)	666,000	171.71
IV (Callisto)	1,170,000	400.54

Use the following figure to solve Problems 51–54.

pH Scale

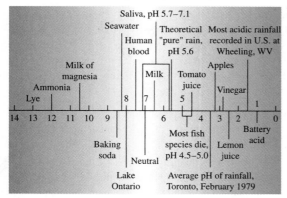

51. pH Find the pH of orange juice if the concentration of the hydrogen ion in the juice is $[H^+] = 6.50 \times 10^{-4}$.

52. pH Find the pH of milk if the concentration of the hydrogen ions in milk is $[H^+] = 1.88 \times 10^{-6}$.

53. pH Find the concentration of hydrogen ions in a glass of wine if the pH is 4.75.

54. pH Find the concentration of hydrogen ions in a bottle of vinegar if the pH is 5.75.

The Richter Scale Find the relative size T of the shock wave of earthquakes with the following magnitudes, as measured on the Richter scale.

55. 5.5   **56.** 6.6   **57.** 8.3   **58.** 8.7

59. Shock Wave How much larger is the shock wave of an earthquake that measures 6.5 on the Richter scale than one that measures 5.5 on the same scale?

60. Shock Wave How much larger is the shock wave of an earthquake that measures 8.5 on the Richter scale than one that measures 5.5 on the same scale?

Depreciation The annual rate of depreciation r on a car that is purchased for P dollars and is worth W dollars t years later can be found from the formula

$$\log(1 - r) = \frac{1}{t} \log \frac{W}{P}$$

61. Find the annual rate of depreciation on a car that is purchased for $9,000 and sold 5 years later for $4,500.

62. Find the annual rate of depreciation on a car that is purchased for $9,000 and sold 4 years later for $3,000.

Two cars depreciate in value according to the following depreciation tables. In each case, find the annual rate of depreciation.

63.

Age in Years	Value in Dollars
New	7,550
5	5,750

64.

Age in Years	Value in Dollars
New	7,550
3	5,750

Simplify each of the following expressions.

65. $\ln e$ **66.** $\ln 1$ **67.** $\ln e^5$

68. $\ln e^{-3}$ **69.** $\ln e^x$ **70.** $\ln e^y$

Use the properties of logarithms to expand each of the following expressions.

71. $\ln 10e^{3t}$ **72.** $\ln 10e^{4t}$

73. $\ln Ae^{-2t}$ **74.** $\ln Ae^{-3t}$

If $\ln 2 = 0.6931$, $\ln 3 = 1.0986$, and $\ln 5 = 1.6094$, find each of the following.

75. $\ln 15$ **76.** $\ln 10$ **77.** $\ln \dfrac{1}{3}$

78. $\ln \dfrac{1}{5}$ **79.** $\ln 9$ **80.** $\ln 25$

81. $\ln 16$ **82.** $\ln 81$

Review Problems

The following problems review material we covered in Section 10.4.

Solve each equation.

83. $x^4 - 2x^2 - 8 = 0$ **84.** $x^4 - 8x^2 - 9 = 0$

85. $x^{2/3} - 5x^{1/3} + 6 = 0$

86. $x^{2/3} - 3x^{1/3} + 2 = 0$

87. $2x - 5\sqrt{x} + 3 = 0$ **88.** $3x - 8\sqrt{x} + 4 = 0$

89. $(3x + 1) - 6\sqrt{3x + 1} + 8 = 0$

90. $(2x - 1) - 2\sqrt{2x - 1} - 15 = 0$

91. Solve $kx^2 + 4x - k = 0$ for x.

92. Solve $4x^2 - 4x + k = 0$ for x.

11.6   Exponential Equations and Change of Base

For items involved in exponential growth, the time it takes for a quantity to double is called the *doubling time.* For example, if you invest \$5,000 in an account that pays 5% annual interest, compounded quarterly, you may want to know how long it will take for your money to double in value. You can find this doubling time if you can solve the equation

$$10,000 = 5,000(1.0125)^{4t}$$

As you will see as you progress through this section, logarithms are the key to solving equations of this type.

Logarithms are very important in solving equations in which the variable appears as an exponent. The equation

$$5^x = 12$$

is an example of one such equation. Equations of this form are called *exponential equations.* Since the quantities 5^x and 12 are equal, so are their common logarithms. We begin our solution by taking the logarithm of both sides:

$$\log 5^x = \log 12$$

We now apply Property 3 for logarithms, $\log x^r = r \log x$, to turn x from an exponent into a coefficient:

$$x \log 5 = \log 12$$

Dividing both sides by $\log 5$ gives us

$$x = \frac{\log 12}{\log 5}$$

If we want a decimal approximation to the solution, we can find log 12 and log 5 on a calculator and divide:

$$x = \frac{1.0792}{0.6990}$$

$$= 1.5439$$

The complete problem looks like this:

$$5^x = 12$$

$$\log 5^x = \log 12$$

$$x \log 5 = \log 12$$

$$x = \frac{\log 12}{\log 5}$$

$$= \frac{1.0792}{0.6990}$$

$$= 1.5439$$

Here is another example of solving an exponential equation using logarithms.

EXAMPLE I   Solve for x: $25^{2x+1} = 15$

Solution   Taking the logarithm of both sides and then writing the exponent $(2x + 1)$ as a coefficient, we proceed as follows:

$$25^{2x+1} = 15$$

$$\log 25^{2x+1} = \log 15 \qquad\qquad \text{Take the log of both sides.}$$

$$(2x + 1)\log 25 = \log 15 \qquad\qquad \text{Property 3}$$

$$2x + 1 = \frac{\log 15}{\log 25} \qquad\qquad \text{Divide by log 25.}$$

$$2x = \frac{\log 15}{\log 25} - 1 \qquad\qquad \text{Add } -1 \text{ to both sides.}$$

$$x = \frac{1}{2}\left(\frac{\log 15}{\log 25} - 1\right) \qquad\qquad \text{Multiply both sides by } \tfrac{1}{2}.$$

Using a calculator, we can write a decimal approximation to the answer:

$$x = \frac{1}{2}\left(\frac{1.1761}{1.3979} - 1\right)$$

$$= \frac{1}{2}(0.8413 - 1)$$

$$= \frac{1}{2}(-0.1587)$$

$$= -0.0794$$

If you invest P dollars in an account with an annual interest rate r that is compounded n times a year, then t years later the amount of money in that account will be

$$A = P\left(1 + \frac{r}{n}\right)^{nt}$$

EXAMPLE 2   How long does it take for $5,000 to double if it is deposited in an account that yields 5% interest compounded once a year?

Solution   Substituting $P = 5{,}000$, $r = 0.05$, $n = 1$, and $A = 10{,}000$ into our formula, we have

$$10{,}000 = 5{,}000(1 + 0.05)^t$$

$$10{,}000 = 5{,}000(1.05)^t$$

$$2 = (1.05)^t \qquad\qquad \text{Divide by 5,000.}$$

This is an exponential equation. We solve by taking the logarithm of both sides:

$$\log 2 = \log(1.05)^t$$

$$= t \log 1.05$$

Dividing both sides by $\log 1.05$, we have

$$t = \frac{\log 2}{\log 1.05}$$

$$= 14.2 \text{ to the nearest tenth}$$

It takes a little over 14 years for $5,000 to double if it earns 5% interest per year, compounded once a year.

There is a fourth property of logarithms we have not yet considered. This last property allows us to change from one base to another and is therefore called the *change-of-base property.*

PROPERTY 4 (CHANGE OF BASE)

If a and b are both positive numbers other than 1, and if $x > 0$, then

$$\log_a x = \frac{\log_b x}{\log_b a}$$

$$\underset{\text{Base } a}{\uparrow} \qquad\qquad \underset{\text{Base } b}{\uparrow}$$

The logarithm on the left side has a base of a, and both logarithms on the right side have a base of b. This allows us to change from base a to any other base b that is a positive number other than 1. Here is a proof of Property 4 for logarithms.

Proof We begin by writing the identity

$$a^{\log_a x} = x$$

Taking the logarithm base b of both sides and writing the exponent $\log_a x$ as a coefficient, we have

$$\log_b a^{\log_a x} = \log_b x$$

$$\log_a x \log_b a = \log_b x$$

Dividing both sides by $\log_b a$, we have the desired result:

$$\frac{\log_a x \log_b a}{\log_b a} = \frac{\log_b x}{\log_b a}$$

$$\log_a x = \frac{\log_b x}{\log_b a}$$

We can use this property to find logarithms we could not otherwise compute on our calculators—that is, logarithms with bases other than 10 or e. The next example illustrates the use of this property.

EXAMPLE 3 Find $\log_8 24$.

Solution Since we do not have base-8 logarithms on our calculators, we can change this expression to an equivalent expression that contains only base-10 logarithms:

$$\log_8 24 = \frac{\log 24}{\log 8} \qquad \text{Property 4}$$

Don't be confused. We did not just drop the base, we changed to base 10. We could have written the last line like this:

$$\log_8 24 = \frac{\log_{10} 24}{\log_{10} 8}$$

From our calculators, we write

$$\log_8 24 = \frac{1.3802}{0.9031}$$

$$= 1.5283$$

APPLICATION

EXAMPLE 4 Suppose that the population in a small city is 32,000 in the beginning of 1994 and that the city council assumes that the population size t

years later can be estimated by the equation

$$P = 32{,}000e^{0.05t}$$

Approximately when will the city have a population of 50,000?

Solution   We substitute 50,000 for P in the equation and solve for t:

$$50{,}000 = 32{,}000e^{0.05t}$$

$$1.56 = e^{0.05t} \qquad \frac{50{,}000}{32{,}000} \text{ is approximately 1.56.}$$

To solve this equation for t, we can take the natural logarithm of each side:

$$\ln 1.56 = \ln e^{0.05t}$$

$$ = 0.05t \ln e \qquad \text{Property 3 for logarithms}$$

$$ = 0.05t \qquad \text{Because } \ln e = 1$$

$$t = \frac{\ln 1.56}{0.05} \qquad \text{Divide each side by 0.05.}$$

$$ = \frac{0.4447}{0.05}$$

$$ = 8.89 \text{ years}$$

We can estimate that the population will reach 50,000 toward the end of 2002.

 Using **TECHNOLOGY**

GRAPHING CALCULATOR

We can evaluate many logarithmic expressions on a graphing calculator by using the fact that logarithmic functions and exponential functions are inverses.

(continued)

(Using Technology, continued)

EXAMPLE 5   Evaluate the logarithmic expression $\log_3 7$ from the graph of an exponential function.

Solution   First, we let $\log_3 7 = x$. Next, we write this expression in exponential form as $3^x = 7$. We can solve this equation graphically by finding the intersection of the graphs $Y_1 = 3^x$ and $Y_2 = 7$, as shown in Figure 1.

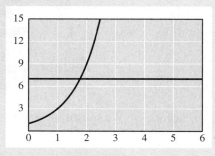

FIGURE 1

Using the calculator, we find the two graphs intersect at $(1.77, 7)$. Therefore, $\log_3 7 = 1.77$ to the nearest hundredth. We can check our work by evaluating the expression $3^{1.77}$ on our calculator with the key strokes

$$3 \boxed{\wedge} 1.77 \boxed{\text{ENTER}}$$

The result is 6.99 to the nearest hundredth, which seems reasonable since 1.77 is accurate to the nearest hundredth. To get a result closer to 7, we would need to find the intersection of the two graphs more accurately.

 ## Getting Ready for Class

After reading through the preceding section, respond in your own words and in complete sentences.

A. What is an exponential equation?

B. How do logarithms help you solve exponential equations?

C. What is the change-of-base property?

D. Write an application modeled by the equation $A = 10{,}000\left(1 + \dfrac{0.08}{2}\right)^{2 \cdot 5}$.

PROBLEM SET 11.6

Solve each exponential equation. Use a calculator to write the answer in decimal form.

1. $3^x = 5$ **2.** $4^x = 3$ **3.** $5^x = 3$

4. $3^x = 4$ **5.** $5^{-x} = 12$ **6.** $7^{-x} = 8$

7. $12^{-x} = 5$ **8.** $8^{-x} = 7$ **9.** $8^{x+1} = 4$

10. $9^{x+1} = 3$ **11.** $4^{x-1} = 4$ **12.** $3^{x-1} = 9$

13. $3^{2x+1} = 2$ **14.** $2^{2x+1} = 3$ **15.** $3^{1-2x} = 2$

16. $2^{1-2x} = 3$ **17.** $15^{3x-4} = 10$

18. $10^{3x-4} = 15$ **19.** $6^{5-2x} = 4$

20. $9^{7-3x} = 5$

Use the change-of-base property and a calculator to find a decimal approximation to each of the following logarithms.

21. $\log_8 16$ **22.** $\log_9 27$ **23.** $\log_{16} 8$

24. $\log_{27} 9$ **25.** $\log_7 15$ **26.** $\log_3 12$

27. $\log_{15} 7$ **28.** $\log_{12} 3$ **29.** $\log_8 240$

30. $\log_6 180$ **31.** $\log_4 321$ **32.** $\log_5 462$

Find a decimal approximation to each of the following natural logarithms.

33. $\ln 345$ **34.** $\ln 3,450$ **35.** $\ln 0.345$

36. $\ln 0.0345$ **37.** $\ln 10$ **38.** $\ln 100$

39. $\ln 45,000$ **40.** $\ln 450,000$

Applying the Concepts

41. Compound Interest How long will it take for $500 to double if it is invested at 6% annual interest compounded 2 times a year?

42. Compound Interest How long will it take for $500 to double if it is invested at 6% annual interest compounded 12 times a year?

43. Compound Interest How long will it take for $1,000 to triple if it is invested at 12% annual interest compounded 6 times a year?

44. Compound Interest How long will it take for $1,000 to become $4,000 if it is invested at 12% annual interest compounded 6 times a year?

45. Doubling Time How long does it take for an amount of money P to double itself if it is invested at 8% interest compounded 4 times a year?

46. Tripling Time How long does it take for an amount of money P to triple itself if it is invested at 8% interest compounded 4 times a year?

47. Tripling Time If a $25 investment is worth $75 today, how long ago must that $25 have been invested at 6% interest computed twice a year?

48. Doubling Time If a $25 investment is worth $50 today, how long ago must that $25 have been invested at 6% interest computed twice a year?

Recall from Section 11.1 that if P dollars are invested in an account with annual interest rate r, compounded continuously, then the amount of money in the account after t years is given by the formula

$$A(t) = Pe^{rt}$$

49. Continuously Compounded Interest Repeat Problem 41 if the interest is compounded continuously.

50. Continuously Compounded Interest Repeat Problem 44 if the interest is compounded continuously.

51. Continuously Compounded Interest How long will it take $500 to triple if it is invested at 6% annual interest, compounded continuously?

52. Continuously Compounded Interest How long will it take $500 to triple if it is invested at 12% annual interest, compounded continuously?

53. Continuously Compounded Interest How long will it take for $1,000 to be worth $2,500 at 8% interest, compounded continuously?

54. Continuously Compounded Interest How long will it take for $1,000 to be worth $5,000 at 8% interest, compounded continuously?

55. Exponential Growth Suppose that the population in a small city is 32,000 at the beginning of 1994 and that the city council assumes that the population size t years later can be estimated by the equation

$$P(t) = 32,000e^{0.05t}$$

Approximately when will the city have a population of 64,000?

56. Exponential Growth Suppose the population of a city is given by the equation

$$P(t) = 100,000e^{0.05t}$$

where t is the number of years from the present time. How large is the population now? (*Now* corresponds to a certain value of t. Once you realize what that value of t is, the problem becomes very simple.)

57. Exponential Growth   Suppose the population of a city is given by the equation

$$P(t) = 15,000e^{0.04t}$$

where t is the number of years from the present time. How long will it take for the population to reach 45,000?

58. Exponential Growth   Suppose the population of a city is given by the equation

$$P(t) = 15,000e^{0.08t}$$

where t is the number of years from the present time. How long will it take for the population to reach 45,000?

Review Problems

The following problems review material we covered in Section 10.5.

Find the vertex for each of the following parabolas, and then indicate if it is the highest or lowest point on the graph.

59. $y = 2x^2 + 8x - 15$ **60.** $y = 3x^2 - 9x - 10$

61. $y = 12x - 4x^2$ **62.** $y = 18x - 6x^2$

63. Maximum Height   An object is projected into the air with an initial upward velocity of 64 feet per second. Its height h at any time t is given by the formula $h = 64t - 16t^2$. Find the time at which

the object reaches its maximum height. Then, find the maximum height.

64. Maximum Height   An object is projected into the air with an initial upward velocity of 64 feet per second from the top of a building 40 feet high. If the height h of the object t seconds after it is projected into the air is $h = 40 + 64t - 16t^2$, find the time at which the object reaches its maximum height. Then, find the maximum height it attains.

Extending the Concepts

65. Solve the formula $A = Pe^{rt}$ for t.

66. Solve the formula $A = Pe^{-rt}$ for t.

67. Solve the formula $A = P2^{-kt}$ for t.

68. Solve the formula $A = P2^{kt}$ for t.

69. Solve the formula $A = P(1 - r)^t$ for t.

70. Solve the formula $A = P(1 + r)^t$ for t.

Use Example 5 as a model to evaluate the following logarithmic expressions by using the graph of an exponential function.

71. $\log_6 23$ **72.** $\log_4 14$

73. $\log_7 29$ **74.** $\log_5 34$

75. In Problem 1 you solved $3^x = 5$ algebraically. Now graph $y = 3^x$ and $y = 5$, and approximate their points of intersection. How do these points of intersection compare to your answer in Problem 1?

76. Now try approximating a solution(s) to $x^2 = 2^x$, which cannot be solved with traditional algebraic techniques (other than "guess and check"). Graph $y = x^2$ and $y = 2^x$, and approximate the points of intersection. What would be the estimate of your solution(s) to $x^2 = 2^x$?

CHAPTER 11 SUMMARY

Examples

1. For the exponential function $f(x) = 2^x$,

$$f(0) = 2^0 = 1$$
$$f(1) = 2^1 = 2$$
$$f(2) = 2^2 = 4$$
$$f(3) = 2^3 = 8$$

Exponential Functions [11.1]

Any function of the form

$$f(x) = b^x$$

where $b > 0$ and $b \neq 1$, is an **exponential function.**

2. The function $f(x) = x^2$ is not one-to-one because 9, which is in the range, comes from both 3 and -3 in the domain.

One-to-One Functions [11.2]

A function is a **one-to-one function** if every element in the range comes from exactly one element in the domain.

3. The inverse of $f(x) = 2x - 3$ is

$$f^{-1}(x) = \frac{x + 3}{2}$$

Inverse Functions [11.2]

The **inverse** of a function is obtained by reversing the order of the coordinates of the ordered pairs belonging to the function. Only one-to-one functions have inverses that are also functions.

4. The definition allows us to write expressions like

$$y = \log_3 27$$

equivalently in exponential form as

$$3^y = 27$$

which makes it apparent that y is 3.

Definition of Logarithms [11.3]

If b is a positive number not equal to 1, then the expression

$$y = \log_b x$$

is equivalent to $x = b^y$. That is, in the expression $y = \log_b x$, y is the number to which we raise b in order to get x. Expressions written in the form $y = \log_b x$ are said to be in *logarithmic form*. Expressions like $x = b^y$ are in *exponential form*.

5. Examples of the two special identities are

$$5^{\log_5 12} = 12$$

and

$$\log_8 8^3 = 3$$

Two Special Identities [11.3]

For $b > 0$, $b \neq 1$, the following two expressions hold for all positive real numbers x:

$$(1) \quad b^{\log_b x} = x$$

$$(2) \quad \log_b b^x = x$$

6. We can rewrite the expression

$$\log_{10} \frac{45^6}{273}$$

using the properties of logarithms, as

$$6 \log_{10} 45 - \log_{10} 273$$

7. $\log_{10} 10{,}000 = \log 10{,}000$
$$= \log 10^4$$
$$= 4$$

8. $\ln e = 1$

$\ln 1 = 0$

9. $\log_6 475 = \dfrac{\log 475}{\log 6}$

$$= \frac{2.6767}{0.7782}$$

$$= 3.44$$

Properties of Logarithms [11.4]
If x, y, and b are positive real numbers, $b \neq 1$, and r is any real number, then

1. $\log_b(xy) = \log_b x + \log_b y$

2. $\log_b \left(\dfrac{x}{y} \right) = \log_b x - \log_b y$

3. $\log_b x^r = r \log_b x$

Common Logarithms [11.5]
Common logarithms are logarithms with a base of 10. To save time in writing, we omit the base when working with common logarithms. That is,

$$\log x = \log_{10} x$$

Natural Logarithms [11.5]
Natural logarithms, written **ln x**, are logarithms with a base of e, where the number e is an irrational number (like the number π). A decimal approximation for e is 2.7183. All the properties of exponents and logarithms hold when the base is e.

Change of Base [11.6]
If x, a, and b are positive real numbers, $a \neq 1$ and $b \neq 1$, then

$$\log_a x = \frac{\log_b x}{\log_b a}$$

COMMON MISTAKES

The most common mistakes that occur with logarithms come from trying to apply the three properties of logarithms to situations in which they don't apply. For example, a very common mistake looks like this:

$$\frac{\log 3}{\log 2} = \log 3 - \log 2 \qquad \text{Mistake}$$

This is not a property of logarithms. In order to write the expression $\log 3 - \log 2$, we would have to start with

$$\log \frac{3}{2} \qquad NOT \qquad \frac{\log 3}{\log 2}$$

There is a difference.

CHAPTER 11   REVIEW

Let $f(x) = 2^x$ and $g(x) = (\frac{1}{3})^x$, and find the following. [11.1]

1. $f(4)$ **2.** $f(-1)$

3. $g(2)$ **4.** $f(2) - g(-2)$

5. $f(-1) + g(1)$ **6.** $g(-1) + f(2)$

7. The graph of $y = f(x)$

8. The graph of $y = g(x)$

For each relation that follows, sketch the graph of the relation and its inverse, and write an equation for the inverse. [11.2]

9. $y = 2x + 1$ **10.** $y = x^2 - 4$

For each of the following functions, find the equation of the inverse. Write the inverse using the notation $f^{-1}(x)$ if the inverse is itself a function. [11.2]

11. $f(x) = 2x + 3$ **12.** $f(x) = x^2 - 1$

13. $f(x) = \frac{1}{2}x + 2$ **14.** $f(x) = 4 - 2x^2$

Write each expression in logarithmic form. [11.3]

15. $3^4 = 81$ **16.** $7^2 = 49$

17. $0.01 = 10^{-2}$ **18.** $2^{-3} = \frac{1}{8}$

Write each expression in exponential form. [11.3]

19. $\log_2 8 = 3$ **20.** $\log_3 9 = 2$

21. $\log_4 2 = \frac{1}{2}$ **22.** $\log_4 4 = 1$

Solve for x. [11.3]

23. $\log_5 x = 2$ **24.** $\log_{16} 8 = x$

25. $\log_x 0.01 = -2$

Graph each equation. [11.3]

26. $y = \log_2 x$ **27.** $y = \log_{1/2} x$

Simplify each expression. [11.3]

28. $\log_4 16$ **29.** $\log_{27} 9$

30. $\log_4(\log_3 3)$

Use the properties of logarithms to expand each expression. [11.4]

31. $\log_2 5x$ **32.** $\log_{10} \dfrac{2x}{y}$

33. $\log_a \dfrac{y^3\sqrt{x}}{z}$ **34.** $\log_{10} \dfrac{x^2}{y^3z^4}$

Write each expression as a single logarithm. [11.4]

35. $\log_2 x + \log_2 y$ **36.** $\log_3 x - \log_3 4$

37. $2\log_a 5 - \frac{1}{2}\log_a 9$

38. $3\log_2 x + 2\log_2 y - 4\log_2 z$

Solve each equation. [11.4]

39. $\log_2 x + \log_2 4 = 3$

40. $\log_2 x - \log_2 3 = 1$

41. $\log_3 x + \log_3(x - 2) = 1$

42. $\log_4(x + 1) - \log_4(x - 2) = 1$

43. $\log_6(x - 1) + \log_6 x = 1$

44. $\log_4(x - 3) + \log_4 x = 1$

Evaluate each expression. [11.5]

45. $\log 346$ **46.** $\log 0.713$

Find x. [11.5]

47. $\log x = 3.9652$ **48.** $\log x = -1.6003$

Simplify. [11.5]

49. $\ln e$ **50.** $\ln 1$

51. $\ln e^2$ **52.** $\ln e^{-4}$

Use the formula $pH = -\log[H^+]$ to find the pH of a solution with the given hydrogen ion concentration. [11.5]

53. $[H^+] = 7.9 \times 10^{-3}$

54. $[H^+] = 8.1 \times 10^{-6}$

Find $[H^+]$ for a solution with the given pH. [11.5]

55. $pH = 2.7$ **56.** $pH = 7.5$

Solve each equation. [11.6]

57. $4^x = 8$ **58.** $4^{3x+2} = 5$

Use the change-of-base property and a calculator to evaluate each expression. Round your answers to the nearest hundredth. [11.6]

59. $\log_{16} 8$ **60.** $\log_{12} 421$

Use the formula $A = P\left(1 + \dfrac{r}{n}\right)^{nt}$ to solve each of the following problems. [11.6]

61. Investing How long does it take $5,000 to double if it is deposited in an account that pays 16% annual interest compounded once a year?

62. Investing How long does it take $10,000 to triple if it is deposited in an account that pays 12% annual interest compounded 6 times a year?

If the current price of an item is P_0 dollars and the annual inflation rate is r, then in t years the price of the item will be $P(t) = P_0(1 + r)^t$ dollars. [11.1]

63. Inflation If the price of a new home is currently $100,000, what will it sell for in 8 years if the annual inflation rate is 4%?

64. Inflation Suppose tuition at a college in Ohio is currently $1,980 per year. If the inflation rate is a constant 3% per year, what will tuition at that school cost in 10 years?

CHAPTER 11 PROJECTS

EXPONENTIAL AND LOGARITHMIC FUNCTIONS

GROUP PROJECT

TWO DEPRECIATION MODELS

Number of People: 3

Time Needed: 20 minutes

Equipment: Paper, pencils, and graphing calculator

Background: Recently, one of the consumer magazines contained an article on leasing a computer. The original price of the computer was $2,500. The term of the lease was 24 months. At the end of the lease, the computer could be purchased for its residual value, $188. This is enough information to find an equation that will give us the value of the computer t months after it has been purchased.

Procedure: We will find models for two types of depreciation: linear depreciation and exponential depreciation. Here are the general equations:

Linear Depreciation	*Exponential Depreciation*
$V = mt + b$	$V = V_0 e^{-kt}$

1. Let t represent time in months and V represent the value of the computer at time t; then find two ordered pairs (t, V) that corresponds to the initial price of the computer, and another ordered pair that corresponds to the residual value of $188 when the computer is 2 years old.

2. Use the two ordered pairs to find m and b in the linear depreciation model; then write the equation that gives us linear depreciation.

NEW DELL DIMENSION XPS R400
400MHz PENTIUM II PROCESSOR

- 96MB 100MHz SDRAM
- 11.5GB Ultra ATA Hard Drive (9ms)
- 1,200HS 19″ (17.9″ vis, .26dp) Monitor
- Diamond Permedia 2 8MB 3D AGP Video Card
- 40X Max™ Variable CD-ROM Drive
- Crystal 3D Wavetable Sound
- Altec Lansing ACS-90 Speakers
- 56K Capable™ U.S. Roboticx × 2 Data/Fax WinModem
- ★ *Upgrade to 128MB 100MHz SDRAM, add $2.65 per Month.*

$99 Mo.◊ Business Lease
36 Mos.
or buy today for $2,579
Order Code #590710

3. Use the two ordered pairs to find V_0 and k in the exponential depreciation model; then write the equation that gives us exponential depreciation.

4. Graph each of the equations on your graphing calculator; then sketch the graphs on the following templates.

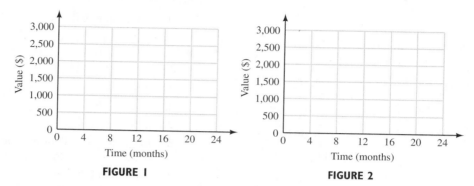

FIGURE 1   **FIGURE 2**

5. Find the value of the computer after 1 year, using both models.

6. Which of the two models do you think best describes the depreciation of a computer?

RESEARCH PROJECT

DRAG RACING

The movie *Heart Like a Wheel* is based on the racing career of drag racer Shirley Muldowney. The movie includes a number of races. Choose four races as the basis for your report. For each race you choose, give a written description of the events leading to the race, along with the details of the race itself. Then draw a graph similar to the one for Problem 42 in Section 11.1. (For each graph, label the horizontal axis from 0 to 12 seconds, and the vertical axis from 0 to 200 miles per hour.) Follow each graph with a description of any significant events that happen during the race (such as a motor malfunction or a crash) and their correlation to the graph.

(The Kobal Collection/20th Century Fox)

CHAPTER 11 TEST

Graph each exponential function. [11.1]

1. $f(x) = 2^x$ **2.** $g(x) = 3^{-x}$

Sketch the graph of each function and its inverse. Find $f^{-1}(x)$ for Problem 3. [11.2]

3. $f(x) = 2x - 3$ **4.** $f(x) = x^2 - 4$

Solve for x. [11.3]

5. $\log_4 x = 3$ **6.** $\log_x 5 = 2$

Graph each of the following. [11.3]

7. $y = \log_2 x$ **8.** $y = \log_{1/2} x$

Evaluate each of the following. [11.3, 11.4, 11.5]

9. $\log_8 4$ **10.** $\log_7 21$
11. $\log 23{,}400$ **12.** $\log 0.0123$
13. $\ln 46.2$ **14.** $\ln 0.0462$

Use the properties of logarithms to expand each expression. [11.4]

15. $\log_2 \dfrac{8x^2}{y}$ **16.** $\log \dfrac{\sqrt{x}}{(y^4)\sqrt[5]{z}}$

Write each expression as a single logarithm. [11.4]

17. $2 \log_3 x - \dfrac{1}{2} \log_3 y$

18. $\dfrac{1}{3} \log x - \log y - 2 \log z$

Use a calculator to find x. [11.5]

19. $\log x = 4.8476$ **20.** $\log x = -2.6478$

Solve for x. [11.4, 11.6]

21. $5 = 3^x$ **22.** $4^{2x-1} = 8$
23. $\log_5 x - \log_5 3 = 1$
24. $\log_2 x + \log_2(x - 7) = 3$
25. **pH** Find the pH of a solution in which $[H^+] = 6.6 \times 10^{-7}$. [11.5]

26. **Compound Interest** If $400 is deposited in an account that earns 10% annual interest compounded twice a year, how much money will be in the account after 5 years? [11.1]

27. **Compound Interest** How long will it take $600 to become $1,800 if the $600 is deposited in an account that earns 8% annual interest compounded 4 times a year? [11.6]

28. **Depreciation** If a car depreciates in value 20% per year for the first 5 years after it is purchased for P_0 dollars, then its value after t years will be $V(t) = P_0(1 - r)^t$ for $0 \le t \le 5$. To the nearest dollar, find the value of a car 4 years after it is purchased for $18,000. [11.1]

Simplify.

1. $-8 + 2[5 - 3(-2 - 3)]$

2. $6(2x - 3) + 4(3x - 2)$

3. $\left(\dfrac{3}{5}\right)^{-2} - \left(\dfrac{3}{13}\right)^{-2}$ **4.** $\dfrac{3}{4} - \dfrac{1}{8} + \dfrac{3}{2}$

5. $\sqrt[3]{27x^4y^3}$

6. $3\sqrt{48} - 3\sqrt{75} + 2\sqrt{27}$

7. $[(6 + 2i) - (3 - 4i)] - (5 - i)$

8. $\log_5[\log_2(\log_3 9)]$ **9.** $1 + \dfrac{x}{1 + \dfrac{1}{x}}$

10. Reduce to lowest terms: $\dfrac{452}{791}$

Multiply.

11. $\left(3t^2 + \dfrac{1}{4}\right)\left(4t^2 - \dfrac{1}{3}\right)$

12. $(\sqrt{6} + 3\sqrt{2})(2\sqrt{6} + \sqrt{2})$

13. Divide: $\dfrac{9x^2 + 9x - 18}{3x - 4}$

14. Rationalize the denominator: $\dfrac{5\sqrt{6}}{2\sqrt{6} + 7}$

15. Subtract: $\dfrac{7}{4x^2 - x - 3} - \dfrac{1}{4x^2 - 7x + 3}$

Solve.

16. $6 - 3(2x - 4) = 2$

17. $\dfrac{2}{3}(6x - 5) + \dfrac{1}{3} = 13$

18. $28x^2 = 3x + 1$ **19.** $|3x - 5| + 6 = 2$

20. $\dfrac{2}{x + 1} = \dfrac{4}{5}$

21. $\dfrac{1}{x + 3} + \dfrac{1}{x - 2} = 1$

22. $\dfrac{1}{x^2 + 3x - 4} + \dfrac{3}{x^2 - 1} = \dfrac{-1}{x^2 + 5x + 4}$

23. $2x - 1 = x^2$

24. $3(4y - 1)^2 + (4y - 1) - 10 = 0$

25. $\sqrt{7x - 4} = -2$

26. $\sqrt{y - 3} - \sqrt{y} = -1$ **27.** $x - 3\sqrt{x} + 2 = 0$

28. $8x^2 + 10x = 3$ **29.** $\log_3 x = 3$

30. $\log_x 0.1 = -1$

31. $\log_3(x - 3) - \log_3(x + 2) = 1$

Solve each system.

32. $-9x + 3y = 1$
$5x - 2y = -2$

33. $4x + 7y = -3$ **34.** $x + y = 4$
$x = -2y - 2$ $x + z = 1$
 $y - 2z = 5$

Solve each inequality and graph the solution.

35. $3y - 6 \geq 3$ or $3y - 6 \leq -3$

36. $|4x + 3| - 6 > 5$

37. $x^3 + 2x^2 - 9x - 18 < 0$

38. Graph the line: $5x + 4y = 20$

39. Write the inequality of the line in slope-intercept form if the slope is $-\dfrac{5}{3}$ and $(3, -3)$ is a point on the line.

40. Write in symbols: The difference of $9a$ and $4b$ is less than their sum.

41. Write 0.0000972 in scientific notation.

42. Factor completely: $50a^4 + 10a^2 - 4$

43. If $f(x) = \dfrac{1}{2}x + 3$, find $f^{-1}(x)$.

44. If $f(x) = 4 \cdot 2^{-x}$, find $f(-2)$.

45. Specify the domain and range for the relation $\{(2, -3), (2, -1), (-3, 3)\}$. Is this relation also a function?

46. State the domain: $y = \sqrt{3 - x}$

47. Give the next number of the sequence, and state whether arithmetic or geometric: $2, -6, 18, \ldots$

48. **Direct Variation** w varies directly with the square root of c. If w is 8 when c is 16, find w when c is 9.

49. **Mixture** How many gallons of 25% alcohol solution and 50% alcohol solution should be mixed to get 20 gallons of 42.5% alcohol solution?

50. **Compound Interest** A $10,000 Treasury bill earns 6% compounded twice a year. How much is the Treasury bill worth after 4 years?

Sequences and Series

(FPG International/Telegraph Colour Library)

INTRODUCTION

Suppose you run up a balance of $1,000 on a credit card that charges 1.65% interest each month (i.e., an annual rate of 19.8%). If you stop using the card and make the minimum payment of $20 each month, how long will it take you to pay off the balance on the card? The answer can be found by using the formula

$$U_n = (1.0165)U_{n-1} - 20$$

where U_n stands for the current unpaid balance on the card, and U_{n-1} is the previous month's balance. Table 1 and Figure 1 were created from this formula and a graphing calculator. As you can see from the table, the balance on the credit card decreases very little each month.

TABLE 1 Monthly Credit Card Balances

Previous Balance $U_{(n-1)}$	Monthly Interest Rate	Payment Number n	Monthly Payment	New Balance $U_{(n)}$
$1,000.00	1.65%	1	$20	$996.50
$996.50	1.65%	2	$20	$992.94
$992.94	1.65%	3	$20	$989.32
$989.32	1.65%	4	$20	$985.64
$985.64	1.65%	5	$20	$981.90
⋮	⋮	⋮	⋮	⋮

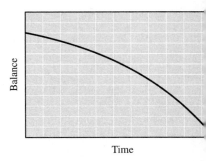

FIGURE 1

In the group project at the end of the chapter, you will use a graphing calculator to continue this table and in so doing find out just how many months it will take to pay off this credit card balance.

Many of the sequences in this chapter will be familiar to you on an intuitive level because you have worked with them for some time now. Here are some of those sequences:

The sequence of odd numbers

$$1, 3, 5, 7, \ldots$$

The sequence of even numbers

$$2, 4, 6, 8, \ldots$$

The sequence of squares

$$1^2, 2^2, 3^2, 4^2, \ldots = 1, 4, 9, 16, \ldots$$

The numbers in each of these sequences can be found from the formulas that define functions. For example, the sequence of even numbers can be found from the function

$$f(x) = 2x$$

by finding $f(1), f(2), f(3), f(4)$, and so forth. This gives us justification for the formal definition of a sequence.

> **DEFINITION** A **sequence** is a function whose domain is the set of positive integers $\{1, 2, 3, 4, \ldots\}$.

As you can see, sequences are simply functions with a specific domain. If we want to form a sequence from the function $f(x) = 3x + 5$, we simply find $f(1)$, $f(2), f(3)$, and so on. Doing so gives us the sequence

$$8, 11, 14, 17, \ldots$$

because $f(1) = 3(1) + 5 = 8$, $f(2) = 3(2) + 5 = 11$, $f(3) = 3(3) + 5 = 14$, and $f(4) = 3(4) + 5 = 17$.

Notation Since the domain for a sequence is always the set $\{1, 2, 3, \ldots\}$, we can simplify the notation we use to represent the terms of a sequence. Using the letter a instead of f, and subscripts instead of numbers enclosed by parentheses, we can represent the sequence from the previous discussion as follows:

$$a_n = 3n + 5$$

Instead of $f(1)$ we write a_1 for the *first term* of the sequence.

Instead of $f(2)$ we write a_2 for the *second term* of the sequence.

Instead of $f(3)$ we write a_3 for the *third term* of the sequence.

Instead of $f(4)$ we write a_4 for the *fourth term* of the sequence.

Instead of $f(n)$ we write a_n for the *nth term* of the sequence.

The *n*th term is also called the **general term** of the sequence. The general term is used to define the other terms of the sequence. That is, if we are given the formula for the general term a_n, we can find any other term in the sequence. The following examples illustrate.

EXAMPLE 1 Find the first four terms of the sequence whose general term is given by $a_n = 2n - 1$.

Solution The subscript notation a_n works the same way function notation works. To find the first, second, third, and fourth terms of this sequence, we simply substitute 1, 2, 3, and 4 for *n* in the formula $2n - 1$:

If   the general term is $a_n = 2n - 1$

then   the first term is $a_1 = 2(1) - 1 = 1$

the second term is $a_2 = 2(2) - 1 = 3$

the third term is $a_3 = 2(3) - 1 = 5$

the fourth term is $a_4 = 2(4) - 1 = 7$

The first four terms of this sequence are the odd numbers 1, 3, 5, and 7. The whole sequence can be written as

$$1, 3, 5, \ldots, 2n - 1, \ldots$$

Since each term in this sequence is larger than the preceding term, we say the sequence is an **increasing sequence.**

EXAMPLE 2 Write the first four terms of the sequence defined by

$$a_n = \frac{1}{n + 1}$$

Solution Replacing *n* with 1, 2, 3, and 4, we have, respectively, the first four terms:

$$\text{First term} = a_1 = \frac{1}{1 + 1} = \frac{1}{2}$$

$$\text{Second term} = a_2 = \frac{1}{2+1} = \frac{1}{3}$$

$$\text{Third term} = a_3 = \frac{1}{3+1} = \frac{1}{4}$$

$$\text{Fourth term} = a_4 = \frac{1}{4+1} = \frac{1}{5}$$

The sequence defined by

$$a_n = \frac{1}{n+1}$$

can be written as

$$\frac{1}{2}, \frac{1}{3}, \frac{1}{4}, \cdots, \frac{1}{n+1}, \cdots$$

Since each term in the sequence is smaller than the term preceding it, the sequence is said to be a **decreasing sequence**.

E X A M P L E   3   Find the fifth and sixth terms of the sequence whose general term is given by $a_n = \dfrac{(-1)^n}{n^2}$.

Solution   For the fifth term, we replace n with 5. For the sixth term, we replace n with 6:

$$\text{Fifth term} = a_5 = \frac{(-1)^5}{5^2} = \frac{-1}{25}$$

$$\text{Sixth term} = a_6 = \frac{(-1)^6}{6^2} = \frac{1}{36}$$

The sequence in Example 3 can be written as

$$-1, \frac{1}{4}, -\frac{1}{9}, \frac{1}{16}, \cdots, \frac{(-1)^n}{n^2}, \cdots$$

Since the terms alternate in sign—if one term is positive, then the next term is negative—we call this an **alternating sequence**. The first three examples all illustrate how we work with a sequence in which we are given a formula for the general term. The next example gives us another way to write the general term. Before we work our next example, let's look at sequences on the graphing calculator.

Using TECHNOLOGY

FINDING SEQUENCES ON A GRAPHING CALCULATOR

Method 1: Using a Table

We can use the table function on a graphing calculator to view the terms of a sequence. To view the terms of the sequence $a_n = 3n + 5$, we set $Y_1 = 3X + 5$. Then we use the table setup feature on the calculator to set the table minimum to 1, and the table increment to 1 also. Here is the setup and result for a TI-83.

Table Setup	Y Variables Setup	Resulting Table	
Table minimum = 1	$Y_1 = 3X + 5$		
Table increment = 1		**X**	**Y**
Independent variable: Auto		1	8
Dependent variable: Auto		2	11
		3	14
		4	17
		5	20

To find any particular term of a sequence, we change the dependent variable setting to Ask, and then input the number of the term of the sequence we want to find. For example, if we want term a_{100}, we input 100 for the independent variable, and the table gives us the value of 305 for that term.

Method 2: Using the Built-in seq(Function

Using this method, first find the seq(function. On a TI-83 it is found in the LIST OPS menu. To find terms a_1 through a_7 for $a_n = 3n + 5$, we first bring up the seq(function on our calculator, then we input the following four items, in order, separated by commas: 3X+5, X, 1, 7. Then we close the parentheses. Our screen will look like this:

$$\text{seq}(3X+5, X, 1, 7)$$

Pressing $\boxed{\text{ENTER}}$ displays the first five terms of the sequence. Pressing the right arrow key repeatedly brings the remaining members of the sequence into view.

Method 3: Using the Built-in Seq Mode

Press the $\boxed{\text{MODE}}$ key on your TI-83 and then select Seq (it's next to Func Par and Pol). Go to the Y variables list and set nMin = 1 and $u(n) = 3n+5$. Then go to the $\boxed{\text{TBLSET}}$ key to set up your table like the one shown in Method 1. Pressing $\boxed{\text{TABLE}}$ will display the sequence you have defined.

RECURSION FORMULAS

Let's go back to one of the first sequences we looked at in this section:

$$8, 11, 14, 17, \ldots$$

Each term in the sequence can be found by simply substituting positive integers for n in the formula $a_n = 3n + 5$. Another way to look at this sequence, however, is to notice that each term can be found by adding 3 to the preceding term; so, we could give all the terms of this sequence by simply saying

Start with 8, and then add 3 to each term to get the next term.

The same idea, expressed in symbols, looks like this:

$$a_1 = 8 \qquad \text{and} \qquad a_n = a_{n-1} + 3 \qquad \text{for } n > 1$$

This formula is called a **recursion formula** because each term is written *recursively* in terms of the term or terms that precede it.

EXAMPLE 4   Write the first four terms of the sequence given recursively by

$$a_1 = 4 \qquad \text{and} \qquad a_n = 5a_{n-1} \qquad \text{for } n > 1$$

Solution   The formula tells us to start the sequence with the number 4, and then multiply each term by 5 to get the next term. Therefore,

$$a_1 = 4$$
$$a_2 = 5a_1 = 5(4) = 20$$
$$a_3 = 5a_2 = 5(20) = 100$$
$$a_4 = 5a_3 = 5(100) = 500$$

The sequence is 4, 20, 100, 500, ...

Using TECHNOLOGY

RECURSION FORMULAS ON A GRAPHING CALCULATOR

We can use a TI-83 graphing calculator to view the sequence defined recursively as

$$a_1 = 8, a_n = a_{n-1} + 3$$

First, put your TI-83 calculator in sequence mode by pressing the $\boxed{\text{MODE}}$ key, and then selecting Seq (it's next to Func Par and Pol). Go to the Y variables list and set $n\text{Min} = 1$ and $u(n) = u(n-1)+3$. (The u is above the 7, and the n is on the $\boxed{\text{X, T, }\theta\text{, }n}$ and is automatically displayed if that key is pressed when the calculator is in the Seq mode. Pressing $\boxed{\text{TABLE}}$ will display the sequence you have defined.

FINDING THE GENERAL TERM

In the first four examples, we found some terms of a sequence after being given the general term. In the next two examples, we will do the reverse. That is, given some terms of a sequence, we will find the formula for the general term.

EXAMPLE 5 Find a formula for the nth term of the sequence 2, 8, 18, 32, ...

Solution Solving a problem like this involves some guessing. Looking over the first four terms, we see each is twice a perfect square:

$$2 = 2(1)$$
$$8 = 2(4)$$
$$18 = 2(9)$$
$$32 = 2(16)$$

If we write each square with an exponent of 2, the formula for the nth term becomes obvious:

$$a_1 = 2 \ = 2(1)^2$$
$$a_2 = 8 \ = 2(2)^2$$
$$a_3 = 18 = 2(3)^2$$
$$a_4 = 32 = 2(4)^2$$
$$\vdots$$
$$a_n = \qquad 2(n)^2 = 2n^2$$

The general term of the sequence 2, 8, 18, 32, . . . is $a_n = 2n^2$.

EXAMPLE 6 Find the general term for the sequence $2, \frac{3}{8}, \frac{4}{27}, \frac{5}{64}, \ \cdot \ \cdot \ \cdot$

Solution The first term can be written as $\frac{2}{1}$. The denominators are all perfect cubes. The numerators are all 1 more than the base of the cubes in the denominators:

$$a_1 = \frac{2}{1} = \frac{1+1}{1^3}$$

$$a_2 = \frac{3}{8} = \frac{2+1}{2^3}$$

$$a_3 = \frac{4}{27} = \frac{3+1}{3^3}$$

$$a_4 = \frac{5}{64} = \frac{4+1}{4^3}$$

Observing this pattern, we recognize the general term to be

$$a_n = \frac{n + 1}{n^3}$$

Note: Finding the nth term of a sequence from the first few terms is not always automatic. That is, it sometimes takes awhile to recognize the pattern. Don't be afraid to guess at the formula for the general term. Many times an incorrect guess leads to the correct formula.

 # Getting Ready for Class

After reading through the preceding section, respond in your own words and in complete sentences.

A. How are subscripts used to denote the terms of a sequence?

B. What is the relationship between the subscripts used to denote the terms of a sequence and function notation?

C. What is a decreasing sequence?

D. What is meant by a recursion formula for a sequence?

PROBLEM SET 12.1

Write the first five terms of the sequences with the following general terms.

1. $a_n = 3n + 1$ **2.** $a_n = 2n + 3$

3. $a_n = 4n - 1$ **4.** $a_n = n + 4$

5. $a_n = n$ **6.** $a_n = -n$

7. $a_n = n^2 + 3$ **8.** $a_n = n^3 + 1$

9. $a_n = \dfrac{n}{n + 3}$ **10.** $a_n = \dfrac{n}{n + 2}$

11. $a_n = \dfrac{n + 1}{n + 2}$ **12.** $a_n = \dfrac{n + 3}{n + 4}$

13. $a_n = \dfrac{1}{n^2}$ **14.** $a_n = \dfrac{1}{n^3}$

15. $a_n = 2^n$ **16.** $a_n = 3^n$

17. $a_n = 3^{-n}$ **18.** $a_n = 2^{-n}$

19. $a_n = 1 + \dfrac{1}{n}$ **20.** $a_n = 1 - \dfrac{1}{n}$

21. $a_n = n - \dfrac{1}{n}$ **22.** $a_n = n + \dfrac{1}{n}$

23. $a_n = (-2)^n$ **24.** $a_n = (-3)^n$

Write the first five terms of the sequences defined by the following recursion formulas.

25. $a_1 = 3$ $a_n = -3a_{n-1}$ $n > 1$

26. $a_1 = -3$ $a_n = 3a_{n-1}$ $n > 1$

27. $a_1 = 3$ $a_n = a_{n-1} - 3$ $n > 1$

28. $a_1 = -3$ $a_n = a_{n-1} - 3$ $n > 1$

29. $a_1 = 1$ $a_n = 2a_{n-1} + 3$ $n > 1$

30. $a_1 = 1$ $a_n = 3a_{n-1} + 2$ $n > 1$

31. $a_1 = 1$ $a_n = a_{n-1} + n$ $n > 1$

32. $a_1 = 2$ $a_n = a_{n-1} - n$ $n > 1$

Determine the general term for each of the following sequences.

33. 2, 3, 4, 5, . . .   **34.** 3, 6, 9, 12, . . .

35. 4, 8, 12, 16, 20, . . .   **36.** 3, 4, 5, 6, . . .

37. 7, 10, 13, 16, . . .   **38.** 4, 9, 14, 19, . . .

39. $1, 4, 9, 16, \ldots$

40. $1, 8, 27, 64, \ldots$

41. $3, 12, 27, 48, \ldots$

42. $2, 16, 54, 128, \ldots$

43. $4, 8, 16, 32, \ldots$

44. $3, 9, 27, 81, \ldots$

45. $-2, 4, -8, 16, \ldots$

46. $-3, 9, -27, 81, \ldots$

47. $\frac{1}{4}, \frac{1}{8}, \frac{1}{16}, \frac{1}{32}, \ldots$

48. $\frac{1}{3}, \frac{1}{9}, \frac{1}{27}, \frac{1}{81}, \ldots$

49. $\frac{1}{4}, \frac{2}{9}, \frac{3}{16}, \frac{4}{25}, \ldots$

50. $\frac{1}{4}, \frac{2}{10}, \frac{3}{28}, \frac{4}{82}, \ldots$

Applying the Concepts

51. Salary Increase   The entry level salary for a teacher is $28,000 with 4% increases after every year of service.
  (a) Write a sequence for this teacher's salary for the first 5 years.
  (b) Find the general term of the sequence in part (a).

52. Holiday Account   To save money for holiday presents, a person deposits $5 in a savings account on January 1, and then deposits an additional $5 every week thereafter until Christmas.
  (a) Write a sequence for the money in that savings account for the first 10 weeks of the year.
  (b) Write the general term of the sequence in part (a).
  (c) If there are 50 weeks from January 1 to Christmas, how much money will be available for spending on Christmas presents?

53. Saving for College   To save money for his son's college education, a person deposits $100 in a savings account on the first day of each month after his son was born.
  (a) Write a sequence to show the money in this college education account at the end of the first, second, third, fourth, and fifth years.
  (b) Write the general term for this sequence in terms of m, if m represents the number of months.
  (c) Write the general term for this sequence in terms of y, if y represents the number of years.

54. Saving for College   To save money for his daughter's college education, a person deposits $100 in a savings account on the first day of each month after his daughter is born. After every year, this monthly deposit is increased by $10—that is, in the second year $110 is deposited each month, in the third year $120 is deposited each month, and so on.

  (a) Write a sequence to show the money in this college education account at the end of the first, second, third, fourth, and fifth years.
  (b) Refer to the preceding exercise and determine what percent more money has been saved for the daughter than for the son.

Review Problems

The problems that follow review material we covered in Section 11.3.

Find x in each of the following.

55. $\log_9 x = \frac{3}{2}$

56. $\log_x \frac{1}{4} = -2$

Simplify each expression.

57. $\log_2 32$

58. $\log_{10} 10,000$

59. $\log_3[\log_2 8]$

60. $\log_5[\log_6 6]$

Extending the Concepts

61. As n increases, the terms in the sequence

$$a_n = \left(1 + \frac{1}{n}\right)^n$$

get closer and closer to the number e (that's the same e we used in defining natural logarithms). It takes some fairly large values of n, however, before we can see this happening. Use a calculator to find a_{100}, $a_{1,000}$, $a_{10,000}$, and $a_{100,000}$, and compare them to the decimal approximation we gave for the number e.

62. The sequence

$$a_n = \left(1 + \frac{1}{n}\right)^{-n}$$

gets close to the number $1/e$ as n becomes large. Use a calculator to find approximations for a_{100} and $a_{1,000}$, and then compare them to $\dfrac{1}{2.7183}$.

63. Write the first ten terms of the sequence defined by the recursion formula

$$a_1 = 1, a_2 = 1, a_n = a_{n-1} + a_{n-2} \qquad n > 2$$

64. Write the first ten terms of the sequence defined by the recursion formula

$$a_1 = 2, a_2 = 2, a_n = a_{n-1} + a_{n-2} \qquad n > 2$$

65. Simplify each complex fraction in the following sequence, and then compare this sequence with the sequence you wrote in Problem 63.

$$1 + \dfrac{1}{1+1}, 1 + \dfrac{1}{1 + \dfrac{1}{1+1}}, 1 + \dfrac{1}{1 + \dfrac{1}{1 + \dfrac{1}{1+1}}}, \ldots$$

66. Write the first five terms of the sequence given by the following recursion formula.

$$a_1 = \dfrac{3}{2}, a_n = 1 + \dfrac{1}{a_{n-1}} \qquad n > 1$$

12.2   Series

There is an interesting relationship between the sequence of odd numbers and the sequence of squares that is found by adding the terms in the sequence of odd numbers.

$$\begin{aligned}
1 &= 1 \\
1 + 3 &= 4 \\
1 + 3 + 5 &= 9 \\
1 + 3 + 5 + 7 &= 16
\end{aligned}$$

When we add the terms of a sequence the result is called a series.

DEFINITION   The sum of a number of terms in a sequence is called a **series.**

A sequence can be finite or infinite depending on whether or not the sequence ends at the nth term. For example,

$$1, 3, 5, 7, 9$$

is a finite sequence, but

$$1, 3, 5, \ldots$$

is an infinite sequence. Associated with each of the preceding sequences is a series found by adding the terms of the sequence:

$$1 + 3 + 5 + 7 + 9 \qquad \text{Finite series}$$

$$1 + 3 + 5 + \ldots \qquad \text{Infinite series}$$

In this section we will consider only finite series. We can introduce a new kind of notation here that is a compact way of indicating a finite series. The notation is called **summation notation,** or **sigma notation** since it is written using the Greek letter sigma. The expression

$$\sum_{i=1}^{4} (8i - 10)$$

is an example of an expression that uses summation notation. The summation notation in this expression is used to indicate the sum of all the expressions $8i - 10$ from $i = 1$ up to and including $i = 4$. That is,

$$\sum_{i=1}^{4} (8i - 10) = (8 \cdot 1 - 10) + (8 \cdot 2 - 10) + (8 \cdot 3 - 10) + (8 \cdot 4 - 10)$$

$$= -2 + 6 + 14 + 22$$

$$= 40$$

The letter i as used here is called the **index of summation,** or just **index** for short.

Here are some examples illustrating the use of summation notation.

EXAMPLE 1 Expand and simplify $\sum_{i=1}^{5} (i^2 - 1)$.

Solution We replace i in the expression $i^2 - 1$ with all consecutive integers from 1 up to 5, including 1 and 5:

$$\sum_{i=1}^{5} (i^2 - 1) = (1^2 - 1) + (2^2 - 1) + (3^2 - 1) + (4^2 - 1) + (5^2 - 1)$$

$$= 0 + 3 + 8 + 15 + 24$$

$$= 50$$

EXAMPLE 2 Expand and simplify $\sum_{i=3}^{6} (-2)^i$.

Solution We replace i in the expression $(-2)^i$ with the consecutive integers beginning at 3 and ending at 6:

$$\sum_{i=3}^{6} (-2)^i = (-2)^3 + (-2)^4 + (-2)^5 + (-2)^6$$

$$= -8 + 16 + (-32) + 64$$

$$= 40$$

Using TECHNOLOGY

SUMMING SERIES ON A GRAPHING CALCULATOR

A TI-83 graphing calculator has a built-in sum(function that, when used with the seq(function, allows us to add the terms of a series. Let's repeat Example 1 using our graphing calculator. First, we go to LIST and select MATH. The fifth option in that list is sum(, which we select. Then we go to LIST again and select OPS. From that list we select seq(. Next we enter X^2−1, X, 1, 5, and then we close both sets of parentheses. Our screen shows the following:

$$\text{sum(seq(X\^2−1, X, 1, 5))} \qquad \text{which will give us } \sum_{i=1}^{5} (i^2 - 1)$$

When we press ENTER the calculator displays 50, which is the same result we obtained in Example 1.

E X A M P L E   3   Expand $\displaystyle\sum_{i=2}^{5} (x^i - 3)$.

Solution   We must be careful not to confuse the letter x with i. The index i is the quantity we replace by the consecutive integers from 2 to 5, not x:

$$\sum_{i=2}^{5} (x^i - 3) = (x^2 - 3) + (x^3 - 3) + (x^4 - 3) + (x^5 - 3)$$

In the first three examples, we were given an expression with summation notation and asked to expand it. The next examples in this section illustrate how we can write an expression in expanded form as an expression involving summation notation.

E X A M P L E   4   Write with summation notation $1 + 3 + 5 + 7 + 9$.

Solution   A formula that gives us the terms of this sum is

$$a_i = 2i - 1$$

where i ranges from 1 up to and including 5. Notice we are using the subscript i in exactly the same way we used the subscript n in the previous section—to indicate the general term. Writing the sum

$$1 + 3 + 5 + 7 + 9$$

with summation notation looks like this:

$$\sum_{i=1}^{5} (2i - 1)$$

EXAMPLE 5   Write with summation notation $3 + 12 + 27 + 48$.

Solution   We need a formula, in terms of i, that will give each term in the sum. Writing the sum as

$$3 \cdot 1^2 + 3 \cdot 2^2 + 3 \cdot 3^2 + 3 \cdot 4^2$$

we see the formula

$$a_i = 3 \cdot i^2$$

where i ranges from 1 up to and including 4. Using this formula and summation notation, we can represent the sum

$$3 + 12 + 27 + 48$$

as

$$\sum_{i=1}^{4} 3i^2$$

EXAMPLE 6   Write with summation notation

$$\frac{x + 3}{x^3} + \frac{x + 4}{x^4} + \frac{x + 5}{x^5} + \frac{x + 6}{x^6}$$

Solution   A formula that gives each of these terms is

$$a_i = \frac{x + i}{x^i}$$

where i assumes all integer values between 3 and 6, including 3 and 6. The sum can be written as

$$\sum_{i=3}^{6} \frac{x + i}{x^i}$$

 # Getting Ready for Class

After reading through the preceding section, respond in your own words and in complete sentences.

A. What is the difference between a sequence and a series?

B. Explain the summation notation $\sum_{i=1}^{4}$ in the series $\sum_{i=1}^{4} (2i + 1)$.

C. When will a finite series result in a numerical value versus an algebraic expression?

D. Determine for what values of n the series $\sum\limits_{i=1}^{n} (-1)^i$ will be equal to 0. Explain your answer.

PROBLEM SET 12.2

Expand and simplify each of the following.

1. $\sum\limits_{i=1}^{4} (2i + 4)$ **2.** $\sum\limits_{i=1}^{5} (3i - 1)$

3. $\sum\limits_{i=1}^{3} (2i - 1)$ **4.** $\sum\limits_{i=1}^{4} (2i - 1)$

5. $\sum\limits_{i=2}^{3} (i^2 - 1)$ **6.** $\sum\limits_{i=3}^{6} (i^2 + 1)$

7. $\sum\limits_{i=1}^{4} \dfrac{i}{1 + i}$ **8.** $\sum\limits_{i=1}^{4} \dfrac{i^2}{1 + i}$

9. $\sum\limits_{i=1}^{3} \dfrac{i^2}{2i - 1}$ **10.** $\sum\limits_{i=3}^{5} (i^3 + 4)$

11. $\sum\limits_{i=1}^{4} (-3)^i$ **12.** $\sum\limits_{i=1}^{4} \left(-\dfrac{1}{3}\right)^i$

13. $\sum\limits_{i=3}^{6} (-2)^i$ **14.** $\sum\limits_{i=4}^{6} \left(-\dfrac{1}{2}\right)^i$

Expand the following.

15. $\sum\limits_{i=1}^{5} (x + i)$ **16.** $\sum\limits_{i=3}^{6} (x - i)$

17. $\sum\limits_{i=2}^{7} (x + 1)^i$ **18.** $\sum\limits_{i=1}^{4} (x + 3)^i$

19. $\sum\limits_{i=1}^{5} \dfrac{x + i}{x - 1}$ **20.** $\sum\limits_{i=1}^{6} \dfrac{x - 3i}{x + 3i}$

21. $\sum\limits_{i=3}^{8} (x + i)^i$ **22.** $\sum\limits_{i=4}^{7} (x - 2i)^i$

23. $\sum\limits_{i=1}^{5} (x + i)^{i+1}$ **24.** $\sum\limits_{i=2}^{6} (x + i)^{i-1}$

Write each of the following sums with summation notation.

25. $2 + 4 + 8 + 16$

26. $3 + 5 + 7 + 9 + 11$

27. $4 + 8 + 16 + 32 + 64$

28. $1 + 3 + 5$

29. $5 + 10 + 17 + 26 + 37$

30. $3 + 8 + 15 + 24$

31. $\dfrac{3}{4} + \dfrac{4}{5} + \dfrac{5}{6} + \dfrac{6}{7} + \dfrac{7}{8}$ **32.** $\dfrac{1}{2} + \dfrac{2}{3} + \dfrac{3}{4} + \dfrac{4}{5}$

33. $\dfrac{1}{3} + \dfrac{2}{5} + \dfrac{3}{7} + \dfrac{4}{9}$ **34.** $\dfrac{3}{1} + \dfrac{5}{3} + \dfrac{7}{5} + \dfrac{9}{7}$

35. $(x - 3) + (x - 4) + (x - 5) + (x - 6)$

36. $x^2 + x^3 + x^4 + x^5 + x^6$

37. $\dfrac{x}{x + 3} + \dfrac{x}{x + 4} + \dfrac{x}{x + 5}$

38. $\dfrac{x - 3}{x^3} + \dfrac{x - 4}{x^4} + \dfrac{x - 5}{x^5} + \dfrac{x - 6}{x^6}$

39. $x^2(x + 2) + x^3(x + 3) + x^4(x + 4)$

40. $x(x + 2)^2 + x(x + 3)^3 + x(x + 4)^4$

41. Repeating Decimals Any repeating, nonterminating decimal may be viewed as a series. For instance, $\dfrac{2}{3} = 0.6 + 0.06 + 0.006 + 0.0006 + \cdots$. Write the following fractions as series.
(a) $\dfrac{1}{3}$
(b) $\dfrac{2}{9}$
(c) $\dfrac{3}{11}$

42. Repeating Decimals Refer to the previous exercise, and express the following repeating decimals as fractions.
(a) $0.55555 \cdots$
(b) $1.33333 \cdots$
(c) $0.29292929 \cdots$

Applying the Concepts

43. Skydiving A skydiver jumps from a plane and falls 16 feet the first second, 48 feet the second second, and 80 feet the third second. If he continues to fall in the same manner, how far will he fall the seventh second? What is the distance he falls in 7 seconds?

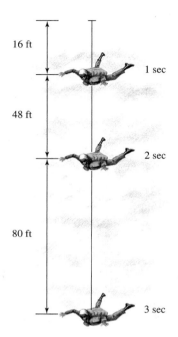

16 ft

1 sec

48 ft

2 sec

80 ft

3 sec

44. Bacterial Growth After 1 day, a colony of 50 bacteria reproduces to become 200 bacteria. After 2 days, they reproduce to become 800 bacteria. If they continue to reproduce at this rate, how many bacteria will be present after 4 days?

45. Converging Series Given the series

$$\frac{1}{1 \cdot 3} + \frac{1}{3 \cdot 5} + \frac{1}{5 \cdot 7} + \cdots$$

(a) Write the first six terms of this series.
(b) Use a graphing calculator to find the sum of the first six terms of this series.
(c) What do you think is the exact sum of this series? (You may need to compute some more terms.)

46. Converging Series Given the series

$$\frac{1}{1 \cdot 3} + \frac{1}{2 \cdot 4} + \frac{1}{3 \cdot 5} + \cdots$$

(a) Write the first six terms of this series.
(b) Use a graphing calculator to find the sum of the first six terms of this series.
(c) What do you think is the exact sum of this series? (You may need to compute some more terms.)

Review Problems

The following problems review material we covered in Section 11.4.

Use the properties of logarithms to expand each of the following expressions.

47. $\log_2 x^3 y$ **48.** $\log_7 \dfrac{x^2}{y^4}$

49. $\log_{10} \dfrac{\sqrt[3]{x}}{y^2}$ **50.** $\log_{10} \sqrt[3]{\dfrac{x}{y^2}}$

Write each expression as a single logarithm.

51. $\log_{10} x - \log_{10} y^2$ **52.** $\log_{10} x^2 + \log_{10} y^2$
53. $2 \log_3 x - 3 \log_3 y - 4 \log_3 z$
54. $\frac{1}{2} \log_6 x + \frac{1}{3} \log_6 y + \frac{1}{4} \log_6 z$

Solve each equation.

55. $\log_4 x - \log_4 5 = 2$
56. $\log_3 6 + \log_3 x = 4$
57. $\log_2 x + \log_2(x - 7) = 3$
58. $\log_5(x + 1) + \log_5(x - 3) = 1$

Extending the Concepts

59. Solve for x by first expanding:

$$\sum_{i=1}^{3} (2^i \cdot x - 4) = 20$$

In this and the following section, we will review and extend two major types of sequences, which we have worked with previously—arithmetic sequences and geometric sequences.

DEFINITION An **arithmetic sequence** is a sequence of numbers in which each term is obtained from the preceding term by adding the same amount each time. An arithmetic sequence is also called an **arithmetic progression.**

The sequence

$$2, 6, 10, 14, \ldots$$

is an example of an arithmetic sequence, since each term is obtained from the preceding term by adding 4 each time. The amount we add each time—in this case, 4—is called the **common difference,** since it can be obtained by subtracting any two consecutive terms. (The term with the larger subscript must be written first.) The common difference is denoted by d.

EXAMPLE 1 Give the common difference d for the arithmetic sequence 4, 10, 16, 22, ...

Solution Since each term can be obtained from the preceding term by adding 6, the common difference is 6. That is, $d = 6$.

EXAMPLE 2 Give the common difference for 100, 93, 86, 79, ...

Solution The common difference in this case is $d = -7$, since adding -7 to any term always produces the next consecutive term.

EXAMPLE 3 Give the common difference for $\frac{1}{2}, 1, \frac{3}{2}, 2, \ldots$.

Solution The common difference is $d = \frac{1}{2}$.

THE GENERAL TERM

The general term a_n of an arithmetic progression can always be written in terms of the first term a_1 and the common difference d. Consider the sequence from Example 1:

$$4, 10, 16, 22, \ldots$$

We can write each term in terms of the first term 4 and the common difference 6:

$$4, \qquad 4 + (1 \cdot 6), \qquad 4 + (2 \cdot 6), \qquad 4 + (3 \cdot 6), \ldots$$

$$a_1, \qquad a_2, \qquad a_3, \qquad a_4, \qquad \ldots$$

Observing the relationship between the subscript on the terms in the second line and the coefficients of the 6's in the first line, we write the general term for the sequence as

$$a_n = 4 + (n - 1)6$$

We generalize this result to include the general term of any arithmetic sequence.

ARITHMETIC SEQUENCES

The **general term** of an arithmetic progression with first term a_1 and common difference d is given by

$$a_n = a_1 + (n - 1)d$$

EXAMPLE 4   Find the general term for the sequence

$$7, 10, 13, 16, \ldots$$

Solution   The first term is $a_1 = 7$, and the common difference is $d = 3$. Substituting these numbers into the formula given earlier, we have

$$a_n = 7 + (n - 1)3$$

which we can simplify, if we choose, to

$$a_n = 7 + 3n - 3$$
$$= 3n + 4$$

EXAMPLE 5   Find the general term of the arithmetic progression whose third term a_3 is 7 and whose eighth term a_8 is 17.

Solution   According to the formula for the general term, the third term can be written as $a_3 = a_1 + 2d$, and the eighth term can be written as $a_8 = a_1 + 7d$. Since these terms are also equal to 7 and 17, respectively, we can write

$$a_3 = a_1 + 2d = 7$$
$$a_8 = a_1 + 7d = 17$$

To find a_1 and d, we simply solve the system:

$$a_1 + 2d = 7$$
$$a_1 + 7d = 17$$

We add the opposite of the top equation to the bottom equation. The result is

$$5d = 10$$

$$d = 2$$

To find a_1, we simply substitute 2 for d in either of the original equations and get

$$a_1 = 3$$

The general term for this progression is

$$a_n = 3 + (n - 1)2$$

which we can simplify to

$$a_n = 2n + 1$$

The sum of the first n terms of an arithmetic sequence is denoted by S_n. The following theorem gives the formula for finding S_n, which is sometimes called the **nth partial sum.**

THEOREM 12.1

The sum of the first n terms of an arithmetic sequence whose first term is a_1 and whose nth term is a_n is given by

$$S_n = \frac{n}{2}(a_1 + a_n)$$

Proof   We can write S_n in expanded form as

$$S_n = a_1 + [a_1 + d] + [a_1 + 2d] + \cdots + [a_1 + (n - 1)d]$$

We can arrive at this same series by starting with the last term a_n and subtracting d each time. Writing S_n this way, we have

$$S_n = a_n + [a_n - d] + [a_n - 2d] + \cdots + [a_n - (n - 1)d]$$

If we add the preceding two expressions term by term, we have

$$2S_n = (a_1 + a_n) + (a_1 + a_n) + (a_1 + a_n) + \cdots + (a_1 + a_n)$$

$$2S_n = n(a_1 + a_n)$$

$$S_n = \frac{n}{2}(a_1 + a_n)$$

E X A M P L E   6   Find the sum of the first ten terms of the arithmetic progression 2, 10, 18, 26, ...

Solution The first term is 2, and the common difference is 8. The tenth term is

$$a_{10} = 2 + 9(8)$$
$$= 2 + 72$$
$$= 74$$

Substituting $n = 10$, $a_1 = 2$, and $a_{10} = 74$ into the formula

$$S_n = \frac{n}{2}(a_1 + a_n)$$

we have

$$S_{10} = \frac{10}{2}(2 + 74)$$
$$= 5(76)$$
$$= 380$$

The sum of the first 10 terms is 380.

 # Getting Ready for Class

After reading through the preceding section, respond in your own words and in complete sentences.

A. Explain how to determine if a sequence is arithmetic.

B. What is a common difference?

C. Suppose the value of a_5 is given. What other possible pieces of information could be given in order to have enough information to obtain the first 10 terms of the sequence?

D. Explain the formula $a_n = a_1 + (n - 1)d$ in words so that someone who wanted to find the nth term of an arithmetic sequence could do so from your description.

PROBLEM SET 12.3

Determine which of the following sequences are arithmetic progressions. For those that are arithmetic progressions, identify the common difference d.

1. $1, 2, 3, 4, \ldots$

2. $4, 6, 8, 10, \ldots$

3. $1, 2, 4, 7, \ldots$

4. $1, 2, 4, 8, \ldots$

5. $50, 45, 40, \ldots$

6. $1, \frac{1}{2}, \frac{1}{4}, \frac{1}{8}, \ldots$

7. $1, 4, 9, 16, \ldots$

8. $5, 7, 9, 11, \ldots$

9. $\frac{1}{3}, 1, \frac{5}{3}, \frac{7}{3}, \ldots$

10. $5, 11, 17, \ldots$

Each of the following problems refers to arithmetic sequences.

11. If $a_1 = 3$ and $d = 4$, find a_n and a_{24}.

12. If $a_1 = 5$ and $d = 10$, find a_n and a_{100}.

13. If $a_1 = 6$ and $d = -2$, find a_{10} and S_{10}.

14. If $a_1 = 7$ and $d = -1$, find a_{24} and S_{24}.

15. If $a_6 = 17$ and $a_{12} = 29$, find the term a_1, the common difference d, and then find a_{30}.

16. If $a_5 = 23$ and $a_{10} = 48$, find the first term a_1, the common difference d, and then find a_{40}.

17. If the third term is 16 and the eighth term is 26, find the first term, the common difference, and then find a_{20} and S_{20}.

18. If the third term is 16 and the eighth term is 51, find the first term, the common difference, and then find a_{50} and S_{50}.

19. Find the sum of the first 100 terms of the sequence 5, 9, 13, 17, ...

20. Find the sum of the first 50 terms of the sequence 8, 11, 14, 17, ...

21. Find a_{35} for the sequence 12, 7, 2, -3, ...

22. Find a_{45} for the sequence 25, 20, 15, 10, ...

23. Find the tenth term and the sum of the first ten terms of the sequence $\frac{1}{2}, 1, \frac{3}{2}, 2, \ldots$.

24. Find the 15th term and the sum of the first 15 terms of the sequence $-\frac{1}{3}, 0, \frac{1}{3}, \frac{2}{3}, \ldots$.

Applying the Concepts

Straight-Line Depreciation   Recall from Section 8.5 that straight-line depreciation is an accounting method used to help spread the cost of new equipment over a number of years. The value at any time during the life of the machine can be found with a linear equation in two variables. For income tax purposes, however, it is the value at the end of the year that is most important, and for this reason sequences can be used.

25. Value of a Copy Machine   A large copy machine sells for $18,000 when it is new. Its value decreases $3,300 each year after that. We can use an arithmetic sequence to find the value of the machine at the end of each year. If we let a_0 represent the value when it is purchased, then a_1 is the value after 1 year, a_2 is the value after 2 years, and so on.
  (a) Write the first 5 terms of the sequence.
  (b) What is the common difference?
  (c) Construct a line graph for the first 5 terms of the sequence using the template that follows.
  (d) Use the line graph to estimate the value of the copy machine 2.5 years after it is purchased.
  (e) Write the sequence from part (a) using a recursive formula.

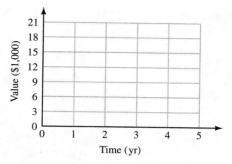

(John Turner/Tony Stone Images)

26. Value of a Forklift   An electric forklift sells for $125,000 when new (see top of p. 764). Each year after that, it decreases $16,500 in value.
  (a) Write an arithmetic sequence that gives the value of the forklift at the end of each of the first 5 years after it is purchased.
  (b) What is the common difference for this sequence?
  (c) Construct a line graph for this sequence using the template below.
  (d) Use the line graph to estimate the value of the forklift 3.5 years after it is purchased.
  (e) Write the sequence from part (a) using a recursive formula

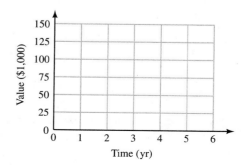

(FPG International LLC/Telegraph Colour Library)

27. Distance A rocket travels vertically 1,500 feet in its first second of flight, and then about 40 feet less each succeeding second. Use these estimates to answer the following questions.
(a) Write a sequence of the vertical distance traveled by a rocket in each of its first 6 seconds.
(b) Is the sequence in part (a) an arithmetic sequence? Explain why or why not.
(c) What is the general term of the sequence in part (a)?

28. Depreciation Suppose an automobile sells for N dollars new, and then depreciated 40% each year.
(a) Write a sequence for the value of this automobile (in terms of N) for each year.
(b) What is the general term of the sequence in part (a)?
(c) Is the sequence in part (a) an arithmetic sequence? Explain why it is or is not.

29. Triangular Numbers The first four triangular numbers are {1, 3, 6, 10, . . .}, and are illustrated in the following diagram.

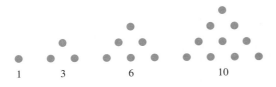

1   3   6   10

(a) Write a sequence of the first 15 triangular numbers.

(b) Write the recursive general term for the sequence of triangular numbers.
(c) Is the sequence of triangular numbers an arithmetic sequence? Explain why it is or is not.

30. Arithmetic Means Three (or more) arithmetic means between two numbers may be found by forming an arithmetic sequence using the original two numbers and the arithmetic means. For example, three arithmetic means between 10 and 34 may be found by examining the sequence {10, a, b, c, 34}. For the sequence to be arithmetic, the common difference must be 6; therefore, $a = 16$, $b = 22$, and $c = 28$. Use this idea to answer the following questions.
(a) Find four arithmetic means between 10 and 35.
(b) Find three arithmetic means between 2 and 62.
(c) Find five arithmetic means between 4 and 28.

31. Increasing Salary Suppose a woman earns $28,000 the first year she works and then gets a raise of $850 every year after that. Write a sequence that gives her salary for each of the first 5 years she works. What is the general term of this sequence? At this rate, how much will she be making the tenth year she works?

32. Increasing Salary Suppose a schoolteacher makes $26,500 the first year he works and then gets a $900 raise every year after that. Write a sequence that gives his salary for the first 5 years he works. What is the general term of this sequence? How much will he be making the 20th year he works?

Review Problems

The problems that follow review material we covered in Section 11.5.

Find the following logarithms.

33. log 576   **34.** log 57,600
35. log 0.0576   **36.** log 0.000576

Find x.

37. log $x = 2.6484$ **38.** log $x = 7.9832$
39. log $x = -7.3516$ **40.** log $x = -2.0168$

This section is concerned with the second major classification of sequences, called geometric sequences. The problems in this section are very similar to the problems in the preceding section.

DEFINITION A sequence of numbers in which each term is obtained from the previous term by multiplying by the same amount each time is called a **geometric sequence**. Geometric sequences are also called **geometric progressions**.

The sequence

$$3, 6, 12, 24, \ldots$$

is an example of a geometric progression. Each term is obtained from the previous term by multiplying by 2. The amount by which we multiply each time—in this case, 2—is called the **common ratio.** The common ratio is denoted by r and can be found by taking the ratio of any two consecutive terms. (The term with the larger subscript must be in the numerator.)

EXAMPLE 1 Find the common ratio for the geometric progression.

$$\frac{1}{2}, \frac{1}{4}, \frac{1}{8}, \frac{1}{16}, \ldots$$

Solution Since each term can be obtained from the term before it by multiplying by $\frac{1}{2}$, the common ratio is $\frac{1}{2}$. That is, $r = \frac{1}{2}$.

EXAMPLE 2 Find the common ratio for $\sqrt{3}, 3, 3\sqrt{3}, 9, \ldots$

Solution If we take the ratio of the third term to the second term, we have

$$\frac{3\sqrt{3}}{3} = \sqrt{3}$$

The common ratio is $r = \sqrt{3}$.

GEOMETRIC SEQUENCES

The **general term** a_n of a geometric sequence with first term a_1 and common ratio r is given by

$$a_n = a_1 r^{n-1}$$

To see how we arrive at this formula, consider the following geometric progression whose common ratio is 3:

$$2, 6, 18, 54, \ldots$$

We can write each term of the sequence in terms of the first term 2 and the common ratio 3:

$$2 \cdot 3^0, \qquad 2 \cdot 3^1, \qquad 2 \cdot 3^2, \qquad 2 \cdot 3^3, \ldots$$
$$a_1, \qquad a_2, \qquad a_3, \qquad a_4, \quad \ldots$$

Observing the relationship between the two preceding lines, we find we can write the general term of this progression as

$$a_n = 2 \cdot 3^{n-1}$$

Since the first term can be designated by a_1 and the common ratio by r, the formula

$$a_n = 2 \cdot 3^{n-1}$$

coincides with the formula

$$a_n = a_1 r^{n-1}$$

EXAMPLE 3   Find the general term for the geometric progression

$$5, 10, 20, \ldots$$

Solution   The first term is $a_1 = 5$, and the common ratio is $r = 2$. Using these values in the formula

$$a_n = a_1 r^{n-1}$$

we have

$$a_n = 5 \cdot 2^{n-1}$$

EXAMPLE 4   Find the tenth term of the sequence $3, \frac{3}{2}, \frac{3}{4}, \frac{3}{8}, \ldots$

Solution   The sequence is a geometric progression with first term $a_1 = 3$ and common ratio $r = \frac{1}{2}$. The tenth term is

$$a_{10} = 3 \left(\frac{1}{2} \right)^9 = \frac{3}{512}$$

EXAMPLE 5   Find the general term for the geometric progression whose fourth term is 16 and whose seventh term is 128.

Solution   The fourth term can be written as $a_4 = a_1 r^3$, and the seventh term can be written as $a_7 = a_1 r^6$.

$$a_4 = a_1 r^3 = 16$$
$$a_7 = a_1 r^6 = 128$$

We can solve for r by using the ratio a_7/a_4.

$$\frac{a_7}{a_4} = \frac{a_1 r^6}{a_1 r^3} = \frac{128}{16}$$

$$r^3 = 8$$

$$r = 2$$

The common ratio is 2. To find the first term, we substitute $r = 2$ into either of the original two equations. The result is

$$a_1 = 2$$

The general term for this progression is

$$a_n = 2 \cdot 2^{n-1}$$

which we can simplify by adding exponents, since the bases are equal:

$$a_n = 2^n$$

As was the case in the preceding section, the sum of the first n terms of a geometric progression is denoted by S_n, which is called the **nth partial sum** of the progression.

THEOREM 12.2

The sum of the first n terms of a geometric progression with first term a_1 and common ratio r is given by the formula

$$S_n = \frac{a_1(r^n - 1)}{r - 1}$$

Proof We can write the sum of the first n terms in expanded form:

$$S_n = a_1 + a_1 r + a_1 r^2 + \cdots + a_1 r^{n-1} \tag{1}$$

Then multiplying both sides by r, we have

$$r S_n = a_1 r + a_1 r^2 + a_1 r^3 + \cdots + a_1 r^n \tag{2}$$

If we subtract the left side of equation (1) from the left side of equation (2) and do the same for the right sides, we end up with

$$r S_n - S_n = a_1 r^n - a_1$$

We factor S_n from both terms on the left side and a_1 from both terms on the right side of this equation:

$$S_n(r - 1) = a_1(r^n - 1)$$

Dividing both sides by $r - 1$ gives the desired result:

$$S_n = \frac{a_1(r^n - 1)}{r - 1}$$

EXAMPLE 6   Find the sum of the first ten terms of the geometric progression 5, 15, 45, 135, ...

Solution   The first term is $a_1 = 5$, and the common ratio is $r = 3$. Substituting these values into the formula for S_{10}, we have the sum of the first ten terms of the sequence:

$$S_{10} = \frac{5(3^{10} - 1)}{3 - 1}$$

$$= \frac{5(3^{10} - 1)}{2}$$

The answer can be left in this form. A calculator will give the result as 147,620.

INFINITE GEOMETRIC SERIES

Suppose the common ratio for a geometric sequence is a number whose absolute value is less than 1—for instance, $\frac{1}{2}$. The sum of the first n terms is given by the formula

$$S_n = \frac{a_1\left[\left(\dfrac{1}{2}\right)^n - 1\right]}{\dfrac{1}{2} - 1}$$

As n becomes larger and larger, the term $(\frac{1}{2})^n$ will become closer and closer to 0. That is, for $n = 10$, 20, and 30, we have the following approximations:

$$\left(\frac{1}{2}\right)^{10} \approx 0.001$$

$$\left(\frac{1}{2}\right)^{20} \approx 0.000001$$

$$\left(\frac{1}{2}\right)^{30} \approx 0.000000001$$

so that for large values of n, there is very little difference between the expression

$$\frac{a_1(r^n - 1)}{r - 1}$$

and the expression

$$\frac{a_1(0 - 1)}{r - 1} = \frac{-a_1}{r - 1} = \frac{a_1}{1 - r} \qquad \text{if} \qquad |r| < 1$$

In fact, the sum of the terms of a geometric sequence in which $|r| < 1$ actually becomes the expression

$$\frac{a_1}{1 - r}$$

as n approaches infinity. To summarize, we have the following:

THE SUM OF AN INFINITE GEOMETRIC SERIES

If a geometric sequence has first term a_1 and common ratio r such that $|r| < 1$, then the following is called an **infinite geometric series:**

$$S = \sum_{i=0}^{\infty} a_1 r^i = a_1 + a_1 r + a_1 r^2 + a_1 r^3 + \cdots$$

Its sum is given by the formula

$$S = \frac{a_1}{1 - r}$$

EXAMPLE 7   Find the sum of the infinite geometric series

$$\frac{1}{5} + \frac{1}{10} + \frac{1}{20} + \frac{1}{40} + \cdots$$

Solution   The first term is $a_1 = \frac{1}{5}$, and the common ratio is $r = \frac{1}{2}$, which has an absolute value less than 1. Therefore, the sum of this series is

$$S = \frac{a_1}{1 - r} = \frac{\frac{1}{5}}{1 - \frac{1}{2}} = \frac{\frac{1}{5}}{\frac{1}{2}} = \frac{2}{5}$$

EXAMPLE 8   Show that 0.999 . . . is equal to 1.

Solution   We begin by writing 0.999 . . . as an infinite geometric series:

$$0.999 \ldots = 0.9 + 0.09 + 0.009 + 0.0009 + \cdots$$

$$= \frac{9}{10} + \frac{9}{100} + \frac{9}{1,000} + \frac{9}{10,000} + \cdots$$

$$= \frac{9}{10} + \frac{9}{10}\left(\frac{1}{10}\right) + \frac{9}{10}\left(\frac{1}{10}\right)^2 + \frac{9}{10}\left(\frac{1}{10}\right)^3 + \cdots$$

As the last line indicates, we have an infinite geometric series with $a_1 = \frac{9}{10}$ and $r = \frac{1}{10}$. The sum of this series is given by

$$S = \frac{a_1}{1 - r} = \frac{\frac{9}{10}}{1 - \frac{1}{10}} = \frac{\frac{9}{10}}{\frac{9}{10}} = 1$$

 Getting Ready for Class

After reading through the preceding section, respond in your own words and in complete sentences.

A. What is a common ratio?

B. Explain the formula $a_n = a_1 r^{n-1}$ in words so that someone who wanted to find the nth term of a geometric sequence could do so from your description.

C. When is the sum of an infinite geometric series a finite number?

D. Explain how a repeating decimal can be represented as an infinite geometric series.

PROBLEM SET 12.4

Identify those sequences that are geometric progressions. For those that are geometric, give the common ratio r.

1. 1, 5, 25, 125, . . .

2. 6, 12, 24, 48, . . .

3. $\frac{1}{2}, \frac{1}{6}, \frac{1}{18}, \frac{1}{54}, \ldots$

4. 5, 10, 15, 20, . . .

5. 4, 9, 16, 25, . . .

6. $-1, \frac{1}{3}, -\frac{1}{9}, \frac{1}{27}, \ldots$

7. $-2, 4, -8, 16, \ldots$

8. 1, 8, 27, 64, . . .

9. 4, 6, 8, 10, . . .

10. 1, $-3, 9, -27, \ldots$

Each of the following problems gives some information about a specific geometric progression.

11. If $a_1 = 4$ and $r = 3$, find a_n.

12. If $a_1 = 5$ and $r = 2$, find a_n.

13. If $a_1 = -2$ and $r = -\frac{1}{2}$, find a_6.

14. If $a_1 = 25$ and $r = -\frac{1}{5}$, find a_6.

15. If $a_1 = 3$ and $r = -1$, find a_{20}.

16. If $a_1 = -3$ and $r = -1$, find a_{20}.

17. If $a_1 = 10$ and $r = 2$, find S_{10}.

18. If $a_1 = 8$ and $r = 3$, find S_5.

19. If $a_1 = 1$ and $r = -1$, find S_{20}.

20. If $a_1 = 1$ and $r = -1$, find S_{21}.

21. Find a_8 for $\frac{1}{5}, \frac{1}{10}, \frac{1}{20}, \ldots$

22. Find a_8 for $\frac{1}{2}, \frac{1}{10}, \frac{1}{50}, \ldots$

23. Find S_5 for $-\frac{1}{2}, -\frac{1}{4}, -\frac{1}{8}, \ldots$

24. Find S_6 for $-\frac{1}{2}, 1, -2, \ldots$

25. Find a_{10} and S_{10} for $\sqrt{2}, 2, 2\sqrt{2}, \ldots$

26. Find a_8 and S_8 for $\sqrt{3}, 3, 3\sqrt{3}, \ldots$

27. Find a_6 and S_6 for 100, 10, 1, ...

28. Find a_6 and S_6 for 100, -10, 1, ...

29. If $a_4 = 40$ and $a_6 = 160$, find r.

30. If $a_5 = \frac{1}{8}$ and $a_8 = \frac{1}{64}$, find r.

Find the sum of each geometric series.

31. $\frac{1}{2} + \frac{1}{4} + \frac{1}{8} + \cdots$

32. $\frac{1}{3} + \frac{1}{9} + \frac{1}{27} + \cdots$

33. $4 + 2 + 1 + \cdots$

34. $8 + 4 + 2 + \cdots$

35. $\frac{2}{5} + \frac{4}{25} + \frac{8}{125} + \cdots$ **36.** $\frac{3}{4} + \frac{9}{16} + \frac{27}{64} + \cdots$

37. $\frac{3}{4} + \frac{1}{4} + \frac{1}{12} + \cdots$ **38.** $\frac{5}{3} + \frac{1}{3} + \frac{1}{15} + \cdots$

39. Show that $0.444 \ldots$ is the same as $\frac{4}{9}$.

40. Show that $0.333 \ldots$ is the same as $\frac{1}{3}$.

41. Show that $0.272727 \ldots$ is the same as $\frac{3}{11}$.

42. Show that $0.545454 \ldots$ is the same as $\frac{6}{11}$.

Applying the Concepts

Declining-Balance Depreciation   The declining-balance method of depreciation is an accounting method businesses use to deduct most of the cost of new equipment during the first few years of purchase. The value at any time during the life of the machine can be found with a linear equation in two variables. For income tax purposes, however, it is the value at the end of the year that is most important, and for this reason sequences can be used.

43. Value of a Crane   A construction crane sells for $450,000 if purchased new. After that, the value decreases by 30% each year. We can use a geometric sequence to find the value of the crane at the end of each year. If we let a_0 represent the value when it is purchased, then a_1 is the value after 1 year, a_2 is the value after 2 years, and so on.
(a) Write the first 5 terms of the sequence.
(b) What is the common ratio?
(c) Construct a line graph for the first 5 terms of the sequence using the template below.
(d) Use the line graph to estimate the value of the crane 4.5 years after it is purchased.
(e) Write the sequence from part (a) using a recursive formula.

(SuperStock)

44. Value of a Printing Press   A large printing press sells for $375,000 when it is new. After that, its value decreases 25% each year.
(a) Write a geometric sequence that gives the value of the press at the end of each of the first 5 years after it is purchased.
(b) What is the common ratio for this sequence?
(c) Construct a line graph for this sequence using the template below.
(d) Use the line graph to estimate the value of the printing press 1.5 years after it is purchased.
(e) Write the sequence from part (a) using a recursive formula.

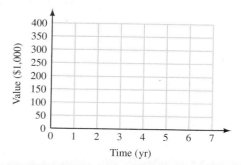

(Jim McCrary/Tony Stone Images)

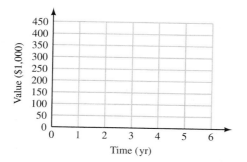

45. Adding Terms Given the geometric series $\frac{1}{3} + \frac{1}{9} + \frac{1}{27} + \cdots$,
(a) Find the sum of all the terms.
(b) Find the sum of the first 6 terms.
(c) Find the sum of all but the first 6 terms.

46. Perimeter Triangle ABC has a perimeter of 40 inches. A new triangle XYZ is formed by connecting the midpoints of the sides of the first triangle, as shown in the following figure. Since midpoints are joined, the perimeter of triangle XYZ will be one-half the perimeter of triangle ABC, or 20 inches.

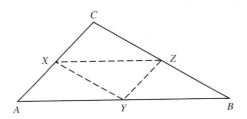

Midpoints of the sides of triangle XYZ are used to form a new triangle RST. If this pattern of using midpoints to draw triangles is continued seven more times, so that there is a total of 10 triangles drawn, what will be the sum of the perimeters of these ten triangles?

47. Bouncing Ball A ball is dropped from a height of 20 feet. Each time it bounces it returns to $\frac{7}{8}$ of the height it fell from. If the ball is allowed to bounce an infinite number of times, find the total vertical distance that the ball travels.

48. Stacking Paper Assume that a thin sheet of paper is 0.002 inch thick. The paper is torn in half, and the two halves placed together.
(a) How thick is the pile of torn paper?
(b) The pile of paper is torn in half again, and then the two halves placed together and torn in half again. The paper is large enough so that this process may be performed a total of 5 times. How thick is the pile of torn paper?
(c) Refer to the tearing and piling process described in part (b). Assuming that somehow the original paper is large enough, how thick is the pile of torn paper if 25 tears are made?

Review Problems

The following problems review material we covered in Section 4.5. Reviewing these problems will help you understand the next section.

Expand and multiply.

49. $(x + 5)^2$

50. $(x + y)^2$

51. $(x + y)^3$

52. $(x - 2)^3$

53. $(x + y)^4$

54. $(x - 1)^4$

Extending the Concepts

Use a calculator to find the given term or partial sum.

55. Find a_{20} if $a_1 = 100$ and $r = 2$.

56. Find a_{20} if $a_1 = 81$ and $r = \frac{1}{3}$.

57. Find S_{18} if $a_1 = 100$ and $r = \frac{1}{2}$.

58. Find S_{22} if $a_1 = 64$ and $r = -\frac{1}{2}$.

59. Find the sum for $a + \frac{a}{2} + \frac{a}{4} + \cdots$ if $a > 1$.

60. Find the sum for $\frac{1}{a} + \frac{1}{a^2} + \frac{1}{a^3} + \cdots$ if $a > 1$.

61. Find the sum for $\frac{a}{b} + \frac{a^2}{b^2} + \frac{a^3}{b^3} + \cdots$ if $\left|\frac{a}{b}\right| < 1$.

62. Sierpinski Triangle In the sequence that follows, the figures are moving toward what is known as the Sierpinski triangle. To construct the figure in stage 2, we remove the triangle formed from the midpoints of the sides of the shaded region in stage 1. Likewise, the figure in stage 3 is found by removing the triangles formed by connecting the midpoints of the sides of the shaded regions in stage 2. If we repeat this process infinitely many times, we arrive at the Sierpinski triangle.
(a) If the shaded region in stage 1 has an area of 1, find the area of the shaded regions in stages 2 through 4.
(b) Do the areas you found in part (a) form an arithmetic sequence or a geometric sequence?

Stage 1   Stage 2   Stage 3   Stage 4

(c) The Sierpinski triangle is the triangle that is formed after the process of forming the stages shown in the figure is repeated infinitely many times. What do you think the area of the shaded region of the Sierpinski triangle will be?

(d) Suppose the perimeter of the shaded region of the triangle in stage 1 is 1. If we were to find the perimeters of the shaded regions in the other stages, would we have an increasing sequence or a decreasing sequence?

12.5   The Binomial Expansion

The purpose of this section is to write and apply the formula for the expansion of expressions of the form $(x + y)^n$, where n is any positive integer. In order to write the formula, we must generalize the information in the following chart:

$$(x + y)^0 = \qquad\qquad 1$$
$$(x + y)^1 = \qquad\qquad x \quad + \quad y$$
$$(x + y)^2 = \qquad x^2 \quad + \quad 2xy \quad + \quad y^2$$
$$(x + y)^3 = \qquad x^3 \quad + \quad 3x^2y \quad + \quad 3xy^2 \quad + \quad y^3$$
$$(x + y)^4 = \quad x^4 \quad + \quad 4x^3y \quad + \quad 6x^2y^2 \quad + \quad 4xy^3 \quad + \quad y^4$$
$$(x + y)^5 = x^5 + 5x^4y \quad + \quad 10x^3y^2 \quad + \quad 10x^2y^3 \quad + \quad 5xy^4 + y^5$$

Note: The polynomials to the right have been found by expanding the binomials on the left — we just haven't shown the work.

There are a number of similarities to notice among the polynomials on the right. Here is a list:

1. In each polynomial, the sequence of exponents on the variable x decreases to 0 from the exponent on the binomial at the left. (The exponent 0 is not shown, since $x^0 = 1$.)
2. In each polynomial, the exponents on the variable y increase from 0 to the exponent on the binomial at the left. (Since $y^0 = 1$, it is not shown in the first term.)
3. The sum of the exponents on the variables in any single term is equal to the exponent on the binomial at the left.

The pattern in the coefficients of the polynomials on the right can best be seen by writing the right side again without the variables. It looks like this:

row 0						1					
row 1					1		1				
row 2				1		2		1			
row 3			1		3		3		1		
row 4		1		4		6		4		1	
row 5	1		5		10		10		5		1

This triangle-shaped array of coefficients is called **Pascal's triangle.** Each entry in the triangular array is obtained by adding the two numbers above it. Each row begins and ends with the number 1. If we were to continue Pascal's triangle, the next two rows would be

| row 6 | | | 1 | 6 | 15 | 20 | 15 | 6 | 1 | |
| row 7 | | 1 | 7 | 21 | 35 | 35 | 21 | 7 | 1 |

The coefficients for the terms in the expansion of $(x + y)^n$ are given in the nth row of Pascal's triangle.

There is an alternative method of finding these coefficients that does not involve Pascal's triangle. The alternative method involves **factorial notation.**

DEFINITION   The expression $n!$ is read "n factorial" and is the product of all the consecutive integers from n down to 1. For example,

$$1! = 1$$
$$2! = 2 \cdot 1 = 2$$
$$3! = 3 \cdot 2 \cdot 1 = 6$$
$$4! = 4 \cdot 3 \cdot 2 \cdot 1 = 24$$
$$5! = 5 \cdot 4 \cdot 3 \cdot 2 \cdot 1 = 120$$

The expression 0! is defined to be 1. We use factorial notation to define binomial coefficients as follows.

DEFINITION   The expression $\binom{n}{r}$ is called a **binomial coefficient** and is defined by

$$\binom{n}{r} = \frac{n!}{r!(n - r)!}$$

EXAMPLE 1   Calculate the following binomial coefficients:

$$\binom{7}{5}, \binom{6}{2}, \binom{3}{0}$$

Solution   We simply apply the definition for binomial coefficients:

$$\binom{7}{5} = \frac{7!}{5!(7 - 5)!}$$

$$= \frac{7!}{5! \cdot 2!}$$

$$= \frac{7 \cdot 6 \cdot \cancel{5 \cdot 4 \cdot 3 \cdot 2 \cdot 1}}{(\cancel{5 \cdot 4 \cdot 3 \cdot 2 \cdot 1})(2 \cdot 1)}$$

$$= \frac{42}{2}$$

$$= 21$$

$$\binom{6}{2} = \frac{6!}{2!(6-2)!}$$

$$= \frac{6!}{2! \cdot 4!}$$

$$= \frac{6 \cdot 5 \cdot \cancel{4 \cdot 3 \cdot 2 \cdot 1}}{(2 \cdot 1)(\cancel{4 \cdot 3 \cdot 2 \cdot 1})}$$

$$= \frac{30}{2}$$

$$= 15$$

$$\binom{3}{0} = \frac{3!}{0!(3-0)!}$$

$$= \frac{3!}{0! \cdot 3!}$$

$$= \frac{\cancel{3 \cdot 2 \cdot 1}}{(1)(\cancel{3 \cdot 2 \cdot 1})}$$

$$= 1$$

If we were to calculate all the binomial coefficients in the following array, we would find they match exactly with the numbers in Pascal's triangle. That is why they are called binomial coefficients—because they are the coefficients of the expansion of $(x + y)^n$.

$$\binom{0}{0}$$

$$\binom{1}{0} \qquad \binom{1}{1}$$

$$\binom{2}{0} \qquad \binom{2}{1} \qquad \binom{2}{2}$$

$$\binom{3}{0} \qquad \binom{3}{1} \qquad \binom{3}{2} \qquad \binom{3}{3}$$

$$\binom{4}{0} \qquad \binom{4}{1} \qquad \binom{4}{2} \qquad \binom{4}{3} \qquad \binom{4}{4}$$

$$\binom{5}{0} \qquad \binom{5}{1} \qquad \binom{5}{2} \qquad \binom{5}{3} \qquad \binom{5}{4} \qquad \binom{5}{5}$$

Using the new notation to represent the entries in Pascal's triangle, we can summarize everything we have noticed about the expansion of binomial powers of the form $(x + y)^n$.

THE BINOMIAL EXPANSION

If x and y represent real numbers and n is a positive integer, then the following formula is known as the **binomial expansion** or **binomial formula**:

$$(x + y)^n = \binom{n}{0} x^n y^0 + \binom{n}{1} x^{n-1} y^1 + \binom{n}{2} x^{n-2} y^2 + \cdots + \binom{n}{n} x^0 y^n$$

It does not make any difference, when expanding binomial powers of the form $(x + y)^n$, whether we use Pascal's triangle or the formula

$$\binom{n}{r} = \frac{n!}{r!(n - r)!}$$

to calculate the coefficients. We will show examples of both methods.

EXAMPLE 2   Expand $(x - 2)^3$.

Solution   Applying the binomial formula, we have

$$(x - 2)^3 = \binom{3}{0} x^3 (-2)^0 + \binom{3}{1} x^2 (-2)^1 + \binom{3}{2} x^1 (-2)^2 + \binom{3}{3} x^0 (-2)^3$$

The coefficients

$$\binom{3}{0}, \binom{3}{1}, \binom{3}{2}, \text{ and } \binom{3}{3}$$

can be found in the third row of Pascal's triangle. They are 1, 3, 3, and 1:

$$(x - 2)^3 = 1x^3 (-2)^0 + 3x^2 (-2)^1 + 3x^1 (-2)^2 + 1x^0 (-2)^3$$
$$= x^3 - 6x^2 + 12x - 8$$

EXAMPLE 3   Expand $(3x + 2y)^4$.

Solution   The coefficients can be found in the fourth row of Pascal's triangle.

$$1, 4, 6, 4, 1$$

Here is the expansion of $(3x + 2y)^4$:

$$(3x + 2y)^4 = 1(3x)^4 + 4(3x)^3(2y) + 6(3x)^2(2y)^2 + 4(3x)(2y)^3 + 1(2y)^4$$
$$= 81x^4 + 216x^3y + 216x^2y^2 + 96xy^3 + 16y^4$$

EXAMPLE 4   Write the first three terms in the expansion of $(x + 5)^9$.

Solution   The coefficients of the first three terms are

$$\binom{9}{0}, \binom{9}{1}, \text{ and } \binom{9}{2}$$

which we calculate as follows:

$$\binom{9}{0} = \frac{9!}{0! \cdot 9!} = \frac{9 \cdot 8 \cdot 7 \cdot 6 \cdot 5 \cdot 4 \cdot 3 \cdot 2 \cdot 1}{(1)(9 \cdot 8 \cdot 7 \cdot 6 \cdot 5 \cdot 4 \cdot 3 \cdot 2 \cdot 1)} = \frac{1}{1} = 1$$

$$\binom{9}{1} = \frac{9!}{1! \cdot 8!} = \frac{9 \cdot 8 \cdot 7 \cdot 6 \cdot 5 \cdot 4 \cdot 3 \cdot 2 \cdot 1}{(1)(8 \cdot 7 \cdot 6 \cdot 5 \cdot 4 \cdot 3 \cdot 2 \cdot 1)} = \frac{9}{1} = 9$$

$$\binom{9}{2} = \frac{9!}{2! \cdot 7!} = \frac{9 \cdot 8 \cdot 7 \cdot 6 \cdot 5 \cdot 4 \cdot 3 \cdot 2 \cdot 1}{(2 \cdot 1)(7 \cdot 6 \cdot 5 \cdot 4 \cdot 3 \cdot 2 \cdot 1)} = \frac{72}{2} = 36$$

From the binomial formula, we write the first three terms:

$$(x + 5)^9 = 1 \cdot x^9 + 9 \cdot x^8(5) + 36x^7(5)^2 + \cdots$$
$$= x^9 + 45x^8 + 900x^7 + \cdots$$

THE KTH TERM OF A BINOMIAL EXPANSION

If we look at each term in the expansion of $(x + y)^n$ as a term in a sequence, $a_1, a_2, a_3, \ldots$, we can write

$$a_1 = \binom{n}{0} x^n y^0$$

$$a_2 = \binom{n}{1} x^{n-1} y^1$$

$$a_3 = \binom{n}{2} x^{n-2} y^2$$

$$a_4 = \binom{n}{3} x^{n-3} y^3 \qquad \text{and so on}$$

To write the formula for the general term, we simply notice that the exponent on y and the number below n in the coefficient are both 1 less than the term number. This observation allows us to write the following:

THE GENERAL TERM OF A BINOMIAL EXPANSION

The kth term in the expansion of $(x + y)^n$ is

$$a_k = \binom{n}{k - 1} x^{n-(k-1)} y^{k-1}$$

EXAMPLE 5 Find the fifth term in the expansion of $(2x + 3y)^{12}$.

Solution Applying the preceding formula, we have

$$a_5 = \binom{12}{4} (2x)^8 (3y)^4$$

$$= \frac{12!}{4! \cdot 8!} (2x)^8 (3y)^4$$

Notice that once we have one of the exponents, the other exponent and the denominator of the coefficient are determined: The two exponents add to 12 and match the numbers in denominator of the coefficient.

Making the calculations from the preceding formula, we have

$$a_5 = 495(256x^8)(81y^4)$$

$$= 10,264,320x^8y^4$$

 Getting Ready for Class

After reading through the preceding section, respond in your own words and in complete sentences.

A. What is Pascal's triangle?

B. Why is $\binom{n}{0} = 1$ for any natural number?

C. State the binomial formula.

D. When is the binomial formula more efficient than multiplying to expand a binomial raised to a whole-number exponent?

PROBLEM SET 12.5

Use the binomial formula to expand each of the following.

1. $(x + 2)^4$

2. $(x - 2)^5$

3. $(x + y)^6$

4. $(x - 1)^6$

5. $(2x + 1)^5$

6. $(2x - 1)^4$

7. $(x - 2y)^5$

8. $(2x + y)^5$

9. $(3x - 2)^4$

10. $(2x - 3)^4$

11. $(4x - 3y)^3$

12. $(3x - 4y)^3$

13. $(x^2 + 2)^4$

14. $(x^2 - 3)^3$

15. $(x^2 + y^2)^3$

16. $(x^2 - 3y)^4$

17. $\left(\frac{x}{2} - 4\right)^3$

18. $\left(\frac{x}{3} + 6\right)^3$

19. $\left(\frac{x}{3} + \frac{y}{2}\right)^4$

20. $\left(\frac{x}{2} - \frac{y}{3}\right)^4$

Write the first four terms in the expansion of the following.

21. $(x + 2)^9$

22. $(x - 2)^9$

23. $(x - y)^{10}$

24. $(x + y)^{10}$

25. $(x + 2y)^{10}$

26. $(x - 2y)^{10}$

Write the first three terms in the expansion of each of the following.

27. $(x + 1)^{15}$

28. $(x - 1)^{15}$

29. $(x - y)^{12}$

30. $(x + y)^{12}$

31. $(x + 2)^{20}$

32. $(x - 2)^{20}$

Write the first two terms in the expansion of each of the following.

33. $(x + 2)^{100}$

34. $(x - 2)^{50}$

35. $(x + y)^{50}$

36. $(x - y)^{100}$

37. Find the ninth term in the expansion of $(2x + 3y)^{12}$.

38. Find the sixth term in the expansion of $(2x + 3y)^{12}$.

39. Find the fifth term of $(x - 2)^{10}$.

40. Find the fifth term of $(2x - 1)^{10}$.

41. Find the fourth term of $(x + 3)^9$.

42. Find the fifth term of $(x + 3)^9$.

43. Write the formula for the 12th term of $(2x + 5y)^{20}$. Do not simplify.

44. Write the formula for the eighth term of $(2x + 5y)^{20}$. Do not simplify.

Applying the Concepts

45. Probability   The third term in the expansion of $(\frac{1}{2} + \frac{1}{2})^7$ will give the probability that in a family with 7 children, 5 will be boys and 2 will be girls. Find the third term.

46. Probability   The fourth term in the expansion of $(\frac{1}{2} + \frac{1}{2})^8$ will give the probability that in a family with 8 children, 3 will be boys and 5 will be girls. Find the fourth term.

47. Multiplication   Without using a calculator, evaluate 1.01^4 by using the binomial formula to expand $(1 + 0.01)^4$.

48. Multiplication   Without using a calculator, evaluate 1.02^5 by using the binomial formula.

Review Problems

The problems that follow review material we covered in Section 11.6.

Solve each equation. Write your answers to the nearest hundredth.

49. $5^x = 7$

50. $10^x = 15$

51. $8^{2x+1} = 16$

52. $9^{3x-1} = 27$

53. Compound Interest   How long will it take $400 to double if it is invested in an account with an annual interest rate of 10% compounded four times a year?

54. Compound Interest   How long will it take $200 to become $800 if it is invested in an account with an annual interest rate of 8% compounded four times a year?

Find each of the following to the nearest hundredth.

55. $\log_4 20$

56. $\log_7 21$

57. $\ln 576$

58. $\ln 5,760$

59. Solve the formula $A = 10e^{5t}$ for t.

60. Solve the formula $A = P2^{-5t}$ for t.

Extending the Concepts

61. Calculate both $\binom{8}{5}$ and $\binom{8}{3}$ to show that they are equal.

62. Calculate both $\binom{10}{8}$ and $\binom{10}{2}$ to show that they are equal.

63. Simplify $\binom{20}{12}$ and $\binom{20}{8}$.

64. Simplify $\binom{15}{10}$ and $\binom{15}{5}$.

65. Show that $\binom{n}{r}$ and $\binom{n}{n-r}$ are equal.

66. Use the results of Problem 65 to find an easy way to compute $\binom{24}{23}$.

67. Choose three values of n, say $n = 3$, $n = 5$, and $n = 7$, and then verify the following formula for each value of n.

$$\binom{n}{0} + \binom{n}{1} + \binom{n}{2} + \cdots + \binom{n}{n} = 2^n$$

68. **Pascal's Triangle** Copy the first eight rows of Pascal's triangle into the eight rows of the following triangular array. (Each number in Pascal's triangle will go into one of the hexagons in the array.) Next, color in each hexagon that contains an odd number. What pattern begins to emerge from this coloring process?

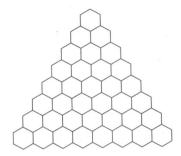

CHAPTER 12 SUMMARY

Examples

1. In the sequence 1, 3, 5, . . . , $2n - 1$, . . . , $a_1 = 1$, $a_2 = 3$, $a_3 = 5$, and $a_n = 2n - 1$.

Sequences [12.1]

A *sequence* is a function whose domain is the set of positive integers. The terms of a sequence are denoted by

$$a_1, a_2, a_3, \ldots, a_n, \ldots$$

where a_1 (read "a sub 1") is the first term, a_2 the second term, and a_n the nth or *general term*.

2. $\displaystyle\sum_{i=3}^{6} (-2)^i$

$= (-2)^3 + (-2)^4 + (-2)^5$
$\quad + (-2)^6$

$= -8 + 16 + (-32) + 64$

$= 40$

Summation Notation [12.2]

The notation

$$\sum_{i=1}^{n} a_i = a_1 + a_2 + a_3 + \cdots + a_n$$

is called *summation notation* or *sigma notation*. The letter i as used here is called the *index of summation* or just *index*.

3. For the sequence 3, 7, 11, 15, . . . , $a_1 = 3$ and $d = 4$. The general term is

$$a_n = 3 + (n - 1)4$$
$$= 4n - 1$$

Using this formula to find the tenth term, we have

$$a_{10} = 4(10) - 1 = 39$$

The sum of the first ten terms is

$$S_{10} = \frac{10}{2}(3 + 39) = 210$$

Arithmetic Sequences [12.3]

An *arithmetic sequence* is a sequence in which each term comes from the preceding term by adding a constant amount each time. If the first term of an arithmetic sequence is a_1 and the amount we add each time (called the *common difference*) is d, then the nth term of the progression is given by

$$a_n = a_1 + (n - 1)d$$

The sum of the first n terms of an arithmetic sequence is

$$S_n = \frac{n}{2}(a_1 + a_n)$$

S_n is called the nth *partial sum*.

4. For the geometric progression 3, 6, 12, 24, . . . , $a_1 = 3$ and $r = 2$. The general term is

$$a_n = 3 \cdot 2^{n-1}$$

The sum of the first ten terms is

$$S_{10} = \frac{3(2^{10} - 1)}{2 - 1} = 3{,}069$$

Geometric Sequences [12.4]

A *geometric sequence* is a sequence of numbers in which each term comes from the previous term by multiplying by a constant amount each time. The constant by which we multiply each term to get the next term is called the *common ratio*. If the first term of a geometric sequence is a_1 and the common ratio is r, then the formula that gives the general term a_n is

$$a_n = a_1 r^{n-1}$$

The sum of the first n terms of a geometric sequence is given by the formula

$$S_n = \frac{a_1(r^n - 1)}{r - 1}$$

5. The sum of the series

$$\frac{1}{3} + \frac{1}{6} + \frac{1}{12} + \cdots$$

is

$$S = \frac{\frac{1}{3}}{1 - \frac{1}{2}} = \frac{\frac{1}{3}}{\frac{1}{2}} = \frac{2}{3}$$

The Sum of an Infinite Geometric Series [12.4]

If a geometric sequence has first term a_1 and common ratio r such that $|r| < 1$, then the following is called an *infinite geometric series:*

$$S = \sum_{i=0}^{\infty} a_1 r^i = a_1 + a_1 r + a_1 r^2 + a_1 r^3 + \cdots$$

Its sum is given by the formula

$$S = \frac{a_1}{1 - r}$$

Factorials [12.5]

The notation $n!$ is called n *factorial* and is defined to be the product of each consecutive integer from n down to 1. That is,

$$0! = 1 \qquad \text{(By definition)}$$
$$1! = 1$$
$$2! = 2 \cdot 1$$
$$3! = 3 \cdot 2 \cdot 1$$
$$4! = 4 \cdot 3 \cdot 2 \cdot 1$$

and so on.

6. $\dbinom{7}{3} = \dfrac{7!}{3!(7 - 3)!}$

$$= \frac{7!}{3! \cdot 4!}$$

$$= \frac{7 \cdot 6 \cdot 5 \cdot \cancel{4 \cdot 3 \cdot 2 \cdot 1}}{(3 \cdot 2 \cdot 1)(\cancel{4 \cdot 3 \cdot 2 \cdot 1})}$$

$$= 35$$

Binomial Coefficients [12.5]

The notation $\dbinom{n}{r}$ is called a *binomial coefficient* and is defined by

$$\binom{n}{r} = \frac{n!}{r!(n - r)!}$$

Binomial coefficients can be found by using the formula above or by *Pascal's triangle,* which is

$$
\begin{array}{ccccccccccc}
 & & & & & 1 & & & & & \\
 & & & & 1 & & 1 & & & & \\
 & & & 1 & & 2 & & 1 & & & \\
 & & 1 & & 3 & & 3 & & 1 & & \\
 & 1 & & 4 & & 6 & & 4 & & 1 & \\
1 & & 5 & & 10 & & 10 & & 5 & & 1
\end{array}
$$

and so on.

7. $(x + 2)^4$

$= x^4 + 4x^3 \cdot 2 + 6x^2 \cdot 2^2$

$+ 4x \cdot 2^3 + 2^4$

$= x^4 + 8x^3 + 24x^2 + 32x + 16$

Binomial Expansion [10.5]

If n is a positive integer, then the formula for expanding $(x + y)^n$ is given by

$$(x + y)^n = \binom{n}{0} x^n y^0 + \binom{n}{1} x^{n-1} y^1 + \binom{n}{2} x^{n-2} y^2 + \cdots + \binom{n}{n} x^0 y^n$$

C H A P T E R   1 2   REVIEW

Write the first four terms of the sequence with the following general terms. [12.1]

1. $a_n = 2n + 5$

2. $a_n = 3n - 2$

3. $a_n = n^2 - 1$

4. $a_n = \dfrac{n + 3}{n + 2}$

5. $a_1 = 4, a_n = 4a_{n-1}, n > 1$

6. $a_1 = \frac{1}{4}, a_n = \frac{1}{4}a_{n-1}, n > 1$

Determine the general term for each of the following sequences. [12.1]

7. $2, 5, 8, 11, \ldots$

8. $-3, -1, 1, 3, 5, \ldots$

9. $1, 16, 81, 256, \ldots$

10. $2, 5, 10, 17, \ldots$

11. $\frac{1}{2}, \frac{1}{4}, \frac{1}{8}, \frac{1}{16}, \ldots$

12. $2, \frac{3}{4}, \frac{4}{9}, \frac{5}{16}, \frac{6}{25}, \ldots$

Expand and simplify each of the following. [12.2]

13. $\displaystyle\sum_{i=1}^{4} (2i + 3)$

14. $\displaystyle\sum_{i=1}^{3} (2i^2 - 1)$

15. $\displaystyle\sum_{i=2}^{3} \dfrac{i^2}{i + 2}$

16. $\displaystyle\sum_{i=1}^{4} (-2)^{i-1}$

17. $\displaystyle\sum_{i=3}^{5} (4i + i^2)$

18. $\displaystyle\sum_{i=4}^{6} \dfrac{i + 2}{i}$

Write each of the following sums with summation notation. [12.2]

19. $3 + 6 + 9 + 12$

20. $3 + 7 + 11 + 15$

21. $5 + 7 + 9 + 11 + 13$

22. $4 + 9 + 16$

23. $\frac{1}{3} + \frac{1}{4} + \frac{1}{5} + \frac{1}{6}$

24. $\frac{1}{3} + \frac{2}{9} + \frac{3}{27} + \frac{4}{81} + \frac{5}{243}$

25. $(x - 2) + (x - 4) + (x - 6)$

26. $\dfrac{x}{x + 1} + \dfrac{x}{x + 2} + \dfrac{x}{x + 3} + \dfrac{x}{x + 4}$

Determine which of the following sequences are arithmetic progressions, geometric progressions, or neither. [12.3, 12.4]

27. $1, -3, 9, -27, \ldots$

28. $7, 9, 11, 13, \ldots$

29. $5, 11, 17, 23, \ldots$

30. $\frac{1}{2}, \frac{1}{3}, \frac{1}{4}, \frac{1}{5}, \ldots$

31. $4, 8, 16, 32, \ldots$

32. $\frac{1}{2}, \frac{1}{4}, \frac{1}{8}, \frac{1}{16}, \ldots$

33. $12, 9, 6, 3, \ldots$

34. $2, 5, 9, 14, \ldots$

Each of the following problems refers to arithmetic progressions. [12.3]

35. If $a_1 = 2$ and $d = 3$, find a_n and a_{20}.

36. If $a_1 = 5$ and $d = -3$, find a_n and a_{16}.

37. If $a_1 = -2$ and $d = 4$, find a_{10} and S_{10}.

38. If $a_1 = 3$ and $d = 5$, find a_{16} and S_{16}.

39. If $a_5 = 21$ and $a_8 = 33$, find the first term a_1, the common difference d, and then find a_{10}.

40. If $a_3 = 14$ and $a_7 = 26$, find the first term a_1, the common difference d, and then find a_9 and S_9.

41. If $a_4 = -10$ and $a_8 = -18$, find the first term a_1, the common difference d, and then find a_{20} and S_{20}.

42. Find the sum of the first 100 terms of the sequence 3, 7, 11, 15, 19, ...

43. Find a_{40} for the sequence 100, 95, 90, 85, 80, ...

Each of the following problems refers to infinite geometric progressions. [12.4]

44. If $a_1 = 3$ and $r = 2$, find a_n and a_{20}.

45. If $a_1 = 5$ and $r = -2$, find a_n and a_{16}.

46. If $a_1 = 4$ and $r = \frac{1}{2}$, find a_n and a_{10}.

47. If $a_1 = -2$ and $r = \frac{1}{3}$, find the sum.

48. If $a_1 = 4$ and $r = \frac{1}{2}$, find the sum.

49. If $a_3 = 12$ and $a_4 = 24$, find the first term a_1, the common ratio r, and then find a_6.

50. Find the tenth term of the sequence 3, $3\sqrt{3}$, 9, $9\sqrt{3}$, ...

Evaluate each of the following. [12.5]

51. $\dbinom{8}{2}$

52. $\dbinom{7}{4}$

53. $\dbinom{6}{3}$

54. $\dbinom{9}{2}$

55. $\dbinom{10}{8}$

56. $\dbinom{100}{3}$

Use the binomial formula to expand each of the following. [12.5]

57. $(x - 2)^4$

58. $(2x + 3)^4$

59. $(3x + 2y)^3$

60. $(x^2 - 2)^5$

61. $\left(\dfrac{x}{2} + 3\right)^4$

62. $\left(\dfrac{x}{3} - \dfrac{y}{2}\right)^3$

Use the binomial formula to write the first three terms in the expansion of the following. [12.5]

63. $(x + 3y)^{10}$

64. $(x - 3y)^9$

65. $(x + y)^{11}$

66. $(x - 2y)^{12}$

Use the binomial formula to write the first two terms in the expansion of the following. [12.5]

67. $(x - 2y)^{16}$

68. $(x + 2y)^{32}$

69. $(x - 1)^{50}$

70. $(x + y)^{150}$

71. Find the sixth term in $(x - 3)^{10}$.

72. Find the fourth term in $(2x + 1)^9$.

CHAPTER 12 PROJECTS

SEQUENCES AND SERIES

GROUP PROJECT

CREDIT CARD PAYMENTS

Number of People: 2

Time Needed: 20–30 minutes

Equipment: Paper, pencil, and graphing calculator

Background: In the beginning of this chapter, you were given a recursive function that you can use to find how long it will take to pay off a credit card. A graphing calculator can be used to model each payment by using the recall function on the graphing calculator. To set up this problem do the following:

(1)   1000   ENTER

(2)   Round (1.0165ANS−20,2)   ENTER

Note: The *Round* function is under the MATH key in the NUM list.

Procedure: Enter the preceding commands into your calculator. While one person in the group hits the ENTER key, another person counts, by keeping a tally of the "payments" made. The credit card is paid off when the calculator displays a negative number.

1. How many months did it take to pay off the credit card?
2. What was the amount of the last payment?
3. What was the total interest paid to the credit card company?
4. How much would you save if you paid $25 per month instead of $20?
5. If the credit card company raises the interest rate to 21.5% annual interest, how long will it take to pay off the balance? How much more would it cost in interest?
6. On larger balances, many credit card companies require a minimum payment of 2% of the outstanding balance. What is the recursion formula for this? How much would this save or cost you in interest?

Detach here and return with check or money order.

Summary of Corporate Card Account
Retain this portion for your files

Corporate Cardmember Name	**Account Number**	**Closing Date**
Leonardo Fibonacci	00000-1000-001	08-01-99

New Balance	**Other Debits**	**Interest Rate**	**Minimum Payment**
$1,000.00	$.00	19.8%	$20.00

RESEARCH PROJECT

BUILDING SQUARES FROM ODD NUMBERS

Leonardo Fibonacci
(Corbis/Bettmann)

A relationship exists between the sequence of squares and the sequence of odd numbers. In *The Book of Squares,* written in 1225, Leonardo Fibonacci has this to say about that relationship:

> I thought about the origin of all square numbers and discovered that they arise out of the increasing sequence of odd numbers.

Work with the sequence of odd numbers until you discover how it can be used to produce the sequence of squares. Then write an essay in which you give a written description of the relationship between the two sequences, along with a diagram that illustrates the relationship. Then see if you can use summation notation to write an equation that summarizes the whole relationship. Your essay should be clear and concise and written so that any of your classmates can read it and understand the relationship you are describing.

CHAPTER 12 TEST

Write the first five terms of the sequences with the following general terms [12.1]

1. $a_n = 3n - 5$

2. $a_1 = 3, a_n = a_{n-1} + 4, n > 1$

3. $a_n = n^2 + 1$ **4.** $a_n = 2n^3$

5. $a_n = \dfrac{n + 1}{n^2}$

6. $a_1 = 4, a_n = -2a_{n-1}, n > 1$

Give the general term for each sequence. [12.1]

7. 6, 10, 14, 18, . . .   **8.** 1, 2, 4, 8, . . .

9. $\dfrac{1}{2}, \dfrac{1}{4}, \dfrac{1}{8}, \dfrac{1}{16}, \ldots$ **10.** $-3, 9, -27, 81, \ldots$

11. Expand and simplify each of the following. [12.2]

  (a) $\displaystyle\sum_{i=1}^{5} (5i + 3)$ (b) $\displaystyle\sum_{i=3}^{5} (2^i - 1)$

  (c) $\displaystyle\sum_{i=2}^{6} (i^2 + 2i)$

12. Find the first term of an arithmetic progression if $a_5 = 11$ and $a_9 = 19$. [12.3]

13. Find the second term of a geometric progression for which $a_3 = 18$ and $a_5 = 162$. [12.4]

Find the sum of the first 10 terms of the following arithmetic progressions. [12.3]

14. 5, 11, 17, . . .   **15.** 25, 20, 15, . . .

16. Write a formula for the sum of the first 50 terms of the geometric progression 3, 6, 12, ... [12.4]

17. Find the sum of $\dfrac{1}{2} + \dfrac{1}{6} + \dfrac{1}{18} + \dfrac{1}{54} + \cdots$. [12.4]

Use the binomial formula to expand each of the following. [12.5]

18. $(x - 3)^4$ **19.** $(2x - 1)^5$

20. Find the first 3 terms in the expansion of $(x - 1)^{20}$. [12.5]

21. Find the sixth term in $(2x - 3y)^8$. [12.5]

CHAPTERS 1-12 CUMULATIVE REVIEW

Simplify.

1. $\dfrac{5(-6) + 3(-2)}{4(-3) + 3}$

2. $9 + 5(4y + 8) + 10y$

3. $\dfrac{18a^7b^{-4}}{36a^2b^{-8}}$

4. $\dfrac{y^2 - y - 6}{y^2 - 4}$

5. $8^{-2/3}$

6. $\log_3 27$

Factor completely.

7. $ab^3 + b^3 + 6a + 6$

8. $8x^2 - 5x - 3$

Solve.

9. $6 - 2(5x - 1) + 4x = 20$

10. $|4x - 3| + 2 = 3$

11. $(x + 1)(x + 2) = 12$

12. $1 - \dfrac{2}{x} = \dfrac{8}{x^2}$

13. $t - 6 = \sqrt{t - 4}$

14. $(4x - 3)^2 = -50$

15. $8t^3 - 27 = 0$

16. $6x^4 - 13x^2 = 5$

Solve and graph the solution on the number line.

17. $-3y - 2 < 7$

18. $|2x + 5| - 2 < 9$

Graph on a rectangular coordinate system.

19. $2x - 3y = 6$

20. $y < \frac{1}{2}x + 3$

21. $y = x^2 - 2x - 3$

22. $y = (\frac{1}{2})^x$

Solve each system.

23. $5x - 3y = -4$
$x + 2y = 7$

24. $x + 2y = 0$
$3y + z = -3$
$2x - z = 5$

Multiply.

25. $\dfrac{3y^2 - 3y}{3y - 12} \cdot \dfrac{y^2 - 2y - 8}{y^2 + 3y + 2}$

26. $(x^{3/5} + 2)(x^{3/5} - 2)$

27. $(2 + 3i)(1 - 4i)$

28. Add: $\dfrac{-3}{x^2 - 2x - 8} + \dfrac{4}{x^2 - 16}$

29. Combine: $4\sqrt{50} + 3\sqrt{8}$

30. Rationalize the denominator:

$$\dfrac{3}{\sqrt{7} - \sqrt{3}}$$

31. If $f(x) = 4x + 1$, find $f^{-1}(x)$.

32. Find x if $\log x = 3.9786$.

33. Find $\log_6 14$ to the nearest hundredth.

Find the general term for each sequence.

34. 3, 14, 25, . . .

35. 16, 8, 4, . . .

36. Solve $S = 2x^2 + 4xy$ for y.

37. Expand and simplify.

$$\sum_{i=1}^{5} (2i - 1)$$

38. Find the slope of the line through $(2, -3)$ and $(-1, -3)$.

39. Find the equation of the line through $(2, 5)$ and $(6, -3)$. Write your answer in slope-intercept form.

40. If $C(t) = 80(\frac{1}{2})^{t/5}$, find $C(5)$ and $C(10)$.

41. Find an equation with solutions $t = -\frac{3}{4}$ and $t = \frac{1}{5}$.

42. If $f(x) = 3x - 7$ and $g(x) = 4 - x$, find $(f \circ g)(x)$.

43. Solve the system.

$$3x - 5y = 2$$
$$2x + 4y = 1$$

44. Find the first term in the expansion of $(2x - y)^5$.

45. Add -8 to the difference of -7 and 4.

46. Factor $x^3 - \frac{1}{8}$.

47. Geometry Two supplementary angles are such that one is 5 times as large as the other. Find the two angles.

48. Variation y varies directly with x. If y is 24 when x is 8, find y when x is 2.

49. Number Problem One number is 3 times another. The sum of their reciprocals is $\frac{4}{3}$. Find the numbers.

50. Mixture How much 30% alcohol solution and 70% alcohol solution must be mixed to get 16 gallons of 60% alcohol solution?

Conic Sections

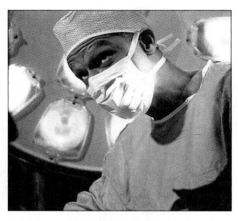

(FPG International/Telegraph Colour Library)

INTRODUCTION

One of the curves we will study in this chapter has interesting reflective properties. Figure 1(a) shows how you can draw one of these curves (an ellipse) using thumbtacks, string, pencil, and paper. Elliptical surfaces will reflect sound waves that originate at one focus through the other focus. This property of ellipses allows doctors to treat patients with kidney stones using a procedure called lithotripsy. A lithotripter is an elliptical device that creates sound waves that crush the kidney stone into small pieces, without surgery. The sound wave originates at one focus of the lithotripter. The energy from it reflects off the surface of the lithotripter and converges at the other focus, where the kidney stone is positioned. Figure 1(b) shows a cross section of a lithotripter, with a patient positioned so that the kidney stone is at the other focus.

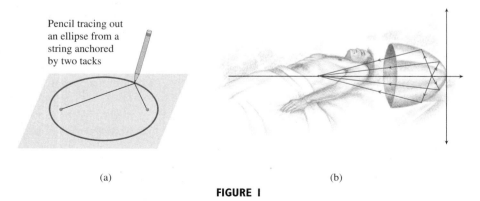

Pencil tracing out an ellipse from a string anchored by two tacks

(a)

(b)

FIGURE I

By studying the conic sections in this chapter, you will be better equipped to understand some of the more technical equipment that exists in the world outside of class.

Conic sections include ellipses, circles, hyperbolas, and parabolas. They are called conic sections because each can be found by slicing a cone with a plane as shown in Figure 1. We begin our work with conic sections by studying circles. Before we find the general equation of a circle, we must first derive what is known as the *distance formula*.

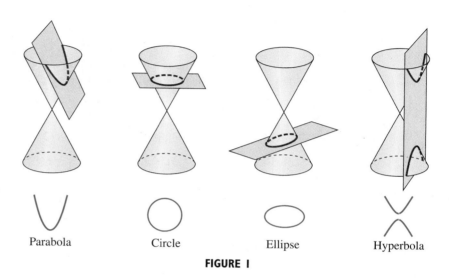

Parabola   Circle   Ellipse   Hyperbola

FIGURE I

Suppose (x_1, y_1) and (x_2, y_2) are any two points in the first quadrant. (Actually, we could choose the two points to be anywhere on the coordinate plane. It is just more convenient to have them in the first quadrant.) We can name the points P_1 and P_2, respectively, and draw the diagram shown in Figure 2.

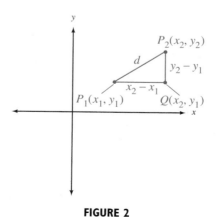

FIGURE 2

Notice the coordinates of point Q. The x-coordinate is x_2 since Q is directly below point P_2. The y-coordinate of Q is y_1 since Q is directly across from point P_1. It is evident from the diagram that the length of P_2Q is $y_2 - y_1$ and the length of P_1Q is $x_2 - x_1$. Using the Pythagorean theorem, we have

$$(P_1P_2)^2 = (P_1Q)^2 + (P_2Q)^2$$

or

$$d^2 = (x_2 - x_1)^2 + (y_2 - y_1)^2$$

Taking the square root of both sides, we have

$$d = \sqrt{(x_2 - x_1)^2 + (y_2 - y_1)^2}$$

We know this is the positive square root, since d is the distance from P_1 to P_2 and must therefore be positive. This formula is called the *distance formula*.

EXAMPLE 1   Find the distance between $(3, 5)$ and $(2, -1)$.

Solution   If we let $(3, 5)$ be (x_1, y_1) and $(2, -1)$ be (x_2, y_2) and apply the distance formula, we have

$$d = \sqrt{(2 - 3)^2 + (-1 - 5)^2}$$
$$= \sqrt{(-1)^2 + (-6)^2}$$
$$= \sqrt{1 + 36}$$
$$= \sqrt{37}$$

EXAMPLE 2   Find x if the distance from $(x, 5)$ to $(3, 4)$ is $\sqrt{2}$.

Solution   Using the distance formula, we have

$$\sqrt{2} = \sqrt{(x - 3)^2 + (5 - 4)^2}$$
$$2 = (x - 3)^2 + 1^2$$
$$2 = x^2 - 6x + 9 + 1$$
$$0 = x^2 - 6x + 8$$
$$0 = (x - 4)(x - 2)$$
$$x = 4 \qquad \text{or} \qquad x = 2$$

The two solutions are 4 and 2, which indicates that two points, $(4, 5)$ and $(2, 5)$, are $\sqrt{2}$ units from $(3, 4)$.

We can use the distance formula to derive the equation of a circle.

THEOREM 13.1

The equation of the circle with center at (a, b) and radius r is given by

$$(x - a)^2 + (y - b)^2 = r^2$$

Proof   By definition, all points on the circle are a distance r from the center (a, b). If we let (x, y) represent any point on the circle, then (x, y) is r units from (a, b). Applying the distance formula, we have

$$r = \sqrt{(x - a)^2 + (y - b)^2}$$

Squaring both sides of this equation gives the equation of the circle:

$$(x - a)^2 + (y - b)^2 = r^2$$

We can use Theorem 13.1 to find the equation of a circle, given its center and radius, or to find its center and radius, given the equation.

EXAMPLE   3   Find the equation of the circle with center at $(-3, 2)$ having a radius of 5.

Solution   We have $(a, b) = (-3, 2)$ and $r = 5$. Applying Theorem 13.1 yields

$$[x - (-3)]^2 + (y - 2)^2 = 5^2$$
$$(x + 3)^2 + (y - 2)^2 = 25$$

EXAMPLE   4   Give the equation of the circle with radius 3 whose center is at the origin.

Solution   The coordinates of the center are $(0, 0)$, and the radius is 3. The equation must be

$$(x - 0)^2 + (y - 0)^2 = 3^2$$
$$x^2 + y^2 = 9$$

We can see from Example 4 that the equation of any circle with its center at the origin and radius r will be

$$x^2 + y^2 = r^2$$

EXAMPLE   5   Find the center and radius, and sketch the graph, of the circle whose equation is

$$(x - 1)^2 + (y + 3)^2 = 4$$

Solution   Writing the equation in the form

$$(x - a)^2 + (y - b)^2 = r^2$$

we have

$$(x - 1)^2 + [y - (-3)]^2 = 2^2$$

The center is at $(1, -3)$, and the radius is 2.

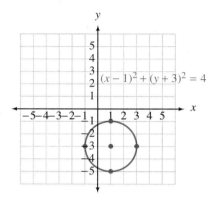

FIGURE 3

EXAMPLE 6   Sketch the graph of $x^2 + y^2 = 9$.

Solution   Since the equation can be written in the form

$$(x - 0)^2 + (y - 0)^2 = 3^2$$

it must have its center at $(0, 0)$ and a radius of 3.

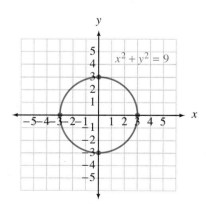

FIGURE 4

EXAMPLE 7   Sketch the graph of $x^2 + y^2 + 6x - 4y - 12 = 0$.

Solution   To sketch the graph, we must find the center and radius. The center and radius can be identified if the equation has the form

$$(x - a)^2 + (y - b)^2 = r^2$$

The original equation can be written in this form by completing the squares on x and y:

$$x^2 + y^2 + 6x - 4y - 12 = 0$$

$$x^2 + 6x \qquad + y^2 - 4y \qquad = 12$$

$$x^2 + 6x + \mathbf{9} + y^2 - 4y + \mathbf{4} = 12 + \mathbf{9} + \mathbf{4}$$

$$(x + 3)^2 + (y - 2)^2 = 25$$

$$(x + 3)^2 + (y - 2)^2 = 5^2$$

From the last line it is apparent that the center is at $(-3, 2)$ and the radius is 5.

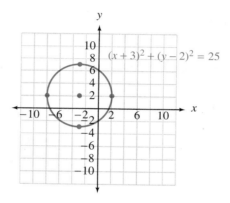

FIGURE 5

Getting Ready for Class

After reading through the preceding section, respond in your own words and in complete sentences.

A. Describe the distance formula in words, as if you were explaining to someone how they should go about finding the distance between two points.

B. What is the mathematical definition of a circle?

C. How are the distance formula and the equation of a circle related?

D. When graphing a circle from its equation, why is completing the square sometimes useful?

PROBLEM SET 13.1

Find the distance between the following points.

1. (3, 7) and (6, 3) **2.** (4, 7) and (8, 1)

3. (0, 9) and (5, 0) **4.** (−3, 0) and (0, 4)

5. (3, −5) and (−2, 1) **6.** (−8, 9) and (−3, −2)

7. (−1, −2) and (−10, 5) **8.** (−3, −8) and (−1, 6)

9. Find x so that the distance between $(x, 2)$ and $(1, 5)$ is $\sqrt{13}$.

10. Find x so that the distance between $(−2, 3)$ and $(x, 1)$ is 3.

11. Find y so that the distance between $(7, y)$ and $(8, 3)$ is 1.

12. Find y so that the distance between $(3, −5)$ and $(3, y)$ is 9.

Write the equation of the circle with the given center and radius.

13. Center (2, 3); $r = 4$

14. Center (3, −1); $r = 5$

15. Center (3, −2); $r = 3$

16. Center (−2, 4); $r = 1$

17. Center (−5, −1); $r = \sqrt{5}$

18. Center (−7, −6); $r = \sqrt{3}$

19. Center (0, −5); $r = 1$

20. Center (0, −1); $r = 7$

21. Center (0, 0); $r = 2$

22. Center (0, 0); $r = 5$

Give the center and radius, and sketch the graph of each of the following circles.

23. $x^2 + y^2 = 4$ **24.** $x^2 + y^2 = 16$

25. $(x − 1)^2 + (y − 3)^2 = 25$

26. $(x − 4)^2 + (y − 1)^2 = 36$

27. $(x + 2)^2 + (y − 4)^2 = 8$

28. $(x − 3)^2 + (y + 1)^2 = 12$

29. $(x + 1)^2 + (y + 1)^2 = 1$

30. $(x + 3)^2 + (y + 2)^2 = 9$

31. $x^2 + y^2 − 6y = 7$ **32.** $x^2 + y^2 − 4y = 5$

33. $x^2 + y^2 + 2x = 1$ **34.** $x^2 + y^2 + 10x = 0$

35. $x^2 + y^2 − 4x − 6y = −4$

36. $x^2 + y^2 − 4x + 2y = 4$

37. $x^2 + y^2 + 2x + y = \dfrac{11}{4}$

38. $x^2 + y^2 − 6x − y = −\dfrac{1}{4}$

Each of the following circles passes through the origin. In each case, find the equation.

39.

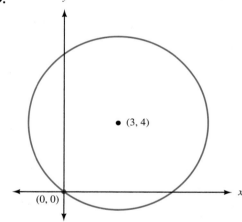

40.

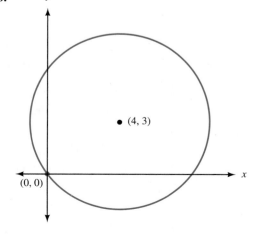

41. Find the equations of circles A, B, and C in the following diagram. The three points are the centers of the three circles.

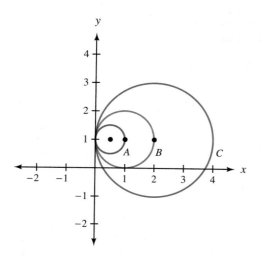

42. Each of the following circles passes through the origin. The centers are as shown. Find the equation of each circle.

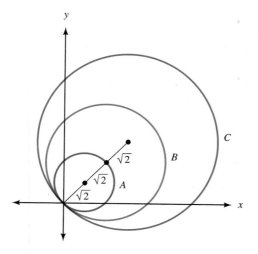

43. Find the equation of the circle with center at the origin that contains the point (3, 4).

44. Find the equation of the circle with center at the origin that contains the point (−5, 12).

45. Find the equation of the circle with center at the origin and x-intercepts 3 and −3.

46. Find the equation of the circle with y-intercepts 4 and −4 and center at the origin.

47. A circle with center at (−1, 3) passes through the point (4, 3). Find the equation.

48. A circle with center at (2, 5) passes through the point (−1, 4). Find the equation.

Review Problems

The problems that follow are a review of material we covered in Sections 12.1 and 12.2.

Find the general term of each sequence. [12.1]

49. 5, 9, 13, 17, . . .

50. 3, 8, 15, 24, . . .

Expand and simplify each series. [12.2]

51. $\displaystyle\sum_{i=2}^{5} \left(\frac{1}{2}\right)^{i}$ **52.** $\displaystyle\sum_{i=3}^{6} (i^2 - 5)$

Write using summation notation. [12.2]

53. $1 + 3 + 5 + 7 + 9$

54. $\frac{2}{3} + \frac{3}{4} + \frac{4}{5} + \frac{5}{6}$

Extending the Concepts

A circle is *tangent to* a line if it touches, but does not cross, the line.

55. Find the equation of the circle with center at (2, 3) if the circle is tangent to the y-axis.

56. Find the equation of the circle with center at (3, 2) if the circle is tangent to the x-axis.

57. Find the equation of the circle with center at (2, 3) if the circle is tangent to the vertical line $x = 4$.

58. Find the equation of the circle with center at (3, 2) if the circle is tangent to the horizontal line $y = 6$.

Find the distance from the origin to the center of each of the following circles.

59. $x^2 + y^2 - 6x + 8y = 144$

60. $x^2 + y^2 - 8x + 6y = 144$

61. $x^2 + y^2 - 6x - 8y = 144$

62. $x^2 + y^2 + 8x + 6y = 144$

This section is concerned with the graphs of ellipses and hyperbolas. To begin, we will consider only those graphs that are centered about the origin.

Suppose we want to graph the equation

$$\frac{x^2}{25} + \frac{y^2}{9} = 1$$

We can find the y-intercepts by letting $x = 0$, and the x-intercepts by letting $y = 0$:

When $x = 0$

$$\frac{0^2}{25} + \frac{y^2}{9} = 1$$

$$y^2 = 9$$

$$y = \pm 3$$

When $y = 0$

$$\frac{x^2}{25} + \frac{0^2}{9} = 1$$

$$x^2 = 25$$

$$x = \pm 5$$

The graph crosses the y-axis at $(0, 3)$ and $(0, -3)$ and the x-axis at $(5, 0)$ and $(-5, 0)$. Graphing these points and then connecting them with a smooth curve gives the graph shown in Figure 1.

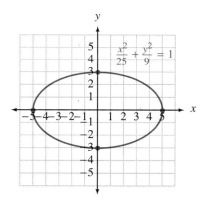

FIGURE 1

We can find other ordered pairs on the graph by substituting in values for x (or y) and then solving for y (or x). For example, if we let $x = 3$, then

$$\frac{3^2}{25} + \frac{y^2}{9} = 1$$

$$\frac{9}{25} + \frac{y^2}{9} = 1$$

$$0.36 + \frac{y^2}{9} = 1$$

$$\frac{y^2}{9} = 0.64$$

$$y^2 = 5.76$$

$$y = \pm 2.4$$

This would give us the two ordered pairs $(3, -2.4)$ and $(3, 2.4)$.

A graph of the type shown in Figure 1 is called an *ellipse*. If we were to find some other ordered pairs that satisfy our original equation, we would find that their graphs lie on the ellipse. Also, the coordinates of any point on the ellipse will satisfy the equation. We can generalize these results as follows.

THE ELLIPSE

The graph of any equation of the form

$$\frac{x^2}{a^2} + \frac{y^2}{b^2} = 1 \qquad \text{Standard form}$$

will be an **ellipse** centered at the origin. The ellipse will cross the x-axis at $(a, 0)$ and $(-a, 0)$. It will cross the y-axis at $(0, b)$ and $(0, -b)$. When a and b are equal, the ellipse will be a circle. Each of the points $(a, 0)$, $(-a, 0)$, $(0, b)$, and $(0, -b)$ is a **vertex** (intercept) of the graph.

The most convenient way to graph an ellipse is to locate the intercepts (vertices).

E X A M P L E   1   Sketch the graph of $4x^2 + 9y^2 = 36$.

Solution   To write the equation in the form

$$\frac{x^2}{a^2} + \frac{y^2}{b^2} = 1$$

we must divide both sides by 36:

$$\frac{4x^2}{36} + \frac{9y^2}{36} = \frac{36}{36}$$

$$\frac{x^2}{9} + \frac{y^2}{4} = 1$$

The graph crosses the x-axis at $(3, 0)$, $(-3, 0)$ and the y-axis at $(0, 2)$, $(0, -2)$. (See Figure 2.)

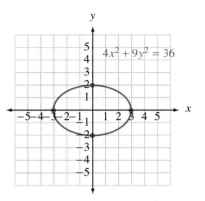

FIGURE 2

Consider the equation

$$\frac{x^2}{9} - \frac{y^2}{4} = 1$$

If we were to find a number of ordered pairs that are solutions to the equation and connect their graphs with a smooth curve, we would have Figure 3.

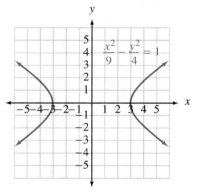

FIGURE 3

This graph is an example of a *hyperbola*. Notice that the graph has x-intercepts at $(3, 0)$ and $(-3, 0)$. The graph has no y-intercepts and hence does not cross the y-axis, since substituting $x = 0$ into the equation yields

$$\frac{0^2}{9} - \frac{y^2}{4} = 1$$

$$-y^2 = 4$$

$$y^2 = -4$$

for which there is no real solution. We can, however, use the number below y^2 to help sketch the graph. If we draw a rectangle that has its sides parallel to the x- and y-axes and that passes through the x-intercepts and the points on the y-axis corresponding to the square roots of the number below y^2, $+2$ and -2, it looks like the rectangle in Figure 4. The lines that connect opposite corners of the rectangle are called *asymptotes*. The graph of the hyperbola

$$\frac{x^2}{9} - \frac{y^2}{4} = 1$$

will approach these lines. Figure 4 shows the graph.

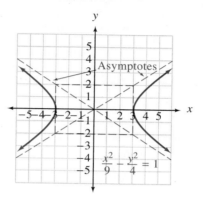

FIGURE 4

E X A M P L E   2   Graph the equation $\dfrac{y^2}{9} - \dfrac{x^2}{16} = 1$.

Solution   In this case the y-intercepts are 3 and -3, and the x-intercepts do not exist. We can use the square roots of the number below x^2, however, to find the asymptotes associated with the graph. The sides of the rectangle used to draw the asymptotes must pass through 3 and -3 on the y-axis, and 4 and -4 on the x-axis. (See Figure 5.)

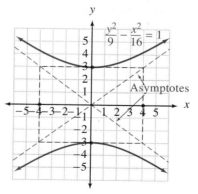

FIGURE 5

Here is a summary of what we have for hyperbolas.

HYPERBOLAS CENTERED AT THE ORIGIN

The graph of the equation

$$\frac{x^2}{a^2} - \frac{y^2}{b^2} = 1$$

will be a **hyperbola centered at the origin.** The graph will have **x-intercepts (vertices)** at $-a$ and a.

The graph of the equation

$$\frac{y^2}{a^2} - \frac{x^2}{b^2} = 1$$

will be a **hyperbola centered at the origin.** The graph will have **y-intercepts (vertices)** at $-a$ and a.

As an aid in sketching either of these equations, the asymptotes can be found by drawing lines through opposite corners of the rectangle whose sides pass through $-a$, a, $-b$, and b on the axes.

ELLIPSES AND HYPERBOLAS NOT CENTERED AT THE ORIGIN

The following equation is that of an ellipse with its center at the point $(4, 1)$:

$$\frac{(x - 4)^2}{9} + \frac{(y - 1)^2}{4} = 1$$

To see why the center is at $(4, 1)$ we substitute x' (read "x prime") for $x - 4$ and y' for $y - 1$ in the equation. That is:

If $\qquad x' = x - 4$

and $\qquad y' = y - 1$

the equation $\quad \dfrac{(x - 4)^2}{9} + \dfrac{(y - 1)^2}{4} = 1$

becomes $\qquad \dfrac{(x')^2}{9} + \dfrac{(y')^2}{4} = 1$

This is the equation of an ellipse in a coordinate system with an x'-axis and a y'-axis. We call this new coordinate system the $x'y'$-**coordinate system.** The center of our ellipse is at the origin in the $x'y'$-coordinate system. The question is this: What are the coordinates of the center of this ellipse in the original xy-coordinate system? To answer this question we go back to our original substitutions:

$$x' = x - 4$$
$$y' = y - 1$$

In the $x'y'$-coordinate system, the center of our ellipse is at $x' = 0, y' = 0$ (the origin of the $x'y'$ system). Substituting these numbers for x' and y', we have

$$0 = x - 4$$
$$0 = y - 1$$

Solving these equations for x and y will give us the coordinates of the center of our ellipse in the xy-coordinate system. As you can see, the solutions are $x = 4$ and

$y = 1$. Therefore, in the xy-coordinate system, the center of our ellipse is at the point $(4, 1)$. Figure 6 shows the graph.

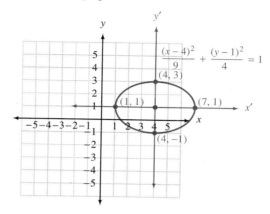

FIGURE 6

The coordinates of all points labeled in Figure 6 are given with respect to the xy-coordinate system. The x'- and y'-axes are shown simply for reference in our discussion. Note that the horizontal distance from the center to the vertices is 3—the square root of the denominator of the $(x - 4)^2$ term. Likewise, the vertical distance from the center to the other vertices is 2—the square root of the denominator of the $(y - 1)^2$ term.

We summarize the information above with the following:

AN ELLIPSE WITH CENTER AT (h, k)

The graph of the equation

$$\frac{(x - h)^2}{a^2} + \frac{(y - k)^2}{b^2} = 1$$

will be an **ellipse with center at (h, k)**. The vertices of the ellipse will be at the points $(h + a, k)$, $(h - a, k)$, $(h, k + b)$, and $(h, k - b)$.

EXAMPLE 3   Graph the ellipse: $x^2 + 9y^2 + 4x - 54y + 76 = 0$

Solution   To identify the coordinates of the center, we must complete the square on x and also on y. To begin, we rearrange the terms so that those containing x are together, those containing y are together, and the constant term is on the other side of the equal sign. Doing so gives us the following equation:

$$x^2 + 4x + 9y^2 - 54y = -76$$

Before we can complete the square on y, we must factor 9 from each term containing y:

$$x^2 + 4x + 9(y^2 - 6y) = -76$$

To complete the square on x, we add 4 to each side of the equation. To complete the square on y, we add 9 inside the parentheses. This increases the left side of the equation by 81 since each term within the parentheses is multiplied by 9. Therefore, we must add 81 to the right side of the equation also.

$$x^2 + 4x + \mathbf{4} + 9(y^2 - 6y + \mathbf{9}) = -76 + \mathbf{4} + \mathbf{81}$$

$$(x + 2)^2 + 9(y - 3)^2 = 9$$

To identify the distances to the vertices, we divide each term on both sides by 9:

$$\frac{(x + 2)^2}{9} + \frac{9(y - 3)^2}{9} = \frac{9}{9}$$

$$\frac{(x + 2)^2}{9} + \frac{(y - 3)^2}{1} = 1$$

The graph is an ellipse with center at $(-2, 3)$, as shown in Figure 7.

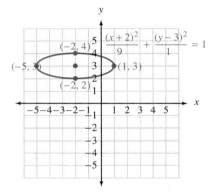

FIGURE 7

The ideas associated with graphing hyperbolas whose centers are not at the origin parallel the ideas just presented about graphing ellipses whose centers have been moved off the origin. Without showing the justification for doing so, we state the following guidelines for graphing hyperbolas:

HYPERBOLAS WITH CENTERS AT (h, k)

The graphs of the equations

$$\frac{(x - h)^2}{a^2} - \frac{(y - k)^2}{b^2} = 1 \quad \text{and} \quad \frac{(y - k)^2}{b^2} - \frac{(x - h)^2}{a^2} = 1$$

will be hyperbolas with their centers at (h, k). The vertices of the graph of the first equation will be at the points $(h + a, k)$ and $(h - a, k)$, and the vertices for the graph of the second equation will be at $(h, k + b)$ and $(h, k - b)$. In either case, the asymptotes can be found by connecting opposite corners of the rectangle that contains the four points $(h + a, k)$, $(h - a, k)$, $(h, k + b)$, and $(h, k - b)$.

E X A M P L E   4   Graph the hyperbola: $4x^2 - y^2 + 4y - 20 = 0$

Solution   To identify the coordinates of the center of the hyperbola, we need to complete the square on y. (Since there is no linear term in x, we do not need to complete the square on x. The x-coordinate of the center will be $x = 0$.)

$$4x^2 - y^2 + 4y - 20 = 0$$

$$4x^2 - y^2 + 4y = 20 \qquad \text{Add 20 to each side.}$$

$$4x^2 - 1(y^2 - 4y) = 20 \qquad \text{Factor } -1 \text{ from each term containing } y.$$

To complete the square on y, we add 4 to the terms inside the parentheses. Doing so adds -4 to the left side of the equation since everything inside the parentheses is multiplied by -1. To keep from changing the equation we must add -4 to the right side also.

$$4x^2 - 1(y^2 - 4y + \mathbf{4}) = 20 - \mathbf{4}$$

$$4x^2 - 1(y - 2)^2 = 16$$

$$\frac{4x^2}{16} - \frac{(y - 2)^2}{16} = \frac{16}{16}$$

$$\frac{x^2}{4} - \frac{(y - 2)^2}{16} = 1$$

This is the equation of a hyperbola with center at $(0, 2)$. The graph opens to the right and left as shown in Figure 8.

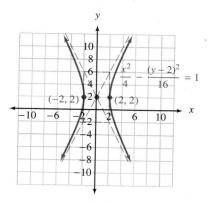

FIGURE 8

 # Getting Ready for Class

After reading through the preceding section, respond in your own words and in complete sentences.

A. How do we find the x-intercepts of a graph from the equation?

B. What is an ellipse?

C. How can you tell by looking at an equation, if its graph will be an ellipse or a hyperbola?

D. Are the points on the asymptotes of a hyperbola in the solution set of the equation of the hyperbola? Explain. (That is, are the asymptotes actually part of the graph?)

PROBLEM SET 13.2

Graph each of the following. Be sure to label both the x- and y-intercepts.

1. $\dfrac{x^2}{9} + \dfrac{y^2}{16} = 1$ **2.** $\dfrac{x^2}{25} + \dfrac{y^2}{4} = 1$

3. $\dfrac{x^2}{16} + \dfrac{y^2}{9} = 1$ **4.** $\dfrac{x^2}{4} + \dfrac{y^2}{25} = 1$

5. $\dfrac{x^2}{3} + \dfrac{y^2}{4} = 1$ **6.** $\dfrac{x^2}{4} + \dfrac{y^2}{3} = 1$

7. $4x^2 + 25y^2 = 100$ **8.** $4x^2 + 9y^2 = 36$

9. $x^2 + 8y^2 = 16$ **10.** $12x^2 + y^2 = 36$

Graph each of the following. Show the intercepts and the asymptotes in each case.

11. $\dfrac{x^2}{9} - \dfrac{y^2}{16} = 1$ **12.** $\dfrac{x^2}{25} - \dfrac{y^2}{4} = 1$

13. $\dfrac{x^2}{16} - \dfrac{y^2}{9} = 1$ **14.** $\dfrac{x^2}{4} - \dfrac{y^2}{25} = 1$

15. $\dfrac{y^2}{9} - \dfrac{x^2}{16} = 1$ **16.** $\dfrac{y^2}{25} - \dfrac{x^2}{4} = 1$

17. $\dfrac{y^2}{36} - \dfrac{x^2}{4} = 1$ **18.** $\dfrac{y^2}{4} - \dfrac{x^2}{36} = 1$

19. $x^2 - 4y^2 = 4$ **20.** $y^2 - 4x^2 = 4$

21. $16y^2 - 9x^2 = 144$ **22.** $4y^2 - 25x^2 = 100$

Find the x- and y-intercepts, if they exist, for each of the following. Do not graph.

23. $0.4x^2 + 0.9y^2 = 3.6$

24. $1.6x^2 + 0.9y^2 = 14.4$

25. $\dfrac{x^2}{0.04} - \dfrac{y^2}{0.09} = 1$ **26.** $\dfrac{y^2}{0.16} - \dfrac{x^2}{0.25} = 1$

27. $\dfrac{25x^2}{9} + \dfrac{25y^2}{4} = 1$ **28.** $\dfrac{16x^2}{9} + \dfrac{16y^2}{25} = 1$

Graph each of the following ellipses. In each case, label the coordinates of the center and the vertices.

29. $\dfrac{(x - 4)^2}{4} + \dfrac{(y - 2)^2}{9} = 1$

30. $\dfrac{(x - 2)^2}{4} + \dfrac{(y - 4)^2}{9} = 1$

31. $4x^2 + y^2 - 4y - 12 = 0$

32. $4x^2 + y^2 - 24x - 4y + 36 = 0$

33. $x^2 + 9y^2 + 4x - 54y + 76 = 0$

34. $4x^2 + y^2 - 16x + 2y + 13 = 0$

Graph each of the following hyperbolas. In each case, label the coordinates of the center and the vertices and show the asymptotes.

35. $\dfrac{(x - 2)^2}{16} - \dfrac{y^2}{4} = 1$

36. $\dfrac{(y - 2)^2}{16} - \dfrac{x^2}{4} = 1$

37. $9y^2 - x^2 - 4x + 54y + 68 = 0$

38. $4x^2 - y^2 - 24x + 4y + 28 = 0$

39. $4y^2 - 9x^2 - 16y + 72x - 164 = 0$

40. $4x^2 - y^2 - 16x - 2y + 11 = 0$

Find the equation for the following ellipses and hyperbolas.

41.

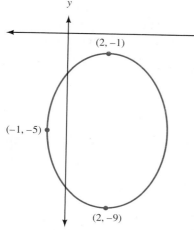

42.

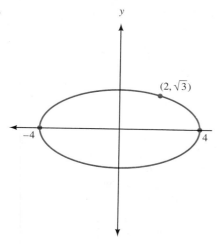

43.

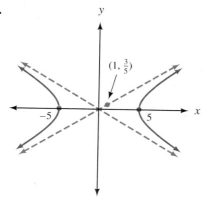

44.

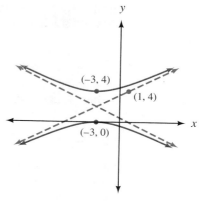

45. Give the equations of the two asymptotes in the graph you found in Problem 15.

46. Give the equations of the two asymptotes in the graph you found in Problem 16.

47. The longer line segment connecting opposite vertices of an ellipse is called the **major axis** of the ellipse. Give the length of the major axis of the ellipse you graphed in Problem 3.

48. The shorter line segment connecting opposite vertices of an ellipse is called the **minor axis** of the ellipse. Give the length of the minor axis of the ellipse you graphed in Problem 3.

Review Problems

The following problems review material we covered in Sections 12.1, 12.3, and 12.4.

Find the general term of each sequence.

49. 5, 11, 17, 23, . . .

50. $-3, 9, -27, 81, \ldots$

51. An arithmetic sequence has a first term of $a_1 = 4$ and a common difference of $d = 5$. Find the sum of the first 20 terms, S_{20}.

52. An arithmetic sequence is such that $a_4 = 23$ and $a_9 = 48$. Find a_{40}.

53. A geometric sequence has a first term of $a_1 = 8$ and a common ratio of $r = \frac{1}{2}$. Find the sum of the first 6 terms.

54. Find the sum: $1 + \frac{1}{2} + \frac{1}{4} + \frac{1}{8} + \cdots$

In Section 8.3 we graphed linear inequalities by first graphing the boundary and then choosing a test point not on the boundary to indicate the region used for the solution set. The problems in this section are very similar. We will use the same general methods for graphing the inequalities in this section that we used in Section 8.3.

EXAMPLE 1 Graph $x^2 + y^2 < 16$.

Solution The boundary is $x^2 + y^2 = 16$, which is a circle with center at the origin and a radius of 4. Since the inequality sign is $<$, the boundary is not included in the solution set and must therefore be represented with a broken line. The graph of the boundary is shown in Figure 1.

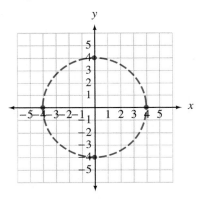

FIGURE 1

The solution set for $x^2 + y^2 < 16$ is either the region inside the circle or the region outside the circle. To see which region represents the solution set, we choose a convenient point not on the boundary and test it in the original inequality. The origin $(0, 0)$ is a convenient point. Since the origin satisfies the inequality $x^2 + y^2 < 16$, all points in the same region will also satisfy the inequality. The graph of the solution set is shown in Figure 2.

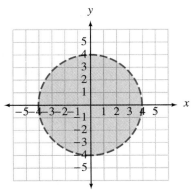

FIGURE 2

EXAMPLE 2   Graph the inequality $y \le x^2 - 2$.

Solution   The parabola $y = x^2 - 2$ is the boundary and is included in the solution set. Using $(0, 0)$ as the test point, we see that $0 \le 0^2 - 2$ is a false statement, which means that the region containing $(0, 0)$ is not in the solution set. (See Figure 3.)

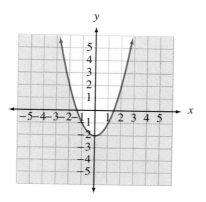

FIGURE 3

EXAMPLE 3   Graph $4y^2 - 9x^2 < 36$.

Solution   The boundary is the hyperbola $4y^2 - 9x^2 = 36$ and is not included in the solution set. Testing $(0, 0)$ in the original inequality yields a true statement, which means that the region containing the origin is the solution set. (See Figure 4.)

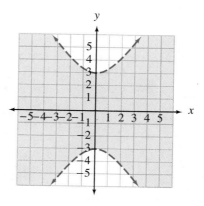

FIGURE 4

EXAMPLE 4   Solve the system.

$$x^2 + y^2 = 4$$
$$x - 2y = 4$$

Solution　In this case the substitution method is the most convenient. Solving the second equation for x in terms of y, we have

$$x - 2y = 4$$
$$x = 2y + 4$$

We now substitute $2y + 4$ for x in the first equation in our original system and proceed to solve for y:

$$(2y + 4)^2 + y^2 = 4$$
$$4y^2 + 16y + 16 + y^2 = 4$$
$$5y^2 + 16y + 12 = 0$$
$$(5y + 6)(y + 2) = 0$$

$$5y + 6 = 0 \qquad \text{or} \qquad y + 2 = 0$$

$$y = -\frac{6}{5} \qquad \text{or} \qquad y = -2$$

These are the y-coordinates of the two solutions to the system. Substituting $y = -\frac{6}{5}$ into $x - 2y = 4$ and solving for x gives us $x = \frac{8}{5}$. Using $y = -2$ in the same equation yields $x = 0$. The two solutions to our system are $(\frac{8}{5}, -\frac{6}{5})$ and $(0, -2)$. Although graphing the system is not necessary, it does help us visualize the situation. (See Figure 5.)

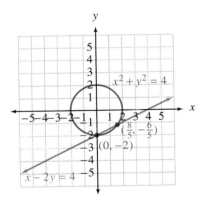

FIGURE 5

EXAMPLE 5　Solve the system.

$$16x^2 - 4y^2 = 64$$
$$x^2 + y^2 = 9$$

Solution　Since each equation is of the second degree in both x and y, it is easier to solve this system by eliminating one of the variables by addition. To eliminate y, we multiply the bottom equation by 4 and add the result to the top equation:

$$
\begin{array}{rl}
16x^2 - 4y^2 = & 64 \\
4x^2 + 4y^2 = & 36 \\
\hline
20x^2 \quad\quad = & 100
\end{array}
$$

$$x^2 = 5$$

$$x = \pm\sqrt{5}$$

The x-coordinates of the points of intersection are $\sqrt{5}$ and $-\sqrt{5}$. We substitute each back into the second equation in the original system and solve for y:

$$\text{When} \qquad\qquad\qquad x = \sqrt{5}$$

$$(\sqrt{5})^2 + y^2 = 9$$

$$5 + y^2 = 9$$

$$y^2 = 4$$

$$y = \pm 2$$

$$\text{When} \qquad\qquad\qquad x = -\sqrt{5}$$

$$(-\sqrt{5})^2 + y^2 = 9$$

$$5 + y^2 = 9$$

$$y^2 = 4$$

$$y = \pm 2$$

The four points of intersection are $(\sqrt{5}, 2)$, $(\sqrt{5}, -2)$, $(-\sqrt{5}, 2)$, and $(-\sqrt{5}, -2)$. Graphically the situation is as shown in Figure 6.

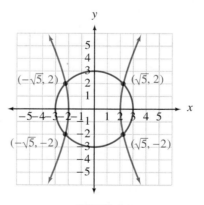

FIGURE 6

E X A M P L E   6   Solve the system.

$$x^2 - 2y = 2$$

$$y = x^2 - 3$$

Solution   We can solve this system using the substitution method. Replacing y in the first equation with $x^2 - 3$ from the second equation, we have

$$x^2 - 2(x^2 - 3) = 2$$
$$-x^2 + 6 = 2$$
$$x^2 = 4$$
$$x = \pm 2$$

Using either $+2$ or -2 in the equation $y = x^2 - 3$ gives us $y = 1$. The system has two solutions: $(2, 1)$ and $(-2, 1)$.

EXAMPLE 7   The sum of the squares of two numbers is 34. The difference of their squares is 16. Find the two numbers.

Solution   Let x and y be the two numbers. The sum of their squares is $x^2 + y^2$, and the difference of their squares is $x^2 - y^2$. (We can assume here that x^2 is the larger number.) The system of equations that describes the situation is

$$x^2 + y^2 = 34$$
$$x^2 - y^2 = 16$$

We can eliminate y by simply adding the two equations. The result of doing so is

$$2x^2 = 50$$
$$x^2 = 25$$
$$x = \pm 5$$

Substituting $x = 5$ into either equation in the system gives $y = \pm 3$. Using $x = -5$ gives the same results, $y = \pm 3$. The four pairs of numbers that are solutions to the original problem are

$$(5, 3) \qquad (-5, 3) \qquad (5, -3) \qquad (-5, -3)$$

We now turn our attention to systems of inequalities. To solve a system of inequalities by graphing, we simply graph each inequality on the same set of axes. The solution set for the system is the region common to both graphs—the intersection of the individual solution sets.

EXAMPLE 8   Graph the solution set for the system

$$x^2 + y^2 \leq 9$$
$$\frac{x^2}{4} + \frac{y^2}{25} \geq 1$$

Solution   The boundary for the top inequality is a circle with center at the origin and a radius of 3. The solution set lies inside the boundary. The boundary for the second inequality is an ellipse. In this case the solution set lies outside the boundary. (See Figure 7.)

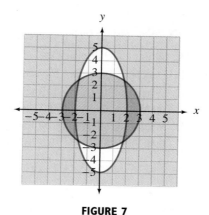

FIGURE 7

The solution set is the intersection of the two individual solution sets.

E X A M P L E   9   Graph the solution set for the following system.

$$x - 2y \leq 4$$

$$x + \ y \leq 4$$

$$x \geq -1$$

Solution   We have three linear inequalities, representing three sections of the coordinate plane. The graph of the solution set for this system will be the intersection of these three sections. The graph of $x - 2y \leq 4$ is the section above and including the boundary $x - 2y = 4$. The graph of $x + y \leq 4$ is the section below and including the boundary line $x + y = 4$. The graph of $x \geq -1$ is all the points to the right of, and including, the vertical line $x = -1$. The intersection of these three graphs is shown in Figure 8.

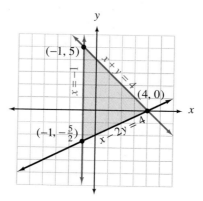

FIGURE 8

Getting Ready for Class

After reading through the preceding section, respond in your own words and in complete sentences.

A. What is the significance of a broken line when graphing inequalities?

B. Describe, in words, the set of points described by $(x - 3)^2 + (y - 2)^2 < 9$.

C. When solving the nonlinear systems whose graphs are a line and a circle, how many possible solutions can you expect?

D. When solving the nonlinear systems whose graphs are both circles, how many possible solutions can you expect?

PROBLEM SET 13.3

Graph each of the following inequalities.

1. $x^2 + y^2 \le 49$ **2.** $x^2 + y^2 < 49$

3. $(x - 2)^2 + (y + 3)^2 < 16$

4. $(x + 3)^2 + (y - 2)^2 \ge 25$

5. $y < x^2 - 6x + 7$ **6.** $y \ge x^2 + 2x - 8$

7. $\dfrac{x^2}{25} - \dfrac{y^2}{9} \ge 1$ **8.** $\dfrac{x^2}{25} - \dfrac{y^2}{9} \le 1$

9. $4x^2 + 25y^2 \le 100$ **10.** $25x^2 - 4y^2 > 100$

Graph the solution sets to the following systems.

11. $x^2 + y^2 < 9$
$\quad\ \ y \ge x^2 - 1$

12. $x^2 + y^2 \le 16$
$\quad\ \ y < x^2 + 2$

13. $\dfrac{x^2}{9} + \dfrac{y^2}{25} \le 1$
$\quad \dfrac{x^2}{4} - \dfrac{y^2}{9} > 1$

14. $\dfrac{x^2}{4} + \dfrac{y^2}{16} \ge 1$
$\quad \dfrac{x^2}{9} - \dfrac{y^2}{25} < 1$

15. $4x^2 + 9y^2 \le 36$
$\quad\quad y > x^2 + 2$

16. $9x^2 + 4y^2 \ge 36$
$\quad\quad y < x^2 + 1$

17. $x + \ \ y \le \ \ 3$
$\quad x - 3y \le \ \ 3$
$\quad\quad\ \ x \ge -2$

18. $x - \ \ y \le \ \ 4$
$\quad x + 2y \le \ \ 4$
$\quad\quad\ \ x \ge -1$

19. $\quad x + y \le \ \ 2$
$\quad -x + y \le \ \ 2$
$\quad\quad\quad y \ge -2$

20. $\quad x - y \le \ \ 3$
$\quad -x - y \le \ \ 3$
$\quad\quad\quad y \le -1$

21. $x + y \le 4$
$\quad\quad x \ge 0$
$\quad\quad y \ge 0$

22. $x - y \le 2$
$\quad\quad x \ge 0$
$\quad\quad y \le 0$

Solve each of the following systems of equations.

23. $x^2 + y^2 = 9$
$\quad 2x + y = 3$

24. $x^2 + \ \ y^2 = 9$
$\quad\ \ x + 2y = 3$

25. $x^2 + \ \ y^2 = 16$
$\quad\ \ x + 2y = \ \ 8$

26. $x^2 + \ \ y^2 = 16$
$\quad\ \ x - 2y = \ \ 8$

27. $x^2 + y^2 = 25$
$\quad x^2 - y^2 = 25$

28. $x^2 + y^2 = 4$
$\quad 2x^2 - y^2 = 5$

29. $x^2 + y^2 = 9$
$\quad\quad y = x^2 - 3$

30. $x^2 + y^2 = 4$
$\quad\quad y = x^2 - 2$

31. $x^2 + y^2 = 16$
$\quad\quad y = x^2 - 4$

32. $x^2 + y^2 = 1$
$\quad\quad y = x^2 - 1$

33. $3x + 2y = 10$
$\quad\quad y = x^2 - 5$

34. $4x + 2y = 10$
$\quad\quad y = x^2 - 10$

35. $y = x^2 + 2x - 3$
$\quad y = \quad\quad -x + 1$

36. $y = -x^2 - 2x + 3$
$\quad y = \quad\quad\quad x - 1$

37. $y = x^2 - 6x + 5$
$\quad y = \quad\quad x - 5$

38. $y = x^2 - 2x - 4$
$\quad y = \quad\quad x - 4$

39. $4x^2 - 9y^2 = 36$
$\quad 4x^2 + 9y^2 = 36$

40. $4x^2 + 25y^2 = 100$
$\quad 4x^2 - 25y^2 = 100$

41. $x \ - y = \ \ 4$
$\quad x^2 + y^2 = 16$

42. $x \ + y = \ \ 2$
$\quad x^2 - y^2 = 4$

Applying the Concepts

43. Number Problem The sum of the squares of two numbers is 89. The difference of their squares is 39. Find the numbers.

44. Number Problem The difference of the squares of two numbers is 35. The sum of their squares is 37. Find the numbers.

45. Number Problem One number is 3 less than the square of another. Their sum is 9. Find the numbers.

46. Number Problem The square of one number is 2 less than twice the square of another. The sum of the squares of the two numbers is 25. Find the numbers.

Review Problems

The following problems review material we covered in Section 12.5.

Expand and simplify.

47. $(x + 2)^4$

48. $(x - 2)^4$

49. $(2x + y)^3$

50. $(x - 2y)^3$

51. Find the first two terms in the expansion of $(x + 3)^{50}$.

52. Find the first two terms in the expansion of $(x - y)^{75}$.

CHAPTER 13   SUMMARY

Examples

1. The distance between $(5, 2)$ and $(-1, 1)$ is

$$d = \sqrt{(5 + 1)^2 + (2 - 1)^2}$$
$$= \sqrt{37}$$

Distance Formula [13.1]

The distance between the two points (x_1, y_1) and (x_2, y_2) is given by the formula

$$d = \sqrt{(x_2 - x_1)^2 + (y_2 - y_1)^2}$$

2. The graph of the circle

$$(x - 3)^2 + (y + 2)^2 = 25$$

has its center at $(3, -2)$ and the radius is 5.

The Circle [13.1]

The graph of any equation of the form

$$(x - a)^2 + (y - b)^2 = r^2$$

is a circle having its center at (a, b) and a radius of r.

3.

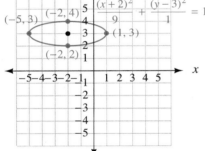

An Ellipse With Center at (h, k) [13.2]

The graph of the equation

$$\frac{(x - h)^2}{a^2} + \frac{(y - k)^2}{b^2} = 1$$

is an ellipse with center at (h, k). The vertices of the ellipse are at the points $(h + a, k)$, $(h - a, k)$, $(h, k + b)$, and $(h, k - b)$.

4.

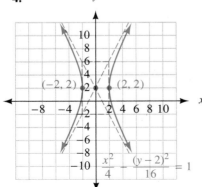

Hyperbolas With Centers at (h, k) [13.2]

The graphs of the equations

$$\frac{(x - h)^2}{a^2} - \frac{(y - k)^2}{b^2} = 1 \quad \text{and} \quad \frac{(y - k)^2}{b^2} - \frac{(x - h)^2}{a^2} = 1$$

are hyperbolas with their centers at (h, k). The vertices of the graph of the first equation are at the points $(h + a, k)$ and $(h - a, k)$, and the vertices for the graph of the second equation are at $(h, k + b)$ and $(h, k - b)$. In either case, the asymptotes can be found by connecting opposite corners of the rectangle that contains the points $(h + a, k)$, $(h - a, k)$, $(h, k + b)$, and $(h, k - b)$.

5. The graph of the inequality

$$x^2 + y^2 < 9$$

is all points inside the circle with center at the origin and radius 3. The circle itself is not part of the solution and is therefore shown with a broken curve.

6. We can solve the system

$$x^2 + y^2 = 4$$

$$x = 2y + 4$$

by substituting $2y + 4$ from the second equation for x in the first equation:

$$(2y + 4)^2 + y^2 = 4$$
$$4y^2 + 16y + 16 + y^2 = 4$$
$$5y^2 + 16y + 12 = 0$$
$$(5y + 6)(y + 2) = 0$$
$$y = -\frac{6}{5} \quad \text{or} \quad y = -2$$

Substituting these values of y into the second equation in our system gives $x = \frac{8}{5}$ and $x = 0$. The solutions are $(\frac{8}{5}, -\frac{6}{5})$ and $(0, -2)$.

Second-Degree Inequalities in Two Variables [13.3]

We graph second-degree inequalities in two variables in much the same way that we graphed linear inequalities. That is, we begin by graphing the boundary, using a solid curve if the boundary is included in the solution (this happens when the inequality symbol is $\geq$ or $\leq$), or a broken curve if the boundary is not included in the solution (when the inequality symbol is $>$ or $<$). After we have graphed the boundary, we choose a test point that is not on the boundary and try it in the original inequality. A true statement indicates we are in the region of the solution. A false statement indicates we are not in the region of the solution.

Systems of Nonlinear Equations [13.3]

A system of nonlinear equations is two equations, at least one of which is not linear, considered at the same time. The solution set for the system consists of all ordered pairs that satisfy both equations. In most cases we use the substitution method to solve these systems; however, the addition method can be used if like variables are raised to the same power in both equations. It is sometimes helpful to graph each equation in the system on the same set of axes in order to anticipate the number and approximate positions of the solutions.

CHAPTER 13   REVIEW

Find the distance between the following points. [13.1]

1. $(2, 6), (-1, 5)$

2. $(3, -4), (1, -1)$

3. $(0, 3), (-4, 0)$

4. $(-3, 7), (-3, -2)$

5. Find x so that the distance between $(x, -1)$ and $(2, -4)$ is 5. [13.1]

6. Find y so that the distance between $(3, -4)$ and $(-3, y)$ is 10. [13.1]

Write the equation of the circle with the given center and radius. [13.1]

7. Center $(3, 1), r = 2$

8. Center $(3, -1)$, $r = 4$

9. Center $(-5, 0)$, $r = 3$

10. Center $(-3, 4)$, $r = 3\sqrt{2}$

Find the equation of each circle. [13.1]

11. Center at the origin, x-intercepts ± 5

12. Center at the origin, y-intercepts ± 3

13. Center at $(-2, 3)$ and passing through the point $(2, 0)$

14. Center at $(-6, 8)$ and passing through the origin

Give the center and radius of each circle, and then sketch the graph. [13.1]

15. $x^2 + y^2 = 4$

16. $(x - 3)^2 + (y + 1)^2 = 16$

17. $x^2 + y^2 - 6x + 4y = -4$

18. $x^2 + y^2 + 4x - 2y = 4$

Graph each of the following. Label the x- and y-intercepts. [13.2]

19. $\dfrac{x^2}{4} + \dfrac{y^2}{9} = 1$ **20.** $4x^2 + y^2 = 16$

Graph the following. Show the asymptotes. [13.2]

21. $\dfrac{x^2}{4} - \dfrac{y^2}{9} = 1$ **22.** $4x^2 - y^2 = 16$

Graph each equation. [13.2]

23. $\dfrac{(x + 2)^2}{9} + \dfrac{(y - 3)^2}{1} = 1$

24. $\dfrac{(x - 2)^2}{16} - \dfrac{y^2}{4} = 1$

25. $9y^2 - x^2 - 4x + 54y + 68 = 0$

26. $9x^2 + 4y^2 - 72x - 16y + 124 = 0$

Graph each of the following inequalities. [13.3]

27. $x^2 + y^2 < 9$

28. $(x + 2)^2 + (y - 1)^2 \le 4$

29. $y \ge x^2 - 1$ **30.** $9x^2 + 4y^2 \le 36$

Graph the solution set for each system. [13.3]

31. $x^2 + y^2 < 16$ **32.** $\quad x + y \le \quad 2$
     $\quad\quad y > x^2 - 4$ $\quad -x + y \le \quad 2$
     $\quad\quad\quad\quad\quad\quad\quad\quad\quad\quad\quad y \ge -2$

Solve each system of equations. [13.3]

33. $\quad x^2 + y^2 = 16$ **34.** $x^2 + y^2 = 4$
     $\quad 2x + y = 4$ $\quad\quad y = x^2 - 2$

35. $9x^2 - 4y^2 = 36$ **36.** $2x^2 - 4y^2 = \quad 8$
     $9x^2 + 4y^2 = 36$ $\quad x^2 + 2y^2 = 10$

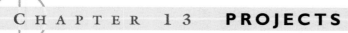

CHAPTER 13 PROJECTS

CONIC SECTIONS

CONSTRUCTING ELLIPSES

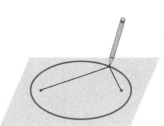

FIGURE 1

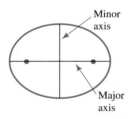

Minor axis

Major axis

FIGURE 2

Number of People: 4

Time Needed: 20 minutes

Equipment: Graph paper, pencils, string, and thumbtacks

Background: The geometric definition for an ellipse is the set of points the sum of whose distances from two fixed points (called foci) is a constant. We can use this definition to draw an ellipse using thumbtacks, string, and a pencil.

Procedure:

1. Start with a piece of string 7 inches long. Place thumbtacks through the string $\frac{1}{2}$ inch from each end, then tack the string to a pad of graph paper so that the tacks are 4 inches apart. Pull the string tight with the tip of a pencil, then trace all the way around the two tacks. (See Figure 1.) The resulting diagram will be an ellipse.

2. The line segment that passes through the tacks (these are the foci) and connects opposite ends of the ellipse is called the major axis. The line segment perpendicular to the major axis that passes through the center of the ellipse and connects opposites ends of the ellipse is called the minor axis. (See Figure 2.) Measure the length of the major axis and the length of the minor axis. Record your results in Table 1.

3. Explain how drawing the ellipse as you have in step 2 shows that the geometric definition of an ellipse given at the beginning of this project is, in fact, correct.

4. Next, move the tacks so that they are 3 inches apart. Trace out that ellipse. Measure the length of the major axis and the length of the minor axis, and record your results in Table 1.

5. Repeat step 4 with the tacks 2 inches apart.

TABLE 1 Ellipses (All Lengths Are Inches)			
Length of String	Distance Between Foci	Length of Major Axis	Length of Minor Axis
6	4		
6	3		
6	2		

6. If the length of the string between the tacks stays at 6 inches, and the tacks were placed 6 inches apart, then the resulting ellipse would be a _____. If the tacks were placed 0 inches apart, then the resulting ellipse would be a _____.

RESEARCH PROJECT

HYPATIA OF ALEXANDRIA

The first woman mentioned in the history of mathematics is Hypatia of Alexandria. Research the life of Hypatia, and then write an essay that begins with a description of the time and place in which she lived and then goes on to give an indication of the type of person she was, her accomplishments in areas other than mathematics, and how she was viewed by her contemporaries.

(Corbis-Bettmann)

CHAPTER 13   TEST

1. Find x so that $(x, 2)$ is $2\sqrt{5}$ units from $(-1, 4)$. [13.1]

2. Give the equation of the circle with center at $(-2, 4)$ and radius 3. [13.1]

3. Give the equation of the circle with center at the origin that contains the point $(-3, -4)$. [13.1]

4. Find the center and radius of the circle. [13.1]

$$x^2 + y^2 - 10x + 6y = 5$$

Graph each of the following. [13.2, 13.3]

5. $4x^2 - y^2 = 16$

6. $\dfrac{x^2}{25} + \dfrac{y^2}{4} = 1$

7. $(x - 2)^2 + (y + 1)^2 \le 9$

8. $9x^2 + 4y^2 - 72x - 16y + 124 = 0$

Solve the following systems. [13.3]

9. $\begin{aligned} x^2 + y^2 &= 25 \\ 2x + y &= 5 \end{aligned}$

10. $\begin{aligned} x^2 + y^2 &= 16 \\ y &= x^2 - 4 \end{aligned}$

Simplify.

1. $2^3 + 3(2 + 20 \div 4)$ **2.** $-|-5|$

3. $-5(2x + 3) + 8x$

4. $4 - 2[3x - 4(x + 2)]$

5. $(3y + 2)^2 - (3y - 2)^2$

6. $\dfrac{\frac{1}{5} - \frac{1}{4}}{\frac{1}{2} + \frac{3}{4}}$ **7.** $x^{2/3} \cdot x^{1/5}$

8. $\sqrt{48x^5y^3}$ (Assume x and y are positive.)

Solve.

9. $5y - 2 = -3y + 6$

10. $3 - 2(3x - 4) = -1$

11. $|3x - 1| - 2 = 6$ **12.** $2x^2 = 5x + 3$

13. $x^3 - 3x^2 - 4x + 12 = 0$

14. $\dfrac{6}{a + 2} = \dfrac{5}{a - 3}$ **15.** $x - 2 = \sqrt{3x + 4}$

16. $(x - 3)^2 = -3$ **17.** $4x^2 + 6x = -5$

18. $\log_2 x = 3$ **19.** $\log_2 x + \log_2 5 = 1$

20. $8^{x+3} = 4$ **21.** $4x + 2y = 4$
$$y = -3x + 1$$

22. $2x - 5y = 3$
$$3x + 2y = -5$$

23. $x + 2y - z = 4$
$$2x - y - 3z = -1$$
$$-x + 2y + 2z = 3$$

Multiply.

24. $\dfrac{x^2 - 16}{x^2 + 5x + 6} \cdot \dfrac{x^2 + 6x + 9}{x^3 + 4x^2}$

25. $(x^{1/5} + 3)(x^{1/5} - 3)$

Divide.

26. $\dfrac{12x^2y^3 - 16x^2y + 8xy^3}{4xy}$

27. $\dfrac{3 - 2i}{1 + 2i}$

Graph.

28. $2x - 3y = 12$ **29.** $3x - y < -2$

30. $y = 2^x$ **31.** $y = \log_2 x$

32. $x^2 + 4x + y^2 - 6y = 12$

33. $9x^2 - 4y^2 = 36$

34. $x^2 + y^2 < 4$ **35.** $y = (x - 2)^2 - 3$

Solve.

36. Find the next number in the sequence: $1, -4, 9, \ldots$

37. If $f(x) = 7x - 8$ find $f^{-1}(x)$.

38. Solve $mx + 2 = nx - 3$ for x.

39. If $f(x) = -\frac{3}{2}x + 1$, find $f(4)$.

40. Find the slope and y-intercept for $2x - 3y = 12$.

41. Find the equation of the line through $(-6, -1)$ and $(-3, -5)$.

42. y varies inversely with the square of x. If y is 3 when x is 3, find y when x is 6.

43. Find the distance between $(-3, 1)$ and $(4, 5)$.

44. Mixture How many gallons of 20% alcohol solution and 60% alcohol solution must be mixed to get 16 gallons of 30% alcohol solution? Be sure to show the system of equations used to solve the problem.

45. If $f(x) = 4 - x^2$ and $g(x) = 5x - 1$, find $(g \circ f)(x)$.

46. Find the second term in the expansion of $(x - 2y)^5$.

47. Projectile Motion An object projected upward with an initial velocity of 48 feet per second will rise and fall according to the equation $s = 48t - 16t^2$, where s is its distance above the ground at the time t. At what times will the object be 20 feet above the ground?

48. Find an equation that has solutions $x = 1$ and $x = \frac{2}{3}$.

49. Graph the solution set for $x^2 + x - 6 > 0$.

50. What is the domain for $f(x) = \dfrac{x + 3}{x - 2}$?

51. Find the general term of the sequence 5, 15, 45, 135, ...

52. Expand and simplify.

$$\sum_{i=1}^{5} (2i + 1)$$

Appendix A

Synthetic Division

*S*ynthetic division is a short form of long division with polynomials. We will consider synthetic division only for those cases in which the divisor is of the form $x + k$, where k is a constant.

Let's begin by looking over an example of long division with polynomials as done in Section 4.8:

$$
\begin{array}{r}
3x^2 - 2x + 4 \\
x + 3\overline{)3x^3 + 7x^2 - 2x - 4} \\
\underline{3x^3 + 9x^2} \\
-2x^2 - 2x \\
\underline{-2x^2 - 6x} \\
4x - 4 \\
\underline{4x + 12} \\
-16
\end{array}
$$

We can rewrite the problem without showing the variable, since the variable is written in descending powers and similar terms are in alignment. It looks like this:

$$
\begin{array}{r}
3 \quad -2 \quad +4 \\
1 + 3\overline{)3 \quad\ \ 7 \quad -2 \quad -4} \\
\underline{(3) \quad +9} \\
-2 \ (-2) \\
\underline{(-2) \ -6} \\
4 \ (-4) \\
\underline{(4) \quad 12} \\
-16
\end{array}
$$

We have used parentheses to enclose the numbers that are repetitions of the numbers above them. We can compress the problem by eliminating all repetitions, except the first one:

$$
\begin{array}{r}
3 \quad -2 \quad\ \ 4 \\
1 + 3\overline{)3 \quad\ \ 7 \quad -2 \quad -4} \\
\underline{9 \quad -6 \quad\ \ 12} \\
3 \quad -2 \quad\ \ 4 \ -16
\end{array}
$$

The top line is the same as the first three terms of the bottom line, so we eliminate the top line. Also, the 1 that was the coefficient of x in the original problem can be eliminated, since we will consider only division problems where the divisor is of the form $x + k$. The following is the most compact form of the original division problem:

$$
\begin{array}{r}
+3\overline{)3 \quad 7 \quad -2 \quad -4} \\
9 \quad -6 \quad 12 \\
\hline
3 \quad -2 \quad 4 \quad -16
\end{array}
$$

If we check over the problem, we find that the first term in the bottom row is exactly the same as the first term in the top row—and it always will be in problems of this type. Also, the last three terms in the bottom row come from multiplication by $+3$ and then subtraction. We can get an equivalent result by multiplying by -3 and adding. The problem would then look like this:

$$
\begin{array}{r}
-3\,\big| \quad 3 \quad\;\; 7 \quad -2 \quad\; -4 \\
\downarrow \quad -9 \quad\;\; 6 \quad -12 \\
\hline
3 \quad -2 \quad\;\; 4 \quad \boxed{-16}
\end{array}
$$

We have used the brackets $\rfloor\lfloor$ to separate the divisor and the remainder. This last expression is synthetic division. It is an easy process to remember. Simply change the sign of the constant term in the divisor, then bring down the first term of the dividend. The process is then just a series of multiplications and additions, as indicated in the following diagram by the arrows:

$$
\begin{array}{r}
-3\,\big| \quad 3 \quad\;\; 7 \quad -2 \quad\; -4 \\
\downarrow \quad -9 \quad\;\; 6 \quad -12 \\
\hline
3 \quad -2 \quad\;\; 4 \quad \boxed{-16}
\end{array}
$$

The last term of the bottom row is always the remainder.

Here are some additional examples of synthetic division with polynomials.

E X A M P L E 1   Divide $x^4 - 2x^3 + 4x^2 - 6x + 2$ by $x - 2$.

Solution We change the sign of the constant term in the divisor to get $+2$ and then complete the procedure:

$$
\begin{array}{r}
+2\,\big| \quad 1 \quad -2 \quad 4 \quad -6 \quad 2 \\
\downarrow \quad\;\; 2 \quad 0 \quad\;\; 8 \quad 4 \\
\hline
1 \quad\;\; 0 \quad 4 \quad\;\; 2 \quad \boxed{6}
\end{array}
$$

From the last line we have the answer:

$$
1x^3 + 0x^2 + 4x + 2 + \frac{6}{x - 2}
$$

E X A M P L E 2   Divide $\dfrac{3x^3 - 4x + 5}{x + 4}$.

Solution   Since we cannot skip any powers of the variable in the polynomial $3x^3 - 4x + 5$, we rewrite it as $3x^3 + 0x^2 - 4x + 5$ and proceed as we did in Example 1:

$$
\begin{array}{r|rrrr}
-4 & 3 & 0 & -4 & 5 \\
 & \downarrow & -12 & 48 & -176 \\
\hline
 & 3 & -12 & 44 & \boxed{-171}
\end{array}
$$

From the synthetic division, we have

$$\frac{3x^3 - 4x + 5}{x + 4} = 3x^2 - 12x + 44 - \frac{171}{x + 4}$$

EXAMPLE 3   Divide $\dfrac{x^3 - 1}{x - 1}$.

Solution   Writing the numerator as $x^3 + 0x^2 + 0x - 1$ and using synthetic division, we have

$$
\begin{array}{r|rrrr}
+1 & 1 & 0 & 0 & -1 \\
 & \downarrow & 1 & 1 & 1 \\
\hline
 & 1 & 1 & 1 & \boxed{0}
\end{array}
$$

which indicates

$$\frac{x^3 - 1}{x - 1} = x^2 + x + 1$$

PROBLEM SET A

Use synthetic division to find the following quotients.

1. $\dfrac{x^2 - 5x + 6}{x + 2}$

2. $\dfrac{x^2 + 8x - 12}{x - 3}$

3. $\dfrac{3x^2 - 4x + 1}{x - 1}$

4. $\dfrac{4x^2 - 2x - 6}{x + 1}$

5. $\dfrac{x^3 + 2x^2 + 3x + 4}{x - 2}$

6. $\dfrac{x^3 - 2x^2 - 3x - 4}{x - 2}$

7. $\dfrac{3x^3 - x^2 + 2x + 5}{x - 3}$

8. $\dfrac{2x^3 - 5x^2 + x + 2}{x - 2}$

9. $\dfrac{2x^3 + x - 3}{x - 1}$

10. $\dfrac{3x^3 - 2x + 1}{x - 5}$

11. $\dfrac{x^4 + 2x^2 + 1}{x + 4}$

12. $\dfrac{x^4 - 3x^2 + 1}{x - 4}$

13. $\dfrac{x^5 - 2x^4 + x^3 - 3x^2 - x + 1}{x - 2}$

14. $\dfrac{2x^5 - 3x^4 + x^3 - x^2 + 2x + 1}{x + 2}$

15. $\dfrac{x^2 + x + 1}{x - 1}$

16. $\dfrac{x^2 + x + 1}{x + 1}$

17. $\dfrac{x^4 - 1}{x + 1}$

18. $\dfrac{x^4 + 1}{x - 1}$

19. $\dfrac{x^3 - 1}{x - 1}$

20. $\dfrac{x^3 - 1}{x + 1}$

Appendix B

Introduction to Determinants

In this appendix we will expand and evaluate *determinants*. The purpose of this appendix is simply to be able to find the value of a given determinant. As we will see in the next appendix, determinants are very useful in solving systems of linear equations. Before we apply determinants to systems of linear equations, however, we must practice calculating the value of some determinants.

DEFINITION The value of the **2 × 2** (2 by 2) **determinant**

$$\begin{vmatrix} a & c \\ b & d \end{vmatrix}$$

is given by

$$\begin{vmatrix} a & c \\ b & d \end{vmatrix} = ad - bc$$

From the preceding definition we see that a determinant is simply a square array of numbers with two vertical lines enclosing it. The value of a 2×2 determinant is found by cross-multiplying on the diagonals and then subtracting, a diagram of which looks like

$$\begin{vmatrix} a & c \\ b & d \end{vmatrix} = ad - bc$$

EXAMPLE 1 Find the value of the following 2×2 determinants:

(a) $\begin{vmatrix} 1 & 2 \\ 3 & 4 \end{vmatrix} = 1(4) - 3(2) = 4 - 6 = -2$

(b) $\begin{vmatrix} 3 & 5 \\ -2 & 7 \end{vmatrix} = 3(7) - (-2)5 = 21 + 10 = 31$

EXAMPLE 2 Solve for x if

$$\begin{vmatrix} x^2 & 2 \\ x & 1 \end{vmatrix} = 8$$

Solution We expand the determinant on the left side to get

$$x^2(1) - x(2) = 8$$
$$x^2 - 2x = 8$$

$$x^2 - 2x - 8 = 0$$
$$(x - 4)(x + 2) = 0$$
$$x - 4 = 0 \quad \text{or} \quad x + 2 = 0$$
$$x = 4 \quad \text{or} \quad x = -2$$

We now turn our attention to 3×3 determinants. A 3×3 determinant is also a square array of numbers, the value of which is given by the following definition.

DEFINITION   The value of the **3×3 determinant**

$$\begin{vmatrix} a_1 & b_1 & c_1 \\ a_2 & b_2 & c_2 \\ a_3 & b_3 & c_3 \end{vmatrix}$$

is given by

$$\begin{vmatrix} a_1 & b_1 & c_1 \\ a_2 & b_2 & c_2 \\ a_3 & b_3 & c_3 \end{vmatrix} = a_1b_2c_3 + a_3b_1c_2 + a_2b_3c_1 - a_3b_2c_1 - a_1b_3c_2 - a_2b_1c_3$$

At first glance, the expansion of a 3×3 determinant looks a little complicated. There are actually two different methods used to find the six products in the preceding definition that simplify matters somewhat.

Method 1   We begin by writing the determinant with the first two columns repeated on the right:

$$\begin{vmatrix} a_1 & b_1 & c_1 \\ a_2 & b_2 & c_2 \\ a_3 & b_3 & c_3 \end{vmatrix} \begin{matrix} a_1 & b_1 \\ a_2 & b_2 \\ a_3 & b_3 \end{matrix}$$

The positive products in the definition come from multiplying down the three full diagonals:

The negative products come from multiplying up the three full diagonals:

EXAMPLE 3   Find the value of

$$\begin{vmatrix} 1 & 3 & -2 \\ 2 & 0 & 1 \\ 4 & -1 & 1 \end{vmatrix}$$

Solution   Repeating the first two columns and then finding the products down the diagonals and the products up the diagonals as given in Method 1, we have

$$= 1(0)(1) + 3(1)(4) + (-2)(2)(-1)$$
$$-4(0)(-2) - (-1)(1)(1) - 1(2)(3)$$
$$= 0 + 12 + 4 - 0 - (-1) - 6$$
$$= 11$$

Method 2   The second method of evaluating a 3×3 determinant is called *expansion by minors*.

DEFINITION   The **minor** for an element in a **3 × 3** determinant is the determinant consisting of the elements remaining when the row and column to which the element belongs are deleted. For example, in the determinant

$$\begin{vmatrix} a_1 & b_1 & c_1 \\ a_2 & b_2 & c_2 \\ a_3 & b_3 & c_3 \end{vmatrix}$$

Minor for element $a_1 = \begin{vmatrix} b_2 & c_2 \\ b_3 & c_3 \end{vmatrix}$

Minor for element $b_2 = \begin{vmatrix} a_1 & c_1 \\ a_3 & c_3 \end{vmatrix}$

Minor for element $c_3 = \begin{vmatrix} a_1 & b_1 \\ a_2 & b_2 \end{vmatrix}$

Before we can evaluate a 3×3 determinant by Method 2, we must first define what is known as the sign array for a 3×3 determinant.

> **DEFINITION**   The **sign array** for a **3 × 3** determinant is a **3 × 3** array of signs in the following pattern:
>
> $$\begin{vmatrix} + & - & + \\ - & + & - \\ + & - & + \end{vmatrix}$$
>
> The sign array begins with a + sign in the upper left-hand corner. The signs then alternate between + and − across every row and down every column.

TO EVALUATE A 3 × 3 DETERMINANT BY EXPANSION OF MINORS

We can evaluate a 3 × 3 determinant by expanding across any row or down any column as follows:

Step 1:   Choose a row or column to expand about.

Step 2:   Write the product of each element in the row or column chosen in step 1 with its minor.

Step 3:   Connect the three products in step 2 with the signs in the corresponding row or column in the sign array.

To illustrate the procedure, we will use the same determinant we used in Example 3.

EXAMPLE 4   Expand across the first row:

$$\begin{vmatrix} 1 & 3 & -2 \\ 2 & 0 & 1 \\ 4 & -1 & 1 \end{vmatrix}$$

Solution   The products of the three elements in row 1 with their minors are

$$1\begin{vmatrix} 0 & 1 \\ -1 & 1 \end{vmatrix} \qquad 3\begin{vmatrix} 2 & 1 \\ 4 & 1 \end{vmatrix} \qquad (-2)\begin{vmatrix} 2 & 0 \\ 4 & -1 \end{vmatrix}$$

Connecting these three products with the signs from the first row of the sign array, we have

$$+1\begin{vmatrix} 0 & 1 \\ -1 & 1 \end{vmatrix} - 3\begin{vmatrix} 2 & 1 \\ 4 & 1 \end{vmatrix} + (-2)\begin{vmatrix} 2 & 0 \\ 4 & -1 \end{vmatrix}$$

We complete the problem by evaluating each of the three 2 × 2 determinants and then simplifying the resulting expression:

$$+1[0 - (-1)] - 3(2 - 4) + (-2)(-2 - 0)$$
$$= 1(1) - 3(-2) + (-2)(-2)$$

$$= 1 + 6 + 4$$
$$= 11$$

The results of Examples 3 and 4 match. It makes no difference which method we use—the value of a 3×3 determinant is unique.

Note: This method of evaluating a determinant is actually more valuable than our first method, because it works with any size determinant from 3×3 to 4×4 to any higher order determinant. Method 1 works only on 3×3 determinants. It cannot be used on a 4×4 determinant.

EXAMPLE 5   Expand down column 2:

$$\begin{vmatrix} 2 & 3 & -2 \\ 1 & 4 & 1 \\ 1 & 5 & -1 \end{vmatrix}$$

Solution   We connect the products of elements in column 2 and their minors with the signs from the second column in the sign array:

$$\begin{vmatrix} 2 & 3 & -2 \\ 1 & 4 & 1 \\ 1 & 5 & -1 \end{vmatrix} = -3\begin{vmatrix} 1 & 1 \\ 1 & -1 \end{vmatrix} + 4\begin{vmatrix} 2 & -2 \\ 1 & -1 \end{vmatrix} - 5\begin{vmatrix} 2 & -2 \\ 1 & 1 \end{vmatrix}$$

$$= -3(-1 - 1) + 4[-2 - (-2)] - 5[2 - (-2)]$$
$$= -3(-2) + 4(0) - 5(4)$$
$$= 6 + 0 - 20$$
$$= -14$$

A Note on the History of Determinants   Determinants were originally known as resultants, a name given to them by Pierre Simon Laplace; however, the work of Gottfried Wilhelm Leibniz contains the germ of the original idea of resultants, or determinants.

PROBLEM SET B

Find the value of the following 2×2 determinants.

1. $\begin{vmatrix} 1 & 0 \\ 2 & 3 \end{vmatrix}$

2. $\begin{vmatrix} 5 & 4 \\ 3 & 2 \end{vmatrix}$

3. $\begin{vmatrix} 2 & 1 \\ 3 & 4 \end{vmatrix}$

4. $\begin{vmatrix} 4 & 1 \\ 5 & 2 \end{vmatrix}$

5. $\begin{vmatrix} 0 & 1 \\ 1 & 0 \end{vmatrix}$

6. $\begin{vmatrix} 1 & 0 \\ 0 & 1 \end{vmatrix}$

7. $\begin{vmatrix} -3 & 2 \\ 6 & -4 \end{vmatrix}$

8. $\begin{vmatrix} 8 & -3 \\ -2 & 5 \end{vmatrix}$

9. $\begin{vmatrix} -3 & -1 \\ 4 & -2 \end{vmatrix}$

10. $\begin{vmatrix} 5 & 3 \\ 7 & -6 \end{vmatrix}$

Solve each of the following for x.

11. $\begin{vmatrix} 2x & 1 \\ x & 3 \end{vmatrix} = 10$

12. $\begin{vmatrix} 3x & -2 \\ 2x & 3 \end{vmatrix} = 26$

13. $\begin{vmatrix} 1 & 2x \\ 2 & -3x \end{vmatrix} = 21$ 14. $\begin{vmatrix} -5 & 4x \\ 1 & -x \end{vmatrix} = 27$

15. $\begin{vmatrix} 2x & -4 \\ 2 & x \end{vmatrix} = -8x$ 16. $\begin{vmatrix} 3x & 2 \\ 2 & x \end{vmatrix} = -11x$

17. $\begin{vmatrix} x^2 & 3 \\ x & 1 \end{vmatrix} = 10$ 18. $\begin{vmatrix} x^2 & -2 \\ x & 1 \end{vmatrix} = 35$

Find the value of each of the following 3 × 3 determinants by using Method 1 of this section.

19. $\begin{vmatrix} 1 & 2 & 0 \\ 0 & 2 & 1 \\ 1 & 1 & 1 \end{vmatrix}$ 20. $\begin{vmatrix} -1 & 0 & 2 \\ 3 & 0 & 1 \\ 0 & 1 & 3 \end{vmatrix}$

21. $\begin{vmatrix} 1 & 2 & 3 \\ 3 & 2 & 1 \\ 1 & 1 & 1 \end{vmatrix}$ 22. $\begin{vmatrix} -1 & 2 & 0 \\ 3 & -2 & 1 \\ 0 & 5 & 4 \end{vmatrix}$

Find the value of each determinant by using Method 2 and expanding across the first row.

23. $\begin{vmatrix} 0 & 1 & 2 \\ 1 & 0 & 1 \\ -1 & 2 & 0 \end{vmatrix}$ 24. $\begin{vmatrix} 3 & -2 & 1 \\ 0 & -1 & 0 \\ 2 & 0 & 1 \end{vmatrix}$

25. $\begin{vmatrix} 3 & 0 & 2 \\ 0 & -1 & -1 \\ 4 & 0 & 0 \end{vmatrix}$ 26. $\begin{vmatrix} 1 & 1 & 1 \\ 1 & -1 & 1 \\ 1 & 1 & -1 \end{vmatrix}$

Find the value of each of the following determinants.

27. $\begin{vmatrix} 2 & -1 & 0 \\ 1 & 0 & -2 \\ 0 & 1 & 2 \end{vmatrix}$ 28. $\begin{vmatrix} 5 & 0 & -4 \\ 0 & 1 & 3 \\ -1 & 2 & -1 \end{vmatrix}$

29. $\begin{vmatrix} 1 & 3 & 7 \\ -2 & 6 & 4 \\ 3 & 7 & -1 \end{vmatrix}$ 30. $\begin{vmatrix} 2 & 1 & 5 \\ 6 & -3 & 4 \\ 8 & 9 & -2 \end{vmatrix}$

Applying the Concepts

31. Slope-Intercept Form Show that the following determinant equation is another way to write the slope-intercept form of the equation of a line.

$$\begin{vmatrix} y & x \\ m & 1 \end{vmatrix} = b$$

32. Temperature Conversions Show that the following determinant equation is another way to write the equation $F = \frac{9}{5}C + 32$.

$$\begin{vmatrix} C & F & 1 \\ 5 & 41 & 1 \\ -10 & 14 & 1 \end{vmatrix} = 0$$

33. Amusement Park Income From 1986 to 1990, the annual income of amusement parks was linearly increasing, after which time it remained fairly constant. The annual income y, in billions of dollars, may be found for one of these years by evaluating the following determinant equation, in which x represents the number of years past 1986.

$$\begin{vmatrix} x & -1.7 \\ 2 & 0.3 \end{vmatrix} = y$$

(a) Write the determinant equation in slope-intercept form.
(b) Use the equation from part (a) to find the approximate income for amusement parks in the year 1988.

34. College Enrollment From 1981 the enrollment of women in the United States armed forces was linearly increasing until 1990, after which it declined. The approximate number of women, w, enrolled in the armed forces from 1981 to 1990 may be found by evaluating the following determinant equation, in which x represents the number of years past 1981.

$$\begin{vmatrix} 6,509 & -2 \\ 85,709 & x \end{vmatrix} = w$$

Use this equation to determine the number of women enrolled in the armed forces in 1985.

Extending the Concepts

A 4 × 4 determinant can be evaluated only by using Method 2, expansion by minors; Method 1 will not work. Below is a 4 × 4 determinant and its associated sign array.

$$\begin{vmatrix} 2 & 0 & 1 & -3 \\ -1 & 2 & 0 & 1 \\ -3 & 0 & 1 & 0 \\ 1 & 1 & 0 & 0 \end{vmatrix} \qquad \begin{vmatrix} + & - & + & - \\ - & + & - & + \\ + & - & + & - \\ - & + & - & + \end{vmatrix}$$

4 × 4 determinant   4 × 4 sign array

35. Use expansion by minors to evaluate the preceding 4 × 4 determinant by expanding it across row 1.

36. Evaluate the preceding determinant by expanding it down column 4.

37. Use expansion by minors down column 3 to evaluate the preceding determinant.

38. Evaluate the preceding determinant by expanding it across row 4.

Appendix C

Cramer's Rule

We begin this appendix with a look at how determinants can be used to solve a system of linear equations in two variables. The method we use is called *Cramer's rule*. We state it here as a theorem without proof.

THEOREM (CRAMER'S RULE)

The solution to the system

$$a_1 x + b_1 y = c_1$$
$$a_2 x + b_2 y = c_2$$

is given by

$$x = \frac{D_x}{D}, \qquad y = \frac{D_y}{D}$$

where

$$D = \begin{vmatrix} a_1 & b_1 \\ a_2 & b_2 \end{vmatrix} \qquad D_x = \begin{vmatrix} c_1 & b_1 \\ c_2 & b_2 \end{vmatrix} \qquad D_y = \begin{vmatrix} a_1 & c_1 \\ a_2 & c_2 \end{vmatrix} \qquad (D \neq 0)$$

The determinant D is made up of the coefficients of x and y in the original system. The determinants D_x and D_y are found by replacing the coefficients of x or y by the constant terms in the original system. Notice also that Cramer's rule does not apply if $D = 0$. In this case the equations are either inconsistent or dependent.

EXAMPLE 1 Use Cramer's rule to solve

$$2x - 3y = 4$$
$$4x + 5y = 3$$

Solution We begin by calculating the determinants D, D_x, and D_y.

$$D = \begin{vmatrix} 2 & -3 \\ 4 & 5 \end{vmatrix} = 2(5) - 4(-3) = 22$$

$$D_x = \begin{vmatrix} 4 & -3 \\ 3 & 5 \end{vmatrix} = 4(5) - 3(-3) = 29$$

$$D_y = \begin{vmatrix} 2 & 4 \\ 4 & 3 \end{vmatrix} = 2(3) - 4(4) = -10$$

$$x = \frac{D_x}{D} = \frac{29}{22} \quad \text{and} \quad y = \frac{D_y}{D} = \frac{-10}{22} = -\frac{5}{11}$$

The solution set for the system is $\{(\frac{29}{22}, -\frac{5}{11})\}$.

Cramer's rule can also be applied to systems of linear equations in three variables.

THEOREM (ALSO CRAMER'S RULE)

The solution set to the system

$$a_1x + b_1y + c_1z = d_1$$
$$a_2x + b_2y + c_2z = d_2$$
$$a_3x + b_3y + c_3z = d_3$$

is given by

$$x = \frac{D_x}{D}, \quad y = \frac{D_y}{D}, \quad \text{and} \quad z = \frac{D_z}{D}$$

where

$$D = \begin{vmatrix} a_1 & b_1 & c_1 \\ a_2 & b_2 & c_2 \\ a_3 & b_3 & c_3 \end{vmatrix} \quad D_x = \begin{vmatrix} d_1 & b_1 & c_1 \\ d_2 & b_2 & c_2 \\ d_3 & b_3 & c_3 \end{vmatrix} \quad (D \neq 0)$$

$$D_y = \begin{vmatrix} a_1 & d_1 & c_1 \\ a_2 & d_2 & c_2 \\ a_3 & d_3 & c_3 \end{vmatrix} \quad D_z = \begin{vmatrix} a_1 & b_1 & d_1 \\ a_2 & b_2 & d_2 \\ a_3 & b_3 & d_3 \end{vmatrix}$$

Again, the determinant D consists of the coefficients of x, y, and z in the original system. The determinants D_x, D_y, and D_z are found by replacing the coefficients of x, y, and z, respectively, with the constant terms from the original system. If $D = 0$, there is no unique solution to the system.

EXAMPLE 2   Use Cramer's rule to solve

$$x + y + z = 6$$
$$2x - y + z = 3$$
$$x + 2y - 3z = -4$$

Solution   This is the same system used in Example 1 in Section 7.7, so we can compare Cramer's rule with our previous methods of solving a system in three variables. We begin by setting up and evaluating D, D_x, D_y, and D_z. (Recall that there are a number of ways to evaluate a 3×3 determinant. Since we have four of these determinants, we can use both Methods 1 and 2 from the previous section.) We evaluate D using Method 1 from Appendix B.

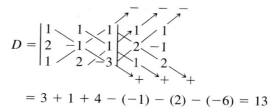

$$= 3 + 1 + 4 - (-1) - (2) - (-6) = 13$$

We evaluate D_x using Method 2 from Appendix B and expanding across row 1:

$$D_x = \begin{vmatrix} 6 & 1 & 1 \\ 3 & -1 & 1 \\ -4 & 2 & -3 \end{vmatrix} = 6 \begin{vmatrix} -1 & 1 \\ 2 & -3 \end{vmatrix} - 1 \begin{vmatrix} 3 & 1 \\ -4 & -3 \end{vmatrix} + 1 \begin{vmatrix} 3 & -1 \\ -4 & 2 \end{vmatrix}$$

$$= 6(1) - 1(-5) + 1(2)$$

$$= 13$$

Find D_y by expanding across row 2:

$$D_y = \begin{vmatrix} 1 & 6 & 1 \\ 2 & 3 & 1 \\ 1 & -4 & -3 \end{vmatrix} = -2 \begin{vmatrix} 6 & 1 \\ -4 & -3 \end{vmatrix} + 3 \begin{vmatrix} 1 & 1 \\ 1 & -3 \end{vmatrix} - 1 \begin{vmatrix} 1 & 6 \\ 1 & -4 \end{vmatrix}$$

$$= -2(-14) + 3(-4) - 1(-10)$$

$$= 26$$

Find D_z by expanding down column 1:

$$D_z = \begin{vmatrix} 1 & 1 & 6 \\ 2 & -1 & 3 \\ 1 & 2 & -4 \end{vmatrix} = 1 \begin{vmatrix} -1 & 3 \\ 2 & -4 \end{vmatrix} - 2 \begin{vmatrix} 1 & 6 \\ 2 & -4 \end{vmatrix} + 1 \begin{vmatrix} 1 & 6 \\ -1 & 3 \end{vmatrix}$$

$$= 1(-2) - 2(-16) + 1(9)$$

$$= 39$$

$$x = \frac{D_x}{D} = \frac{13}{13} = 1 \qquad y = \frac{D_y}{D} = \frac{26}{13} = 2 \qquad z = \frac{D_z}{D} = \frac{39}{13} = 3$$

The solution set is $\{(1, 2, 3)\}$.

Note: We are solving each of these determinants by expanding about different rows or columns just to show the different ways these determinants can be evaluated.

E X A M P L E   3   Use Cramer's rule to solve

$$x + y = -1$$

$$2x - z = 3$$

$$y + 2z = -1$$

Solution   It is helpful to rewrite the system using zeros for the coefficients of those variables not shown:

$$x + y + 0z = -1$$
$$2x + 0y - z = 3$$
$$0x + y + 2z = -1$$

The four determinants used in Cramer's rule are

$$D = \begin{vmatrix} 1 & 1 & 0 \\ 2 & 0 & -1 \\ 0 & 1 & 2 \end{vmatrix} = -3$$

$$D_x = \begin{vmatrix} -1 & 1 & 0 \\ 3 & 0 & -1 \\ -1 & 1 & 2 \end{vmatrix} = -6$$

$$D_y = \begin{vmatrix} 1 & -1 & 0 \\ 2 & 3 & -1 \\ 0 & -1 & 2 \end{vmatrix} = 9$$

$$D_z = \begin{vmatrix} 1 & 1 & -1 \\ 2 & 0 & 3 \\ 0 & 1 & -1 \end{vmatrix} = -3$$

$$x = \frac{D_x}{D} = \frac{-6}{-3} = 2 \qquad y = \frac{D_y}{D} = \frac{9}{-3} = -3 \qquad z = \frac{D_z}{D} = \frac{-3}{-3} = 1$$

The solution set is $\{(2, -3, 1)\}$.

Finally, we should mention the possible situations that can occur when the determinant D is 0, when we are using Cramer's rule.

If $D = 0$ and at least one of the other determinants, D_x or D_y (or D_z), is not 0, then the system is inconsistent. In this case there is no solution to the system.

On the other hand, if $D = 0$ and both D_x and D_y (and D_z in a system of three equations in three variables) are 0, then the system is a dependent one.

A Note on the History of Cramer's Rule   Cramer's rule is named after the Swiss mathematician *Gabriel Cramer* (1704–1752). Cramer's rule appeared in the appendix of an algebraic work of his classifying curves, but the basic idea behind his now-famous rule was formulated earlier by Leibniz and Chinese mathematicians. It was actually Cramer's superior notation that helped to popularize the technique.

Cramer has a respectable reputation as a mathematician, but he does not rank with the great mathematicians of his time, although through his extensive travels he met many of them, such as the Bernoullis, Euler, and D'Alembert.

Cramer had very broad interests. He wrote on philosophy, law, and government, as well as mathematics; served in public office; and was an expert on cathedrals, often instructing workers about their repair and coordinating excavations to recover cathedral archives. Cramer never married, and a fall from a carriage eventually led to his death.

PROBLEM SET C

Solve each of the following systems using Cramer's rule.

1. $2x - 3y = 3$
$\quad 4x - 2y = 10$

2. $3x + y = -2$
$\quad -3x + 2y = -4$

3. $5x - 2y = 4$
$\quad -10x + 4y = 1$

4. $-4x + 3y = -11$
$\quad 5x + 4y = 6$

5. $4x - 7y = 3$
$\quad 5x + 2y = -3$

6. $3x - 4y = 7$
$\quad 6x - 2y = 5$

7. $9x - 8y = 4$
$\quad 2x + 3y = 6$

8. $4x - 7y = 10$
$\quad -3x + 2y = -9$

9. $x + y + z = 4$
$\quad x - y - z = 2$
$\quad 2x + 2y - z = 2$

10. $-x + y + 3z = 6$
$\quad x + y + 2z = 7$
$\quad 2x + 3y + z = 4$

11. $x + y - z = 2$
$\quad -x + y + z = 3$
$\quad x + y + z = 4$

12. $-x - y + z = 1$
$\quad x - y + z = 3$
$\quad x + y - z = 4$

13. $3x - y + 2z = 4$
$\quad 6x - 2y + 4z = 8$
$\quad x - 5y + 2z = 1$

14. $2x - 3y + z = 1$
$\quad 3x - y - z = 4$
$\quad 4x - 6y + 2z = 3$

15. $2x - y + 3z = 4$
$\quad x - 5y - 2z = 1$
$\quad -4x - 2y + z = 3$

16. $4x - y + 5z = 1$
$\quad 2x + 3y + 4z = 5$
$\quad x + y + 3z = 2$

17. $-x - 7y = 1$
$\quad x + 3z = 11$
$\quad 2y + z = 0$

18. $x + y = 2$
$\quad -x + 3z = 0$
$\quad 2y + z = 3$

19. $x - y = 2$
$\quad 3x + z = 11$
$\quad y - 2z = -3$

20. $4x + 5y = -1$
$\quad 2y + 3z = -5$
$\quad x + 2z = -1$

Applying the Concepts

21. Break-Even Point If a company has fixed costs of $100 per week and each item it produces costs $10 to manufacture, then the total cost y per week to produce x items is

$$y = 10x + 100$$

If the company sells each item it manufactures for $12, then the total amount of money y the company brings in for selling x items is

$$y = 12x$$

Use Cramer's rule to solve the system

$$y = 10x + 100$$
$$y = 12x$$

for x to find the number of items the company must sell per week in order to break even.

22. Break-Even Point Suppose a company has fixed costs of $200 per week and each item it produces costs $20 to manufacture.
(a) Write an equation that gives the total cost per week y to manufacture x items.
(b) If each item sells for $25, write an equation that gives the total amount of money y the company brings in for selling x items.
(c) Use Cramer's rule to find the number of items the company must sell each week to break even.

23. Health Insurance For years between 1980 and 1991, the number (in millions) of U.S. residents without health insurance, y, may be approximated by the equation

$$y = 0.98x - 1{,}915.8$$

where x represents the year, and $1980 \leq x \leq 1991$. To determine the year in which 30 million U.S. residents were without health insurance, we solve the system of equations made up of the equation above and the equation $y = 30$. Solve this system using Cramer's rule. (When you obtain an answer, you will need to round it to the nearest year.)

24. Price Index From 1970 to 1990, the price index of dental care, d, may be closely approximated by the equation

$$d = 6x - 11,780$$

where x is the year, and $1970 \le x \le 1990$. Determine when the price index for dental care reached 120 by forming a system of equations using the equation above along with the equation $d = 120$. Solve this system using Cramer's rule.

Extending the Concepts

Solve for x and y using Cramer's rule. Your answers will contain the constants a and b.

25. $ax + by = -1$
    $bx + ay = 1$

26. $ax + y = b$
    $bx + y = a$

27. $a^2x + by = 1$
    $b^2x + ay = 1$

28. $ax + by = a$
    $bx + ay = a$

29. Name the system of equations for which Cramer's rule yields the following determinants.

$$D = \begin{vmatrix} 1 & 2 \\ 3 & 4 \end{vmatrix} \qquad D_x = \begin{vmatrix} 1 & 2 \\ 0 & 4 \end{vmatrix}$$

30. Name the system of equations for which Cramer's rule yields the following determinants.

$$D = \begin{vmatrix} 1 & 3 & 2 \\ -1 & 0 & 4 \\ 2 & 5 & -1 \end{vmatrix} \qquad D_y = \begin{vmatrix} 1 & 1 & 2 \\ -1 & 3 & 4 \\ 2 & 5 & -1 \end{vmatrix}$$

Appendix D

Matrix Solutions to Linear Systems

In mathematics, a **matrix** is a rectangular array of elements considered as a whole. We can use matrices to represent systems of linear equations. To do so, we write the coefficients of the variables and the constant terms in the same position in the matrix as they occur in the system of equations. To show where the coefficients end and the constant terms begin, we use vertical lines instead of equal signs. For example, the system

$$2x + 5y = -4$$
$$x - 3y = 9$$

can be represented by the matrix

$$\left[\begin{array}{cc|c} 2 & 5 & -4 \\ 1 & -3 & 9 \end{array}\right]$$

which is called an **augmented matrix** because it includes both the coefficients of the variables and the constant terms.

To solve a system of linear equations by using the augmented matrix for that system, we need the following row operations as the tools of that solution process. The row operations tell us what we can do to an augmented matrix that may change the numbers in the matrix, but will always produce a matrix that represents a system of equations with the same solution as that of our original system.

ROW OPERATIONS

1. We can interchange any two rows of a matrix.
2. We can multiply any row by a nonzero constant.
3. We can add to any row a constant multiple of another row.

The three row operations are simply a list of the properties we use to solve systems of linear equations, translated to fit an augmented matrix. For instance, the second operation in our list is actually just another way to state the multiplication property of equality.

We solve a system of linear equations by transforming the augmented matrix into a matrix that has 1's down the diagonal of the coefficient matrix, and 0's below it. For instance, we will have solved the system

$$2x + 5y = -4$$
$$x - 3y = 9$$

When the matrix

$$\begin{bmatrix} 2 & 5 & | & -4 \\ 1 & -3 & | & 9 \end{bmatrix}$$

has been transformed, using the row operations listed earlier, we get a matrix of the form

$$\begin{bmatrix} 1 & - & | & - \\ 0 & 1 & | & - \end{bmatrix}$$

To accomplish this, we begin with the first column and try to produce a 1 in the first position and a 0 below it. Interchanging rows 1 and 2 gives us a 1 in the top position of the first column:

↓ Interchange rows 1 and 2.

$$\begin{bmatrix} 1 & -3 & | & 9 \\ 2 & 5 & | & -4 \end{bmatrix}$$

Multiplying row 1 by -2 and adding the result to row 2 gives us a 0 where we want it.

↓ Multiply row 1 by -2 and
add the result to row 2.

$$\begin{bmatrix} 1 & -3 & | & 9 \\ 0 & 11 & | & -22 \end{bmatrix}$$

↓ Multiply row 2 by $\frac{1}{11}$.

$$\begin{bmatrix} 1 & -3 & | & 9 \\ 0 & 1 & | & -2 \end{bmatrix}$$

Taking this last matrix and writing the system of equations it represents, we have

$$x - 3y = 9$$
$$y = -2$$

Substituting -2 for y in the top equation gives us

$$x = 3$$

The solution to our system is $(3, -2)$.

EXAMPLE I   Solve the following system using an augmented matrix:

$$x + y - z = 2$$
$$2x + 3y - z = 7$$
$$3x - 2y + z = 9$$

Solution We begin by writing the system in terms of an augmented matrix:

$$\left|\begin{array}{ccc|c} 1 & 1 & -1 & 2 \\ 2 & 3 & -1 & 7 \\ 3 & -2 & 1 & 9 \end{array}\right|$$

Next, we want to produce 0's in the second two positions of column 1:

$$\downarrow \quad \begin{array}{l}\text{Multiply row 1 by } -2 \text{ and} \\ \text{add the result to row 2.}\end{array}$$

$$\left[\begin{array}{ccc|c} 1 & 1 & -1 & 2 \\ 0 & 1 & 1 & 3 \\ 3 & -2 & 1 & 9 \end{array}\right]$$

$$\downarrow \quad \begin{array}{l}\text{Multiply row 1 by } -3 \text{ and} \\ \text{add the result to row 3.}\end{array}$$

$$\left[\begin{array}{ccc|c} 1 & 1 & -1 & 2 \\ 0 & 1 & 1 & 3 \\ 0 & -5 & 4 & 3 \end{array}\right]$$

Note that we could have done these two steps in one single step. As you become more familiar with this method of solving systems of equations, you will do just that.

$$\downarrow \quad \begin{array}{l}\text{Multiply row 2 by 5 and} \\ \text{add the result to row 3.}\end{array}$$

$$\left[\begin{array}{ccc|c} 1 & 1 & -1 & 2 \\ 0 & 1 & 1 & 3 \\ 0 & 0 & 9 & 18 \end{array}\right]$$

$$\downarrow \quad \text{Multiply row 3 by } \tfrac{1}{9}.$$

$$\left[\begin{array}{ccc|c} 1 & 1 & -1 & 2 \\ 0 & 1 & 1 & 3 \\ 0 & 0 & 1 & 2 \end{array}\right]$$

Converting back to a system of equations, we have

$$x + y - z = 2$$

$$y + z = 3$$

$$z = 2$$

This system is equivalent to our first one, but much easier to solve.

Substituting $z = 2$ into the second equation, we have

$$y = 1$$

Substituting $z = 2$ and $y = 1$ into the first equation, we have

$$x = 3$$

The solution to our original system is (3, 1, 2). It satisfies each of our original equations. You can check this, if you like.

PROBLEM SET D

Solve the following systems of equations by using matrices.

1. $x + y = 5$
$3x - y = 3$

2. $x + y = -2$
$2x - y = -10$

3. $3x - 5y = 7$
$-x + y = -1$

4. $2x - y = 4$
$x + 3y = 9$

5. $2x - 8y = 6$
$3x - 8y = 13$

6. $3x - 6y = 3$
$-2x + 3y = -4$

7. $x + y + z = 4$
$x - y + 2z = 1$
$x - y - z = -2$

8. $x - y - 2z = -1$
$x + y + z = 6$
$x + y - z = 4$

9. $x + 2y + z = 3$
$2x - y + 2z = 6$
$3x + y - z = 5$

10. $x - 3y + 4z = -4$
$2x + y - 3z = 14$
$3x + 2y + z = 10$

11. $x + 2y = 3$
$y + z = 3$
$4x - z = 2$

12. $x + y = 2$
$3y - 2z = -8$
$x + z = 5$

13. $x + 3y = 7$
$3x - 4z = -8$
$5y - 2z = -5$

14. $x + 4y = 13$
$2x - 5z = -3$
$4y - 3z = 9$

Solve each system using matrices. Remember, multiplying a row by a nonzero constant will not change the solution to the system.

15. $\frac{1}{3}x + \frac{1}{5}y = 2$
$\frac{1}{3}x - \frac{1}{2}y = -\frac{1}{3}$

16. $\frac{1}{2}x + \frac{1}{3}y = 13$
$\frac{1}{5}x + \frac{1}{8}y = 5$

The systems that follow are inconsistent systems. In both cases, the lines are parallel. Try solving each system using matrices and see what happens.

17. $2x - 3y = 4$
$4x - 6y = 4$

18. $10x - 15y = 5$
$-4x + 6y = -4$

The systems that follow are dependent systems. In each case, the lines coincide. Try solving each system using matrices and see what happens.

19. $-6x + 4y = 8$
$-3x + 2y = 4$

20. $x + 2y = 5$
$-x - 2y = -5$

Appendix E

Answers to Odd-Numbered Problems, Chapter Reviews, Chapter Tests, and Cumulative Reviews

CHAPTER 1

PROBLEM SET 1.1

1. $x + 5 = 14$ **3.** $5y < 30$ **5.** $3y \leq y + 6$ **7.** $\dfrac{x}{3} = x + 2$ **9.** 9 **11.** 49 **13.** 8 **15.** 64
17. 16 **19.** 100 **21.** 121 **23.** 11 **25.** 16 **27.** 17 **29.** 42 **31.** 30 **33.** 30 **35.** 24
37. 80 **39.** 27 **41.** 35 **43.** 13 **45.** 4 **47.** 37 **49.** 37 **51.** 16 **53.** 16 **55.** 81
57. 41 **59.** 345 **61.** 2,345 **63.** 2 **65.** 148 **67.** 36 **69.** 36 **71.** 58 **73.** 62 **75.** 5
77. 10 **79.** 25 **81.** 10 **83.** (a) 600 mg (b) 231 mg **85.**

Activity	Calories Burned in 1 Hour by a 150-Pound Person
Bicycling	374
Bowling	265
Handball	680
Jogging	680
Skiing	544

87. $2(10 + 4)$ **89.** $3(50) - 14$ **91.** 10 cookies **93.** 420 calories **95.** Approximately 224 chips
97. 95 grams

PROBLEM SET 1.2

1–7.

9. $\frac{18}{24}$ **11.** $\frac{12}{24}$ **13.** $\frac{15}{24}$ **15.** $\frac{36}{60}$ **17.** $\frac{22}{60}$ **19.** $-10, \frac{1}{10}, 10$ **21.** $-\frac{3}{4}, \frac{4}{3}, \frac{3}{4}$ **23.** $-\frac{11}{2}, \frac{2}{11}, \frac{11}{2}$
25. $3, -\frac{1}{3}, 3$ **27.** $\frac{2}{5}, -\frac{5}{2}, \frac{2}{5}$ **29.** $-x, \frac{1}{x}, |x|$ **31.** < **33.** > **35.** > **37.** > **39.** < **41.** <
43. 6 **45.** 22 **47.** 3 **49.** 7 **51.** 3 **53.** $\frac{8}{15}$ **55.** $\frac{3}{2}$ **57.** $\frac{5}{4}$ **59.** 1 **61.** 1 **63.** 1
65. $\frac{9}{16}$ **67.** $\frac{8}{27}$ **69.** $\frac{1}{10,000}$ **71.** $\frac{1}{9}$ **73.** $\frac{1}{25}$ **75.** 4 inches; 1 inch^2 **77.** 4.5 inches; 1.125 inches^2
79. 10.25 centimeters; 5 centimeters^2 **81.** $-8, -2$ **83.** $-64°F; -54°F$ **85.** $-15°F$

87.

Month	Temperature (°F)
January	−36
February	−30
March	−14
April	−2
May	19
June	22
July	35
August	30
September	19
October	15
November	−11
December	−26

89. −100 ft; −105 ft

91. Area = 93.5 inches^2: perimeter = 39 inches   **93.** 1,387 calories   **95.** 654 more calories

PROBLEM SET 1.3

1. 3 + 5 = 8; 3 + (−5) = −2; −3 + 5 = 2; −3 + (−5) = −8
3. 15 + 20 = 35; 15 + (−20) = −5; −15 + 20 = 5; −15 + (−20) = −35   **5.** 3   **7.** −7   **9.** −14
11. −3   **13.** −25   **15.** −12   **17.** −19   **19.** −25   **21.** −8   **23.** −4   **25.** 6   **27.** 6   **29.** 8
31. −4   **33.** −14   **35.** −17   **37.** 4   **39.** 3   **41.** 15   **43.** −8   **45.** 12   **47.** 23, 28
49. 30, 35   **51.** 0, −5   **53.** −12, −18   **55.** −4, −8   **57.** Yes   **59.** 5 + 9 = 14
61. [−7 + (−5)] + 4 = −8   **63.** [−2 + (−3)] + 10 = 5   **65.** 3   **67.** −3   **69.** −12 + 4
71. 10 + (−6) + (−8) = −4   **73.** −$30 + $40 = $10   **75.** $6.50, $6.75, $7.00, $7.25, $7.50, $7.75, $8.00; yes
77. (a)

Professions	Salaries
Engineering	$42,862
Computer Science	$40,920
Math/Statistics	$40,523
Chemistry	$36,036
Business Admin.	$34,831
Accounting	$33,702
Sales/Marketing	$33,252
Teaching	$25,735

(b) $78,898

PROBLEM SET 1.4

1. −3   **3.** −6   **5.** 0   **7.** −10   **9.** −16   **11.** −12   **13.** −7   **15.** 35   **17.** 0   **19.** −4
21. 4   **23.** −24   **25.** −28   **27.** 25   **29.** 4   **31.** 7   **33.** 17   **35.** 8   **37.** 4   **39.** 18
41. 10   **43.** 17   **45.** 1   **47.** 1   **49.** 27   **51.** −26   **53.** −2   **55.** 68   **57.** −7 − 4 = −11

59. $12 - (-8) = 20$ **61.** $-5 - (-7) = 2$ **63.** $[4 + (-5)] - 17 = -18$ **65.** $8 - 5 = 3$
67. $-8 - 5 = -13$ **69.** $8 - (-5) = 13$ **71.** 10 **73.** -2 **75.** $1,500 - 730$
77. $-\$35 + \$15 - \$20 = -\$40$ **79.** $\$98 - \$65 - \$53 = -\$20$
81. $\$4,500, \$3,950, \$3,400, \$2,850, \$2,300$; yes   **83.** 769 feet   **85.** 439 feet   **87.** 2 seconds
89. (a)   (b) 84 million tons

Year	Garbage (millions of tons)
1960	88
1970	121
1980	152
1990	205
1997	217

91. (a)   (b) 5 cents/minute   **93.** $35°$ **95.** $60°$

Year	Cents/Minute
1998	33
1999	28
2000	25
2001	23
2002	22
2003	20

PROBLEM SET 1.5

1. Commutative   **3.** Multiplicative inverse   **5.** Commutative   **7.** Distributive   **9.** Commutative, associative
11. Commutative, associative   **13.** Commutative   **15.** Commutative, associative   **17.** Commutative
19. Additive inverse   **21.** $3x + 6$ **23.** $9a + 9b$ **25.** 0 **27.** 0 **29.** 10 **31.** $(4 + 2) + x = 6 + x$
33. $x + (2 + 7) = x + 9$ **35.** $(3 \cdot 5)x = 15x$ **37.** $(9 \cdot 6)y = 54y$ **39.** $\left(\frac{1}{2} \cdot 3\right)a = \frac{3}{2}a$ **41.** $\left(\frac{1}{3} \cdot 3\right)x = x$
43. $\left(\frac{1}{2} \cdot 2\right)y = y$ **45.** $\left(\frac{3}{4} \cdot \frac{4}{3}\right)x = x$ **47.** $\left(\frac{6}{5} \cdot \frac{5}{6}\right)a = a$ **49.** $8x + 16$ **51.** $8x - 16$ **53.** $4y + 4$
55. $18x + 15$ **57.** $6a + 14$ **59.** $54y - 72$ **61.** $\frac{3}{2}x - 3$ **63.** $x + 2$ **65.** $3x + 3y$ **67.** $8a - 8b$
69. $12x + 18y$ **71.** $12a - 8b$ **73.** $3x + 2y$ **75.** $4a + 25$ **77.** $6x + 12$ **79.** $14x + 38$ **81.** No
83. No, not commutative   **85.** $8 \div 4 \neq 4 \div 8$ **87.** $12(2,400 - 480) = \$23,040$; $12(2,400) - 12(480) = \$23,040$
89. $P = 2(l + w)$

PROBLEM SET 1.6

1. -42 **3.** -21 **5.** -16 **7.** 3 **9.** 121 **11.** 6 **13.** -60 **15.** 24 **17.** 49 **19.** -27
21. 6 **23.** 10 **25.** 9 **27.** 45 **29.** 14 **31.** -2 **33.** 216 **35.** -2 **37.** -18 **39.** 29
41. 38 **43.** -5 **45.** 37 **47.** 80 **49.** $-\frac{10}{21}$ **51.** -4 **53.** 1 **55.** $\frac{9}{16}$ **57.** $-\frac{8}{27}$ **59.** $-8x$
61. $42x$ **63.** x **65.** x **67.** $-4a - 8$ **69.** $-\frac{3}{2}x + 3$ **71.** $-6x + 8$ **73.** $-15x - 30$ **75.** -25
77. $2(-4x) = -8x$ **79.** -26 **81.** 8 **83.** -80 **85.** $\frac{1}{8}$ **87.** 1 **89.** -24 **91.** $\$60$ **93.** $1°F$
95. $\$500, \$1,000, \$2,000, \$4,000, \$8,000, \$16,000$; yes   **97.** 465 calories

PROBLEM SET 1.7

1. -2 **3.** -3 **5.** $-\frac{1}{3}$ **7.** 3 **9.** $\frac{1}{7}$ **11.** 0 **13.** 9 **15.** -15 **17.** -36 **19.** $-\frac{1}{4}$ **21.** $\frac{16}{15}$

23. $\frac{4}{3}$ **25.** $-\frac{8}{13}$ **27.** -1 **29.** 1 **31.** $\frac{3}{5}$ **33.** $-\frac{5}{3}$ **35.** -2 **37.** -3 **39.** Undefined

41. Undefined   **43.** 5 **45.** $-\frac{7}{3}$ **47.** -1 **49.** -7 **51.** $\frac{15}{17}$ **53.** $-\frac{32}{17}$ **55.** $\frac{1}{3}$ **57.** 1 **59.** 1

61. -2 **63.** $\frac{9}{7}$ **65.** $\frac{16}{11}$ **67.** -1 **69.** 3 **71.** -10 **73.** -3 **75.** -8 **77.** \$350

79. Drops 3.5°F each hour.

PROBLEM SET 1.8

1. $\{0, 1, 2, 3, 4, 5, 6\}$ **3.** $\varnothing$ **5.** $\{0, 1, 2, 3, 4, 5, 6\}$ **7.** $\{0, 2\}$ **9.** $\{0, 6\}$ **11.** $\{0, 1, 2, 3, 4, 5, 6, 7\}$

13. $0, 1$ **15.** $-3, -2.5, 0, 1, \frac{3}{2}$ **17.** All   **19.** $-10, -8, -2, 9$ **21.** π **23.** True   **25.** False

27. False   **29.** True   **31.** Composite, $2^4 \cdot 3$ **33.** Prime   **35.** Composite, $3 \cdot 11 \cdot 31$ **37.** $2^4 \cdot 3^2$

39. $2 \cdot 19$ **41.** $3 \cdot 5 \cdot 7$ **43.** $2^2 \cdot 3^2 \cdot 5$ **45.** $5 \cdot 7 \cdot 11$ **47.** 11^2 **49.** $2^2 \cdot 3 \cdot 5 \cdot 7$ **51.** $2^2 \cdot 5 \cdot 31$ **53.** $\frac{7}{11}$

55. $\frac{5}{7}$ **57.** $\frac{11}{13}$ **59.** $\frac{14}{15}$ **61.** $\frac{5}{9}$ **63.** $\frac{5}{8}$ **65.** $6^3 = (2 \cdot 3)^3 = 2^3 \cdot 3^3$ **67.** $9^4 \cdot 16^2 = (3^2)^4(2^4)^2 = 2^8 \cdot 3^8$

69. $3 \cdot 8 + 3 \cdot 7 + 3 \cdot 5 = 24 + 21 + 15 = 60 = 2^2 \cdot 3 \cdot 5$ **71.** Irrational numbers   **73.** $8, 21, 34$

PROBLEM SET 1.9

1. $\frac{2}{3}$ **3.** $-\frac{1}{4}$ **5.** $\frac{1}{2}$ **7.** $\frac{x-1}{3}$ **9.** $\frac{3}{2}$ **11.** $\frac{x+6}{2}$ **13.** $-\frac{3}{5}$ **15.** $\frac{10}{a}$ **17.** $\frac{7}{8}$ **19.** $\frac{1}{10}$ **21.** $\frac{7}{9}$

23. $\frac{7}{3}$ **25.** $\frac{1}{4}$ **27.** $\frac{7}{6}$ **29.** $\frac{19}{24}$ **31.** $\frac{13}{60}$ **33.** $\frac{29}{35}$ **35.** $\frac{949}{1,260}$ **37.** $\frac{13}{420}$ **39.** $\frac{41}{24}$ **41.** $\frac{5}{4}$ **43.** $\frac{160}{63}$

45. $\frac{5}{8}$ **47.** $-\frac{2}{3}$ **49.** $\frac{7}{3}$ **51.** $\frac{1}{125}$

CHAPTER 1 REVIEW

1. $-7 + (-10) = -17$ **2.** $(-7 + 4) + 5 = 2$ **3.** $(-3 + 12) + 5 = 14$ **4.** $4 - 9 = -5$

5. $9 - (-3) = 12$ **6.** $-7 - (-9) = 2$ **7.** $(-3)(-7) - 6 = 15$ **8.** $5(-6) + 10 = -20$

9. $2[(-8)(3x)] = -48x$ **10.** $\frac{-25}{-5} = 5$ **11.** $\frac{-40}{8} - 7 = -12$ **12.** $\frac{-45}{15} + 9 = 6$

13–18.

19. 12 **20.** 3 **21.** $\frac{4}{5}$ **22.** $\frac{7}{10}$ **23.** 1.8 **24.** -10 **25.** $-6, \frac{1}{6}$ **26.** $-\frac{3}{10}, \frac{10}{3}$ **27.** $9, -\frac{1}{9}$

28. $\frac{12}{5}, -\frac{5}{12}$ **29.** $\frac{6}{35}$ **30.** -5 **31.** $-\frac{5}{4}$ **32.** 3 **33.** -38 **34.** 0 **35.** -22 **36.** -25

37. 1 **38.** -3 **39.** 22 **40.** -4 **41.** -20 **42.** -15 **43.** 14 **44.** 12 **45.** 28 **46.** -30

47. -12 **48.** -24 **49.** 12 **50.** -4 **51.** $-\frac{1}{4}$ **52.** $-\frac{2}{3}$ **53.** 23 **54.** 47 **55.** -3 **56.** -35

57. 32 **58.** 2 **59.** 30 **60.** -3 **61.** -98 **62.** 70 **63.** 2 **64.** $\frac{17}{2}$ **65.** Undefined   **66.** 8

67. 3 **68.** Associative   **69.** Multiplicative identity   **70.** Commutative   **71.** Additive inverse

72. Multiplicative inverse   **73.** Additive identity   **74.** Commutative, associative   **75.** Distributive

76. $12 + x$ **77.** $28a$ **78.** x **79.** y **80.** $14x + 21$ **81.** $6a - 12$ **82.** $\frac{5}{2}x - 3$ **83.** $-\frac{3}{2}x + 3$

84. $-\frac{1}{3}, 0, 5, -4.5, \frac{2}{5}, -3$ **85.** $0, 5$ **86.** $\sqrt{7}, \pi$ **87.** $0, 5, -3$ **88.** $2 \cdot 3^2 \cdot 5$ **89.** $2^3 \cdot 3 \cdot 5 \cdot 7$ **90.** $\frac{173}{210}$

91. $\frac{109}{420}$ **92.** -2 **93.** 810 **94.** 8 **95.** 12 **96.** -1 **97.** $\frac{1}{16}$

CHAPTER 1 TEST

1. $x + 3 = 8$ **2.** $5y = 15$ **3.** 40   **4.** 16   **5.** Opposite: 4; reciprocal: $-\frac{1}{4}$; absolute value: 4
6. Opposite: $-\frac{3}{4}$; reciprocal: $\frac{4}{3}$; absolute value: $\frac{3}{4}$ **7.** -4 **8.** 17   **9.** -12 **10.** 0   **11.** c   **12.** e
13. d   **14.** a   **15.** -21 **16.** 64   **17.** -2 **18.** $-\frac{8}{27}$ **19.** 4   **20.** 204   **21.** 25   **22.** 52
23. 2   **24.** 2   **25.** $8 + 2x$ **26.** $10x$ **27.** $6x + 10$ **28.** $-2x + 1$ **29.** $1, -8$ **30.** $1, 1.5, \frac{3}{4}, -8$
31. $\sqrt{2}$ **32.** All of them   **33.** $2^4 \cdot 37$ **34.** $2^2 \cdot 5 \cdot 67$ **35.** $\frac{25}{42}$ **36.** $\frac{8}{x}$ **37.** $8 + (-3) = 5$
38. $-24 - 2 = -26$ **39.** $-5(-4) = 20$ **40.** $\frac{-24}{-2} = 12$ **41.** 12   **42.** $\frac{1}{2}$

CHAPTER 2

PROBLEM SET 2.1

1. $-3x$ **3.** $-a$ **5.** $12x$ **7.** $6a$ **9.** $6x - 3$ **11.** $7a + 5$ **13.** $5x - 5$ **15.** $4a + 2$
17. $-9x - 2$ **19.** $12a + 3$ **21.** $10x - 1$ **23.** $21y + 6$ **25.** $-6x + 8$ **27.** $-2a + 3$
29. $-4x + 26$ **31.** $4y - 16$ **33.** $-6x - 1$ **35.** $2x - 12$ **37.** $10a + 33$ **39.** $4x - 9$ **41.** $7y - 39$
43. $-19x - 14$ **45.** 5   **47.** -9 **49.** 4   **51.** 4   **53.** -37 **55.** -41 **57.** 64   **59.** 64
61. 144   **63.** 144   **65.** 3   **67.** 0   **69.** 15   **71.** 6   **73.** 5, 7, 9, 11   **75.** 2, 5, 10, 17
77. (a)

n	1	2	3	4
$3n$	3	6	9	12

(b)

n	1	2	3	4
n^3	1	8	27	64

79. (a) 42°F   (b) 28°F   (c) -14°F

81. (a) \$37.50   (b) \$40.00   (c) \$42.50   **83.** $0.71G$; \$887.50   **85.** $x - 5; -7$ **87.** $2(x + 10)$; 16
89. $\frac{10}{x}; -5$ **91.** $[3x + (-2)] - 5; -13$ **93.** $-\frac{7}{2}$ **95.** $\frac{51}{40}$

PROBLEM SET 2.2

1. 11   **3.** 4   **5.** $-\frac{3}{4}$ **7.** -5.8 **9.** -17 **11.** $-\frac{1}{8}$ **13.** -4 **15.** -3.6 **17.** 1   **19.** $-\frac{7}{45}$
21. 3   **23.** $\frac{11}{8}$ **25.** 21   **27.** 7   **29.** 3.5   **31.** 22   **33.** -6 **35.** 0   **37.** -2 **39.** -16
41. -3 **43.** 10   **45.** -12 **47.** -1 **49.** 4   **51.** 2   **53.** -5 **55.** -1 **57.** -3 **59.** 8
61. -8 **63.** 2   **65.** 11   **67.** $x = 3, y = -1, z = 0$
69. (a) 6% of the light is reflected.   (b) 5% is reflected.   (c) 2% is absorbed.   (d) 75% is reflected.
71. 70°   **73.** 11   **75.** 2   **77.** $18x$ **79.** x **81.** y **83.** x **85.** a

PROBLEM SET 2.3

1. 2   **3.** 4   **5.** $-\frac{1}{2}$ **7.** -2 **9.** 3   **11.** 4   **13.** 0   **15.** 0   **17.** 6   **19.** -50 **21.** $\frac{3}{2}$
23. 12   **25.** -3 **27.** 32   **29.** -8 **31.** $\frac{1}{2}$ **33.** 4   **35.** 8   **37.** -4 **39.** 4   **41.** -15
43. $-\frac{1}{2}$ **45.** 3   **47.** 1   **49.** $\frac{1}{4}$ **51.** -3 **53.** 3   **55.** 2   **57.** $-\frac{3}{2}$ **59.** $-\frac{3}{2}$ **61.** 1   **63.** 1
65. -2 **67.** -2 **69.** 200 tickets   **71.** 2 three-pointers   **73.** (a) 9   (b) 15   (c) 18
75. $\frac{17}{3}; 3x + 2 = 19$ **77.** $10; 2(x + 10) = 40$ **79.** $10x - 43$ **81.** $-3x - 13$ **83.** $-6y + 4$
85. $-5x + 7$

PROBLEM SET 2.4

1. 3   **3.** -2 **5.** -1 **7.** 2   **9.** -4 **11.** -2 **13.** 0   **15.** 1   **17.** $\frac{1}{2}$ **19.** 7   **21.** 8
23. $-\frac{1}{3}$ **25.** $\frac{3}{4}$ **27.** 75   **29.** 2   **31.** 6   **33.** 8   **35.** 0   **37.** $\frac{3}{7}$ **39.** 1   **41.** $\frac{3}{2}$ **43.** 4
45. 1   **47.** $6x - 10$ **49.** $\frac{3}{2}x + 3$ **51.** $-x + 2$

PROBLEM SET 2.5

1. 100 feet   **3.** 0   **5.** 2   **7.** 15   **9.** 10   **11.** 4   **13.** 2   **15.** $l = \dfrac{A}{w}$ **17.** $r = \dfrac{d}{t}$ **19.** $h = \dfrac{V}{lw}$
21. $P = \dfrac{nRT}{V}$ **23.** $a = P - b - c$ **25.** $x = 3y - 1$ **27.** $y = 3x + 6$ **29.** $y = -\frac{2}{3}x + 2$
31. $y = -2x + 4$ **33.** $y = \frac{5}{2}x - \frac{3}{2}$ **35.** $w = \dfrac{P - 2l}{2}$ **37.** $v = \dfrac{h - 16t^2}{t}$ **39.** $h = \dfrac{A - \pi r^2}{2\pi r}$
41. $y = -\frac{3}{2}x + 3$ **43.** $y = \frac{3}{7}x - 3$ **45.** $y = 2x + 8$ **47.** $60°; 150°$ **49.** $45°; 135°$ **51.** 10
53. 240   **55.** 25%   **57.** 35%   **59.** 64   **61.** 2,000
63. (a) $68,840,000 = 0.628T$ (b) 109,617,834 people   **65.** (a) 60%   (b) 40%   **67.** 100°C; yes
69. 20°C; yes   **71.** $C = \frac{5}{9}(F - 32)$ **73.** 60%   **75.** 26.5%
77.   **79.** (a) 7 meters   (b) $\frac{3}{2}$ or 1.5 inches

Altitude (feet)	Temperature (°F)
0	56
9,000	24.5
15,000	3.5
21,000	-17.5
28,000	-42
40,000	-84

81. The sum of 4 and 1 is 5.   **83.** The difference between 6 and 2 is 4.   **85.** $2(6 + 3)$ **87.** $2 \cdot 5 + 3 = 13$

PROBLEM SET 2.6

Along with the answers to the odd-numbered problems in this problem set and the next, we are including some of the equations used to solve the problems. Be sure that you try the problems on your own before looking here to see the correct equations.
1. 8   **3.** $2x + 4 = 14$; 5   **5.** -1 **7.** The two numbers are x and $x + 2$; $x + (x + 2) = 8$; 3 and 5.
9. 6 and 14   **11.** Barney's age is x; Fred's age is $x + 4$; $(x - 5) + (x - 1) = 48$; Barney is 27; Fred is 31.
13. Lacy is 16; Jack is 32.   **15.** Patrick is 18; Pat is 38.   **17.** $l = 11$ inches; $w = 6$ inches
19. The length of a side is x; $4x = 48$; $x = 12$ meters.   **21.** $l = 17$ inches; $w = 10$ inches
23. If the number of nickels is x, then the number of dimes is $x + 9$; $5x + 10(x + 9) = 210$; 8 nickels, 17 dimes.
25. 15 dimes, 30 quarters   **27.** If she has x nickels, then she has $x + 3$ dimes and $x + 5$ quarters: $5x + 10(x + 3) + 25(x + 5) = 435$; 7 nickels, 10 dimes, 12 quarters.   **29.** 4 is less than 10.   **31.** 9 is greater than or equal to -5.
33. $<$ **35.** $<$ **37.** 2   **39.** 12

PROBLEM SET 2.7

1. $4,000 invested at 8%, $6,000 invested at 9%   **3.** If $x = $ the amount invested at 10%, then $x + 500$ is the amount invested at 12%; $0.10x + 0.12(x + 500) = 214$; $700 invested at 10%, $1,200 invested at 12%   **5.** $500 at 8%, $1,000 at 9%, $1,500 at 10%   **7.** $45°, 45°, 90°$ **9.** $22.5°, 45°, 112.5°$ **11.** $53°, 90°$ **13.** Let $x = $ the number of minutes after the first minute; $41 + 32x = 521$; $x = 15$, so the call was 16 minutes.

15. Let x = the number of hours past 35 hours; $35(12) + 18x = 492$; $x = 4$, so she worked 39 hours that week.
17. Let x = the number of children's tickets; then $2x$ = the number of adult tickets; $4.5x + 6(2x) = 115.5$; $x = 7$; she sold 7 children's tickets and 14 adult tickets.   **19.** Jeff   **21.** $10.38   **23.** Yes   **25.** $54.00   **27.** Yes
29. $79,626   **31.** -8 **33.** $\frac{3}{4}$ **35.** 25   **37.** 0

PROBLEM SET 2.8

1. $x < 12$

3. $a \le 12$

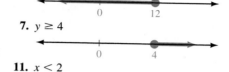

5. $x > 13$

7. $y \ge 4$

9. $x > 9$

11. $x < 2$

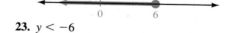

13. $a \le 5$

15. $x > 15$

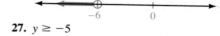

17. $x < -3$

19. $x \le 6$

21. $x \ge -50$

23. $y < -6$

25. $x < 6$

27. $y \ge -5$

29. $x < 3$

31. $x \le 18$

33. $a < -20$

35. $y < 25$

37. $a \le 3$

39. $x \ge \frac{15}{2}$

41. $x < -1$

43. $y \ge -2$

45. $x < -1$

47. $m \le -6$

49. $x \le -5$

51. $y < -\frac{3}{2}x + 3$ **53.** $y < \frac{2}{5}x - 2$ **55.** $y \le \frac{3}{7}x + 3$

57. $y \le \frac{1}{2}x + 1$ **59.** $x < 3$ **61.** $x \ge 3$ **63.** At least 291   **65.** $2x + 6 < 10; x < 2$
67. $4x > x - 8; x > -\frac{8}{3}$ **69.** $2x + 2(3x) \ge 48; x \ge 6$; the width is at least 6 meters.
71. $x + (x + 2) + (x + 4) > 24; x > 6$; the shortest side is even and greater than 6 inches.   **73.** $t \ge 100$
75. Lose money if they sell less than 200 tickets; make a profit if they sell more than 200 tickets   **77.** b   **79.** a
81. b and c

CHAPTER 2 REVIEW

1. $-3x$　**2.** $-2x - 3$　**3.** $4a - 7$　**4.** $-2a + 3$　**5.** $-6y$　**6.** $-6x + 1$　**7.** 19　**8.** -11
9. -18　**10.** -13　**11.** 8　**12.** 6　**13.** -8　**14.** $\frac{15}{14}$　**15.** 2　**16.** -5　**17.** -5　**18.** 0
19. 12　**20.** -8　**21.** -1　**22.** 1　**23.** 5　**24.** $\frac{5}{2}$　**25.** 0　**26.** 0　**27.** 10　**28.** 2　**29.** 2
30. $-\frac{8}{3}$　**31.** 0　**32.** -5　**33.** 0　**34.** -4　**35.** -8　**36.** 4　**37.** $y = \frac{2}{5}x - 2$　**38.** $y = \frac{5}{2}x - 5$
39. $h = \dfrac{V}{\pi r^2}$　**40.** $w = \dfrac{P - 2l}{2}$　**41.** 206.4　**42.** 9%　**43.** 11　**44.** 5 meters; 25 meters
45. $500 at 9%; $800 at 10%　**46.** 10 nickels; 5 dimes　**47.** $x > -2$　**48.** $x < 2$　**49.** $a \geq 6$
50. $a < -15$
51.

52.

53.

CHAPTER 2 TEST

1. $-4x + 5$　**2.** $3a - 4$　**3.** $-3y - 12$　**4.** $11x + 28$　**5.** 22　**6.** 25　**7.** 6　**8.** $\frac{4}{3}$　**9.** 2
10. $-\frac{1}{2}$　**11.** -3　**12.** 70　**13.** -3　**14.** -1　**15.** 5.7　**16.** 2,000　**17.** 3　**18.** $\frac{28}{3}$ inches
19. $y = -\frac{2}{5}x + 4$　**20.** $v = \dfrac{h - x - 16t^2}{t}$　**21.** Rick is 20; Dave is 40.
22. Width is 10 inches; length is 20 inches.　**23.** 8 quarters, 15 dimes　**24.** $800 at 7%; $1,400 at 9%
25. $x < 1$

26. $a < -4$

27. $x \leq -3$

28. $m \geq -2$

CHAPTERS 1 & 2 CUMULATIVE REVIEW

1. 30　**2.** -7　**3.** -14　**4.** -12　**5.** $2x$　**6.** 8　**7.** $-\frac{8}{27}$　**8.** $-72x$　**9.** $-\frac{4}{5}$　**10.** $\frac{8}{35}$
11. $-\frac{8}{3}$　**12.** 0　**13.** $\frac{1}{4}$　**14.** $\frac{8}{9}$　**15.** $-\frac{1}{15}$　**16.** $11x + 7$　**17.** 4　**18.** -16　**19.** -50　**20.** 5
21. $\frac{3}{2}$　**22.** 1　**23.** $\frac{1}{3}$　**24.** 12　**25.** 48　**26.** $c = P - a - b$　**27.** $y = -\frac{3}{4}x + 3$, or $\dfrac{-3x + 12}{4}$
28. $x \geq 5$　**29.** $x > -10$　**30.** $x \leq 6$

31. $x > 3$

32. $x - 5 = 12$　**33.** $\frac{2}{3}, -\frac{3}{2}, \frac{2}{3}$　**34.** $P = 12$ inches, $A = 9$ inches^2　**35.** -8　**36.** Additive inverse
37. $2x - 1$　**38.** $-5, -3, -1.7, 2.3, \frac{12}{7}$　**39.** $\frac{3}{4}$　**40.** 11　**41.** 25　**42.** 15　**43.** 12　**44.** She lost $9.
45. 48° and 90°　**46.** Width = 5 inches, length = 9 inches　**47.** 65°　**48.** 4 quarters and 12 dimes
49. $400 at 5%; $600 at 6%　**50.** 18 hours

CHAPTER 3

PROBLEM SET 3.1

1–17.

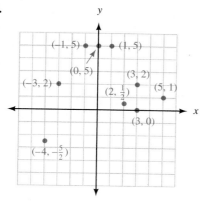

19. $(-4, 4)$ **21.** $(-4, 2)$ **23.** $(-3, 0)$ **25.** $(2, -2)$
27. $(-5, -5)$

29. Yes   **31.** No

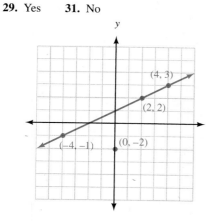

33. Yes   **35.** No

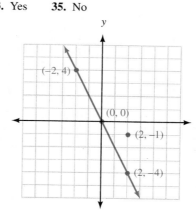

37. Yes   **39.** No

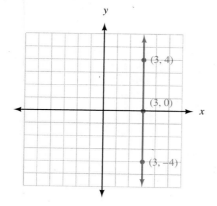

41. No   **43.** No

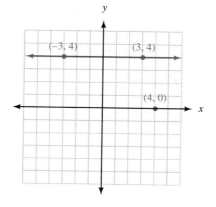

45. Every point on this line has a *y*-coordinate of −3.

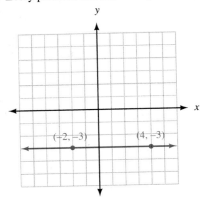

47. Along the *y*-axis

49. Any three:

(0, 0), (5, 40), (10, 80),
(15, 120), (20, 160), (25, 200),
(30, 240), (35, 280), (40, 320)

51.

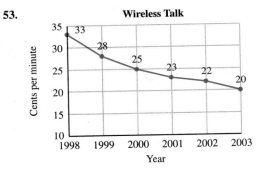

53.

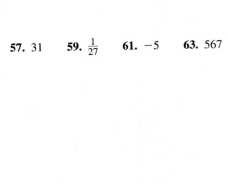

55.

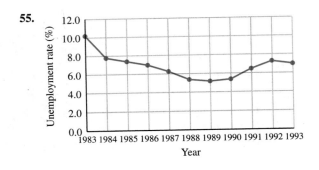

57. 31   **59.** $\frac{1}{27}$ **61.** −5   **63.** 567

65. 15

PROBLEM SET 3.2

1. $(0, 6), (3, 0), (6, -6)$ **3.** $(0, 3), (4, 0), (-4, 6)$ **5.** $(1, 1), (\frac{3}{4}, 0)(5, 17)$ **7.** $(2, 13), (1, 6), (0, -1)$

9. $(-5, 4), (-5, -3), (-5, 0)$

11.

x	y
1	3
-3	-9
4	12
6	18

13.

x	y
0	0
$-\frac{1}{2}$	-2
-3	-12
3	12

15.

x	y
2	3
3	2
5	0
9	-4

17.

x	y
2	0
3	2
1	-2
-3	-10

19.

x	y
0	-1
-1	-7
-3	-19
$\frac{3}{2}$	8

21. $(0, -2)$ **23.** $(1, 5), (0, -2), (-2, -16)$ **25.** $(1, 6), (-2, -12), (0, 0)$

27. $(2, -2)$ **29.** $(3, 0), (3, -3)$ **31.** 12 inches

33.

Minutes	Cost
0	$ 3.00
10	$ 4.00
20	$ 5.00
30	$ 6.00
40	$ 7.00
50	$ 8.00
60	$ 9.00
70	$10.00
80	$11.00
90	$12.00
100	$13.00

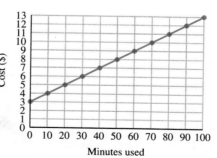

35.

Hours	Cost
0	$10
1	$13
2	$16
3	$19
4	$22
5	$25
6	$28
7	$31
8	$34
9	$37
10	$40

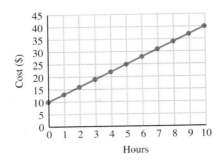

37.

Bags Collected	Cost
5	$20.50
7	$23.50
9	$26.50
11	$29.50
13	$32.50

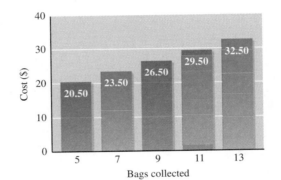

39.

Gallons per Month	Monthly Cost ($)
15	23.50
20	29.00
25	34.50
30	40.00
35	45.50

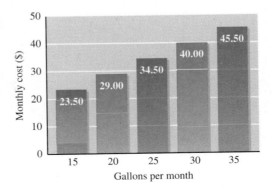

41. -3 **43.** 2 **45.** 0 **47.** $y = -5x + 4$ **49.** $y = \frac{3}{2}x - 3$

PROBLEM SET 3.3

1.

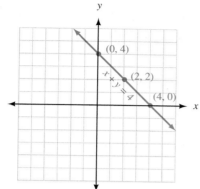

3.

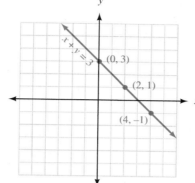

5.

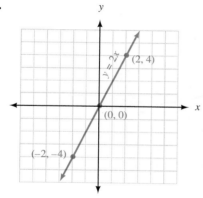

7.

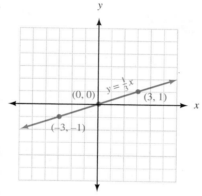

9.

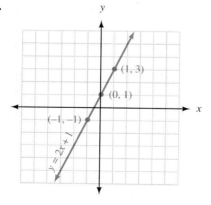

11.

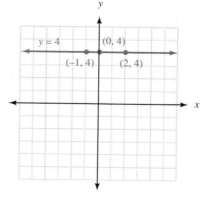

13.

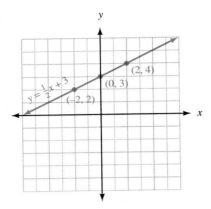

15.

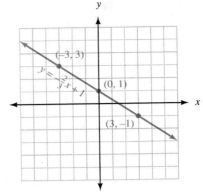

17.

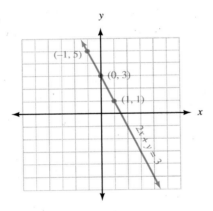

19.

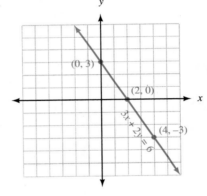

21.

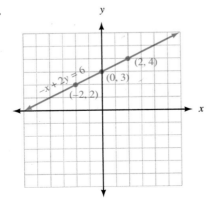

23.

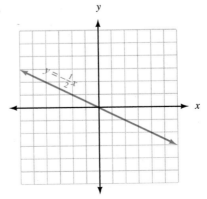

25.

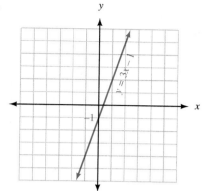

27.

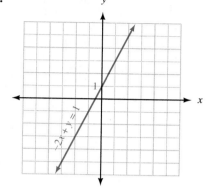

29.

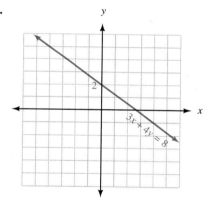

31.

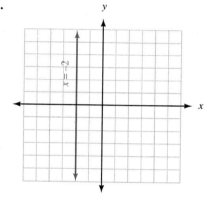

33.

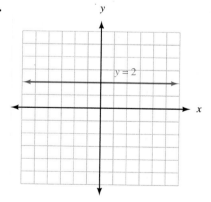

35.

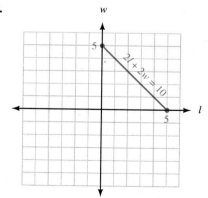

Note: The graph does not go beyond the x- and y-axes because l and w are the length and width of a rectangle and therefore cannot be negative.

37.

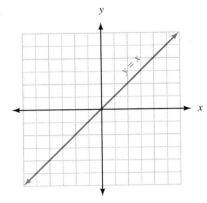

39.

x	y
−4	−3
−2	−2
0	−1
2	0
6	2

41.

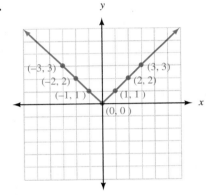

43.

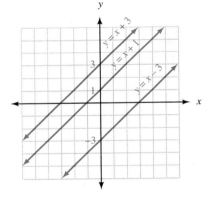

45.

Distance From Airport (miles)	Height (miles)
0	0
15	1.1
30	2.2
45	3.3
60	4.4
75	5.5
90	6.6

47.

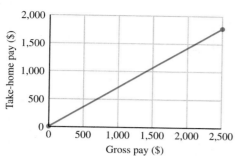

49. 5   **51.** −2   **53.** 3   **55.** −6

PROBLEM SET 3.4

1.

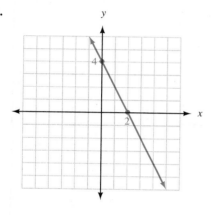

3.

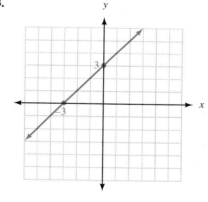

5.

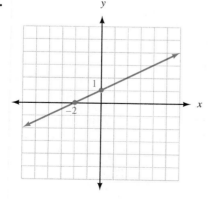

7.

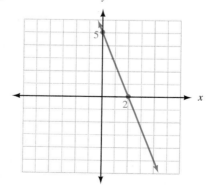

9.

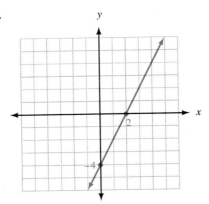

11.

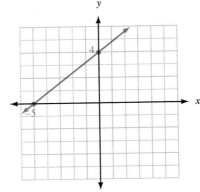

13.

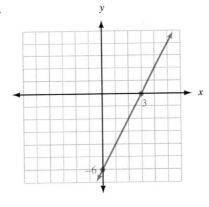

15.

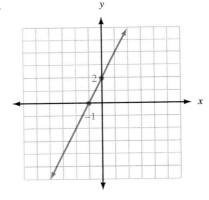

17.

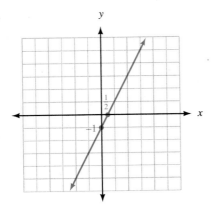

19.

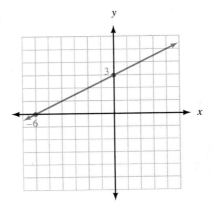

21.

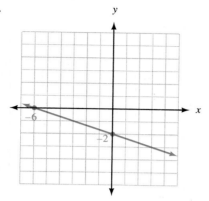

23.

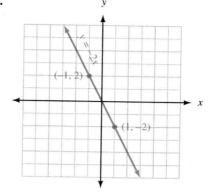

25.

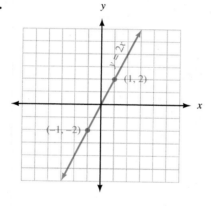

27.

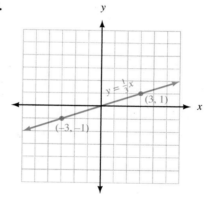

29.

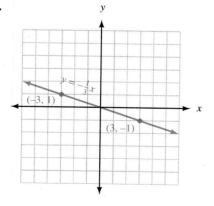

31.

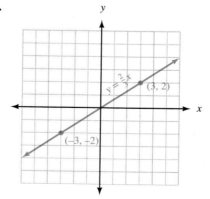

33.

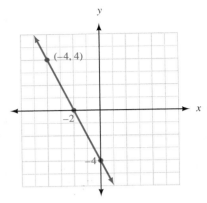

35.

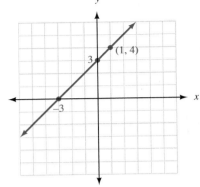

37.

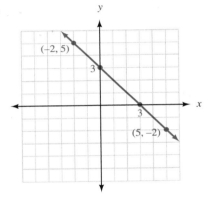

39.

x	y
−2	1
0	−1
−1	0
1	−2

41.

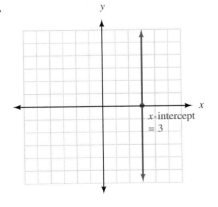

43.

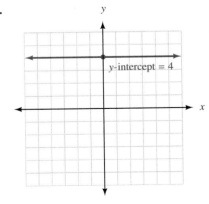

45.

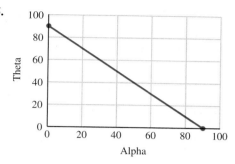

47. $x \leq -4$ **49.** $x \geq 3$ **51.** $x > -4$

PROBLEM SET 3.5

1.

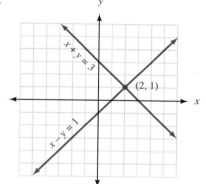

3.

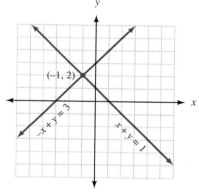

5.

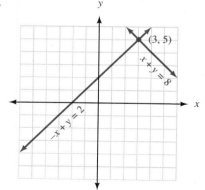

7.

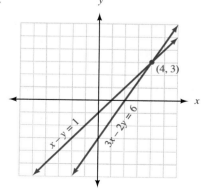

9.

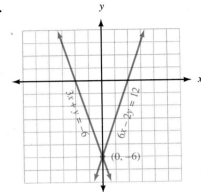

11.

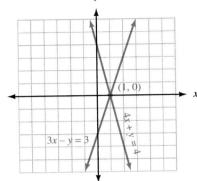

13.

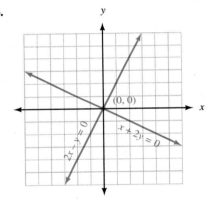

15.

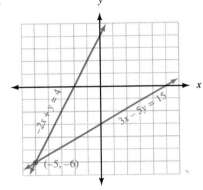

17.

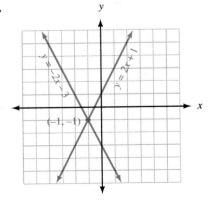

19.

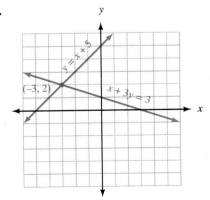

21.

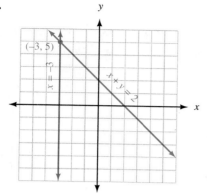

23.

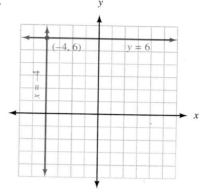

25. No solution

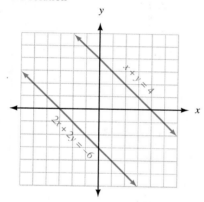

27. Infinitely many solutions

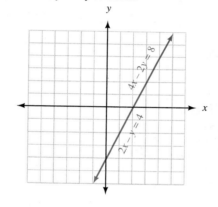

29.

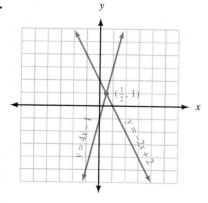

31.

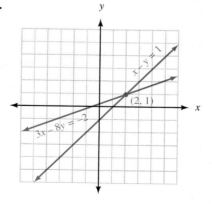

33. (a) 25 hours   (b) Gigi's   (c) Marcy's

35. (a) 4 hours   (b) Computer Service   (c) ICM World

37. -15 **39.** 27   **41.** -27 **43.** 13

PROBLEM SET 3.6

1. (2, 1)   **3.** (3, 7)   **5.** (2, −5)   **7.** (−1, 0)   **9.** Lines coincide.   **11.** (4, 8)   **13.** ($\frac{1}{5}$, 1)   **15.** (1, 0)

17. (−1, −2)   **19.** (−5, $\frac{3}{4}$)   **21.** (−4, 5)   **23.** (−3, −10)   **25.** (3, 2)   **27.** (5, $\frac{1}{3}$)   **29.** (−2, $\frac{2}{3}$)

31. (2, 2)   **33.** Lines are parallel.   **35.** (1, 1)   **37.** Lines are parallel.   **39.** (7, 5)   **41.** (10, 12)

43. 75   **45.** 400

PROBLEM SET 3.7

1. (4, 7)   **3.** (3, 17)   **5.** ($\frac{3}{2}$, 2)   **7.** (2, 4)   **9.** (0, 4)   **11.** (−1, 3)   **13.** (1, 1)   **15.** (2, −3)

17. (−2, $\frac{3}{5}$)   **19.** (−3, 5)   **21.** Lines are parallel.   **23.** (3, 1)   **25.** ($\frac{1}{2}$, $\frac{3}{4}$)   **27.** (2, 6)   **29.** (4, 4)

31. (5, −2)   **33.** (18, 10)   **35.** Lines coincide.   **37.** (10, 12)

39. (a) 1,000 miles   (b) Car   (c) Truck   (d) Miles ≥ 0   **41.** 40 gallons   **43.** l = 9 meters; w = 3 meters

45. 12 nickels, 15 dimes   **47.** 480   **49.** $800 at 8%, $1,600 at 10%

PROBLEM SET 3.8

1. $x + y = 25$ The two numbers   **3.** 3 and 12   **5.** $x − y = 5$ The two numbers   **7.** 6 and 29
$y = x + 5$ are 10 and 15. $x = 2y + 1$ are 9 and 4.

9. Let x = the amount invested at 6% and y = the amount invested at 8%.   **11.** $2,000 at 6%, $8,000 at 5%
$x + y = 20{,}000$ He had $9,000 at 8%
$0.06x + 0.08y = 1{,}380$ and $11,000 at 6%.

13. 6 nickels, 8 quarters   **15.** Let x = the number of dimes and y = the number of quarters.
$x + y = 21$ He has 12 dimes
$0.10x + 0.25y = 3.45$ and 9 quarters.

17. Let x = the number of liters of 50% solution and y = the number of liters of 20% solution.
$x + y = 18$ 6 liters of 50% solution
$0.50x + 0.20y = 0.30(18)$ 12 liters of 20% solution

19. 10 gallons of 10% solution, 20 gallons of 7% solution   **21.** 20 adults, 50 kids   **23.** 16 feet wide, 32 feet long

25. 33 $5 chips, 12 $25 chips   **27.** 50 at $11, 100 at $20   **29.** $−6x + 11$ **31.** $−8$ **33.** $−4$ **35.** $w = \dfrac{P − 2l}{2}$

37.   **39.** $y \geq \frac{3}{2}x − 6$ **41.** Width 2 inches, length 11 inches

CHAPTER 3 REVIEW

1. (4, −6), (0, 6), (1, 3), (2, 0)   **2.** (5, −2), (0, −4), (15, 2), (10, 0)   **3.** (4, 2), (2, −2), ($\frac{9}{2}$, 3)

4. (2, 13), (−$\frac{3}{5}$, 0), (−$\frac{6}{5}$, −3)   **5.** (2, −3), (−1, −3), (−3, −3)   **6.** (6, 5), (6, 0), (6, −1)   **7.** (2, −$\frac{3}{2}$)

8. $\left(-\frac{8}{3}, -1\right), (-3, -2)$ **9-14.**

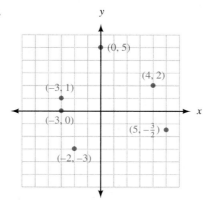

15. $(-2, 0), (0, -2), (1, -3)$

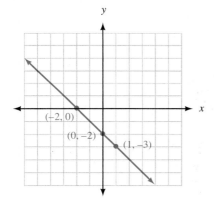

16. $(-1, -3), (1, 3), (0, 0)$

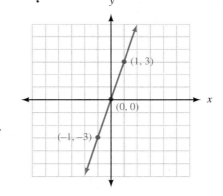

17. $(1, 1), (0, -1), (-1, -3)$

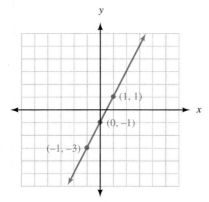

18.

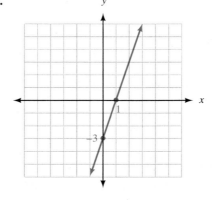

19.

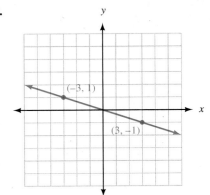

20.

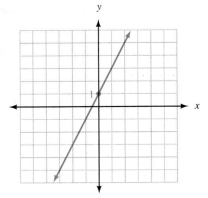

21.

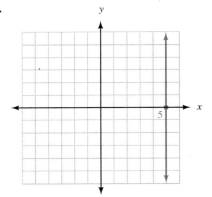

22.

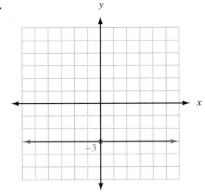

23. x-intercept = 2   **24.** x-intercept = 12   **25.** x-intercept = 3   **26.** x-intercept = 2

  y-intercept = -6 y-intercept = -4 y-intercept = -3 y-intercept = -6

27. $(4, -2)$ **28.** $(-3, 2)$ **29.** $(3, -2)$ **30.** $(2, 0)$ **31.** $(2, 1)$ **32.** $(-1, -2)$ **33.** $(1, -3)$

34. $(-2, 5)$ **35.** Lines coincide.   **36.** $(2, -2)$ **37.** $(1, 1)$ **38.** Lines are parallel.   **39.** $(-2, -3)$

40. $(5, -1)$ **41.** $(-2, 7)$ **42.** $(-4, -2)$ **43.** $(-1, 4)$ **44.** $(2, -6)$ **45.** Lines are parallel.

46. $(-3, -10)$ **47.** $(1, -2)$ **48.** Lines coincide.   **49.** 10, 8   **50.** 24, 8   **51.** \$4,000 at 4%, \$8,000 at 5%

52. \$3,000 at 6%, \$11,000 at 8%   **53.** 10 dimes, 7 nickels   **54.** 9 dimes, 6 quarters

55. 40 liters of 10% solution, 10 liters of 20% solution   **56.** 20 liters of 25% solution, 20 liters of 15% solution

CHAPTER 3 TEST

1. $(0, -2), (5, 0), (10, 2), (-\frac{5}{2}, -3)$ **2.** $(2, 5), (0, -3)$ **3.**

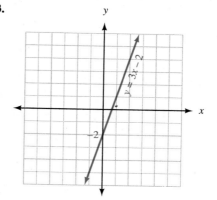

4.

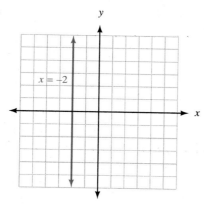

5. y-intercept $= -3$
x-intercept $= 5$

6. y-intercept $= 1$
x-intercept $= -\frac{2}{3}$

7.

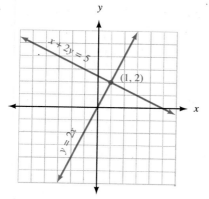

8.

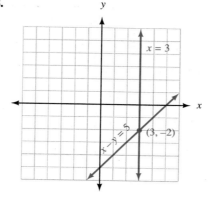

9. $(-3, -4)$ **10.** $(5, -3)$ **11.** $(2, -2)$ **12.** Lines coincide.   **13.** $(2, 7)$ **14.** $(4, 3)$ **15.** $(2, 9)$
16. $(-5, -1)$ **17.** $5, 7$ **18.** $3, 12$ **19.** \$6,000   **20.** 7 nickels, 5 quarters

CHAPTERS 1–3 CUMULATIVE REVIEW

1. 17 **2.** 48 **3.** 36 **4.** $2x - 2$ **5.** 32 **6.** 16 **7.** $\frac{5}{4}$ **8.** $-10a + 29$ **9.** -8 **10.** 0
11. 7 **12.** -6 **13.** $x \le -9$

14. $x \ge 12$

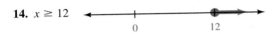

15.

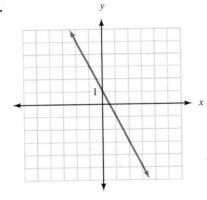

16.

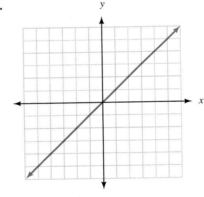

17.

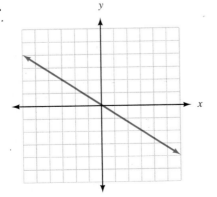

18. $\{1, 2\}$

19.

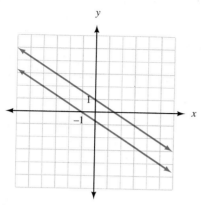

20. (1, 0)

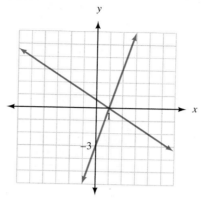

21. Lines coincide.   **22.** (1, −3)   **23.** (2, 3)   **24.** (6, 7)   **25.** (9, 3)   **26.** (4, −1)   **27.** (4, 0)
28. (3, −1)   **29.** (−1, −3)   **30.** Lines coincide.   **31.** Commutative property of addition   **32.** 32
33. −5   **34.** $2^2 \cdot 3^2 \cdot 5$ **35.** 25%   **36.** 15.5   **37.** 13   **38.** −21   **39.** −13   **40.** (−2, −4)
41. $5 − (−8) = 13$ **42.**

x	y
2	$-\frac{4}{5}$
5	1

43. (4, 0), (0, −3)   **44.** {−3, 0, 5}   **45.** −3   **46.** 14
47. Width is $\frac{17}{3}$ cm; length is $\frac{49}{3}$ cm.   **48.** 3 ounces

49. $1,200 at 6%, $2,100 at 8%   **50.** 8 nickels, 7 dimes

CHAPTER 4

PROBLEM SET 4.1

1. Base 4; exponent 2; 16   **3.** Base 0.3; exponent 2; 0.09   **5.** Base 4; exponent 3; 64
7. Base −5; exponent 2; 25   **9.** Base 2; exponent 3; −8   **11.** Base 3; exponent 4; 81   **13.** Base $\frac{2}{3}$; exponent 2; $\frac{4}{9}$
15. Base $\frac{1}{2}$; exponent 4; $\frac{1}{16}$ **17.** (a)   (b) Either *larger* or *greater* will work.

Number x	Square x^2
1	1
2	4
3	9
4	16
5	25
6	36
7	49

19. x^9 **21.** y^{30} **23.** 2^{12} **25.** x^{28} **27.** x^{10} **29.** 5^{12} **31.** y^9 **33.** 2^{50} **35.** a^{3x} **37.** b^{xy}
39. $16x^2$ **41.** $32y^5$ **43.** $81x^4$ **45.** $0.25a^2b^2$ **47.** $64x^3y^3z^3$ **49.** $8x^{12}$ **51.** $16a^6$ **53.** x^{14}

55. a^{11} **57.** $128x^7$ **59.** $432x^{10}$ **61.** $16x^4 y^6$ **63.** $\frac{8}{27}a^{12} b^{15}$

65.

Number x	Square x^2
-3	9
-2	4
-1	1
0	0
1	1
2	4
3	9

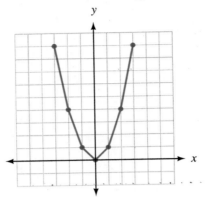

69. 4.32×10^4 **71.** 5.7×10^2 **73.** 2.38×10^5 **75.** 2,490   **77.** 352

67.

Number x	Square x^2
-2.5	6.25
-1.5	2.25
-0.5	0.25
0	0
0.5	0.25
1.5	2.25
2.5	6.25

79. 28,000   **81.** 27 inches^3 **83.** 15.6 inches^3 **85.** 36 inches^3
87. Yes, the dimensions could be 2 feet $\times$ 3 feet $\times$ 7 feet.   **89.** 6.5×10^8 seconds   **91.** $740,000
93. 180,000   **95.** 219 inches^3 **97.** 182 inches^3
99. (a) Length of Stage 1 is $\frac{4}{3}$ feet.   (b) Length of Stage 2 is $(\frac{4}{3})^2 = \frac{16}{9}$ feet.

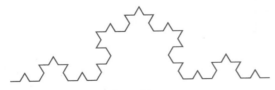

(c) Length of Stage 3 is $(\frac{4}{3})^3 = \frac{64}{27}$ feet.   (d) Length of Stage 10 is $(\frac{4}{3})^{10}$ feet.
101. -3 **103.** 11   **105.** -5 **107.** 5

PROBLEM SET 4.2

1. $\frac{1}{9}$ **3.** $\frac{1}{36}$ **5.** $\frac{1}{64}$ **7.** $\frac{1}{125}$ **9.** $\frac{2}{x^3}$ **11.** $\frac{1}{8x^3}$ **13.** $\frac{1}{25y^2}$ **15.** $\frac{1}{100}$

17.

Number x	Square x^2	Power of 2 2^x
-3	9	$\frac{1}{8}$
-2	4	$\frac{1}{4}$
-1	1	$\frac{1}{2}$
0	0	1
1	1	2
2	4	4
3	9	8

19. $\frac{1}{25}$ **21.** x^6 **23.** 64   **25.** $8x^3$ **27.** 6^{10}

29. $\frac{1}{6^{10}}$ **31.** $\frac{1}{2^8}$ **33.** 2^8 **35.** $27x^3$ **37.** $81x^4y^4$ **39.** 1   **41.** $2a^2b$ **43.** $\frac{1}{49y^6}$ **45.** $\frac{1}{x^8}$

47. $\frac{1}{y^3}$ **49.** x^2 **51.** a^6 **53.** $\frac{1}{y^9}$ **55.** y^{40} **57.** $\frac{1}{x}$ **59.** x^9 **61.** a^{16} **63.** $\frac{1}{a^4}$

65.

Number x	Power of 2 2^x
-3	$\frac{1}{8}$
-2	$\frac{1}{4}$
-1	$\frac{1}{2}$
0	1
1	2
2	4
3	8

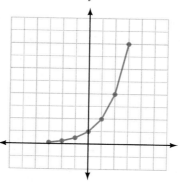

67. 4.8×10^{-3} **69.** 2.5×10^1 **71.** 9×10^{-6}

73.

Expanded Form	Scientific Notation $n \times 10^r$
0.000357	3.57×10^{-4}
0.00357	3.57×10^{-3}
0.0357	3.57×10^{-2}
0.357	3.57×10^{-1}
3.57	3.57×10^0
35.7	3.57×10^1
357	3.57×10^2
3,570	3.57×10^3
35,700	3.57×10^4

75. 0.00423   **77.** 0.00008   **79.** 4.2   **81.** 0.002 seconds

83. 6×10^{-3} inches **85.** 2.5×10^4 **87.** 2.35×10^5 **89.** 8.2×10^{-4} **91.** 100 inches^2; 400 inches^2; 4
93. x^2; $4x^2$; 4 **95.** 216 inches^3; 1,728 inches^3; 8 **97.** x^3; $8x^3$; 8 **99.** $7x$ **101.** $2a$ **103.** $10y$

PROBLEM SET 4.3

1. $12x^7$ **3.** $-16y^{11}$ **5.** $32x^2$ **7.** $200a^6$ **9.** $-24a^3b^3$ **11.** $24x^6y^8$ **13.** $3x$ **15.** $\dfrac{6}{y^3}$ **17.** $\dfrac{1}{2a}$

19. $-\dfrac{3a}{b^2}$ **21.** $\dfrac{x^2}{9z^2}$ **23.**

25. 6×10^8

a	b	ab	$\dfrac{a}{b}$	$\dfrac{b}{a}$
10	$5x$	$50x$	$\dfrac{2}{x}$	$\dfrac{x}{2}$
$20x^3$	$6x^2$	$120x^5$	$\dfrac{10x}{3}$	$\dfrac{3}{10x}$
$25x^5$	$5x^4$	$125x^9$	$5x$	$\dfrac{1}{5x}$
$3x^{-2}$	$3x^2$	9	$\dfrac{1}{x^4}$	x^4
$-2y^4$	$8y^7$	$-16y^{11}$	$-\dfrac{1}{4y^3}$	$-4y^3$

27. 1.75×10^{-1} **29.** 1.21×10^{-6} **31.** 4.2×10^3 **33.** 3×10^{10} **35.** 5×10^{-3} **37.** $8x^2$
39. $-11x^5$ **41.** 0 **43.** $4x^3$ **45.** $31ab^2$ **47.** **49.** $4x^3$

a	b	ab	$a + b$
$5x$	$3x$	$15x^2$	$8x$
$4x^2$	$2x^2$	$8x^4$	$6x^2$
$3x^3$	$6x^3$	$18x^6$	$9x^3$
$2x^4$	$-3x^4$	$-6x^8$	$-x^4$
x^5	$7x^5$	$7x^{10}$	$8x^5$

51. $\dfrac{1}{b^2}$ **53.** $\dfrac{6y^{10}}{x^4}$ **55.** 2×10^6 **57.** 1×10^1 **59.** 4.2×10^{-6} **61.** $9x^3$ **63.** $-20a^2$ **65.** $6x^5y^2$
67. 2 **69.** 4 **71.** $(4 + 5)^2 = 9^2 = 81$; $4^2 + 5^2 = 16 + 25 = 41$
73. $(3 + 4)^2 = 7^2 = 49$; $3^2 + 2(3)(4) + 4^2 = 49$ **75.** $P = 6x$; $A = 2x^2$ **77.** $V = 8x^2$ inches^3
79. (a) 4.7×10^3 feet^3 (b) 3.1×10^4 boxes **81.** $(-4, 16), (-2, 4), (-1, 1), (0, 0), (1, 1), (2, 4), (4, 16)$

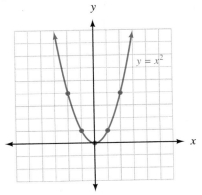

83. $(-3, 18), (-2, 8), (-1, 2), (0, 0), (1, 2), (2, 8), (3, 18)$

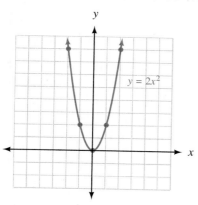

85. $(-4, 8), (-2, 2), (-1, \frac{1}{2}), (0, 0), (1, \frac{1}{2}), (2, 2), (4, 8)$

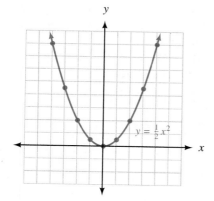

87. $(-4, 4), (-2, 1), (-1, \frac{1}{4}), (0, 0), (1, \frac{1}{4}), (2, 1), (4, 4)$

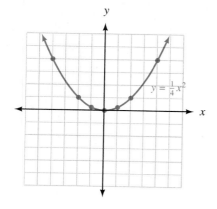

89. 9   **91.** 0

93.

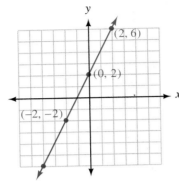

95.

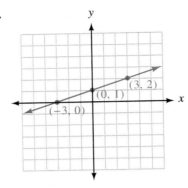

PROBLEM SET 4.4

1. Trinomial, 3   **3.** Trinomial, 3   **5.** Binomial, 1   **7.** Binomial, 2   **9.** Monomial, 2   **11.** Monomial, 0
13. $5x^2 + 5x + 9$ **15.** $5a^2 - 9a + 7$ **17.** $x^2 + 6x + 8$ **19.** $6x^2 - 13x + 5$ **21.** $x^2 - 9$
23. $3y^2 - 11y + 10$ **25.** $6x^3 + 5x^2 - 4x + 3$ **27.** $2x^2 - x + 1$ **29.** $2a^2 - 2a - 2$
31. $-\frac{1}{9}x^3 - \frac{2}{3}x^2 - \frac{5}{2}x + \frac{7}{4}$ **33.** $-4y^2 + 15y - 22$ **35.** $x^2 - 33x + 63$ **37.** $8y^2 + 4y + 26$
39. $75x^2 - 150x - 75$ **41.** $12x + 2$ **43.** 4   **45.** 25   **47.** 16   **49.** 18π inches^3 **51.** $-15x^2$
53. $6x^3$ **55.** $6x^4$

PROBLEM SET 4.5

1. $6x^2 + 2x$ **3.** $6x^4 - 4x^3 + 2x^2$ **5.** $2a^3b - 2a^2b^2 + 2ab$ **7.** $3y^4 + 9y^3 + 12y^2$
9. $8x^5y^2 + 12x^4y^3 + 32x^2y^4$ **11.** $x^2 + 7x + 12$ **13.** $x^2 + 7x + 6$ **15.** $x^2 + 2x + \frac{3}{4}$ **17.** $a^2 + 2a - 15$
19. $xy + bx - ay - ab$ **21.** $x^2 - 36$ **23.** $y^2 - \frac{25}{36}$ **25.** $2x^2 - 11x + 12$ **27.** $2a^2 + 3a - 2$
29. $6x^2 - 19x + 10$ **31.** $2ax + 8x + 3a + 12$ **33.** $25x^2 - 16$ **35.** $2x^2 + \frac{5}{2}x - \frac{3}{4}$ **37.** $3 - 10a + 8a^2$
43. $a^3 - 6a^2 + 11a - 6$

39.

	x	3
x	x^2	$3x$
2	$2x$	6

$(x + 2)(x + 3) = x^2 + 2x + 3x + 6$
$= x^2 + 5x + 6$

41.

	x	x	2
x	x^2	x^2	$2x$
1	x	x	2

$(x + 1)(2x + 2) = 2x^2 + 4x + 2$

45. $x^3 + 8$ **47.** $2x^3 + 17x^2 + 26x + 9$ **49.** $5x^4 - 13x^3 + 20x^2 + 7x + 5$ **51.** $2x^4 + x^2 - 15$
53. $6a^6 + 15a^4 + 4a^2 + 10$ **55.** $x^3 + 12x^2 + 47x + 60$ **57.** $A = x(2x + 5) = 2x^2 + 5x$
59. $A = x(x + 1) = x^2 + x$ **61.** $R = (1{,}200 - 100p)p = 1{,}200p - 100p^2$
63. $R = (1{,}700 - 100p)p = 1{,}700p - 100p^2$

65.

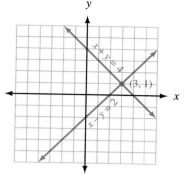

67. $(1, -1)$ **69.** $(-\frac{1}{2}, -\frac{1}{2})$ **71.** 6 dimes, 5 quarters

PROBLEM SET 4.6

1. $x^2 - 4x + 4$ **3.** $a^2 + 6a + 9$ **5.** $x^2 - 10x + 25$ **7.** $a^2 - a + \frac{1}{4}$ **9.** $x^2 + 20x + 100$
11. $a^2 + 1.6a + 0.64$ **13.** $4x^2 - 4x + 1$ **15.** $16a^2 + 40a + 25$ **17.** $9x^2 - 12x + 4$
19. $9a^2 + 30ab + 25b^2$ **21.** $16x^2 - 40xy + 25y^2$ **23.** $49m^2 + 28mn + 4n^2$ **25.** $36x^2 - 120xy + 100y^2$
27. $x^4 + 10x^2 + 25$ **29.** $a^4 + 2a^2 + 1$ **31.**

x	$(x + 3)^2$	$x^2 + 9$	$x^2 + 6x + 9$
1	16	10	16
2	25	13	25
3	36	18	36
4	49	25	49

33.

a	b	$(a + b)^2$	$a^2 + b^2$	$a^2 + ab + b^2$	$a^2 + 2ab + b^2$
1	1	4	2	3	4
3	5	64	34	49	64
3	4	49	25	37	49
4	5	81	41	61	81

35. $a^2 - 25$ **37.** $y^2 - 1$ **39.** $81 - x^2$

41. $4x^2 - 25$ **43.** $16x^2 - \frac{1}{9}$ **45.** $4a^2 - 49$ **47.** $36 - 49x^2$ **49.** $x^4 - 9$ **51.** $a^4 - 16$
53. $25y^8 - 64$ **55.** $2x^2 - 34$ **57.** $-12x^2 + 20x + 8$ **59.** $a^2 + 4a + 6$ **61.** $8x^3 + 36x^2 + 54x + 27$
63. $(50 - 1)(50 + 1) = 2{,}500 - 1 = 2{,}499$ **65.** Both equal 25.   **67.** $x^2 + (x + 1)^2 = 2x^2 + 2x + 1$
69. $x^2 + (x + 1)^2 + (x + 2)^2 = 3x^2 + 6x + 5$ **71.** $a^2 + ab + ba + b^2 = a^2 + 2ab + b^2$

73.

	x	x	1
x	x^2	x^2	x
x	x^2	x^2	x
1	x	x	1

$(2x + 1)^2 = 4x^2 + 4x + 1$ **75.** $5x$ **77.** $\dfrac{a^4}{2b^2}$

79.

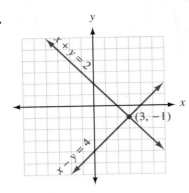

81.

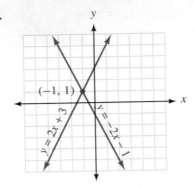

PROBLEM SET 4.7

1. $x - 2$ **3.** $3 - 2x^2$ **5.** $5xy - 2y$ **7.** $7x^4 - 6x^3 + 5x^2$ **9.** $10x^4 - 5x^2 + 1$ **11.** $-4a + 2$

13. $-8a^4 - 12a^3$ **15.** $-4b - 5a$ **17.** $-6a^2b + 3ab^2 - 7b^3$ **19.** $-\dfrac{a}{2} - b - \dfrac{b^2}{2a}$ **21.** $3x + 4y$

23. $-y + 3$ **25.** $5y - 4$ **27.** $xy - x^2 y^2$ **29.** $-1 + xy$ **31.** $-a + 1$ **33.** $x^2 - 3xy + y^2$

35. $2 - 3b + 5b^2$ **37.** $-2xy + 1$ **39.** $xy - \dfrac{1}{2}$ **41.** $\dfrac{1}{4x} - \dfrac{1}{2a} + \dfrac{3}{4}$ **43.** $\dfrac{4x^2}{3} + \dfrac{2}{3x} + \dfrac{1}{x^2}$

45. $3a^{3m} - 9a^m$ **47.** $2x^{4m} - 5x^{2m} + 7$ **49.** $3x^2 - x + 6$ **51.** 4   **53.** $x + 5$ **55.** Both equal 7.

57. $\dfrac{3(10) + 8}{2} = 19; 3(10) + 4 = 34$ **59.** $(7, -1)$ **61.** $(2, 3)$ **63.** $(1, 1)$ **65.** Lines coincide.

PROBLEM SET 4.8

1. $x - 2$ **3.** $a + 4$ **5.** $x - 3$ **7.** $x + 3$ **9.** $a - 5$ **11.** $x + 2 + \dfrac{2}{x + 3}$ **13.** $a - 2 + \dfrac{12}{a + 5}$

15. $x + 4 + \dfrac{9}{x - 2}$ **17.** $x + 4 + \dfrac{-10}{x + 1}$ **19.** $a + 1 + \dfrac{-1}{a + 2}$ **21.** $x - 3 + \dfrac{17}{2x + 4}$

23. $3a - 2 + \dfrac{7}{2a + 3}$ **25.** $2a^2 - a - 3$ **27.** $x^2 - x + 5$ **29.** $x^2 + x + 1$ **31.** $x^2 + 2x + 4$

33. 5, 20   **35.** \$800 at 8%, \$400 at 9%   **37.** Eight \$5 bills, twelve \$10 bills
39. 10 gallons of 20%, 6 gallons of 60%

CHAPTER 4 REVIEW

1. -1 **2.** -64 **3.** $\dfrac{9}{49}$ **4.** y^{12} **5.** x^{30} **6.** x^{35} **7.** 2^{24} **8.** $27y^3$ **9.** $-8x^3 y^3 z^3$ **10.** $\dfrac{1}{49}$

11. $\dfrac{4}{x^5}$ **12.** $\dfrac{1}{27y^3}$ **13.** a^6 **14.** $\dfrac{1}{x^4}$ **15.** x^{15} **16.** $\dfrac{1}{x^5}$ **17.** 1   **18.** -3 **19.** $9x^6 y^4$

20. $64a^{22}b^{20}$ **21.** $\dfrac{-1}{27x^3 y^6}$ **22.** b **23.** $\dfrac{1}{x^{35}}$ **24.** $5x^7$ **25.** $\dfrac{6y^7}{x^5}$ **26.** $4a^6$ **27.** $2x^5 y^2$

28. 6.4×10^7 **29.** 2.3×10^8 **30.** 8×10^3 **31.** $8a^2 - 12a - 3$ **32.** $-12x^2 + 5x - 14$
33. $-4x^2 - 6x$ **34.** 14   **35.** $12x^2 - 21x$ **36.** $24x^5 y^2 - 40x^4 y^3 + 32x^3 y^4$ **37.** $a^3 + 6a^2 + a - 4$
38. $x^3 + 125$ **39.** $6x^2 - 29x + 35$ **40.** $25y^2 - \dfrac{1}{25}$ **41.** $a^4 - 9$ **42.** $a^2 - 10a + 25$
43. $9x^2 + 24x + 16$ **44.** $y^4 + 6y^2 + 9$ **45.** $-2b - 4a$ **46.** $-5x^4 y^3 + 4x^2 y^2 + 2x$ **47.** $x + 9$
48. $2x + 5$ **49.** $x^2 - 4x + 16$ **50.** $x - 3 + \dfrac{16}{3x + 2}$ **51.** $x^2 - 4x + 5 + \dfrac{5}{2x + 1}$ **52.** $V = 3x^3$ **53.** Yes
54. (a) Surface area $= 50.24 \text{ ft}^2$ (b) Volume $= 33.5 \text{ ft}^3$ **55.** The trampoline has a radius of 6 feet.

CHAPTER 4 TEST

1. 81   **2.** $\frac{9}{16}$ **3.** $72x^{18}$ **4.** $\frac{1}{9}$ **5.** 1   **6.** a^2 **7.** x **8.** 2.78×10^{-2} **9.** 243,000   **10.** $\frac{y^2}{2x^4}$

11. $3a$ **12.** $10x^5$ **13.** 9×10^9 **14.** $8x^2 + 2x + 2$ **15.** $3x^2 + 4x + 6$ **16.** $3x - 4$ **17.** 10

18. $6a^4 - 10a^3 + 8a^2$ **19.** $x^2 + \frac{5}{6}x + \frac{1}{6}$ **20.** $8x^2 + 2x - 15$ **21.** $x^3 - 27$ **22.** $x^2 + 10x + 25$

23. $9a^2 - 12ab + 4b^2$ **24.** $9x^2 - 16y^2$ **25.** $a^4 - 9$ **26.** $2x^2 + 3x - 1$ **27.** $4x + 3 + \dfrac{4}{2x - 3}$

28. $3x^2 + 9x + 25 + \dfrac{76}{x - 3}$ **29.** 15.625 centimeters^3 **30.** x^3

CHAPTERS 1–4 CUMULATIVE REVIEW

1. $\frac{3}{4}$ **2.** 24   **3.** 105   **4.** Undefined   **5.** $9a + 15$ **6.** 1   **7.** -12 **8.** 63   **9.** 45   **10.** y^{16}

11. $16y^4$ **12.** $108a^{18}b^{29}$ **13.** $\dfrac{1,152y^{13}}{x^3}$ **14.** 5×10^{-5} **15.** $25x^2 - 10x + 1$ **16.** $-13x + 10$

17. $x^3 - 1$ **18.** 4   **19.** $\frac{5}{2}$ **20.** 4   **21.** -2 **22.**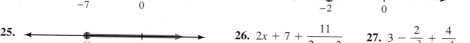

23.

24.

25.    **26.** $2x + 7 + \dfrac{11}{2x - 3}$ **27.** $3 - \dfrac{2}{x^3} + \dfrac{4}{x^4}$

28.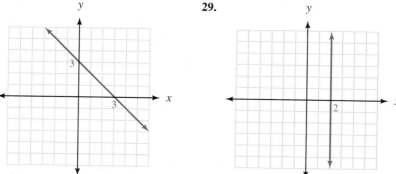

29.

30.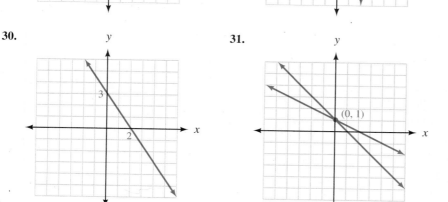

31.

32.

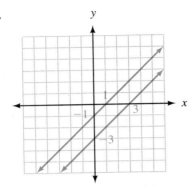

33.

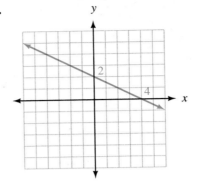

34. (3, 1)   **35.** Lines are parallel.   **36.** (−2, 1)   **37.** (3, 2)   **38.** (4, 5)   **39.** (5, 1)   **40.** −17

41. 21   **42.** $h = \dfrac{2A}{b}$ **43.** $-\sqrt{2}, \pi$ **44.** 3   **45.** $(4, 0), (\frac{16}{3}, 1)$ **46.** x-intercept −2; y-intercept 4

47. 1.86×10^5 **48.** 0.000987   **49.** 6×10^{-3}

50. 5 liters of 40% solution and 15 liters of 16% solution

CHAPTER 5

PROBLEM SET 5.1

1. $5(3x + 5)$ **3.** $3(2a + 3)$ **5.** $4(x - 2y)$ **7.** $3(x^2 - 2x - 3)$ **9.** $3(a^2 - a - 20)$
11. $4(6y^2 - 13y + 6)$ **13.** $x^2(9 - 8x)$ **15.** $13a^2(1 - 2a)$ **17.** $7xy(3x - 4y)$ **19.** $11ab^2(2a - 1)$
21. $7x(x^2 + 3x - 4)$ **23.** $11(11y^4 - x^4)$ **25.** $25x^2(4x^2 - 2x + 1)$ **27.** $8(a^2 + 2b^2 + 4c^2)$
29. $4ab(a - 4b + 8ab)$ **31.** $11a^2b^2(11a - 2b + 3ab)$ **33.** $12x^2y^3(1 - 6x^3 - 3x^2y)$ **35.** $(x + 3)(y + 5)$
37. $(x + 2)(y + 6)$ **39.** $(a - 3)(b + 7)$ **41.** $(a - b)(x + y)$ **43.** $(2x - 5)(a + 3)$ **45.** $(b - 2)(3x - 4)$
47. $(x + 2)(x + a)$ **49.** $(x - b)(x - a)$ **51.** $(x + y)(a + b + c)$ **53.** $(3x + 2)(2x + 3)$
55. $(10x - 1)(2x + 5)$ **57.** $(4x + 5)(5x + 1)$ **59.** $(x + 2)(x^2 + 3)$ **61.** $(3x - 2)(2x^2 + 5)$ **63.** 6
65. $3(4x^2 + 2x + 1)$ **67.** $A = 1{,}000(1 + r)$; when $r = 0.12$, $A = 1{,}000(1.12) = \$1{,}120$.
69. (a) $1{,}000{,}000(1 + r)$ (b) 1,300,000 bacteria   **71.** (a) $A = 7{,}000(1 - r)$ (b) 5,390 names in the year 2000
73. $x^2 - 5x - 14$ **75.** $x^2 - x - 6$ **77.** $x^3 + 27$ **79.** $2x^3 + 9x^2 - 2x - 3$

PROBLEM SET 5.2

1. $(x + 3)(x + 4)$ **3.** $(x + 1)(x + 2)$ **5.** $(a + 3)(a + 7)$ **7.** $(x - 2)(x - 5)$ **9.** $(y - 3)(y - 7)$
11. $(x - 4)(x + 3)$ **13.** $(y + 4)(y - 3)$ **15.** $(x + 7)(x - 2)$ **17.** $(r - 9)(r + 1)$ **19.** $(x - 6)(x + 5)$
21. $(a + 7)(a + 8)$ **23.** $(y + 6)(y - 7)$ **25.** $(x + 6)(x + 7)$ **27.** $2(x + 1)(x + 2)$ **29.** $3(a + 4)(a - 5)$
31. $100(x - 2)(x - 3)$ **33.** $100(p - 5)(p - 8)$ **35.** $x^2(x + 3)(x - 4)$ **37.** $2r(r + 5)(r - 3)$
39. $2y^2(y + 1)(y - 4)$ **41.** $x^3(x + 2)(x + 2)$ **43.** $3y^2(y + 1)(y - 5)$ **45.** $4x^2(x - 4)(x - 9)$
47. $(x + 2y)(x + 3y)$ **49.** $(x - 4y)(x - 5y)$ **51.** $(a + 4b)(a - 2b)$ **53.** $(a - 5b)(a - 5b)$
55. $(a + 5b)(a + 5b)$ **57.** $(x - 6a)(x + 8a)$ **59.** $(x + 4b)(x - 9b)$ **61.** $(x^2 - 3)(x^2 - 2)$
63. $(x - 100)(x + 20)$ **65.** $(x - \frac{1}{2})(x - \frac{1}{2})$ **67.** $(x + 0.2)(x + 0.4)$ **69.** $(x + 16)$ **71.** $4x^2 - x - 3$
73. $6a^2 + 13a + 2$ **75.** $6a^2 + 7a + 2$ **77.** $6a^2 + 8a + 2$ **79.** $2x^2 + 7x - 11$ **81.** $3x + 8$
83. $3x^2 + 4x - 5$

PROBLEM SET 5.3

1. $(2x + 1)(x + 3)$ **3.** $(2a - 3)(a + 1)$ **5.** $(3x + 5)(x - 1)$ **7.** $(3y + 1)(y - 5)$ **9.** $(2x + 3)(3x + 2)$
11. $(2x - 3y)(2x - 3y)$ **13.** $(4y + 1)(y - 3)$ **15.** $(4x - 5)(5x - 4)$ **17.** $(10a - b)(2a + 5b)$
19. $(4x - 5)(5x + 1)$ **21.** $(6m - 1)(2m + 3)$ **23.** $(4x + 5)(5x + 3)$ **25.** $(3a - 4b)(4a - 3b)$
27. $(3x - 7y)(x + 2y)$ **29.** $(2x + 5)(7x - 3)$ **31.** $(3x - 5)(2x - 11)$ **33.** $(5t - 19)(3t - 2)$
35. $2(2x + 3)(x - 1)$ **37.** $2(4a - 3)(3a - 4)$ **39.** $x(5x - 4)(2x - 3)$ **41.** $x^2(3x + 2)(2x - 5)$
43. $2a(5a + 2)(a - 1)$ **45.** $3x(5x + 1)(x - 7)$ **47.** $5y(7y + 2)(y - 2)$ **49.** $a^2(5a + 1)(3a - 1)$
51. $3y(4x + 5)(2x - 3)$ **53.** $2y(2x - y)(3x - 7y)$ **55.** Both equal 25.   **57.** $4x^2 - 9$ **59.** $x^4 - 81$
61. $h = 2(4 - t)(1 + 8t)$
63. (a) $x(9 - 2x)(11 - 2x)$ (b) 9 inches by 11 inches   **65.** $x^2 - 9$
67. $36a^2 - 1$ **69.** $x^2 + 8x + 16$ **71.** $4x^2 + 12x + 9$

Time t (seconds)	Height h (feet)
0	8
1	54
2	68
3	50
4	0

PROBLEM SET 5.4

1. $(x + 3)(x - 3)$ **3.** $(a + 6)(a - 6)$ **5.** $(x + 7)(x - 7)$ **7.** $4(a + 2)(a - 2)$ **9.** Cannot be factored
11. $(5x + 13)(5x - 13)$ **13.** $(3a + 4b)(3a - 4b)$ **15.** $(3 + m)(3 - m)$ **17.** $(5 + 2x)(5 - 2x)$
19. $2(x + 3)(x - 3)$ **21.** $32(a + 2)(a - 2)$ **23.** $2y(2x + 3)(2x - 3)$ **25.** $(a^2 + b^2)(a + b)(a - b)$
27. $(4m^2 + 9)(2m + 3)(2m - 3)$ **29.** $3xy(x + 5y)(x - 5y)$ **31.** $(x - 1)^2$ **33.** $(x + 1)^2$ **35.** $(a - 5)^2$
37. $(y + 2)^2$ **39.** $(x - 2)^2$ **41.** $(m - 6)^2$ **43.** $(2a + 3)^2$ **45.** $(7x - 1)^2$ **47.** $(3y - 5)^2$
49. $(x + 5y)^2$ **51.** $(3a + b)^2$ **53.** $3(a + 3)^2$ **55.** $2(x + 5y)^2$ **57.** $5x(x + 3y)^2$
59. $(x + 3 + y)(x + 3 - y)$ **61.** $(x + y + 3)(x + y - 3)$ **63.** 14   **65.** 25
67. (a) $x^2 - 16$ (b) $(x + 4)(x - 4)$ (c)

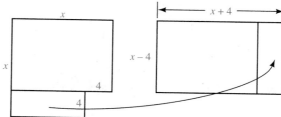

69. $a^2 - b^2 = (a - b)(a + b)$ **71.** $x - 2 + \dfrac{2}{x - 3}$ **73.** $3x - 2 + \dfrac{9}{2x + 3}$

PROBLEM SET 5.5

1. $(x + 9)(x - 9)$ **3.** $(x + 5)(x - 3)$ **5.** $(x + 3)^2$ **7.** $(y - 5)^2$ **9.** $2ab(a^2 + 3a + 1)$
11. Cannot be factored   **13.** $3(2a + 5)(2a - 5)$ **15.** $(3x - 2y)^2$ **17.** $4x(x^2 + 4y^2)$ **19.** $2y(y + 5)^2$
21. $a^4(a^2 + 4b^2)$ **23.** $(x + 4)(y + 3)$ **25.** $(x^2 + 4)(x + 2)(x - 2)$ **27.** $(x + 2)(y - 5)$ **29.** $5(a + b)^2$
31. Cannot be factored   **33.** $3(x + 2y)(x + 3y)$ **35.** $(2x + 19)(x - 2)$ **37.** $100(x - 2)(x - 1)$
39. $(x + 8)(x - 8)$ **41.** $(x + a)(x + 3)$ **43.** $a^5(7a + 3)(7a - 3)$ **45.** Cannot be factored
47. $a(5a + 1)(5a + 3)$ **49.** $(x + y)(a - b)$ **51.** $3a^2b(4a + 1)(4a - 1)$ **53.** $5x^2(2x + 3)(2x - 3)$

55. $(3x + 41y)(x - 2y)$ **57.** $2x^3(2x - 3)(4x - 5)$ **59.** $(2x + 3)(x + a)$ **61.** $(y^2 + 1)(y + 1)(y - 1)$
63. $3x^2y^2(2x + 3y)^2$ **65.** 5 **67.** $-\frac{3}{2}$ **69.** $-\frac{3}{4}$ **71.** x^{15} **73.** $72x^{18}$ **75.** 5.76×10^4

PROBLEM SET 5.6

1. $-2, 1$ **3.** 4, 5 **5.** $0, -1, 3$ **7.** $-\frac{2}{3}, -\frac{3}{2}$ **9.** $0, -\frac{4}{3}, \frac{4}{3}$ **11.** $0, -\frac{1}{3}, -\frac{3}{5}$ **13.** $-1, -2$ **15.** 4, 5
17. $6, -4$ **19.** 2, 3 **21.** -3 **23.** $4, -4$ **25.** $\frac{3}{2}, -4$ **27.** $-\frac{2}{3}$ **29.** 5 **31.** $4, -\frac{5}{2}$ **33.** $\frac{5}{3}, -4$
35. $\frac{7}{2}, -\frac{7}{2}$ **37.** $0, -6$ **39.** $0, 3$ **41.** $0, 4$ **43.** $0, 5$ **45.** $2, 5$ **47.** $\frac{1}{2}, -\frac{4}{3}$ **49.** $4, -\frac{5}{2}$
51. $8, -10$ **53.** 5, 8 **55.** 6, 8 **57.** -4 **59.** 5, 8 **61.** $6, -8$ **63.** $0, -\frac{3}{2}, -4$ **65.** $0, 3, -\frac{5}{2}$
67. $0, \frac{1}{2}, -\frac{5}{2}$ **69.** $0, \frac{3}{5}, -\frac{3}{2}$ **71.** $-3, -2, 2$ **73.** $-4, -1, 4$ **75.** Bicycle, \$75; suit, \$15
77. House, \$2,400; lot, \$600 **79.** $\frac{1}{8}$ **81.** x^8 **83.** x^{18} **85.** 5.6×10^{-3}

PROBLEM SET 5.7

1. Two consecutive even integers are x and $x + 2$; $x(x + 2) = 80$; 8, 10 and $-10, -8$ **3.** 9, 11 and $-11, -9$
5. $x(x + 2) = 5(x + x + 2) - 10$; 8, 10 and 0, 2 **7.** 8, 6
9. The numbers are x and $5x + 2$; $x(5x + 2) = 24$; 2, 12 and $-\frac{12}{5}, -10$ **11.** 5, 20 and 0, 0
13. Let $x = $ the width; $x(x + 1) = 12$; width 3 inches, length 4 inches
15. Let $x = $ the base; $\frac{1}{2}(x)(2x) = 9$; base 3 inches **17.** $x^2 + (x + 2)^2 = 10^2$; 6 inches and 8 inches **19.** 12 meters
21. $1,400 = 400 + 700x - 100x^2$; 200 or 500 **23.** 200 videotapes or 800 videotapes
25. $R = xp = (1,200 - 100p)p = 3,200$; \$4 or \$8 **27.** \$7 or \$10 **29.** (a) 5 feet (b) 12 feet
31. (a) 25 seconds later (b)

t (sec)	h (feet)
0	100
5	1,680
10	2,460
15	2,440
20	1,620
25	0

33. $200x^{24}$ **35.** x^7 **37.** 8×10^1 **39.** $10ab^2$ **41.** $6x^4 + 6x^3 - 2x^2$ **43.** $9y^2 - 30y + 25$
45. $4a^4 - 49$

CHAPTER 5 REVIEW

1. $10(x - 2)$ **2.** $x^2(4x - 9)$ **3.** $5(x - y)$ **4.** $x(7x^2 + 2)$ **5.** $4(2x + 1)$ **6.** $2(x^2 + 7x + 3)$
7. $8(3y^2 - 5y + 6)$ **8.** $15xy^2(2y - 3x^2)$ **9.** $7(7a^3 - 2b^3)$ **10.** $6ab(b + 3a^2b^2 - 4a)$ **11.** $(x + a)(y + b)$
12. $(x - 5)(y + 4)$ **13.** $(2x - 3)(y + 5)$ **14.** $(5x - 4a)(x - 2b)$ **15.** $(y + 7)(y + 2)$
16. $(w + 5)(w + 10)$ **17.** $(a - 6)(a - 8)$ **18.** $(r - 6)(r - 12)$ **19.** $(y + 9)(y + 11)$ **20.** $(y + 6)(y + 2)$
21. $(2x + 3)(x + 5)$ **22.** $(2y - 5)(2y - 1)$ **23.** $(5y + 6)(y + 1)$ **24.** $(5a - 3)(4a - 3)$
25. $(2r + 3t)(3r - 2t)$ **26.** $(2x - 7)(5x + 3)$ **27.** $(n + 9)(n - 9)$ **28.** $(2y + 3)(2y - 3)$
29. Cannot be factored. **30.** $(6y + 11x)(6y - 11x)$ **31.** $(8a + 11b)(8a - 11b)$ **32.** $(8 + 3m)(8 - 3m)$
33. $(y + 10)^2$ **34.** $(m - 8)^2$ **35.** $(8t + 1)^2$ **36.** $(4n - 3)^2$ **37.** $(2r - 3t)^2$ **38.** $(3m + 5n)^2$
39. $2(x + 4)(x + 6)$ **40.** $a(a - 3)(a - 7)$ **41.** $3m(m + 1)(m - 7)$ **42.** $5y^2(y - 2)(y + 4)$

43. $2(2x + 1)(2x + 3)$ **44.** $a(3a + 1)(a - 5)$ **45.** $2m(2m - 3)(5m - 1)$ **46.** $5y(2x - 3y)(3x - y)$
47. $4(x + 5)^2$ **48.** $x(2x + 3)^2$ **49.** $5(x + 3)(x - 3)$ **50.** $3x(2x + 3y)(2x - 3y)$ **51.** $3ab(2a + b)(a + 5b)$
52. $x^3(x + 1)(x - 1)$ **53.** $y^4(4y^2 + 9)$ **54.** $4x^3(x + 2y)(3x - y)$ **55.** $5a^2b(3a - b)(2a + 3b)$
56. $3ab^2(2a - b)(3a + 2b)$ **57.** $-2, 5$ **58.** $-\frac{5}{2}, \frac{5}{2}$ **59.** $2, -5$ **60.** $7, -7$ **61.** $0, 9$ **62.** $-\frac{3}{2}, -\frac{2}{3}$
63. $0, -\frac{5}{3}, \frac{2}{3}$ **64.** $10, 12$ and $-12, -10$ **65.** $10, 11$ and $-11, -10$ **66.** $5, 7$, and $-1, 1$ **67.** $5, 15$
68. $6, 11$ and $-\frac{11}{2}, -12$ **69.** 2 inches

CHAPTER 5 TEST

1. $5(x - 2)$ **2.** $9xy(2x - 1 - 4y)$ **3.** $(x + 2a)(x - 3b)$ **4.** $(x - 7)(y + 4)$ **5.** $(x - 2)(x - 3)$
6. $(x - 3)(x + 2)$ **7.** $(a + 4)(a - 4)$ **8.** Cannot be factored   **9.** $(x^2 + 9)(x + 3)(x - 3)$
10. $3(3x - 5y)(3x + 5y)$ **11.** $(x + 5)(x + 3)(x - 3)$ **12.** $(x - b)(x + 5)$ **13.** $2(2a + 1)(a + 5)$
14. $3(m - 3)(m + 2)$ **15.** $(2y - 1)(3y + 5)$ **16.** $2x(2x + 1)(3x - 5)$ **17.** $-3, -4$ **18.** 2   **19.** $-6, 6$
20. $5, -4$ **21.** $5, 6$ **22.** $0, 4, -4$ **23.** $3, -\frac{5}{2}$ **24.** $0, 1, -\frac{1}{3}$ **25.** $4, 16$ **26.** $3, 5$, and $-3, -1$
27. 3 feet, 14 feet   **28.** 5 meters and 12 meters   **29.** 200 or 300 items   **30.** \$3 or \$6

CHAPTERS 1–5 CUMULATIVE REVIEW

1. -9 **2.** -6 **3.** 29   **4.** 9   **5.** 9   **6.** -64 **7.** 2   **8.** $5a - 5$ **9.** $-8a - 3$ **10.** x^{40}
11. 1   **12.** x^4 **13.** $4x^4y^6$ **14.** $15x^2 + 14x - 8$ **15.** -6 **16.** 10   **17.** -21 **18.** -2
19. $0, 7, \frac{7}{2}$ **20.** $-\frac{9}{4}, \frac{9}{4}$ **21.** $-\frac{1}{4}, \frac{3}{2}$ **22.** $x < 4$ **23.** $x \geq -\frac{6}{5}$ **24.**

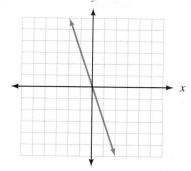

25.

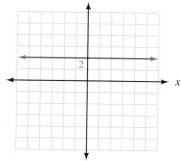

26.

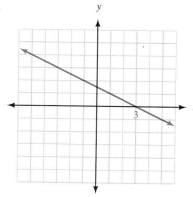

27. x-intercept $= 3$; y-intercept $= -10$ **28.** $\frac{1}{25}$ **29.** $(0, 2), (4, \frac{2}{5})$ **30.** $2x + 9 = 5$

31.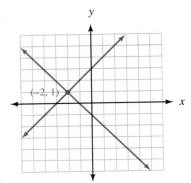

32. (2, 5)   **33.** (2, −4)   **34.** (1, 2)   **35.** (1,500, 3,500)

36. (−2, 1)   **37.** $(n − 9)(n + 4)$ **38.** $(2x + 5y)(7x − 2y)$ **39.** $(4 + a)(4 − a)$ **40.** $(7x − 1)^2$
41. $5y(3x − y)^2$ **42.** $3x(3x + y)(2x − y)$ **43.** $(y + 5)(3x − 2)$ **44.** Opposite is 2; reciprocal is $−\frac{1}{2}$.
45. Commutative   **46.** 1.05×10^9 **47.** $−4x^3y^2 + 2xy − 3$ **48.** $x^2 + 3x + 9$ **49.** 34 inches, 38 inches
50. 6 burgers, 8 fries

CHAPTER 6

PROBLEM SET 6.1

1. $\frac{1}{x − 2}, x \neq 2$ **3.** $\frac{1}{a + 3}, a \neq −3, 3$ **5.** $\frac{1}{x − 5}, x \neq −5, 5$ **7.** $\frac{(x + 2)(x − 2)}{2}$ **9.** $\frac{2(x − 5)}{3(x − 2)}, x \neq 2$

11. 2   **13.** $\frac{5(x − 1)}{4}$ **15.** $\frac{1}{x − 3}$ **17.** $\frac{1}{x + 3}$ **19.** $\frac{1}{a − 5}$ **21.** $\frac{1}{3x + 2}$ **23.** $\frac{x + 5}{x + 2}$

25. $\frac{2m(m + 2)}{m − 2}$ **27.** $\frac{x − 1}{x − 4}$ **29.** $\frac{2(2x − 3)}{2x + 3}$ **31.** $\frac{2x − 3}{2x + 3}$ **33.** $\frac{1}{(x^2 + 9)(x − 3)}$ **35.** $\frac{3x − 5}{(x^2 + 4)(x − 2)}$

37. $\frac{2x(7x + 6)}{2x + 1}$ **39.** $\frac{x + 2}{x + 5}$ **41.** $\frac{x + a}{x + b}$ **43.** $\frac{x + a}{x + 3}$ **45.** $\frac{4}{3}$ **47.** $\frac{4}{5}$ **49.** $\frac{8}{1}$

51.

Checks Written x	Total Cost 2.00 + 0.15x	Cost per Check $\frac{2.00 + 0.15x}{x}$
0	$2.00	Undefined
5	$2.75	$0.55
10	$3.50	$0.35
15	$4.25	$0.28
20	$5.00	$0.25

53. 40.7 miles/hour   **55.** 39.25 feet/minute

57. 12.95 feet/second
59. Level ground: 10 minutes/mile or 0.1 mile/minute; downhill: $\frac{20}{3}$ minutes/mile or $\frac{3}{20}$ mile/minute

61. 48 miles/gallon   **63.** $5 + 4 = 9$ **65.**

x	$\dfrac{x-3}{3-x}$
-2	-1
-1	-1
0	-1
1	-1
2	-1

The entries are all -1 because the numerator and denominator are opposites.
Or, $\dfrac{x-3}{3-x} = \dfrac{-1(x-3)}{3-x} = -1$.

67.

x	$\dfrac{x-5}{x^2-25}$	$\dfrac{1}{x+5}$
0	$\frac{1}{5}$	$\frac{1}{5}$
2	$\frac{1}{3}$	$\frac{1}{3}$
-2	$\frac{1}{7}$	$\frac{1}{7}$
5	Undefined	Undefined
-5	Undefined	$\frac{1}{10}$

69. 0   **71.** $16a^2b^2$ **73.** $19x^4 + 21x^2 - 42$

75. $7a^4b^4 + 9b^3 - 11a^3$

PROBLEM SET 6.2

1. 2   **3.** $\dfrac{x}{2}$ **5.** $\dfrac{3}{2}$ **7.** $\dfrac{1}{2(x-3)}$ **9.** $\dfrac{4a(a+5)}{7(a+4)}$ **11.** $\dfrac{y-2}{4}$ **13.** $\dfrac{2(x+4)}{x-2}$ **15.** $\dfrac{x+3}{(x-3)(x+1)}$

17. 1   **19.** $\dfrac{y-5}{(y+2)(y-2)}$ **21.** $\dfrac{x+5}{x-5}$ **23.** 1   **25.** $\dfrac{a+3}{a+4}$ **27.** $\dfrac{2y-3}{y-6}$ **29.** $\dfrac{x-1}{x-2}$ **31.** $\dfrac{3}{2}$

33. $2(x-3)$ **35.** $2a$ **37.** $(x+2)(x+1)$ **39.** $-2x(x-5)$ **41.** $\dfrac{2(x+5)}{x(y+5)}$ **43.** $\dfrac{2x}{x-y}$ **45.** $\dfrac{1}{x-2}$

47. $\frac{1}{5}$ **49.** $\frac{1}{100}$ **51.** 2.7 miles   **53.** 742 miles/hour   **55.** 0.45 mile/hour   **57.** 8.8 miles/hour

59. Level ground: 6 miles/hour; downhill: 9 miles/hour   **61.** 3   **63.** $\frac{11}{4}$ **65.** $\frac{11}{35}$ **67.** $9x^2$ **69.** $6ab^2$

PROBLEM SET 6.3

1. $\dfrac{7}{x}$ **3.** $\dfrac{4}{a}$ **5.** 1   **7.** $y+1$ **9.** $x+2$ **11.** $x-2$ **13.** $\dfrac{6}{x+6}$ **15.** $\dfrac{(y+2)(y-2)}{2y}$

17. $\dfrac{3+2a}{6}$ **19.** $\dfrac{7x+3}{4(x+1)}$ **21.** $\frac{1}{5}$ **23.** $\frac{1}{3}$ **25.** $\dfrac{3}{x-2}$ **27.** $\dfrac{4}{a+3}$ **29.** $\dfrac{2x-20}{(x+5)(x-5)}$

31. $\dfrac{x+2}{x+3}$ **33.** $\dfrac{a+1}{a+2}$ **35.** $\dfrac{1}{(x+3)(x+4)}$ **37.** $\dfrac{y}{(y+5)(y+4)}$ **39.** $\dfrac{3(x-1)}{(x+4)(x+1)}$ **41.** $\frac{1}{3}$

43.

Number x	Reciprocal $\dfrac{1}{x}$	Sum $1 + \dfrac{1}{x}$	Sum $\dfrac{x + 1}{x}$
1	1	2	2
2	$\dfrac{1}{2}$	$\dfrac{3}{2}$	$\dfrac{3}{2}$
3	$\dfrac{1}{3}$	$\dfrac{4}{3}$	$\dfrac{4}{3}$
4	$\dfrac{1}{4}$	$\dfrac{5}{4}$	$\dfrac{5}{4}$

45.

x	$x + \dfrac{4}{x}$	$\dfrac{x^2 + 4}{x}$	$x + 4$
1	5	5	5
2	4	4	6
3	$\dfrac{13}{3}$	$\dfrac{13}{3}$	7
4	5	5	8

47. $\dfrac{x + 3}{x + 2}$ **49.** $\dfrac{x + 2}{x + 3}$ **51.** $x + \dfrac{2}{x} = \dfrac{x^2 + 2}{x}$ **53.** $\dfrac{1}{x} + \dfrac{1}{2x} = \dfrac{3}{2x}$ **55.** 3 **57.** -2 **59.** $\dfrac{1}{5}$
61. $-2, -3$ **63.** $3, -2$ **65.** $0, 5$

PROBLEM SET 6.4

1. -3 **3.** 20   **5.** -1 **7.** 5   **9.** -2 **11.** 4   **13.** 3, 5   **15.** $-8, 1$ **17.** 2   **19.** 1   **21.** 8
23. 5   **25.** Possible solution 2, which does not check; $\varnothing$ **27.** 3
29. Possible solution -3, which does not check; $\varnothing$ **31.** 0   **33.** Possible solutions 2 and 3, but only 2 checks; 2
35. -4 **37.** -1 **39.** $-6, -7$ **41.** Possible solutions -3 and -1, but only -1 checks; -1 **43.** 1, 4
45. 7   **47.** Length 13 inches, width 4 inches   **49.** 6, 8 or $-8, -6$ **51.** 6 inches, 8 inches

PROBLEM SET 6.5

1. $\dfrac{1}{4}, \dfrac{3}{4}$ **3.** $x + \dfrac{1}{x} = \dfrac{13}{6}; \dfrac{2}{3}$ and $\dfrac{3}{2}$ **5.** -2 **7.** $\dfrac{1}{x} + \dfrac{1}{x + 2} = \dfrac{5}{12}$; 4 and 6
9. Let $x =$ the speed of the boat in still water.

	d	r	t
Upstream	26	$x - 3$	$\dfrac{26}{x - 3}$
Downstream	38	$x + 3$	$\dfrac{38}{x + 3}$

The equation is $\dfrac{26}{x - 3} = \dfrac{38}{x + 3}$; $x = 16$ miles per hour.
11. 300 miles/hour   **13.** 170 miles/hour, 190 miles/hour   **15.** 9 miles/hour   **17.** 8 miles/hour

19. $\dfrac{1}{12} - \dfrac{1}{15} = \dfrac{1}{x}$; 60 hours   **21.** $\dfrac{60}{11}$ minutes   **23.** 12 minutes

25.

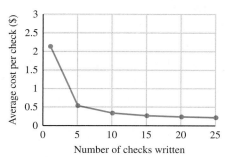

27.

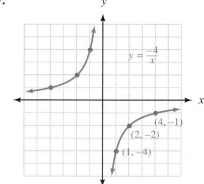

29.

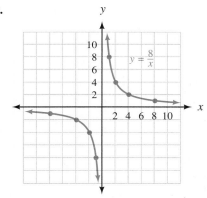

31.

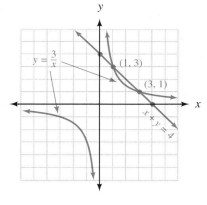

33. $5ab(3a^2b^2 - 4a - 7b)$ **35.** $(x - 6)(x + 2)$ **37.** $(x^2 + 4)(x + 2)(x - 2)$ **39.** $5x(x - 6)(x + 1)$
41. $0, 6$ **43.** $-10, 8$ **45.** 9 inches and 12 inches

PROBLEM SET 6.6

1. 6 **3.** $\frac{1}{6}$ **5.** xy^2 **7.** $\frac{xy}{2}$ **9.** $\frac{y}{x}$ **11.** $\frac{a + 1}{a - 1}$ **13.** $\frac{x + 1}{2(x - 3)}$ **15.** $a - 3$ **17.** $\frac{y + 3}{y + 2}$

19. $x + y$ **21.** $\frac{a + 1}{a}$ **23.** $\frac{1}{x}$ **25.** $\frac{2a + 3}{3a + 4}$ **27.** $\frac{x - 1}{x + 2}$ **29.** $\frac{x + 4}{x + 1}$ **31.** $\frac{7}{3}, \frac{17}{7}, \frac{41}{17}$

33.

Number x	Reciprocal $\frac{1}{x}$	Quotient $\dfrac{x}{\frac{1}{x}}$	Square x^2
1	1	1	1
2	$\frac{1}{2}$	4	4
3	$\frac{1}{3}$	9	9
4	$\frac{1}{4}$	16	16

35.

Number x	Reciprocal $\dfrac{1}{x}$	Sum $1 + \dfrac{1}{x}$	Quotient $\dfrac{1 + \dfrac{1}{x}}{\dfrac{1}{x}}$
1	1	2	2
2	$\frac{1}{2}$	$\frac{3}{2}$	3
3	$\frac{1}{3}$	$\frac{4}{3}$	4
4	$\frac{1}{4}$	$\frac{5}{4}$	5

37. $x < 1$ **39.** $x \geq -7$ **41.** $x < 6$

43. $x \leq 2$

PROBLEM SET 6.7

1. 1 **3.** 10 **5.** 40 **7.** $\frac{5}{4}$ **9.** $\frac{7}{3}$ **11.** $3, -4$ **13.** $4, -4$ **15.** $2, -4$ **17.** $6, -1$ **19.** 15 hits

21. 21 milliliters **23.** 45.5 grams **25.** 14.7 inches **27.** 343 miles **29.** $\dfrac{x + 2}{x + 3}$ **31.** $\dfrac{2(x + 5)}{x - 4}$

33. $\dfrac{1}{x - 4}$

CHAPTER 6 REVIEW

1. $\dfrac{1}{2(x - 2)}, x \neq 2$ **2.** $\dfrac{1}{a - 6}, a \neq -6, 6$ **3.** $\dfrac{2x - 1}{x + 3}, x \neq -3$ **4.** $\dfrac{1}{x + 4}, x \neq 4$ **5.** $\dfrac{3x - 2}{2x - 3}, x \neq \frac{3}{2}, -6, 0$

6. $\dfrac{1}{(x - 2)(x^2 + 4)}, x \neq \pm 2$ **7.** $x - 2, x \neq -7$ **8.** $a + 8, a \neq -8$ **9.** $\dfrac{y + b}{y + 5}, x \neq -a, y \neq -5$ **10.** $\dfrac{x}{2}$

11. $\dfrac{x + 2}{x - 4}$ **12.** $(a - 6)^2$ **13.** $\dfrac{x - 1}{x + 2}$ **14.** 1 **15.** $x - 9$ **16.** $\dfrac{13}{a + 8}$ **17.** $\dfrac{x^2 + 5x + 45}{x(x + 9)}$

18. $\dfrac{4x + 5}{4(x + 5)}$ **19.** $\dfrac{1}{(x + 6)(x + 2)}$ **20.** $\dfrac{a - 6}{(a + 5)(a + 3)}$ **21.** 4 **22.** 9 **23.** 1, 6

24. Possible solution 2, which does not check; $\varnothing$ **25.** 5 **26.** 3 and $\frac{7}{3}$ **27.** 15 miles/hour **28.** 84 hours

29. $\frac{1}{2}$ **30.** $\dfrac{y + 3}{y + 7}$ **31.** $\dfrac{4a - 7}{a - 1}$ **32.** $\frac{2}{5}$ **33.** $\frac{2}{9}$ **34.** 12 **35.** $-6, 6$ **36.** $-6, 8$

CHAPTER 6 TEST

1. $\dfrac{x + 4}{x - 4}$ **2.** $\dfrac{2}{a + 2}$ **3.** $\dfrac{y + 7}{x + a}$ **4.** 3 **5.** $(x - 7)(x - 1)$ **6.** $\dfrac{x + 4}{3x - 4}$ **7.** $(x - 3)(x + 2)$

8. $\dfrac{-3}{x - 2}$ **9.** $\dfrac{2x + 3}{(x + 3)(x - 3)}$ **10.** $\dfrac{3x}{(x + 1)(x - 2)}$ **11.** $\frac{11}{5}$ **12.** 1 **13.** 6 **14.** 15 miles/hour

15. 60 hours   **16.** $\frac{1}{2}$ and $\frac{1}{3}$ **17.** 132   **18.** $\dfrac{x+1}{x-1}$ **19.** $\dfrac{x+4}{x+2}$

CHAPTERS 1–6 CUMULATIVE REVIEW

1. -3 **2.** -6 **3.** -4 **4.** -3 **5.** $-4x-4$ **6.** $4-x$ **7.** $\frac{1}{81}$ **8.** $\dfrac{1}{x^4}$ **9.** 6   **10.** 1

11. $-2a^3 - 10a^2 - 5a + 13$ **12.** $a^4 - 49$ **13.** $x - 7$ **14.** $\dfrac{5x+3}{5x+15}$ **15.** $\dfrac{x-2}{4(x+2)}$ **16.** $\frac{11}{4}$

17. $\frac{19}{12}$ **18.** -2 **19.** $-\frac{3}{7}, \frac{3}{7}$ **20.** $0, \frac{7}{2}, 2$ **21.** 6   **22.** No solution   **23.** $-3, 6$ **24.** $(3, -1)$

25. $(\frac{1}{3}, -\frac{1}{2})$ **26.** $(17, 5)$ **27.** $(6, -12)$

28.

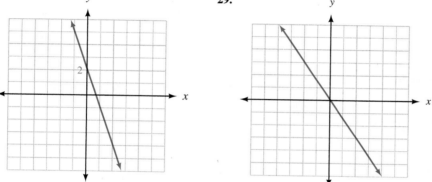

29.

30. $a \geq 6$ **31.** $(x + a)(y + 5)$ **32.** $(a + 7)(a - 5)$ **33.** $(4y - 3)(5y - 3)$ **34.** $(2r - 3t)(2r + 3t)$

35. $(4x + 9y)^2$ **36.**   **37.** $(1, -3)$

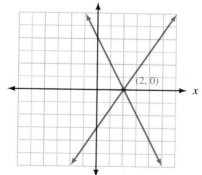

38. x-intercept $= 5$; y-intercept $= 2$ **39.** 72.6   **40.** 1.35×10^{-3} **41.** $-\frac{1}{16}$ **42.** -14

43. Associative property   **44.** $-3, 0, \frac{3}{4}, 1.5$ **45.** $\dfrac{x+1}{4(x+2)}$ **46.** $x - 7$ **47.** $\dfrac{2x+3}{3x+2}$ **48.** $l = \dfrac{P-2w}{2}$

49. 29, 11   **50.** $7,500 at 6%, $10,500 at 8%

CHAPTER 7

PROBLEM SET 7.1

1. 5   **3.** $-\frac{9}{2}$ **5.** $-\frac{4}{3}$ **7.** -10 **9.** -2 **11.** $\frac{3}{4}$ **13.** 3   **15.** -3 **17.** 4   **19.** 2   **21.** 6

23. $-1, 6$ **25.** $0, 2, 3$,   **27.** $\frac{2}{3}, \frac{3}{2}$ **29.** $-4, \frac{5}{2}$ **31.** $-10, 0$ **33.** $-5, 1$ **35.** $-2, 3$ **37.** $-3, -2, 2$

39. $\frac{4}{5}$ **41.** $-5, 5$ **43.** 9   **45.** $-\frac{4}{3}$ **47.** 1   **49.** $-2, \frac{1}{4}$ **51.** $-\frac{4}{3}, 0, \frac{4}{3}$ **53.** $-3, -\frac{3}{2}, \frac{3}{2}$

55. Any method of solution results in a false statement.   **57.** No solution   **59.** All real numbers are solutions.

61. No solution   **63.** 10 feet by 20 feet   **65.** $92.00   **67.** $54

69.

Age (years)	Maximum Heart Rate (beats per minute)
18	202
19	201
20	200
21	199
22	198
23	197

71. For a 20-year-old person

Resting Heart Rate (beats per minute)	Training Heart Rate (beats per minute)
60	144
62	145
64	146
68	147
70	148
72	149

73. $-5, -3$ and $3, 5$ **75.** 24 feet   **77.** 2 ft by 8 ft   **79.** 9   **81.** 2, 3
83. Possible solutions 2 and 4 ; only 2 checks.

PROBLEM SET 7.2

1. $-4, 4$ **3.** $-2, 2$ **5.** $\varnothing$ **7.** $-1, 1$ **9.** $\varnothing$ **11.** $-6, 6$ **13.** $-3, 7$ **15.** $\frac{17}{3}, \frac{7}{3}$ **17.** $2, 4$

19. $-\frac{5}{2}, \frac{5}{6}$ **21.** $-1, 5$ **23.** $\varnothing$ **25.** $-4, 20$ **27.** $-4, 8$ **29.** $-\frac{10}{3}, \frac{2}{3}$ **31.** $\varnothing$ **33.** $-1, \frac{3}{2}$

35. $5, 25$ **37.** $-30, 26$ **39.** $-12, 28$ **41.** $-5, \frac{3}{5}$ **43.** $1, \frac{1}{9}$ **45.** $-\frac{1}{2}$ **47.** 0   **49.** $-\frac{1}{2}$

51. $-\frac{1}{6}, -\frac{7}{4}$ **53.** All real numbers   **55.** All real numbers

57. $t = \frac{d}{r}$ **59.** 25%   **61.** $x > 13$ **63.** $x \geq 4$ **65.** $x > -4$

PROBLEM SET 7.3

1. -1 5

3. -3 0

5. -1 6

7. 2 4

9. -2 4

11. -1 5

13. -1 3

15. -3 -2

17. 2 3 $(-\infty, 2) \cup (3, \infty)$

19. -3 6 $(-3, 6)$

21. 2 3 $(-\infty, 2] \cup [3, \infty)$

23. -3 5 $[-3, 5]$

25. $(-\frac{11}{3}, \frac{11}{2})$

27. $(1, 2)$

29.

31.

33.

35.

37.

39.

41. $-2 < x < 3$ **43.** $x \le -2$ or $x \ge 3$ **45.** (a) $2x + x > 10$; $x + 10 > 2x$; $2x + 10 > x$ (b) $-10 < x < 10$

47.   **49.** $5 < x < 15$ **51.** $4 < x < 5$

53. The width is between 3 inches and $\frac{11}{2}$ inches.   **55.** -5 **57.** 5   **59.** 7   **61.** 9   **63.** 6   **65.** $2x - 3$

67. $-3, 0, 2$

PROBLEM SET 7.4

1. $-3 < x < 3$

3. $x \le -2$ or $x \ge 2$

5. $-3 < x < 3$

7. $t < -7$ or $t > 7$

9. $\varnothing$ **11.** All real numbers   **13.** $-4 < x < 10$

15. $a \le -9$ or $a \ge -1$ **17.** $\varnothing$

19. $-1 < x < 5$

21. $y \le -5$ or $y \ge -1$

23. $k \le -5$ or $k \ge 2$

25. $-1 < x < 7$

27. $a \le -2$ or $a \ge 1$

29. $-6 < x < \frac{8}{3}$

31. $x < 2$ or $x > 8$

33. $x \le -3$ or $x \ge 12$

35. $x < 2$ or $x > 6$

37. $0.99 < x < 1.01$ **39.** $x \le -\frac{3}{5}$ or $x \ge -\frac{2}{5}$ **41.** $-\frac{1}{6} \le x \le \frac{3}{2}$ **43.** $-0.05 < x < 0.25$
45. $|x| \le 4$ **47.** $|x - 5| \le 1$ **49.** The minimum number of channels is 12, and the maximum is 36.
51. $|65 - s| \le 20$ **53.** $4x^2(4x^2 - 5x + 2)$ **55.** $(2x - 3)(a + 4)$ **57.** $(x + 5)(x - 7)$
59. $(x + 2y)(x - 3y)$ **61.** $(x + 3)(x - 3)$ **63.** $\frac{2}{3}$

PROBLEM SET 7.5

1. $(x - 3)^2$ **3.** $(a - 6)^2$ **5.** $(5 - t)^2$ **7.** $(2y^2 - 3)^2$ **9.** $(4a + 5b)^2$ **11.** $(\frac{1}{5} + \frac{1}{4}t^2)^2$
13. $(x + 2 + 3)^2 = (x + 5)^2$ **15.** $(7x - 8y)(7x + 8y)$ **17.** $(2a - \frac{1}{2})(2a + \frac{1}{2})$ **19.** $(x - \frac{3}{5})(x + \frac{3}{5})$
21. $(5 - t)(5 + t)$ **23.** $(4a^2 + 9)(2a - 3)(2a + 3)$ **25.** $(x - 5 + y)(x - 5 - y)$
27. $(a + 4 + b)(a + 4 - b)$ **29.** $(x + 2)(x + 5)(x - 5)$ **31.** $(2x + 3)(x + 2)(x - 2)$
33. $(x + 3)(2x + 3)(2x - 3)$ **35.** $(x - y)(x^2 + xy + y^2)$ **37.** $(a + 2)(a^2 - 2a + 4)$
39. $(y - 1)(y^2 + y + 1)$ **41.** $10(r - 5)(r^2 + 5r + 25)$ **43.** $(4 + 3a)(16 - 12a + 9a^2)$
45. $(t + \frac{1}{3})(t^2 - \frac{1}{3}t + \frac{1}{9})$ **47.** $(x + 9)(x - 9)$ **49.** $(x - 3)(x + 5)$ **51.** $(x^2 + 2)(y^2 + 1)$
53. $2ab(a^2 + 3a + 1)$ **55.** Does not factor   **57.** $3(2a + 5)(2a - 5)$ **59.** $(5 - t)^2$ **61.** $4x(x^2 + 4y^2)$
63. $(x + 5)(x + 3)(x - 3)$ **65.** Does not factor   **67.** $(x - 3)(x - 7)^2$ **69.** $(2 - 5x)(4 + 3x)$
71. $(r + \frac{1}{5})(r - \frac{1}{5})$ **73.** Does not factor   **75.** $100(x - 3)(x + 2)$ **77.** $(3x^2 + 1)(x^2 - 5)$
79. $3a^2b(2a - 1)(4a^2 + 2a + 1)$ **81.** $(4 - r)(16 + 4r + r^2)$ **83.** $5x^2(2x + 3)(2x - 3)$
85. $2x^3(4x - 5)(2x - 3)$ **87.** $(y + 1)(y - 1)(y^2 - y + 1)(y^2 + y + 1)$ **89.** $2(5 + a)(5 - a)$
91. $(x - 2 + y)(x - 2 - y)$ **93.** 30 and -30 **95.** $112.36   **97.** $p^3 - r^3 = (p - r)(p^2 + pr + r^2)$

99.

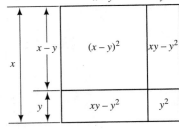

101. $(1, 3)$ **103.** $(1, 3)$ **105.** $(1, 1)$
107. 225 adults, 700 children
109. 3 gallons of 50%, 6 gallons of 20%

PROBLEM SET 7.6

1. $(4, 3)$

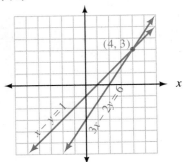

3. $(-5, -6)$

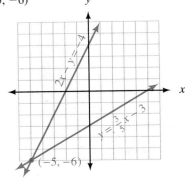

5. (4, 2)   **7.** Lines are parallel; there is no solution.

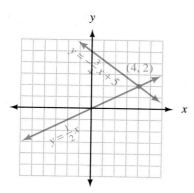

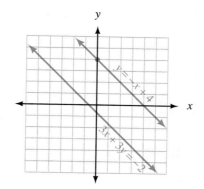

9. Lines coincide; any solution to one of the equations is a solution to the other.   **11.** (2, 3)   **13.** (1, 1)

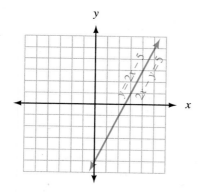

15. Lines coincide: $\{(x, y) \mid 3x - 2y = 6\}$ **17.** $(1, -\frac{1}{2})$ **19.** $(\frac{1}{2}, -3)$ **21.** $(-\frac{8}{3}, 5)$ **23.** (2, 2)

25. Parallel lines; $\varnothing$ **27.** (12, 30)   **29.** (10, 24)   **31.** $(4, \frac{10}{3})$ **33.** (3, −3)   **35.** $(\frac{4}{3}, -2)$ **37.** (6, 2)

39. (2, 4)   **41.** Lines coincide: $\{(x, y) \mid 2x - y = 5\}$ **43.** Lines coincide: $\{(x, y) \mid x = \frac{3}{2}y\}$ **45.** $(-\frac{15}{43}, -\frac{27}{43})$

47. $(\frac{60}{43}, \frac{46}{43})$ **49.** $(\frac{9}{41}, -\frac{11}{41})$ **51.** (6,000, 4,000)   **53.** 2   **55.** 5, 13   **57.** 150 adults, 150 children

59. 5 gallons 20%, 10 gallons 14%   **61.** Boat, 9 mi/hr; current, 3 mi/hr   **63.** Airplane, 270 mi/hr; wind, 30 mi/hr

65. $675x^{14}$ **67.** x^7 **69.** 2×10^{-4} **71.** $2a^2 + 10a + 8$ **73.** $16a^2 - 81$ **75.** $24x^3y^2 - 18x^2y^3 + 12xy^4$

77. $x - 2 + \dfrac{2}{x - 3}$

PROBLEM SET 7.7

1. (1, 2, 1)   **3.** (2, 1, 3)   **5.** (2, 0, 1)   **7.** $(\frac{1}{2}, \frac{2}{3}, -\frac{1}{2})$ **9.** No solution, inconsistent system   **11.** (4, −3, −5)

13. No unique solution   **15.** (4, −5, −3)   **17.** No unique solution   **19.** $(\frac{1}{2}, 1, 2)$ **21.** $(\frac{1}{2}, \frac{1}{3}, \frac{1}{4})$

23. (1, 3, 1)   **25.** (−1, 2, −2)   **27.** 4 amp, 3 amp, 1 amp   **29.** $200 at 6%, $1,400 at 8%, $600 at 9%

31. 3 of each   **33.** $h = -16t^2 + 64t + 80$ **35.** $\dfrac{x + 2}{x + 3}$ **37.** $\dfrac{2(x + 5)}{x - 4}$ **39.** $\dfrac{1}{x - 4}$ **41.** $\dfrac{x + 5}{x - 3}$

43. −2   **45.** 24 hours

CHAPTER 7 REVIEW PROBLEMS

1. -3 **2.** $\frac{10}{13}$ **3.** $-\frac{5}{11}$ **4.** $-\frac{4}{9}$ **5.** $-\frac{3}{2}, 4$ **6.** $-\frac{1}{9}, \frac{1}{9}$ **7.** $1, 4$ **8.** $-2, -\frac{2}{3}, \frac{2}{3}$ **9.** 4 feet by 12 feet

10. $3, 4, 5$ **11.** $2, 4$ **12.** $-1, 5$ **13.** $-1, 4$ **14.** $-\frac{5}{3}, 3$ **15.** $-\frac{3}{2}, 3$ **16.** $-\frac{5}{3}, \frac{7}{3}$ **17.** $(-\infty, 12]$

18. $(-1, \infty)$ **19.** $[2, 6]$ **20.** $[-3, 2]$ **21.** $(-\infty, -\frac{3}{2}] \cup [3, \infty)$ **22.** $(-\infty, \frac{4}{5}] \cup [2, \infty)$

23.   **24.**

25.   **26.**

27. $\varnothing$ **28.** All real numbers   **29.** $(x^2 + 4)(x + 2)(x - 2)$ **30.** $3(a^2 + 3)^2$ **31.** $(a - 2)(a^2 + 2a + 4)$

32. $5x(x + 3y)^2$ **33.** $3ab(a - 3b)(a + 3b)$ **34.** $(x - 5 + y)(x - 5 - y)$ **35.** $(6 - 5a)(6 + 5a)$

36. $(x + 3)(x - 3)(x + 4)$ **37.** $(0, 1)$ **38.** Lines coincide   **39.** $(3, 5)$ **40.** $(4, 8)$ **41.** $(\frac{3}{2}, \frac{1}{2})$

42. $(4, 1)$ **43.** $(-5, 3)$ **44.** Parallel lines   **45.** $(3, -1, 4)$ **46.** $(2, \frac{1}{2}, -3)$ **47.** Dependent system

48. Dependent system   **49.** $(2, -1, 4)$ **50.** Dependent system

CHAPTER 7 TEST

1. 28 **2.** $-\frac{1}{3}, 2$ **3.** $0, 5$ **4.** $-\frac{7}{4}$ **5.** $-5, 2$ **6.** 6 inches, 12 inches   **7.** 0 seconds and 2 seconds

8. $2, 6$ **9.** $\varnothing$ **10.** $(-\infty, 4) \cup [6, \infty)$

11. $[-\frac{2}{5}, 2]$ **12.**

13.   **14.** $(x + 4)(x - 3)$ **15.** $(4a^2 + 9y^2)(2a + 3y)(2a - 3y)$

16. $(t + \frac{1}{2})(t^2 - \frac{1}{2}t + \frac{1}{4})$ **17.** $4a^3b(a - 8b)(a + 2b)$ **18.** $(1, 2)$ **19.** $(15, 12)$ **20.** $(2, -4)$

21. $(3, -2, 1)$ **22.** $5, 9$ **23.** $\$4,000$ at 5%, $\$8,000$ at 6%   **24.** 11 nickels, 3 dimes, 1 quarter

CHAPTERS 1–7 CUMULATIVE REVIEW

1. 6 **2.** -93 **3.** $\frac{25}{4}$ **4.** $18x + 7$ **5.** $6x - 19$ **6.** $24y$ **7.** $2x^3 - 5x^2 - 8x + 6$ **8.** -1

9. 0 **10.** $5, \frac{2}{3}$ **11.** $\varnothing$ **12.** $(4, 3)$ **13.** Parallel lines   **14.** Lines coincide   **15.** $(4, 0, 9)$

16. $(-4, 4)$

17. $[2, 5]$

18. $(-\infty, -2] \cup [9, \infty)$

19. $(-\infty, -4]$ **20.** $\dfrac{13}{a - b}$ **21.** $x < -3$ or $x > 3$ **22.** $7, -\frac{1}{7}$ **23.** -1 **24.** $-5 < x < 5$

25. $-1, 0, 2.35, 4$ **26.** $\dfrac{a + b}{a^2 + ab + b^2}$ **27.** $\frac{5}{3}, -2$ **28.** 36 **29.** $\dfrac{30y^6}{x^8 z^5}$ **30.** 4.69×10^{-4}

31. 4.9×10^2 **32.** 0.000123 **33.** $x \neq 5, x \neq -5$

34.

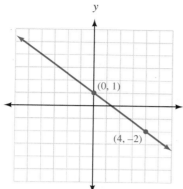

35.

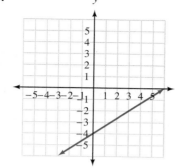

36. $(-1, 2, -1)$

37. $\frac{3}{5}, -3$

38. $y = -\frac{2}{3}x - 3$

39. $y = \frac{7}{3}x - 6$

40. $y = 3x + 1$

41. $(4y + \frac{1}{4})^2$

42. $(2x + y^2)(3a^2 + 1)$

43. $(x - 2)(x^2 + 2x + 4)$ **44.** $(4a^2 + 9b^2)(2a - 3b)(2a - 3b)$ **45.** $2x - 3 + \dfrac{3}{x - 2}$

46. $\dfrac{y}{y + 1}$ **47.** $\dfrac{1}{(x + 3)(x + 1)}$ **48.** $\dfrac{x - 2}{x - 3}$ **49.** 8 mph; 4 mph   **50.** $\pi(a - b)(a + b)$

CHAPTER 8

PROBLEM SET 8.1

1. $\frac{3}{2}$ **3.** No slope   **5.** $\frac{2}{3}$

7.

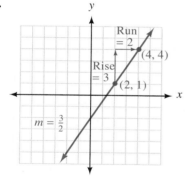

9.

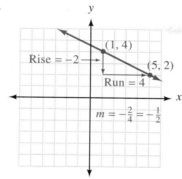

11.

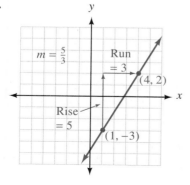

13.

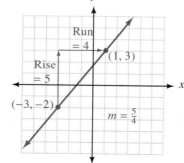

15.

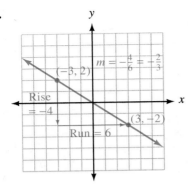

17.

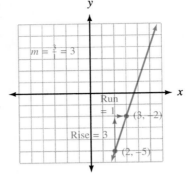

19.

x	y
0	2
3	0

Slope $= -\frac{2}{3}$

21.

x	y
0	-5
3	-3

Slope $= \frac{2}{3}$

23.

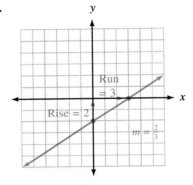

25.

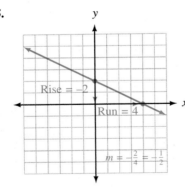

27. $\frac{1}{5}$ **29.** 0   **31.** (a) Yes   (b) No

33. 24 feet   **35.** 10 minutes   **37.** 20; °C per minute

39. 1st minute   **41.** Slope $= -1{,}250$; dollars per year

43. 2 to 3 years   **45.** (a) 15,800 feet   (b) $-\frac{7}{100}$

47. $1,875; $310,000   **49.** 0

51. $y = -\frac{3}{2}x + 6$

53. $t = \dfrac{A - P}{Pr}$ **55.**

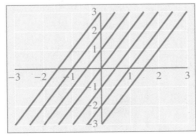

57.

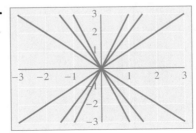

59.

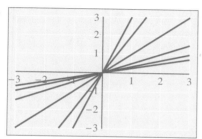

PROBLEM SET 8.2

1. $y = 2x + 3$ **3.** $y = x - 5$ **5.** $y = \frac{1}{2}x + \frac{3}{2}$ **7.** $y = 4$

9. Slope $= 3$,
   y-intercept $= -2$,
   perpendicular slope $= -\frac{1}{3}$

11. Slope $= \frac{2}{3}$,
   y-intercept $= -4$,
   perpendicular slope $= -\frac{3}{2}$

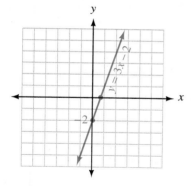

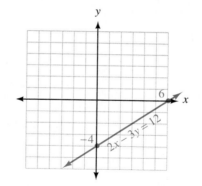

13. Slope $= -\frac{4}{5}$,
   y-intercept $= 4$,
   perpendicular slope $= \frac{5}{4}$

15. Slope $= \frac{1}{2}$, y-intercept $= -4$, $y = \frac{1}{2}x - 4$

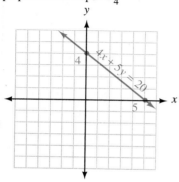

17. Slope $= -\frac{2}{3}$, y-intercept $= 3$, $y = -\frac{2}{3}x + 3$ **19.** $y = 2x - 1$ **21.** $y = -\frac{1}{2}x - 1$ **23.** $y = -3x + 1$

25. $x - y = 2$ **27.** $2x - y = 3$ **29.** $6x - 5y = 3$ **31.** $(0, -4), (2, 0); y = 2x - 4$

33. $(-2, 0), (0, 4); y = 2x + 4$ **35.** Slope = 0, y-intercept = -2 **37.** $y = 3x + 7$ **39.** $y = -\frac{5}{2}x - 13$

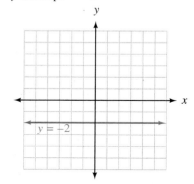

41. $y = \frac{1}{4}x + \frac{1}{4}$ **43.** $y = -\frac{2}{3}x + 2$ **45.** (b) $86°$ **47.** (a) $(y - 1) = 0.00877(x - 98)$ (b) $40 \le x \le 220$

49. (a) $(y - 3{,}000) = 4{,}250(x - 1{,}984)$ or $y = 4{,}250x - 8{,}429{,}000$ (b) $11{,}500$ cases

51. $(1, 2, 3)$ **53.** $(1, 3, 1)$ **55.**

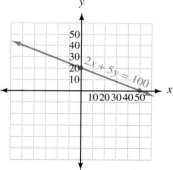

57.

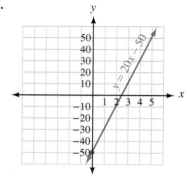

59. $y = -\frac{3}{2}x + 3$; slope = $-\frac{3}{2}$, y-intercept = 3, x-intercept = 2

61. $y = \frac{3}{2}x + 3$; slope = $\frac{3}{2}$, y-intercept = 3, x-intercept = -2 **63.** Slope = $-\frac{b}{a}$, x-intercept = a, y-intercept = b

PROBLEM SET 8.3

1.

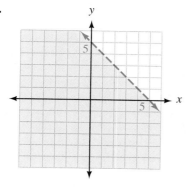

3.

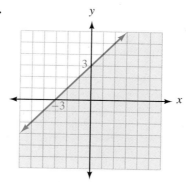

5.

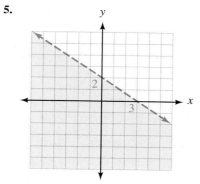

7.

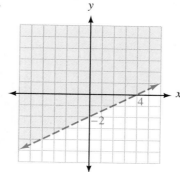

9.

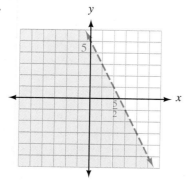

11.

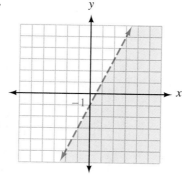

13. $x + y > 4$ **15.** $-x + 2y \le 4$ **17.**

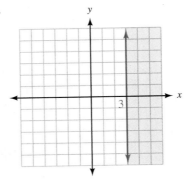

19.

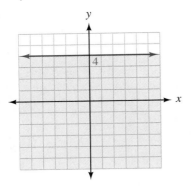

21.

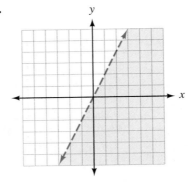

23.

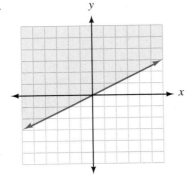

25.

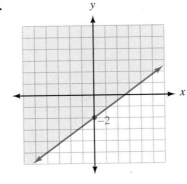

27.

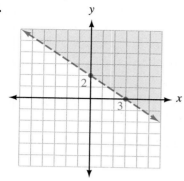

29.

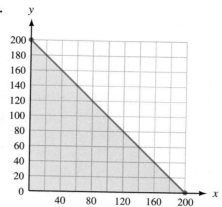

31.

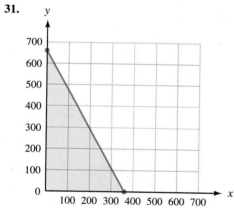

33.

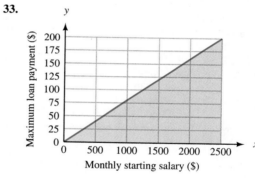

$y \le 0.08x$

35. $t > -2$ **37.** $x < -4$ or $x > 4$ **39.** $-1 < t < 2$ **41.**

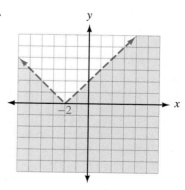

43.

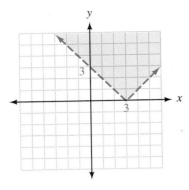

45.

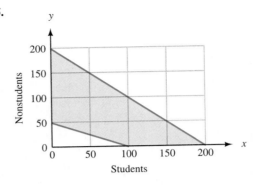

PROBLEM SET 8.4

1. (a) $y = 8.5x$ for $10 \leq x \leq 40$

(b)

Hours Worked	Function Rule	Gross Pay ($)
x	$y = 8.5x$	y
10	$y = 8.5(10) = 85$	85
20	$y = 8.5(20) = 170$	170
30	$y = 8.5(30) = 255$	255
40	$y = 8.5(40) = 340$	340

(c)

(d) Domain $= \{x \mid 10 \leq x \leq 40\}$; Range $= \{y \mid 85 \leq y \leq 340\}$ (e) Minimum $= \$85$; Maximum $= \$340$
3. Domain $= \{1, 2, 4\}$; Range $= \{3, 5, 1\}$; a function   **5.** Domain $= \{-1, 1, 2\}$; Range $= \{3, -5\}$; a function
7. Domain $= \{7, 3\}$; Range $= \{-1, 4\}$; not a function   **9.** Yes   **11.** No   **13.** No   **15.** Yes   **17.** Yes

19. (a)

(b) Domain $= \{t \mid 0 \leq t \leq 1\}$;
Range $= \{h \mid 0 \leq h \leq 4\}$

Time (sec)	Function Rule	Distance (ft)
t	$h = 16t - 16t^2$	h
0	$h = 16(0) - 16(0)^2$	0
0.1	$h = 16(0.1) - 16(0.1)^2$	1.44
0.2	$h = 16(0.2) - 16(0.2)^2$	2.56
0.3	$h = 16(0.3) - 16(0.3)^2$	3.36
0.4	$h = 16(0.4) - 16(0.4)^2$	3.84
0.5	$h = 16(0.5) - 16(0.5)^2$	4
0.6	$h = 16(0.6) - 16(0.6)^2$	3.84
0.7	$h = 16(0.7) - 16(0.7)^2$	3.36
0.8	$h = 16(0.8) - 16(0.8)^2$	2.56
0.9	$h = 16(0.9) - 16(0.9)^2$	1.44
1	$h = 16(1) - 16(1)^2$	0

(c)

21. Domain $= \{x \mid -5 \leq x \leq 5\}$
Range $= \{y \mid 0 \leq y \leq 5\}$

23. Domain $= \{x \mid -5 \leq x \leq 3\}$
Range $= \{y \mid y = 3\}$

25. Domain = All real numbers;
Range = $\{y \mid y \geq -1\}$; a function

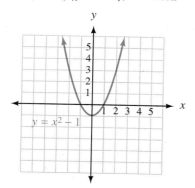

27. Domain = All real numbers;
Range = $\{y \mid y \geq 4\}$; a function

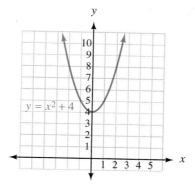

29. Domain = $\{x \mid x \geq -1\}$;
Range = All real numbers; not a function

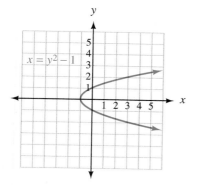

31. Domain = $\{x \mid x \geq 4\}$;
Range = All real numbers; not a function

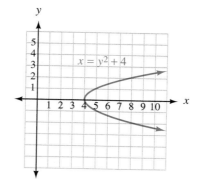

33. Domain = All real numbers;
Range = $\{y \mid y \geq 0\}$; a function

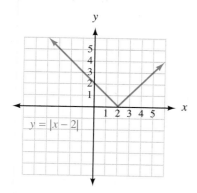

35.

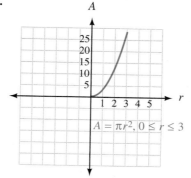

37. x = the width, so $x + 2$ = the length; $P = 2x + 2(x + 2) = 4x + 4$. The variable x must be positive.
39. $A = x(x + 2)$, where $x > 0$ **41.** (a) Yes   (b) Domain = $\{t \mid 0 \leq t \leq 6\}$; Range = $\{h \mid 0 \leq h \leq 60\}$
(c) $t = 3$ (d) $h = 60$ (e) $t = 6$ **43.** (a) F11   (b) F12   (c) F10   (d) F9   **45.** 10   **47.** -14
49. 1   **51.** -3

PROBLEM SET 8.5

1. −1   **3.** −11   **5.** 2   **7.** 4   **9.** 35   **11.** −13   **13.** 1   **15.** −9   **17.** 8   **19.** 19   **21.** 16

23. 0   **25.** $3a^2 - 4a + 1$ **27.** 4   **29.** 0   **31.** 2   **33.** −8   **35.** −1   **37.** $2a^2 - 8$ **39.** $2b^2 - 8$

41. 0   **43.** −2   **45.** −3   **47.**   **49.** $x = 4$

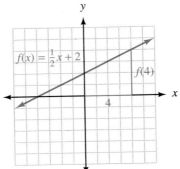

51.

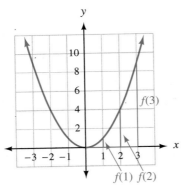

53. $V(3) = 300$, the painting is worth $300 in 3 years; $V(6) = 600$, the painting is worth $600 in 6 years.

55. $P(x) = 2x + 2(2x + 3) = 6x + 6$, where $x > 0$ **57.** $A(2) = 3.14(4) = 12.56$; $A(5) = 3.14(25) = 78.5$; $A(10) = 3.14(100) = 314$

59. (a) $2.49   (b) $1.53 for a 6-minute call   (c) 5 minutes   **61.** (a) $5,625   (b) $1,500

(c) Domain = $\{t \mid 0 \le t \le 5\}$ (d)

(e) Range = $\{V(t) \mid 1{,}500 \le V(t) \le 18{,}000\}$ (f) 2.42 years

63. (a)

Weight (oz)	0.6	1.0	1.1	2.5	3.0	4.8	5.0	5.3
Cost (cents)	32	32	55	78	78	124	124	147

(b) In words: Over 2 ounces, but not over 3 ounces. Inequality: $2 < x \leq 3$ (c) Domain $= \{x \mid 0 < x \leq 6\}$
(d) Range $= \{C(x) \mid C(x) = 32, 55, 78, 101, 124, 147\}$ **65.** $-\frac{2}{3}, 4$ **67.** $-2, 1$ **69.** $\varnothing$
71. (a) 2   (b) 0   (c) 1   (d) 4   (e) 1   (f) 3   (g) 0   (h) 2

PROBLEM SET 8.6

1. $6x + 2$ **3.** $-2x + 8$ **5.** $8x^2 + 14x - 15$ **7.** $\dfrac{2x + 5}{4x - 3}$ **9.** $4x - 7$ **11.** $3x^2 - 10x + 8$

13. $-2x + 3$ **15.** $3x^2 - 11x + 10$ **17.** $9x^3 - 48x^2 + 85x - 50$ **19.** $x - 2$ **21.** $\dfrac{1}{x - 2}$

23. $3x^2 - 7x + 3$ **25.** $6x^2 - 22x + 20$ **27.** 15   **29.** 98   **31.** $\frac{3}{2}$ **23.** 1   **35.** 40   **37.** 147
39. (a) 81   (b) 29   (c) $(x + 4)^2$ (d) $x^2 + 4$ **41.** (a) -2 (b) -1 (c) $16x^2 + 4x - 2$ (d) $4x^2 + 12x - 1$
45. (a) $R(x) = 11.5x - .05x^2$ (b) $C(x) = 2x + 200$ (c) $P(x) = -.05x^2 + 9.5x - 200$ (d) $\overline{C}(x) = 2 + \dfrac{200}{x}$
47. (a) $M(x) = 220 - x$ (b) $M(24) = 196$ (c) 142   (d) 135   (e) 128   **49.** $x = 1, y = 2$
51. $x = \frac{5}{2}, y = -1$ **53.** $x = 0, y = 3$ **55.** $x = 3, y = 4$

PROBLEM SET 8.7

1. 30   **3.** 5   **5.** -6 **7.** $\frac{1}{2}$ **9.** 40   **11.** 225   **13.** $\frac{81}{5}$ **15.** 40.5   **17.** 64   **19.** 8

21. $\frac{50}{3}$ pounds   **23.** (a) $T = 4P$ (b)

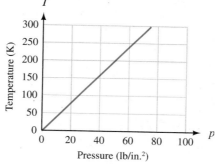

(c) 70 pounds per square inch   **25.** 12 pounds per square inch   **27.** (a) $f = \frac{80}{d}$
(b)   (c) An f-stop of 8

29. $\frac{1,504}{15}$ square inches   **31.** 1.5 ohms

33. (a) $P = 0.21\sqrt{l}$ (b)   (c) 3.15 seconds

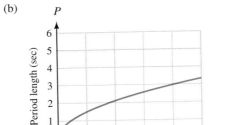

35. 1.28 meters   **37.** $F = \dfrac{G(m_1 \cdot m_2)}{d^2}$ **39.** $x \le -9$ or $x \ge -1$

41. All real numbers   **43.** $-6 < y < \frac{8}{3}$

45. (a)

Height Above Surface (ft)	Illumination (foot-candles)
2	900
4	225
6	100
8	56.25
10	36

Height Above Surface (ft)	Area of Illumination Region (ft²)
2	$\pi \approx 3.1$
4	$4\pi \approx 12.6$
6	$9\pi \approx 28.3$
8	$16\pi \approx 50.3$
10	$25\pi \approx 78.5$

(b)

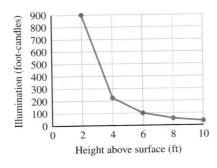

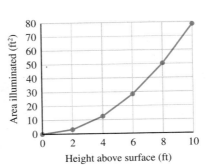

(c) Illumination F is inversely proportional to the square of distance h.
Area A is directly proportional to the square of distance h.

(d) $A = \frac{\pi}{4}h^2$
$F = \frac{3,600}{h^2}$

CHAPTER 8 REVIEW

1. -2 **2.** 0 **3.**

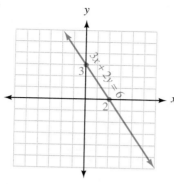

slope $= -\frac{3}{2}$

4.

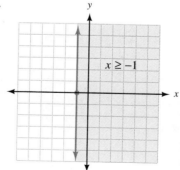

slope $= -\frac{3}{2}$

5. No slope

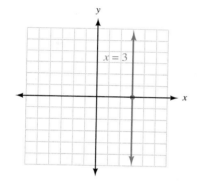

6. 3 **7.** 5 **8.** -5 **9.** 6 **10.** $y = 3x + 5$

11. $y = -2x$ **12.** $m = 3, b = -6$ **13.** $m = \frac{2}{3}, b = -3$ **14.** $y = 2x$ **15.** $y = -\frac{1}{3}x$ **16.** $y = 2x + 1$

17. $y = 7$ **18.** $y = -\frac{3}{2}x - \frac{17}{2}$ **19.** $y = 2x - 7$ **20.** $y = \frac{1}{3}x - \frac{2}{3}$

21.

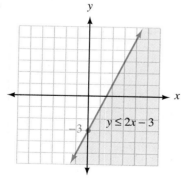

$y \le 2x - 3$

22.

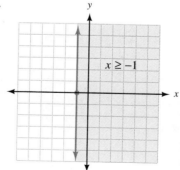

$x \ge -1$

23. Domain $= \{2, 3, 4\}$;
Range $= \{4, 3, 2\}$;
a function

24. Domain $= \{6, -4, -2\}$; Range $= \{3, 0\}$; a function   **25.** 0 **26.** 1 **27.** 1 **28.** $3a + 2$ **29.** 1
30. 31 **31.** $(f + g)(x) = x^2 + 2x - 3$ **32.** 5 **33.** -9 **34.** $(fg)(x) = 2x^3 + x^2 - 8x - 4$ **35.** -5
36. $(g \circ f)(x) = 4x^2 + 4x - 3$ **37.** 24 **38.** 6 **39.** 4 **40.** 25 **41.** 84 pounds   **42.** 16 foot-candles

CHAPTER 8 TEST

1. x-intercept $= 3$,
   y-intercept $= 6$,
   slope $= -2$

2. x-intercept $= -\frac{3}{2}$
   y-intercept $= -3$
   slope $= -2$

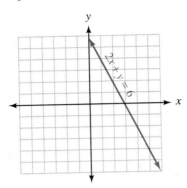

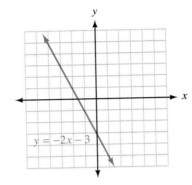

3. x-intercept $= -\frac{8}{3}$,
   y-intercept $= 4$,
   slope $= \frac{3}{2}$

4. x-intercept $= -2$,
   no y-intercept,
   no slope

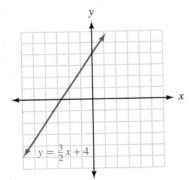

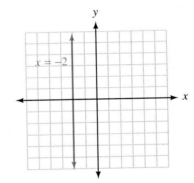

5. $y = 2x + 5$ **6.** $y = -\frac{3}{7}x + \frac{5}{7}$ **7.** $y = \frac{2}{5}x - 5$ **8.** $y = -\frac{1}{3}x - \frac{7}{3}$ **9.** $x = 4$

10.

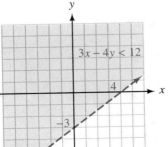

11.

12. Domain = {−2, −3}; Range = {0, 1}; not a function
13. Domain = All real numbers; Range = {$y \mid y \geq -9$}; a function   **14.** 11   **15.** −4   **16.** 8   **17.** 4
18. $0 < x < 4$ **19.** $V(x) = x(8 - 2x)^2$
20. $V(2) = 32$ cubic inches is the volume of the box if a square with 2-inch sides is cut from each corner.
21. 18   **22.** $\frac{81}{4}$ **23.** $\frac{2{,}000}{3}$ pounds

CHAPTERS 1–8 CUMULATIVE REVIEW

1. 18   **2.** $\frac{36}{25}$ **3.** x^3 **4.** $\dfrac{x^3}{y^7}$ **5.** $-3x - 12$ **6.** $\dfrac{1}{a^2 + a + 1}$ **7.** $12x$ **8.** $(f \circ g)(x) = 3x^2 - 5$

9. −2   **10.** $5a - 7b > 5a + 7b$ **11.** $(2x - 3)(4x^2 + 6x + 9)$ **12.** 12, arithmetic

13. Commutative, associative   **14.** $\dfrac{1}{(y + 3)(y + 2)}$ **15.** $-\dfrac{1}{x + y}$ **16.** $12t^4 - \frac{1}{12}$ **17.** $x + 2$

18. $2x^2 - 3x^3$ **19.** $a^3 - a^2 + 2a - 4 + \dfrac{7}{a + 2}$ **20.** −20   **21.** 3   **22.** 1   **23.** −12, 12

24. −5, 3, 5   **25.** 5   **26.** $\frac{3}{2}$, 4   **27.** $(-3, -7)$ **28.** $(0, -\frac{1}{2})$ **29.** Lines are parallel.   **30.** $(0, \frac{6}{5}, \frac{-14}{5})$

31. $(\frac{31}{50}, \frac{13}{50})$ **32.** $(\frac{51}{31}, \frac{10}{31}, \frac{19}{31})$ **33.** $x \geq -1$

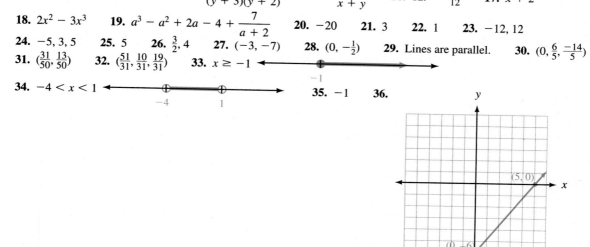

34. $-4 < x < 1$ **35.** −1   **36.**

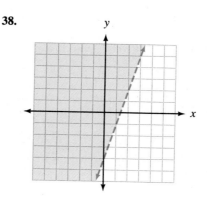

37.

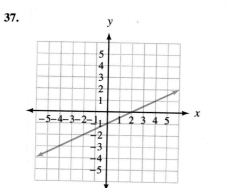

38.

39. $2^3 \times 3 \times 7$ **40.** $(x - 10)(x + 7)$ **41.** $(x + 5 - y)(x + 5 + y)$ **42.** $x + 3$ **43.** 9.27×10^6
44. 225   **45.** 0   **46.** $\frac{3}{4}$ **47.** $x = -2$ **48.** -108 **49.** $b = 10$ feet, $h = 15$ feet   **50.** $120°, 40°, 20°$

CHAPTER 9

PROBLEM SET 9.1

1. 12   **3.** Not a real number   **5.** -7 **7.** -3 **9.** 2   **11.** Not a real number   **13.** 0.2   **15.** 0.2
17. $6a^4$ **19.** $3a^4$ **21.** xy^2 **23.** $2x^2y$ **25.** $2a^3b^5$ **27.** 6   **29.** -3 **31.** 2   **33.** -2 **35.** 2
37. $\frac{9}{5}$ **39.** $\frac{4}{5}$ **41.** 9   **43.** 125   **45.** 8   **47.** $\frac{1}{3}$ **49.** $\frac{1}{27}$ **51.** $\frac{6}{5}$ **53.** $\frac{8}{27}$ **55.** 7   **57.** $\frac{3}{4}$
59. $x^{4/5}$ **61.** a **63.** $\dfrac{1}{x^{2/5}}$ **65.** $x^{1/6}$ **67.** $x^{9/25}y^{1/2}z^{1/5}$ **69.** $\dfrac{b^{7/4}}{a^{1/8}}$ **71.** $y^{3/10}$ **73.** $\dfrac{1}{a^2b^4}$ **75.** $\dfrac{s^{1/2}}{r^{20}}$
77. $10b^3$ **79.** $(9^{1/2} + 4^{1/2})^2 = (3 + 2)^2 = 5^2 = 25 \neq 9 + 4$ **81.** $\sqrt{\sqrt{a}} = (a^{1/2})^{1/2} = a^{1/4} = \sqrt[4]{a}$ **83.** 25 mph
85. 1.618   **87.** $\frac{13}{8}$, numerator and denominator are consecutive members of the Fibonacci sequence.
89. (a) 420 pm   (b) 594 pm   (c) 5.94×10^{-10} m   **91.** (a) $\sqrt{2}$ (b) $\sqrt{3}$
93. (a) B   (b) A   (c) C   (d) (0, 0) and (1, 1)   **95.** $x^6 - x^3$ **97.** $x^2 + 2x - 15$
99. $x^4 - 10x^2 + 25$ **101.** $x^3 - 27$ **103.** When $x = 2$, $y = 1.7$ **105.** When $x = 10$, $y = 5.6$
107. Graphs intersect at $x = 1$, $y = 1$ and $x = 0$, $y = 0$
109. (a) 1.62 micrograms   (b) 0.87 microgram   (c) 0.00293 microgram   (d) 0.0000029 microgram

PROBLEM SET 9.2

1. $x + x^2$ **3.** $a^2 - a$ **5.** $6x^3 - 8x^2 + 10x$ **7.** $12x^2 - 36y^2$ **9.** $x^{4/3} - 2x^{2/3} - 8$
11. $a - 10a^{1/2} + 21$ **13.** $20y^{2/3} - 7y^{1/3} - 6$ **15.** $10x^{4/3} + 21x^{2/3}y^{1/2} + 9y$ **17.** $t + 10t^{1/2} + 25$
19. $x^3 + 8x^{3/2} + 16$ **21.** $a - 2a^{1/2}b^{1/2} + b$ **23.** $4x - 12x^{1/2}y^{1/2} + 9y$ **25.** $a - 3$ **27.** $x^3 - y^3$
29. $t - 8$ **31.** $4x^3 - 3$ **33.** $x + y$ **35.** $a - 8$ **37.** $8x + 1$ **39.** $t - 1$ **41.** $2x^{1/2} + 3$
43. $3x^{1/3} - 4y^{1/3}$ **45.** $3a - 2b$ **47.** $3(x - 2)^{1/2}(4x - 11)$ **49.** $5(x - 3)^{7/5}(x - 6)$
51. $3(x + 1)^{1/2}(3x^2 + 3x + 2)$ **53.** $(x^{1/3} - 2)(x^{1/3} - 3)$ **55.** $(a^{1/5} - 4)(a^{1/5} + 2)$ **57.** $(2y^{1/3} + 1)(y^{1/3} - 3)$
59. $(3t^{1/5} + 5)(3t^{1/5} - 5)$ **61.** $(2x^{1/7} + 5)^2$ **63.** $\dfrac{3 + x}{x^{1/2}}$ **65.** $\dfrac{x + 5}{x^{1/3}}$ **67.** $\dfrac{x^3 + 3x^2 + 1}{(x^3 + 1)^{1/2}}$
69. $\dfrac{-4}{(x^2 + 4)^{1/2}}$ **71.** 2   **73.** 27   **75.** 0.871   **77.** 15.8%   **79.** 5.9%   **81.** 3   **83.** $m = \frac{2}{3}, b = -2$
85. $y = \frac{2}{3}x + 6$

PROBLEM SET 9.3

1. $2\sqrt{2}$ **3.** $7\sqrt{2}$ **5.** $12\sqrt{2}$ **7.** $4\sqrt{5}$ **9.** $4\sqrt{3}$ **11.** $15\sqrt{3}$ **13.** $3\sqrt[3]{2}$ **15.** $4\sqrt[3]{2}$ **17.** $6\sqrt[3]{2}$
19. $2\sqrt[5]{2}$ **21.** $3x\sqrt{2x}$ **23.** $2y\sqrt[4]{2y^3}$ **25.** $2xy^2\sqrt[3]{5xy}$ **27.** $4abc^2\sqrt{3b}$ **29.** $2bc\sqrt[3]{6a^2c}$ **31.** $2xy^2\sqrt[5]{2x^3y^2}$
33. $3xy^2z\sqrt[5]{x^2}$ **35.** $2\sqrt{3}$ **37.** $\sqrt{-20}$, which is not a real number   **39.** $\dfrac{\sqrt{11}}{2}$ **41.** $\dfrac{2\sqrt{3}}{3}$ **43.** $\dfrac{5\sqrt{6}}{6}$

45. $\dfrac{\sqrt{2}}{2}$ **47.** $\dfrac{\sqrt{5}}{5}$ **49.** $2\sqrt[3]{4}$ **51.** $\dfrac{2\sqrt[3]{3}}{3}$ **53.** $\dfrac{\sqrt[4]{24x^2}}{2x}$ **55.** $\dfrac{\sqrt[4]{8y^3}}{y}$ **57.** $\dfrac{\sqrt[3]{36xy^2}}{3y}$ **59.** $\dfrac{\sqrt[3]{6xy^2}}{3y}$

61. $\dfrac{\sqrt[4]{2x}}{2x}$ **63.** $\dfrac{3x\sqrt{15xy}}{5y}$ **65.** $\dfrac{5xy\sqrt{6xz}}{2z}$ **67.** $\dfrac{2ab\sqrt[3]{6ac^2}}{3c}$ **69.** $\dfrac{2xy^2\sqrt[3]{3z^2}}{3z}$ **71.** $5|x|$ **73.** $3|xy|\sqrt{3x}$

75. $|x-5|$ **77.** $|2x+3|$ **79.** $2|a(a+2)|$ **81.** $2|x|\sqrt{x-2}$

83. $\sqrt{9+16} = \sqrt{25} = 5$; $\sqrt{9} + \sqrt{16} = 3 + 4 = 7$ **85.** $5\sqrt{13}$ feet

89. $\sqrt{2}, \sqrt{3}, \sqrt{4}, \sqrt{5}, \sqrt{6}, \sqrt{7}, \ldots$; 10th term $= \sqrt{11}$; 100th term $= \sqrt{101}$ **91.** 3   **93.** 0   **95.** 1   **97.** 3

99. 17   **101.** $\dfrac{\sqrt[10]{a^7}}{a}$ **103.** $\dfrac{\sqrt[20]{a^9}}{a}$ **105.**

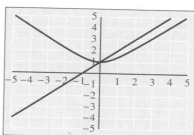

107. About $\dfrac{3}{4}$

109. $x = 0$

111. (a) $s = \dfrac{a+b+c}{2}$ (b) $A \approx 14.70$

PROBLEM SET 9.4

1. $7\sqrt{5}$ **3.** $-x\sqrt{7}$ **5.** $\sqrt[3]{10}$ **7.** $9\sqrt[5]{6}$ **9.** 0   **11.** $\sqrt{5}$ **13.** $-32\sqrt{2}$ **15.** $-3x\sqrt{2}$ **17.** $-2\sqrt[3]{2}$

19. $8x\sqrt[3]{xy^2}$ **21.** $3a^2b\sqrt{3ab}$ **23.** $11ab\sqrt[3]{3a^2b}$ **25.** $10xy\sqrt[4]{3y}$ **27.** $\sqrt{2}$ **29.** $\dfrac{8\sqrt{5}}{15}$ **31.** $\dfrac{(x-1)\sqrt{x}}{x}$

33. $\dfrac{3\sqrt{2}}{2}$ **35.** $\dfrac{5\sqrt{6}}{6}$ **37.** $\dfrac{8\sqrt[3]{25}}{5}$ **39.** $\sqrt{12} \approx 3.464$; $2\sqrt{3} = 2(1.732) \approx 3.464$

41. $\sqrt{8} + \sqrt{18} \approx 2.828 + 4.243 = 7.071$; $\sqrt{50} \approx 7.071$; $\sqrt{26} \approx 5.099$ **43.** $8\sqrt{2x}$ **45.** 5

53.

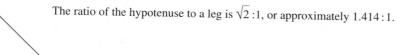

The ratio of the hypotenuse to a leg is $\sqrt{2} : 1$, or approximately $1.414 : 1$.

55. (a) $\sqrt{2} : 1 \approx 1.414 : 1$ (b) $5 : \sqrt{2}$ (c) $5 : 4$

57.   **59.**

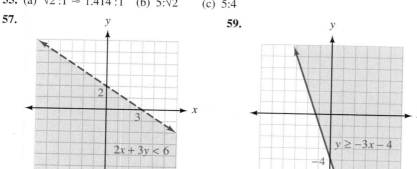

61.

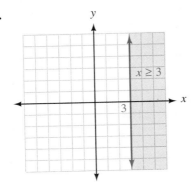

PROBLEM SET 9.5

1. $3\sqrt{2}$ **3.** $10\sqrt{21}$ **5.** 720 **7.** 54 **9.** $\sqrt{6} - 9$ **11.** $24 + 6\sqrt[3]{4}$ **13.** $7 + 2\sqrt{6}$ **15.** $x + 2\sqrt{x} - 15$
17. $34 + 20\sqrt{3}$ **19.** $19 + 8\sqrt{3}$ **21.** $x - 6\sqrt{x} + 9$ **23.** $4a - 12\sqrt{ab} + 9b$ **25.** $x + 4\sqrt{x} - 4$

27. $x - 6\sqrt{x - 5} + 4$ **29.** 1 **31.** $a - 49$ **33.** $25 - x$ **35.** $x - 8$ **37.** $10 + 6\sqrt{3}$ **39.** $\dfrac{\sqrt{3} + 1}{2}$

41. $\dfrac{5 - \sqrt{5}}{4}$ **43.** $\dfrac{x + 3\sqrt{x}}{x - 9}$ **45.** $\dfrac{10 + 3\sqrt{5}}{11}$ **47.** $\dfrac{3\sqrt{x} + 3\sqrt{y}}{x - y}$ **49.** $2 + \sqrt{3}$ **51.** $\dfrac{11 - 4\sqrt{7}}{3}$

53. $\dfrac{a + 2\sqrt{ab} + b}{a - b}$ **55.** $\dfrac{x + 4\sqrt{x} + 4}{x - 4}$ **57.** $\dfrac{5 - \sqrt{21}}{4}$ **59.** $\dfrac{\sqrt{x} - 3x + 2}{1 - x}$ **63.** $10\sqrt{3}$

65. $x + 6\sqrt{x} + 9$ **67.** 75 **69.** $\dfrac{5\sqrt{2}}{4}$ second; $\dfrac{5}{2}$ second **75.** 147 **77.** 50 **79.** 100

PROBLEM SET 9.6

1. 4 **3.** $\varnothing$ **5.** 5 **7.** $\varnothing$ **9.** $\dfrac{39}{2}$ **11.** $\varnothing$ **13.** 5 **15.** 3 **17.** $-\dfrac{32}{3}$ **19.** 3, 4 **21.** $-1, -2$
23. -1 **25.** $\varnothing$ **27.** 7 **29.** 0, 3 **31.** -4 **33.** 8 **35.** 0 **37.** 9 **39.** 0 **41.** 8
43. Possible solution 9, which does not check; $\varnothing$ **45.** Possible solutions 0 and 32; only 0 checks; 0
47. Possible solutions -2 and 6; only 6 checks; 6 **49.** $h = 100 - 16t^2$ **51.** $\dfrac{392}{121} \approx 3.24$ ft

53. $l = \dfrac{10}{3}(\sqrt{x} + 1)$ **55.** 5 meters **57.** 2,500 meters **59.** $0 \le y \le 100$

61.

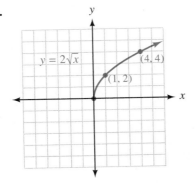

63.

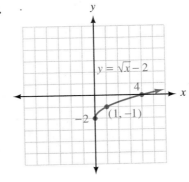

65.

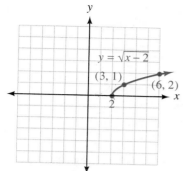

67.

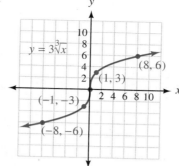

69.

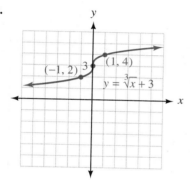

71.

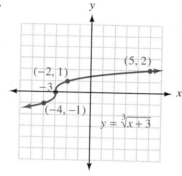

73. $\sqrt{6} - 2$ **75.** $x + 10\sqrt{x} + 25$ **77.** $\dfrac{x - 3\sqrt{x}}{x - 9}$ **79.** Possible solutions 2 and 6; only 6 checks; 6

81. $1, -1, -3$ **83.** $y = \frac{1}{8}x^2$ **85.**

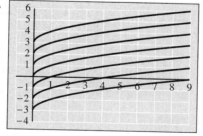

87. The value of b shifts the curve b units along the y-axis.   **89.**

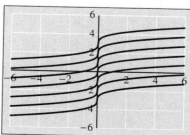

91. The value of b shifts the curve b units along the y-axis.

93.

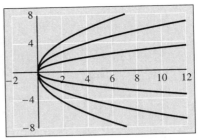

95. The smaller the absolute value of a, the more slowly the graph rises or falls. If a is negative, the graph lies below the x-axis.

PROBLEM SET 9.7

1. $6i$ **3.** $-5i$ **5.** $6i\sqrt{2}$ **7.** $-2i\sqrt{3}$ **9.** 1 **11.** -1 **13.** $-i$ **15.** $x = 3, y = -1$
17. $x = -2, y = -\frac{1}{2}$ **19.** $x = -8, y = -5$ **21.** $x = 7, y = \frac{1}{2}$ **23.** $x = \frac{3}{7}, y = \frac{2}{5}$ **25.** $5 + 9i$
27. $5 - i$ **29.** $2 - 4i$ **31.** $1 - 6i$ **33.** $2 + 2i$ **35.** $-1 - 7i$ **37.** $6 + 8i$ **39.** $2 - 24i$
41. $-15 + 12i$ **43.** $18 + 24i$ **45.** $10 + 11i$ **47.** $21 + 23i$ **49.** $-2 + 2i$ **51.** $2 - 11i$
53. $-21 + 20i$ **55.** $-2i$ **57.** $-7 - 24i$ **59.** 5 **61.** 40 **63.** 13 **65.** 164 **67.** $-3 - 2i$
69. $-2 + 5i$ **71.** $\frac{8}{13} + \frac{12}{13}i$ **73.** $-\frac{18}{13} - \frac{12}{13}i$ **75.** $-\frac{5}{13} + \frac{12}{13}i$ **77.** $\frac{13}{15} - \frac{2}{5}i$
79. $R = -11 - 7i$ ohms   **81.** Domain $= \{1, 3, 4\}$, Range $= \{2, 4\}$, a function
83. Domain $= \{1, 2, 3\}$, Range $= \{1, 2, 3\}$, a function   **85.** A function

CHAPTER 9 REVIEW

1. 7 **2.** -3 **3.** 2 **4.** 27 **5.** $2x^3y^2$ **6.** $\frac{1}{16}$ **7.** x^2 **8.** a^2b^4 **9.** $a^{7/20}$ **10.** $a^{5/12}b^{8/3}$
11. $12x + 11x^{1/2}y^{1/2} - 15y$ **12.** $a^{2/3} - 10a^{1/3} + 25$ **13.** $4x^{1/2} + 2x^{5/6}$ **14.** $2(x - 3)^{1/4}(4x - 13)$
15. $\dfrac{x + 5}{x^{1/4}}$ **16.** $2\sqrt{3}$ **17.** $5\sqrt{2}$ **18.** $2\sqrt[3]{2}$ **19.** $3x\sqrt{2}$ **20.** $4ab^2c\sqrt{5a}$ **21.** $2abc\sqrt[4]{2bc^2}$ **22.** $\dfrac{3\sqrt{2}}{2}$
23. $3\sqrt[3]{4}$ **24.** $\dfrac{4x\sqrt{21xy}}{7y}$ **25.** $\dfrac{2y\sqrt[3]{45x^2z^2}}{3z}$ **26.** $-2x\sqrt{6}$ **27.** $3\sqrt{3}$ **28.** $\dfrac{8\sqrt{5}}{5}$ **29.** $7\sqrt{2}$
30. $11a^2b\sqrt{3ab}$ **31.** $-4xy\sqrt[3]{xz^2}$ **32.** $\sqrt{6} - 4$ **33.** $x - 5\sqrt{x} + 6$ **34.** $3\sqrt{5} + 6$ **35.** $6 + \sqrt{35}$
36. $\dfrac{63 + 12\sqrt{7}}{47}$ **37.** 0 **38.** 3 **39.** 5 **40.** $\varnothing$

41.

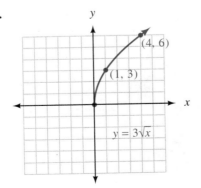

42.

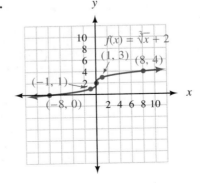

43. 1   **44.** $-i$ **45.** $x = -\frac{3}{2}, y = -\frac{1}{2}$ **46.** $x = -2, y = -4$ **47.** $9 + 3i$ **48.** $-7 + 4i$

49. $-6 + 12i$ **50.** $5 + 14i$ **51.** $12 + 16i$ **52.** 25   **53.** $1 - 3i$ **54.** $-\frac{6}{5} + \frac{3}{5}i$

55. 22.5 ft   **56.** 65 yd

CHAPTER 9 TEST

1. $\frac{1}{9}$ **2.** $\frac{7}{5}$ **3.** $a^{5/12}$ **4.** $\frac{x^{13/12}}{y}$ **5.** $7x^4y^5$ **6.** $2x^2y^4$ **7.** $2a$ **8.** $x^{n^2-n}y^{1-n^3}$ **9.** $6a^2 - 10a$

10. $16a^3 - 40a^{3/2} + 25$ **11.** $(3x^{1/3} - 1)(x^{1/3} + 2)$ **12.** $(3x^{1/3} - 7)(3x^{1/3} + 7)$ **13.** $\frac{x + 4}{x^{1/2}}$

14. $\frac{3}{(x^2 - 3)^{1/2}}$ **15.** $5xy^2\sqrt{5xy}$ **16.** $2x^2y^2\sqrt[3]{5xy^2}$ **17.** $\frac{\sqrt{6}}{3}$ **18.** $\frac{2a^2b\sqrt{15bc}}{5c}$ **19.** $-6\sqrt{3}$ **20.** $-ab\sqrt[3]{3}$

21. $x + 3\sqrt{x} - 28$ **22.** $21 - 6\sqrt{6}$ **23.** $\frac{5 + 5\sqrt{3}}{2}$ **24.** $\frac{x - 2\sqrt{2x} + 2}{x - 2}$

25. Possible solutions 1 and 8; only 8 checks; 8   **26.** -4 **27.** -3

28.   **29.**   **30.** $x = \frac{1}{2}, y = 7$ **31.** $6i$

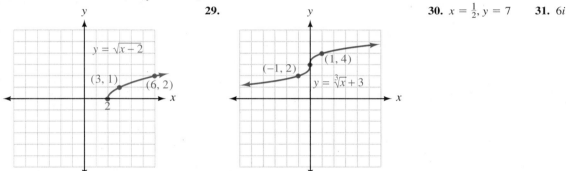

32. $17 - 6i$ **33.** $9 - 40i$ **34.** $-\frac{5}{13} - \frac{12}{13}i$ **35.** $i^{38} = (i^2)^{19} = (-1)^{19} = -1$

CHAPTERS 1–9 CUMULATIVE REVIEW

1. 23   **2.** 1   **3.** 59   **4.** $22x + 11$ **5.** $\frac{8}{y^3}$ **6.** 4   **7.** $6y^2\sqrt{2y}$ **8.** $\frac{4\sqrt{15}}{5}$ **9.** $-12, \frac{1}{12}$

10. $D = \{-1, 3\}, R = \{2, -1, 0\}$; no   **11.** 4.15×10^4 **12.** 1   **13.** $(5a - 2b)(5a + 2b)(25a^2 + 4b^2)$

14. $2(4a^2 + 3)(3a^2 - 1)$ **15.** $\frac{2}{7}$ **16.** $\frac{2z}{x}$ **17.** $\frac{x - 5}{x - 3}$ **18.** $\frac{18}{7}$ **19.** $6x^2 - 19xy + 10y^2$

20. $3x^2 + 20x + 25$ **21.** $x - 4\sqrt{x} + 4$ **22.** $-2a^2 + 1 - 3b^2$ **23.** $\frac{6}{y^2}$ **24.** $\frac{x + 2\sqrt{xy} + y}{x - y}$

25. $\frac{2}{3}x^3 - \frac{1}{12}x^2 + \frac{1}{3}x + \frac{5}{12}$ **26.** $\frac{1}{(x + 3)(x + 5)}$ **27.** 2   **28.** $\frac{5}{6}$ **29.** $-\frac{7}{3}, 3$ **30.** $-4, -2, 4$

31. Possible solution -1, which does not check; $\varnothing$ **32.** $x = 3$ **33.** $-10, 7$ **34.** 2   **35.** 6

36. $y \leq -2$ or $y \geq 3$ **37.** $x \leq -\frac{2}{5}$ or $x \geq 2$ **38.** $(-9, -7)$

39. $(5, -2)$ **40.** $(3, -5, 0)$ **41.** $(9, -11, 1)$ **42.** Both equal 25

43.

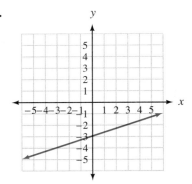

44.

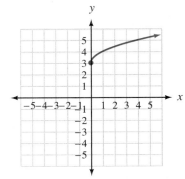

45.

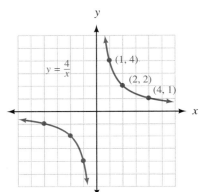

46. 6   **47.** $\frac{4}{5}, -3$ **48.** $y = -\frac{2}{5}x + 5$ **49.** $y = \frac{2}{3}x - 3$

50. $L = 24$ ft, $W = 10$ ft

CHAPTER 10

PROBLEM SET 10.1

1. ± 5 **3.** $\pm 3i$ **5.** $\pm\frac{\sqrt{3}}{2}$ **7.** $\pm 2i\sqrt{3}$ **9.** $\pm\frac{3\sqrt{5}}{2}$ **11.** $-2, 3$ **13.** $\frac{-3 \pm 3i}{2}$ **15.** $\frac{-2 \pm 2i\sqrt{2}}{5}$

17. $-4 \pm 3i\sqrt{3}$ **19.** $\frac{3 \pm 2i}{2}$ **21.** ± 1 **23.** ± 1 **25.** ± 1 **27.** $x^2 + 12x + 36 = (x + 6)^2$

29. $x^2 - 4x + 4 = (x - 2)^2$ **31.** $a^2 - 10a + 25 = (a - 5)^2$ **33.** $x^2 + 5x + \frac{25}{4} = (x + \frac{5}{2})^2$

35. $y^2 - 7y + \frac{49}{4} = (y - \frac{7}{2})^2$ **37.** $-6, 2$ **39.** $-3, -9$ **41.** $1 \pm 2i$ **43.** $4 \pm \sqrt{15}$ **45.** $\frac{5 \pm \sqrt{37}}{2}$

47. $1 \pm \sqrt{5}$ **49.** $\frac{4 \pm \sqrt{13}}{3}$ **51.** $\frac{3 \pm i\sqrt{71}}{8}$ **53.** $\frac{\sqrt{3}}{2}$ inch, 1 inch   **55.** $x\sqrt{3}, 2x$ **57.** $\sqrt{2}$ inches

59. $\frac{\sqrt{2}}{2}$ inch   **61.** $x\sqrt{2}$ **63.** 781 feet   **65.** $\frac{1,170}{5,630} = 0.21$ to the nearest hundredth   **67.** $20\sqrt{2} \approx 28$ feet

69. 7.3% to the nearest tenth   **71.** 19.6 to the nearest tenth

73. $3\sqrt{5}$ **75.** $3y^2\sqrt{3y}$ **77.** $3x^2y\sqrt[3]{2y^2}$ **79.** 13   **81.** $\frac{3\sqrt{2}}{2}$ **83.** $\sqrt[3]{2}$ **85.** $x = \pm 2a$ **87.** $x = -a$

89. $x = 0, -2a$ **91.** $x = \frac{-p \pm \sqrt{p^2 - 4q}}{2}$

PROBLEM SET 10.2

1. $-2, -3$ **3.** $2 \pm \sqrt{3}$ **5.** $1, 2$ **7.** $\dfrac{2 \pm i\sqrt{14}}{3}$ **9.** $0, 5$ **11.** $0, -\dfrac{4}{3}$ **13.** $\dfrac{3 \pm \sqrt{5}}{4}$ **15.** $-3 \pm \sqrt{17}$

17. $\dfrac{-1 \pm i\sqrt{5}}{2}$ **19.** 1 **21.** $\dfrac{1 \pm i\sqrt{47}}{6}$ **23.** $4 \pm \sqrt{2}$ **25.** $\dfrac{1}{2}, 1$ **27.** $-\dfrac{1}{2}, 3$ **29.** $\dfrac{-1 \pm i\sqrt{7}}{2}$

31. $1 \pm \sqrt{2}$ **33.** $\dfrac{-3 \pm \sqrt{5}}{2}$ **35.** $3, -5$ **37.** $2, -1 \pm i\sqrt{3}$ **39.** $-\dfrac{3}{2}, \dfrac{3 \pm 3i\sqrt{3}}{4}$ **41.** $\dfrac{1}{5}, \dfrac{-1 \pm i\sqrt{3}}{10}$

43. $0, \dfrac{-1 \pm i\sqrt{5}}{2}$ **45.** $0, 1 \pm i$ **47.** $0, \dfrac{-1 \pm i\sqrt{2}}{3}$ **49.** $\dfrac{-3 - 2i}{5}$ **51.** 2 seconds

53. $\dfrac{1}{4}$ second and 1 second   **55.** $40 \pm 20 = 20$ or 60 items   **57.** $\dfrac{3.5 \pm 0.5}{0.004} = 750$ or $1,000$ patterns

59. $(10.5 - 2x)(8.2 - 2x) = 0.8(10.5)(8.2)$; $4x^2 - 37.4x + 17.22 = 0$; $x \approx 8.86$ centimeters (impossible) or 0.49 centimeter

61. (a) $21 + 2w = 20$, or $l + w = 10$; $l \cdot w = 15$
    (b) 8.16 or 1.84 to the nearest hundredth
    (c) Two answers are possible because either dimension (long or short) may be considered the length.

63. $4y + 1 + \dfrac{-2}{2y - 7}$ **65.** $x^2 + 7x + 12$ **67.** 5 **69.** $\dfrac{27}{125}$ **71.** $\dfrac{1}{4}$ **73.** $21x^3y$ **75.** $-2\sqrt{3}, \sqrt{3}$

77. $\dfrac{-1 \pm \sqrt{3}}{\sqrt{2}} = \dfrac{-\sqrt{2} \pm \sqrt{6}}{2}$ **79.** $-2i, i$ **81.** $-i, 4i$

PROBLEM SET 10.3

1. $D = 16$; two rational   **3.** $D = 0$; one rational   **5.** $D = 5$; two irrational   **7.** $D = 17$; two irrational
9. $D = 36$; two rational   **11.** $D = 116$; two irrational   **13.** ± 10 **15.** ± 12 **17.** 9 **19.** -16
21. $\pm 2\sqrt{6}$ **23.** $x^2 - 7x + 10 = 0$ **25.** $t^2 - 3t - 18 = 0$ **27.** $y^3 - 4y^2 - 4y + 16 = 0$
29. $2x^2 - 7x + 3 = 0$ **31.** $4t^2 - 9t - 9 = 0$ **33.** $6x^3 - 5x^2 - 54x + 45 = 0$ **35.** $10a^2 - a - 3 = 0$
37. $9x^3 - 9x^2 - 4x + 4 = 0$ **39.** $x^4 - 13x^2 + 36 = 0$ **41.** $(x - 3)(x + 5)^2 = 0$ or $x^3 + 7x^2 - 5x - 75 = 0$
43. $(x - 3)^2(x + 3)^2 = 0$ or $x^4 - 18x^2 + 81 = 0$ **45.** $-3, -2, -1$ **47.** $-3, -4, 2$ **49.** $1 \pm i$
51. $5, 4 \pm 3i$ **53.** $a^{11/2} - a^{9/2}$ **55.** $x^3 - 6x^{3/2} + 9$ **57.** $6x^{1/2} - 5x$ **59.** $5(x - 3)^{1/2}(2x - 9)$
61. $(2x^{1/3} - 3)(x^{1/3} - 4)$ **63.** $-2, 1, i, -i$ **65.** $x = -a$ is a solution of multiplicity 3.   **67.** $-1, 5$
69. $1 + \sqrt{2} \approx 2.41, 1 - \sqrt{2} \approx -0.41$ **71.** $-1, \dfrac{1}{2}, 1$ **73.** $\dfrac{1}{2}, \sqrt{2} \approx 1.41, -\sqrt{2} \approx -1.41$ **75.** $\dfrac{2}{3}, 1 + i, 1 - i$

PROBLEM SET 10.4

1. $1, 2$ **3.** $-8, -\dfrac{5}{2}$ **5.** $\pm 3, \pm i\sqrt{3}$ **7.** $\pm 2i, \pm i\sqrt{5}$ **9.** $\dfrac{7}{2}, 4$ **11.** $-\dfrac{9}{8}, \dfrac{1}{2}$ **13.** $\pm \dfrac{\sqrt{30}}{6}, \pm i$

15. $\pm \dfrac{\sqrt{21}}{3}, \pm \dfrac{i\sqrt{21}}{3}$ **17.** $4, 25$ **19.** Possible solutions 25 and 9; only 25 checks; 25

21. Possible solutions $\dfrac{25}{9}$ and $\dfrac{49}{4}$; only $\dfrac{25}{9}$ checks; $\dfrac{25}{9}$ **23.** $27, 38$ **25.** $4, 12$ **27.** $t = \dfrac{v \pm \sqrt{v^2 + 64h}}{32}$

29. $x = \dfrac{-4 \pm 2\sqrt{4 - k}}{k}$ **31.** $x = -y$ **33.** $t = \dfrac{1 \pm \sqrt{1 + h}}{4}$ **35.** $t = \dfrac{v \pm \sqrt{v^2 + 1,280}}{32}$

37. $2 + 2\sqrt{5}; 4\left(\dfrac{1 + \sqrt{5}}{2}\right) = 2(1 + \sqrt{5}) = 2 + 2\sqrt{5}$ **39.** $1 + \sqrt{5}; \dfrac{BC}{AB} = \dfrac{1 + \sqrt{5}}{2}$

41. (a)

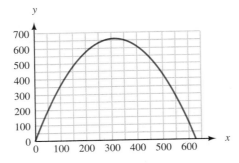

(b) The legs are 630 feet apart.

43. (a) $l + 2w = 160$

(b) $A = (160 - 2w) \cdot w$ or $A = -2w^2 + 160w$

(c) Some possible values are shown below:

w	$A = -2w^2 + 160w$
20	2,400
25	2,750
30	3,000
35	3,150
40	3,200
45	3,150
50	3,000
55	2,750

(d) 3,200 square yards

45. $3\sqrt{7}$ **47.** $5\sqrt{2}$ **49.** $39x^2y\sqrt{5x}$ **51.** $-11 + 6\sqrt{5}$ **53.** $x + 4\sqrt{x} + 4$ **55.** $\dfrac{7 + 2\sqrt{7}}{3}$

57. x-intercepts $= -2, 0, 2$; y-intercept $= 0$ **59.** x-intercepts $= -3, -\frac{1}{3}, 3$; y-intercept $= -9$

61. $\frac{1}{2}$ and -1

PROBLEM SET 10.5

1. x-intercepts $= -3, 1$;
Vertex $= (-1, -4)$

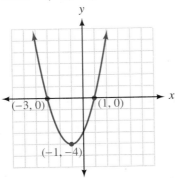

3. x-intercepts $= -5, 1$;
Vertex $= (-2, 9)$

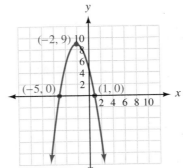

5. x-intercepts $= -1, 1$;
Vertex $= (0, -1)$

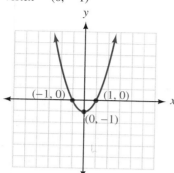

7. x-intercepts $= 3, -3$;
Vertex $= (0, 9)$

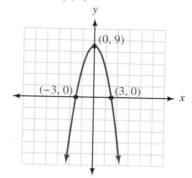

9. x-intercepts $= -1, 3$;
Vertex $= (1, -8)$

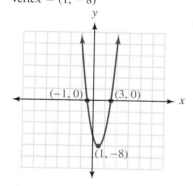

11. x-intercepts $= 1 + \sqrt{5}, 1 - \sqrt{5}$;
Vertex $= (1, -5)$

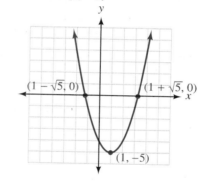

13. Vertex $= (2, -8)$

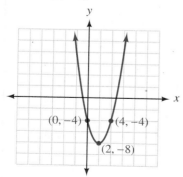

15. Vertex $= (1, -4)$

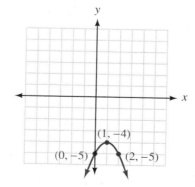

17. Vertex = (0, 1)

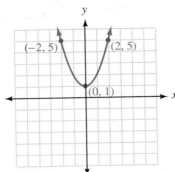

19. Vertex = (0, − 3)

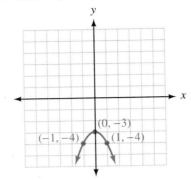

21. Vertex = $(-\frac{2}{3}, -\frac{1}{3})$

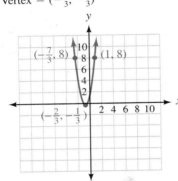

23. (3, − 4) lowest

25. (1, 9) highest

27. (2, 16) highest

29. (− 4, 16) highest

31. 40 items; maximum profit $500

33. 875 patterns; maximum profit $731.25

35. The ball is in her hand when $h(t) = 0$, which means $t = 0$ or $t = 2$ seconds. Maximum height is $h(1) = 16$ feet.

37. 256 feet   **39.** 40 feet by 20 feet

41. Maximum $R = \$3,600$
when $p = \$6.00$

43. Maximum $R = \$7,225$
when $p = \$8.50$

45. $1 - i$

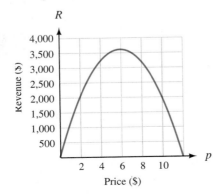

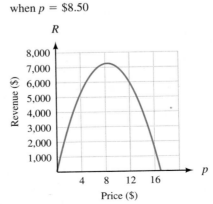

47. $27 + 5i$ **49.** $\frac{1}{10} + \frac{3}{10}i$ **51.** $y = (x - 2)^2 - 4$ **53.** $y = -\frac{1}{135}(x - 90)^2 + 60$

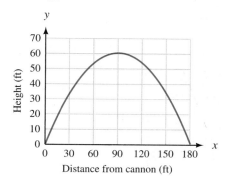

55. Problem 1: $(x - 2)(x + 2)$
Problem 2: $(x - 2)(x + 2) = 0; x = 2, -2$
Problem 3:

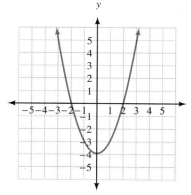

Relationship: If $x - 2$ is a factor of polynomial $P(x)$, then $x = 2$ is a solution to the equation $P(x) = 0$, and the graph of $y = P(x)$ has an x-intercept at 2.

Problem 4: $0 = x^2 - 4; (x - 2)(x + 2) = 0; x = -2, 2$

PROBLEM SET 10.6

1.

3.

5.

7.

9.

11.

13.

15. All real numbers   **17.** No solution; $\varnothing$

19.

21.

23.

25.

27.

29.

31.

33.

35. (a) $-2 < x < 2$ (b) $x < -2$ or $x > 2$ (c) $x = -2$ or $x = 2$

37. (a) $-2 < x < 5$ (b) $x < -2$ or $x > 5$ (c) $x = -2$ or $x = 5$

39. (a) $x < -1$ or $1 < x < 3$ (b) $-1 < x < 1$ or $x > 3$ (c) $x = -1$ or $x = 1$ or $x = 3$

41. $x \geq 4$; the width is at least 4 inches.   **43.** $5 \leq p \leq 8$; charge at least $5 but no more than $8 for each radio.
45. Let x = the number of $10 increases in dues. Then,
Income $= (10,000 - 200x)(100 + 10x) = -2,000x^2 + 80,000x + 1,000,000$. The vertex of this parabola is at
$x = -b/(2a) = (-80,000)/[2(-2,000)] = 20$. Thus, there should be 20 increases of $10, or an increase of $200, making
the new maximum income $= (10,000 - 200 \cdot 20)(100 + 10 \cdot 20) = \$1,800,000$.

47. Let x = the number of $2 increases. Then,
Income $= (20 + 2x)(40 - 2x) = -4x^2 + 40x + 800$. The vertex of this parabola is at $x = -b/(2a) =$
$(-40)/[2(-4)] = 5$. This implies there should be five increases of $2 for maximum income; that is, $\$20 + 5 \cdot \$2 = \$30$
for an oil change.

49. $\frac{5}{3}$ **51.** Possible solutions 1 and 6; only 6 checks; 6

53.

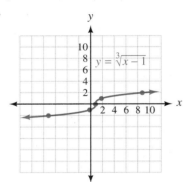

55.

57.

CHAPTER 10 REVIEW

1. $0, 5$ **2.** $0, \frac{4}{3}$ **3.** $\dfrac{4 \pm 7i}{3}$ **4.** $-3 \pm \sqrt{3}$ **5.** $-5, 2$ **6.** $-3, -2$ **7.** 3 **8.** 2 **9.** $\dfrac{-3 \pm \sqrt{3}}{2}$

10. $\dfrac{3 \pm \sqrt{5}}{2}$ **11.** $-5, 2$ **12.** $0, \frac{9}{4}$ **13.** $\dfrac{2 \pm i\sqrt{15}}{2}$ **14.** $1 \pm \sqrt{2}$ **15.** $-\frac{4}{3}, \frac{1}{2}$ **16.** 3 **17.** $5 \pm \sqrt{7}$

18. $0, \dfrac{5 \pm \sqrt{21}}{2}$ **19.** $-1, \frac{3}{5}$ **20.** $3, \dfrac{-3 \pm 3i\sqrt{3}}{2}$ **21.** $\dfrac{1 \pm i\sqrt{2}}{3}$ **22.** $\dfrac{3 \pm \sqrt{29}}{2}$ **23.** 100 or 170 items

24. 20 or 60 items   **25.** $D = 0$; one rational   **26.** $D = 0$; one rational   **27.** $D = 25$; two rational
28. $D = 361$; two rational   **29.** $D = 5$; two irrational   **30.** $D = 5$; two irrational   **31.** $D = -23$; two complex
32. $D = -87$; two complex   **33.** ± 20 **34.** ± 20 **35.** 4   **36.** 4   **37.** 25   **38.** 49

39. $x^2 - 8x + 15 = 0$ **40.** $x^2 - 2x - 8 = 0$ **41.** $2y^2 + 7y - 4 = 0$ **42.** $t^3 - 5t^2 - 9t + 45 = 0$

43. $-4, 12$ **44.** $-\frac{3}{4}, -\frac{1}{6}$ **45.** $\pm 2, \pm i\sqrt{3}$ **46.** Possible solutions 4 and 1; only 4 checks; 4   **47.** $\frac{9}{4}, 16$

48. 4   **49.** 4   **50.** 7   **51.** $t = \dfrac{5 \pm \sqrt{25 + 16h}}{16}$ **52.** $t = \dfrac{v \pm \sqrt{v^2 + 640}}{32}$

53. ![number line with open circles at -1 and 2, segment between]
-1 2

54. ![number line with closed circles at 2/3 and 4, segment between]
$\frac{2}{3}$ 4

55. ![number line with closed circles at -4 and 3/2, segment between]
-4 $\frac{3}{2}$

56. ![number line with open circles at -4, -2 and 3]
$-4\ -2$ 3

57.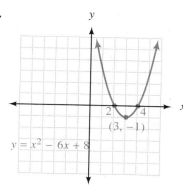

$y = x^2 - 6x + 8$

58.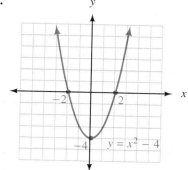

$y = x^2 - 4$

Chapter 10 Test

1. $-\frac{9}{2}, \frac{1}{2}$ **2.** $3 \pm i\sqrt{2}$ **3.** $5 \pm 2i$ **4.** $1 \pm i\sqrt{2}$ **5.** $\frac{5}{2}, \dfrac{-5 \pm 5i\sqrt{3}}{4}$ **6.** $-1 \pm i\sqrt{5}$ **7.** $r = \pm\dfrac{\sqrt{A}}{8} - 1$

8. $2 \pm \sqrt{2}$ **9.** $\frac{1}{2}$ and $\frac{3}{2}$ seconds   **10.** 15 or 100 cups   **11.** 9   **12.** $D = 81$; two rational

13. $3x^2 - 13x - 10 = 0$ **14.** $x^3 - 7x^2 - 4x + 28 = 0$ **15.** $\pm\sqrt{2}, \pm\frac{1}{2}i$ **16.** $1, \frac{1}{2}$ **17.** $\frac{1}{4}, 9$

18. $t = \dfrac{7 \pm \sqrt{49 + 16h}}{16}$

19.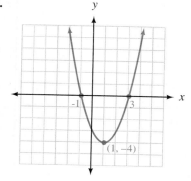

20.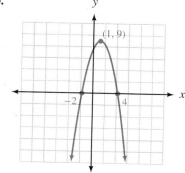

21.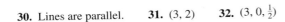
$-2 \qquad 3$

22.
$-3 \qquad \frac{1}{2}$

23. Maximum profit = \$900 by selling 100 items per week

CHAPTERS 1–10: CUMULATIVE REVIEW

1. 0 **2.** $-\frac{8}{27}$ **3.** 88 **4.** $-7x + 10$ **5.** $10x - 37$ **6.** $\frac{x^3}{y^7}$ **7.** $2\sqrt[3]{4}$ **8.** $\frac{9}{20}$ **9.** $\frac{1}{7}$ **10.** $\frac{4}{5}$

11. $x - 6y$ **12.** $x - 3$ **13.** $3x^3 - 11x^2 + 4$ **14.** $2i$ **15.** $\frac{23}{13} + \frac{11}{13}i$ **16.** -2 **17.** 15

18. $-9, 9$ **19.** Possible solutions 3 and 2; only 2 checks. **20.** $-3, -2$ **21.** $\frac{4}{3} - \sqrt{2}$ and $\frac{4}{3} + \sqrt{2}$

22. $-\frac{3}{2}, -1$ **23.** $0, \dfrac{5 + \sqrt{37}}{6}, \dfrac{5 - \sqrt{37}}{6}$ **24.** $-\frac{1}{2}, \frac{2}{3}$ **25.** $\frac{9}{4}$ **26.** $\dfrac{8}{a - b}$

27. $8 \le x \le 20$
$8 \qquad 20$

28. $x \le -\frac{1}{2}$ or $x \ge 2$
$-\frac{1}{2} \qquad 2$
29. Lines coincide.

30. Lines are parallel. **31.** $(3, 2)$ **32.** $(3, 0, \frac{1}{2})$

33.

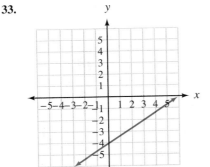

34.

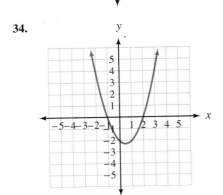

35. $4x - 3y = 2$ **36.** $(x + 4 + y)(x + 4 - y)$ **37.** $\dfrac{7\sqrt[3]{3}}{3}$ **38.** $\dfrac{\sqrt{5} + 1}{2}$ **39.** $100°, 55°, 25°$ **40.** $\frac{25}{16}$

CHAPTER 11

PROBLEM SET 11.1

1. 1   **3.** 2   **5.** $\frac{1}{27}$ **7.** 13

9.

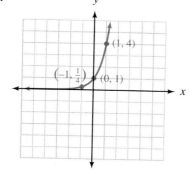

11.

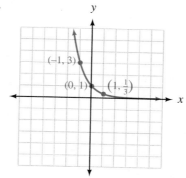

13.

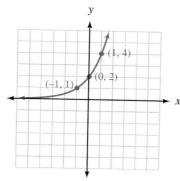

15.

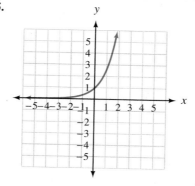

17.

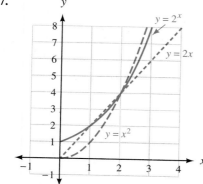

19.

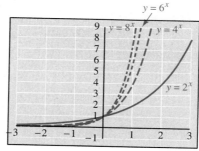

21. $h = 6 \cdot \left(\frac{2}{3}\right)^n$; fifth bounce: $6\left(\frac{2}{3}\right)^5 \approx 0.79$ feet

23.

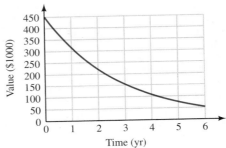

$Q(t) = \frac{1}{2}$ when $t \approx 4.3$ days

$Q(t) = \frac{1}{2}$ when $t \approx 4.3$ days

25. (a) $A(t) = 1,200\left(1 + \dfrac{0.06}{4}\right)^{4t}$ (b) $\$1,932.39$ (c) About 11.64 years   (d) $\$1,939.29$

27. (a) $\$129,138.48$ (b) Domain: $\{t \mid 0 \le t \le 6\}$ (c)
(d) Range: $\{V(t) \mid 52,942.05 < V(t) \le 450,000\}$
(e) After approximately 4 years and 8 months

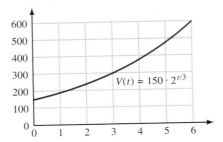

29. $f(1) = 200, f(2) = 800, f(3) = 3,200$

31. $V(t)$

$V(t) = 150 \cdot 2^{t/3}$

33. Domain $= \{-2, 2\}$, range $= \{3, 6, 8\}$, not a function

35. All real numbers except -7 and 5   **37.** -18 **39.** 2

41. (a) 1   (b) $\frac{1}{3}$ (c) 3   (d) 1   (e) $\frac{1}{2}$ (f) 2   (g) 3   (h) $\frac{1}{2}$

43. (a)

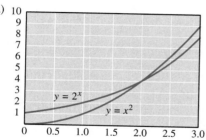

(b) $2^2 = 4; 2^4 = 16$ and $4^2 = 16$

(c)

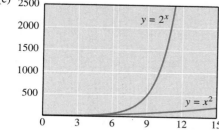

From the graph, $y = 2^x$ grows faster. This is confirmed because 2^{100} is much larger than 100^2, which is only 10,000.

PROBLEM SET 11.2

1. $f^{-1}(x) = \dfrac{x + 1}{3}$ **3.** $f^{-1}(x) = \sqrt[3]{x}$ **5.** $f^{-1}(x) = \dfrac{x - 3}{x - 1}$ **7.** $f^{-1}(x) = 4x + 3$

9. $f^{-1}(x) = 2(x + 3) = 2x + 6$ **11.** $f^{-1}(x) = \dfrac{1 - x}{3x - 2}$ **13.**

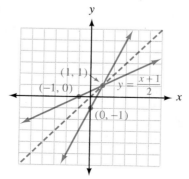

15.

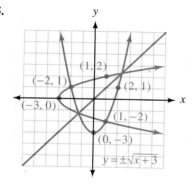

17.

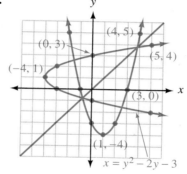

19.

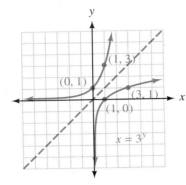

21.

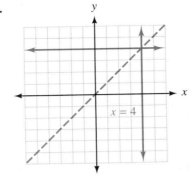

23.

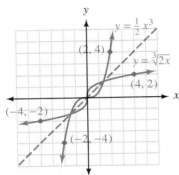

25.

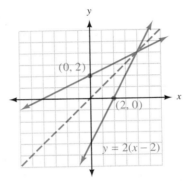

27.

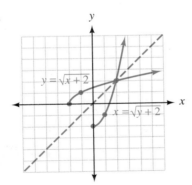

29. (a) Yes (b) No (c) Yes   **31.** (a) 4 (b) $\frac{4}{3}$ (c) 2 (d) 2

33. $f^{-1}(x) = \frac{1}{x}$ **35.** $f^{-1}(x) = 7(x + 2)$

37. (a) -3 (b) -6 (c) 2 (d) 3 (e) -2 (f) 3 (g) Each is the inverse of the other.   **39.** $-2, 3$ **41.** 25

43. $5 \pm \sqrt{17}$ **45.** $1 \pm i$ **47.** $f^{-1}(x) = \dfrac{x - 5}{3}$ **49.** $f^{-1}(x) = \sqrt[3]{x - 1}$ **51.** $f^{-1}(x) = \dfrac{2x - 4}{x - 1}$

53. (a) 1 (b) 2 (c) 5 (d) 0 (e) 1 (f) 2 (g) 2 (h) 5

PROBLEM SET 11.3

1. $\log_2 16 = 4$ **3.** $\log_5 125 = 3$ **5.** $\log_{10} 0.01 = -2$ **7.** $\log_2 \frac{1}{32} = -5$ **9.** $\log_{1/2} 8 = -3$

11. $\log_3 27 = 3$ **13.** $10^2 = 100$ **15.** $2^6 = 64$ **17.** $8^0 = 1$ **19.** $10^{-3} = 0.001$ **21.** $6^2 = 36$

23. $5^{-2} = \frac{1}{25}$ **25.** 9   **27.** $\frac{1}{125}$ **29.** 4   **31.** $\frac{1}{3}$ **33.** 2   **35.** $\sqrt[3]{5}$

37.

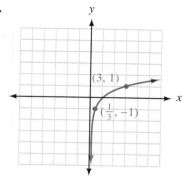

39.

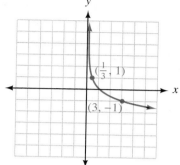

41.

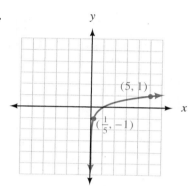

43.

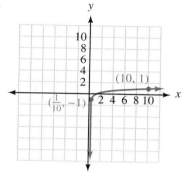

45. $y = 3^x$ **47.** $y = \log_{1/3} x$ **49.** 4   **51.** $\frac{3}{2}$ **53.** 3   **55.** 1   **57.** 0   **59.** 0   **61.** $\frac{1}{2}$

63.

Prefix	Multiplying Factor	$\log_{10}$ (Multiplying Factor)
Nano	0.000 000 001	-9
Micro	0.000 001	-6
Deci	0.1	-1
Giga	1,000,000,000	9
Peta	1,000,000,000,000,000	15

65. 2   **67.** 10^8 times as large   **69.** $\dfrac{-2 \pm \sqrt{10}}{2}$ **71.** $\dfrac{2 \pm i\sqrt{3}}{2}$ **73.** 5, $\dfrac{-5 \pm 5i\sqrt{3}}{2}$ **75.** 0, 5, -1

77. $\dfrac{3 \pm \sqrt{29}}{2}$

79. (a)

x	-1	0	1	2
$f(x)$	$\frac{1}{8}$	1	8	64

(b)

x	$\frac{1}{8}$	1	8	64
$f^{-1}(x)$	-1	0	1	2

(c) $f(x) = 8^x$

(d) $f^{-1}(x) = \log_8 x$

PROBLEM SET 11.4

1. $\log_3 4 + \log_3 x$ **3.** $\log_6 5 - \log_6 x$ **5.** $5 \log_2 y$ **7.** $\frac{1}{3} \log_9 z$ **9.** $2 \log_6 x + 4 \log_6 y$

11. $\frac{1}{2}\log_5 x + 4\log_5 y$ **13.** $\log_b x + \log_b y - \log_b z$ **15.** $\log_{10} 4 - \log_{10} x - \log_{10} y$

17. $2\log_{10} x + \log_{10} y - \frac{1}{2}\log_{10} z$ **19.** $3\log_{10} x + \frac{1}{2}\log_{10} y - 4\log_{10} z$ **21.** $\frac{2}{3}\log_b x + \frac{1}{3}\log_b y - \frac{4}{3}\log_b z$

23. $\log_b xz$ **25.** $\log_3 \dfrac{x^2}{y^3}$ **27.** $\log_{10} \sqrt{x}\sqrt[3]{y}$ **29.** $\log_2 \dfrac{x^3\sqrt{y}}{z}$ **31.** $\log_2 \dfrac{\sqrt{x}}{y^3 z^4}$ **33.** $\log_{10} \dfrac{x^{3/2}}{y^{3/4} z^{4/5}}$ **35.** $\frac{2}{3}$

37. 18   **39.** Possible solutions -1 and 3; only 3 checks; 3   **41.** 3

43. Possible solutions -2 and 4; only 4 checks; 4   **45.** Possible solutions -1 and 4; only 4 checks; 4

47. Possible solutions $-\frac{5}{2}$ and $\frac{5}{3}$; only $\frac{5}{3}$ checks; $\frac{5}{3}$ **51.** (a) 1.602 (b) 2.505 (c) 3.204

53. $\text{pH} = 6.1 + \log_{10} x - \log_{10} y$ **55.** 2.52   **57.** $D = -7$; two complex   **59.** $x^2 - 2x - 15 = 0$

61. $3y^2 - 11y + 6 = 0$

PROBLEM SET 11.5

1. 2.5775   **3.** 1.5775   **5.** 3.5775   **7.** -1.4225 **9.** 4.5775   **11.** 2.7782   **13.** 3.3032   **15.** -2.0128

17. -1.5031 **19.** -0.3990 **21.** 759   **23.** 0.00759   **25.** 1,430   **27.** 0.00000447   **29.** 0.0000000918

31. 10^{10} **33.** 10^{-10} **35.** 10^{20} **37.** $\frac{1}{100}$ **39.** 1,000

41. Possible solutions 1 and 100; both check

43. The calculator gives a "domain error," indicating that -10 is not in the domain of $f(x) = \log x$.

45. The San Francisco earthquake was approximately 2,000 times greater.

47.

x	$(1 + x)^{(1/x)}$
1	2
0.5	2.25
0.1	2.5937
0.01	2.7048
0.001	2.7169
0.0001	2.7181
0.00001	2.7183

$(1 + x)^{(1/x)}$ appears to approach e.

49. $3\log R - 2\log T = \log R^3 - \log T^2$

$$= \log \frac{R^3}{T^2}$$

So if $\dfrac{R^3}{T^2}$ is constant, then its logarithm $\left(\log \dfrac{R^3}{T^2}\right)$ will also be constant.

51. Approximately 3.19   **53.** 1.78×10^{-5} **55.** 3.16×10^5 **57.** 2.00×10^8 **59.** 10 times as large

61. 12.9%   **63.** 5.3%   **65.** 1   **67.** 5   **69.** x **71.** $\ln 10 + 3t$ **73.** $\ln A - 2t$ **75.** 2.7080

77. -1.0986 **79.** 2.1972   **81.** 2.7724   **83.** $\pm 2, \pm i\sqrt{2}$ **85.** 8, 27   **87.** $1, \frac{9}{4}$ **89.** 5, 1

91. $\dfrac{-2 \pm \sqrt{4 + k^2}}{k}$

PROBLEM SET 11.6

1. 1.4650   **3.** 0.6826   **5.** -1.5440 **7.** -0.6477 **9.** -0.3333 **11.** 2.0000   **13.** -0.1845

15. 0.1845   **17.** 1.6168   **19.** 2.1131   **21.** 1.333   **23.** 0.7500   **25.** 1.3917   **27.** 0.7186

29. 2.6356   **31.** 4.1632   **33.** 5.8435   **35.** -1.0642 **37.** 2.3026   **39.** 10.7144   **41.** 11.72 years

43. 9.25 years   **45.** 8.75 years   **47.** 18.58 years   **49.** 11.55 years   **51.** 18.31 years   **53.** 11.45 years

55. 13.9 years later or toward the end of 2007   **57.** 27.5 years   **59.** $(-2, -23)$; lowest   **61.** $(\frac{3}{2}, 9)$; highest

63. 2 seconds, 64 feet   **65.** $t = \dfrac{1}{r}\ln\dfrac{A}{P}$ **67.** $t = \dfrac{1}{k}\dfrac{\log P - \log A}{\log 2}$ **69.** $t = \dfrac{\log A - \log P}{\log(1 - r)}$ **71.** 1.75

73. 1.73   **75.** (1.46, 5); they are nearly the same point

CHAPTER 11 REVIEW

1. 16   **2.** $\frac{1}{2}$ **3.** $\frac{1}{9}$ **4.** -5 **5.** $\frac{5}{6}$ **6.** 7   **7.**

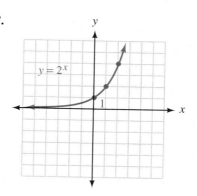

8.

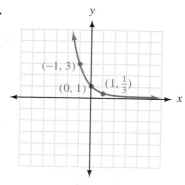

9.

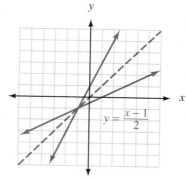

10.

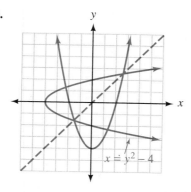

11. $f^{-1}(x) = \dfrac{x - 3}{2}$ **12.** $y = \pm\sqrt{x + 1}$ **13.** $f^{-1}(x) = 2x - 4$

14. $y = \pm\sqrt{\dfrac{4 - x}{2}}$ **15.** $\log_3 81 = 4$ **16.** $\log_7 49 = 2$ **17.** $\log_{10} 0.01 = -2$ **18.** $\log_2 \frac{1}{8} = -3$

19. $2^3 = 8$ **20.** $3^2 = 9$ **21.** $4^{1/2} = 2$ **22.** $4^1 = 4$ **23.** 25   **24.** $\frac{3}{4}$ **25.** 10

26.

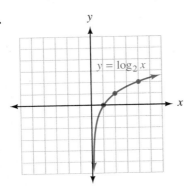

27.

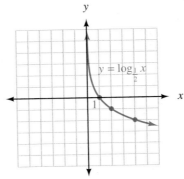

28. 2   **29.** $\frac{2}{3}$ **30.** 0   **31.** $\log_2 5 + \log_2 x$ **32.** $\log_{10} 2 + \log_{10} x - \log_{10} y$

33. $3 \log_a y + \frac{1}{2} \log_a x - \log_a z$ **34.** $2 \log_{10} x - 3 \log_{10} y - 4 \log_{10} z$ **35.** $\log_2 xy$ **36.** $\log_3 \frac{x}{4}$ **37.** $\log_a \frac{25}{3}$

38. $\log_2 \frac{x^3 y^2}{z^4}$ **39.** 2   **40.** 6   **41.** Possible solutions -1 and 3; only 3 checks; 3   **42.** 3

43. Possible solutions -2 and 3; only 3 checks; 3   **44.** Possible solutions -1 and 4; only 4 checks; 4   **45.** 2.5391

46. -0.1469 **47.** 9,230   **48.** 0.0251   **49.** 1   **50.** 0   **51.** 2   **52.** -4 **53.** 2.1   **54.** 5.1

55. 2.0×10^{-3} **56.** 3.2×10^{-8} **57.** $\frac{3}{2}$ **58.** $x = \frac{1}{3} \left(\frac{\log 5}{\log 4} - 2 \right) = -0.28$ **59.** 0.75   **60.** 2.43

61. About 4.67 years   **62.** About 9.25 years   **63.** \$136,856.91   **64.** \$2,660.95 per year

CHAPTER 11 TEST

1.

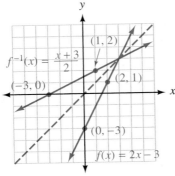

2.

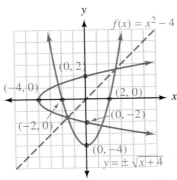

3.

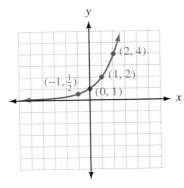

4.

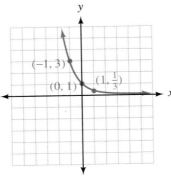

5. 64   **6.** $\sqrt{5}$

7.

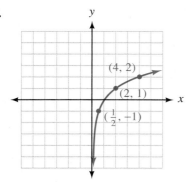

8.

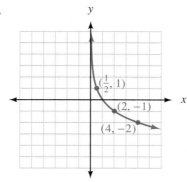

9. $\frac{2}{3}$ **10.** 1.5646   **11.** 4.3692   **12.** -1.9101 **13.** 3.8330   **14.** -3.0748 **15.** $3 + 2\log_2 x - \log_2 y$

16. $\frac{1}{2}\log x - 4\log y - \frac{1}{5}\log z$ **17.** $\log_3 \dfrac{x^2}{\sqrt{y}}$ **18.** $\log \dfrac{\sqrt[3]{x}}{yz^2}$ **19.** 7.04×10^4 **20.** 2.25×10^{-3}

21. 1.46   **22.** $\frac{5}{4}$ or 1.25   **23.** 15   **24.** Possible solutions -1 and 8; only 8 checks; 8
25. 6.2   **26.** \$651.56   **27.** About 13.9 years   **28.** \$7,372.80

CHAPTERS 1–11: CUMULATIVE REVIEW

1. 32   **2.** $24x - 26$ **3.** -16 **4.** $\frac{17}{8}$ **5.** $3xy\sqrt[3]{x}$ **6.** $3\sqrt{3}$ **7.** $-2 + 7i$ **8.** 0   **9.** $\dfrac{x^2 + x + 1}{x + 1}$

10. $\frac{4}{7}$ **11.** $12t^4 - \frac{1}{12}$ **12.** $18 + 14\sqrt{3}$ **13.** $3x + 7 + \dfrac{10}{3x - 4}$ **14.** $\dfrac{-12 + 7\sqrt{6}}{5}$ **15.** $\dfrac{24}{(4x + 3)(4x - 3)}$

16. $\frac{8}{3}$ **17.** 4   **18.** $\frac{1}{4}, -\frac{1}{7}$ **19.** $\varnothing$ **20.** $\frac{3}{2}$ **21.** $\dfrac{1 \pm \sqrt{29}}{2}$ **22.** $-\frac{12}{5}$ **23.** 1   **24.** $-\frac{1}{4}, \frac{2}{3}$ **25.** $\varnothing$

26. 4   **27.** 1, 4   **28.** $-\frac{3}{2}, \frac{1}{4}$ **29.** 27   **30.** 10   **31.** Possible solution $-\frac{9}{2}$; does not check; $\varnothing$ **32.** $(\frac{4}{3}, \frac{13}{3})$

33. $(8, -5)$ **34.** $(3, 1, -2)$ **35.** $y \le 1$ or $y \ge 3$

36. $x < -\frac{7}{2}$ or $x > 2$

37. $x < -3$ or $-2 < x < 3$

38.

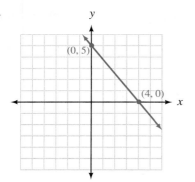

39. $y = -\frac{5}{3}x + 2$ **40.** $9a - 4b < 9a + 4b$ **41.** 9.72×10^{-5} **42.** $2(5a^2 + 2)(5a^2 - 1)$
43. $f^{-1}(x) = 2x - 6$ **44.** 1 **45.** $D = \{2, -3\}, R = \{-3, -1, 3\}$, no **46.** $\{x \mid x \le 3\}$
47. -54, geometric **48.** 6 **49.** 6 gallons 25%; 14 gallons 50% **50.** $12,667.70

CHAPTER 12

PROBLEM SET 12.1

1. 4, 7, 10, 13, 16 **3.** 3, 7, 11, 15, 19 **5.** 1, 2, 3, 4, 5 **7.** 4, 7, 12, 19, 28 **9.** $\frac{1}{4}, \frac{2}{5}, \frac{3}{6}, \frac{4}{7}, \frac{5}{8}$ **11.** $\frac{2}{3}, \frac{3}{4}, \frac{4}{5}, \frac{5}{6}, \frac{6}{7}$
13. $1, \frac{1}{4}, \frac{1}{9}, \frac{1}{16}, \frac{1}{25}$ **15.** 2, 4, 8, 16, 32 **17.** $\frac{1}{3}, \frac{1}{9}, \frac{1}{27}, \frac{1}{81}, \frac{1}{243}$ **19.** $2, \frac{3}{2}, \frac{4}{3}, \frac{5}{4}, \frac{6}{5}$ **21.** $0, \frac{3}{2}, \frac{8}{3}, \frac{15}{4}, \frac{24}{5}$
23. $-2, 4, -8, 16, -32$ **25.** $3, -9, 27, -81, 243$ **27.** $3, 0, -3, -6, -9$ **29.** 1, 5, 13, 29, 61
31. 1, 3, 6, 10, 15 **33.** $a_n = n + 1$ **35.** $a_n = 4n$ **37.** $a_n = 3n + 4$ or recursively as $a_1 = 7, a_n = a_{n-1} + 3$
39. $a_n = n^2$ **41.** $a_n = 3n^2$ **43.** $a_n = 2^{n+1}$ or recursively as $a_1 = 4, a_n = 2a_{n-1}$
45. $a_n = (-2)^n$ or recursively as $a_1 = -2, a_n = -2a_{n-1}$ **47.** $a_n = \frac{1}{2^{n+1}}$ or recursively as $a_1 = \frac{1}{4}, a_n = \frac{1}{2}a_{n-1}$
49. $a_n = \frac{n}{(n + 1)^2}$ **51.** (a) $28,000, $29,120, $30,284.80, $31,496.19, $32,756.04 (b) $a_n = 28,000(1.04)^{n-1}$
53. (a) $1,200, $2,400, $3,600, $4,800, $6,000 (b) $a_m = 100m$ (c) $a_y = 100(12y)$ **55.** 27 **57.** 5 **59.** 1
61. $a_{100} \approx 2.7048; a_{1,000} \approx 2.7169; a_{10,000} \approx 2.7181; a_{100,000} \approx 2.7183$ **63.** 1, 1, 2, 3, 5, 8, 13, 21, 34, 55
65. $\frac{3}{2}, \frac{5}{3}, \frac{8}{5}$

PROBLEM SET 12.2

1. 36 **3.** 9 **5.** 11 **7.** $\frac{163}{60}$ **9.** $\frac{62}{15}$ **11.** 60 **13.** 40 **15.** $5x + 15$ when simplified
17. $(x + 1)^2 + (x + 1)^3 + (x + 1)^4 + (x + 1)^5 + (x + 1)^6 + (x + 1)^7$
19. $\frac{x + 1}{x - 1} + \frac{x + 2}{x - 1} + \frac{x + 3}{x - 1} + \frac{x + 4}{x - 1} + \frac{x + 5}{x - 1}$
21. $(x + 3)^3 + (x + 4)^4 + (x + 5)^5 + (x + 6)^6 + (x + 7)^7 + (x + 8)^8$
23. $(x + 1)^2 + (x + 2)^3 + (x + 3)^4 + (x + 4)^5 + (x + 5)^6$ **25.** $\sum_{i=1}^{4} 2^i$ **27.** $\sum_{i=2}^{6} 2^i$ **29.** $\sum_{i=2}^{6} (i^2 + 1)$
31. $\sum_{i=3}^{7} \frac{i}{i + 1}$ **33.** $\sum_{i=1}^{4} \frac{i}{2i + 1}$ **35.** $\sum_{i=3}^{6} (x - i)$ **37.** $\sum_{i=3}^{5} \frac{x}{x + i}$ **39.** $\sum_{i=2}^{4} x^i(x + i)$
41. (a) $\frac{1}{3} = 0.3 + 0.03 + 0.003 + 0.0003 + \cdots$ (b) $\frac{2}{9} = 0.2 + 0.02 + 0.002 + 0.0002 + \cdots$
(c) $\frac{3}{11} = 0.27 + 0.0027 + 0.000027 + \cdots$ **43.** 208 feet, 784 feet **45.** (a) $\frac{1}{3}, \frac{1}{15}, \frac{1}{35}, \frac{1}{63}, \frac{1}{99}, \frac{1}{143}$ (b) $\frac{6}{13}$ (c) $\frac{1}{2}$
47. $3 \log_2 x + \log_2 y$ **49.** $\frac{1}{3} \log_{10} x - 2 \log_{10} y$ **51.** $\log_{10} \frac{x}{y^2}$ **53.** $\log_3 \frac{x^2}{y^3 z^4}$
55. 80 **57.** Possible solutions -1 and 8, only 8 checks; 8 **59.** $\frac{16}{7}$

PROBLEM SET 12.3

1. 1 **3.** Not an arithmetic progression **5.** -5 **7.** Not an arithmetic progression **9.** $\frac{2}{3}$
11. $a_n = 4n - 1; a_{24} = 95$ **13.** $a_{10} = -12; S_{10} = -30$ **15.** $a_1 = 7; d = 2; a_{30} = 65$
17. $a_1 = 12; d = 2; a_{20} = 50; S_{20} = 620$ **19.** 20,300 **21.** -158 **23.** $a_{10} = 5; S_{10} = \frac{55}{2}$
25. (a) $18,000, $14,700, $11,400, $8,100, $4,800 (b) $-3,300$

(c)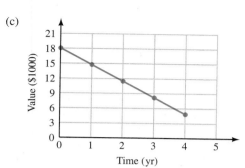

Time (yr)

(d) $9,750   (e) $a_n = 18,000 - (n - 1)3,300$

27. (a) 1,500, 1,460, 1,420, 1,380, 1,340, 1,300
(b) It is arithmetic because the same amount is subtracted from each succeeding term.   (c) $a_n = 1,500 - (n - 1)40$
29. (a) 1, 3, 6, 10, 15, 21, 28, 36, 45, 55, 66, 78, 91, 105, 120   (b) $a_n = n + a_{n-1}$
(c) No, it is not arithmetic because the same amount is not added to each term.
31. $28,000, $28,850, $29,700, $30,550, $31,400; $a_n = 27,150 + 850n$; $a_{10} = $35,650$ **33.** 2.7604
35. -1.2396 **37.** 445   **39.** 4.45×10^{-8}

PROBLEM SET 12.4

1. 5   **3.** $\frac{1}{3}$ **5.** Not geometric   **7.** -2 **9.** Not geometric   **11.** $a_n = 4 \cdot 3^{n-1}$

13. $a_6 = -2\left(-\dfrac{1}{2}\right)^5 = \dfrac{1}{16}$ **15.** $a_{20} = 3(-1)^{19} = -3$ **17.** $S_{10} = \dfrac{10(2^{10} - 1)}{2 - 1} = 10{,}230$

19. $S_{20} = \dfrac{1[(-1)^{20} - 1]}{-1 - 1} = 0$ **21.** $a_8 = \dfrac{1}{5}\left(\dfrac{1}{2}\right)^7 = \dfrac{1}{640}$ **23.** $S_5 = \dfrac{-\dfrac{1}{2}\left[\left(\dfrac{1}{2}\right)^5 - 1\right]}{\dfrac{1}{2} - 1} = -\dfrac{31}{32}$

25. $a_{10} = \sqrt{2}(\sqrt{2})^9 = (\sqrt{2})^{10} = 32$; $S_{10} = \dfrac{\sqrt{2}[(\sqrt{2})^{10} - 1]}{\sqrt{2} - 1} = \dfrac{31\sqrt{2}}{\sqrt{2} - 1} = 62 + 31\sqrt{2}$

27. $a_6 = 100\left(\dfrac{1}{10}\right)^5 = \dfrac{1}{10^3} = \dfrac{1}{1{,}000}$; $S_6 = \dfrac{100\left[\left(\dfrac{1}{10}\right)^6 - 1\right]}{\dfrac{1}{10} - 1} = \dfrac{100(-0.999999)}{-0.9} = 111.111$ **29.** $r = \pm 2$

31. $S = \dfrac{\dfrac{1}{2}}{1 - \dfrac{1}{2}} = 1$ **33.** $S = \dfrac{4}{1 - \dfrac{1}{2}} = 8$ **35.** $S = \dfrac{\dfrac{2}{5}}{1 - \dfrac{2}{5}} = \dfrac{2}{3}$ **37.** $S = \dfrac{\dfrac{3}{4}}{1 - \dfrac{1}{3}} = \dfrac{9}{8}$

43. (a) $450,000, $315,000, $220,500, $154,350, $108,045   (b) 0.7
(c)   (d) $90,000   (e) $a_n = 0.7a_{n-1}$

Time (yr)

45. (a) $\frac{1}{2}$ (b) $\frac{364}{729}$ (c) $\frac{1}{1,458}$ **47.** 300 feet   **49.** $x^2 + 10x + 25$ **51.** $x^3 + 3x^2y + 3xy^2 + y^3$

53. $x^4 + 4x^3y + 6x^2y^2 + 4xy^3 + y^4$ **55.** 52,428,800   **57.** 199.99924   **59.** $S = 2a$ **61.** $S = \dfrac{a}{b - a}$

PROBLEM SET 12.5

1. $x^4 + 8x^3 + 24x^2 + 32x + 16$ **3.** $x^6 + 6x^5y + 15x^4y^2 + 20x^3y^3 + 15x^2y^4 + 6xy^5 + y^6$
5. $32x^5 + 80x^4 + 80x^3 + 40x^2 + 10x + 1$ **7.** $x^5 - 10x^4y + 40x^3y^2 - 80x^2y^3 + 80xy^4 - 32y^5$
9. $81x^4 - 216x^3 + 216x^2 - 96x + 16$ **11.** $64x^3 - 144x^2y + 108xy^2 - 27y^3$

13. $x^8 + 8x^6 + 24x^4 + 32x^2 + 16$ **15.** $x^6 + 3x^4y^2 + 3x^2y^4 + y^6$ **17.** $\dfrac{x^3}{8} - 3x^2 + 24x - 64$

19. $\dfrac{x^4}{81} + \dfrac{2x^3y}{27} + \dfrac{x^2y^2}{6} + \dfrac{xy^3}{6} + \dfrac{y^4}{16}$ **21.** $x^9 + 18x^8 + 144x^7 + 672x^6$ **23.** $x^{10} - 10x^9y + 45x^8y^2 - 120x^7y^3$
25. $x^{10} + 20x^9y + 180x^8y^2 + 960x^7y^3$ **27.** $x^{15} + 15x^{14} + 105x^{13}$ **29.** $x^{12} - 12x^{11}y + 66x^{10}y^2$
31. $x^{20} + 40x^{19} + 760x^{18}$ **33.** $x^{100} + 200x^{99}$ **35.** $x^{50} + 50x^{49}y$

37. $a_9 = \dfrac{12!}{8!4!}(2x)^4(3y)^8 = 495(16x^4)(6,561y^8) = 51,963,120x^4y^8$ **39.** $a_5 = \dfrac{10!}{4!6!}x^6(-2)^4 = 210x^6(16) = 3,360x^6$

41. $a_4 = \dfrac{9!}{3!6!}x^6(3)^3 = 84x^6(27) = 2,268x^6$ **43.** $a_{12} = \dfrac{20!}{11!9!}(2x)^9(5y)^{11}$ **45.** $\frac{21}{128}$ **47.** 1.04060401

49. $x = \dfrac{\log 7}{\log 5} \approx 1.21$ **51.** $\frac{1}{6}$ or 0.17   **53.** Approximately 7 years   **55.** 2.16   **57.** 6.36   **59.** $t = \dfrac{1}{5}\ln\dfrac{A}{10}$

61. 56   **63.** 125,970
67. As one example, if $n = 5$, $\binom{5}{0} + \binom{5}{1} + \binom{5}{2} + \binom{5}{3} + \binom{5}{4} + \binom{5}{5} = 1 + 5 + 10 + 10 + 5 + 1 = 32 = 2^5$.

CHAPTER 12 REVIEW

1. 7, 9, 11, 13   **2.** 1, 4, 7, 10   **3.** 0, 3, 8, 15   **4.** $\frac{4}{3}, \frac{5}{4}, \frac{6}{5}, \frac{7}{6}$ **5.** 4, 16, 64, 256   **6.** $\frac{1}{4}, \frac{1}{16}, \frac{1}{64}, \frac{1}{256}$

7. $3n - 1$ **8.** $2n - 5$ **9.** n^4 **10.** $n^2 + 1$ **11.** 2^{-n} **12.** $\dfrac{n+1}{n^2}$ **13.** 32   **14.** 25   **15.** $\frac{14}{5}$

16. -5 **17.** 98   **18.** $\frac{127}{30}$ **19.** $\displaystyle\sum_{i=1}^{4} 3i$ **20.** $\displaystyle\sum_{i=1}^{4} (4i - 1)$ **21.** $\displaystyle\sum_{i=1}^{5} (2i + 3)$ **22.** $\displaystyle\sum_{i=2}^{4} i^2$ **23.** $\displaystyle\sum_{i=1}^{4} \dfrac{1}{i + 2}$

24. $\displaystyle\sum_{i=1}^{5} \dfrac{i}{3^i}$ **25.** $\displaystyle\sum_{i=1}^{3} (x - 2i)$ **26.** $\displaystyle\sum_{i=1}^{4} \dfrac{x}{x + i}$ **27.** Geometric   **28.** Arithmetic   **29.** Arithmetic
30. Neither   **31.** Geometric   **32.** Geometric   **33.** Arithmetic   **34.** Neither   **35.** $a_n = 3n - 1: 59$
36. $a_n = 8 - 3n: -40$ **37.** 34: 160   **38.** 78: 648   **39.** $a_1 = 5, d = 4, a_{10} = 41$
40. $a_1 = 8, d = 3, a_9 = 32, S_9 = 180$ **41.** $a_1 = -4, d = -2, a_{20} = -42, S_{20} = -460$ **42.** 20,100
43. -95 **44.** $a_n = 3(2)^{n-1}, a_{20} = 3(2)^{19}$ **45.** $a_n = 5(-2)^{n-1}, a_{16} = 5(-2)^{15}$
46. $a_n = 4(\frac{1}{2})^{n-1}, a_{10} = 4(\frac{1}{2})^9 = \frac{1}{128}$ **47.** -3 **48.** 8   **49.** $a_1 = 3, r = 2, a_6 = 96$ **50.** $243\sqrt{3}$
51. 28   **52.** 35   **53.** 20   **54.** 36   **55.** 45   **56.** 161,700   **57.** $x^4 - 8x^3 + 24x^2 - 32x + 16$
58. $16x^4 + 96x^3 + 216x^2 + 216x + 81$ **59.** $27x^3 + 54x^2y + 36xy^2 + 8y^3$
60. $x^{10} - 10x^8 + 40x^6 - 80x^4 + 80x^2 - 32$ **61.** $\frac{1}{16}x^4 + \frac{3}{2}x^3 + \frac{27}{2}x^2 + 54x + 81$
62. $\frac{1}{27}x^3 - \frac{1}{6}x^2y + \frac{1}{4}xy^2 - \frac{1}{8}y^3$ **63.** $x^{10} + 30x^9y + 405x^8y^2$ **64.** $x^9 - 27x^8y + 324x^7y^2$
65. $x^{11} + 11x^{10}y + 55x^9y^2$ **66.** $x^{12} - 24x^{11}y + 264x^{10}y^2$ **67.** $x^{16} - 32x^{15}y$ **68.** $x^{32} + 64x^{31}y$

69. $x^{50} - 50x^{49}$ **70.** $x^{150} + 150x^{149}y$ **71.** $\dfrac{10!}{5!5!}x^5(-3)^5 = 252x^5(-243) = -61,236x^5$

72. $\dfrac{9!}{3!6!}(2x)^6 = 84(64x^6) = 5,376x^6$

CHAPTER 12 TEST

1. $-2, 1, 4, 7, 10$ **2.** $3, 7, 11, 15, 19$ **3.** $2, 5, 10, 17, 26$ **4.** $2, 16, 54, 128, 250$ **5.** $2, \frac{3}{4}, \frac{4}{9}, \frac{5}{16}, \frac{6}{25}$

6. $4, -8, 16, -32, 64$ **7.** $a_n = 4n + 2$ **8.** $a_n = 2^{n-1}$ **9.** $a_n = (\frac{1}{2})^n = 1/2^n$ **10.** $a_n = (-3)^n$

11. (a) 90 (b) 53 (c) 130 **12.** 3 **13.** ± 6 **14.** 320 **15.** 25 **16.** $S_{50} = 3(2^{50} - 1)$ **17.** $\frac{3}{4}$

18. $x^4 - 12x^3 + 54x^2 - 108x + 81$ **19.** $32x^5 - 80x^4 + 80x^3 - 40x^2 + 10x - 1$ **20.** $x^{20} - 20x^{19} + 190x^{18}$

21. $\frac{8!}{5!3!} (2x)^3(-3y)^5 = 56(8x^3)(-243y^5) = -108{,}864x^3y^5$

CHAPTERS 1–12: CUMULATIVE REVIEW

1. 4 **2.** $30y + 49$ **3.** $\frac{a^5b^4}{2}$ **4.** $\frac{y - 3}{y - 2}$ **5.** $\frac{1}{4}$ **6.** 3 **7.** $(a + 1)(b^3 + 6)$ **8.** $(x - 1)(8x + 3)$

9. -2 **10.** $1, \frac{1}{2}$ **11.** $-5, 2$ **12.** $-2, 4$ **13.** 8 **14.** $\frac{3 \pm 5i\sqrt{2}}{4}$ **15.** $\frac{3}{2}, \frac{-3 \pm 3i\sqrt{3}}{4}$

16. $\pm \frac{\sqrt{10}}{2}$ and $\pm \frac{\sqrt{3}}{3}i$ **17.** $-3 \quad 0$ **18.**  $-8 \quad 3$

19.

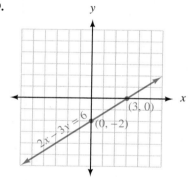

20.

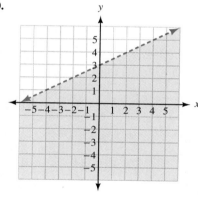

21.

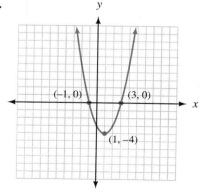

22.

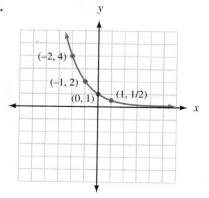

23. (1, 3)   **24.** (4, −2, 3)   **25.** $\dfrac{y(y-1)}{y+1}$ **26.** $x^{6/5} - 4$ **27.** $14 - 5i$ **28.** $\dfrac{1}{(x+4)(x+2)}$

29. $26\sqrt{2}$ **30.** $\dfrac{3\sqrt{7} + 3\sqrt{3}}{4}$ **31.** $f^{-1}(x) = \dfrac{x-1}{4}$ **32.** 9,519   **33.** 1.47   **34.** $a_n = a_{n-1} + 11$

35. $a_n = \frac{1}{2}(a_{n-1})$ **36.** $y = \dfrac{S - 2x^2}{4x}$ **37.** 25   **38.** 0   **39.** $y = -2x + 9$ **40.** $C(5) = 40, C(10) = 20$

41. $20t^2 + 11t - 3$ **42.** $(f \circ g)(x) = 5 - 3x$ **43.** $(\frac{13}{22}, -\frac{1}{22})$ **44.** $32x^5$ **45.** -19

46. $(x - \frac{1}{2})(x^2 + \frac{1}{2}x + \frac{1}{4})$ **47.** 30°, 150°   **48.** 6   **49.** 1, 3   **50.** 4 gal. of 30%, 12 gal. of 70%

CHAPTER 13

PROBLEM SET 13.1

1. 5   **3.** $\sqrt{106}$ **5.** $\sqrt{61}$ **7.** $\sqrt{130}$ **9.** 3 or −1   **11.** 3   **13.** $(x - 2)^2 + (y - 3)^2 = 16$
15. $(x - 3)^2 + (y + 2)^2 = 9$ **17.** $(x + 5)^2 + (y + 1)^2 = 5$ **19.** $x^2 + (y + 5)^2 = 1$ **21.** $x^2 + y^2 = 4$
23. Center = (0, 0)
  Radius = 2

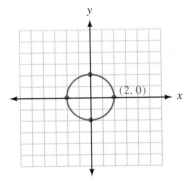

25. Center = (1, 3)
  Radius = 5

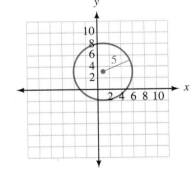

27. Center = (−2, 4)
  Radius = $2\sqrt{2}$

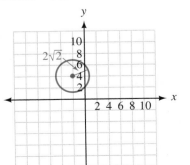

29. Center = (−1, −1)
  Radius = 1

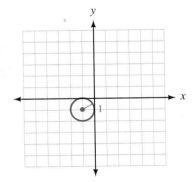

31. Center = $(0, 3)$
Radius = 4

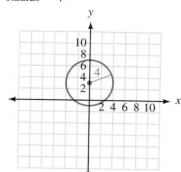

33. Center = $(-1, 0)$
Radius = $\sqrt{2}$

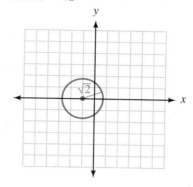

35. Center = $(2, 3)$,
Radius = 3

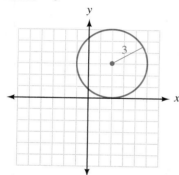

37. Center = $(-1, -\frac{1}{2})$,
Radius = 2

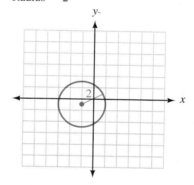

39. $(x - 3)^2 + (y - 4)^2 = 25$

41. (a) $(x - \frac{1}{2})^2 + (y - 1)^2 = \frac{1}{4}$ (b) $(x - 1)^2 + (y - 1)^2 = 1$ (c) $(x - 2)^2 + (y - 1)^2 = 4$ **43.** $x^2 + y^2 = 25$

45. $x^2 + y^2 = 9$ **47.** $(x + 1)^2 + (y - 3)^2 = 25$ **49.** $a_n = 4n + 1$ **51.** $\frac{15}{32}$ **53.** $\sum\limits_{i=1}^{5} (2i - 1)$

55. $(x - 2)^2 + (y - 3)^2 = 4$ **57.** $(x - 2)^2 + (y - 3)^2 = 4$ **59.** 5   **61.** 5

PROBLEM SET 13.2

1.

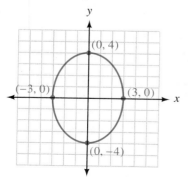

3.

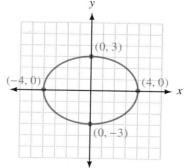

5.

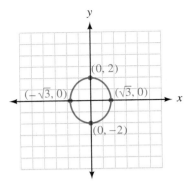

7.

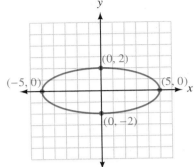

9.

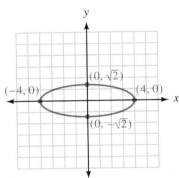

11.

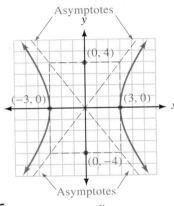

13.

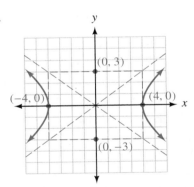

15.

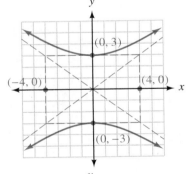

17.

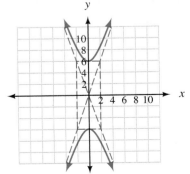

19.

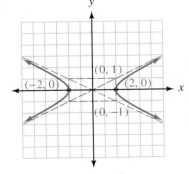

21.

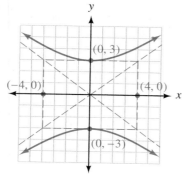

23. x-intercepts $= \pm 3$, y-intercepts $= \pm 2$

25. x-intercepts $= \pm 0.2$, no y-intercepts

27. x-intercepts $= \pm \frac{3}{5}$, y-intercepts $= \pm \frac{2}{5}$

29.

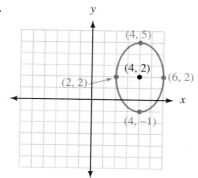

31.

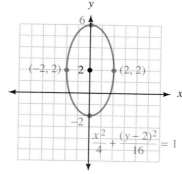

33.

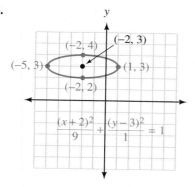

35.

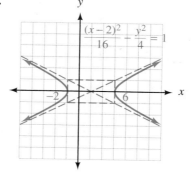

37.

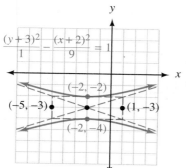

39.

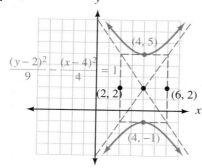

41. $\dfrac{(x-2)^2}{9} + \dfrac{(y+5)^2}{16} = 1$ **43.** $\dfrac{x^2}{25} - \dfrac{y^2}{9} = 1$ **45.** $y = \frac{3}{4}x,\ y = -\frac{3}{4}x$ **47.** 8   **49.** $a_n = 6n - 1$

51. 1,030   **53.** $\dfrac{63}{4}$

PROBLEM SET 13.3

1.

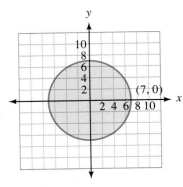

3.

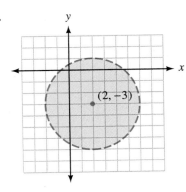

5.

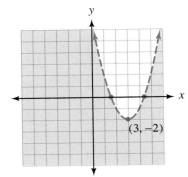

7.

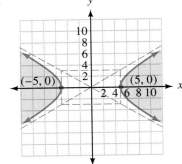

9.

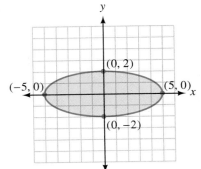

11.

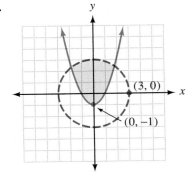

13.

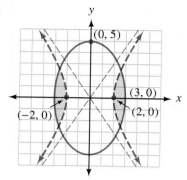

15. No intersection

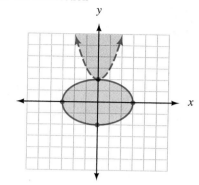

17.

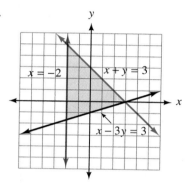

19.

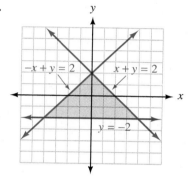

21.

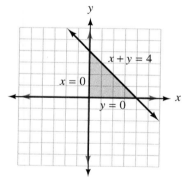

23. $(0, 3), (\frac{12}{5}, -\frac{9}{5})$ **25.** $(0, 4), (\frac{16}{5}, \frac{12}{5})$ **27.** $(5, 0), (-5, 0)$ **29.** $(0, -3), (\sqrt{5}, 2), (-\sqrt{5}, 2)$

31. $(0, -4), (\sqrt{7}, 3), (-\sqrt{7}, 3)$ **33.** $(-4, 11), (\frac{5}{2}, \frac{5}{4})$ **35.** $(-4, 5), (1, 0)$ **37.** $(2, -3), (5, 0)$

39. $(3, 0), (-3, 0)$ **41.** $(4, 0), (0, -4)$ **43.** 8, 5 or $-8, -5$ or 8, -5 or $-8, 5$ **45.** 6, 3 or 13, -4

47. $x^4 + 8x^3 + 24x^2 + 32x + 16$ **49.** $8x^3 + 12x^2y + 6xy^2 + y^3$ **51.** $x^{50} + 150x^{49}$

CHAPTER 13 REVIEW

1. $\sqrt{10}$ **2.** $\sqrt{13}$ **3.** 5   **4.** 9   **5.** $-2, 6$ **6.** $-12, 4$ **7.** $(x - 3)^2 + (y - 1)^2 = 4$

8. $(x - 3)^2 + (y + 1)^2 = 16$ **9.** $(x + 5)^2 + y^2 = 9$ **10.** $(x + 3)^2 + (y - 4)^2 = 18$ **11.** $x^2 + y^2 = 25$

12. $x^2 + y^2 = 9$ **13.** $(x + 2)^2 + (y - 3)^2 = 25$ **14.** $(x + 6)^2 + (y - 8)^2 = 100$

15. $(0, 0); r = 2$ **16.** $(3, -1); r = 4$

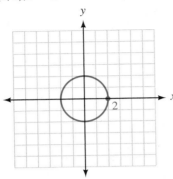

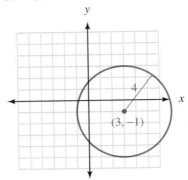

17. $(3, -2); r = 3$ **18.** $(-2, 1); r = 3$

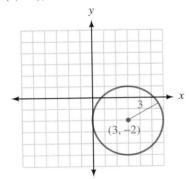

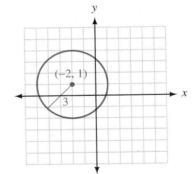

19.

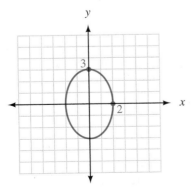

20.

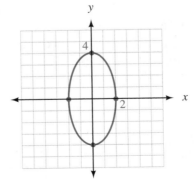

21.

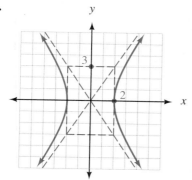

22.

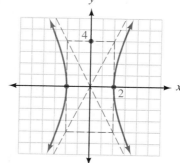

23.

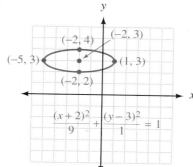

24.

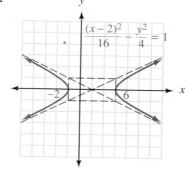

25.

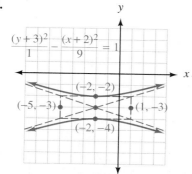

26.

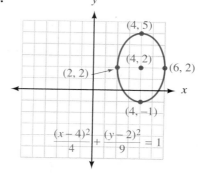

27.

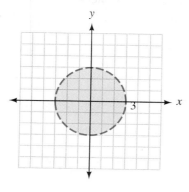

28.

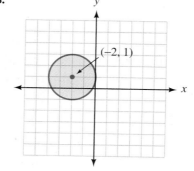

29.

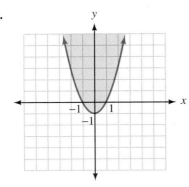

30.

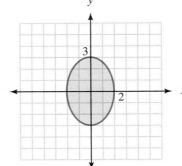

31.

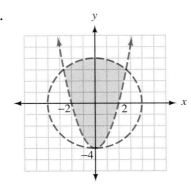

32.

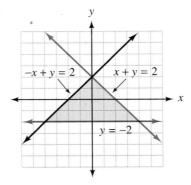

33. $(0, 4)$, $(\frac{16}{5}, -\frac{12}{5})$ **34.** $(0, -2)$, $(\sqrt{3}, 1)$, $(-\sqrt{3}, 1)$ **35.** $(-2, 0)$, $(2, 0)$

36. $\left(-\sqrt{7}, -\frac{\sqrt{6}}{2}\right)$, $\left(-\sqrt{7}, \frac{\sqrt{6}}{2}\right)$, $\left(\sqrt{7}, -\frac{\sqrt{6}}{2}\right)$, $\left(\sqrt{7}, \frac{\sqrt{6}}{2}\right)$

CHAPTER 13 TEST

1. -5 and 3 **2.** $(x + 2)^2 + (y - 4)^2 = 9$ **3.** $x^2 + y^2 = 25$ **4.** Center $= (5, -3)$, Radius $= \sqrt{39}$

5.

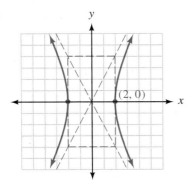

6.

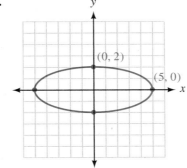

7.

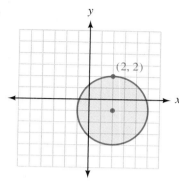

8. $\dfrac{(x-4)^2}{4} + \dfrac{(y-2)^2}{9} = 1$

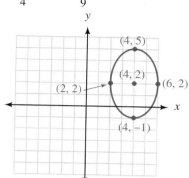

9. $(0, 5), (4, -3)$ **10.** $(0, -4), (\sqrt{7}, 3), (-\sqrt{7}, 3)$

CHAPTERS 1–13: CUMULATIVE REVIEW

1. 29   **2.** -5 **3.** $-2x - 15$ **4.** $2x + 20$ **5.** $24y$ **6.** $-\dfrac{1}{25}$ **7.** $x^{13/15}$ **8.** $4x^2y\sqrt{3xy}$ **9.** 1

10. 2   **11.** $-\dfrac{7}{3}, 3$ **12.** $-\dfrac{1}{2}, 3$ **13.** $-2, 2, 3$ **14.** 28   **15.** Possible solutions 0 and 7; only 7 checks.

16. $3 \pm i\sqrt{3}$ **17.** $-\dfrac{3}{4} \pm \dfrac{i\sqrt{11}}{4}$ **18.** 8   **19.** $\dfrac{2}{5}$ **20.** $-\dfrac{7}{3}$ **21.** $(-1, 4)$ **22.** $(-1, -1)$

23. $(-1, 2, -1)$ **24.** $\dfrac{x^2 - x - 12}{x^3 + 2x^2}$ **25.** $x^{2/5} - 9$ **26.** $3xy^2 - 4x + 2y^2$ **27.** $-\dfrac{1}{5} - \dfrac{8}{5}i$

28.

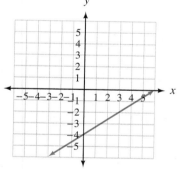

29.

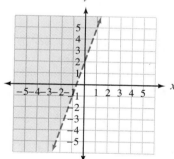

30.

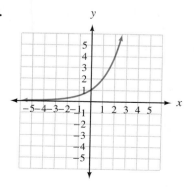

31.

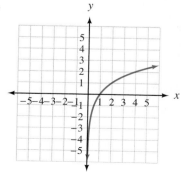

32.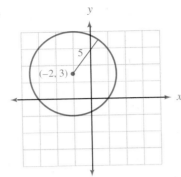

33.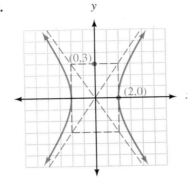

34.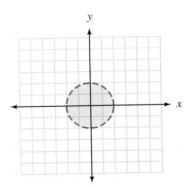

35.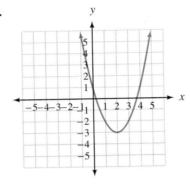

36. -16 **37.** $f^{-1}(x) = \dfrac{x + 8}{7}$ **38.** $-\dfrac{5}{m - n}$ **39.** -5

40. Slope $= \frac{2}{3}$; y-intercept $= -4$ **41.** $y = -\frac{4}{3}x - 9$ **42.** $\frac{3}{4}$ **43.** $\sqrt{65}$

44. 4 gallons of 60%, 12 gallons of 20%; the equations are $x + y = 16$ and $0.2x + 0.6y = 0.3(16)$

45. $(g \circ f)(x) = 19 - 5x^2$ **46.** $-10x^4 y$ **47.** 0.5 second and 2.5 seconds   **48.** $3x^2 - 5x + 2 = 0$

49. $x < -3$ or $x > 2$

50. Domain $= \{x \mid x \neq 2\}$ **51.** $a_n = 5 \cdot 3^{n-1}$ **52.** 35

APPENDIX A

1. $x - 7 + \dfrac{20}{x + 2}$ **3.** $3x - 1$ **5.** $x^2 + 4x + 11 + \dfrac{26}{x - 2}$ **7.** $3x^2 + 8x + 26 + \dfrac{83}{x - 3}$ **9.** $2x^2 + 2x + 3$

11. $x^3 - 4x^2 + 18x - 72 + \dfrac{289}{x + 4}$ **13.** $x^4 + x^2 - x - 3 - \dfrac{5}{x - 2}$ **15.** $x + 2 + \dfrac{3}{x - 1}$

17. $x^3 - x^2 + x - 1$ **19.** $x^2 + x + 1$

APPENDIX B

1. 3   **3.** 5   **5.** -1 **7.** 0   **9.** 10   **11.** 2   **13.** -3 **15.** -2 **17.** $-2, 5$ **19.** 3   **21.** 0

23. 3   **25.** 8   **27.** 6   **29.** -228 **31.** $\begin{vmatrix} y & x \\ m & 1 \end{vmatrix} = y - mx = b; \ y = mx + b$

33. (a) $y = 0.3x + 3.4$ (b) $y = 4$ billion dollars   **35.** 4   **37.** 4

APPENDIX C

1. $(3, 1)$ **3.** Lines are parallel; $\varnothing$ **5.** $(-\frac{15}{43}, -\frac{27}{43})$ **7.** $(\frac{60}{43}, \frac{46}{43})$ **9.** $(3, -1, 2)$ **11.** $(\frac{1}{2}, \frac{5}{2}, 1)$

13. No unique solution   **15.** $(-\frac{10}{91}, -\frac{9}{13}, \frac{107}{91})$ **17.** $(\frac{71}{13}, -\frac{12}{13}, \frac{24}{13})$ **19.** $(3, 1, 2)$ **21.** $x = 50$ items

23. 1986   **25.** $x = -\dfrac{1}{a-b} = \dfrac{1}{b-a}, y = \dfrac{1}{a-b}$ **27.** $x = \dfrac{1}{a^2 + ab + b^2}, y = \dfrac{a+b}{a^2 + ab + b^2}$

29. $x + 2y = 1$
   $3x + 4y = 0$

APPENDIX D

1. $(2, 3)$ **3.** $(-1, -2)$ **5.** $(7, 1)$ **7.** $(1, 2, 1)$ **9.** $(2, 0, 1)$ **11.** $(1, 1, 2)$ **13.** $(4, 1, 5)$

15. $(4, \frac{10}{3})$

Index